Multilevel Converters

Multilevel Converters: Analysis, Modulation, Topologies, and Applications

Special Issue Editors

Gabriele Grandi
Alex Ruderman

MDPI • Basel • Beijing • Wuhan • Barcelona • Belgrade

Special Issue Editors
Gabriele Grandi
University of Bologna
Italy

Alex Ruderman
Nazarbayev University
Kazakhstan

Editorial Office
MDPI
St. Alban-Anlage 66
4052 Basel, Switzerland

This is a reprint of articles from the Special Issue published online in the open access journal *Energies* (ISSN 1996-1073) from 2018 to 2019 (available at: https://www.mdpi.com/journal/energies/special_issues/multilevel_converters)

For citation purposes, cite each article independently as indicated on the article page online and as indicated below:

LastName, A.A.; LastName, B.B.; LastName, C.C. Article Title. *Journal Name* **Year**, *Article Number*, Page Range.

ISBN 978-3-03921-481-5 (Pbk)
ISBN 978-3-03921-482-2 (PDF)

Contents

About the Special Issue Editors

Gabriele Grandi received his M.Sc. (cum laude) and Ph.D. degrees in Electrical Engineering from the University of Bologna, Bologna, Italy, in 1990 and 1994, respectively. He has been with the Department of Electrical, Electronic, and Information Engineering, University of Bologna as Research Associate (since 1995), Associate Professor (since 2005), and most recently as Full Professor (since 2016) in Electrical Engineering. He is the Founder and Leader of the research laboratory "SolarTronic-Lab" at University of Bologna, dealing with power electronic circuits, multiphase and multilevel converters, photovoltaics, electric vehicle chargers, and circuit modeling. He has authored or co-authored more than 160 papers in conference proceedings and international journals, mainly with the IEEE. Dr. Grandi serves as Editor-at-Large for *IET Power Electronics*, Academic Editor for MDPI *Energies* and MDPI *Electronics*, and Associate Editor for *IEEE Trans. on Industrial Electronics*.

Alex Ruderman obtained his M.Sc. (cum laude) and Ph.D. degrees from the Electrotechnical University and Polytechnic University (former Leningrad, USSR; now St. Petersburg, Russia) in 1980 and 1987 respectively. In 1995–2003, he worked as an R&D engineer for Intel Microprocessor Development Center, Haifa, Israel. In 2004–2013, Alex taught electronics-related courses in Bar Ilan University, Ariel University and Holon Institute of Technology as Adjunct Professor. In 2006, he joined Elmo Motion Control, Petach Tikva, Israel, the makers of compact intelligent servo drives, as a Chief Scientist (Elmo drives are used by NASA in Mars Curiosity and InSight missions). Since 2013, Alex has been Associate Professor at Nazarbayev University School of Engineering and Digital Sciences, Electrical and Computer Engineering Department. Alex is a regular reviewer for *IEEE Transactions on Industrial Electronics* and *Power Electronics* and a program committee member for several international Power Electronics Conferences. His major research focus is multilevel power converters—he has authored over 60 conference and journal papers on the subject. Alex is an IEEE Senior Member and Associate Editor for the *IET Journal of Power Electronics*.

Preface to "Multilevel Converters: Analysis, Modulation, Topologies, and Applications"

Multilevel inverters (MLIs) have been widely used for medium- and high-voltage power applications in the recent decades, mainly for grid-connected applications to interface with renewable energy sources. Compared to basic two-level inverters, MLIs offer many advantages, such as reduced voltage rating of power switches, reduced voltage and current harmonic distortion, reduced electromagnetic interference, as well as flexibility and modularity.

MLIs became a standard for applications such as in medium voltage drives and HVDC grids, and are promising for lower voltage applications such as in battery chargers, active filters, static compensators, dynamic voltage restorers, rectifiers, grid-tied inverters, and many more. Increased efficiency and reduced harmonic distortion are beneficial for photovoltaic systems and uninterruptible power supplies. The introduction of multilevel topologies has shifted the power converter design paradigm, including control and modulation strategies, component selection and requirements, reliability aspects, amongst others. While relatively low-power applications employ high-frequency PWM, for high-voltage/current applications, the switching frequency of the power semiconductors is limited to few kHz by switching loss considerations, and the use of multilevel converters becomes mandatory.

Gabriele Grandi, Alex Ruderman
Special Issue Editors

Article

A Full-bridge Director Switches based Multilevel Converter with DC Fault Blocking Capability and Its Predictive Control Strategy

Jin Zhu, Tongzhen Wei *, Qunhai Huo and Jingyuan Yin

Institute of Electrical Engineering, Chinese Academy of Sciences, Haidian District, Beijing 100190, China; zhujin@mail.iee.ac.cn (J.Z.); huoqunhai@mail.iee.ac.cn (Q.H.); yinjingyuan@mail.iee.ac.cn (J.Y.)
* Correspondence: tzwei@mail.iee.ac.cn; Tel.: +86-182-1116-1108

Received: 23 October 2018; Accepted: 21 December 2018; Published: 28 December 2018

Abstract: Voltage source converter-based high-voltage direct current transmission system (VSC-HVDC) technology has been widely used. However, traditional half-bridge sub module (HBSM)-based module multilevel converter (MMC) cannot block a DC fault current. This paper proposes that a full-bridge director switches based multi-level converter can offer features such as DC side fault blocking capability and is more compact and lower cost than other existing MMC topologies. A suitable predictive control strategy is proposed to minimize the error of the output AC current and the capacitor voltage of the sub-module while the director switches are operated in low-frequency mode. The validity of the proposed topology and control method is demonstrated based on simulation and experimental studies.

Keywords: multilevel converter; DC side fault blocking; predictive control

1. Introduction

The modular multilevel converter (MMC) has been accepted as a suitable solution for high-voltage and high-power application fields due to several inherent features [1–8]. However, blocking the DC fault current becomes a difficult problem because the anti-parallel diodes are still conducting after the insulated-gate bipolar transistors (IGBTs) of HBSM are turned off [9].

To solve this problem, recent research has highlighted a number of interesting converter topologies which combine the features of the multilevel output AC voltage waveform and DC fault blocking capacity [9–22]. Full-bridge sub-module (FBSM) based MMC (F-MMC) is a basic configuration with DC fault current blocking capacity [10,11]. However, the DC fault current blocking capability comes at a cost of nearly doubling power losses and number of semiconductor devices. Some other type of sub-module is proposed instead of FBSM to make a further optimization in reducing the number of IGBTs, such as a clamp double sub-module (CDSM) proposed in [14,15] and a three-level cross-connected sub-module (TCSM) proposed in [16]. Several hybrid MMC topologies are also proposed, based on HBSM and those various types of sub-module [9,12,13,16–19,23], for further reducing the cost and loss on the premise of having the DC fault blocking capacity, such as hybrid MMC based on CDSM and HBSM (CH-MMC). However, there are still some drawbacks, for example, as they are composed of a large number of sub-modules, the system needs to be more complicated and the converter station bulkier.

The alternate-arm multilevel converter (AAMC) based on the hybrid topology of HBSM and director switches is proposed in [20–22]. The AAMC further improves the traditional MMC topology by cutting the number of sub-modules, reducing DC bus voltage, and gaining the ability to block DC fault currents [22]. However, some features still have the possibility for further optimization, such as the size and cost of the overall system. One of the main technical challenges associated with the control of such a director switches based multilevel converter is to simultaneously keep the capacitor voltages balanced and provide good output current tracking performance, while the director switches keep switching in low frequency.

In order to further optimize the size and cost of the voltage source converter-based high-voltage direct current transmission system (VSC-HVDC) converter with the blocking ability of DC faults, this paper proposes a full bridge director switches based alternate-arm multi-level converter (FA-MMC) and a corresponding control strategy:

1. The size and cost of the overall system can be significantly reduced by reducing the number of SM capacitors, IGBTs, and other related devices. In addition, an FA-MMC retains the ability to block DC-side faults since it uses H-bridge SMs as the AAMC.

2. Similar to AAMC, a systematic multi-objective control method is needed for this kind of topology to minimize the error of output AC current and the capacitor voltage of the sub-module while the director switches are operated in low-frequency mode. A suitable predictive control strategy for this kind of topology is presented in this paper to achieve the flexibility to include the previously mentioned multiple system requirements.

2. Proposed Topology

2.1. Structure and Basic Operation

The basic circuit configuration of full-bridge director switches based modular multi-Level converter (FA-MMC) proposed in this paper is shown in Figure 1b. The proposed topology consists of a stack of H-bridge SMs and four director switches (S1–S4) made of series IGBTs or IGCTs. The ability of DC-side fault blocking is still retained since the H-bridge SMs structure is the same as the AAMC.

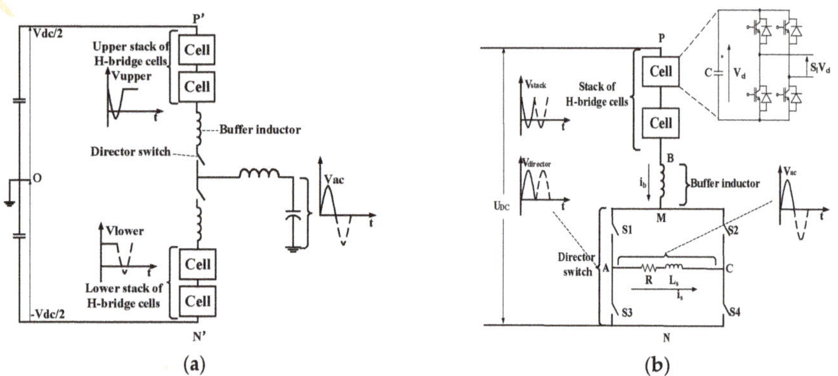

Figure 1. Schematic representation of the two topologies: (a) alternate-arm multilevel converter (AAMC); (b) full bridge director switches based alternate-arm multi-level converter (FA-MMC).

The voltage of the director switches ($U_{director}$ in Figure 1b) is equal to the DC voltage (U_{DC} in Figure 2) plus the voltage produced by the stack of H-bridge SMs which can be considered as only one controllable voltage source. Therefore, the voltage of director switches can be adjusted flexibly so that the switching of S_1–S_4 can switch at near to zero voltage. The ideal voltage waveform is shown in Figure 2a.

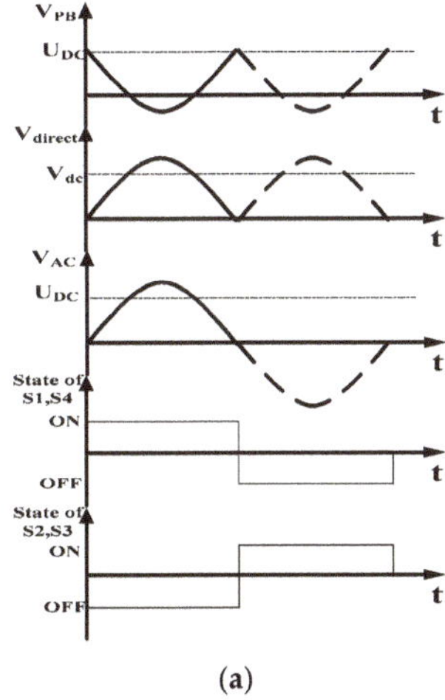
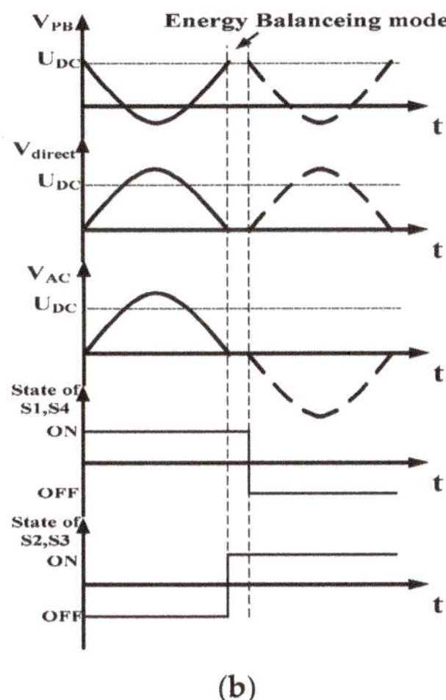

Figure 2. The voltage waveform and state of S1–S4: (**a**) without energy balance mode; (**b**) with energy balance mode.

The working cycle of S_1–S_4 is synchronized with the output AC voltage. S_1 and S_4 are conducting and S_2 and S_3 are turned off while the output AC voltage (U_{ac} in Figure 1b is in its positive half-cycle, in contrast, S_2 and S_3 are conducting and S_1 and S_4 are turned off while the output AC voltage is in its negative half-cycle. This ensures that the four director switches can switch at low-frequency and at the point of zero-voltage-crossing as shown in Figure 2a. These features lead to low switching losses, and low demand for dynamic voltage sharing at the switching instant of the series switches, so that the system design has been simplified.

2.2. Energy Balance

When the AC current flows through the stack of H-bridge sub module. In order to ensure the continuous operation of the system, the energy balance of the stack of H-bridges should be guaranteed. The amount of energy transferred from the AC side (E_{AC}) and going to the DC side (E_{DC}) should be equal over half the fundamental period and is given as

$$E_{AC} = \frac{3}{2} \frac{\hat{V}_{AC}\hat{I}_{AC}}{\omega} \pi \cos(\varphi), \tag{1}$$

$$E_{DC} = \frac{6 U_{DC} \hat{I}_{AC}}{\omega} \cos(\varphi), \tag{2}$$

For E_{AC} to equal E_{DC}, the relationship between the DC voltage magnitudes and AC voltage magnitudes mentioned in Equations (1) and (2) can be given as

$$U_{DC} = \frac{\pi}{4} \hat{V}_{AC}, \tag{3}$$

However, since the converters can't operate in the perfect given by Equation (3), an energy balancing strategy should be used.

Reference [22] presented two methods to achieve energy balance for AAMCs that can also be used in this topology: Overlap current and third harmonic current injection. In this paper the overlap current method is used to extend the period when the current directed from S_1 and S_4 to S_2 and S_3 is extended and S_1–S_4 are all conducting. The overlap current is used to exchange power between the sub module capacitors and the DC bus. The load current is only slightly affected, since the overlap time is very short and the inductance can smooth the change in current. Considering its effect on the grid current, the overlap time is determined to be less than 0.8 ms.

3. Predictive Control Strategy

The control strategy of the proposed topology requires minimizing the error of the output current and DC voltage in each sub module, and, meanwhile, the director switches switching should be operated in low-frequency and zero-voltage switching mode.

3.1. Dynamic Modeling

Based on Figure 2, the governing equations of the single-phase FA-MMC can be shown as follows:

$$U_{DC} - V_{PB} - L_b\frac{di_b}{dt} = V_{MN} \tag{4}$$

$$V_{AC} = L_s\frac{di_s}{dt} + Ri_s \tag{5}$$

$$V_{AC} = S_d V_{MN} \tag{6}$$

As presented in Section 2, the value of L_b is small and the voltage on it can be ignored; S_1–S_4 have five switching state combinations depending on a switching function S_d as shown in Table 1.

Table 1. Switching states of director switches.

Mode	S_d	S_1	S_2	S_3	S_4	Output Voltage (V_{AC})
Basic Operating Mode	1	ON	OFF	OFF	ON	V_{MN}
	−1	OFF	ON	ON	OFF	$-V_{MN}$
	0	ON	ON	ON	ON	0
Energy Balancing Mode	0	ON	ON	OFF	OFF	0
	0	OFF	OFF	ON	ON	0

The output voltage of each H-bridge sub module is equal to V_{ci} (capacitor voltage of the ith sub module ($i = 1, 2, \cdots, n$)), $-V_{ci}$, or zero, depending on the switching states, and depends on a switching function Si

$$S_i = \begin{cases} 1 \\ 0 \\ -1 \end{cases} \quad (i = 1, 2, \ldots, n). \tag{7}$$

The relationship between V_{ci} and U_{DC} is formalized as

$$\sum_{i=1}^{n} V_{ci} \approx U_{DC} \tag{8}$$

Based on Equation (16) and the basic principle, V_{PB} is formalized as

$$\sum_{i=1}^{n} S_i V_{ci} = V_{PB} \tag{9}$$

The dynamic capacitor voltage of the cells of the H-bridge sub module in Figure 1b is formalized as

$$S_i i_b = C\frac{dV_{ci}}{dt} \tag{10}$$

The relationship of currents is and i_b in Figure 1b, which was also indicated by the switching function S_d according to Table 1, is expressed as

$$i_b = S_d i_s \ (S_d = 1 \text{ or } -1) \tag{11a}$$

$$L\frac{di_b}{dt} = U_{DC} - \sum_{i=1}^{n} S_i V_{ci} \ (S_d = 0) \tag{11b}$$

The switching states of director switches operate in an energy balancing mode, as mentioned in Table 1. As discussed previously, the current i_b flows through the stack of H-bridge sub modules, buffer inductor, and director switch to the DC side, charging or discharging the capacitor of the H-bridge sub modules.

Only considering the basic operating mode, substituting Equations (5), (6), (9), and (11a) into (4), a dynamic model of the single-phase proposed topology in basic operating mode can be expressed as

$$S_d(U_{DC} - \sum_{i=0}^{n} S_i V_{ci} - S_d L_b \frac{di_s}{dt}) = L_s \frac{di_s}{dt} + R i_s \tag{12}$$

where S_d = 1 or -1. Equation (12) can be simplified as

$$L\frac{di_s}{dt} = S_d(U_{DC} - \sum_{i=0}^{n} S_i V_{ci}) - R i_s \tag{13}$$

where $L = L_b + L_s$.

3.2. Proposed Predictive Control

The predictive control strategy is proposed in this section based on the dynamic model of the FA-MMC presented above, the three primary targets of the predictive control strategy is achieved as follows:

3.2.1. AC-Side Current Control

Assuming a sampling period of T_s, a discrete-time model of the FA-MMC AC-side current in basic operating mode based on Equation (3) is calculated by

$$\frac{L}{T_s}(i_s(k+1) - i_s(k)) = S_d(k)(U_{DC}(k) - \sum_{i=0}^{n} S_i(k) V_{ci}(k)) - R i_s(k) \tag{14}$$

the value of S_d could be assumed as a constant value during a short sampling period of T_s. $i_s(k)$ is the actual AC current at time k and $i_s(k + 1)$ is the predicted AC current at time $k + 1$, $U_{DC}(k)$ can be considered as a constant value if the DC side voltage is controlled. Finally, $V_{ci}(k)$ is the capacitor voltage of the sub module i at time k.

To reduce the error between the predicted current and the reference current, a cost function associated with the current error is defined as

$$J_i = \left| i_{sref}(k+1) - i_s(k+1) \right| \tag{15}$$

where i_{sref} is the reference current and $i_s(k + 1)$ is the predicted current obtained from Equation (14). Ideally, J_i will be equal to its minimum value of (J_{min} = 0 in Figure 4) if the AC-side current is controlled well.

3.2.2. Capacitor Voltage Balancing

Based on Equations (10) and (11), $V_{ci}(k+1)$ can be deduced as

$$V_{ci}(k+1) = V_{ci}(k) + \frac{S_i(k)S_d(k)i_s(k)}{C} T_s \tag{16}$$

where $V_{ci}(k)$ can be measured in real time. Another cost function for balancing the capacitor voltage of sub modules is given as

$$J_{vc} = \sum_{i=1}^{n} \left| V_{ci}(k+1) - V_{ciref}(k+1) \right| \tag{17}$$

where $V_{ciref}(k+1)$ is the reference DC capacitor voltage of sub module i (with i between 1 and n), which can be equal to the average voltage of all cells (given as $\frac{\sum_{i=1}^{n} V_{ci}(k+1)}{n}$), and $V_{ci}(k+1)$ is a predicted value, which can be obtained from Equation (16).

Consequently, by adding the above cost function together a combined cost function, which can simultaneously achieve the two main control objectives mentioned above is given as the linear combination

$$J_{all} = \alpha J_i + \beta J_{vc} \tag{18}$$

where α and β are weighing factors, α is adjusted based on the cost contribution allocated to the error of AC-Side current, and β is adjusted based on the cost contribution allocated to the voltage deviations of sub module capacitors. The empirical method to determine the value of cost function is presented in [24].

Within each sampling and computing period T_s, the combined cost function J_{all} is re-calculated, and the best switching indicated to the minimum value for Equation (18) will be adopted for the current control cycle.

3.2.3. Director Switch Control

As presented in Figure 2a in Section 2.1, the state of director switches S_1–S_4 at the next step should depend on the value of V_{AC}. According to Equation (5), the necessary value of V_{AC} at the current step can be expressed as

$$V_{AC}(k+1) = Ri_{sref}(k+1) + \frac{L_s}{T_s}(i_{sref}(k+1) - i_s(k)) \tag{19}$$

However, the fluctuation of $V_{AC}(k+1)$ due to differences between $i_{sref}(k+1)$ and $i_s(k)$ during zero voltage crossings will lead to high frequency repeated switching of S_1–S_4, resulting in an increase of switching losses.

Therefore, a director switch control strategy should be taken considering the need to

1. Add the energy-balancing mode ($S_d = 0$ in Table 1) in to achieve energy balancing of the stack of H-bridge by exchanging power with DC bus.
2. Avoid repeated switching of the director switches.

Replacing $i_s(k)$ by $i_{sref}(k)$, the necessary value of V_{AC} at the current step can be expressed as

$$V_{ACref}(k+1) = Ri_{sref}(k+1) + \frac{L}{T_s}(i_{sref}(k+1) - i_{sref}(k)) \tag{20}$$

where $i_{sref}(k)$ is the reference value of the current of the current step. Voltage $V_{ACref}(k+1)$, obtained by Equation (20), is a standard sine wave, which can avoid the fluctuation of $V_{AC}(k+1)$ due to differences between $i_{sref}(k+1)$ and $i_s(k)$ during zero voltage crossings. Finally, the implementation procedure of the proposed director switch control strategy is summarized in Part I of Figure 4. The schematic diagram of the control system is shown in Figure 3.

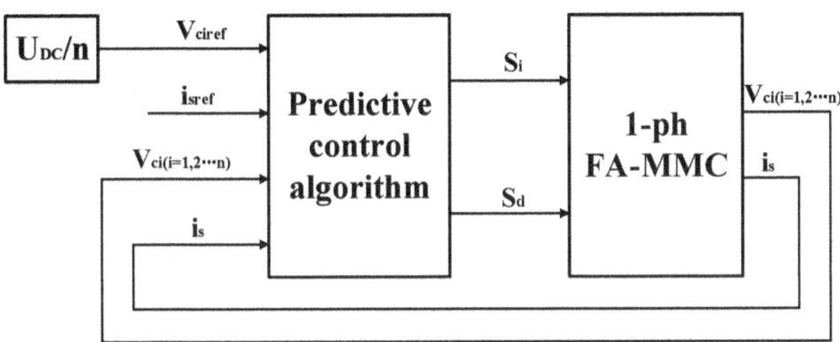

Figure 3. Schematic diagram of the control system.

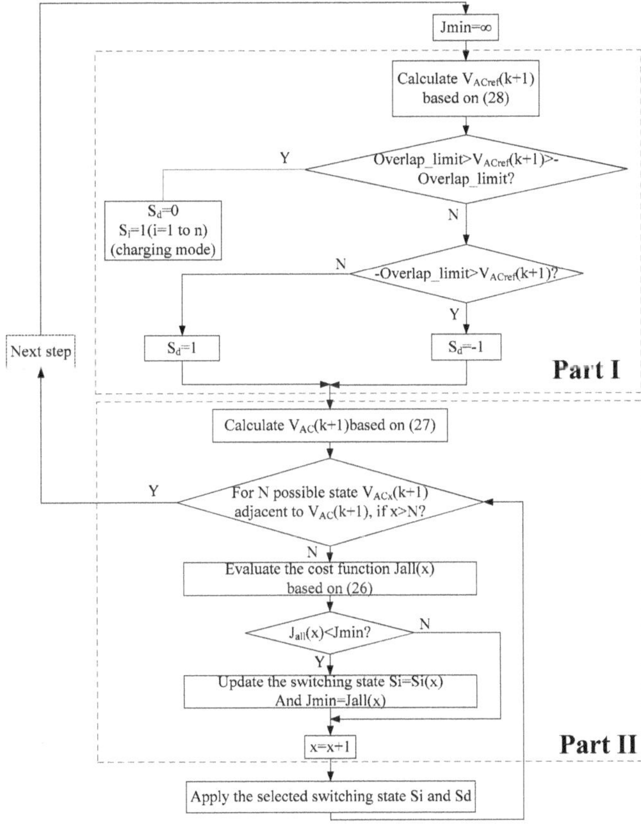

Figure 4. Block diagram of the predictive control strategy

4. Simulation Results

This section evaluates the performance of the proposed FA-MMC and control method with a simulation. The simulation parameters are given in Table 2.

Table 2. Parameters of the study system of Figure 1b.

DC voltage U_{DC}	3000 V
Submodule capacitor C	3300 μF
Load inductance L_s	3 mH
Buffer inductors L_b	0.1 mH
Load inductance R	6
Sampling period T_s	100 μs
Nominal frequencies f	50 Hz
No. of cell in the stack of H-bridge cells	2

4.1. Operating Performance under a Steady-State Condition

Figure 5 shows the voltage of the stack of H-bridges cells, the voltage across the director switches S_1–S_4, and the AC output voltage while the load current tracks the reference in steady-state operation. The simulation results are consistent with the working principle of the topology described in Figure 1b of Section 2. The voltage waveforms appear staircased because there are only two cells, while they would more closely resemble a sine curve with an increase in the number of cells. Figure 6 shows that the capacitor voltages in the two cells are averaged well and mostly under the control of MPC in basic operating mode. Further, they get closer to the given value U_{DC}/n_{cell_FA} in energy-balancing mode.

Figure 7 shows the director switch control signal of S_1–S_4. It can be seen in Figure 7a that they all operated at a frequency of 100 Hz and achieved zero voltage switching under the director switch control strategy described in Part I of Figure 4. In contrast, when Part I of Figure 4 is removed, the director switch control signal, which is only determined by $V_{AC}(k + 1)$, is shown in Figure 6b. The difference in responses occurs because $V_{ACref}(k + 1)$ in Equation (20) is obviously a standard sine wave while the $V_{AC}(k + 1)$ is repeatedly crossing the zero voltage point as shown in Figure 8. This demonstrates the effectiveness of the director switch control strategy.

Figure 9 reveals that the relation of the current across L_b (I_b in Figure 9) and the load current (Is in Figure 9) is similar to Equation (11a) in basic operating mode. The current across L_b (I_b in Figure 9) becomes an overlap current that charges or discharges the capacitor of the cells when S1–S4 are all conducting in energy balancing mode.

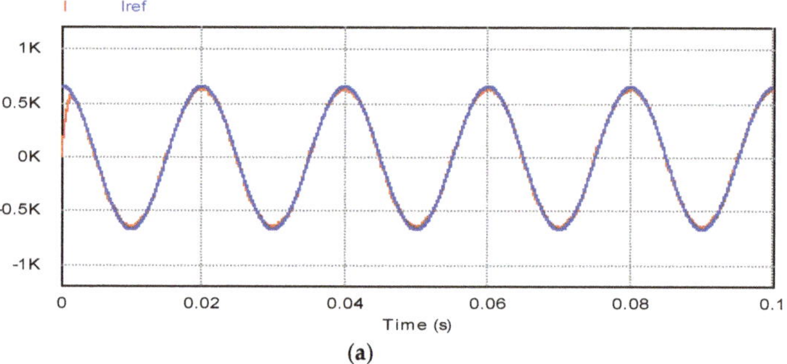

Figure 5. *Cont.*

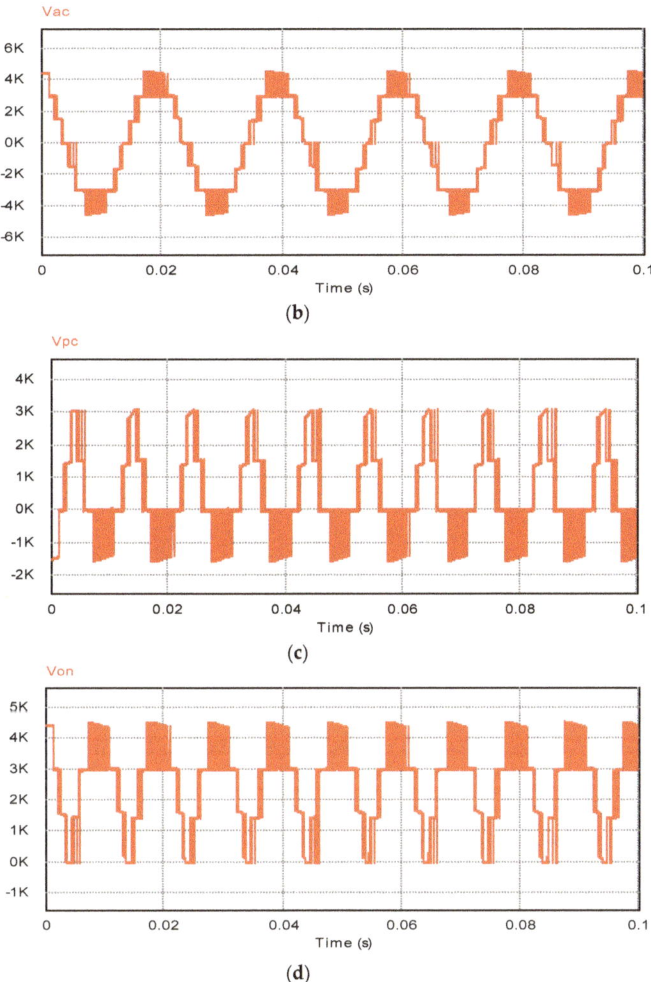

Figure 5. Simulation waveform of the single-phase FA-MMC in steady state operation: (**a**) Load current and reference current; (**b**) output AC voltage; (**c**) voltage of the stack of H-bridges; (**d**) voltage across the director switches S_1–S_4.

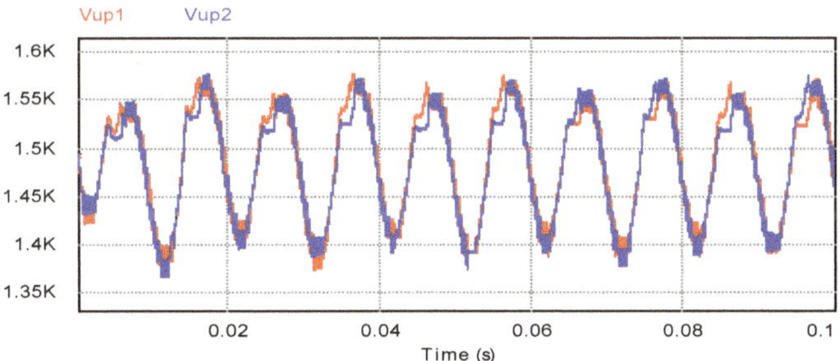

Figure 6. Capacitor voltages of the cells.

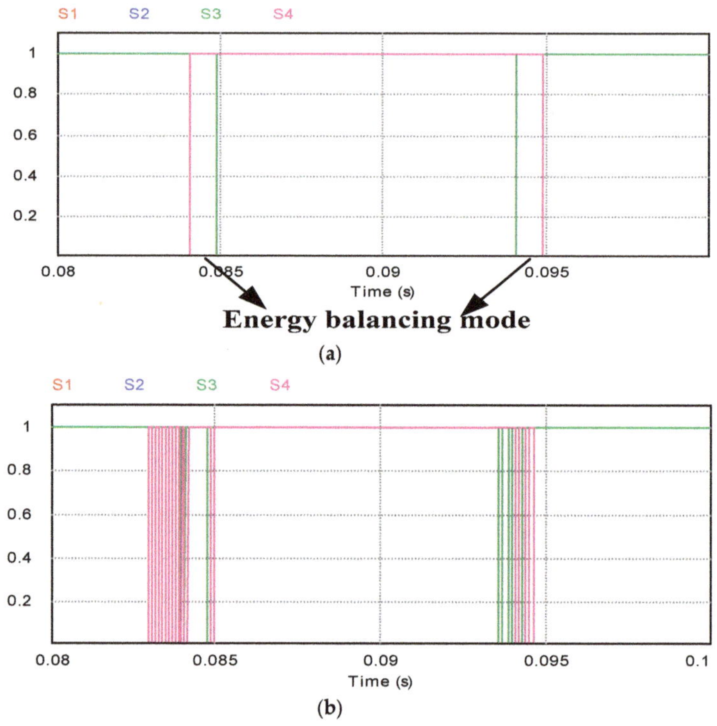

(a)

(b)

Figure 7. Director switch control signal of S_1–S_4: (**a**) Control signal based on $V_{acref(t+Ts)}$; (**b**) control signal based on $V_{ac(t+Ts)}$.

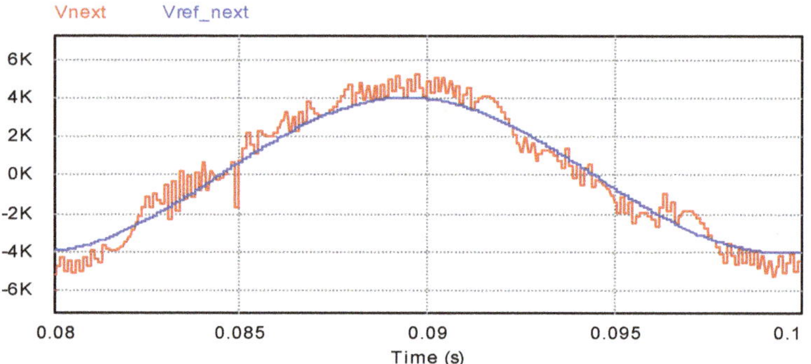

Figure 8. The waveforms of V_{ac} and V_{acref}.

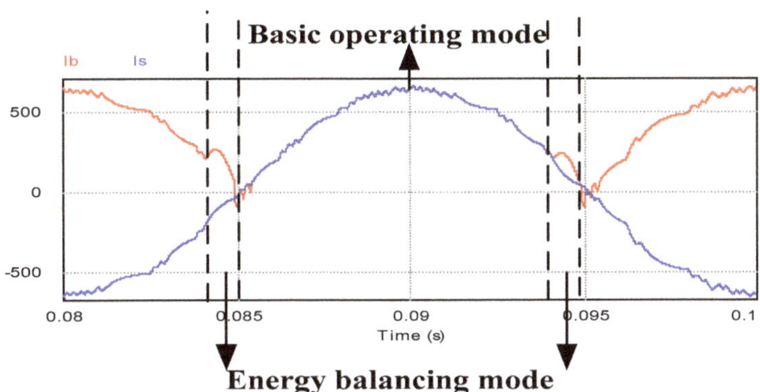

Figure 9. The waveforms of V_{ac} and V_{acref}.

4.2. Operating Performance under a Transient-State Condition

To test the dynamic performance, a sudden change in the reference current is set at 0.04 s, and the behavior of the system is shown in Figure 10. It can be seen that the current tracked the reference value well. The time of reference tracking (from 600 A to −600 A) is less than 0.01 ms as shown in Figure 10b, and the output AC voltage waveform is shown in Figure 11.

The capacitor voltage is shown in Figure 12. It can be seen that the capacitor voltage of the two sub-module remains balanced after a sudden change of load current. Figure 12b shows that there is a deviation in the beginning, but is averaged well immediately by the predictive control strategy after 1 ms.

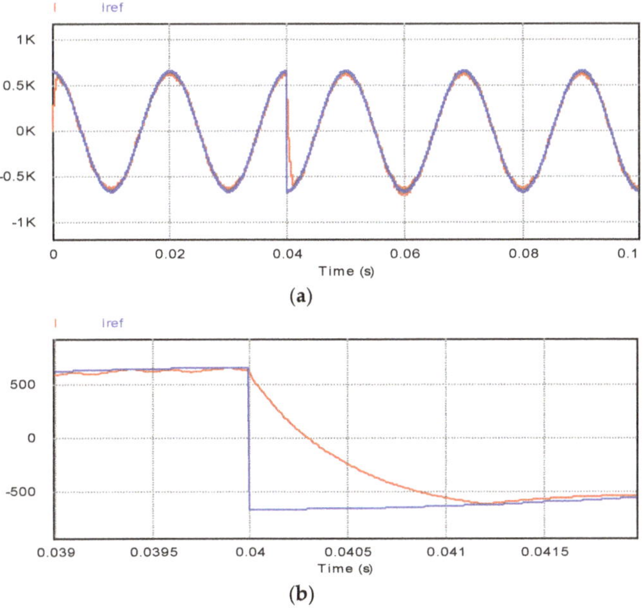

Figure 10. Load current for a sudden change: (**a**) reference current and actual load current; (**b**) Detail of the reference current and actual load current at the instant.

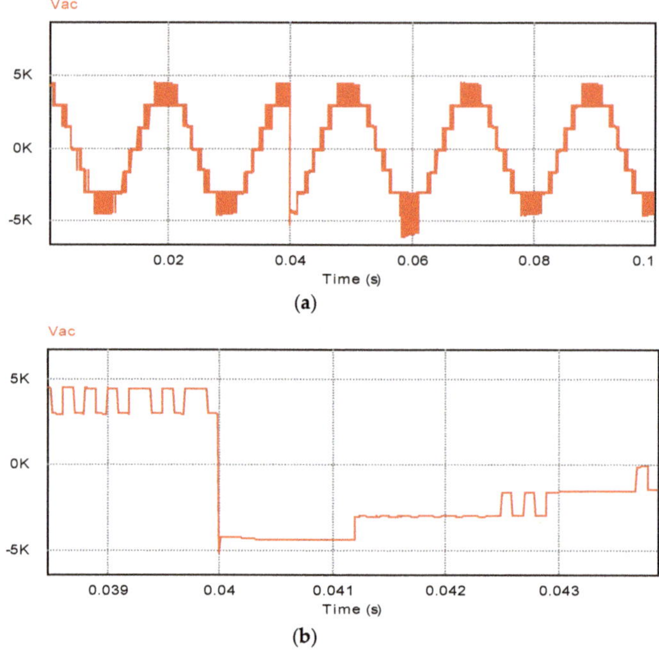

(a)

(b)

Figure 11. Output AC voltage waveforms: (**a**) Output AC voltage ; (**b**) detail of the output AC voltage at the instant.

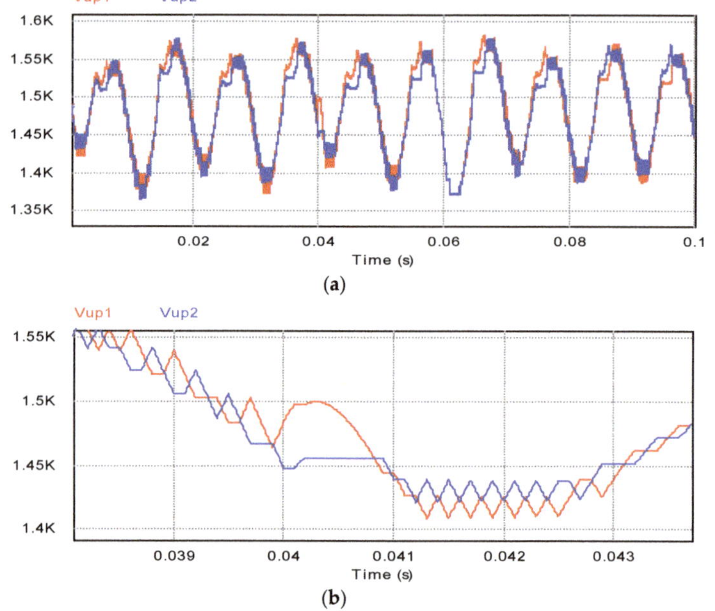

(a)

(b)

Figure 12. Capacitor voltages of the sub-module.

Figure 13 shows the states of the director switches at the instant of the sudden change of load current, demonstrating that the director switches are controlled well and operated in low-frequency mode.

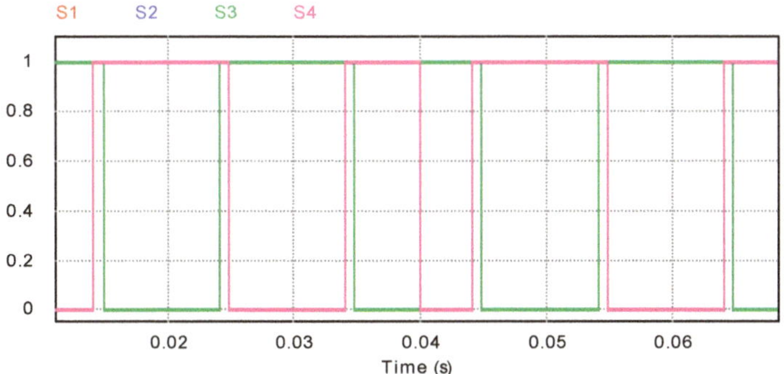

Figure 13. Control signal of S_1–S_4.

4.3. Operating Performance under a DC Fault

Having verified the normal operation of the converter, the model was tested under a DC fault. A three-phase model was built, and a DC fault was induced at 0.04 s. The blocking time is set to be 3 ms after the fault current is detected considering the sensor delay time. Figure 14 shows that the voltage of the cell capacitor is kept at 1.5 kV and the AC current follows the given value before 0.04 s. When a DC short-circuit happens at 0.04 s, the direction of current is reversed and the AC side current rises at first because during the sensor delay, the capacitors discharge and current flows from the AC side to the DC side. After 3 ms, when the converter station is blocked, the DC and AC side currents gradually reduce to zero along with the charging of the capacitor.

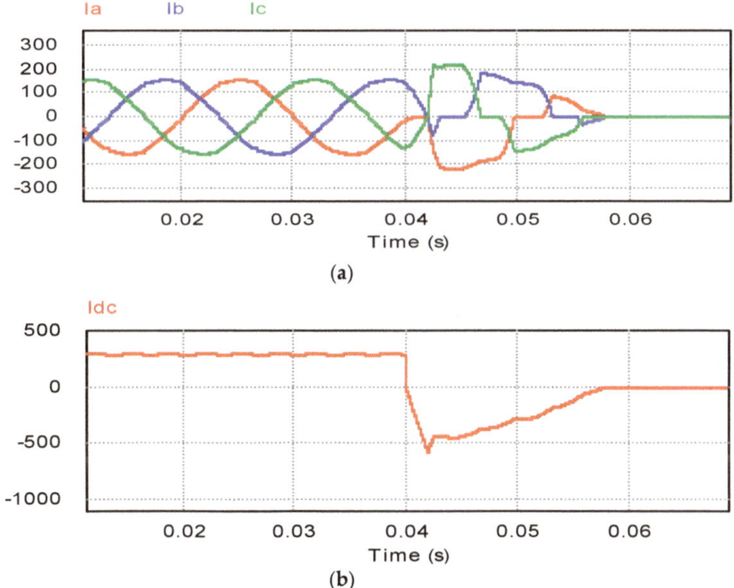

Figure 14. *Cont.*

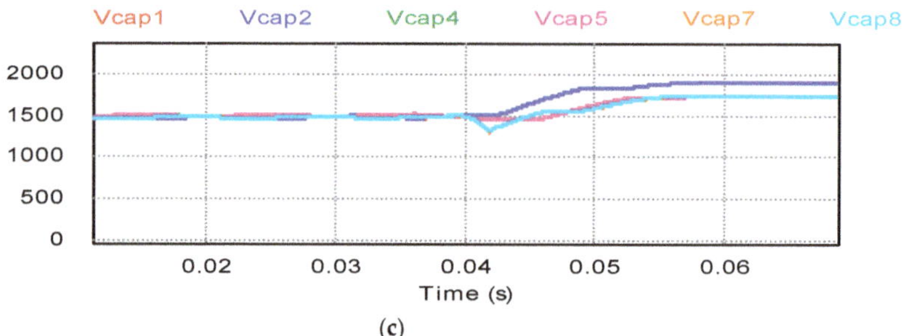

Figure 14. Current and voltage simulation waveforms of a DC fault: (**a**) AC current; (**b**) DC current; (**c**) capacitor voltages.

5. Experimental Results

Experiments on an FA-MMC-based inverter were also carried out to verify the proposed topology and test the predictive control strategy. The parameters for the experiment are listed in Table 3. A photo of the inverter is shown in Figure 15 and an IGBT is utilized as the power switch. The main control algorithms were implemented in a combination of a DSP and FPGA. The DC-link voltage was obtained via a three-phase autotransformer.

Table 3. Experiment parameters.

DC voltage U_{DC}	100 V
Submodule capacitor C	3300 μF
Load inductance L_s	3 mH
Buffer inductors L_b	0.1 mH
Load inductance R	6
Sampling period T_s	100 μs
Nominal frequencies f	50 Hz
No. of cell in the stack of H-bridge cells	2

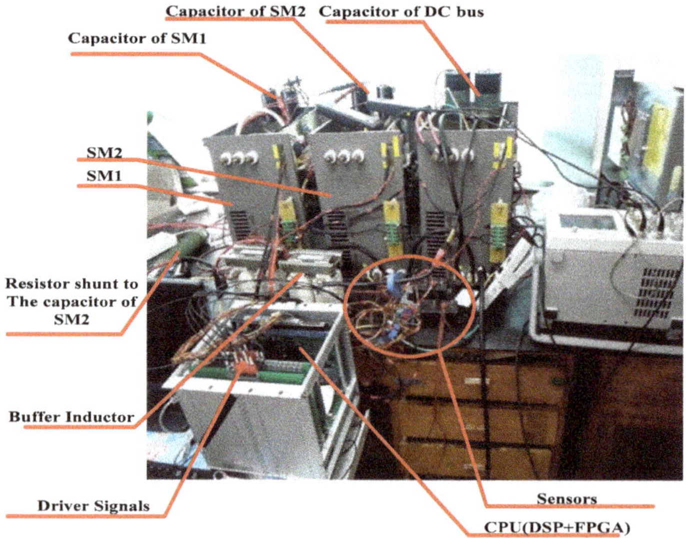

Figure 15. Photo of the modular multilevel converter (MMC)-based inverter for the experiment.

To test the system balance ability, a 100 Ω resistor was shunted to capacitor SM2. The topology worked in this unbalanced condition by appropriately setting the value of the weighting factor β, which is used to balance the capacitor voltage of the two SMs to zero. In this paper, we set the weighting factor α to a fixed value of 50, and set weighting factor β to 0 or 100 to compare the waveforms. Figure 16 shows the capacitor voltages of the two SMs. At first, the capacitor voltage of SM1 is lower than SM2, and the fluctuation is larger due to the unbalanced condition. After giving a suitable value to weighting factor β, each cell capacitor voltage is well regulated to their reference value and the fluctuation of the two cells is also the same.

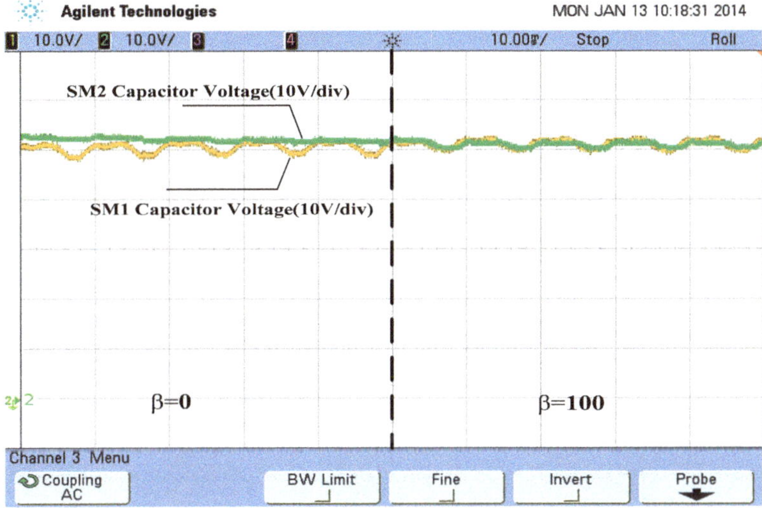

Figure 16. weighting factor β's effects on the capacitor voltages.

Figures 17 and 18 show the output current and voltage of this topology, which are both measured during balanced and unbalanced operation. From Figure 18 we can see that the voltage ripple of the two cell capacitors does not affect the current, apparently due to the robustness of the predictive control. When the weighting factor β is set to 100, meaning that the capacitor voltage balance is considered as a control goal, only a slight distortion is introduced into the output current.

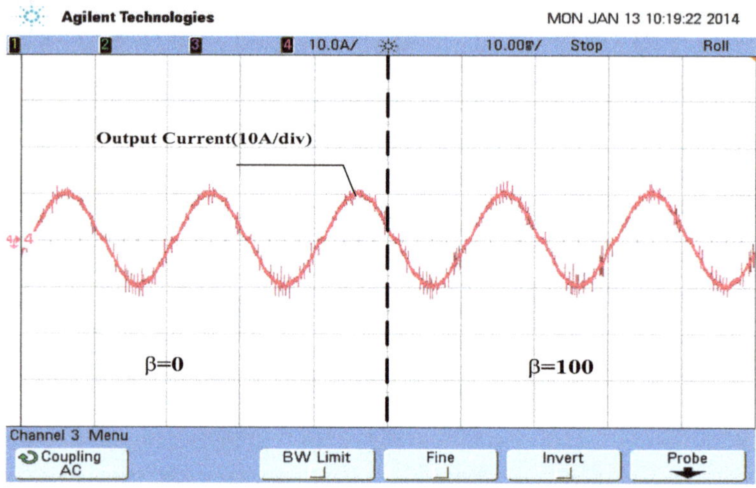

Figure 17. The waveform of the output current under balanced and unbalanced conditions.

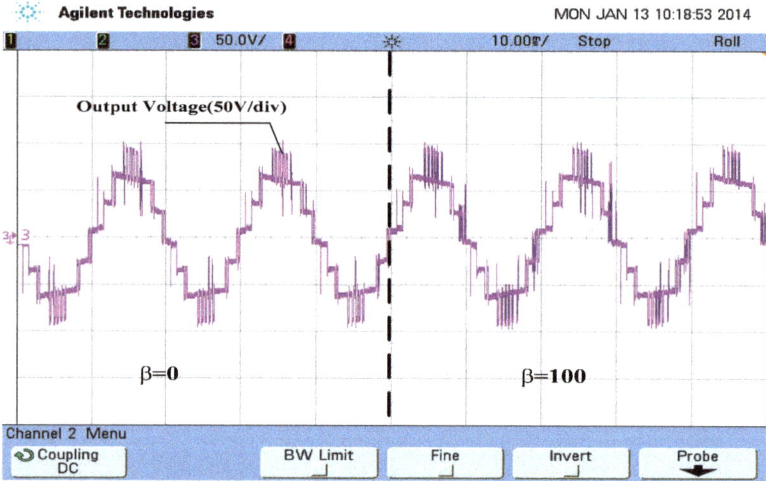

Figure 18. The waveform of the output voltage under balanced and unbalanced conditions.

To test the dynamic performance, the behavior of the FA-MMC and corresponding control method for a step in the angle of the reference current is shown in Figure 19. The waveforms show that the voltage changed quickly to drive the current to its new reference value and that the current is well-tracked. Figure 19 also shows that the dynamic capacitor voltage waveform is not influenced by the step in the angle of the reference current. Figure 20 shows a detailed view of the output voltage and current for a step in the angle of the reference current. The reference tracking of the proposed method that considered the possible switching states adjacent to $V_{AC}(k + 1)$ is fast, because extreme voltage changes are possible. The results are similar to simulation results.

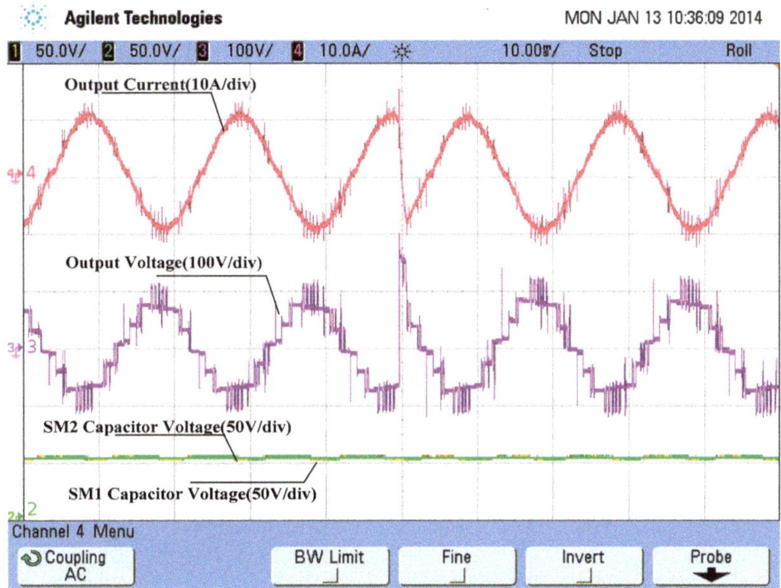

Figure 19. Waveform of the voltage and current for a step in the angle of the reference current.

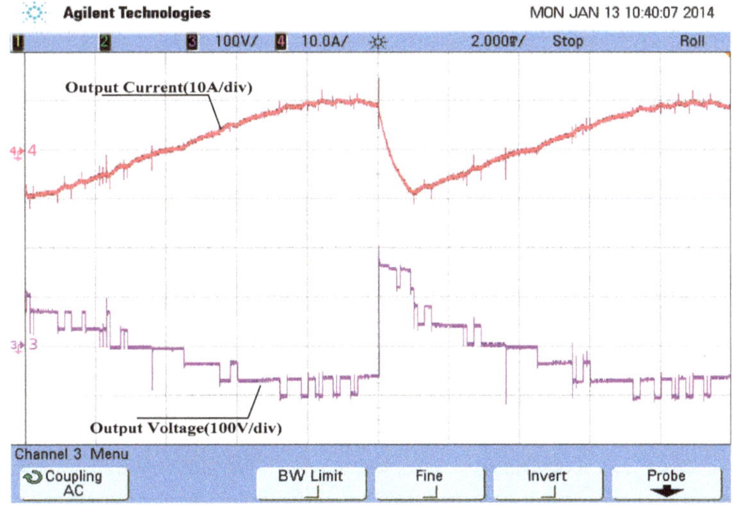

Figure 20. Detail of the output voltage and current for a step in the angle of the reference current.

6. Characteristic Analysis and Comparison with Other Topologies

6.1. DC Fault Blocking Capacity

When a DC-side short-circuit happens, with all IGBTs turned off, the director switches and stack of H-bridges behave as a number of uncontrolled diodes connecting with all DC capacitors in the cells connected in series, as shown in Figure 21. The equivalent capacitor value in Figure 21 can be expressed as

$$C_e = C / n_{cell\_FA} \tag{21}$$

where C is the capacitance of the capacitor in each cell and ncell_FA is the number of cells in one phase of the FA-MMC. The AC source charges the equivalent capacitor and inductors (including the L_s, L_b, L_d) through the DC fault current, thus, limiting the rising rate of the fault current. Consequently, the value of U_{ce} will rise rapidly, and the DC fault current will be blocked.

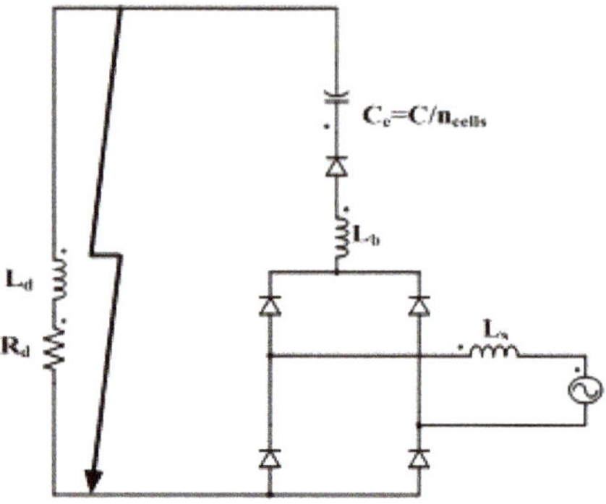

Figure 21. The equivalent circuit of the insulated-gate bipolar transistors (IGBTs) blocking when a DC fault occurs.

6.2. Number of Sub-module and IGBTs

Equation (3) shows that the DC bus voltage is lower than the peak value of output AC voltage by 27% in FA-MMC topology. This implies that the voltage rating of the director switches should be at least equal to the peak value of output AC voltage since they have to support higher voltages.

Assuming that the maximum allowable working voltage rating of the IGBTs is equal to the voltage rating of the DC capacitors in the sub-modules, the number of sub-modules of the proposed FA-MMC is given by

$$n_{cell\_FA} = \frac{\hat{U}_{AC}}{U_{RATED}} \tag{22}$$

where U_{RATED} is the voltage rating of the IGBTs. The number of IGBTs of each phase of the FA-MMC is given by

$$n_{IGBT1} = 4n_{cell\_FA} + 4n_S = \frac{8\hat{U}_{AC}}{U_{RATED}} \tag{23}$$

where n_S, the number of IGBT in S_1–S_4, is given by

$$n_S = \frac{\hat{V}_{MN}}{U_{RATED}} = \frac{\hat{U}_{AC}}{U_{RATED}} \tag{24}$$

Given the same AC output voltage, we also can deduce the DC voltage, number of cells, and IGBTs needed in an AAMC. The relationship between DC and AC voltage magnitudes in an AAMC, which has been derived in [22], can be expressed as

$$V_{dc} = \frac{\pi}{2}\hat{V}_{AC} \tag{25}$$

It can be seen from Equation (3) and (25) that the FA-MMC can reduce the DC bus voltage by half with the same AC output voltage and same active/reactive power flow compared with an AAMC.

Considering the sum of the sub-module capacitor voltage must be greater than the peak value of the line-to-line voltage to achieve DC current blocking capability, the number of sub-modules of an AAMC can be expressed as

$$n_{cell\_AA} = \frac{\sqrt{3}\hat{U}_{AC}}{U_{RATED}} \tag{26}$$

But, it only has two direct switches, so considering Equation (24) and (26), the number of IGBTs of a single-phase AAMC is given by

$$n_{IGBT2} = 4n_{cell\_AA} + 2n_S = \frac{(4\sqrt{3}+2)\hat{U}_{AC}}{U_{RATED}} \tag{27}$$

To summarize, the number of IGBTs of the FA-MMC is less than that of the AAMC, and the DC voltage and number of sub-modules of the FA-MMC is nearly half those of the AAMC, leading to smaller size, less need for insulation, and lower cost. And the comparison results of the number of IGBTs and sub-module capacitors between FA-MMC, AAMC and various MMC topologies mentioned in the introduction is shown in Table 4.

Table 4. Number of semiconductor devices and sub-module capacitors.

Topology	Number of Sub-Module Capacitor	Number of IGBTs	Number of Diodes
H-MMC	N	2 N	0
F-MMC	N	4 N	0
CH-MMC	N	2.35 N	0.7 N
AAMC	0.34 N	1.8 N	0
FA-MMC	0.2 N	1.6 N	0

6.3. Efficiency Analysis

To evaluate the power losses of the FA-MMC and AAMC, a simple loss calculation method for module multilevel converter is adopted [25]. And the result is shown in Table 5. To summarize, the losses of FA2MC increases slightly compared with AAMC, but it is still significantly lower than other types of MMC topologies.

Table 5. Losses calculation results

Topology	Switching Losses	Conduction Losses	Total Losses
H-MMC	0.29%	0.82%	1.11%
F-MMC	0.29%	1.88%	2.18%
CH-MMC	0.29%	1.19%	1.48%
AAMC	0.16%	0.47%	0.63%
FA-MMC	0.16%	0.66%	0.82%

6.4. Comprehensive Comparison with Other Topological Structures

According to the above analysis, a comprehensive comparison between the full-bridge MMC, half-bridge MMC, CH-MMC, A2MC, and FA-MMC is summarized in Table 6, where more "+" means the corresponding topology performs better in the corresponding characteristic. It can be seen in Table 6 that the FA-MMC has advantages in several aspects compared with the other topologies.

Table 6. Comprehensive comparison with other various topology

Topology	Economy	Efficiency	Volume	DC Fault Blocking Capacity	Demand for Insulation
H-MMC	+++	+++	+	+	+
F-MMC	+	+	+	++	+
CH-MMC	++	++	+	++	+
AAMC	++++	+++++	++	++	++
FA-MMC	+++++	++++	+++	++	+++

7. Conclusions

In this paper, a FA-MMC topology and its predictive control scheme have been proposed. The effectiveness of the proposed topology and proposed control strategy under various operating conditions are evaluated based on simulation studies in the PowerSIM environment and experiments, and the comparisons with other topological structures are also given. Through the analysis and demonstration mentioned above, the characteristics of this topology and its predictive control strategy are summarized as follows:

(1) The sub-module capacitor number of FA-MMC reduce significantly while connecting to the same AC voltage level and power level, results in a more compact structure;

(2) Further, it reduces the number of needed IGBTs while retaining the ability to block a DC-side fault compared with other topologies, so that the cost of the system is reduced;

(3) The algorithm the algorithm has been proved to be able to achieve multiple control objectives of FA-MMC simultaneously (i.e., capacitor voltages balancing and ac-side currents control). The developed control strategy also contains a director switch control function so that the director switch maintains operation in low-frequency and zero voltage switching mode.

Author Contributions: Conceptualization, J.Z.; methodology, T.W.; software, J.Y.; validation, J.Z., Q.H. and J.Y.; writing—original draft preparation, J.Z.

Funding: This work was supported by National Key R&D Program of China (2016YFB0900900), the National Natural Science Fund of China (No. 51607171).

Glossary of Terms

E_{AC}	The amount of energy transferred from AC side over half the fundamental period
E_{DC}	The amount of energy going to DC side over half the fundamental period.
V_{AC}	Output ac voltage
I_{AC}	AC current
U_{DC}	DC voltage of FA MMC
C_e	Equivalent capacitance while all capacitors in cells connected in series
U_{ce_int}	Initial voltage of Ce when dc fault blocking
φ	Angular position of AC current
n_{cell_FA}	Number of cells of FA-MMC each phase
U_{RATED}	The rated voltage of IGBT
n_{IGBT1}	The needed number of IGBTs of FA-MMC each phase
n	The needed number of IGBTs of S1–S4
n_{cell_AA}	The needed number of cells of AAMC each phase
n_{IGBT2}	The needed number of IGBTs of AAMC each phase
V_{dc}	DC voltage of AAMC while the output AC voltage is equal to FA MMC
V_{PB}	The voltage produced by the stack of H-bridge cells of FA MMC
S_i	Switching function of the ith cell
V_{ci}	The capacitor voltage of the ith cell

L_b	Buffer inductor
I_b	The current through Lb
C	The capacitor value of cell
S_d	Switching function of director switch
L_s	Filter inductor
I_s	The current through Ls
T_s	Sampling period
J_{all}, J_i, J_{vc}	Cost function
α, β	Weighting factor

References

1. Wang, Y.; Yang, B.; Zuo, H.; Liu, H.; Yan, H. A DC Short-Circuit Fault Ride Through Strategy of MMC-HVDC Based on the Cascaded Star Converter. *Energies* **2018**, *11*, 2079. [CrossRef]
2. Xiao, L.; Li, Y.; Xiao, H.; Zhang, Z.; Xu, Z. Electromechanical Transient Modeling of Line Commutated Converter-Modular Multilevel Converter-Based Hybrid Multi-Terminal High Voltage Direct Current Transmission Systems. *Energies* **2018**, *11*, 2102. [CrossRef]
3. Wu, Z.; Chu, J.; Gu, W.; Huang, Q.; Chen, L.; Yuan, X. Hybrid Modulated Model Predictive Control in a Modular Multilevel Converter for Multi-Terminal Direct Current Systems. *Energies* **2018**, *11*, 1861. [CrossRef]
4. Li, Z.; Tan, G. A Black Start Scheme Based on Modular Multilevel Control-High Voltage Direct Current. *Energies* **2018**, *11*, 1715. [CrossRef]
5. Pouresmaeil, E.; Mehrasa, M.; Rodrigues, E.; Godina, R.; Catalão, J.P.S. Control of Modular Multilevel Converters Under Loading Variations in Distributed Generation Applications. In Proceedings of the 2018 IEEE International Conference on Environment and Electrical Engineering and 2018 IEEE Industrial and Commercial Power Systems Europe (EEEIC/I CPS Europe), Palermo, Italy, 12–15 June 2018; pp. 1–6.
6. Hakimi, S.M.; Hajizadeh, A. Integration of Photovoltaic Power Units to Power Distribution System through Modular Multilevel Converter. *Energies* **2018**, *11*, 2753. [CrossRef]
7. Jafarishiadeh, M.; Ahmadi, R. Two-and-One Set of Arms MMC-Based Multilevel Converter with Reduced Submodule Counts. In Proceedings of the 2018 IEEE Transportation Electrification Conference and Expo (ITEC), Long Beach, CA, USA, 13–15 June 2018; pp. 779–782.
8. Yang, D.; Yin, L.; Xu, S.; Wu, N. Power and Voltage Control for Single-Phase Cascaded H-Bridge Multilevel Converters under Unbalanced Loads. *Energies* **2018**, *11*, 2435. [CrossRef]
9. Zhang, J.; Zhao, C. The Research of SM Topology with DC Fault Tolerance in MMC-HVDC. *IEEE Trans. Power Deliv.* **2015**, *30*, 1561–1568. [CrossRef]
10. Dong., S.; He, H.; Yuan, Z.; Yang, Y.; He, J.; He, H. Transient over-voltage of the full-bridge MMC HVDC system with overhead line fault. In Proceedings of the 2016 Power Systems Computation Conference, Chengdu, China, 19–22 September 2016; pp. 1–4.
11. Lin, W.; Jovcic, D.; Nguefeu, S.; Saad, H. Protection of full bridge MMC DC grid employing mechanical DC circuit breakers. In Proceedings of the 2017 IEEE Power Energy Society General Meeting, Chicago, IL, USA, 16–20 July 2017; pp. 1–5.
12. Zeng, R.; Xu, L.; Yao, L.; Williams, B.W. Design and Operation of a Hybrid Modular Multilevel Converter. *IEEE Trans. Power Electron.* **2015**, *30*, 1137–1146. [CrossRef]
13. Li, R.; Adam, G.P.; Holliday, D.; Fletcher, J.E.; Williams, B.W. Hybrid Cascaded Modular Multilevel Converter with DC Fault Ride-Through Capability for the HVDC Transmission System. *IEEE Trans. Power Deliv.* **2015**, *30*, 1853–1862. [CrossRef]
14. Marquardt, R. Modular Multilevel Converter: An universal concept for HVDC-Networks and extended DC-Bus-applications. In Proceedings of the 2010 International Power Electronics Conference, Sapporo, Japan, 21–24 June 2010; pp. 502–507.
15. Marquardt, R. Modular Multilevel Converter topologies with DC-Short circuit current limitation. In Proceedings of the 8th International Conference on Power Electronics, Jeju, South Korea, 30 May–3 June 2011; pp. 1425–1431.

16. Qin, J.; Saeedifard, M.; Rockhill, A.; Zhou, R. Hybrid Design of Modular Multilevel Converters for HVDC Systems Based on Various Submodule Circuits. *IEEE Trans. Power Deliv.* **2015**, *30*, 385–394. [CrossRef]
17. Li, R.; Fletcher, J.E.; Xu, L. A Hybrid Modular Multilevel Converter with Novel Three-level Cells for DC Fault Blocking Capability. *IEEE Trans. Power Deliv.* **2015**, *30*, 2017–2026. [CrossRef]
18. Zeng, R.; Xu, L.; Yao, L. An improved modular multilevel converter with DC fault blocking capability. In Proceedings of the 2014 IEEE PES General Meeting, National Harbor, MD, USA, 27–31 July 2014; pp. 1–5.
19. Xu, J.; Zhao, P.; Zhao, C. Reliability Analysis and Redundancy Configuration of MMC with Hybrid Sub-Module Topologies. *IEEE Trans. Power Electron.* **2016**, *31*, 2020–2029. [CrossRef]
20. Davidson, C.C.; Trainer, D.R. Innovative concepts for hybrid multi-level converters for HVDC power transmission. In Proceedings of the 9th IET International Conference an AC and DC Power Transmission, London, UK, 19–21 October 2010; pp. 1–5.
21. Cross, A.M.; Trainer, D.R.; Crookes, R.W. Chain-link based HVDC Voltage Source Converter using current injection. In Proceedings of the 9th IET International Conference an AC and DC Power Transmission, London, UK, 19–21 October 2010; pp. 1–5.
22. Merlin, M.M.C.; Green, T.C.; Mitcheson, P.D.; Trainer, D.R.; Critchley, D.R.; Crookes, R.W. A new hybrid multi-level Voltage-Source Converter with DC fault blocking capability. In Proceedings of the 9th IET International Conference an AC and DC Power Transmission, London, UK, 19–21 October 2010; pp. 1–5.
23. Lee, H.Y.; Asif, M.; Park, K.H.; Lee, B.W. Assessment of Appropriate MMC Topology Considering DC Fault Handling Performance of Fault Protection Devices. *Appl. Sci.* **2018**, *8*, 1834. [CrossRef]
24. Cortes, P.; Kouro, S.; Rocca, B.L.; Vargas, R.; Rodriguez, J.; Leon, J.I.; Vazquez, S.; Franquelo, L.G. Guidelines for weighting factors design in Model Predictive Control of power converters and drives. In Proceedings of the 2009 IEEE International Conference on Industrial Technology, Gippsland, Australia, 10–13 February 2009; pp. 1–7.
25. Zhang, Z.; Xu, Z.; Xue, Y. Valve Losses Evaluation Based on Piecewise Analytical Method for MMC–HVDC Links. *IEEE Trans. Power Deliv.* **2014**, *29*, 1354–1362. [CrossRef]

Article

A Harmonic Voltage Injection Based DC-Link Imbalance Compensation Technique for Single-Phase Three-Level Neutral-Point-Clamped (NPC) Inverters

Kyoung-Pil Kang [1], Younghoon Cho [1,*], Myung-Hyo Ryu [2] and Ju-Won Baek [2]

[1]   Department of Electrical Engineering, Konkuk University, Seoul 05029, Korea; kkpsp@konkuk.ac.kr
[2]   Industry Application Research Laboratory, Korea Electrotechnology Research Institute,
      Changwon 123456, Korea; mhryu@keri.re.kr (M.-H.R.); jwbaek@keri.re.kr (J.-W.B.)
*   Correspondence: yhcho98@konkuk.ac.kr; Tel.: +82-10-6207-0431

Received: 15 May 2018; Accepted: 11 July 2018; Published: 19 July 2018

Abstract: In three-level neutral-point-clamped (NPC) inverters, the voltage imbalance problem between the upper and lower dc-link capacitors is one of the major concerns. This paper proposed a dc-link capacitor voltage balancing method where a common offset voltage was injected. The offset voltage consists of harmonic components and a voltage difference between the upper and the lower capacitors. Here, both the second-order harmonics and the half-wave of the second-order component were injected to compensate for the unbalanced voltage between the capacitors. In order to show the effectiveness of the proposed voltage injection, the theoretical analyses, simulations, and experimental results are provided. Since the proposed method does not require any hardware modifications, it can be easily adapted. Both the simulations and the experiments validated that the voltage difference of the dc-link could be effectively reduced with the proposed method.

Keywords: neutral-point-clamped (NPC) inverter; dc-link capacitor voltage balance; offset voltage injection; harmonic component

---

1. Introduction

Recently, multilevel power inverters have been popularly employed in many electronic applications [1,2]. For example, solid-state transformers (SST) and dc distribution systems, which are high voltage (HV) or medium voltage (MV) applications, essentially require the use of multilevel topologies [3–7]. In multilevel topologies, three-level neutral-point-clamped (NPC) inverters have been widely used in MV and HV applications. Compared to two-level inverters, three-level NPC inverters have some advantages, as follows. NPC inverters have more output voltage levels than two-level inverters. Therefore, the output voltages of an NPC inverter are more similar to sinusoidal waves than other topologies and NPC inverters have less of a harmonic component on output voltage. Additionally, in NPC inverters, the voltage rating of the switching device can be half of the one used in two-level inverters. In addition, NPC inverters generate relatively less leakage current flowing through the ground paths, so electromagnetic interference (EMI) induced problems are relatively lower than the two-level inverters.

However, the NPC inverter has a major drawback associated with the neutral-point voltage located between the upper and the lower dc-link capacitors. The voltage between the positive dc-link rail and the neutral-point should be identical to the voltage across the neutral-point and the negative dc-link rail. Unfortunately, there is a voltage imbalance between the upper and lower capacitors. This voltage imbalance harms the stability of the system, and limits the switching operation of the power stage [8–11]. In order to mitigate the voltage imbalance, many strategies that are based on additional hardware configurations or control algorithms have been proposed, and have been successfully adapted in some applications [12–25]. In [12,13], additional circuits for dc-link balancing were proposed. Although these methods achieved the dc-link voltage balancing successfully, the increase in the cost and the losses were major defects. To avoid these disadvantages, several modulation techniques for single-phase three-level NPC inverters have been presented in [14–21]. Among these modulation techniques, the carrier-based pulse width modulation (CB-PWM) approaches have been extensively preferred due to their simplicity of implementation. In [14], the offset voltage injection with the zero-sequence component in the reference voltage was presented. The zero-sequence component is calculated at every switching period based on the dc-link link voltage and the grid current. Another type of offset voltage injection method was discussed in [15]. In this paper, the offset voltage with a distribution factor was added into the modulation signal. However, these strategies face difficulties in being implemented because they are a burden on the prediction of the line current and the avoidance of nonlinearity in the injection signal. Additionally, the exact parameter information is essential to implement these methods as the algorithms are highly dependent on the system parameters. In [16], a simple signal injection method was proposed to balance out the dc-link capacitor voltages by utilizing the harmonic signal consisting of the dc-link voltage difference and the double frequency of the utility grid. The method can easily be implemented as well as reducing the harmonic distortion in the input current of the NPC inverter.

In this paper, the method proposed in [16] was further extended and detailed. In the proposed method, an even harmonic signal was added to the reference signal, which is generated by the current controller. Compared to other harmonic injection methods, the proposed method showed less voltage distortion on the synthesized output voltage. Furthermore, fast voltage balancing performance was obtained with the proposed strategy. A 10-kW single-phase three-level NPC inverter was built and tested. Here, the input grid voltage was 943 V in root mean square (RMS) and the output dc-link voltage was 1.8 kV. To artificially create voltage imbalance conditions, an unbalanced load bank was attached to the individual capacitors in the dc-link. The proposed method was compared with the method suggested in [14] through simulations. The experimental results are presented to validate the effectiveness of the proposed method. This paper is organized as follows. In Section 2, the pole voltage of the NPC inverter is analyzed with the proposed offset voltage injection method. The theoretical analysis of control performance with the offset voltage is discussed in Section 3. Simulations and experimental results with the proposed method are shown in Section 4. Finally, Section 5 concludes this paper.

2. The Operation of the Single-Phase NPC Inverter and Its Neutral Current

Figure 1 illustrates a switching leg of the three-level NPC inverter and its conduction states. As shown in Figure 1, the switching leg consists of four switching devices, Q_{x1}, Q_{x2}, Q_{x3}, and Q_{x4}, two clamping diodes, D_1 and D_2, and two dc-link capacitors, C_{CH} and C_{CL}. The pole voltage v_{x0} has three different levels, V_{CH}, 0, and $-V_{CL}$ according to the values of the switch function S_x during the conduction periods, as shown in Figure 1b–d. All parameters used in this paper are defined in Table 1.

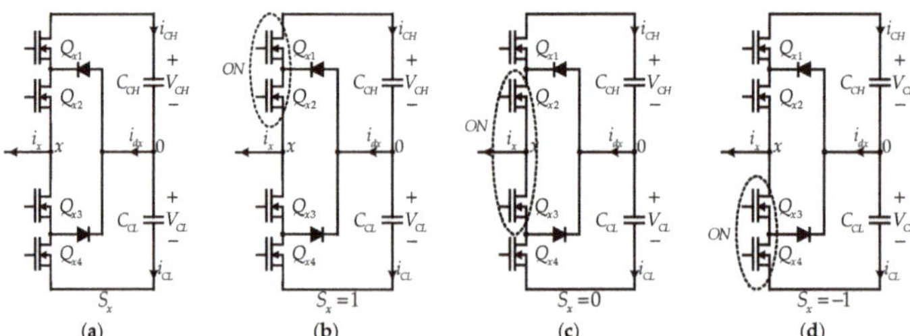

Figure 1. The switching leg of the three-level neutral-point-clamped (NPC) inverter and its switching states. (**a**) The circuit structure; (**b**) the conduction state with $S_x = 1$; (**c**) the conduction state with $S_x = 0$; and (**d**) the conduction state with $S_x = -1$.

Table 1. Nomenclature of the hardware and controller parameters.

Parameters	Description	Parameters	Description
Q_{xj}	Power switch "j" in leg "x".	$\bar{v}_{x0}$	Average pole voltage.
D_x	Clamped diode in leg "x".	$v_Z{}^*$	Injection voltage reference.
i_x	Instantaneous current from leg "x" to grid.	$V_C{}^+; V_C{}^-$	Triangular carrier signals, a positive ($V_C{}^+$); and negative ($V_C{}^-$) one.
C_{CH}, C_{CL}	Individuals capacitances of dc-link capacitors, the upper (CH); and the lower (CL) one.	V_{CH}, V_{CL}	Individuals capacitor voltages of dc-link capacitors, the upper (CH); and the lower (CL) one.
$R_o; R_{add}$	Resistive output load; and additional resistive load.	q_{add}	Additional switch to control dc-link capacitor unbalance circuit.
e_g	Instantaneous voltage of the grid utility.	$i_g; i_g{}^*$	Instantaneous phase current of NPC inverter; and its reference value.
$V_{DC}; V_{DC}{}^*$	dc-link capacitor voltage and its reference value.	$v_g; v_g{}^*$	Instantaneous phase voltage of NPC inverter; and its reference value.
i_{dx}	Instantaneous current of clamped diode.	$u_{x0}{}^*; \bar{u}_{x0}{}^*$	Reference signal of leg "x" and its average value.
$\bar{i}_d$	Average current of clamped diode.	K	Coefficient of injection voltage
L_g	Input inductance of NPC inverter.	ω	Angular frequency of phase voltage.
τ_x	Pulse width of leg "x".	δ_g	Phase angle of grid voltage.
T_c	One switching period.	m	Modulation index.
S_x	Switch conduction state of leg "x".	f_{sw}	Switching frequency of NPC inverter.
$v_{x0}; v_{x0}{}^*$	Instantaneous pole voltage of leg "x" and its reference value.	$u_{offset}{}^*$	Offset signal for dc-link balancing control.

When $S_x = 1$, the upper two devices, Q_{x1} and Q_{x2}, are turned on. At this condition, shown in Figure 1b, the output power is supplied by the upper dc-link capacitor C_{CH}, so that v_{x0} becomes V_{CH}. If S_x is 0, the middle switches, Q_{x2} and Q_{x3}, conduct, and the output voltage is also clamped by the clamping diodes. Here, the amplitude of v_{x0} is 0, and this state is illustrated in Figure 1c. In Figure 1d, S_x is defined as -1, and the lower switching devices, Q_{x3} and Q_{x4}, are closed. The output voltage is fed by the lower dc-link capacitor C_{CL}, and v_{x0} becomes $-V_{CL}$.

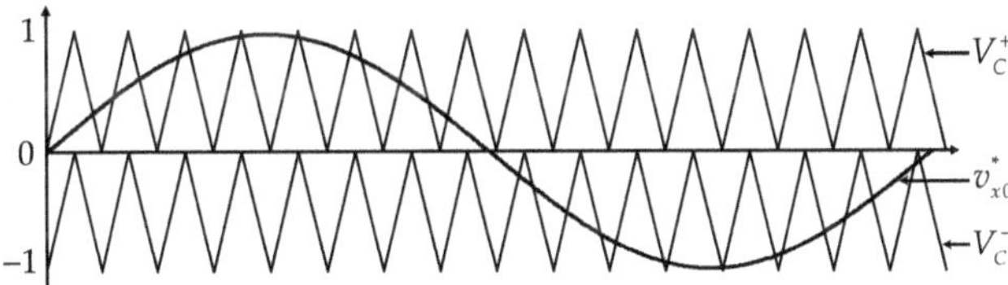

Figure 2. The normalized pole voltage reference and the carrier signals.

Figure 2 shows the normalized pole voltage reference u_{x0}^* and two carrier signals V_C^+ and V_C^-. Here, u_{x0}^* is defined as Equation (1):

$$u_{x0}^* = \frac{2v_{x0}^*}{V_{CH} + V_{CL}} \tag{1}$$

where v_{x0}^* is the reference of the pole voltage. Equation (2) defines the relationships between the magnitudes of the carrier signals and the normalized pole voltage reference and the values of the switching function.

$$\text{if} \begin{cases} V_C^+ \leq u_{x0}^* \\ V_C^- \leq u_{x0}^* \leq V_C^+ \\ u_{x0}^* \leq V_C^- \end{cases}, \text{ then } \begin{cases} S_x = 1 \\ S_x = 0 \\ S_x = -1 \end{cases} \tag{2}$$

Let us consider that the switching frequency f_{sw} is much higher than the frequency of u_{x0}^*. By doing so, u_{x0}^* in the single switching period T_c can be assumed as a constant value. Consequently, the on-time duration of the switch, τ_x, is given as:

$$\tau_x = |u_{x0}^*| T_c \tag{3}$$

By using Equation (3), the average output pole voltage, $\bar{v}_{x0}$, over one switching period is simply written as follows:

$$\bar{v}_{x0} = \begin{cases} \tau_x V_{CH}/T_c, & u_{x0}^* \geq 0 \\ -\tau_x V_{CL}/T_c, & u_{x0}^* \leq 0 \end{cases} \tag{4}$$

If the upper and the lower capacitors have the same voltage, $V_{CH} = V_{CL} = 0.5V_{DC}$, the pole voltage shown in Equation (4) can be rewritten as:

$$\bar{v}_{x0} = \frac{\tau_x}{T_c} \frac{V_{DC}}{2} \tag{5}$$

In order to be placed in the linear modulation range, the following conditions should be satisfied.

$$|u_{x0}^*| \leq 1, \ |v_{x0}| \leq \frac{V_{DC}}{2} \tag{6}$$

Figure 3 represents the configuration of the single-phase three-level NPC inverter dealt with in this paper. Here, two three-level switching legs were employed. In each switching leg, the middle points of the clamping diodes were connected to the neutral point of the dc-link. By referring the notations in Figure 3, the pole voltage references of the individual switching legs are written as follows:

$$v_{A0}^* = v_g^* + v_z^* \tag{7}$$

$$v_{B0}^* = v_z^* \tag{8}$$

where v_g^* and v_z^* are the line-to-line voltage reference and the virtual offset voltage between the switching pole B and the neutral point of the dc-link, respectively. It should be noticed that both v_{A0}^* and v_{B0}^* should be operated in the linear modulation region. This means that they should satisfy the conditions expressed in Equation (6). In addition, v_z^* should be also placed in the linear modulation region, because v_z^* is identical to v_{B0}^*. From this analysis, the following condition can be derived:

$$-0.5V_{DC} - \min(v_g^*, 0) \leq v_z^* \leq 0.5V_{DC} - \max(v_g^*, 0) \tag{9}$$

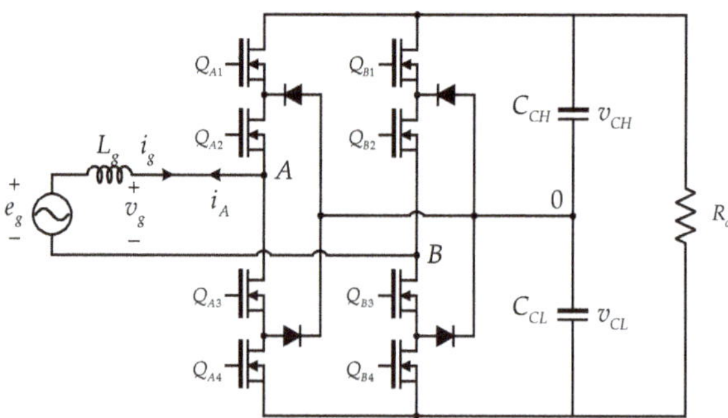

Figure 3. The single-phase NPC inverter topology.

Equation (9) offers that various virtual offset voltages can be selected with various control purposes. The object of the approach taken in this paper was to balance the dc-link capacitor voltages. To do this, the second-order harmonic injection approach was proposed in this paper. Here, v_z^* is selected as Equation (10):

$$v_z^* = -\frac{1}{2}v_g^* + K(V_{CH} - V_{CL})\sin{(2\omega t)} \tag{10}$$

where ω and K are the fundamental electrical angular frequency of the grid voltage e_g and the injection gain of the second harmonic voltage, respectively. By substituting Equation (10) into Equations (7) and (8), the pole voltages are expressed as follows:

$$v_{A0}^* = \frac{1}{2}v_g^* + K(V_{CH} - V_{CL})\sin{(2\omega t)} \tag{11}$$

$$v_{B0}^* = \frac{1}{2}v_g^* + K(V_{CH} - V_{CL})\sin{(2\omega t)} \tag{12}$$

In Figure 3, the voltage of the lower capacitor was adjusted by injecting the neutral current i_{dA} and i_{dB}, which are represented with the phase current from the switching pole A to the grid i_A and the switching functions of each switching leg, S_{A0} and S_{B0}, as follows.

$$i_{dA}(t) = [1 - S_A^2(t)]i_A(t) \tag{13}$$

$$i_{dB}(t) = -[1 - S_B^2(t)]i_A(t) \tag{14}$$

The entire neutral current flowing into the neutral point is simply obtained as:

$$i_d(t) = i_{dA}(t) + i_{dB}(t) = [S_B^2(t) - S_A^2(t)]i_A(t) \tag{15}$$

The average value of the neutral current over a single switching period is calculated as below:

$$\overline{i_d} = \frac{1}{T_c}\int_{T_c} i_d(t)dt = (|u_{B0}^*| - |u_{A0}^*|)\overline{i_A} \tag{16}$$

where $\overline{i_A}$ is the average of $i_A(t)$ in the switching period. By applying the pole voltages, Equation (16) is rewritten as Equation (17).

$$\overline{i_d} = \frac{2}{V_{DC}}(|v_{B0}^*| - |v_{A0}^*|)\overline{i_A} \tag{17}$$

By substituting Equations (11) and (12) into Equation (17), $\overline{i_d}$ is obtained, and is expressed in two ways according to the polarities of v_{A0}^* and v_{B0}^*. When the polarities of the pole voltage references are the same, the average neutral current is written as Equation (18).

$$\overline{i_d} = \mp\frac{2v_g^*}{V_{DC}} \tag{18}$$

If the polarities of v_{A0}^* and v_{B0}^* are different, Equation (19) is obtained.

$$\overline{i_d} = \mp\frac{4K}{V_{DC}}\sin(2\omega t) \tag{19}$$

By adjusting the pole voltage references, the average current expressed by Equations (18) or (19) is controlled to balance out the dc-link capacitors' voltages.

3. Analysis of the Injected Offset Voltage

This section compares the proposed second-order harmonic injection method above-mentioned with the partially rectified wave injection method. Figure 4 shows the pole voltage range which consisted of $0.5V_{DC} - \max(v_g{}^*,0)$ to $-0.5V_{DC} - \min(v_g{}^*,0)$.

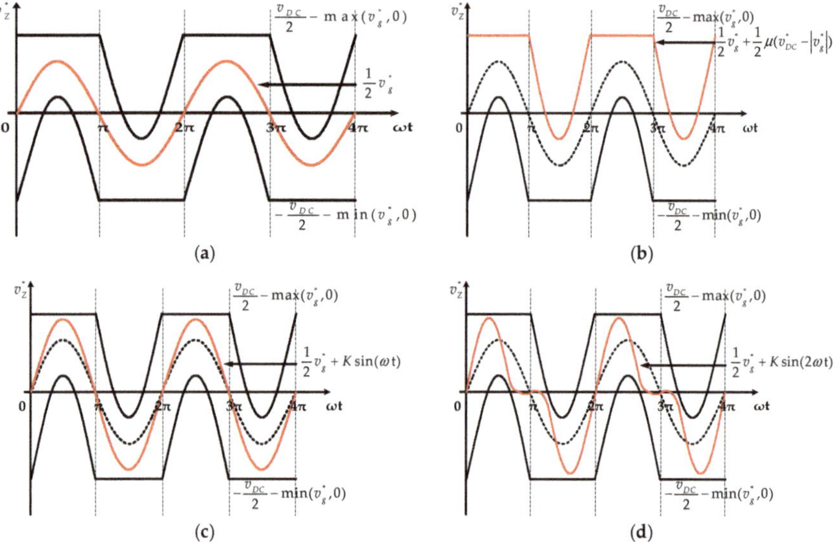

Figure 4. Limitation pole voltage $v_{A0}{}^*$ with the offset voltage $v_z{}^*$. (**a**) $v_z{}^* = 0$; (**b**) using distribution factor μ; (**c**) fundamental component; and (**d**) second-order harmonic component.

Figure 4a shows the pole voltage reference when there was no difference between upper and lower capacitor voltage in Equation (8). In this case, the pole voltage reference always satisfies the range of the pole voltage (Equation (9)). Figure 4b shows the synthesized pole voltage reference where this method was proposed in [15]. In this case, the offset voltage was composed of distribution factor μ and the absolute value of the phase voltage reference. This method can cover the full range of dc-link capacitor voltage difference, but this offset signal injection method requires repetitive calculation since the absolute values are used in Equation (17). Figure 4c shows the synthesized pole voltage where the offset voltage composed of the dc-link capacitor voltage difference and the fundamental component was injected instead of the second-order harmonic component in the pole voltage reference (Equation (9)). In this case, the maximum value of the pole voltage reference was $0.5\,mV_{DC} + K$ at $\pi/2$, and the minimum value of the pole voltage reference was $-0.5\,mV_{DC} - K$ at $3\pi/2$. Therefore, the voltage difference that can be injected for the dc-link capacitor voltage balance is restricted by modulation index m. Figure 4d shows the pole voltage where the offset voltage is composed of the dc-link capacitor voltage difference and the second-order harmonic component. In this case, the maximum value can be found by calculating the divergence of the pole voltage $v_{A0}{}^*$. Solving Equation (20) equal to zero, the roots are as follows:

$$\frac{dv^*_{A0}}{d\omega t} = \frac{1}{2}mV_{DC}\cos(\omega t) + 2K\cos(2\omega t) \tag{20}$$

$$
\omega t = \begin{cases}
\pi - \mathrm{acos}\left(\left(m - \sqrt{(32K^2 + m^2)} \right)/8K \right) \\
\pi + \mathrm{acos}\left(\left(m + \sqrt{(32K^2 + m^2)} \right)/8K \right) \\
\pi + \mathrm{acos}\left(\left(m - \sqrt{(32K^2 + m^2)} \right)/8K \right) \\
\pi - \mathrm{acos}\left(\left(m + \sqrt{(32K^2 + m^2)} \right)/8K \right)
\end{cases}
\tag{21}
$$

Among these roots, the maximum value and the minimum value exist at:

$$
\begin{aligned}
\omega t_{\min} &= \pi + \mathrm{acos}\left(\left(m + \sqrt{(32K^2 + m^2)} \right)/8K \right) \\
\omega t_{\max} &= \pi - \mathrm{acos}\left(\left(m + \sqrt{(32K^2 + m^2)} \right)/8K \right)
\end{aligned}
\tag{22}
$$

The other roots are imaginary roots when the value of K is below the specific value determined by the modulation index m. In addition, the offset voltage $v_z{}^*$ can be considered, which is composed with the voltage difference K and the half-wave rectified by the second-order harmonic component. In this case, the maximum value of the pole voltage $v_{A0}{}^*$ is the same as the condition that injected the second-order harmonic component, but the minimum value was the same as the condition where the dc-link voltage difference is zero. The reason each injection voltage was inserted in a subdivided way as follows: for the 0 to $\pi/2$ region, the voltage difference was reduced, but within the next $\pi/2$ to π region, the offset voltage signal made switching operations for each leg to diverge. In this region, the switching state of leg A was increased to a 0 state and the switching state of leg B was increased to a -1 state by a synthesized reference signal. Consequently, a new reference signal increase the voltage difference, which turned on more low-side switches than the high-side ones. Therefore, the injection voltage had to be inserted in a subdivided way.

Figure 5 shows the waveforms which are reference signal for legs A, B, and the offset signal that has double the frequency of the reference signal. In addition, it also shows that the switching operation changed every $\pi/2$ cycle, when the reference signal and offset signal were synthesized. In Figure 5a, Region 1, the switching operation for leg A was increased to a 1 state by the synthesized reference signal with the offset signal ($S_A = 1$). On the other hand, the switching operation for leg B was increased to the 0 state ($S_B = 0$). In Region 2, the offset signal had a negative value. Therefore, the switching operation for leg A was increased to the 0 state ($S_A = 0$). In the same manner, the switching operation for leg B was increased to the -1 state ($S_B = -1$). In Region 3, the reference signal had a negative value and the offset signal had a positive value. The switching operation for leg A was increased to the 0 state ($S_A = 0$), and for leg B it was increased to the 1 state ($S_B = 1$). In Region 4, the switching operation for leg A was increased to the -1 state ($S_A = -1$) and for leg B, it was increased to the 0 state ($S_B = 0$). In Figure 5b, the switching operation states for leg A and leg B could be easily observed by comparing it with Figure 5a. In Regions 1 and 3, the increased switching state for each leg, (S_A,S_B), were (1,0) and (0,1), respectively. In Regions 2 and 4, there was no offset signal, so the synthesized reference signals were the same as the reference signal.

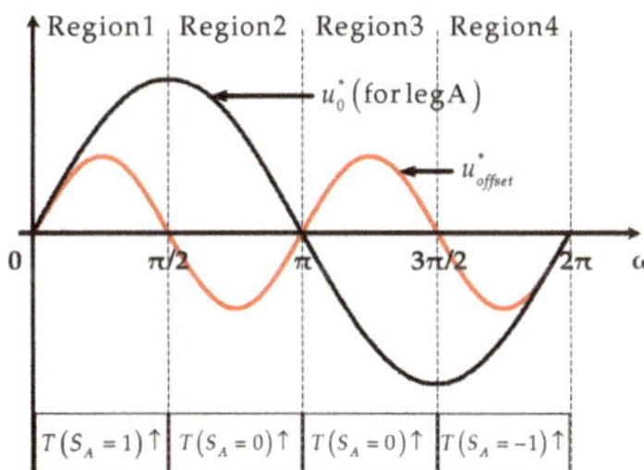

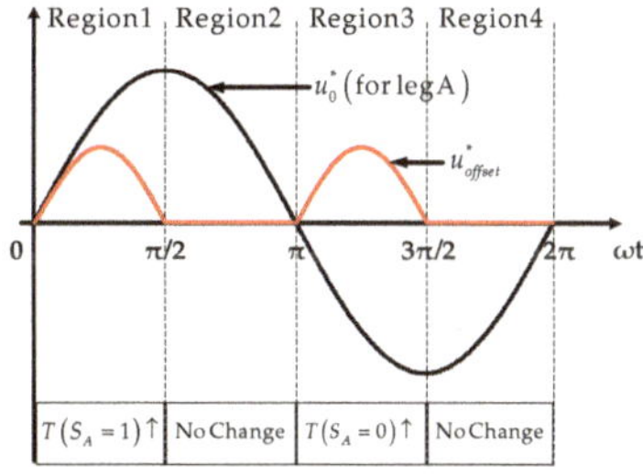

Figure 5. *Cont.*

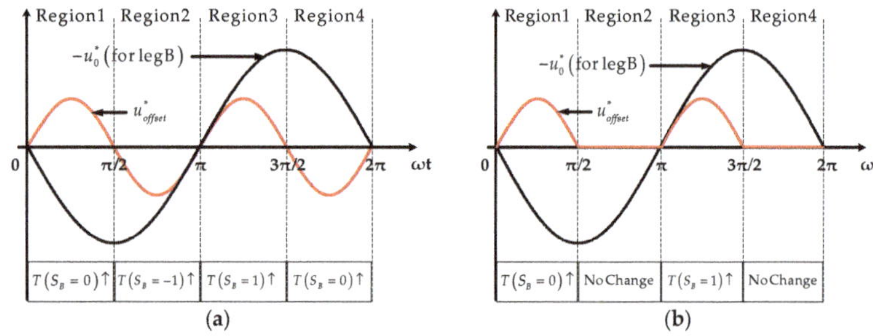

Figure 5. Waveforms pole voltage reference signal $u_{x0}*$ and offset signal $u_{offset}*$. (**a**) Second-order harmonic signal; and (**b**) half-wave signal of second-order harmonic.

The maximum and minimum values obtained from the previous equations were determined by the voltage difference K and modulation index m. With the same modulation index m, the offset voltage injection method with distribution factor μ could be adopted for the full range of capacitor voltage difference such as from 0 V to V_{DC}, but there were voltage oscillations on the dc-link voltage when the balancing control was adopted [15]. In contrast, the offset voltage injection method with the second-order harmonic component or the half-wave rectified can be adopted for smooth control. Furthermore, the proposed method could easily configure the controller using Equations (11), (12), and (22), and the PWM modulator described in Figure 6. However, the proposed method has limitations on the range of capacitor voltage difference given the effects of the voltage difference of the dc-link capacitor on the injection signal. This could be a larger reference signal than the previous reference signal for dc-link voltage control. In this case, an over-distorted reference signal could not control either the dc-link voltage control or dc-link capacitor voltage balancing control. The voltage range of the proposed method is up to the point where the remaining two roots other than the maximum and minimum are zero when the modulation index m is constant in Equation (21).

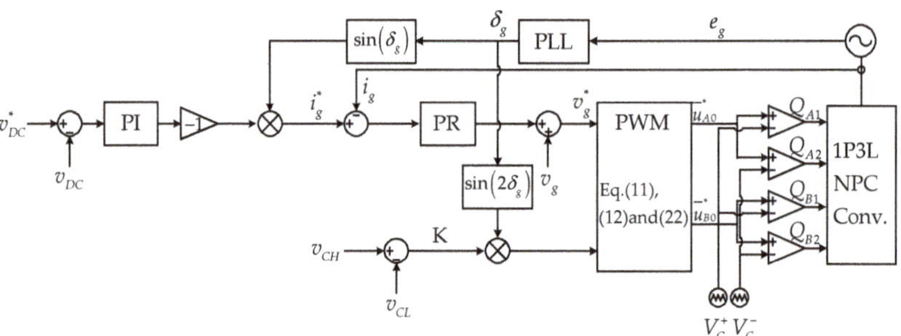

Figure 6. Control block diagram for capacitor voltage balance.

4. Simulation and Experimental Results

Simulations and experiments for the single-phase three-level NPC inverter were performed to verify the effectiveness of the proposed half-wave of the even-harmonics voltage injection method. In addition, these simulations and experiments were done in a single-phase NPC circuit structure with an additional resistive load circuit on top of the dc-link capacitor, as shown in Figure 2. The proposed dc-link capacitor voltage balancing control and the conventional method were carried out in a PSIM simulation, and the simulation scheme is shown in Figure 7. The controller contained three blocks to generate the modulation signal: the PI controller for dc-link capacitor voltage control, PR controller for input current control, and the proposed capacitor voltage balancing controller. The control sequence for the dc-link capacitor voltage balance is as follows. The NPC inverter operates as a general PWM rectifier that traces the dc-link capacitor voltage reference with the PI and PR controller before the capacitor voltage imbalance occurs. In this case, the offset signal $v_z{}^*$ in Equations (10) and (11) is zero because there is no voltage difference on the dc-link capacitor, so the coefficient of $v_z{}^*$, K, is zero. When a voltage difference occurs, the balancing control operates to reduce the voltage difference. The parameters for balancing control are obtained as follows: the capacitor voltage (V_{CH}, V_{CL}) and input current i_g are measured by the voltage and current sensor, respectively. The grid voltage phase angle, δ_g, is calculated from the phased locked loop (PLL) scheme by measuring grid voltage e_g. At this point, the PWM modulator generates a new pole voltage reference by synthesizing the reference signal and offset signal.

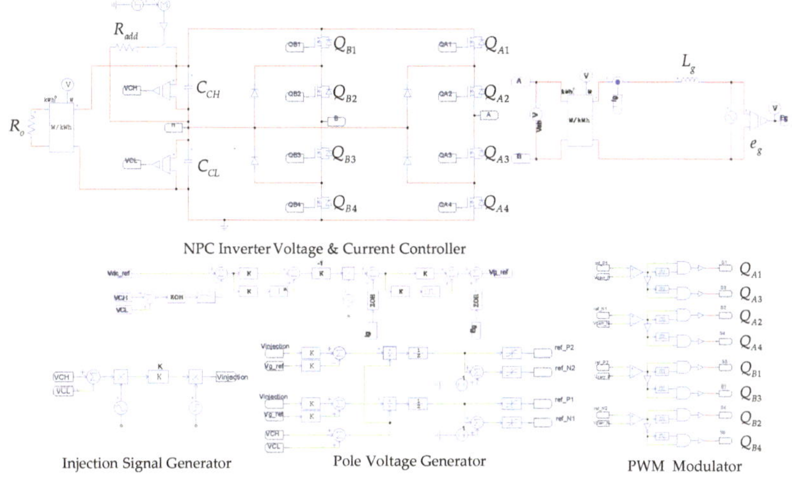

Figure 7. PSIM scheme for capacitor voltage balancing control.

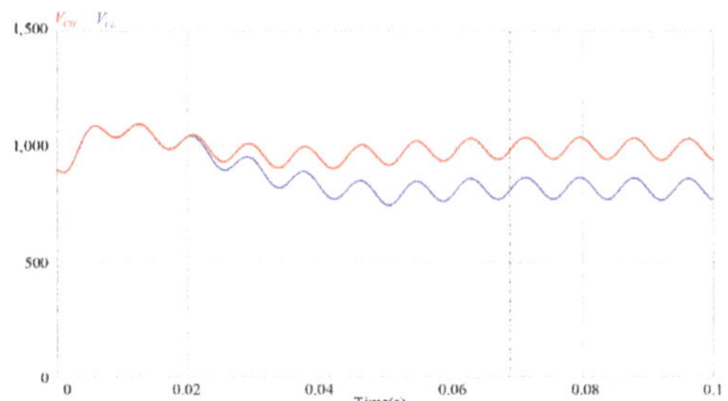

Figure 8. Unbalanced voltage waveforms of upper capacitor V_{CH} and V_{CL}.

In the simulation, the dc-link capacitor voltage difference between the upper capacitor V_{CH} and lower capacitor V_{CL} was intentionally made to generate an imbalance condition on the dc-link capacitor by using the attached additional circuit on top of the dc-link capacitor in Figure 3. Figure 8 shows the voltage levels of the upper and lower capacitors, and the NPC inverter controls the dc-link voltage V_{DC} to the dc-link voltage reference $V_{DC}{}^*$ when the switch of the additional circuit was closed at t = 0.02 s. If there were no additional controls for the dc-link capacitor voltage balancing, the upper capacitor voltage value remained at a lower value than the lower capacitor voltage V_{CH}.

The simulation parameters of a single-phase three-level NPC inverter adopting the proposed balancing control were as follows. Capacitance of the dc-link capacitor C_{CH}, C_{CL} was 250 µF, inductance of the filter inductor L_g was 14 mH, resistance of the load R_o and the additional load were 540 Ω, the switching frequency f_{SW} was 10 kHz, the grid side voltage v_g was 943 V/60 Hz in the root mean square (rms) value, and the controlled dc-link voltage reference $V_{DC}{}^*$ was 1.8 kV.

Figure 9 shows the simulation results for the dc-link capacitor voltage balancing control with the type of injection voltage under the same modulation index. In the simulation, the voltage difference between the upper and lower capacitor of the dc-link was enforced at about 334 V. Figure 9a shows the simulation results of the dc-link capacitor voltage balancing control using distribution factor μ. From the simulation results, each capacitor voltage V_{CH} and V_{CL} reached a balanced point at $t = 0.4941$ s. Figure 9b shows the simulation results when injecting a second-order harmonic signal into the modulation signal. In this case, each capacitor voltage V_{CH} and V_{CL} reached a balanced point at $t = 0.4249$ s. Figure 9c shows the simulation result when injecting the half-wave of a second-order harmonic signal into the modulation signal. In this case, the upper and lower capacitor voltage reached a balanced point at 0.3875 s. From the simulation results, the half-wave of the second-order harmonic signal injection method was better than other offset voltage injection methods.

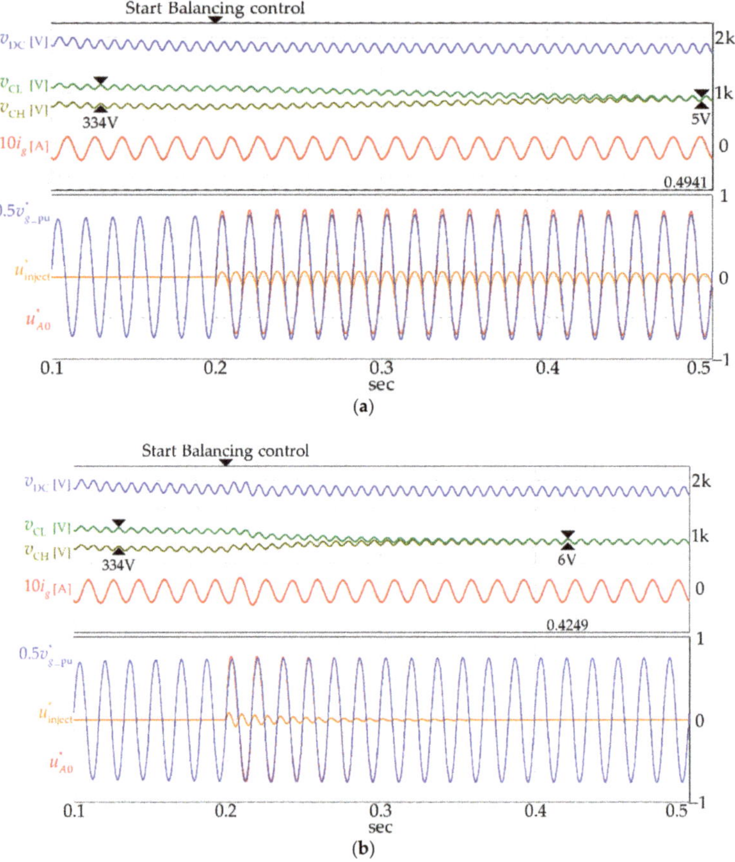

Figure 9. *Cont.*

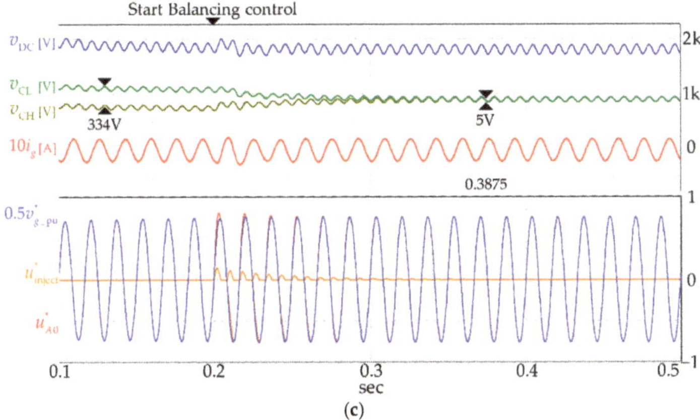

Figure 9. Simulation results for the single-phase NPC inverter with V_{DC}, V_{CH}, V_{CL} and i_g: (**a**) using distribution factor μ; (**b**) using second-order harmonic component; and (**c**) using half-wave of second-order harmonic component.

Figure 10 shows a photograph of the experimental setup for a single-phase three-level NPC inverter used to verify the proposed method. The NPC inverter module consisted of two NPC half-bridges, a series filter inductor, and two electrolytic capacitors. For the test, a 6 kW resistive load was connected to the dc-link capacitor of the NPC inverter. The same system parameters for the NPC inverter were applied as in the previous simulation. A 1:6 transformer was connected to the output of the variac to obtain 943 V AC voltage. Each leg of the NPC inverter consisted of four MOSFETs and two clamping diodes. SiC MOSFETs, Cree's C2M0040120D, and a SiC Schottky diode, Cree's C4D20120D, were utilized for each leg. By adapting silicon carbide devices, the switching frequency can be higher than for silicon-based devices. When using a higher switching frequency, the physical size of the magnetic component for the NPC inverter can be reduced. Furthermore, a lower switching loss is expected than with silicon-based devices. Component specifications of the NPC inverter are shown in Table 2. In order to measure the dc-link capacitor voltage and input current, a differential probe and a current probe were used. For each probe of the dc-link capacitor, PINTEK's high voltage differential probe DP-50 was used. In addition, for input current, Lecroy's current probe CP150 was used. The control structure of the NPC inverter contained a current controller for input current, a voltage controller for the dc-link capacitor voltage, and a voltage balance controller for the dc-ink capacitor voltage balance. For these controllers, a digital control board based on Texas Instruments' TMS320F28335 was used, which was made in-laboratory. The phase voltage reference signal for the NPC inverter was generated by measuring the dc-link capacitor voltage and input current, and the injection signal for the dc-link capacitor voltage balance was calculated from the measured dc-link capacitor voltage and the phase angle of the grid. The NPC inverter regulated the AC voltage to DC voltage without the dc-link capacitor voltage imbalance caused by the generated phase voltage reference signal and injection signal.

Table 2. Parameter specifications of the NPC inverter.

Parameter	Value	Quantity
Switches	1.2 kV/60 A	8
Clamped Diodes	1.2 kV/33 A	4
Filter Inductor	14 mH	1
Capacitors	250 uF	2

In the above experimental setup, the experimental process was as follows: first, the AC input voltage was increased to 943 V under dc-link voltage control conditions for 1.8 kV dc-link voltage. Second, the dc-link capacitor voltage imbalance was enforced at about 100 V by operating an additional resistive load circuit, which was attached on the upper capacitor. At this time, the voltage difference should be suppressed by the voltage rating of the electrolytic capacitor. In addition, then, the adopting proposed method, dc-link voltage, separated capacitor voltage, and current distortion were observed intensively under unbalanced dc-link capacitor voltage.

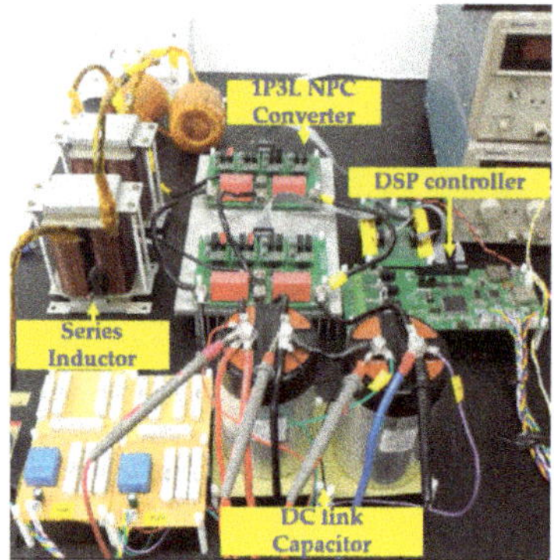

Figure 10. Hardware configuration for the dc-link capacitor voltage balancing test.

The experimental results when adopting the proposed method for dc-link capacitor voltage balance are presented in Figure 11. As shown in Figure 11a,b, the capacitor voltage difference was enforced at about 100 V. Before adopting the balancing control in Figure 11a, the upper and lower capacitor voltage levels were measured as 828 V and 924 V, respectively. After adopting the proposed second-order harmonic injection method, the upper capacitor voltage level increased from 828 V to 869 V, and the lower capacitor voltage level decreased from 924 V to 887 V. In Figure 11b, the other experimental result, the capacitor voltage difference was measured as 106 V. After adopting the proposed half-wave rectified second-order harmonic injection, the capacitor voltage difference decreased from 106 V to 9 V. At the beginning of the capacitor voltage balancing control, the line current i_g was instantaneously distorted, but this distortion disappeared within 50 ms as the upper and the lower capacitor voltage level became equal.

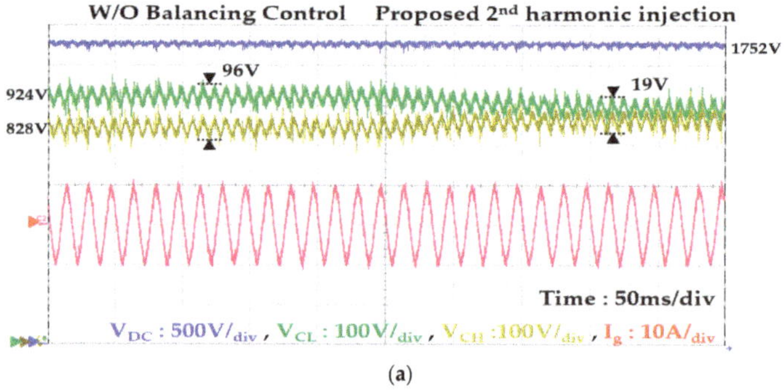

(a)

Figure 11. *Cont.*

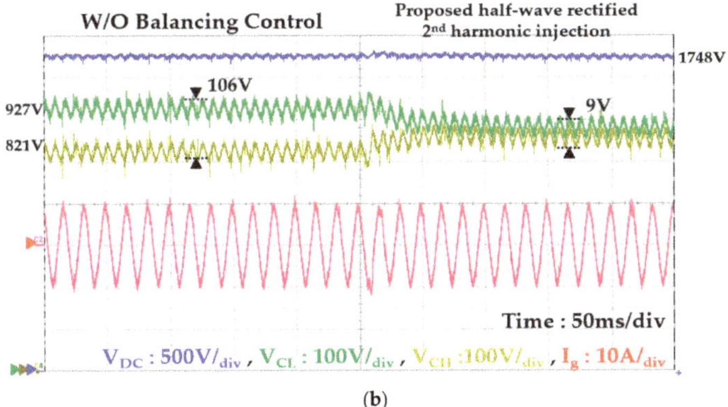

(**b**)

Figure 11. Experimental results for the single-phase NPC inverter with V_{DC}, V_{CH}, V_{CL} and i_g (**a**) using second-order harmonic component; and (**b**) using subdivided wave of second-order harmonic component.

From the simulations and experimental results, the proposed capacitor voltage balancing control strategies of injecting the second-order harmonic signal and the half-wave of the second-order harmonic signal were effective for dc-link capacitor voltage balancing. Among these balancing methods, the method of injecting the half-wave signal into the modulation signal was faster, at about 50 ms, at achieving capacitor voltage balance than the injection of the full-wave signal of the second-order harmonic. However, in the case of input current i_g, the full-wave injection method of the second-order harmonic had a lower distortion than the half-wave signal injection method as it seems that the synthesized pole voltage reference signal was distorted when the half cycle of the second-order harmonic signal had a negative polarity.

5. Conclusions

This paper described an offset voltage injection method for dc-link capacitor voltage balance on a single-phase three-level NPC inverter. The operations and the balancing strategies were explained. The proposed offset voltage consisted of double the frequency of the grid and the voltage difference between the upper and lower capacitors of the dc-link. In addition, the partial offset voltage injection method of the second-order harmonic signal was proposed to achieve dc-link voltage balance. The proposed method does not require hard calculations and additional hardware setup for dc-link balancing control; it is simpler and more intuitive to implement than the conventional offset voltage injection method. However, the proposed method can operate only in a narrower voltage difference than the conventional method. This limitation is caused by the maximum and minimum values of the offset voltage, which consists of the capacitor voltage difference and the phase angle of the grid. Simulations and experiments were performed at 60% load of the NPC inverter. In addition, the results based on a single-phase NPC inverter application verified the validity and feasibility of the proposed method. The proposed method of reducing dc-link capacitor voltage difference can be adopted for other topologies that have separated-dc link capacitors. Furthermore, it seems that using a variable weight factor on the offset voltage for the dc-link balance could be possible.

Author Contributions: K.-P.K. implemented the system, and performed the experiments. Y.C. managed the project. M.-H.R. and J.-W.B. assisted with the idea development and paper writing.

Funding: This research was funded by the Human Resources Program in Energy Technology of the Korea Institute of Energy Technology Evaluation and Planning (KETEP), granted financial resources from the Ministry of Trade, Industry & Energy, Republic of Korea (No. 20174030201660), and by the Korea Electric Power Corporation

(KEPCO) under project entitled Demonstration study for Low Voltage Direct Current Distribution Network in Island (D3080).

Acknowledgments: This research was funded by the Human Resources Program in Energy Technology of the Korea Institute of Energy Technology Evaluation and Planning (KETEP), granted financial resource from the Ministry of Trade, Industry & Energy, Republic of Korea (No. 20174030201660), and by the Korea Electric Power Corporation (KEPCO) under project entitled Demonstration study for Low Voltage Direct Current Distribution Network in Island (D3080).

Conflicts of Interest: The authors declare no conflict no interest.

References

1. Olivares, D.E.; Mehrizi-Sani, A.; Etemadi, A.H.; Cañizares, C.A.; Iravani, R.; Kazerani, M.; Hajimiragha, A.H.; Gomis-Bellmunt, O.; Saeedifard, M.; Palma-Behnke, R.; et al. Trends in Microgrid Control. *IEEE Trans. Smart Grid* **2014**, *5*, 1905–1919. [CrossRef]
2. Asmus, P. Microgrids, virtual power plants and our distributed energy future. *Electr. J.* **2010**, *23*, 72–82. [CrossRef]
3. She, X.; Burgos, R.; Wang, G.; Wang, F.; Huang, A.Q. Review of solid state transformer in the distribution system: From components to field application. In Proceedings of the 2012 IEEE Energy Conversion Congress and Exposition (ECCE), Raleigh, NC, USA, 15–20 September 2012; pp. 4077–4084.
4. Kolar, J.W.; Ortiz, G. Solid-state-transformers: Key components of future traction and smart grid systems. In Proceedings of the International Power Electronics Conference (IPEC), Hiroshima, Japan, 18–21 May 2014.
5. Huber, J.E.; Kolar, J.W. Optimum number of cascaded cells for high-power medium-voltage multilevel converters. In Proceedings of the 2013 IEEE Energy Conversion Congress and Exposition (ECCE), Denver, CO, USA, 15–19 September 2013; pp. 359–366.
6. Guillod, T.; Huber, J.E.; Ortiz, G.; De, A.; Franck, C.M.; Kolar, J.W. Characterization of the voltage and electric field stresses in multi-cell solid-state transformers. In Proceedings of the 2014 IEEE Energy Conversion Congress and Exposition (ECCE), Pittsburgh, PA, USA, 14–18 September 2014; pp. 4726–4734.
7. Huber, J.E.; Kolar, J.W. Volume/weight/cost comparison of a 1MVA 10 kV/400 V solid-state against a conventional low-frequency distribution transformer. In Proceedings of the 2014 IEEE Energy Conversion Congress and Exposition (ECCE), Pittsburgh, PA, USA, 14–18 September 2014; pp. 4545–4552.
8. Jih-Sheng, L.; Zheng, P.F. Multilevel converters-a new breed of power converters. *IEEE Trans. Ind. Appl.* **1996**, *32*, 509–517. [CrossRef]
9. Rodriguez, J.; Jih-Sheng, L.; Zheng, P.F. Multilevel inverters: A survey of topologies, controls, and applications. *IEEE Trans. Ind. Electron.* **2002**, *49*, 724–738. [CrossRef]
10. Nabae, A.; Takahashi, I.; Akagi, H. A New Neutral-Point-Clamped PWM Inverter. *IEEE Trans. Ind. Appl* **1981**, *IA-17*, 518–523. [CrossRef]
11. Schweizer, M.; Friedli, T.; Kolar, J.W. Comparative Evaluation of Advanced Three-Phase Three-Level Inverter/Converter Topologies Against Two-Level Systems. *IEEE Trans. Ind. Electron.* **2013**, *60*, 5515–5527. [CrossRef]
12. Stala, R. Application of Balancing Circuit for DC-Link Voltages Balance in a Single-Phase Diode-Clamped Inverter with Two Three-Level Legs. *IEEE Trans. Ind. Electron.* **2011**, *58*, 4185–4195. [CrossRef]
13. Boora, A.; Nami, A.; Zare, F.; Ghosh, A.; Balaabjerg, F. Voltage-sharing converter to supply single-phase asymmetrical four-level diode-clamped inverter with high power factor loads. *IEEE Trans. Power Electorn.* **2010**, *25*, 2507–2520. [CrossRef]
14. Song, W.; Feng, X.; Smedley, K.M. A Carrier-Based PWM Strategy with the Offset Voltage Injection for Single-Phase Three-Level Neutral-Point-Clamped Converters. *IEEE Trans. Power Electron.* **2013**, *28*, 1083–1095. [CrossRef]
15. De Freitas, I.S.; Bandeira, M.M.; Barros, L.D.M.; Jacobina, C.B.; dos Santos, E.C.; Salvadori, F.; de Silva, S.A. A Carrier-Based PWM Technique for Capacitor Voltage Balancing of Single-Phase Three-Level Neutral-Point-Clamped Converters. *IEEE Trans. Ind. Appl.* **2015**, *51*, 3227–3235. [CrossRef]
16. Kang, K.-P.; Ryu, M.-H.; Baek, J.-W.; Kim, J.-Y.; Cho, Y. A carrier-based pwm method with the double frequency voltage injection for three-level neutral-point clamped (NPC) converters. In Proceedings of the 2017 Asian Conference on Energy, Power and Transportation Electrification (ACEPT), Singapore, 24–26 October 2017; pp. 1–6.

17. Du, Z.; Tolbert, L.M.; Ozpineci, B.; Chiasson, J.N. Fundamental Frequency Switching Strategies of a Seven-Level Hybrid Cascaded H-Bridge Multilevel Inverter. *IEEE Trans. Power Electron.* **2009**, *24*, 25–33. [CrossRef]
18. Ye, Z.; Xu, Y.; Wu, X.; Tan, G.; Deng, X.; Wang, Z. A Simplified PWM Strategy for a Neutral-Point-Clamped (NPC) Three-Level Converter with Unbalanced DC Links. *IEEE Trans. Power Electron.* **2016**, *31*, 3227–3238. [CrossRef]
19. Kou, L.; Wang, H.; Liu, Y.F.; Sen, P.C. DC-link capacitor voltage balancing technique for phase-shifted PWM-based seven-switch five-level ANPC inverter. In Proceedings of the 2017 IEEE Applied Power Electronics Conference and Exposition (APEC), Tampa, FL, USA, 26–30 March 2017; pp. 671–678.
20. WenSheng, S.; XiaoYun, F.; Chenglin, X. A neutral point voltage regulation method with SVPWM control for single-phase three-level NPC converters. In Proceedings of the 2008 IEEE Vehicle Power and Propulsion Conference, Harbin, China, 3–5 September 2008; pp. 1–4.
21. Celanovic, N.; Boroyevich, D. A comprehensive study of neutral-point voltage balancing problem in three-level neutral-point-clamped voltage source PWM inverters. *IEEE Trans. Power Electron.* **2000**, *15*, 242–249. [CrossRef]
22. Hong-Seok, S.; Keil, R.; Mutschler, P.; Weem, J.v.d.; Kwanghee, N. Advanced control scheme for a single-phase PWM rectifier in traction applications. In Proceedings of the 38th IAS Annual Meeting on Conference Record of the Industry Applications Conference 2003, Salt Lake City, UT, USA, 12–16 October 2003; pp. 1558–1565.
23. Abdelsalam, M.; Marei, M.; Tennakoon, S.; Griffiths, A. Capacitor voltage balancing strategy based on sub-module capacitor voltage estimation for modular multilevel converters. *CSEE J. Power Energy Syst.* **2016**, *2*, 65–73. [CrossRef]
24. Ghias, A.M.Y.M.; Pou, J.; Acuna, P.; Ceballos, S.; Heidari, A.; Agelidis, V.G.; Merabet, A. Elimination of Low-Frequency Ripples and Regulation of Neutral-Point Voltage in Stacked Multicell Converters. *IEEE Trans. Power Electron.* **2017**, *32*, 164–175. [CrossRef]
25. Yang, R.; Li, B.; Wang, G.; Cecati, C.; Zhou, S.; Xu, D. Asymmetric Mode Control of MMC to Suppress Capacitor Voltage Ripples in Low-Frequency, Low-Voltage Conditions. *IEEE Trans. Power Electron.* **2017**, *32*, 4219–4230. [CrossRef]

Article

A Highly Efficient Single-Phase Three-Level Neutral Point Clamped (NPC) Converter Based on Predictive Control with Reduced Number of Commutations

Eun-Su Jun and Sangshin Kwak *

School of Electrical and Electronics Engineering, Chung-ang University, Seoul 06974, Korea;
elizame3@naver.com
* Correspondence: sskwak@cau.ac.kr; Tel.: +82-2-820-5346

Received: 16 November 2018; Accepted: 14 December 2018; Published: 18 December 2018

Abstract: This paper proposes a highly efficient single-phase three-level neutral point clamped (NPC) converter operated by a model predictive control (MPC) method with reduced commutations of switches. The proposed method only allows switching states with none or a single commutation at the next step as candidates for future switching states for the MPC method. Because the proposed method preselects switching states with reduced commutations when selecting an optimal state at a future step, the proposed method can reduce the number of switchings and the corresponding switching losses. Although the proposed method slightly increases the peak-to-peak variations of the two dc capacitor voltages, the developed method does not deteriorate the input current quality and input power factor despite the reduced number of switching numbers and losses. Thus, the proposed method can reduce the number of switching losses and lead to high efficiency, in comparison with the conventional MPC method.

Keywords: model predictive control; single-phase three-level NPC converter; commutation

---

1. Introduction

Recently, multilevel converters have become popular in a variety of high-power systems owing to their low voltage stress, improved waveform qualities, and low electromagnetic interference (EMI) compared to two-level converters [1,2]. Among several kinds of multilevel converters, three-level neutral point clamped (NPC) converters with relatively simple configurations have been realized for many application areas. In addition to three-phase NPC converters, single-phase three-level NPC converters have been employed for high-speed traction systems as well. In order to control single-phase three-level NPC converters, traditional carrier-based pulse width modulation (CBPWM) methods combined with linear proportional and integral (PI) controllers have been investigated to synthesize ac sinusoidal current waveforms with three-level NPC converters. Aside from their adjustable ac voltage and current synthesis, the NPC converters require balancing of the two dc capacitor voltages because of their structure, which has two split dc capacitors in the dc link. As a result, CBPWM methods with offset voltage injection to remove imbalance of neutral point (NP) voltage in the NPC converters have been often used [3–7].

Recently, model predictive control (MPC) methods have been studied for numerous power converters including three-level NPC converters [8–10]. There have been several studies on MPC algorithms for single-phase NPC converters as well as three-phase NPC converters [11–13]. In the MPC methods for single-phase NPC converters, a cost function to determine an optimal switching state generally consists of two terms combined with a weighting factor not only to control both the ac sinusoidal currents but also to balance the two capacitor voltages. The ac sinusoidal current is controlled by changing the converter voltage levels, whereas the NP voltage balance is adjusted by using redundant switching states that yield the same voltage level.

The conventional MPC method selects an optimal switching state among nine switching states allowed by the single-phase three-level NPC converter on the basis of a cost function considering the ac source current and the NP voltage balance. Consideration of all possible switching states in the conventional method can choose a switching state involved in many commutations as an optimal switching state for the next step, which can lead to an increased number of switchings and corresponding switching losses [14–18]. In addition, the conventional MPC method changes the optimal switching state by evaluating the capacitor voltage balance term using the redundant switching states to equal the two capacitor voltages [19,20] owing to a slight voltage difference even when the converter does not require a change in the voltage level. Thus, this operation can increase the number of switchings and switching losses as well [21,22]. Several trials to reduce switching losses based on the model predictive control methods have been addressed for a variety of power converters in literatures. In [23,24], approaches to reduce switching losses of matrix converters have been addressed. Ref. [23] proposed a switching loss reduction technique by adding an additional term related with a number of future commutations to a cost function used to control the matrix converter. In [24], a trial to decrease switching losses of the matrix converter has been presented, where a cost function includes an extra term directly representing switching losses at next step by calculating switch currents and switch voltages. In [25], a model predictive control method for modular multilevel converters (MMCs) has been developed with a cost function which is aimed at the elimination of the MMC circulating currents, regulating the arm voltages, and controlling the ac-side currents. In addition, this strategy tried to reduce power losses by decreasing the submodule switching frequency. In [26,27], reduction techniques of switching losses for two-level voltage source inverters have been presented. Ref. [26] proposed a switching strategy based on the model predictive control method to clamp one phase with the largest load current among the three legs in the voltage source inverter every sampling period, which can successfully reduce switching losses of the voltage source inverter. In addition, a model predictive control method for the voltage source inverter has been developed to reduce switching losses by injecting future zero-sequence voltage [27]. This approach decreased the switching losses by implementing optimal discontinuous pulse patterns to stop switching operations at vicinity of peak values of load currents. However, there has not been, to the authors' best knowledge, tried to reduce switching losses, using a trade-off between switching losses and capacitor voltage balancing in the three-level NPC converters, although several trials to reduce switching losses based on the model predictive control methods have been addressed for a variety of power converters.

In this paper, a highly efficient algorithm with a reduced number of switching and low switching losses for single-phase three-level neutral point clamped (NPC) converters is proposed based on a model predictive control (MPC) method with a decreased number of commutations of switches. The proposed method pre-excludes, from the candidates for possible future switching states, the switching states that yield more than two commutations in the next sampling period. As a result, the proposed technique can reduce the number of switchings and switching losses by utilizing switching states involving no commutation or only one commutation during every sampling instant for single-phase three-level NPC converters. In addition, the developed method does not deteriorate the input current quality or input power factor despite the reduced switching numbers and losses. Although the proposed method slightly increases the peak-to-peak variations of the two dc capacitor voltages at the expense of reduced commutation, the increased voltage variation is not high. Thus, the proposed method can obtain high efficiency and low switching losses at the expense of a slightly increased peak-to-peak variation of the NP voltage. The performance of the proposed method with a reduced number of switchings and higher efficiency is evaluated in terms of the total harmonic distortion (THD) and peak-to-peak variations of the capacitor voltages. Simulations and experimental results are presented to verify the effectiveness of the proposed method.

2. Single-Phase Three-Level NPC Converter and Model Predictive Control Method

Figure 1 shows a circuit diagram for the single-phase three-level NPC converter. As shown in Figure 1, the single-phase three-level NPC converter has an input inductor L_s and resistor R_s as an ac side filter, as well as two capacitors C_1 and C_2 at the dc side. In addition, v_{c1} and v_{c2} are the dc voltage of each capacitor, and R_L is a load resistor. Switches S_{aj} and S_{bj} ($j = 1, 2, 3, 4$) are Insulated Gate Bipolar Transistors (IGBTs) at the a-phase and b-phase, respectively. The switch states at each phase, produced by the converter, can be defined as a function of the switching status of the two upper devices as:

$$S_x = \begin{cases} 1 & (S_{x1}, S_{x2} : ON) \\ 0 & (S_{x2}, S_{x3} : ON) \\ -1 & (S_{x3}, S_{x4} : ON) \end{cases} (x = a \text{ or } b) \tag{1}$$

The two switches S_{x1} and S_{x3} operate complementarily. Similarly, S_{x2} and S_{x4} work in a complementary manner. As a result, the switching status of the two lower devices is automatically determined by the upper switches. Owing to possible combinations of the switching states of (1) in the a and b phases, a total of nine operating states can be generated by the single-phase three-level NPC converter. On the basis of nine operating states, the phase switching state, upper device switching status, and converter input voltage v_{ab} are listed in Table 1. The nine operating states yield five voltage levels for the converter input voltage v_{ab}, which provides the single-phase NPC converter with redundancy.

As shown in Table 1, the two states (1, 0) and (0, −1) for (S_a, S_b) are redundant switching states that apply the same voltage level to the converter input terminal of the single-phase three-level NPC converter by assuming that the two capacitor voltages are well balanced. Likewise, the states (0, 1) and (−1, 0) for (S_a, S_b) are also redundant because they yield the equal converter input voltage v_{ab}. These redundancies can be utilized to balance the two capacitor voltages v_{c1} and v_{c2}. The currents i_u and i_l shown in Figure 1 can be expressed using the switching status and the source current i_s as in (2) and (3). Thus, they can be obtained without additional measurements [16]:

$$i_u = \frac{S_a(S_a + 1) - S_b(S_b + 1)}{2} i_s \tag{2}$$

$$i_l = -\frac{S_a(S_a - 1) - S_b(S_b - 1)}{2} i_s \tag{3}$$

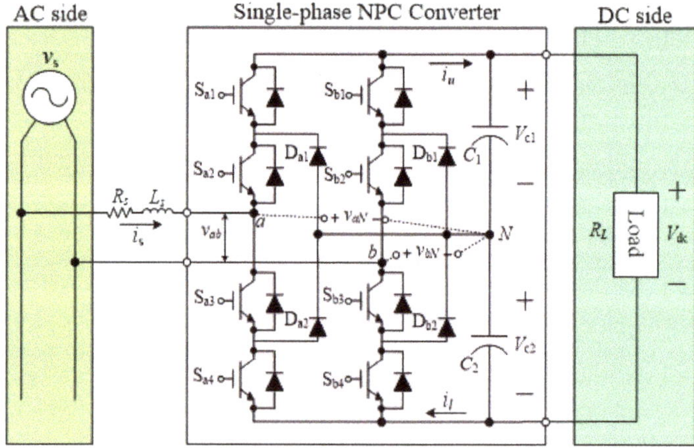

Figure 1. Single-phase three-level NPC converter.

Table 1. Nine operating states, phase switching state, upper device switching status, and converter input voltage of single-phase three-level NPC converter.

#	Operating Status		Phase Switching State				Converter input Voltage	Capacitor Voltage	
	S_a	S_b	S_{a1}	S_{a2}	S_{b1}	S_{b2}	v_{ab}	v_{c1}	v_{c2}
1	0	0	OFF	ON	OFF	ON	0	-	-
2	1	1	ON	ON	ON	ON	0	-	-
3	−1	−1	OFF	OFF	OFF	OFF	0	-	-
4	1	−1	ON	ON	OFF	OFF	V_{dc}	↑	↑
5	1	0	ON	ON	OFF	ON	$V_{dc}/2$	↑	↓
6	0	−1	OFF	ON	OFF	OFF	$V_{dc}/2$	↓	↑
7	0	1	OFF	ON	ON	ON	$-V_{dc}/2$	↑	↓
8	−1	0	OFF	OFF	OFF	ON	$-V_{dc}/2$	↓	↑
9	−1	1	OFF	OFF	ON	ON	$-V_{dc}$	↓	↓

The capacitor voltage dynamics of the dc link are calculated by using differential equations:

$$\frac{dv_{c1}}{dt} = \frac{1}{C_1} i_u \tag{4}$$

$$\frac{dv_{c2}}{dt} = \frac{1}{C_2} i_l \tag{5}$$

Using a constant sampling period T_s, the capacitor voltage dynamics in the discrete-time domain are described as:

$$\frac{dv_{cm}}{dt} \approx \frac{v_{cm}(k+1) - v_{cm}(k)}{T_s} \quad (m = 1,\ 2) \tag{6}$$

Using (6), Equations (4) and (5) can be expressed in the discrete time domain as:

$$v_{c1}(k+1) = v_{c1}(k) + \frac{T_s}{C_1}\left(\frac{S_a(S_a+1) - S_b(S_b+1)}{2} \right) i_s \tag{7}$$

$$v_{c2}(k+1) = v_{c2}(k) - \frac{T_s}{C_2}\left(\frac{S_a(S_a-1) - S_b(S_b-1)}{2} \right) i_s \tag{8}$$

The input current of the ac side shown in Figure 1 is expressed in the continuous time domain as:

$$v_s = R i_s + L\frac{di_s}{dt} + v_{ab} \tag{9}$$

Equation (9) is expressed in the discrete time domain as:

$$i_s(k+1) = \left(1 - \frac{RT_s}{L}\right) i_s(k) + \frac{T_s}{L}\left(v_s(k) - v_{ab}(k)\right) \tag{10}$$

The ac source current at the next step, $i_s(k+1)$, in (10) can have five possible movements owing to the five possible voltage levels for $v_{ab}(k)$. The single-phase three-level NPC converter needs to balance the two capacitor voltages by manipulating the phase switching states $S_a(k)$ and $S_b(k)$ shown in (7) and (8) as well as control the source current by changing the converter input voltage $v_{ab}(k)$ in (10). As a result, the cost function with two terms for the ac source current control part and the neutral point (NP) voltage control part of the two capacitor voltages is:

$$g = |i_s^*(k+1) - i_s(k+1)| + \lambda_c|v_{c1}(k+1) - v_{c2}(k+1)| \tag{11}$$

where λ_c represents the weighting factor of the capacitor voltage balancing term in the cost function. Moreover, the future ac reference current can be expressed with past and present currents from a Lagrange extrapolation as [28–31]:

$$i_s^*(k+1) = 3i_s^*(k) - 3i_s^*(k-1) + i_s^*(k-2) \tag{12}$$

where $i_s^*(k)$ is the present current, and $i_s^*(k-1)$ and $i_s^*(k-2)$ are the reference value of the one-step and two-step past ac source currents, respectively. The ac sinusoidal current is controlled by changing the converter voltage levels, whereas the NP voltage balance is adjusted by using redundant switching states that yield the same voltage level. As a result, the conventional MPC method changes the optimal switching state by evaluating the capacitor voltage balance term using the redundant switching states to equal the two capacitor voltages owing to a slight voltage difference even when the converter does not require a change in the voltage level. Thus, this operation can increase the number of switchings and the switching losses as well.

3. Proposed MPC Method Based on Voltage Tolerance Band

The conventional MPC method selects an optimal switching state among nine switching states allowed by the single-phase three-level NPC converter on the basis of a cost function considering the source current and the NP voltage balance. Consideration of all possible switching states in the conventional method can help to choose a switching state involved in many commutations as an optimal switching state for the next step. In addition, the conventional MPC method changes the optimal switching state by evaluating the capacitor voltage balance term using the redundant switching states to equal the two capacitor voltages owing to a slight voltage difference even when the converter does not require a change in the voltage level. Table 2 illustrates the number of commutations involved in switch transitions from the current step to the next step, which vary from zero to four.

Table 2. Number of commutations involved in switch transitions from current step to next step in conventional MPC method.

Current Operating Status		Next Possible Operating Status								
(S_a,S_b)	(0,0)	(0,0)	(0,1)	(1,0)	(−1,0)	(0,−1)	(1,−1)	(−1,1)	(1,1)	(−1,−1)
number of commutations		0	1	1	1	1	2	2	2	2
(S_a,S_b)	(1.1)	(1,1)	(1,0)	(0,1)	(−1,1)	(0,0)	(1,−1)	(0,−1)	(−1,0)	(−1,−1)
number of commutations		0	1	1	2	2	2	3	3	4
(S_a,S_b)	(−1.−1)	(−1,−1)	(−1,0)	(0,−1)	(−1,1)	(0,0)	(1,−1)	(0,1)	(1,0)	(1,1)
number of commutations		0	1	1	2	2	2	3	3	4
(S_a,S_b)	(1.−1)	(1,−1)	(0,−1)	(1,0)	(−1,−1)	(0,0)	(1,1)	(−1,0)	(0,1)	(−1,1)
number of commutations		0	1	1	2	2	2	3	3	4
(S_a,S_b)	(1.0)	(1,0)	(1,1)	(1,−1)	(0,0)	(0,1)	(0,−1)	(−1,0)	(−1,1)	(−1,−1)
number of commutations		0	1	1	1	2	2	2	3	3
(S_a,S_b)	(0.−1)	(0,−1)	(1,−1)	(−1,−1)	(0,0)	(0,1)	(1,0)	(−1,0)	(−1,1)	(1,1)
number of commutations		0	1	1	1	2	2	2	3	3
(S_a,S_b)	(0.1)	(0,1)	(1,1)	(−1,1)	(0,0)	(0,−1)	(1,0)	(−1,0)	(1,−1)	(−1,−1)
number of commutations		0	1	1	1	2	2	2	3	3
(S_a,S_b)	(−1.0)	(−1,0)	(−1,1)	(−1,−1)	(0,0)	(0,1)	(0,−1)	(1,0)	(1,−1)	(1,1)
number of commutations		0	1	1	1	2	2	2	3	3
(S_a,S_b)	(−1.1)	(−1,1)	(0,1)	(−1,0)	(−1,−1)	(0,0)	(1,1)	(1,0)	(0,−1)	(1,−1)
number of commutations		0	1	1	2	2	2	3	3	4

A switching operation with a number of commutations equal to that shown in Table 2, for example, implies that one switch turns off and another switch turns on at a switching instant. Likewise, two switches are off and two are on at a switching moment when the switching operation corresponds to a number of commutations equal to two in Table 2. The conventional method, which selects a next-step switching state depending on the cost function, does not consider the number of commutations.

Thus, the number of switchings can increases in a case where an optimal switching state with many commutations is chosen at the next step. Figure 2 shows simulation waveforms obtained by the conventional MPC method for a single-phase three-level NPC converter.

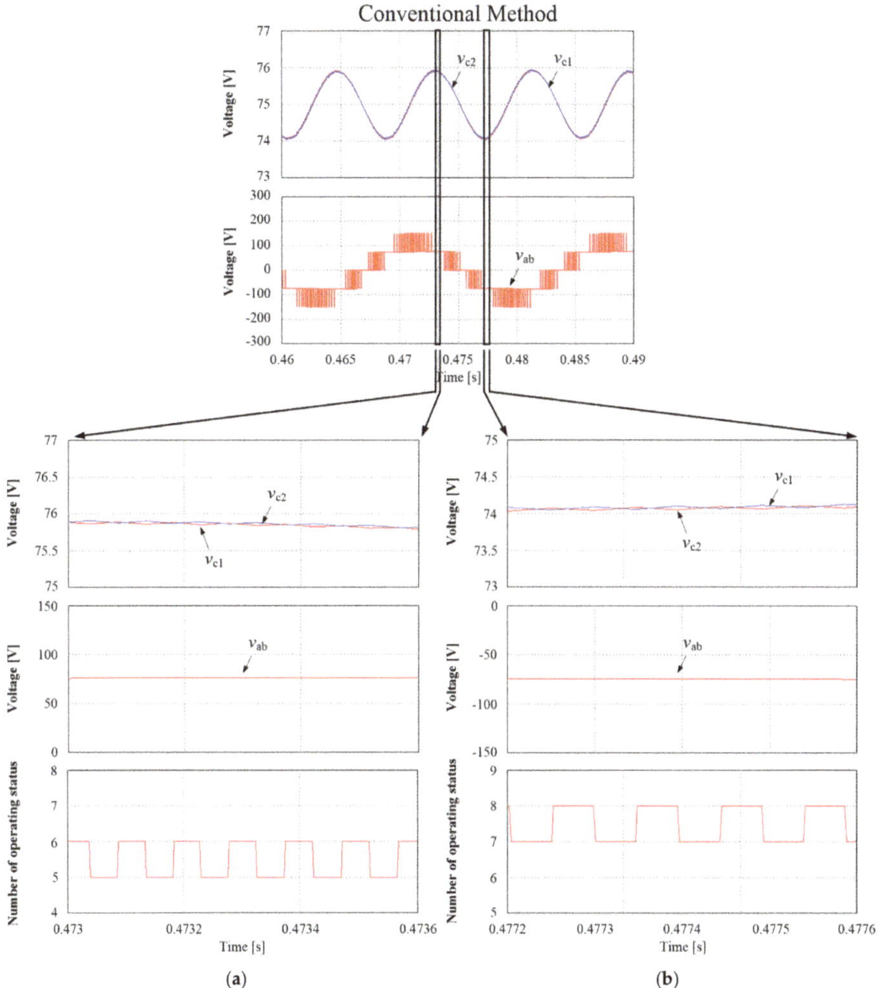

Figure 2. Simulation waveforms of conventional MPC method (**a**) during period with converter voltage v_{ab} fixed to $V_{dc}/2$ and (**b**) during period with converter input voltage v_{ab} fixed to $-V_{dc}/2$.

It is seen from Figure 2a that the conventional method, during the period with the converter voltage v_{ab} fixed to $V_{dc}/2$, repeatedly changes the switching states corresponding to an operating status between $(1, 0)$ and $(0, -1)$. This is the redundant state with respect to each other, although the switch transitions do not be required in terms of the ac source current control. These switching operations involve two commutations at every switching instant, as shown in Table 2. As a result, the number of switching operations substantially increases, whereas the two capacitor voltages perfectly match. Similarly, the simulation waveforms obtained by the conventional MPC method, especially during the period with the converter input voltage v_{ab} fixed to $-V_{dc}/2$, are depicted in Figure 2b. It is seen that the switch transition repeatedly occurs between $(0, 1)$ and $(-1, 0)$ in terms of the operating status, which also involves two commutations at every switching instant, as shown in Table 2. Therefore, it is noted that the two capacitor voltages are tightly balanced by repeatedly using the redundant switching states, at the expense of an increased number of switchings in the conventional MPC method. The proposed method pre-excludes, from the candidates for possible future switching states, the switching states that yield more than two commutations in the next sampling period. As a result, the proposed technique can reduce the number of switchings and the switching losses by utilizing switching states involving no commutation or only one commutation at every sampling instant for single-phase three-level NPC converters. Table 3 shows the switching states allowed in the proposed method, which are states with the number of commutations restricted to zero or one

Table 3. Switching states allowed in proposed method.

Current Operating Status	Next Possible Operating Status
(0,0)	(0,0) (0,1) (1,0) (−1,0) (0,−1)
(1,1)	(1,1) (1,0) (0,1)
(−1,−1)	(−1,−1) (−1,0) (0,−1)
(1,−1)	(1,−1) (0,−1) (1,0)
(1,0)	(1,0) (1,1) (1,−1) (0,0)
(0,−1)	(0,−1) (1,−1) (−1,−1) (0,0)
(0,1)	(0,1) (1,1) (−1,1) (0,0)
(−1,0)	(−1,0) (−1,1) (−1,−1) (0,0)
(−1,1)	(−1,1) (0,1) (−1,0)

Figure 3 shows simulation waveforms obtained by the proposed MPC method for a single-phase three-level NPC converter. It is seen from Figure 3a that the proposed method, during the period with the converter voltage v_{ab} fixed to $V_{dc}/2$, does not change the switching states. This is because the operating status $(0, -1)$ corresponding to the redundant status of $(1, 0)$ is not a possible state for the next state when the current operating status is $(0, -1)$. Similarly, simulation waveforms obtained by the proposed MPC method, especially during the period with the converter input voltage v_{ab} fixed to $-V_{dc}/2$, are depicted in Figure 3b. It is also seen that there is no switch transition because the operating status $(-1, 0)$ corresponding to the redundant status of $(0, 1)$ is not a possible state for the next state when the current operating status is $(-1, 0)$. As a result, the proposed method can reduce the number of switchings and the corresponding switching losses, whereas an NP voltage imbalance between the two capacitor voltages occurs. The NP voltage imbalance that occurs when the periods of the converter voltage are fixed at $V_{dc}/2$ or $-V_{dc}/2$ is resolved by selecting switching states to eliminate the imbalance afterward. Figure 4 depicts simulation waveforms obtained by the proposed MPC method during the period when the converter voltage v_{ab} oscillates between $V_{dc}/2$ and V_{dc}. It is seen that the NP voltage imbalance is solved by the proposed algorithm, where the optimal states are 5, 4, 6, 4, 5, and so on, as shown in Figure 4. Figure 5 shows the capacitor voltage behavior in the switching states. Because switching states 4, 5, and 6 can increase or decrease the upper and lower capacitor voltages, the proposed method can successfully eliminate the NP voltage imbalance quickly.

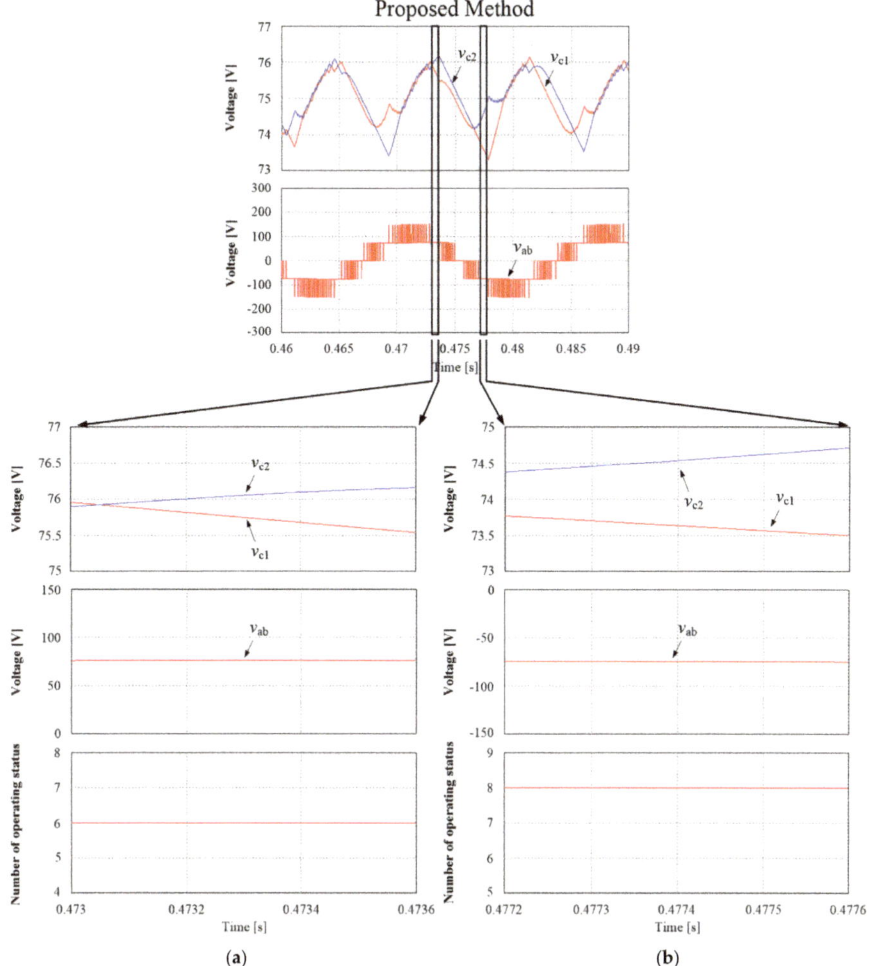

Figure 3. Simulation waveforms of proposed MPC method (**a**) during period with converter voltage v_{ab} fixed to $V_{dc}/2$ and (**b**) during period with converter input voltage v_{ab} fixed to $-V_{dc}/2$.

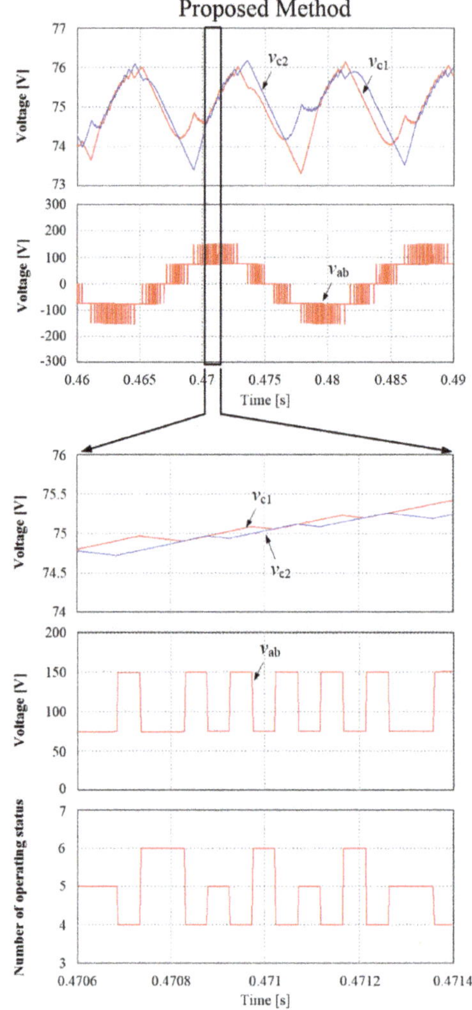

Figure 4. Simulation waveforms of proposed MPC method during period with converter voltage v_{ab} between $V_{dc}/2$ and V_{dc}.

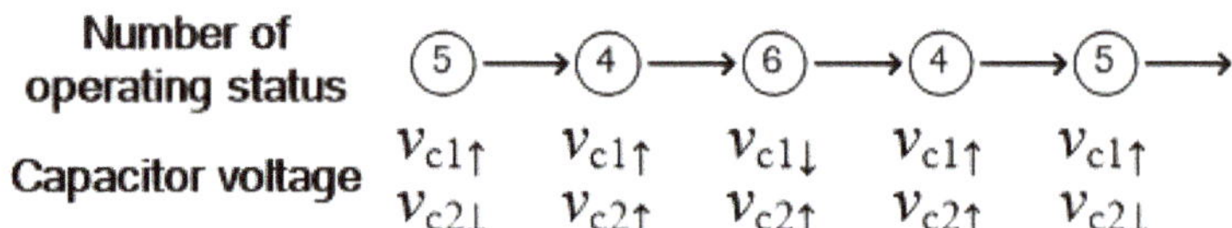

Figure 5. Number of operating status and capacitor voltage behavior of proposed MPC method during period with converter voltage v_{ab} between $V_{dc}/2$ and V_{dc}.

The performance of the MPC methods owing to its inherent operational principle is strongly influenced by the sampling frequency. The number of switchings by the conventional MPC and proposed methods as functions of the sampling periods are shown in Figure 6. It is seen that the number of switchings of the proposed method is lower than that of the conventional MPC method for all considered sampling periods. Increasing the sampling frequency increases the number of switchings, leading to an increasing difference in the number of switchings obtained from the two methods. The conventional MPC and the proposed methods are compared in terms of the THDs of the source current and the peak-to-peak capacitor voltages vs. the sampling periods shown in Figure 6. It is observed that the proposed method results in almost the same THDs in the source current as those in the conventional MPC method. In addition, the peak-to-peak capacitor voltages of the proposed method are slightly higher than those of the conventional method, at the expense of a decreased number of switchings. Thus, it can be concluded that compared to the conventional MPC method, the proposed method can lead to a reduced number of switchings, which can lead to lower switching losses and a nearly equal THD of the source currents.

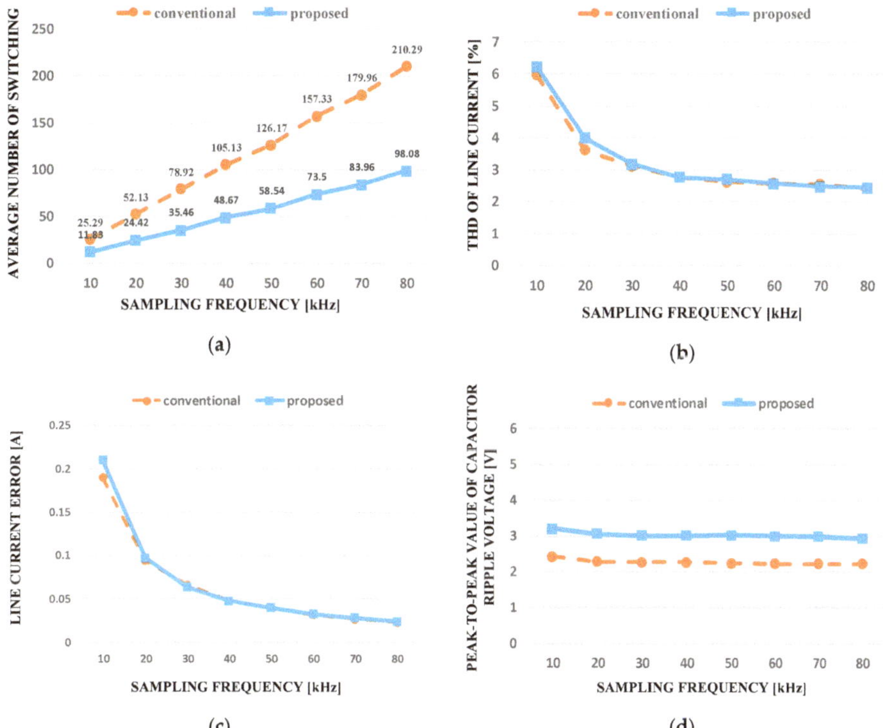

Figure 6. Comparison results obtained by conventional MPC method and proposed method vs. sampling frequency: (**a**) number of switchings; (**b**) THD values of source currents; (**c**) current errors; and (**d**) peak-to-peak values of capacitor ripple voltages.

Loss analysis and stress distribution among the switching devices were further conducted on conditions with v_s = 730 V, V_{dc} = 1000 V, and P_{in} = 10 kW. Losses resulted in each switching component by the conventional and the proposed methods are depicted in Figure 7. The proposed method yields reduced losses in all the switching component, including the IGBTs and the clamping diodes, in comparison with the conventional method. By comparing the conduction loss and the switching loss in Figure 7, the conduction losses generated by the two methods are almost the same. On the other hand, the switching losses of the proposed method are lower than those of the conventional method for all the components. Total efficiency of the conventional and the proposed method was 98% and

98. 7%, respectively. Regarding loss distribution shown in Figure 7, the two methods lead to more losses in the inner switches, S_{a2} and S_{b2}, than the outer switches, which is general in the three-level NPC converters. However, it is seen that the losses by the proposed method are less concentrated on the inner switches than the conventional method, as shown in Figure 7.

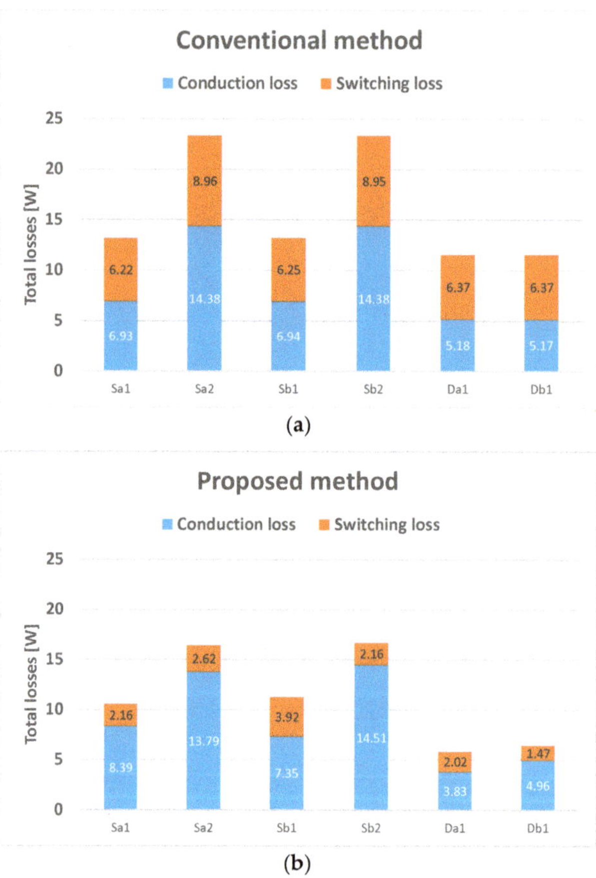

Figure 7. Loss comparison of (**a**) conventional method (**b**) proposed method.

4. Simulation and Experimental Results

In order to demonstrate the proposed method, a single-phase three-level NPC converter with the proposed method was operated at v_s = 110 V, V_{dc} = 150 V, T_s = 50 μs, R_L = 100 Ω, R_s = 1 Ω, and L_s = 10 mH. The weighting factor λ_c = 0.5 in (11) was used for both the conventional and the proposed methods. Figure 8 shows simulation waveforms of the source current (i_s), source voltage (v_s), line-to-line converter input voltage (v_{ab}), converter pole voltages (t), and frequency spectrum of the input current (i_s) obtained by the conventional and the proposed methods.

It is seen that the proposed method, operated with only a consideration of the reduced number of commutations, and the conventional method, using all possible switching states, make the source voltage and the source current in phase. This yields a unity power factor. It is noted that the source current and the ac line-to-line converter voltage generated by both methods are almost the same. On the other hand, the converter pole voltages of the proposed method are different from those of the conventional method because of the reduced number of switchings. It is seen that the pole voltage of the proposed method has a lower number of commutations than the conventional method owing to the reduced switching operations of the proposed method. From the frequency spectrum waveforms, it can be shown that the two methods represent almost the same current THD values. Therefore, the proposed method can reduce the number of switchings and the switching losses without deteriorating the quality of the ac current waveform in comparison with the conventional method.

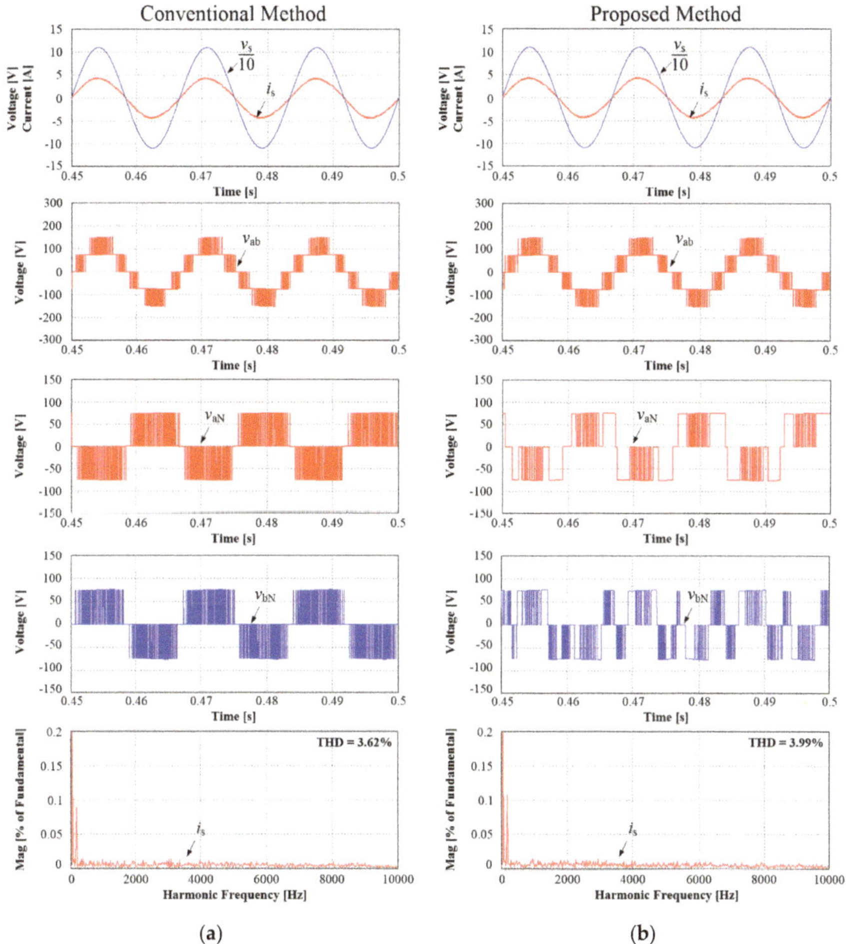

Figure 8. Simulation results of ac source current (i_s), source voltage (v_s), converter line-to-line voltage (v_{ab}), pole voltages (v_{aN}, v_{bN}), and frequency spectrum of input current (i_s) obtained by (**a**) conventional and (**b**) proposed methods.

Figure 9 shows simulation waveforms of the upper and the lower capacitor voltages (v_{c1} and v_{c2}) and switching patterns of the four upper switches (S_{a1}, S_{a2}, S_{b1}, and S_{b2}) during the steady state as obtained by the conventional and proposed methods. In Figure 9a, obtained by the conventional MPC method using all possible switching states, the two capacitor voltages with the NP voltage controlled by the redundant switching states are almost equal with an avoidable oscillation at a certain voltage boundary ΔV_C.

In the proposed method, as shown in Figure 9b, the converter is operated with only a consideration of the reduced number of commutations. It is clearly seen from the switching patterns that the proposed method yields a reduced number of switchings compared with the conventional method. This can lead to a decreased number of switching losses and higher efficiency. In addition, in the proposed method of Figure 9b, the NP voltage balance is well regulated without a continuous increase or decrease in the capacitor voltages, whereas the peak-to-peak ripple voltages of the two capacitors obtained by the proposed method are slightly increased compared with those of the conventional method. The number of switchings and switching losses of the proposed method were reduced by almost half in comparison with the conventional method.

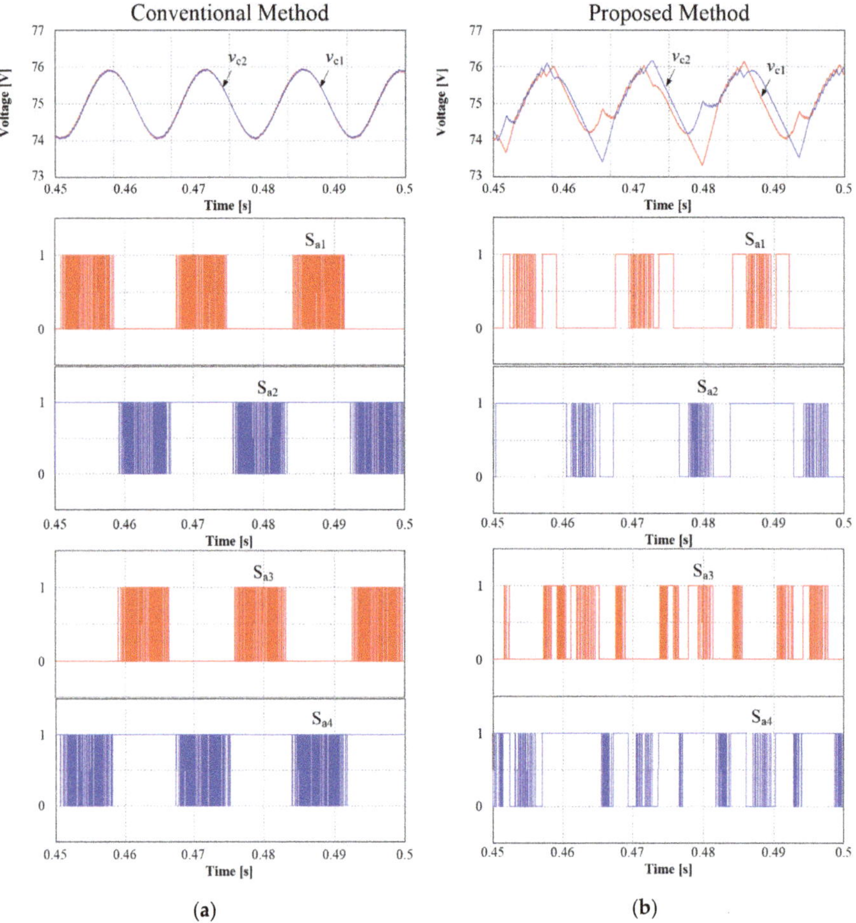

Figure 9. Simulation results of upper and lower capacitor voltages (v_{c1}, v_{c2}) and switching patterns of four upper switches (S_{a1}, S_{a2}, S_{b1}, S_{b2}) during steady state in (**a**) conventional MPC method and (**b**) proposed MPC method.

Figure 10 shows simulation waveforms of the two methods when imbalance conditions of the capacitor voltages, which were intentionally generated, occur. Both the conventional and proposed methods can balance the capacitor voltages, as shown in Figure 10. It is seen that the proposed method, using a reduced number of possible switching states for a reduced number of commutations, can yield an NP voltage balance at almost the same speed as the conventional method.

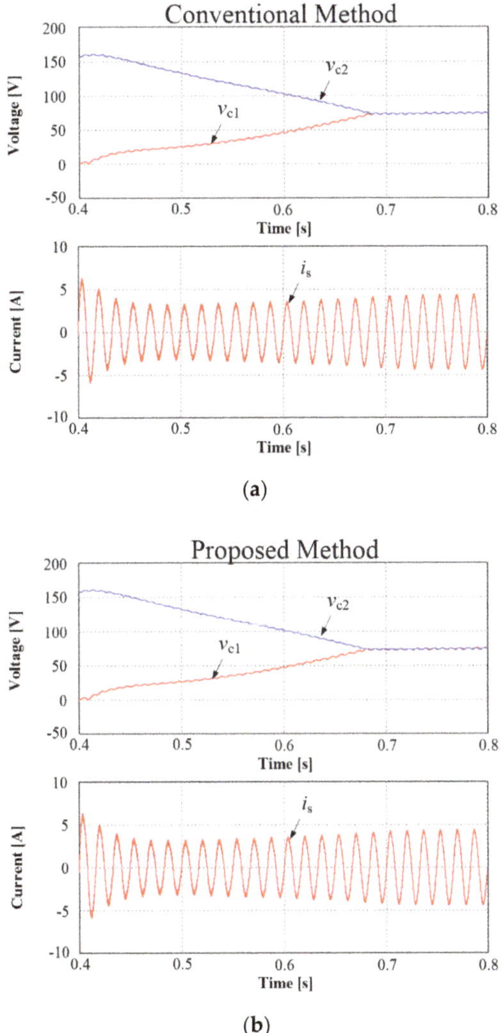

Figure 10. Simulation results of capacitor voltages (v_{c1}, v_{c2}) and source current during imbalanced NP voltage conditions obtained by (**a**) conventional method and (**b**) proposed method.

Figures 11 and 12 show simulation waveforms of step changes of the load resistance and the dc load voltage obtained by the two methods. It is seen that the proposed method achieves dynamic responses as quickly as the conventional method despite the reduced number of possible switching states to decrease the number of switching losses.

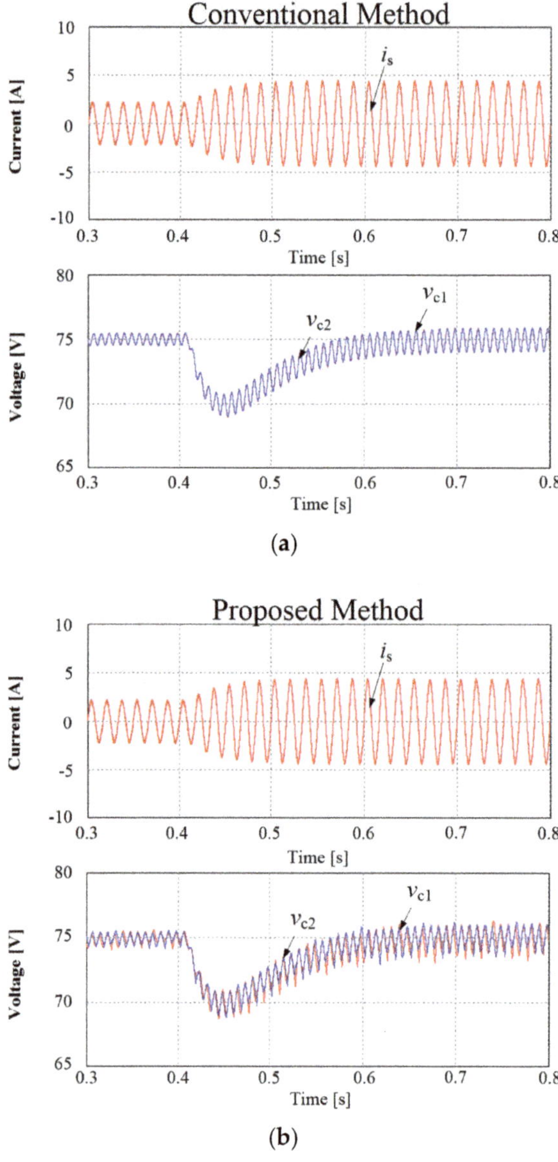

Figure 11. Simulation results of capacitor voltages (v_{c1}, v_{c2}) and source current with step change of load resistor from 200 Ω to 100 Ω obtained by (**a**) conventional method and (**b**) proposed method.

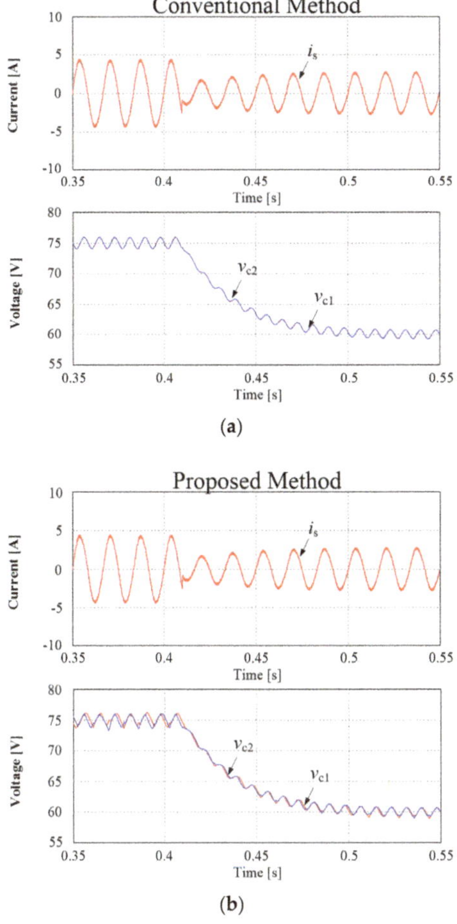

Figure 12. Simulation results of capacitor voltages (v_{c1}, v_{c2}) and source current with step change of dc voltage from 150 V to 120 V obtained by (**a**) conventional method and (**b**) proposed method.

Effects of the control parameter λ_c on performance were investigated, where Figure 13 shows the average number of switching, the THD values of line current, and the peak-to-peak values of capacitor ripple voltage of the conventional and the proposed methods, as a function of the weighting factor λ_c varying from 0.05 to 2. It is shown from Figure 13 that the proposed method results in much lower average number of switching and almost same THD values of the line currents in comparison with the conventional method, over the range of the varying weighting factor. Figure 14. depicts simulation results of ac source current, frequency spectrum of source current, and two capacitor voltages obtained by the conventional and the proposed methods with weighting factor $\lambda_c = 0.05$, $\lambda_c = 0.5$, and $\lambda_c = 2$, respectively. The peak-to-peak value of the two capacitor ripple voltages of the proposed method is slightly increased compared with that of the conventional method, with the trade-off with the reduced number of switching and the consequently decreased switching losses. It is seen that the proposed method with the three different weighting factors regulates the sinusoidal input current well and maintains the two capacitor voltage balancing, even with the lower switching operations than the conventional method.

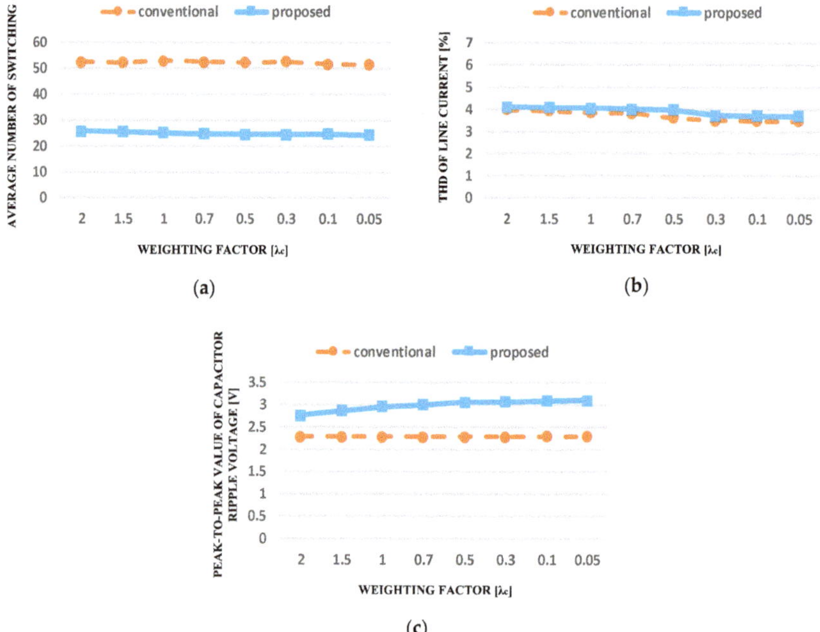

Figure 13. Effects of weighting factor λ_c varying from 0.05 to 2 on (**a**) average number of switching (**b**) THD of line current (**c**) peak-to-peak value of capacitor ripple voltage of the conventional and the proposed methods.

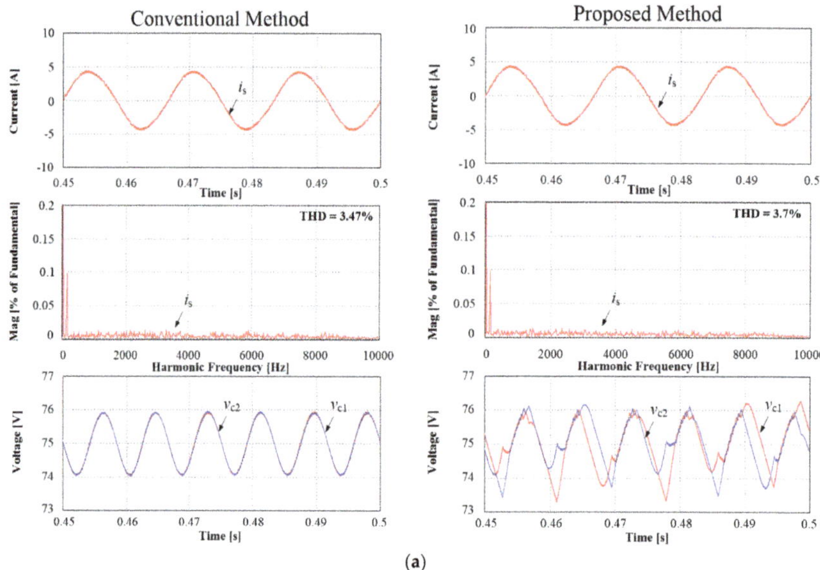

Figure 14. *Cont.*

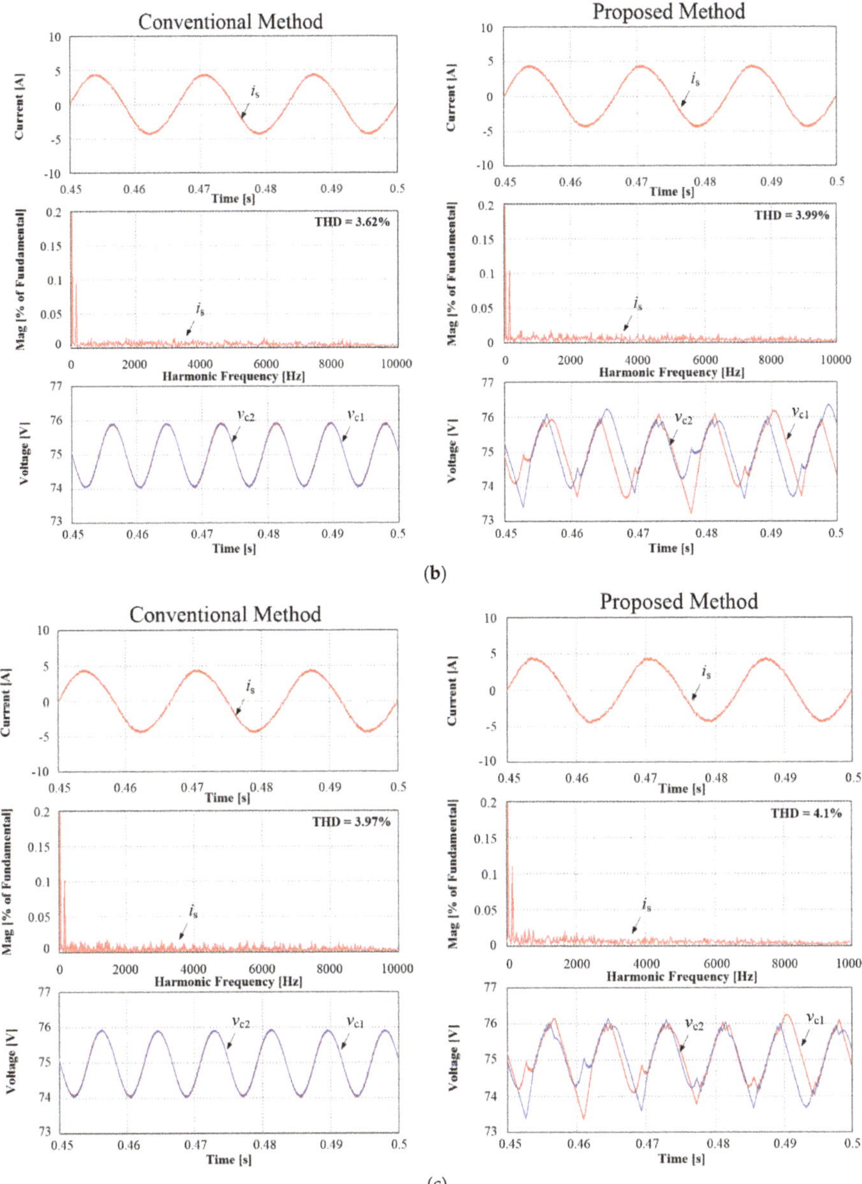

Figure 14. Simulation results of ac source current (i_s), frequency spectrum of input current (i_s), and capacitor voltages (v_{aN}, v_{bN}) obtained by the conventional and the proposed methods with weighting factor (**a**) $\lambda_c = 0.05$, (**b**) $\lambda_c = 0.5$, and (**c**) $\lambda_c = 2$.

Performances with larger input resistance and input inductance were investigated. Figure 15 shows the average number of switching, the THD values of line current, and the peak-to-peak values of capacitor ripple voltage of the conventional and the proposed methods, for several values of the input resistance and the input inductance. It is shown from Figure 15 that the proposed method results in much lower average number of switching and almost same THD values of the line currents in comparison with the conventional method, for the different input parameters. Figure 16. depicts simulation results of ac source current, frequency spectrum of source current, and two capacitor voltages obtained by the conventional and the proposed methods with the three different input resistances and input inductances. The peak-to-peak value of the two capacitor ripple voltages of the proposed method is slightly increased compared with that of the conventional method, with the trade-off with the reduced number of switching and the consequently decreased switching losses. It is seen that the proposed method with the three different input parameters regulates the sinusoidal input current well and maintains the two capacitor voltage balancing, even with the lower switching operations than the conventional method.

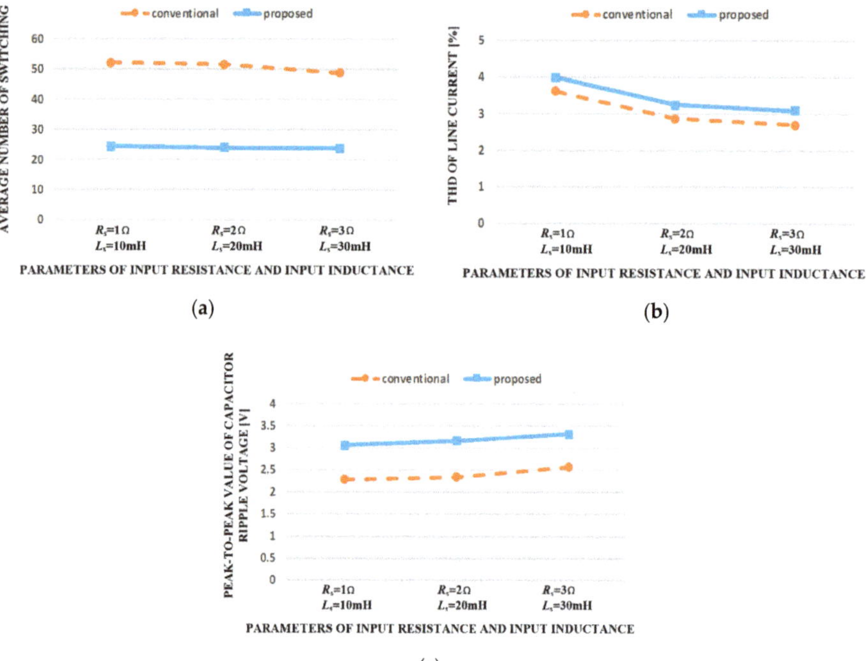

Figure 15. Effects of input resistance and input inductance on (**a**) average number of switching (**b**) THD of line current (**c**) peak-to-peak value of capacitor ripple voltage of the conventional and the proposed methods.

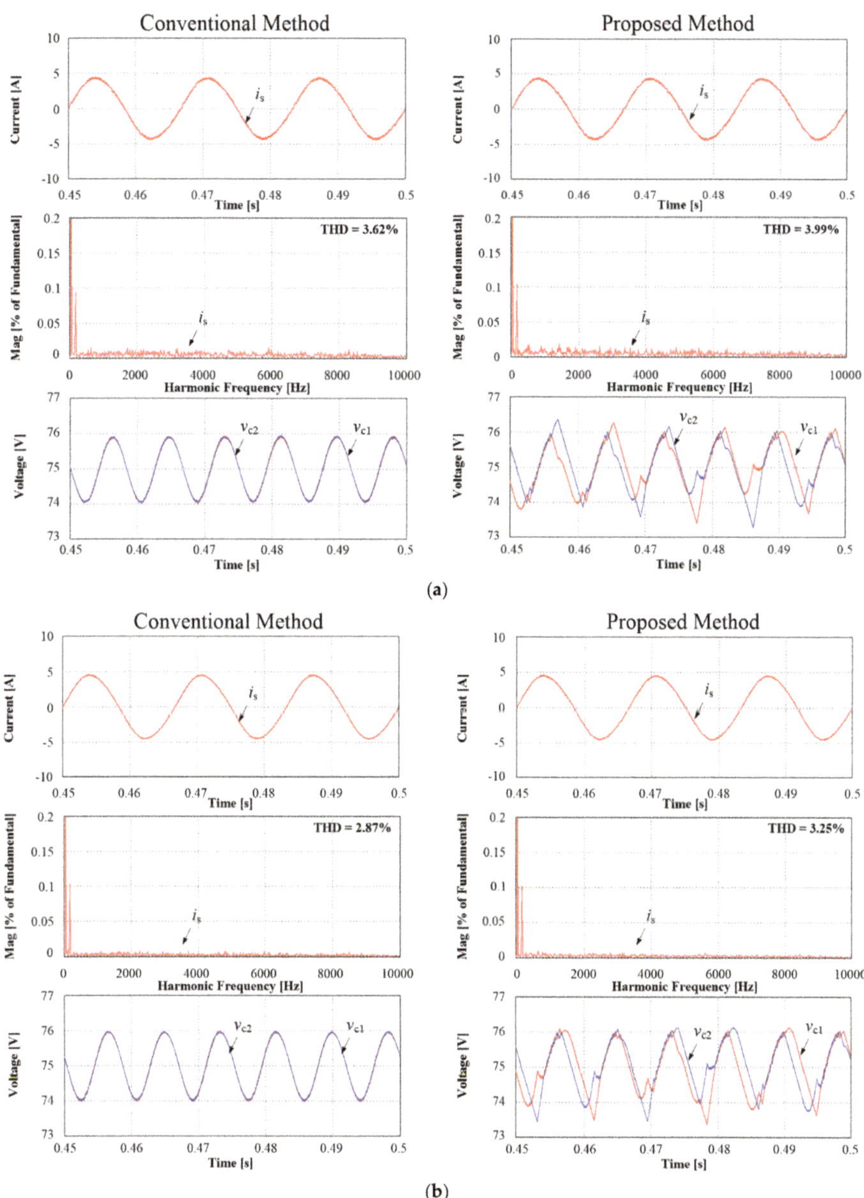

(a)

(b)

Figure 16. *Cont.*

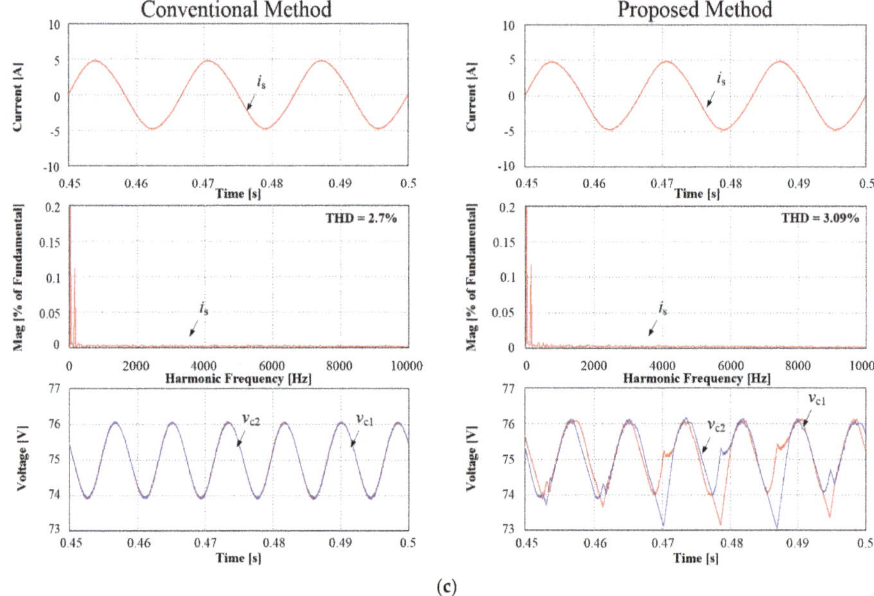

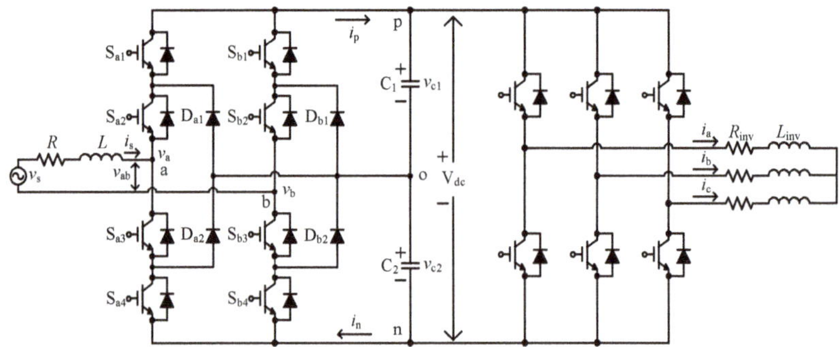

Figure 16. Simulation results of ac source current (i_s), frequency spectrum of input current (i_s), and capacitor voltages (v_{aN}, v_{bN}) obtained by the conventional and the proposed methods with input parameters (**a**) $R_s = 1\ \Omega$ and $L_s = 10$ mH (**b**) $R_s = 2\ \Omega$ and $L_s = 20$ mH, and (**c**) $R_s = 3\ \Omega$ and $L_s = 30$ mH.

The single-phase three-level NPC converter operated with the proposed method was tested with a nonlinear load, which is a three-phase voltage source inverter with a fundamental frequency of 80 Hz as shown in Figure 17. For the purpose of comparison, the simulation results obtained by the conventional method were also included. It is seen from Figure 18 that the single-phase three-level NPC converter with the proposed method well regulates the sinusoidal source current in phase with the source voltage with a low THD value, even with a nonlinear load. In addition, the two capacitor voltages of the proposed method are balanced in a case of the nonlinear load as the same as the linear load, as shown in Figure 18.

Figure 17. Schematic with a three-phase voltage source inverter as a nonlinear load.

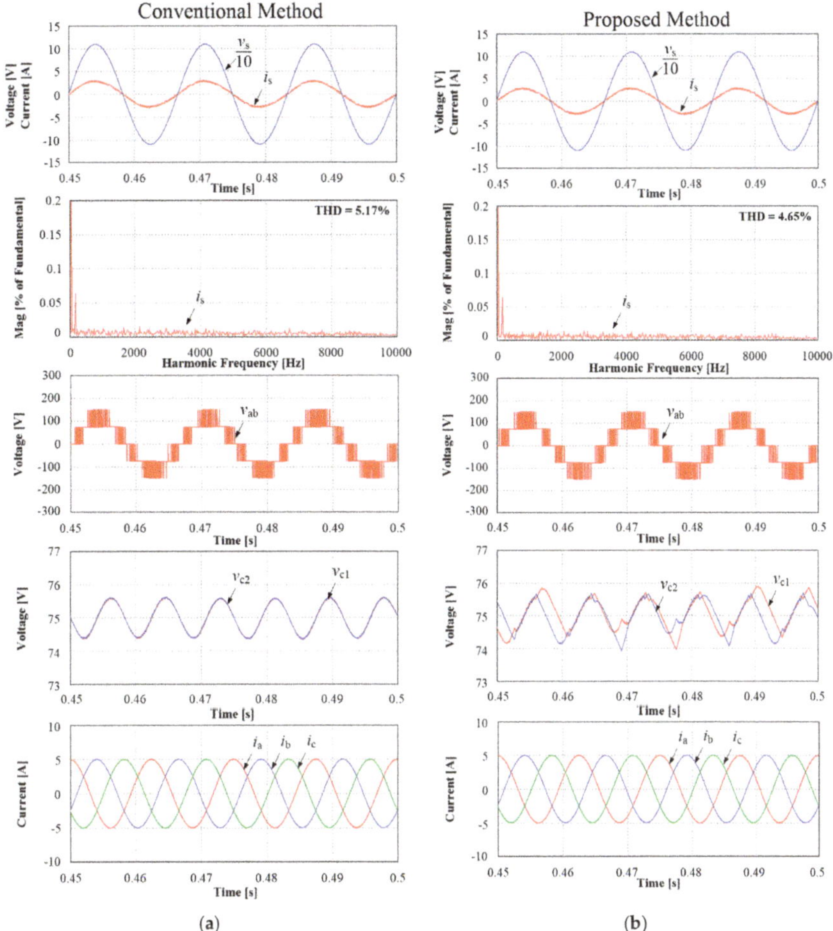

Figure 18. Simulation results with the three-phase voltage source inverter as a nonlinear load: ac source current (i_s), frequency spectrum of input current (i_s), line to line source voltages (v_{ab}), capacitor voltages (v_{c1}, v_{c2}), and three-phase load currents of the voltage source inverter (from top to bottom) obtained by (**a**) the conventional method (**b**) the proposed methods.

A prototype of a single-phase three-level NPC converter, shown in Figure 19, was fabricated in a laboratory to prove the proposed method. The conventional and proposed methods were implemented using a DSP board (TMS320F28335). To compare the performance of the two methods, experiments were conducted under the same conditions as the simulation. Figure 20 shows experimental waveforms of the source voltage/current, converter input voltage, each pole voltage, and an FFT analysis of the source current for the conventional method and the proposed method during steady-state conditions.

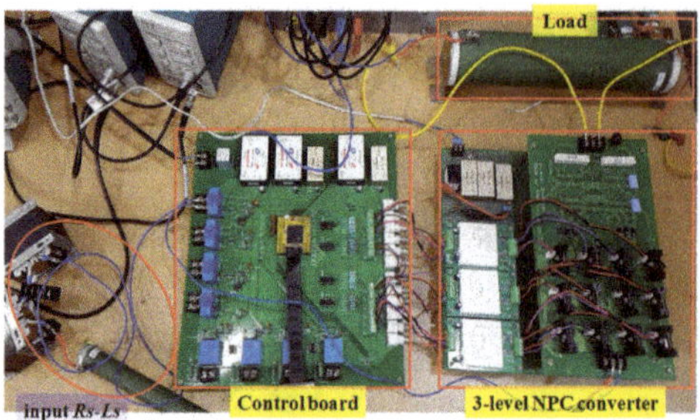

Figure 19. Photograph of prototype setup for single-phase three-level NPC converter.

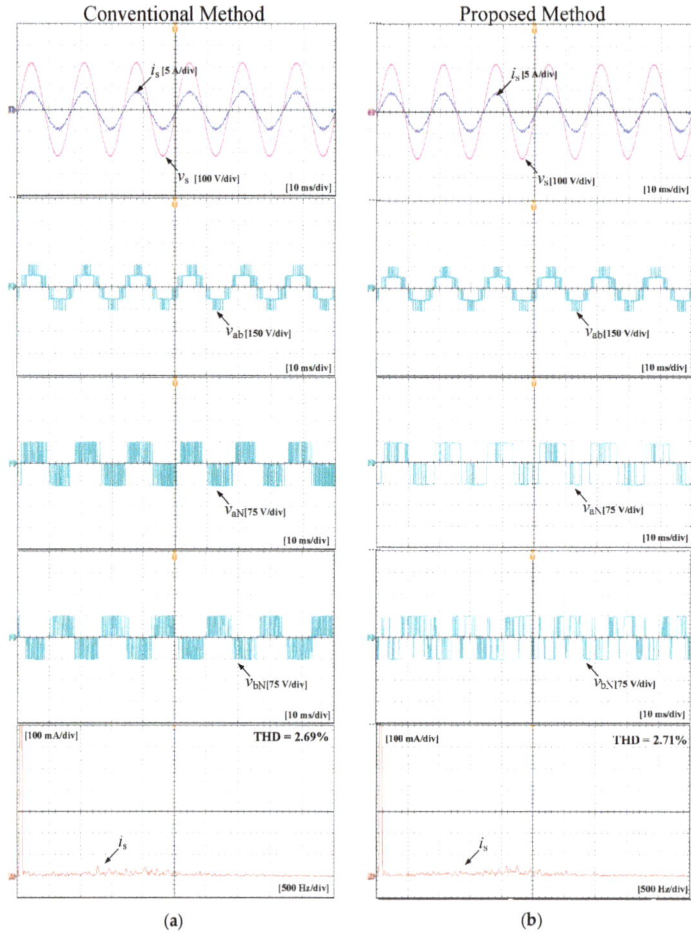

Figure 20. Experimental results of ac source current (i_s) and source voltage (v_s), converter input voltage (v_{ab}), pole voltage (v_{aN}, v_{bN}), and FFT analysis of source current (i_s) in (**a**) conventional and (**b**) proposed methods.

As in the simulation, the proposed method shows almost the same source current and converter input voltage waveforms as the conventional method. In addition, the proposed method through the FFT analysis shows performance that is very similar to that of the conventional method. On the other hand, as shown in Figure 20, the proposed method has a quite different pole voltage from the conventional method owing to the reduced number of switchings.

Figure 21 shows the upper and lower capacitor voltages, source current, and switching state in the steady state. As shown in Figure 21, the proposed method reduces the number of switchings in comparison with the conventional method. In addition, the NP voltage balance in the proposed method is well regulated without a continuous increase or decrease in the capacitor voltages, whereas the peak-to-peak ripple voltages of the two capacitors obtained by the proposed method is slightly increased compared with the conventional method.

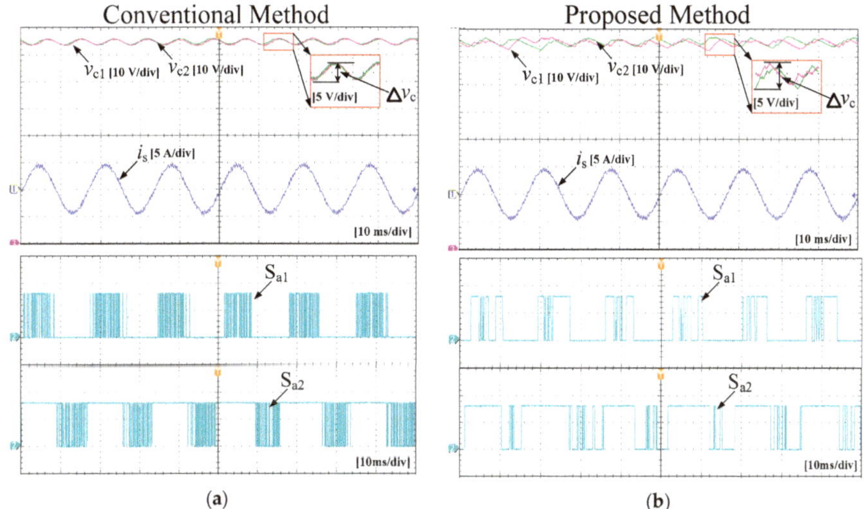

Figure 21. Experimental results of capacitor voltages (v_{c1}, v_{c2}), source current (i_s), and upper switch of *a* leg (S_{a1}, S_{a2}) during steady state in (**a**) conventional method and (**b**) proposed method.

In Figure 22, experimental waveforms of the two methods are shown when imbalanced NP voltage conditions of the capacitor voltages, which are intentionally generated, occur. Both the conventional and proposed methods can balance the capacitor voltages, as shown in Figure 22. This is the same as the simulation results of Figure 10. It is seen that the proposed method, using a reduced number of possible switching states for a reduced number of commutations, can yield an NP voltage balance at almost the same speed as the conventional method. Figures 23 and 24 show experimental waveforms of step changes of the load resistance and the dc load voltage obtained by the two methods. It is seen that the proposed method achieves dynamic responses as quickly as the conventional method despite the reduced number of possible switching states to decrease the number of switching losses.

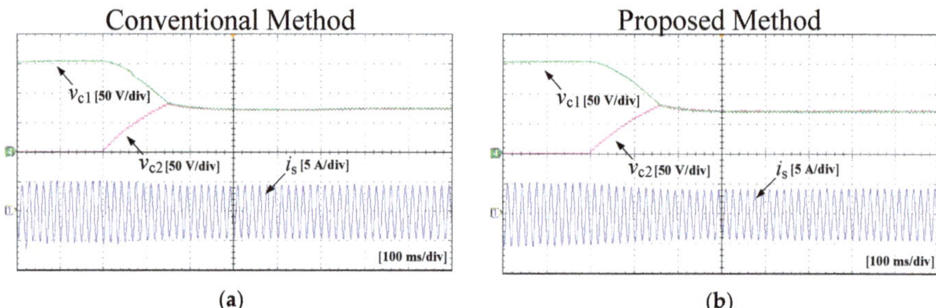

Figure 22. Experimental results of capacitor voltages (v_{c1}, v_{c2}) and source current during imbalanced NP voltage conditions obtained by (**a**) conventional method and (**b**) proposed method.

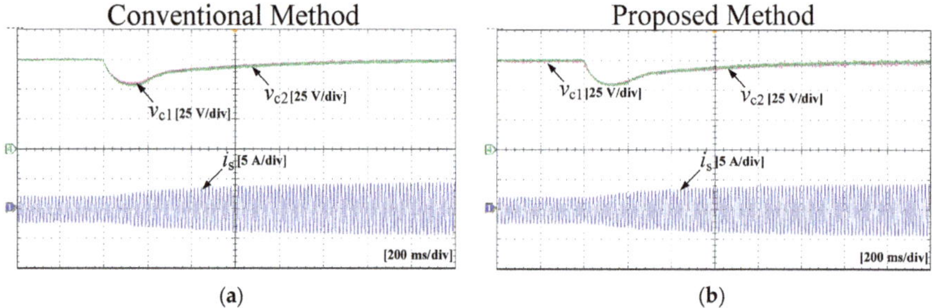

Figure 23. Experimental results of capacitor voltages (v_{c1}, v_{c2}) and source current with step change of load resistor from 200 Ω to 100 Ω obtained by (**a**) conventional method and (**b**) proposed method.

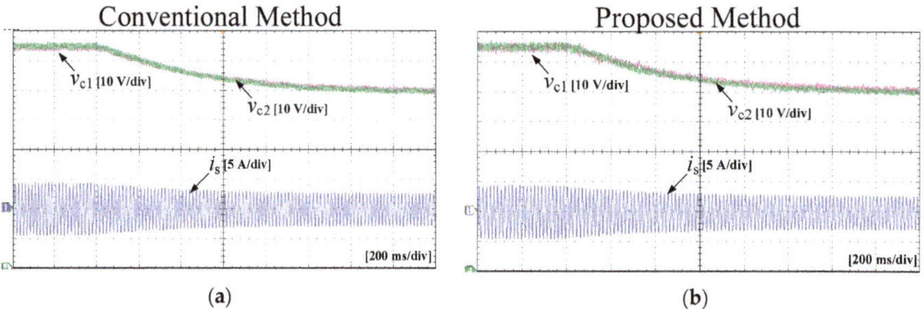

Figure 24. Experimental results of capacitor voltages (v_{c1}, v_{c2}) and source current with step change of dc voltage from 150 V to 120 V obtained by (**a**) conventional method and (**b**) proposed method.

5. Conclusions

This paper proposed a highly efficient algorithm with a reduced number of switchings and low switching losses for single-phase three-level NPC converters based on an MPC method with a decreased number of commutations of switches. The proposed method pre-excludes, from the candidates for possible future switching states, the switching states that yield more than two commutations in the next sampling period. As a result, the proposed technique can reduce the number of switchings and the switching losses by utilizing switching states involving no commutation or only one commutation at every sampling instant for single-phase three-level NPC converters. In addition, the developed method does not deteriorate the input current quality or input power factor despite the reduced switching numbers and losses. Although the proposed method slightly increases the peak-to-peak variations of the two dc capacitor voltages at the expense of reduced commutation, the increased voltage variation is not high. Thus, the proposed method can obtain high efficiency and low switching losses at the expense of slightly increased peak-to-peak variations of the NP voltage. Simulations and experimental results were presented to verify the effectiveness of the proposed method.

Author Contributions: All authors contributed to this work by collaboration.

Acknowledgments: This research was supported by the National Research Foundation of Korea (NRF) grant funded by the Korean government (MSIP) (2017R1A2B4011444) and the Human Resources Development (No.20174030201810) of the Korea Institute of Energy Technology Evaluation and Planning (KETEP) grant funded by the Korea government Ministry of Trade, Industry and Energy.

Conflicts of Interest: The authors declare no conflict of interest.

Nomenclature

Abbreviations

EMI	Electromagnetic interference
NPC	Neutral point clamped
CBPWM	Carrier-based pulse width modulation
PI	Proportional and integral
NP	Neutral point
MPC	Model predictive control
THD	Total harmonic distortion
IGBT	Insulated gate bipolar transistor

Variables

S_a	Operating status of a-phase
S_b	Operating status of b-phase
v_a	Source voltage
i_a	Source current
i_a^*	Reference source current
R_s	Input resistance
L_s	Input inductance
v_{ab}	Line-to-line source voltage
v_{aN}	Phase voltage of a-phase
v_{bN}	Phase voltage of b-phase
C_1	Output capacitance in upper capacitor
C_2	Output capacitance in lower capacitor
i_u	Current of upper dc-bus bar
i_l	Current of lower dc-bus bar
R_L	Output load resistance
V_{dc}	Output dc voltage
g	Cost function
λ_c	Weighting factor
T_s	Sampling period

References

1. Portillo, R.; Prats, M.M.; Leon, J.I.; Sanchez, J.A.; Carrasco, J.M.; Galvan, E.; Franquelo, L.G. Modeling strategy for back-to-back three-level converters applied to high-power wind turbines. *IEEE Trans. Ind. Electron.* **2006**, *53*, 1483–1491. [CrossRef]

2. Alepuz, S.; Busquets-Monge, S.; Bordonau, J.; Gago, J.; Gonzalez, D.; Balcells, J. Interfacing renewable energy sources to the utility grid using a three-level inverter. *IEEE Trans. Ind. Electron.* **2006**, *53*, 1504–1511. [CrossRef]

3.   Stala, R. Application of balancing circuit for DC-link voltages balance in a single-phase diode-clamped inverter with two three-level legs. *IEEE Trans. Ind. Electron.* **2011**, *58*, 4185–4195. [CrossRef]

4.   Song, W.; Feng, X.; Smedley, K. A carrier-based PWM strategy with the offset voltage injection for single-phase three-level neutral-point-clamped converters. *IEEE Trans. Power Electron.* **2013**, *28*, 1083–1095. [CrossRef]

5.   Choi, U.-M.; Lee, H.-H.; Lee, K.-B. Simple neutral-point voltage control for three-level inverters using a discontinuous pulse width modulation. *IEEE Trans. Energy Convers.* **2013**, *28*, 434–443. [CrossRef]

6.   Song, W.; Wang, S.; Xiong, C.; Ge, X.; Feng, X. Single-phase three-level space vector pulse width modulation algorithm for grid-side railway tractionconverter and its relationship of carrier-based pulse width modulation. *IET Electr. Syst. Transp.* **2014**, *14*, 78–87. [CrossRef]

7.   Freitas, I.S.D.; Bandeira, M.M.; Barros, L.D.M.; Jacobina, C.B.; Santos, E.C.D.; Salvadori, F.; Silva, S.A.D. A carrier-based PWM technique for capacitor voltage balancing of single-phase three-level neutral-point-clamped converters. *IEEE Trans. Ind. Appl.* **2015**, *51*, 3227–3235. [CrossRef]

8.   Habibullah, M.; Lu, D.D.; Xiao, D.; Fletcher, J.E.; Rahman, M.F. Predictive torque control of induction motor sensorless drive fed by a 3L-NPC inverter. *IEEE Trans. Ind. Inf.* **2017**, *13*, 60–70. [CrossRef]

9.   Yaramasu, V.; Wu, B. Predictive control of a three-level boost converter and an NPC inverter for high-power PMSG-based medium voltage wind energy conversion systems. *IEEE Trans. Power Electron.* **2014**, *29*, 5308–5322. [CrossRef]

10. Acuna, P.; Moran, L.; Rivera, M.; Aguilera, R.; Burgos, R.; Agelidis, V.G. A single-objective predictive control method for a multivariable single-phase three-level NPC converter-based active power filter. *IEEE Trans. Ind. Electron.* **2015**, *62*, 4598–4607. [CrossRef]

11. Song, W.; Ma, J.; Zhou, L.; Feng, X. Deadbeat predictive power control of single-phase three-level neutral-point-clamped converters using space-vector modulation for electric railway traction. *IEEE Trans. Power Electron.* **2016**, *31*, 1083–1095. [CrossRef]

12. Acuna, P.; Aguilera, R.P.; Ghias, A.M.Y.M.; Rivera, M.; Baier, C.R.; Agelidis, V.G. Cascade-free model predictive control for single-phase grid-connected power converters. *IEEE Trans. Ind. Electron.* **2017**, *64*, 721–732. [CrossRef]

13. Deng, Y.; Li, J.; Shin, K.H.; Viitanen, T.; Saeedifard, M.; Harley, R.G. Improved modulation scheme for loss balancing of three-level active NPC converters. *IEEE Trans. Power Electron.* **2017**, *32*, 2521–2532. [CrossRef]

14. Jing, X.; He, J.; Demerdash, N.A.O. Loss balancing SVPWM for active NPC converters. In Proceedings of the 2014 IEEE Applied Power Electronics Conference and Exposition—APEC 2014, Fort Worth, TX, USA, 16–20 March 2014.

15. Pulikanti, S.; Dahidah, M.S.A.; Agelidis, V. Voltage balancing control of three-level active NPC converter using SHE-PWM. *IEEE Trans. Power Deliv.* **2011**, *26*, 258–267. [CrossRef]

16. Bruckner, T.; Bernet, S.; Guldner, H. The active NPC converter and its loss-balancing control. *IEEE Trans. Ind. Electron.* **2005**, *52*, 855–868. [CrossRef]

17. Floricau, D.; Gateau, G.; Leredde, A.; Teodorescu, R. The efficiency of three-level active NPC converter for different PWM strategies. In Proceedings of the 2009 13th European Conference on Power Electronics and Applications, Barcelona, Spain, 8–10 September 2009.

18. Bruckner, T.; Bernet, S.; Steimer, P. Feedforward loss control of three-level active NPC converters. *IEEE Trans. Ind. Appl.* **2007**, *43*, 1588–1596. [CrossRef]

19. Brückner, T.; Bernet, S. Loss balancing in three-level voltage source inverters applying active NPC switches. In Proceedings of the IEEE Power Electronics Specialists Conference (PESC), Vancouver, BC, Canada, 17–21 June 2001; pp. 1135–1140.

20. Ma, L.; Kerekes, T.; Rodriguez, P.; Jin, X.; Teodorescu, R.; Liserre, M. A new PWM strategy for grid-connected half-bridge active NPC converters with losses distribution balancing mechanism. *IEEE Trans. Power Electron.* **2015**, *30*, 5331–5340. [CrossRef]

21. Jiao, Y.; Lee, F.C. New modulation scheme for three-level active neutral-point-clamped converter with loss and stress reduction. *IEEE Trans. Ind. Electron.* **2015**, *62*, 5468–5479. [CrossRef]

22. Habibullah, M.; Lu, D.D.; Xiao, D.; Rahman, M.F. Finite-state predictive torque control of induction motor supplied from a three-level NPC voltage source inverter. *IEEE Trans. Power Electron.* **2017**, *32*, 479–489. [CrossRef]

23. Vargas, R.; Ammann, U.; Rodriguez, J. Predictive approach to increase efficiency and reduce switching losses on matrix converters. *IEEE Trans. Power Electron.* **2009**, *24*, 894–902. [CrossRef]
24. Vargas, R.; Rodriguez, J.; Rojas, C.A.; Rivera, M. Predictive control of an induction machine fed by a matrix converter with increased efficiency and reduced common-mode voltage. *IEEE Trans. Energy Convers.* **2014**, *29*, 473–485.
25. Vatani, M.; Bahrani, B.; Saeedifard, M.; Hovd, M. Indirect finite control set model predictive control of modular multilevel converters. *IEEE Trans. Smart Grid* **2015**, *6*, 1520–1529. [CrossRef]
26. Kwak, S.; Park, J. Switching strategy based on model predictive control of VSI to obtain high efficiency and balanced loss distribution. *IEEE Trans. Power Electron.* **2014**, *29*, 4551–4567. [CrossRef]
27. Kwak, S.; Park, J. Predictive control method with future zero-sequence voltage to reduce switching losses in three-phase voltage source inverters. *IEEE Trans. Power Electron.* **2015**, *30*, 1558–1566. [CrossRef]
28. Kukrer, O. Discrete-time current control of voltage-fed three-phase PWM inverters. *IEEE Trans. Power Electron.* **1996**, *11*, 260–269. [CrossRef]
29. Hildetirand, F.B. *Introduction to Numerical Analysis*; McGraw-Hill: New York, NY, USA, 1956.
30. Venkatraman, K.; Selvan, M.P.; Moorthi, S. Predictive current control of distribution static compensator for load compensation in distribution system. *IET Gener. Transm. Distrib.* **2016**, *10*, 2410–2423. [CrossRef]
31. Kumar, C.; Mishra, M.K. Operation and control of an improved performance interactive DSTATCOM. *IEEE Trans. Ind. Electron.* **2015**, *62*, 6024–6034. [CrossRef]

Article

A Method for the Simultaneous Suppression of DC Capacitor Fluctuations and Common-Mode Voltage in a Five-Level NPC/H Bridge Inverter

Ming Wu [1], Zhenhao Song [1,*], Zhipeng Lv [1], Kai Zhou [2] and Qi Cui [2]

[1] China Electric Power Research Institute, Haidian District, Beijing 100192, China; wuming@epri.sgcc.com.cn (M.W.); lvzhipeng@epri.sgcc.com.cn (Z.L.)

[2] Jiangsu Province Laboratory of Mining Electric and Automation, China University of Mining and Technology, Xuzhou 221008, China; zhoukai659274@163.com (K.Z.); kuangdarencuiqi@sina.com (Q.C.)

* Correspondence: songzhenhao@epri.sgcc.com.cn; Tel.: +86-1821-054-0873

Received: 7 December 2018; Accepted: 20 February 2019; Published: 26 February 2019

Abstract: To suppress the direct current (DC) capacitor voltage fluctuations and the common-mode voltage (CMV) in a three-phase, five-level, neutral-point-clamped (NPC)/H-bridge inverter, this paper analyzes the influence of all voltage vectors on the neutral point potential of each phase under different pulse mappings in detail with an explanation of the CMV distribution. Then, based on the traditional space vector pulse width modulation (SVPWM) algorithm, a dual-pulse-mapping algorithm is proposed to suppress the DC capacitor fluctuations and the CMV simultaneously. In the algorithm, the reference voltage synthesis selects the voltage vector that has the smallest CMV value as the priority. In addition, the two kinds of pulse mappings that have opposite effects on the neutral point potential are switched to output. At the same time, regulating factors are introduced to adjust the working time of each voltage vector under the two pulse mappings; then, the capacitor voltages can be balanced. Both the simulation and experiment demonstrate the algorithm's effectiveness.

Keywords: NPC/H Bridge; five-level; Balance of capacitor voltage; Suppression of CMV; SVPWM

---

1. Introduction

In the past several decades, multilevel power topologies have increasingly been used in high-power, medium-voltage drives [1,2]. Due to the advantages of a higher voltage capability, lower switching frequencies, better power quality, and smaller voltage jumps (dv/dt), multilevel converters have become more popular and are applied to high-voltage variable frequencies, flexible alternating current (AC) transmissions, high-voltage direct current (HVDC), and so on [3]. Several strategies have been proposed for multilevel converters, such as sinusoidal pulse width modulation (SPWM), selective harmonic elimination pulse width modulation (SHEPWM), and space vector pulse width modulation (SVPWM). Compared to SPWM, SVPWM has lower total harmonic distortion (THD) and a high utilization of direct current (DC)-side voltage [4]. SHEPWM is kind of offline strategy that requires excellent storage performance [5]. Moreover, various multilevel converter topologies have been studied, such as neutral-point-clamped (NPC) [6], cascaded H-Bridge (CHB) [7], modular multilevel converter (MMC) [8], and Matrix converter [9]. Among all of the topologies, the neutral-point-clamped and the cascaded H-Bridge are the two most widely used, and they are suitable for medium-voltage, high-power drives [10,11]. By combining the NPC and H-bridge topologies, a five-level NPC/H-bridge topology was proposed in 1990 [12] and since then has been applied in industrial drive applications. Figure 1 shows a schematic diagram of a three-phase, five-level NPC/H bridge inverter's main circuit with RL (resistance and inductance) load. Each phase has an independent DC power V_{dc}, and the DC-side is connected by two capacitors in series. Under ideal conditions, the voltage of each is $V_{dc}/2$.

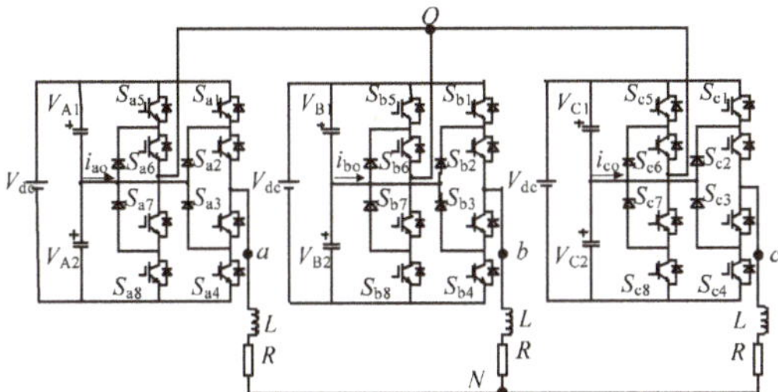

Figure 1. A diagram of a five-level, neutral-point-clamped (NPC)/H-bridge inverter's main circuit.

However, the NPC/H bridge topology has two crucial problems: an imbalance in capacitor voltages in three phases and a requirement that the common-mode voltage be suppressed. In fact, there are some reports about solving the imbalance in capacitor voltages of the NPC/H bridge topology. Most use two six-pulse rectifying devices to obtain two independent DC powers to supply each phase, which not only increases the complexity of the topology, but also increases the cost of the system circuit [13,14].

The best way to solve the problem is to utilize software balancing techniques because of the cost. In other five-level topologies, the object function optimization, zero sequence voltage injection, and virtual space-vector PWM methods are common techniques that mainly utilize the redundant vectors with a different combination of switching sequences [15,16]. In [15], for a five-level, diode-clamped converter (DCC), the duty cycles are calculated for the redundant states and adjusted between two zero vectors to control the capacitor voltages. This method is not suitable for the NPC/H bridge topology because the redundant states have the same influences on the neutral potential of capacitors. In [16], the authors propose to solve the capacitor voltage balancing issue of the five-level DCC based on a Model Predictive Control strategy. Although this strategy balances the capacitor voltages, it causes voltage jumps that could damage the switches.

In addition, the common-mode voltage in the NPC/H topology also deserves more attention. High common-mode voltage and its common-mode current will produce electromagnetic interference that will damage the motor. There are many methods to solve the problem. In [17], a five-level NPC/H inverter is considered to be equivalent to two three-level NPC inverters. One equivalent inverter always operates with zero common-mode voltage to suppress CMV, and the other one operates with the conventional three-level SVPWM. This measure can inhibit the CMV to $V_{dc}/3$. Another author proposes a strategy called hierarchical model predictive voltage control (HMPVC) [18], which can suppress CMV effectively.

In this paper, a dual-pulse-mapping algorithm is proposed for the simultaneous suppression of DC-side capacitor midpoint potentials and CMV in five-level NPC/H bridge inverter. In fact, the algorithm is suitable for NPC/H-Bridge converters with any number of voltage levels. In the analysis in Section 2, each voltage vector is found to correspond to a variety of pulse mappings, so the two pulse mappings (A) and (C), which have completely opposite effects on the midpoint potential of capacitors in each phase, are selected. In the proposed algorithm, the first step is to select the voltage vectors with the smallest CMV based on a line-voltage coordinate system. In order to ensure the output performance, the target of the algorithm is to suppress the CMV to $V_{dc}/6$. The second step is to synthesize the reference voltage based on space vector modulation (SVM) with six segments. The first three segments work under pulse mapping (A) and the other three segments work under pulse mapping (C). The key of the algorithm is to choose regulating factor for each vector. The value of the factor decides the working time of each vector that influences the charge or discharge of capacitors under the two pulse mappings. So, adjusting the regulating factor can balance the capacitor voltages effectively. The algorithm proposed in this paper can perfectly solve the two crucial problems of the

NPC/H bridge topology, which makes the NPC/H bridge topology play a better role in high-voltage and high-power applications.

2. Switching States of the Five-Level NPC/H Bridge Converter

In Figure 1, each phase has four pairs of complementary switches. They are (S_{x1}, S_{x3}), (S_{x2}, S_{x4}), (S_{x5}, S_{x7}), and (S_{x6}, S_{x8}), x = (a, b, c). Each phase has five switching states, such as $S_x = \{2, 1, 0, -1, -2\}$, where every state has different pulse mappings, which are shown in Table 1.

Table 1. The switching states and pulse mappings.

Switching State S_x	Pulse Mapping Number	$S_{x1}, S_{x2}, S_{x3}, S_{x4}, S_{x5}, S_{x6}, S_{x7}, S_{x8}$	Phase Voltage u_{xo}
2	1	11000011	V_{dc}
1	2	11000110	$V_{dc}/2$
	3	01100011	
0	4	11001100	0
	5	01100110	
	6	00110011	
−1	7	01101100	$-V_{dc}/2$
	8	00110110	
−2	9	00111100	$-V_{dc}$

Based on the main circuit diagram of the system shown in Figure 1 and the corresponding switching states in Table 1, the output voltage of each phase of the inverter is

$$\begin{cases} u_{ao} = V_{dc} * S_a/2 \\ u_{bo} = V_{dc} * S_b/2 \\ u_{co} = V_{dc} * S_c/2 \end{cases} \tag{1}$$

3. Analysis of the Midpoint Potential and CMV

3.1. Mechanism of DC-Side Capacitor Voltage Imbalance

We take the phase A of the five-level NPC/H Bridge inverter as an example to analyze the effect of nine pulse mappings on the midpoint potential.

Figure 2 shows the circuit of pulse mapping 1, where S_{x1}, S_{x2}, S_{x7}, S_{x8} are equal to 1. The phase voltage is V_{dc} and the midpoint potential has no change because the current of the load is not linked to the midpoint potential. Similar to pulse mapping 1, the circuits of pulse mapping 4, 5, 6, and 9 also have no influence on the midpoint potential.

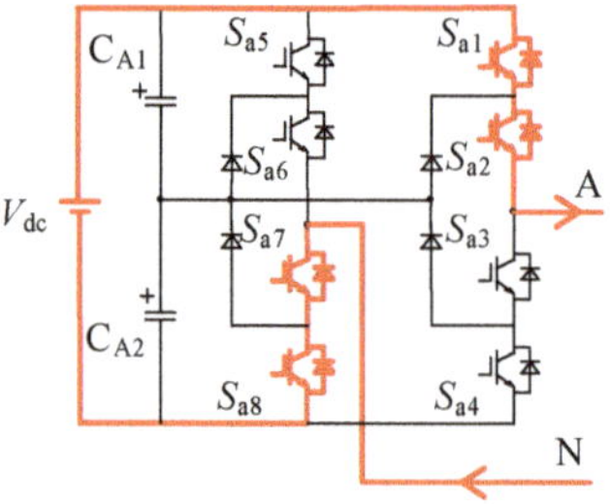

Figure 2. The circuit of pulse mapping 1.

Figure 3 shows the circuit of pulse mapping 2. Assuming that the direction of phase current i_a is the same as in Figure 3, the current flows through S_{a1}, S_{a2}, S_{a7}, and D_{a4} and then arrives at the midpoint potential. Due to the effect of i_a, the midpoint potential increases and the voltage of capacitor C_{A1} decreases. If the current direction is opposite to that shown in Figure 3, the current flows through D_{a3}, S_{a6} and the freewheeling diodes of S_{a2}, S_{a1}, then arrives at the positive pole. At this time, the midpoint potential decreases and the voltage of capacitor C_{A1} increases. In these two situations, the current has an opposite influence on the midpoint potential. Figure 4 shows the circuit of pulse mapping 3. If the current direction is same as the definition of Figure 4, the current flows through D_{a1}, S_{a2}'s freewheeling diode, S_{a7} and S_{a8}, then arrives at the negative pole. The midpoint potential decreases and the voltage of capacitor C_{A1} increases. If the current direction is opposite to the definition, the current flows through the freewheeling diodes of S_{a8} and S_{a7}, S_{a3}, and D_{a2}. At this time, the midpoint potential increases and the voltage of capacitor C_{A1} decreases. We can conclude that different directions of i_a have different effects on the midpoint potential. The changes caused by pulse mapping 7 and 8 are same as those caused by pulse mapping 2 and 3. Table 2 shows the changes in midpoint potential when using pulse mapping 2, 3, 7, and 8 with opposite direction currents.

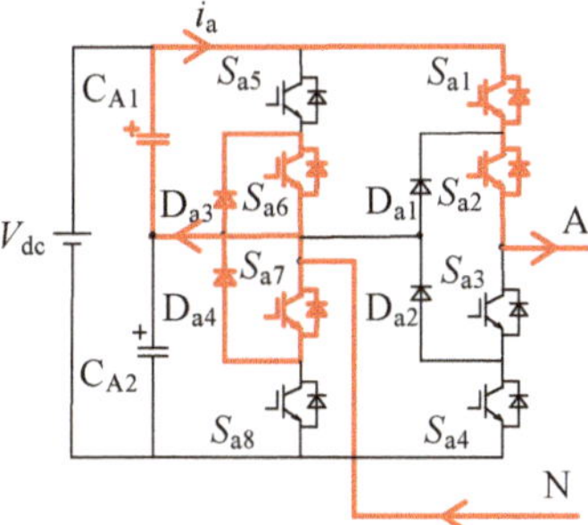

Figure 3. The circuit of pulse mapping 2.

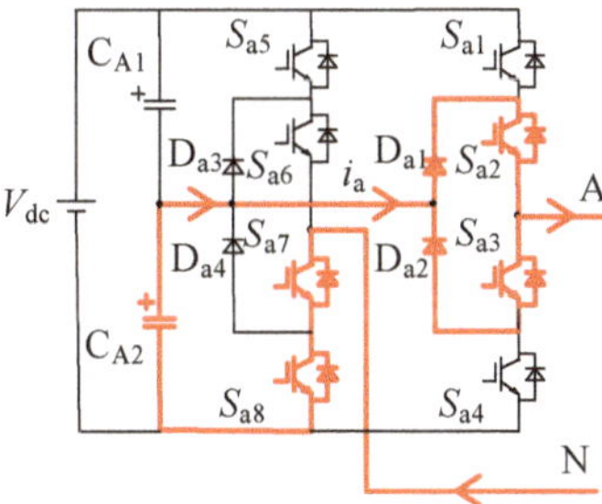

Figure 4. The circuit of pulse mapping 3.

Table 2. Changes in midpoint potential in pulse mapping 2, 3, 7, and 8 with different direction currents.

Switching State S_x	Pulse Mapping Number	$S_{x1}, S_{x2}, S_{x3}, S_{x4},$ $S_{x5}, S_{x6}, S_{x7}, S_{x8}$	Direction of i_a	Midpoint Potential
1	2	11000110	+ / −	↑ / ↓
1	3	01100011	+ / −	↓ / ↑
−1	7	01101100	+ / −	↑ / ↓
−1	8	00110110	+ / −	↓ / ↑

As shown in Table 2, the changes caused by pulse mapping 2 are the total opposite of those caused by pulse mapping 3. This situation also exists between pulse mapping 7 and 8. So, we select pulse mapping 2 and 7, which have the same influence on the midpoint potential, to form pulse mapping (A), and pulse mapping 3 and 8 constitute pulse mapping (B), as listed in Table 3.

Based on the analysis of Tables 2 and 3, we know that pulse mapping (A) and (B) have a completely opposite effect on the midpoint potential whatever the direction of i_a is. If the pulse mapping (A) and (B) are used properly, the midpoint potential can be balanced.

Table 3. The two pulse mappings of the five-level NPC/H-Bridge inverter.

State	(A)	(B)
2	11000011	11000011
1	11000110	01100011
0	01100110	01100110
−1	01101100	00110110
−2	00111100	00111100

3.2. Distribution of CMV

In a five-level NPC/H bridge inverter, the common-mode voltage U_{CMV} is expressed as Equation (2):

$$U_{CMV} = \frac{u_{ao} + u_{bo} + u_{co}}{3} \tag{2}$$

where u_{ao}, u_{bo}, and u_{co} are the output voltage of each phase, which are obtained by Equation (1). Their values are: $\pm V_{dc}$, $\pm V_{dc}/2$, 0. So, all of the values of U_{CMV} can be calculated as: $\pm V_{dc}$, $\pm 5V_{dc}/6$, $\pm 2V_{dc}/3$, $\pm V_{dc}/2$, $\pm V_{dc}/3$, $\pm V_{dc}/6$, and 0. The common-mode voltage distribution of the five-level NPC/H bridge inverter is shown in Figure 5 (where the V_{dc} is omitted).

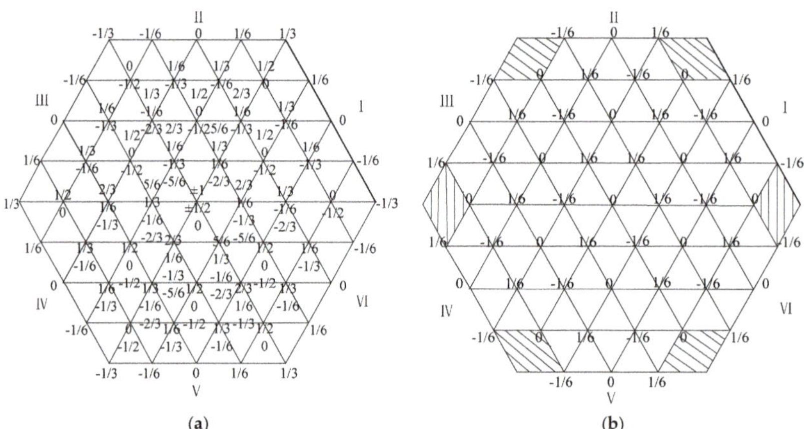

Figure 5. The common-mode voltage distribution. (**a**) Common-mode-voltage distribution diagram of traditional strategy; and (**b**) Common-mode-voltage distribution diagram of proposed strategy.

Figure 5a shows the traditional common-mode voltage distribution of the five-level NPC/H bridge inverter. In order to suppress the common-mode voltage of the system while not affecting the output performance of the inverter, this paper retains the vector that has the smallest common-mode voltage among all the redundant vectors. Additionally, the six vertex vectors of the outer hexagon are eliminated. So, the common-mode voltage distribution in Figure 5a can be simplified as shown in Figure 5b.

When the reference voltage is located in the shaded quadrangle of Figure 5b, the closest three voltage vectors are utilized in the synthesis. Thus, the common-mode voltage of the five-level NPC/H bridge inverter is suppressed to $\pm V_{dc}/6$ and 0 while the output performance of the inverter is ensured.

4. The Proposed Dual-Pulse-Mapping Algorithm

In order to balance the DC-side capacitor voltages of a five-level NPC/H bridge inverter and suppress the CMV effectively at the same time, a dual-pulse-mapping algorithm is proposed. From the analysis above, we found that mapping (A) and (B) have the completely opposite influence on the DC-side capacitor voltages. When mapping (A) and (B) work, the influences of each vector on the three-phase midpoint potentials of capacitors are opposite. Thus, this paper selects mapping (A) and (B) to control the voltage alternately based on the six-segment SVPWM.

4.1. Vector Selection and Vector Duration Time

In this paper, the vector selection and calculation are based on the line-voltage coordinate system [6]. Assuming the three-phase reference voltages are U_a, U_b, and U_c, the *a-b-c* coordinate is converted to the *ab-bc-ca* line-voltage coordinate and unitized as Equation (3). The components of the reference voltage V_{ref} on the axis *ab*, *bc*, and *ca* are U_{ab}, U_{bc}, and U_{ca}, respectively.

$$\begin{bmatrix} U_{ab} \\ U_{bc} \\ U_{ca} \end{bmatrix} = \frac{2}{V_{dc}} \begin{bmatrix} 1 & -1 & 0 \\ 0 & 1 & -1 \\ -1 & 0 & 1 \end{bmatrix} \begin{bmatrix} U_a \\ U_b \\ U_c \end{bmatrix} \tag{3}$$

The per unit values of components are rounded up and down to an integer. In the equation, "floor" means rounding down to an integer and "ceil" represents rounding up to an integer.

$$\begin{cases} F_{ab} = floor(U_{ab}),\ C_{ab} = ceil(U_{ab})\ ; \\ F_{bc} = floor(U_{bc}),\ C_{bc} = ceil(U_{bc})\ ; \\ F_{ca} = floor(U_{ca}),\ C_{ca} = ceil(U_{ca})\ ; \end{cases} \tag{4}$$

All of the triangles can be separated into regular and inverted ones. If $F_{ab} + F_{bc} + F_{ca} = -1$, the end of the reference voltage vector is located in the regular triangle, and the coordinates of the three triangle vertexes (U_A, U_B, and U_C) and the time duty ratio of three basic vectors are as follows.

$$\begin{cases} U_A = (C_{ab}, F_{bc}, C_{ca}),\ d_A = U_{ab} - F_{ab}\ ; \\ U_B = (F_{ab}, C_{bc}, F_{ca}),\ d_B = U_{bc} - F_{bc}\ ; \\ U_C = (F_{ab}, F_{bc}, C_{ca}),\ d_C = U_{ca} - F_{ca}\ ; \end{cases} \tag{5}$$

If $F_{ab} + F_{bc} + F_{ca} = -1$, the end of the reference voltage vector is located in the inverted triangle. The coordinates and the time duty ratio are as follows.

$$\begin{cases} U_A = (C_{ab}, F_{bc}, C_{ca}),\ d_A = C_{ab} - U_{ab}\ ; \\ U_B = (F_{ab}, C_{bc}, C_{ca}),\ d_B = C_{bc} - U_{bc}\ ; \\ U_C = (C_{ab}, C_{bc}, F_{ca}),\ d_C = C_{ca} - U_{ca}\ ; \end{cases} \tag{6}$$

After selecting the three basic vectors, it is essential to find all the switching states corresponding to each basic vector. (S_a, S_b, S_c) represents the switching states of basic vectors in the $\alpha - \beta$ coordinate system. $U_X = (a, b, c)$ represents the coordinates of triangle-vertex vectors in the line-voltage coordinate system (where X = A, B, or C).

$$\begin{cases} S_a = S_a \\ S_b = S_a - a \\ S_c = S_a + c \\ 0 \leq S_a, S_b, S_c \leq 4 \end{cases} \tag{7}$$

Equations (5)–(7) can quickly determine the triangle-vertex vectors and their time duty ratios. The relationship between common-mode voltage and switching states is obtained by Equations (1) and (2) as the following:

$$U_{CMV} = \frac{S_a + S_b + S_c}{6} V_{dc} \tag{8}$$

The dual-pulse-mapping algorithm proposed in this paper utilizes Equation (8) to select the vectors that have the smallest CMV among redundant vectors.

4.2. Voltage Balancing Algorithm

Based on the traditional SVPWM, this paper presents a new algorithm, called the dual-pulse-mapping algorithm, which is composed of six segments. The first three segments are controlled by pulse mapping (A), and the other segments are controlled by mapping (B). At the same time, the regulating factor k is adopted for each vector. According to the midpoint potentials of the three phases, adjusting the working time of each vector under the two pulse mappings can balance the capacitor voltages effectively and stably. Taking sector I as example, the vectors with the smallest value of CMV are shown in Figure 6.

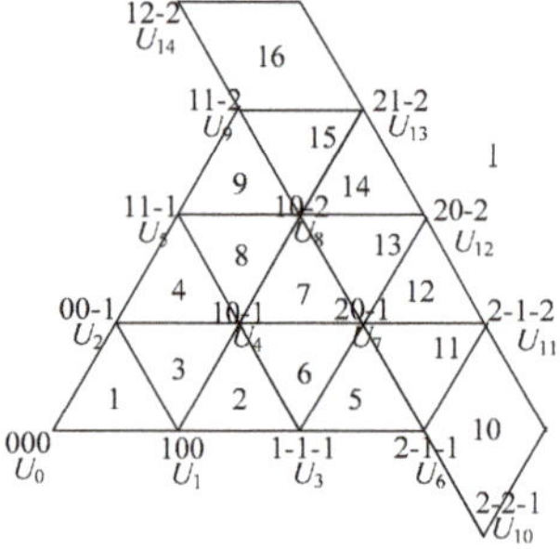

Figure 6. The vectors with the minimum common-mode voltage (CMV) in Sector I.

We take the 13th triangle in Figure 6 as an example to explain how the dual-pulse mapping works. Because the reference voltage is in the 13th triangle, the vector synthesizing sequence is showing as Figure 7.

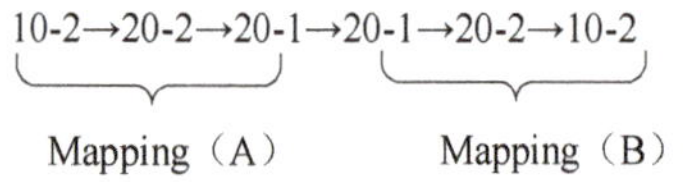

$$10\text{-}2 \rightarrow 20\text{-}2 \rightarrow 20\text{-}1 \rightarrow 20\text{-}1 \rightarrow 20\text{-}2 \rightarrow 10\text{-}2$$

Mapping（A）　　　　Mapping（B）

Figure 7. The vector synthesizing sequence of dual-pulse mapping strategy.

As shown in Tables 2 and 3, the pulse mappings (A) and (B) have the opposite effect on the midpoint potential. Based on this, the regulating factors k_a and k_c are adopted. Assuming that the working time of the three vectors are T_1, T_2, and T_3, the working time of the six segments is assigned as Figure 8.

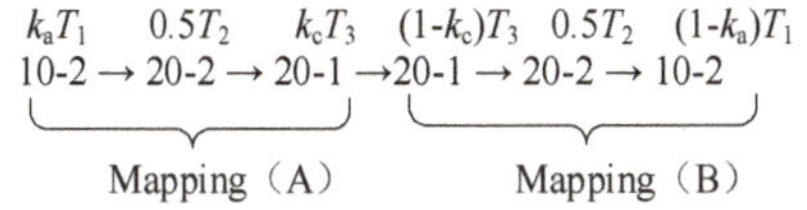

$$k_aT_1 \quad 0.5T_2 \quad k_cT_3 \quad (1-k_c)T_3 \quad 0.5T_2 \quad (1-k_a)T_1$$
$$10\text{-}2 \rightarrow 20\text{-}2 \rightarrow 20\text{-}1 \rightarrow 20\text{-}1 \rightarrow 20\text{-}2 \rightarrow 10\text{-}2$$

Mapping（A）　　　　　Mapping（B）

Figure 8. The assigned working time of sex segments.

If the midpoint potential of phase A is equal to $V_{dc}/2$, then k_a is 0.5. If the midpoint potential of phase A is higher than $V_{dc}/2$, we only need to reduce the value of k_a to <0.5, for example 0.4, in order to increase the working time of 10-2 under mapping (B) (because the time if the follow current's state is much smaller than the sampling time and a follow current also exists when using pulse mapping (B), so this state will not be taken into consideration). It is all the same when the midpoint potential of phase A is lower than $V_{dc}/2$. By adjusting the operating time of 10-2 under the two kinds of pulse mappings, the midpoint potential of phase A will be balanced.

5. Simulation Verification and Analysis

In order to show the performance of the proposed method, simulation studies were carried out in the MATLAB/Simulink software environment. The condition were set as follows: the DC-side voltage equals 1000 V, the rated DC-side capacitor voltages V_{dc} equal 500 V, the DC capacitances are 3300 μF, the output voltage frequency f is 50 Hz, the load power factor $\cos\varphi$ is 0.985, and its impedance Z equals 26.43.

5.1. Simulation Results of the Conventional SVPWM Strategy

Figure 9a shows the three-phase midpoint potentials (which are defined as the difference values between the midpoint potentials of the unbalanced three-phase capacitors and the midpoint potentials of the balanced capacitors) when using the conventional SVPWM, which only adopts mapping (A). The DC-side voltage equals 1000 V, so the midpoint potentials of the balanced capacitors are 500 V. The midpoint potentials of the three-phase capacitors are continuously increased from 0 V to 500 V, which is consistent with the theoretical analysis. Figure 9b shows the line voltage between phase A and phase B; the nine-level line voltage attenuates to five-level as time goes by. When the modulation index m has high values, the midpoint current of the inflow capacitor is large, and the voltage of the capacitor changes greatly. Therefore, the midpoint potential of the capacitor rises rapidly. On the contrary, when m has low values, the midpoint potential of the capacitor rises slowly.

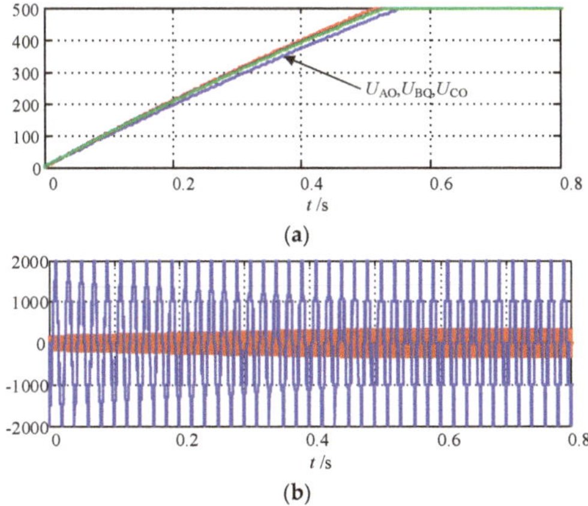

Figure 9. The inverter without capacitor voltage control. (**a**) Midpoint Potential of three phases; and (**b**) Line voltage and common mode voltage.

The line voltage between phase A and phase B is degraded to five-level from the original nine-level. The imbalanced DC-side voltages inject harmonics into the output voltage of the inverter, thereby distorting the output voltage waveform. If the inverter continues to work in an unbalanced capacitor voltage state, the power electronics in the system will be unevenly pressed or will even break down, and eventually the entire system will not work properly.

5.2. Simulation Results with the Proposed Dual-Pulse-Mapping Strategy

The proposed dual-pulse-mapping strategy was also analyzed using simulations, and the simulation parameters were the same as those for the simulations of the conventional SVPWM. Figure 10 shows the three-phase current, the midpoint potentials of the three-phase DC-side capacitors, the line voltage between phase A and phase B, and the common-mode voltage under the modulation index of 0.4 and 0.9 when the power factor of the load equals 0.984.

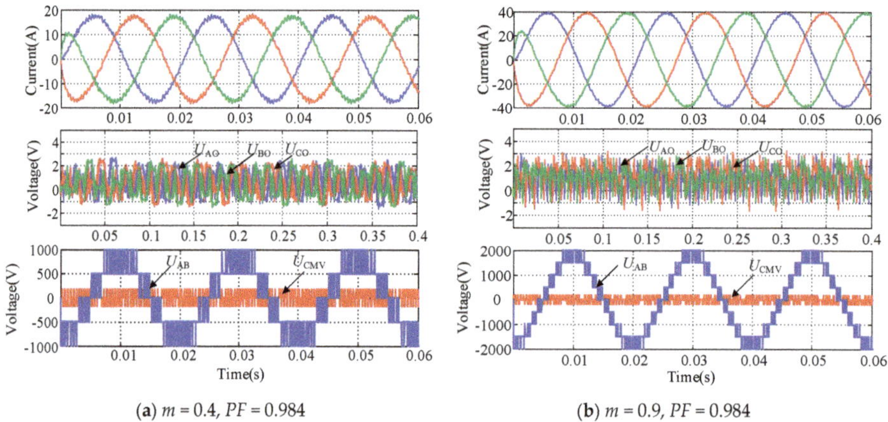

(**a**) $m = 0.4$, $PF = 0.984$ (**b**) $m = 0.9$, $PF = 0.984$

Figure 10. The simulation waveforms under the algorithm. (**a**) The three-phase current, midpoint potentials, line voltage between phase A and phase B, and common-mode voltage when m is 0.4; and (**b**) The three-phase current, midpoint potentials, line voltage between phase A and phase B, and common-mode voltage when m is 0.9.

According to the Figure 10, under the modulation index of 0.4 and 0.9, the proposed dual-pulse-mapping strategy could control the operation of the inverter stably. The midpoint potentials of the three-phase DC-side capacitors fluctuate within ± 3 V, which indicates that the proposed algorithm can effectively balance the midpoint potentials of DC-side capacitors. The largest ripple in the common voltage is $1/6V_{dc}$, which means that the common voltage has effectively been suppressed.

5.3. Simulation Results of Dynamic Performance

We performed a simulation with the conventional strategy in which the modulation index was 0.9, the power factor was 0.984, and the pre-charged voltage of the DC-side capacitor was 500 V. When the midpoint potentials of the DC-side capacitors reached $V_{dc}/2$, the proposed dual-pulse-mapping strategy was used to balance the voltage. Figure 11 shows the simulation diagram of the midpoint potentials of the DC-side capacitors, the voltages of phase A's two capacitors, and the line voltage between phase A and phase B. Under the pulse mapping (A), the midpoint potentials of the three-phase capacitors rise rapidly from 0 V to 500 V and the two capacitor voltages of phase A also change from 500 V to 0 V and 1000 V, respectively. The line voltage between phase A and phase B degrades to five-level. Then, the dual-pulse-mapping algorithm was adopted. The midpoint potentials of three-phase capacitors return to a balanced state and the line voltage also returns to nine-level. The two capacitor voltages of phase A change from 0 V and 100 V to 50 V. These simulation results show that the proposed algorithm has intelligent dynamic performance.

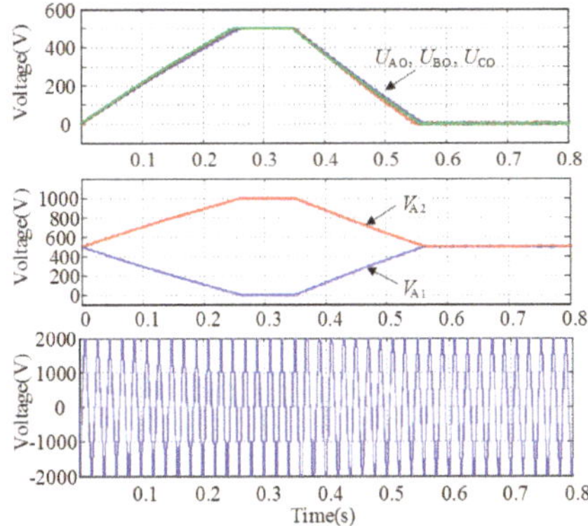

Figure 11. The simulation waveform of dynamic performance.

6. Experimental Results

In order to further verify the effectiveness of the proposed control algorithm for balancing DC-side (Direct Current) capacitor voltages and the suppression of the common-mode voltage and its dynamic response capability, a three-phase, five-level NPC/H Bridge inverter with an *RL* load experimental platform was built as shown in Figure 12. The structure is composed of a 12-pulse uncontrolled rectifier and a five-level NPC/H-bridge-type inverter. The basic parameters of the experiment are shown in Table 4.

Table 4. The experimental parameters of the system.

Parameters	Values
DC-side voltage V_{dc}	100 V
Output voltage frequency f_{AC}	50 Hz
DC-side capacitance C	3300 μF
RL load	26 Ω + 15 mH
Modulation index m	0.9 and 0.4
Switching frequency f_s	800 Hz

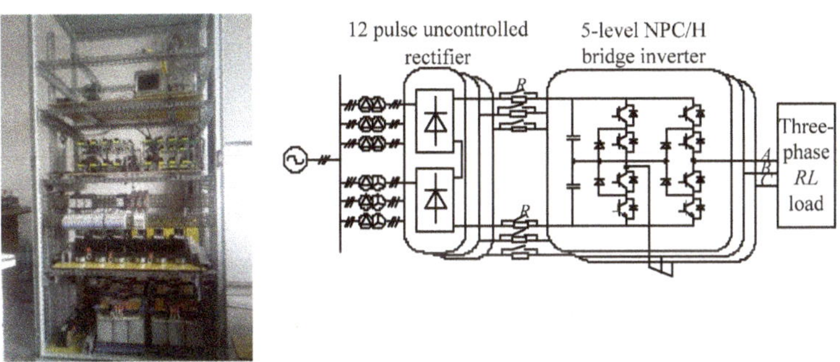

Figure 12. The experimental platform.

In Figure 13a,b, the DC-side capacitor voltages' fluctuations are very small regardless of whether the reference voltage locates in the high- or low-modulation index region, and the common-mode voltage of the system is suppressed at the $-V_{dc}/6$, 0, $V_{dc}/6$. This proves that the dual-pulse-mapping algorithm controls the stable operation of the three-phase, five-level NPC/H-bridge inverter. Figure 13c,d show the harmonic analysis of the line voltage. It can be seen that, when the modulation index m is 0.9 and 0.4, the THD is 17.8% and 35.0%, respectively, which means that the output performance of the algorithm is the same as that of traditional SVPWM.

The pulse-width modulations (PWMs) of S_{a1}, S_{a2}, S_{a3}, and S_{a4} in phase A are shown in Figure 14. The switching frequencies of the four switches in phase A were calculated approximately during 250~350 Hz, with a modulation index m of 0.4 and 0.9. The higher the modulation index is, the lower the switching frequency of each switch is. Phase B and C are both the same. Compared with the traditional SVPWM, the switching frequency of each switch just a little higher.

Figure 15a,b show the experimental waveforms of the dynamic response when the reference voltage modulation index suddenly decreases. The modulation index changes from $m = 0.9$ to $m = 0.4$. The line voltage and phase currents respond immediately during a very short time. The capacitor voltage is still balanced but has a little increase because of the uncontrolled rectifier.

In order to verify the strong self-balancing property of the algorithm, the conditions were set as follows: in the beginning, the modulation index was $m = 0.9$, and the values of regulating factors were all 0.5, which means that the dual-pulse-mapping algorithm was equivalent to a common six-segment SVPWM. Under the mapping (A), the midpoint potentials V_{AO}, V_{BO}, and V_{CO} rise quickly and achieve the maximum $V_{dc}/2$ after some time. The line voltage also degenerates to five-level from nine-level. The inverter works under an unstable condition. Then, the dual-pulse-mapping algorithm was adopted. In Figure 15c, the midpoint potentials of the three-phase capacitors return to a balanced state after a short time. The line voltage also returns to nine-level. The two capacitor voltages in phase A change from 0 V and 100 V to 50 V in Figure 15d.

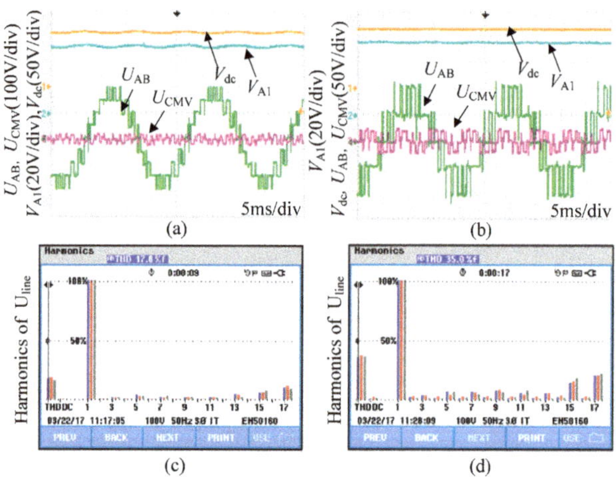

Figure 13. The experimental results of steady-state operation. (**a**) The DC-side voltage, the capacitance voltage of phase A, line voltage between phase A and phase B, and common-mode voltage when m is 0.9; (**b**) The DC-side voltage, the capacitance voltage of phase A, line voltage between phase A and phase B, and common-mode voltage when m is 0.4; (**c**) The harmonic analysis of the line voltage when m is 0.9; and (**d**) The harmonic analysis of the line voltage when m is 0.4.

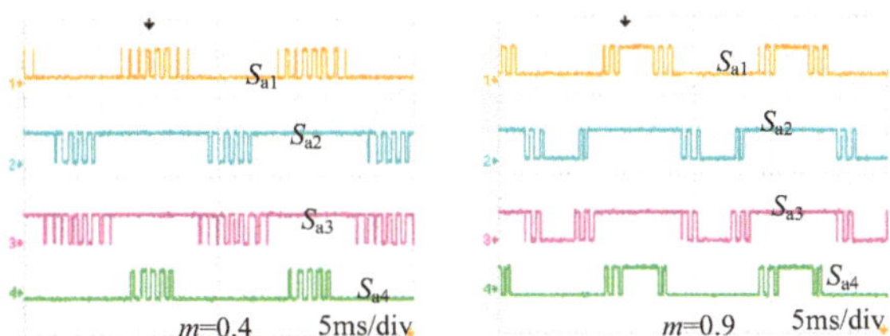

Figure 14. The pulse width modulations (PWMs) of phase *A* when $m = 0.4$ and $m = 0.9$.

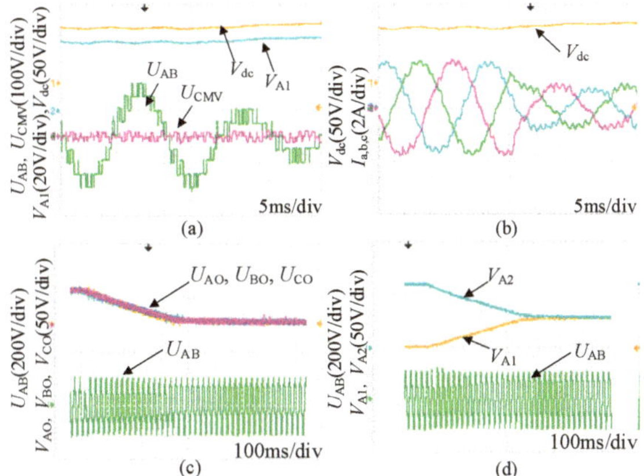

Figure 15. The experimental results of dynamic performance. (**a**) The DC-side voltage, the capacitance voltage of phase A, line voltage between phase A and phase B, and common-mode voltage with changing *m*; (**b**) The DC-side voltage and the currents of three-phase with changing *m*; (**c**) The midpoint potentials of three-phase and the line voltage between phase A and phase B after dual-pulse mapping strategy started to work; and (**d**) The two capacitor voltages in phase A after dual-pulse mapping strategy started to work.

7. Conclusions

In this paper, the balance of the midpoint potentials of three-phase DC-side capacitors and the suppression of common-mode voltage were studied in a three-phase, five-level NPC/H-bridge inverter. A dual-pulse-mapping algorithm was proposed based on the characteristics of this topology. In combination with the traditional six-segment SVPWM, the proposed algorithm selectively uses the two pulse mappings that have the opposite effects on the midpoint potentials of the capacitors. The simulation and experimental results show that the proposed algorithm can balance the midpoint potentials perfectly and can suppress the common-mode voltage effectively. Based on the dynamic experiments, we can conclude that the proposed algorithm has a good capability to respond and the output performance is as good as that of the traditional SVPWM. Compared with the hardware balancing method, the volume and the cost of the system are both reduced. So, the proposed algorithm has a certain significance for cascaded NPC/H-bridge multilevel topology research. In high-voltage and high-power applications, the NPC/H bridge topology will be more suitable with the proposed algorithm.

Author Contributions: Z.S. put forward the main idea of this paper. K.Z. and Q.C. performed the data analysis and the simulations/experiments. M.W. and Z.L. assisted with the idea's development and the paper's writing.

Funding: This research was funded by National Key R&D Program of China, grant number is 2016YFB 0900400.

Conflicts of Interest: The authors declare no conflict of interest.

References

1. Wu, B. *High Power Converters and AC Drives*; IEEE Press/Wiley: New York, NY, USA, 2006.
2. Rodgriguez, J.; Lai, J.S.; Peng, F.Z. Multilevel inverters: A survey of topologies, controls and applications. *IEEE Trans. Ind. Appl.* **2002**, *49*, 724–738.
3. Saeedifard, M.; Barbosa, P.M.; Steimer, P.K. Operation and control of a hybrid seven-level converter. *IEEE Trans. Power Electron.* **2012**, *27*, 652–660. [CrossRef]
4. Kokila, A.; Kumar, V.S. Comparison of SPWM and SVPWM control techniques for diode clamped multilevel inverter with photovoltaic system. In Proceedings of the 2018 4th International Conference on Electrical Energy Systems (ICEES), Chennai, India, 7–9 February 2018; pp. 234–241.
5. Yang, P.; Li, C.; Lan, Z. Investigation on the subsection synchronous modulation strategy for high power NPC/H-bridge inverter. In Proceedings of the IEEE 8th International Power Electronics and Motion Control Conference (IPEMC-ECCE Asia), Hefei, China, 22–26 May 2016.
6. Chang, H.; Wei, R.; Ge, Q.; Wang, X.; Zhu, H. Comparison of SVPWM for five level NPC/H-Bridge inverter and traditional five level NPC inverter based on line-voltage coordinate system. In Proceedings of the International Conference on Electrical Machines and Systems, Pattaya, Thailand, 25–28 October 2015; pp. 574–577.
7. Gong, Z.; Wu, X.; Dai, P.; Zhu, R. Modulated model predictive control for MMC-based active frontend rectifiers under unbalanced grid conditions. *IEEE Trans. Ind. Electron.* **2019**, *66*, 2398–2409. [CrossRef]
8. Rendusara, D.A.; Cengelci, E.; Enjeti, P.N.; Stefanovic, V.R.; Victor, R.; Gray, J.W. Analysis of common mode voltage — neutral shift in medium voltage PWM adjustable speed drive (MV-ASD) systems. *IEEE Trans. Power Electron.* **2014**, *29*, 6281–6292.
9. Gong, Z.; Zhang, H.; Dai, P.; Sun, N.; Li, M. A low-cost phase-angle compensation method for the indirect matrix converters operating at the unity grid power factor. *IEEE Trans. Power Electron.* **2019**. [CrossRef]
10. Sun, L.; Wu, L.Z.; Ma, W.; Fei, X.; Cai, X.; Zhou, L. Analysis of the DC-link capacitor current of power cells in cascaded H-Bridge inverters for high-voltage drives. *IEEE Trans. Power Electron.* **2014**, *29*, 6281–6292. [CrossRef]
11. Wu, C.M.; Lau, W.H.; Chung, H. A five-level neutral-point-clamped H-bridge PWM inverter with superior harmonic suppression: A theoretical analysis. In Proceedings of the 1999 IEEE International Symposium on Cirsuits and Systems, Orlando, FL, USA, 30 May–2 June 1999; Volume 5, pp. 198–201.
12. Chen, Z.Y.; Wu, B. A Novel Switching Sequence Design for Five-Level NPC/H-Bridge Inverters With Improved Output Voltage Spectrum and Minimized Device Switching Frequency. *IEEE Trans. Power Electron.* **2017**, *22*, 2138–2145. [CrossRef]
13. Shen, J.; Schroder, S.; Gao, J.P.; Qu, B. Impact of DC-Link Voltage Ripples on the Machine-Side Performance in NPC H-Bridge Topology. *IEEE Trans. Ind. Appl.* **2016**, *52*, 3212–3223. [CrossRef]
14. Rojas, C.A.; Kouro, S.; Edwards, D.; Wu, B. Five-level H-bridge NPC central photovoltaic inverter with open-end winding grid connection. In Proceedings of the 40th Annual Conference of the IEEE Industrial Electronics Society (IECON), Dallas, TX, USA, 30 October–1 November 2014; pp. 4622–4627.
15. Hotait, H.A.; Massoud, A.M.; Finney, S.J.; Williams, B.W. Capacitor Voltage Balancing Using Redundant States of Space Vector Modulation for Five-level Diode Clamped Inverter. *IET Power Electron.* **2010**, *3*, 292–313. [CrossRef]
16. Qin, J.C.; Saeedifard, M. Capacitor Voltage Balancing of a Five-level Diode-Clamped Converter Based on a Predictive Current Control Strategy. In Proceedings of the 26th Annual IEEE Applied Power Electronics Conference and Exposition, Fort Worth, TX, USA, 6–10 March 2011; pp. 1656–1660.

17. Long, L.Z.; Zhang, Y.G.; Kuang, G.J. A Modified Space Vector Modulation Scheme to Reduce Common Mode Voltage for Cascaded NPC/H-bridge Inverter. In Proceedings of the IEEE 7th International Power Electronics and Motion Control Conference (ECCE), Harbin, China, 25 June 2012; pp. 1837–1841.
18. Gong, Z.; Dai, P.; Wu, X.J.; Deng, F.J.; Liu, D.; Chen, Z. A Hierarchical Model Predictive Voltage Control for NPC/H-Bridge Converters with a Reduced Computational Burden. *J. Power Electron.* **2017**, *17*, 136–148. [CrossRef]

Article

A Novel Fault-Tolerant Control of Modular Multilevel Converter under Sub-Module Faults Based on Phase Disposition PWM

Jingyuan Yin [1,*], Wen Wu [2], Tongzhen Wei [1], Xuezhi Wu [2] and Qunhai Huo [1]

[1] Institute of Electrical Engineering, Chinese Academy of Sciences, Beijing 100044, China; tzwei@mail.iee.ac.cn (T.W.); huoqunhai@mail.iee.ac.cn (Q.H.)

[2] National Active Distribution Network Technology Research Center, Beijing Jiaotong University, Beijing 100044, China; 17117417@bjtu.edu.cn (W.W.); xzhwu@bjtu.edu.cn (X.W.)

* Correspondence: yinjingyuan@mail.iee.ac.cn; Tel.: +86-138-1151-8632

Received: 14 November 2018; Accepted: 18 December 2018; Published: 21 December 2018

Abstract: Each arm of modular multilevel converter (MMC) consists of a large number of sub-module (SM) units. However, it also increases the probability of SM failure during the long-term system operation. Focusing on the fault-tolerant operation issue for the MMC under SM faults, the traditional zero-sequence voltage injection fault-tolerant control algorithm is analyzed detailed and its disadvantages are concluded. Based on this, a novel fault-tolerant control strategy based on phase disposition pulse-width modulation (PD-PWM) is proposed in this paper, which has three main benefits: (i) it has carrier and modulation wave dual correction mechanism, which control ability is more higher and flexible; (ii) it only needs to inject zero-sequence voltage in half a cycle of the modulation wave, which simplifies the complexity of traditional zero-sequence voltage injection control algorithms and much easier for implement; (iii) furthermore, the zero-sequence voltage can even be avoided injecting under the symmetrical fault conditions. Finally, the effectiveness of the proposed control strategy is verified with the simulation and experiment studies under different fault conditions.

Keywords: modular multilevel converter (MMC); Sub-module (SM) fault; fault-tolerant control; Phase Disposition PWM

PACS: J0101

---

1. Introduction

Compared with the traditional two-level converter, modular multilevel converter (MMC) enjoys the advantage of modular design [1,2], which allows MMC good scalability and facilitates the improvement of voltage level, as well as other advantages as many output levels and good harmonic characteristics. Therefore, in recent years, MMC topology converters have been widely studied and applied in high-voltage direct current transmission (HVDC), staticsynchronous compensator (STATCOM), and medium-voltage motor drive [3–6].

In actual transmission projects, in order to adapt to higher transmission voltage level, a large number of sub-modules (SMs) are often cascaded. For example, Trans Bay Cable Project is reported to have 216 SMs per arm [7], and the Zhou Shan multi-terminal flexible HVDC transmission project completed in 2014 in China is larger cascaded reach 250 SMs per arm [8].

As the number of cascaded SMs increasing, the security risks of MMC converters become larger, where the SM faults are prone to occur. When a SM fault occurs and is not processed in time, it may cause system shutdown and even endanger the security of the power grid [9,10]. Therefore, fault-tolerant control approach under SM faults is a problem that needs to be studied.

Generally, the common fault-tolerant control strategies are often designed based on equipping redundant SMs [8]. According to the operation state of the redundant SMs, two main schemes are developed, which are cold-reserve mode and hot-reserve mode, respectively. For the cold-reserve scheme [11,12], it means that the redundant SMs are all in bypassed state when the system operating normally and will replace the same number of faulty SMs when SMs malfunction. For the hot-reserve scheme [13–15], it means that the redundant SMs are all operating identically as the rated SMs. However, there are two further detailed modes under this scheme. We call it hot-reserve mode-a, b, respectively, where hot-reserve mode-a means that simultaneously bypass the same number of faulty SMs in each arm to keep the system strictly symmetrical [13]; and hot-reserve mode-b means that only bypass the faulty SMs [14,15]. Compared to the mode-a, the hot-reserve mode-b scheme can ensure full utilizing of all SMs during normal operation period and can avoid the long changing time for redundant SM capacitors to rated values. Furthermore, some extra healthy SMs may need be bypassed in hot-reserve mode-a, which might induce a greater transient impact on the system when large number of SMs are simultaneously bypassed. Therefore, hot-reserve mode-b is a more preferable redundant option for the MMC since its economic and reliability. However, this scheme also has a potential problem, where the system may operate asymmetric under asymmetrical faults. For ensuring the subsequent stable operation, the corresponding fault-tolerant control approach should be further considered, and this is also what this paper deals with.

Based on the background of hot-reserve mode-b, an energy-balancing-based fault-tolerant control strategy is proposed in [16]. Its main control idea is to increase the SM capacitor voltages on the faulty arms to ensure the energy equalization of each arm, thereby achieving the subsequent balance operation of the system. However, the reference command of the SM capacitor voltage needs to be calculated, and the increased value of SM capacitor voltage will become larger as the number of malfunction SMs increases. The neutral-point shift control method has been applied in the H-bridge cascaded multilevel converter system, which can well solve the problems caused by the module faults [17–20]. Inspired by this idea, the corresponding neutral-point shift control approaches suitable for the MMC system are researched in [15,21]. With injecting the zero-sequence voltage to modify the modulation waves, the converter line-voltages in the grid side are ensured balanced to maintain the system stable operation under SM faults. The advantage of these methods in [15,21] is that they do not need to adjust the SM capacitor voltages in fault arms. Similar to the methods in [15,21], an improved zero-sequence voltage injection fault-tolerant control strategy is proposed in [22]. It simplifies the calculation process of the required zero-sequence voltage. However, the common shortcoming in [15–22] is that they main only consider one arm occurs SM faults. When the upper and lower arm occur SM faults simultaneously, multiple zero-sequence voltages need to be calculated, which increase the complexity of the control method. It should be further improved.

Aiming at this problem and for better deal with the multiple SM fault conditions, also under the background of the redundant SMs equipped with hot reserve scheme and only bypassing the faulty SMs when SMs malfunction, a novel fault-tolerant control strategy is proposed in this paper. Compared to the current approaches, it has three main benefits: (i) it has carrier and modulation wave dual correction mechanism, which control ability is more higher and flexible; (ii) it only needs to inject zero-sequence voltage in half a cycle of the modulation wave, which simplifies the complexity of traditional zero-sequence voltage injection control algorithms and more easier for implement; (iii) furthermore, the zero-sequence voltage can even be avoided injecting under the symmetrical fault conditions. Finally, the effectiveness of the proposed control strategy is verified with the simulation and experiment studies under different fault conditions.

2. Basics Principles of MMC

2.1. MMC Topology

A typical three-phase MMC topology is shown in Figure 1. It consists of six symmetrical arms, which can be divided into three units: phase, arm, and SM. N identical SM units and one arm inductance for suppressing the circulating current and the fault current rising rate are connected in series on each arm, and the AC output terminal is taken out from the midpoint of the upper and lower arm inductances connected to each other. In order to enhance the reliability of the system, there are a certain number of redundant SMs.

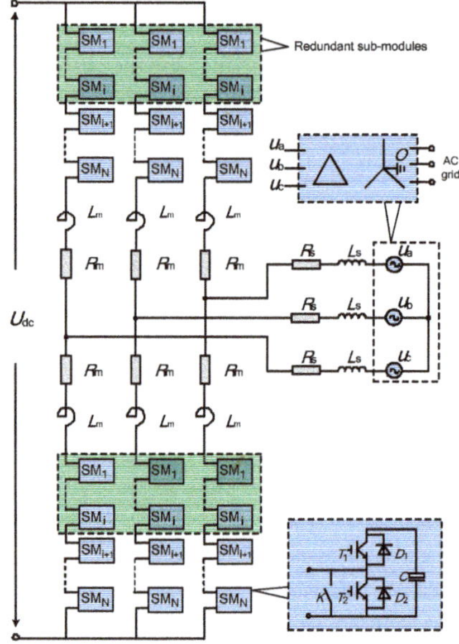

Figure 1. Topology structure of three-phase modular multilevel converter (MMC).

Among them, the internal structure of the SM unit has various forms [23–25], and the more common half-bridge SM structure is selected as the research object. Specifically, it is composed of an insulated gate bipolar transistor (IGBT) half bridge, a DC capacitor C, and a bypass switch K. In the normal working state of the system, the SM capacitor voltage U_{SM} has two levels of output states of 0 and DC capacitor voltage U_C. When T_1 is turned on and T_2 is off, the SM is inserted and outputs U_C. When T_1 is turned off and T_2 is on, the SM is bypassed and output zero. Thus, the AC side multi-level output of the MMC converter can be realized by properly controlling the on-off state of the SM.

2.2. Phase Disposition PWM and Voltage Balancing Strategy

Currently, MMC's commonly used modulation methods include carrier phase shift modulation, phase disposition modulation, and nearest level modulation. Because the phase disposition modulation (PD-PWM) method has better harmonic characteristics and is easier to implement [26,27], in this paper, PD-PWM is selected as the basic modulation strategy in subsequent analysis and experiment.

In addition, in order to reduce the fluctuation of SM capacitor voltage, this paper selects the more common capacitor voltage balancing control based on the sorting algorithm [16,28]. The schematic diagram of the relevant control method is shown in Figure 2. Where v_j^* is the sampled phase voltage of phase j after standard (j = a, b, c). i_{jp} and i_{jl} are the arm current of the upper and lower arm of phase j, respectively. N_{jsum} is the rated total number of the inserted SMs of phase j (i.e., as shown in Figure 1, it is equal to N if no SM faults occur). n_{jp} is the number of the SMs need to be inserted in the upper arm of phase j, which is obtained from the PD-PWM modulation. n_{jl} is the number of the SMs need to be inserted in the lower arm of phase j, which can be obtained directly by calculating $N_{\mathrm{jsum}} - n_{\mathrm{jp}}$.

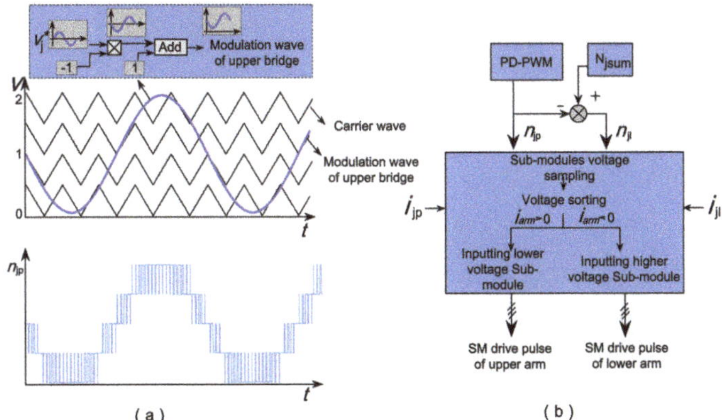

Figure 2. Schematic diagram basic control: (**a**) Diagram of the pulse-width modulation (PD-PWM); (**b**) Diagram of the capacitor voltage sorting algorithm.

When the system is operating, the numbers of the SMs need to be inserted in the upper arm n_{jp} and the lower arm n_{jl} are obtained according to the PD-PWM modulation firstly, and then the voltage of each SM in the arm is sorted by the capacitor voltage sorting algorithm. According to the direction of the arm current, the SM with lower capacitor voltage is preferentially charged, and the SM with higher capacitor voltage is preferentially bypassed and discharged, thereby ensuring a relatively balanced voltage of each module during the whole dynamic operation.

3. Analysis of the Traditional Zero-Sequence Voltage Injection Control Method

After the SM occur faults, it is easy to know that the phase voltage output ability of the fault phase will be changed, and the neutral-point of the phase voltage in the ac side will be offset accordingly. In order to ensure that the MMC converter can continue stable operation with connected to the grid, the zero-sequence voltages can be injected to ensure the line voltage of the MMC converter remains unchanged before and after the SM faults, thereby realizing the fault-crossing of the SM [22]. The basic modulation diagram of the zero-sequence voltage injection control method is shown in Figure 3. Where Figure 3a depicts the schematic diagram of three-phase modulation waves when only the upper arm of phase occur faults, Figure 3b depicts the schematic diagram of three-phase modulation waves when the upper and lower arms of phase j simultaneously occur faults.

With comparing the Figure 3a,b, it can be observed that if the SM faults only occur only in the upper arm, on the basis of bypassing the faulty SMs, the fault-crossing can be realized by only adding one corresponding zero-sequence voltage component on the modulation wave of each phase. It is easy to achieve. However, if the SM simultaneously occur faults in the upper and lower arms, two injected zero-sequence voltages and the four corresponding correction times t_1, t_1, t_1, and t_4 are need to be calculated. In addition, if the numbers of the faulty SMs in the upper and lower arms are not identical, the two required zero-sequence voltages are needed to be calculated respectively. This will increase the complexity of the fault-tolerant control algorithm.

Considering that the location and the time of occurring SM faults are random, the MMC might with (i) the signal faulty arm; (ii) the upper and lower symmetrical faulty arms; and (iii) the upper and lower asymmetrical faulty arms after SMs malfunction. Furthermore, the faulty arms state may be re-changed between the mentioned three states when the SMs fault again. Therefore, in order to ensure the stable operation of MMC system and flexibly deal with the multiple possible arms faulty states after SMs malfunction, this paper improves the traditional zero-sequence voltage injection control method, and proposes a new fault-tolerant control strategy based on PD-PWM.

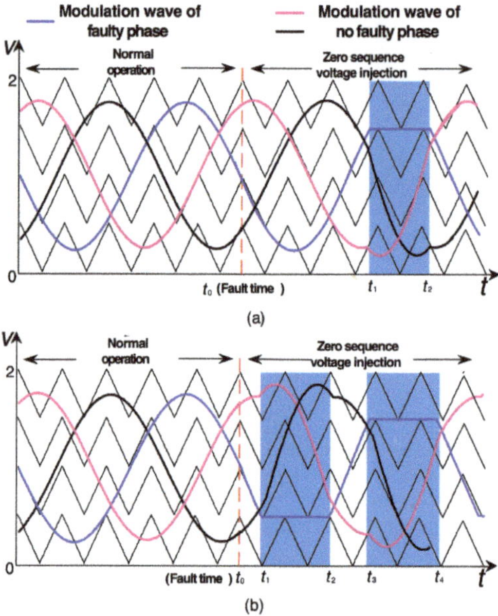

Figure 3. Diagram of zero sequence voltage injection modulation: (**a**) Schematic diagram of fault modulation wave of upper arm; (**b**) Schematic diagram of fault modulation wave of upper arm and lower arm.

4. Novel Fault-Tolerant Control Strategy

4.1. Overall Control Process of the Proposed Fault-Tolerant Control Strategy

In order to facilitate the later analysis, we define the faulty SM numbers in the upper and lower arms of phase j are n_{jp_fau} and n_{jl_fau}, respectively. At the same time, we assume the modulation ratio of the system is m under the normal operation. When the number of the faulty SMs in the arm satisfies Equation (1), the fault will not affect the normal operation of the system because of itself modulation margin of the system. While when the Equation (1) is not valid, the fault-tolerant control strategy must be inserted to ensure the subsequent stable operation of the MMC converter.

$$\begin{cases} m \leq 1 - \dfrac{2n_{jp\_fau}}{N} \\ m \leq 1 - \dfrac{2n_{jl\_fau}}{N} \end{cases} \tag{1}$$

Figure 4 shows the overall control flow chart of the fault-tolerant control strategy proposed in this paper. It can be divided into three main parts in the process of implementation:

1. Sub module fault handling. Its main task is to block the trigger pulse of the faulty SMs and isolate the faulty SMs through closing the bypass switch.
2. Correction of the modulation state. The main task of this part is that with combining the faulty SM numbers of the fault phase, achieving the correction of the original carrier and modulation waves by adopting the proposed correction algorithm (where its specific implementation method will be discussed in detail in Section 4.2, thereby obtaining the SM numbers that need to be inserted in

each arms after SMs malfunction. In addition, it should be note that this step is the also core part of the fault-tolerant control strategy. In the traditional zero-sequence voltage injection control method, it only with correcting the modulation waves to help achieve the SM fault-crossing. Compared to it, we add the correction algorithm of the carrier waves on the basis of the correction algorithm of the modulation waves, and realized their combination. This effectively simplifies the complexity of traditional control algorithm when deal with multiple arms occurring SM faults.

3. Generation of the SM drive pulses. In this part, with combining the SM numbers that need to be inserted in each arms obtained from the step (2), the drive pulses of the remained healthy SMs are generated by using SM voltage sorting control algorithm, and finally completes the fault-tolerant control.

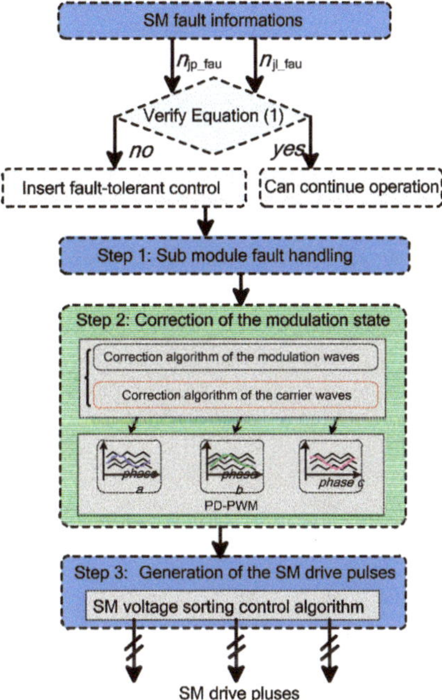

Figure 4. Overall flow chart of the proposed fault-tolerant control strategy.

4.2. Specific Implementation Method of the Correction of the Carrier and Modulation Waves

When the system is operation under the normal state, we can depict the modulation diagram of one phase as Figure 5.

As shown in Figure 5, when the MMC system is operating normally, the carrier waves of the N SMs in the upper arm equally distributed in the interval [0, 2] according to the ratio of γ, where it is easy to know:

$$\gamma = \frac{2}{N} \tag{2}$$

However, if the SMs occurring faults, the number of SMs that can normally be switched in the faulty phase will be change as the faulty SMs are bypassed. Therefore, the original modulation link must be revised and divided into two parts: 'Correction algorithm of the carrier waves' and 'Correction algorithm of the modulation waves'. We will describe the implementation method detailed in Sections 4.2.1 and 4.2.2, respectively.

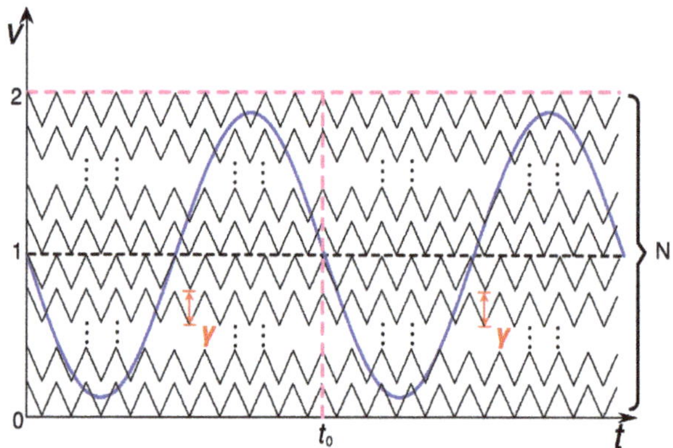

Figure 5. Modulation diagram of one phase under the normal operation.

4.2.1. Correction Algorithm of the Carrier Waves

Assuming the SMs occurring faults at t_0, and to ensure the generality of the analysis process, supposing the SMs fault in both the upper and lower arms of phase j. Firstly, with comparing the upper arm faulty SM numbers n_{jp_fau} and the lower arm faulty SM numbers of n_{jl_fau} of phase j, we define the smaller of the two is χ_j, and the larger of the two is τ_j. Thereby we obtain:

$$\chi_j = \min\left[n_{jp\_fau}, n_{jl\_fau}\right] \tag{3}$$

$$\tau_j = \max\left[n_{jp\_fau}, n_{jl\_fau}\right] \tag{4}$$

In order to ensure the maximum utilization of the remained healthy SMs, the number of SMs that need to participate in switching operation in the two arms of the fault phase should respectively be:

$$N_{j\chi} = N - \chi_j \tag{5}$$

$$N_{j\tau} = N - \tau_j \tag{6}$$

Then, for correcting carrier wave conveniently, the correction coefficient K_{jc} ($K_{jc} \geq 1$) about the amplitude of the carrier waves is introduced. Simultaneously, combing the smaller faulty SM number χ_j, we establish a relationship between them, where:

$$K_{jc} \cdot \gamma = \frac{2}{N - \chi_j} \tag{7}$$

Further, we can obtain:

$$K_{jc} = \frac{2}{\gamma(N - \chi_j)} = \frac{N}{N - \chi_j} \tag{8}$$

Based on the Equation (8), enlarging the amplitude of the carrier waves with the coefficient K_{jc}, we can obtain the initially corrected modulation diagram of the faulty phase as shown in Figure 6. It can be observed that the carrier waves in the interval [0, 2] are re-distributed equally according to the ratio of $K_{jc}*\gamma$ at this time.

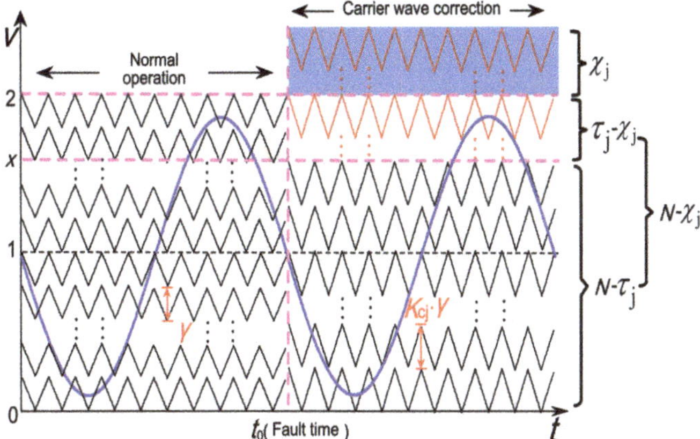

Figure 6. Modulation diagram of the faulty phase under carrier wave correction.

Therefore, through the correction algorithm of the carrier waves, we firstly guarantee the SMs on the smaller faulty SM numbers arm can all normally participate in the switching operation. In other words, we can equivalently consider that the MMC is under the 'signal arm fault' state, and its rated SM number is $N_j\chi = N - \chi_j$; its faulty SM number is $\tau_j - \chi_j$. Especially, we can found that when the $\tau_j - \chi_j = 0$ (e.g., the faulty SM numbers in the upper and the lower arm are equal), the SM fault-crossing can be achieved only with the carrier wave correction algorithm.

In addition, if the influence of the different faulty SM numbers $\tau_j - \chi_j$ is not considered at present, it can be considered that the fault phase will operate normally according to the newly rated total number of the inserted SMs of phase $N_{j\text{sum}} = N_{j\chi}$. Under the control of sorting algorithm, the SM capacitor voltage of the fault phase will automatically be average raised and operating stable [16,28].

4.2.2. Correction Algorithm of the Modulation Waves

As analyzed in the previous section, the faulty phase can be equivalently converted to 'signal arm fault' state with the carrier wave correction algorithm. For further eliminating the effects of the different faulty SM numbers $\tau_j - \chi_j$, we establish the modulation wave correction algorithm, where it is achieved by the zero-sequence voltage injection control method.

Firstly, according to the Figure 6, we can calculate:

$$x = K_{jc} \cdot \gamma \cdot (N - \tau_j) = 2 \cdot \frac{N - \tau_j}{N - \chi_j} \tag{9}$$

Then, according the traditional zero-sequence voltage injection control method [23], the zero sequence voltage that needed to be injected into the initial modulation wave $v_j{}^*$ of each phase can be obtained by (9):

$$k_{jm}(t) = -v_j{}^* + (x - 1) \tag{10}$$

Substituting Equations (3), (4) and (9) to (10), we can obtain:

$$k_{jm}(t) = -v_j{}^* + (2 \cdot \frac{N - \tau_j}{N - \chi_j} - 1) = -v_j{}^* + (2 \cdot \frac{N - \max\left[n_{jp\_fau},\, n_{jl\_fau}\right]}{N - \min\left[n_{jp\_fau},\, n_{jl\_fau}\right]} - 1) \tag{11}$$

Based on the Figure 6, injecting the zero-sequence voltage $k_{jm}(t)$ in the interval $[t_1, t_2]$, we can obtain the final corrected modulation diagram of the faulty phase as shown in Figure 7.

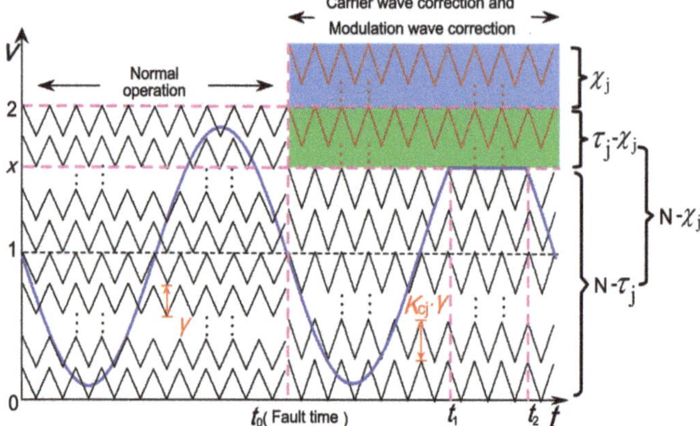

Figure 7. Final modulation diagram of the faulty phase under fault-tolerant control.

In addition, assuming the sampled phase voltage of phase j after the standard is:

$$v_j{}^* = m\sin(\omega t + \varphi_j) \tag{12}$$

where, φ_j denotes the initial phase angle of the phase voltage. Combining Equations (9) and (11), the corresponding times t_1 and t_2 in one modulation period can be calculated respectively:

$$\begin{cases} t_1 = \dfrac{\arcsin[(1-2\cdot\frac{N-\tau_j}{N-\chi_j})/m]-\varphi_j}{\omega} \\[4mm] t_2 = \pi - \dfrac{\arcsin[(1-2\cdot\frac{N-\tau_j}{N-\chi_j})/m]-\varphi_j}{\omega} \end{cases} \tag{13}$$

Based on the above analysis, the overall implementation block diagram of the fault-tolerant control strategy in the MMC system can be obtained as shown in Figure 8.

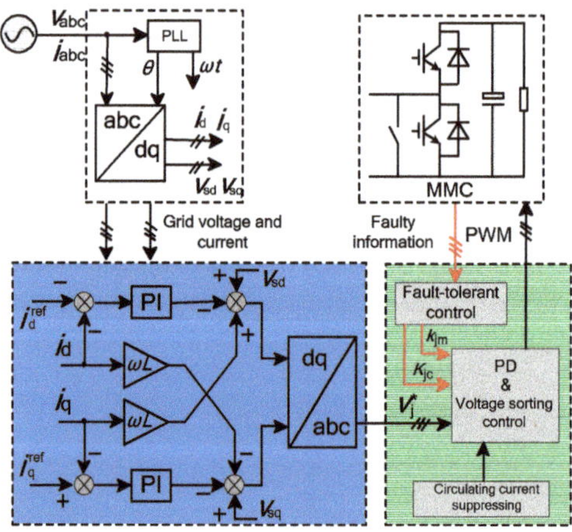

Figure 8. Schematic diagram of control system.

4.3. Maximum Control Range of the Proposed Control Strategy

Restricting condition 1: meet the voltage pressure requirements. Because when the faulty arm is asymmetric, the carrier wave of each SM will be revised through the increasing carrier wave amplitude control link. At this time, the steady-state operating voltage of the SM will be raised. In order to ensure that the working voltage of the SM is less than the withstand voltage of the SM, it is necessary to satisfy the following requirements:

$$\frac{U_{dc}}{N - \chi_j} < U_m \tag{14}$$

With

$$\min\left[n_{jp\_fau}, n_{jl\_fau}\right] < N\left(1 - \frac{U_{dc}}{U_m}\right) \tag{15}$$

Restricting condition 2: ensure that the system does not appear over modulation. According to the analysis of the previous section, combining the Equations (9)–(13), the set of the modulation amplitudes in the upper arm of each phase in the limiting modulation wave range [t1, t2] can be calculated as follows:

$$U_j^* = \left\{2 \cdot \frac{N - \tau_j}{N - \chi_j}, 1 - m, 1 + m, \left(1 - \sqrt{3}m - 2\frac{\tau_j - \chi_j}{N - \chi_j}\right)\right\} \tag{16}$$

As the system works normally, there must be:

$$\begin{cases} 0 \leq m \leq 1 \\ 0 \leq \chi_j < \tau_j \leq N \end{cases} \tag{17}$$

Therefore, to ensure that the system does not appear over modulation after the SM fault, it is necessary to ensure that:

$$0 \leq 1 - \sqrt{3}m - 2 \cdot \frac{\tau_j - \chi_j}{N - \chi_j} \leq 2 \tag{18}$$

with

$$\max\left[n_{jp\_fau}, n_{jl\_fau}\right] \leq \frac{1 - \sqrt{3}m}{2}N - \frac{1 + \sqrt{3}m}{2}\chi_j \tag{19}$$

5. Simulation Studies

A three-phase MMC simulation model is built in MATLAB/Simulink simulation software as shown in Figure 1. The main parameters of the simulation system are shown in Table 1. In the simulation process, the circulating current suppression strategy is always put into operation when the grid is connected [29].

Table 1. Main parameters of the simulation system.

Parameters	Value
Ac system nominal voltage	10 kV
Ac System inductance L_s	5 mH
Fundamental frequency	50 Hz
Ac system power losses R_s	0.03 Ω
Arm inductance L_m	5 mH
Series arm resistance R_m	0.01 Ω
Dc bus voltage U_{dc}	20 kV
Number of SMs per arm N	20
Number of redundant SMs per arm	5
Sub-module capacitor C	2000 μF
Transformer ratio	1:1 (Y/Δ)

5.1. Case 1

Fault condition 1: Single arm fault to symmetrical arm fault. When the system operates to t = 0.4 s, five SMs in the upper arm of phase a occur fault; at t = 0.5 s, the fault-tolerant control strategy is enabled; at t = 0.9 s, five SMs in the lower arm of phase a are failed again. Figure 9 is the related simulation results.

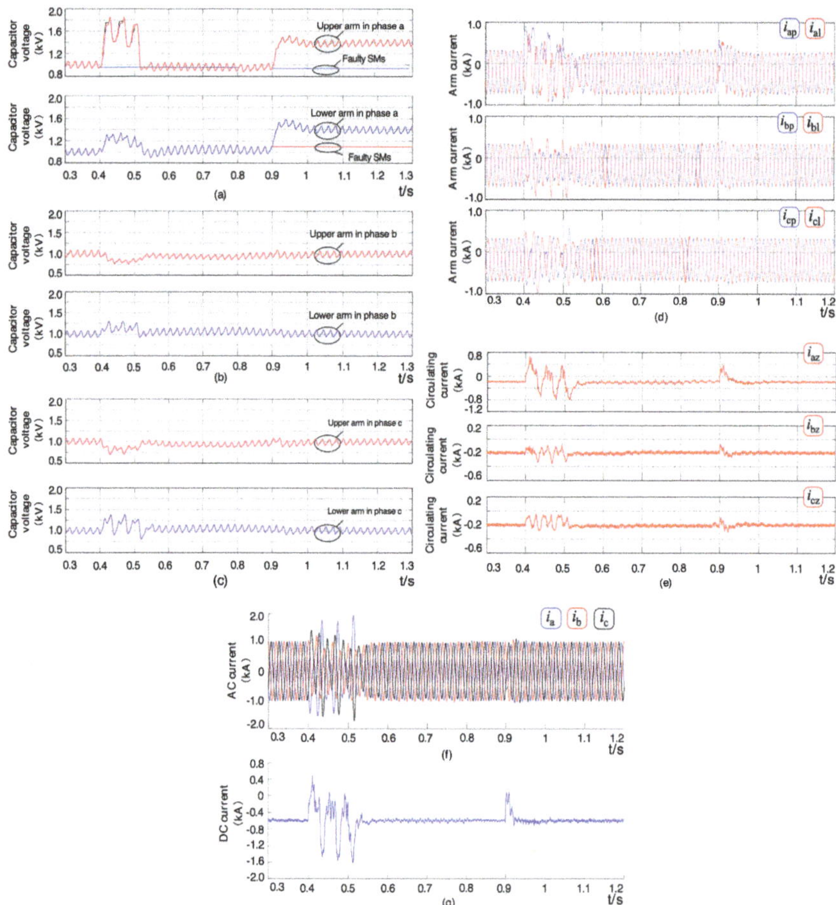

Figure 9. Simulation results of fault condition 1: (**a**) Capacitor voltage of phase a; (**b**) Capacitor voltage of phase b; (**c**) Capacitor voltage of phase c; (**d**) Arm current of phase j (j = a,b,c); (**e**) Circulating current of phase j (j = a,b,c); (**f**) AC current; (**g**) DC-link current.

From Figure 9a–g, it can be seen that: (1) When the system is in normal operation, the voltage and current of the system are stable. (2) When the SMs fail at t = 0.4 s, the SM capacitor voltage of each phase begins to oscillate irregularly, the arm current and AC grid-connected current show an asymmetric state, and the DC current and circulating current of the system also appear relatively fluctuant. (3) When fault-tolerant control strategy is enabled at t = 0.5 s, the system SM voltage immediately starts to return to normal value, the fluctuation of circulating current and DC current is obviously suppressed, and the arm current and AC grid-connected current are quickly restored to symmetry and stable operation. (4) When the SMs occur fault again at t = 0.9 s, through the control of the increasing carrier wave amplitude, the SM voltage of fault phase is in the boost operation (up to about 1.3 kV) and the SM voltage of the non-fault phase continues to maintain the stable operation of the original rated value (1.0 kV), and all the current can continue to maintain stable operation through rapid transient regulation.

The above results show that the proposed fault-tolerant control strategy can quickly realize fault-tolerant operation after the SM fault, and can respond quickly when the single arm SM fault transforms to symmetrical fault.

5.2. Case 2

Fault condition 2: Single arm fault to arm asymmetric fault. When the system operates to t = 0.4 s, five SMs in the upper arm of phase a occur fault; at t = 0.5 s, the fault-tolerant control strategy is enabled; at t = 0.9 s, three SMs in the lower arm of phase a are failed again. Figures 10 and A1 are the related simulation results.

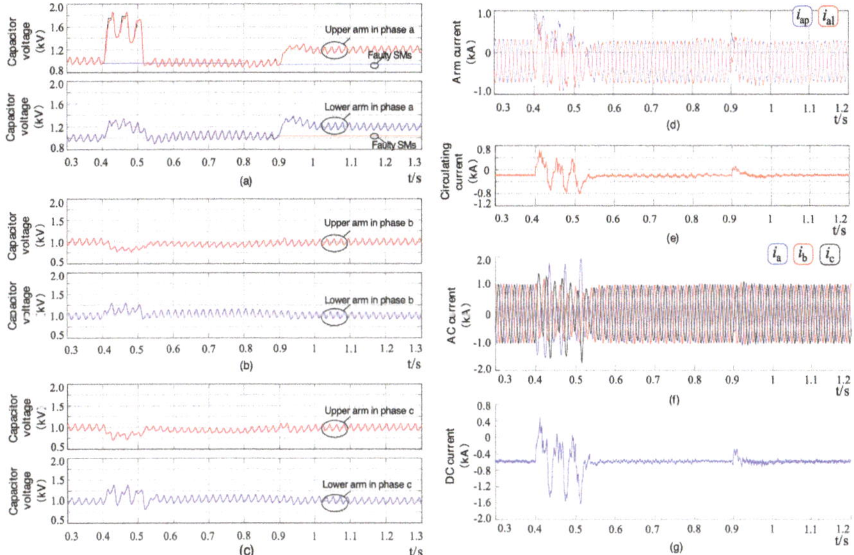

Figure 10. Simulation results of fault condition 2: (**a**) Capacitor voltage of phase a; (**b**) Capacitor voltage of phase b; (**c**) Capacitor voltage of phase c; (**d**) Arm current of phase a; (**e**) Circulating current of phase a; (**f**) AC current; (**g**) DC-link current.

It can be seen that: (1) Since the operation conditions of the system before t = 0.9 s are the same as the fault condition 1, the simulation of the system in 0~0.9 s is basically the same as the fault condition 1. (2) When the SMs occur fault again at t = 0.9 s, the system is operating asymmetrically in the upper and lower arms. The SM voltage of the fault phase is increased to about 1.2 kV by increasing carrier wave amplitude, while the SM voltage of the non-fault phase keeps the stable operation with the original rated value, and all the current can continue to maintain stable operation through rapid transient regulation.

Because the χ_a (the smaller number of faulty SMs of phase a) is smaller than the fault condition 1, the boosting amplitude of SM voltage of the fault phase is smaller than the fault condition 1, and the impact of the AC and DC current of the system is relatively small at t = 0.9 s. The above results also show that the proposed fault-tolerant control strategy can effectively achieve fault-tolerant operation after the SM fault, and can also respond quickly when the single arm fault transforms to the asymmetric arm fault.

5.3. Case 3

Fault condition 3: Arm symmetrical fault to arm asymmetric fault. When the system operates to t = 0.4 s, three SMs in the upper arm and lower arm of phase occur fault simultaneously; at t = 0.5 s, the fault-tolerant control strategy is enabled; at t = 0.9 s, two SMs in the upper arm of phase a are failed again. Figures 11 and A2 are the related simulation results.

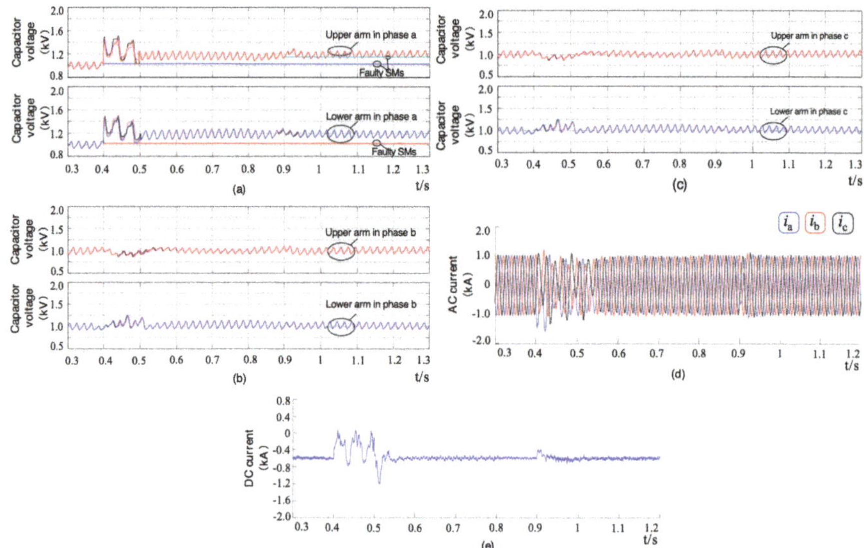

Figure 11. Simulation results of fault condition 3: (**a**) Capacitor voltage of phase a; (**b**) Capacitor voltage of phase b; (**c**) Capacitor voltage of phase c; (**d**) AC current; (**e**) DC-link current.

It can be seen that: (1) When the system is in normal operation, the voltage and current of the system are stable. (2) When the SMs fail at t = 0.4 s, the capacitor voltage of each phase SM begins to oscillate irregularly, the arm current and AC grid-connected current show an asymmetric state. But at this time, the arm structure is still in symmetrical state and the number of faulty SMs is smaller than fault conditions 1 and 2, so the oscillation amplitude of voltage and current is slight compared with the fault conditions 1 and 2. (3) When fault-tolerant control strategy is enabled at t = 0.5 s, the SM voltage of the fault phase is increased to about 1.2 kV, while the SM voltage of the non-fault phase keeps the stable operation with the original rated value, and all the current can continue to maintain stable operation rapidly. (4) When the SMs occur fault again at t = 0.9 s, through the control of the limiting modulation wave amplitude, the system can continue to operate stably.

Based on the simulation results of fault conditions 1–3, it can be seen that the fault-tolerant control strategy of the SM proposed in this paper can effectively realize the fault-tolerant operation control of the system after the SM fault, and can respond flexibly to different fault conditions of the arm, and can respond quickly when the fault state of the arm changes.

6. Experimental Studies

In order to verify the fault-tolerant control strategy, a 2-terminal MMC-based experimental platform is built in the laboratory. The experimental platform is shown in Figure 12. Among them, MMC1 is the DC voltage source and MMC2 is to verify the control strategy. The basic parameters of MMC prototype are shown in Table 2.

In the course of the experiment, the SM1 in the upper arm of phase a occurs fault at t1, fault-tolerant control strategy is enabled at t2 and the SM4 in the lower arm of phase a occurs fault again at t3. Figure 13 shows the relevant experimental waveform.

Figure 12. The photograph of MMC experiment prototype.

Table 2. Experimental parameters of the MMC prototype.

Parameters	Value
AC System inductance L_s	5 mH
Arm inductance L_m	5 mH
DC bus voltage U_{dc}	20 kV
Number of SMs per arm N	4
Number of redundant SMs per arm	1
Sub-module capacitor C	2000 μF
Transformer ratio	380 V/380 V (Y/Δ)

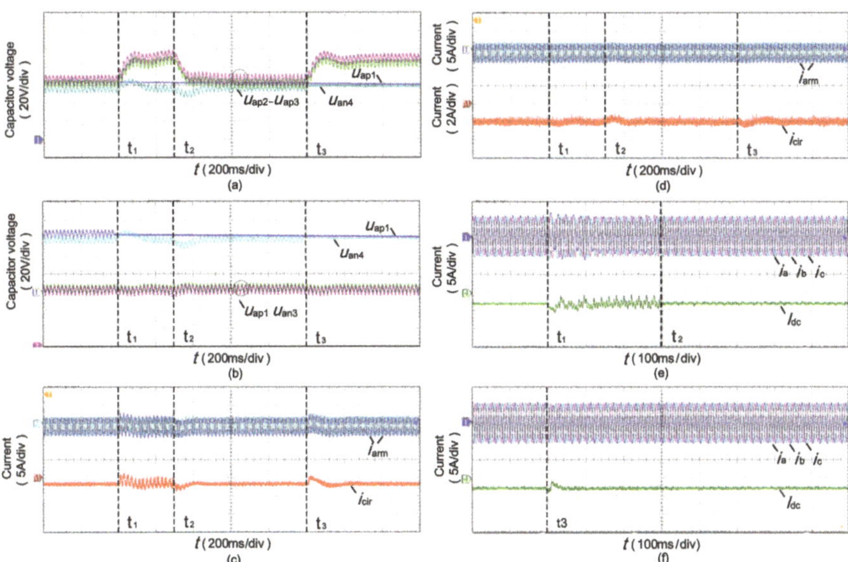

Figure 13. Waveforms of experiment: (**a**) Capacitor voltage of phase a; (**b**) Capacitor voltage of phase b; (**c**) Circulating current of phase a; (**d**) Circulating current of phase b; (**e**) AC current and DC current from t1 to t3; (**f**) AC current and DC current after t3.

It can be seen from Figure 13 that: (1) Before the fault, the system operates steadily. When the SM1 fails at t1, the capacitor voltage of SM1 begins to oscillate irregularly, the arm current and AC grid-connected current show an asymmetric state, and the DC current and circulating current of the system also appear relatively fluctuant. (2) When the fault-tolerant control strategy is put into operation at t2, SM1 voltage starts to return to normal value and the arm current and AC grid-connected current are restored to symmetry and stable operation. (3) When the SM4 occurs fault again at t3, the SM4 voltage of the fault phase is increased to 80 V, while the SM voltage of the non-fault phase keeps the stable operation with the original rated value (60 V). There is no major impact on the system in the process, and the system current is still in a stable state.

Limited by the number of SMs of the experimental platform, the experiment only validates the fault condition 1, but the experimental results are basically consistent with the simulation. The results of simulation and experiment show that the design of fault-tolerant control strategy proposed in this paper is reasonable and effective, and it is helpful to realize fault-crossing of SMs under different operating conditions.

7. Conclusions

The fault-tolerant operation issue of MMC under SM faults is studied in this paper. The main works and contributions can be summarized as:

(1) The traditional zero-sequence voltage injection fault-tolerant control algorithm is analyzed detailed. It reveals that the traditional method is easy to implement under the signal arm faulty state. However, if the SM simultaneously occurring faults in the upper and lower arms or appearing multiple arm failures, the required zero-sequence voltages will be calculated difficult. The SM fault-crossing is complicated to realize.

(2) A novel fault-tolerant control strategy based on PD-PWM is proposed, which has three main benefits: (i) it has carrier and modulation wave dual correction mechanism, which control ability is more higher and flexible; (ii) it only needs to inject zero-sequence voltage in half a cycle of the modulation wave, which simplifies the complexity of traditional zero-sequence voltage injection control algorithms and much easier for implement; (iii) furthermore, the zero-sequence voltage can even be avoided injecting under the symmetrical fault conditions.

(3) The simulations in the MATLAB/SIMULINK and experiments with 2-terminal a MMC-based prototype are all studied with the proposed control strategy under different fault conditions. The results confirm the efficiency of the control strategy.

Author Contributions: Conceptualization, W.W., T.W. and Q.H.; Formal analysis, J.Y.; Writing—original draft, J.Y.; Writing—review & editing, J.Y.

Acknowledgments: This research work was supported by National Key R&D Program of China (2016YFB0900900), National Natural Science Foundation of China (51607171), Chinese Academy of Sciences Youth Innovation Association (2017180).

Conflicts of Interest: The authors declare no conflict of interest.

Appendix A

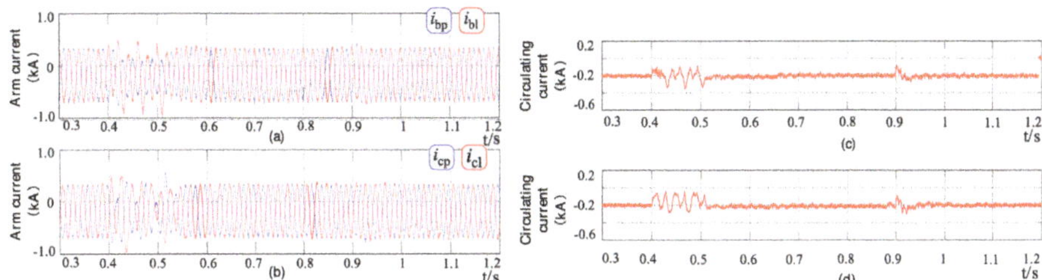

Figure A1. Simulation results of fault condition 2: (**a**) Arm current of phase b; (**b**) Arm current of phase c; (**c**) Circulating current of phase b; (**d**) Circulating current of phase c.

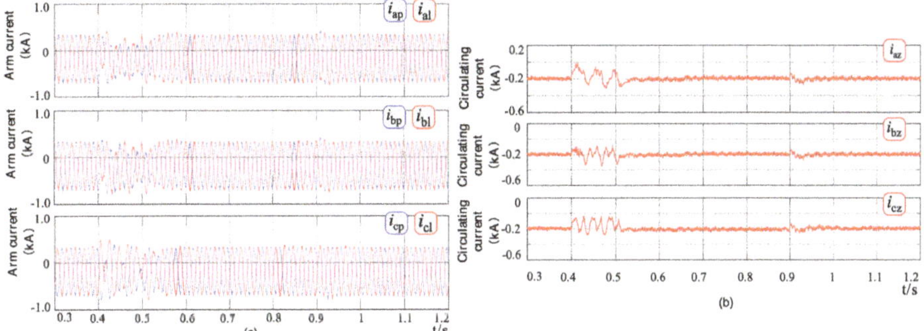

Figure A2. Simulation results of fault condition 3: (**a**) Arm current of phase j (j = a,b,c); (**b**) Circulating current of phase j (j = a,b,c).

References

1. Popova, L.; Pyrhonen, J.; Ma, K.; Blaabjerg, F. Device loading of modular multilevel converter MMC in wind power application. In Proceedings of the 2014 International Power Electronics Conference (IPEC-Hiroshima 2014—ECCE-ASIA), Hiroshima, Japan, 18–21 May 2014; pp. 548–554.
2. Gowaid, I.A.; Adam, G.P.; Ahmed, S.; Holliday, D.; Williams, B.W. Analysis and design of a modular multilevel converter with trapezoidal modulation for medium and high voltage DC-DC transformers. *IEEE Trans. Power Electron.* **2015**, *30*, 5439–5457. [CrossRef]
3. Perez, M.A.; Bernet, S.; Rodriguez, J.; Kouro, S.; Lizana, R. Circuit topologies, modeling, control schemes, and applications of modular multilevel converters. *IEEE Trans. Power Electron.* **2018**, *30*, 4–17. [CrossRef]
4. Nami, A.; Liang, J.; Dijkhuizen, F.; Demetriades, G. Modular multilevel converters for HVDC applications: Review on converter cells and functionalities. *IEEE Trans. Power Electron.* **2015**, *30*, 18–36. [CrossRef]
5. Lesnicar, A.; Marquardt, R. An innovative modular multilevel converter topology suitable for a wide power range. In Proceedings of the IEEE Bologna Power Tech Conference Proceedings, Bologna, Italy, 23–26 June 2003; pp. 23–26.
6. Debnath, S.; Saeedifard, M. Simulation-based gradient-descent optimization of modular multilevel converter controller parameters. *IEEE Trans Ind. Electron.* **2016**, *63*, 102–112. [CrossRef]
7. Li, Y.; Jiang, W.; Yu, S.; Zou, X. Design of zhoushan multi-terminal VSC-HVDC transmission project. *High Voltage Eng.* **2014**, *40*, 2490–2496.
8. Wu, W.; Wu, X.; Yin, J.; Jing, L.; Wang, S.; Li, J. Characteristic analysis and fault-tolerant control of circulating current for modular multilevel converters under SM faults. *Energies* **2017**, *10*, 1827. [CrossRef]

9.	Abdelsalam, M.; Tennakoon, S.; Griffiths, A.L.; Marei, M. Investigation of sub-module fault types of modular multi-level converters in HVDC networks. In Proceedings of the International Universities Power Engineering Conference (UPEC), Stoke, UK, 1–4 September 2015; pp. 1–6.

10.	Jiang, W.; Wang, C.; Wang, M. Fault detection and remedy of multilevel inverter based on BP neural network. In Proceedings of the Asia-Pacific Power and Energy Engineering Conference (APPEEC), Shanghai, China, 27–29 March 2012; pp. 1–4.

11.	Moon, J.W.; Kim, C.S.; Park, J.W. Circulating current control in MMC under the unbalanced voltage. *IEEE Trans. Power Deliv.* **2013**, *28*, 1952–1959. [CrossRef]

12.	Son, G.; Lee, H.J.; Nam, T. Design and control of a modular multilevel HVDC converter with redundant power modules for noninterruptible energy transfer. *IEEE Trans. Power Deliv.* **2012**, *27*, 1611–1619.

13.	Konstantinou, G.; Pou, J.; Ceballos, S. Active redundant sub-module configuration in modular multilevel converters. *IEEE Trans. Power Deliv.* **2013**, *28*, 2333–2341. [CrossRef]

14.	Liu, G.; Xu, Z.; Xue, Y. Optimized control strategy based on dynamic redundancy for the modular multilevel converter. *IEEE Trans. Power Electron.* **2015**, *30*, 339–348. [CrossRef]

15.	Li, S.; Wang, Z.; Wang, G. Fluctuation volt age control and fault-tolerant operation of modular multilevel converters with zero-sequence injection. *Int. Trans. Electr. Energy Syst.* **2014**, *24*, 944–959. [CrossRef]

16.	Hu, P.; Jiang, D.; Zhou, Y. Energy-balancing control strategy for modular multilevel converters under submodule Fault Conditions. *IEEE Trans. Power Electron.* **2014**, *29*, 5021–5029. [CrossRef]

17.	Peter, W. Enhancing the reliability of modular medium-voltage drives. *IEEE Trans Ind. Electron.* **2002**, *49*, 948–954.

18.	Jose, R.; Peter, W.; Jorge, P.; Rodrigo, M.; Maria, J. Operation of a medium-voltage drive under faulty conditions. *IEEE Trans Ind. Electron.* **2005**, *52*, 1080–1085.

19.	Pablo, L.; Gabriel, O. Extended operation of cascade multicell converters under fault condition. *IEEE Trans Ind. Electron.* **2009**, *56*, 2697–2703.

20.	Fernanda, C.; Humberto, P.; Cassiano, R. Generalized carrier-based modulation strategy for cascaded multilevel converters operating under fault conditions. *IEEE Trans. Ind. Electron.* **2012**, *59*, 679–689.

21.	Yang, Q.; Qin, J. A post-fault strategy to control the modular multilevel converter under submodule failure. *IEEE Trans. Power Deliv.* **2012**, *27*, 1538–1547.

22.	Shen, K.; Xiao, B.; Mei, J. A modulation reconfiguration based fault-tolerant control scheme for modular multilevel converters. In Proceedings of the IEEE Applied Power Electronics Conference and Exposition (APEC), Long Beach, CA, USA, 17–31 March 2013; pp. 3251–3255.

23.	Yang, W.; Song, Q.; Xu, S.; Rao, H.; Liu, W. An MMC topology Based on unidirectional current H-bridge submodule with active circulating current injection. *IEEE Trans. Power Electron.* **2018**, *33*, 3870–3883. [CrossRef]

24.	Ning, L.; Venkata, D. Dynamic electro-magnetic-thermal modeling of MMC-based DC–DC converter for real-time simulation of MTDC grid. *IEEE Trans. Power Deliv.* **2018**, *33*, 1337–1347.

25.	Xu, J.; Penghao, Z.; Zhao, C. Reliability analysis and redundancy configuration of MMC with hybrid submodule topologies. *IEEE Trans. Power Electron.* **2016**, *31*, 2720–2729. [CrossRef]

26.	Lu, C. *Research on the Control Strategy of Modular Multilevel Converter*; China University of Ming and Technology: Beijing, China, 2014.

27.	Wang, D.; Yang, J.; Chen, Z.; Mao, C.; Lu, J. A Transformerless medium voltage multiphase motor drive system. *Energies* **2016**, *9*, 323. [CrossRef]

28.	Meshram, P.M.; Borghate, V.B. A Novel Voltage Balancing Method of Modular Multilevel Converter (MMC). In Proceedings of the International Conference on Energy, Automation and Signal, Bhubaneswar, Odisha, India, 28–30 December 2011; pp. 1–5.

29.	Li, J.; Jin, X.; Wu, X. Analysis of Fault Circulating current without redundant SM and its suppression strategy. *Power Syst. Technol.* **2016**, *40*, 32–39.

Article

A Novel Multilevel Bidirectional Topology for On-Board EV Battery Chargers in Smart Grids

Rafael S. Leite, João L. Afonso and Vítor Monteiro *

Industrial Electronics Department, School of Engineering, University of Minho, 4800-058 Guimarães, Portugal; a73417@alunos.uminho.pt (R.S.L.); jla@dei.uminho.pt (J.L.A.)
* Correspondence: vmonteiro@dei.uminho.pt; Tel.: +351-253-510-392

Received: 31 October 2018; Accepted: 6 December 2018; Published: 10 December 2018

Abstract: This paper proposes a novel on-board electric vehicle (EV) battery charger (EVBC) based on a bidirectional multilevel topology. The proposed topology is formed by an AC-DC converter for the grid-side interface and by a DC-DC converter for the battery-side interface. Both converters are interfaced by a split DC-link used to achieve distinct voltage levels in both converters. Characteristically, the proposed EVBC operates with sinusoidal grid-side current, unitary power factor, controlled battery-side current or voltage, and controlled DC-link voltages. The grid-side converter operates with five voltage levels, while the battery-side operates with three voltage levels. An assessment, for comparison with classical multilevel converters for EVBCs is considered along the paper, illustrating the key benefits of the proposed topology. As the proposed EVBC is controlled in bidirectional mode, targeting the EV incorporation into smart grids, the grid-to-vehicle (G2V) and vehicle-to-grid (V2G) operation modes are discussed and evaluated. Both converters of the proposed EVBC use discrete-time predictive control algorithms, which are described in the paper. An experimental validation was performed under real operating conditions, employing a developed laboratory prototype.

Keywords: multilevel converter; electric vehicle; on-board battery charger; power factor correction; power quality; smart grid

---

1. Introduction

The electric vehicle (EV) is considered as the central element to support electric mobility in smart grids, serving to help to address major energy concerns. From a global perspective, different options of EVs can be considered distinguished by the energy storage system, as battery EVs (BEVs) or fuel cell EVs (FCEVs), and by the external interface for the charging process, as plug-in EVs (PHEV) [1–3]. Within the scope of this paper, the final application is for EVs using batteries as the energy storage system, where the main advantage is the capacity of the energy storage system and the main drawback is the required charging time. The relevance of the EV for this purpose is carefully addressed and evaluated in [4–6] in terms of power electronics and control methodologies for the grid-side. As demonstrated in [7,8], since the EV batteries are charged from the power grid (independently of the on-board or off-board technology), power quality is an imperative feature for assuring the grid stability. In this perspective, advanced contributions for the EV controlled action in smart grids, and bearing in mind power quality issues, are presented in [9]. Additionally, the opportunity to operate in bidirectional mode, as well as to operate in the four quadrants in terms of power quality will also be decisive for contributing to establish energy management strategies in a smart grid perspective. These new contributions for the EV operation in four-quadrants and framed in smart grids, are examined in [10,11]. The flexible incorporation of an EV into the energy management of a smart home is presented in [12], perspective an advanced communication toward to control the charging and discharging processes.

Classically, on-board EV battery chargers (EVBC) are projected with two- or three-level topologies for the grid-side coupling converter [13], however, by increasing the levels, the size of the passive filters can be reduced, as demonstrated in [14,15] for other types of applications, different from EVBC. Nonetheless, the levels cannot be augmented indeterminately. By establishing a tradeoff between power density and required hardware and software, five-level topologies are identified as interesting solutions for different purposes. For instance, considering the grid-side converter, a five-level topology with reduced switching devices and based on the active neutral point clamped in presented in [16] for grid-tied solar photovoltaic applications; a five-level Vienna-type structure is proposed in [17] for active rectifiers; a unidirectional five-level based on the T-rectifier topology is analyzed in [18] for high-speed gen-set applications; an improved five-level topology is presented in [19] for active rectifiers or grid-tied applications; a unidirectional five-level topology is proposed in [20] for power factor correction converters, including applications of EVBC; a symmetric cascade five-level topology is proposed in [21] for grid-tied inverters; based on the previous structure, a bidirectional five-level topology is proposed in [19], allowing the operation as active converter or as grid-tied converter; a modular unidirectional five-level topology is offered in [22] for applications of renewable energy sources; a five-level topology for renewables applications is proposed in [23]; a five-level topology based on the neutral point clamped arrangement is presented in [24] for motor drivers; a novel five-level topology is proposed in [25] for unidirectional EVBC; and a five-level topology with reduced switching devices is proposed in [20] for active rectifier applications.

In terms of the battery-side converter, two-level topologies are usually employed for the battery-side coupling converter. However, it should be noted that DC-DC multi-level topologies can be applied for other applications, for instance, in [26] is presented a novel multilevel boost converter for applications of photovoltaics or fuel cell generation systems, in [27] a review of DC-DC four- and three-level topologies is presented, and in [28] a two-level interleaved and intercoupled boost converter for high power applications is analyzed. The three-level topology is presented in [29], however, the application is for the voltage balancing of series connected batteries. On the other hand, in this paper, the split DC-link is used as interface for the grid-side converter, where the voltage of the capacitors is always controlled by the grid-side power converter. Moreover, the control algorithm is completely different, in order to obtain a reduced current ripple, where the proposed approach consists in using the converter controlled by current controlling only two switching devices during each operation mode. It should be noted that in [29] are controlled four switching devices, therefore, controlling only two it is possible to reduce the switching losses and the control complexity (e.g., it is not necessary to deal with any type of dead-time for the switching devices).

The above-mentioned five-level topologies for the grid-side converter were validated in the scope of diverse applications, but not all were validated neither compared in the scope of an EVBC. The same occurs with the battery-side converter. Moreover, no EVBC was validated employing multilevel topologies in the grid-side and the battery-side converters. In this sense, this paper proposes an on-board EVBC multilevel topology. The internal constitution of a conventional on-board EVBC is presented in Figure 1, where is highlighted the bidirectional power flow, between the batteries and the grid, in order to accomplish with the grid-to-vehicle and vehicle-to-grid operations in smart grids. The key contributions of this paper are: (a) a novel EVBC based on a multilevel topology; (b) an analysis of the proposed EVBC in terms of operation targeting smart grids; and (c) an experimental validation using a dedicated developed on-board EVBC.

The paper is outlined as follows: A description of the hardware topology of the proposed EVBC is presented in Section 2. The discrete-time predictive control algorithms used for the grid-side converter and for the battery-side converter are presented in Section 3. The foremost experimental results considering diverse operating states for smart grids are presented in Section 4. The main conclusions are discussed in Section 5.

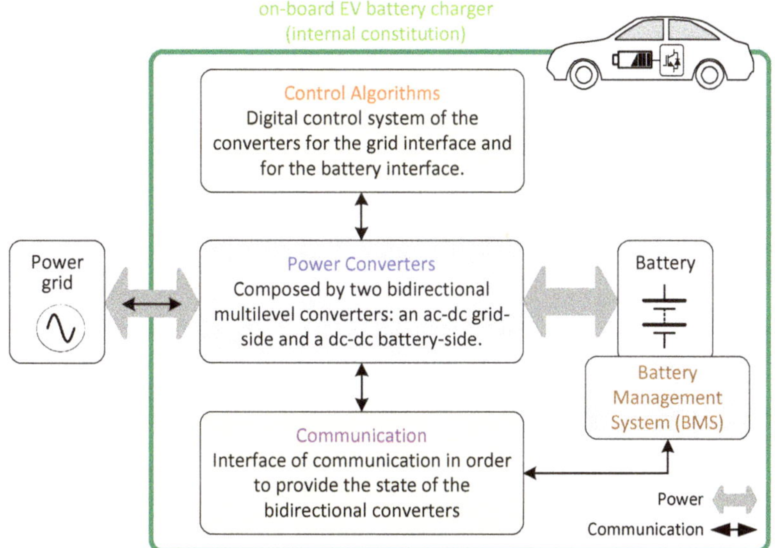

Figure 1. Internal constitution of a conventional on-board electric vehicle battery charger (EVBC).

2. EV Battery Charger: Topology Description

Figure 2 shows the global electrical schematic of the proposed EVBC. This topology consists of a grid-side converter and a battery-side converter, both with a multilevel characteristic supported by a split DC-link formed by two sets of capacitors (C_1 and C_2). In terms of other components, the EVBC consists of twelve insulated-gate bipolar transistors (IGBTs) (used as controlled switching devices, eight for the grid-side converter and four the battery-side converter), an inductive coupling filter (L_1, L_2), and a LC passive filter interfacing the batteries (L_3, L_4 and C_3).

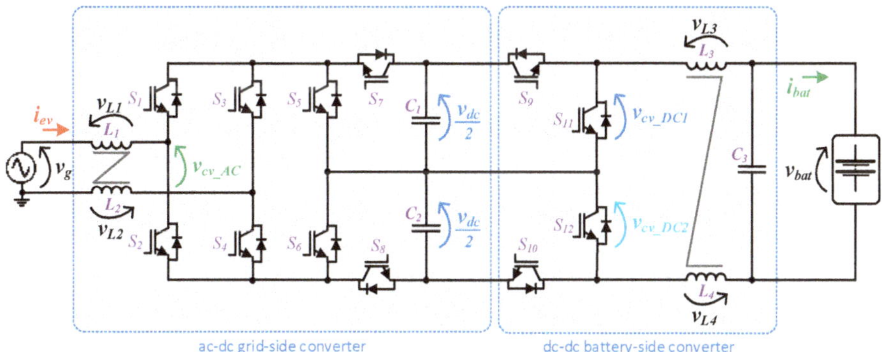

Figure 2. Topology of the proposed on-board electric vehicle battery charger (EVBC).

The values of the parameters constituting the on-board EVBC, as well as the specifications of the system that was taken into account when choosing components, are given in Table 1.

Table 1. Parameters and specifications of the on-board EVBC.

Parameter		Value	Unit
Inductive Passive Filter	L_1, L_2	5	mH
Inductive Passive Filter	L_3, L_4	2.5	mH
Capacitive Passive Filter	C_1, C_2	2.24	mF
Capacitive Passive Filter	C_3	20	µF
Nominal Input Voltage (grid-side)	v_g	$230 \pm 10\%$	V
Nominal Grid Frequency	f	$50 \pm 1\%$	Hz
Nominal Input Current (grid-side)	i_{ev}	16	A
Total Harmonic Distortion (grid current)	-	<5%	-
Total Power Factor	-	1	-
Nominal DC-link Voltage	v_{dc}	400	V
Nominal Output Voltage (battery-side)	v_{bat}	200–400	V
Nominal Output Current (battery-side)	i_{bat}	10	A
Switching Frequency	f_{sw}	20	kHz
Sampling Frequency	f_s	40	kHz

2.1. Topology Description: Grid-Side Converter

In the development of power electronics systems, electrical grid power quality issues are, more than ever, a major concern. Since the voltage levels produced by grid-side multilevel converters are directly proportional to the quality of the obtained grid current, multilevel converters have emerged as contributors for this concern. The circuit topology of the grid-side multilevel converter proposed for the on-board EVBC is presented in Figure 2. The proposed topology emerged as a derivation of the traditional full-bride rectifier with four devices connected to the split DC-link as the power factor correction (PFC) three-level DC-DC converter. This topology can produce five distinct voltage levels ($+v_{dc}$, $+v_{dc/2}$, 0, $-v_{dc/2}$, $-vdc$) at the terminals of the converter (v_{cv_AC}). As can be seen, each IGBT is applied to a maximum voltage of $+v_{dc}$. The configuration of the topology allows to switch necessarily only six of the eight IGBTs, during the full operation as active rectifier or grid-tied inverter, hence, decreasing the switching losses. When the grid side converter operates as active rectifier, Figure 3, during the positive half cycle of the power grid voltage, the IGBTs (S_1, S_4) are always switched on and the IGBTs (S_2, S_3) are always switched off. When the IGBT S_6 is switched on, the IGBT S_5 is switched to state the voltage levels 0 and $+v_{dc/2}$. On the other hand, when the IGBT S_5 is switched off, the IGBT S_6 is switched to state the voltage levels $+v_{dc/2}$ and $+vdc$. During the negative half cycle of the power grid voltage, the IGBTs (S_2, S_3) are always switched on and the IGBTs (S_1, S_4) are always switched off. When the IGBT S_5 is switched on, the IGBT S_6 is switched to state the voltage levels 0 and $-v_{dc/2}$. Finally, when the IGBT S_6 is switched off, the IGBT S_5 is switched to state the voltage levels $-v_{dc/2}$ and $-v_{dc}$. When the grid side converter operates as an inverter, Figure 4, during the positive half cycle of the power grid voltage, the IGBTs (S_1, S_4) are always switched on and the IGBTs (S_2, S_3) are always switched off. When the IGBT S_8 is switched off, the IGBT S_7 is switched to state the voltage levels 0 and $+v_{dc/2}$. On the other hand, when the IGBT S_7 is switched on, the IGBT S_8 is switched to state the voltage levels $+v_{dc/2}$ and $+v_{dc}$. During the negative half cycle of the power grid voltage, the IGBTs (S_2, S_3) are always switched on and the IGBTs (S_1, S_4) are always switched off. When the IGBT S_7 is switched off, the IGBT S_8 is switched to state the voltage levels 0 and $-v_{dc/2}$. Finally, when the IGBT S_8 is switched on, the IGBT S_7 is switched to state the voltage levels $-v_{dc/2}$ and $-v_{dc}$.

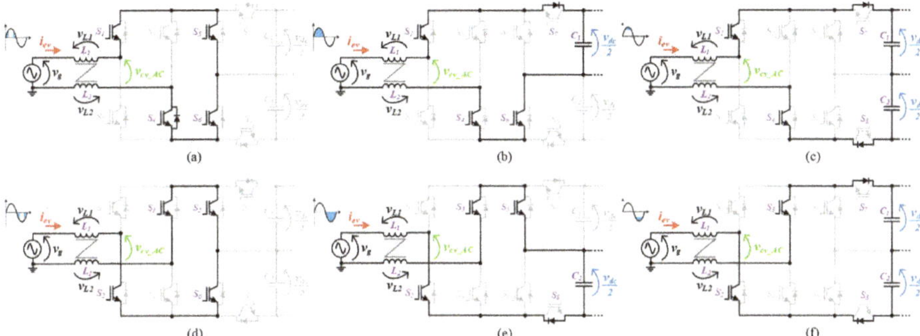

Figure 3. Operation stages for the grid-side converter during the operation as active rectifier: (**a**) v_{cv_AC} = 0; (**b**) v_{cv_AC} = +v_{dc}/2; (**c**) v_{cv_AC} = +v_{dc}; (**d**) v_{cv_AC} = 0; (**e**) v_{cv_AC} = −v_{dc}/2; (**f**) v_{cv_AC} = −v_{dc}.

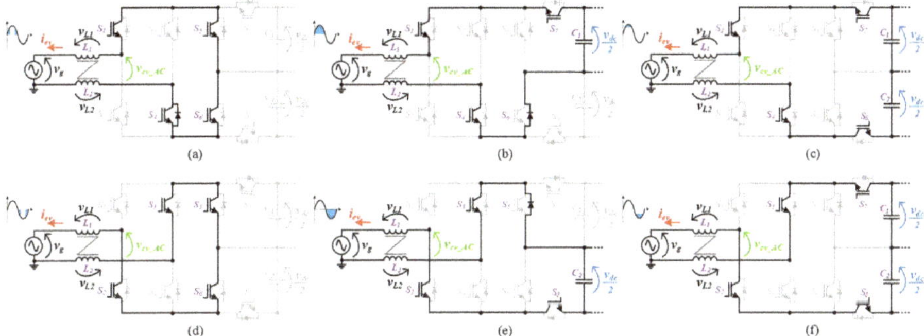

Figure 4. Operation stages for the grid-side converter during the operation as grid-tied inverter: (**a**) v_{cv_AC} = 0; (**b**) v_{cv_AC} = +v_{dc}/2; (**c**) v_{cv_AC} = +v_{dc}; (**d**) v_{cv_AC} = 0; (**e**) v_{cv_AC} = −v_{dc}/2; (**f**) v_{cv_AC} = −v_{dc}.

A predictive current control technique, with a fixed switching frequency of 20 kHz, was applied to obtain a sinusoidal EVBC current and in phase with the grid voltage (or phase opposition in vehicle-to-grid mode). Since this current control is identified as a linear current control, the modulation technique is applied individually. In this sense, Figure 5 shows the pulse-width modulation (PWM) modulation technique arrangement used for the grid-side converter. The proposed PWM modulation technique requires only one carrier signal and two reference signals to control eight IGBTs. This figure shows the adapted voltage reference (ref_1, ref_2), the triangular carrier signal ($v_{carrier}$), the PWM signals of the IGBTs, the voltage levels produced by the converter (v_{cv_AC}) and the power grid voltage (v_g). The pulses signal of the IGBTs (S_1, S_2, S_3, S_4) are exclusively dependent of the instantaneous value of the power grid voltage. During the operation as grid-tied inverter, the IGBTs (S_7, S_8) have an opposite a command signal as the IGBTs (S_5, S_6), respectively. In the modulation strategy, the voltages references (ref_1, ref_2) are adapted, based on a digital codification, from the modulating signal s_M established by the Equation (1), where v_{cv_AC} are the voltage levels produced by the converter and m_a the amplitude modulation index. The voltage levels produced by the converter are obtained in the control algorithm, explained in Section 3.1.

$$s_M = 2v_{cv\_AC}m_a. \tag{1}$$

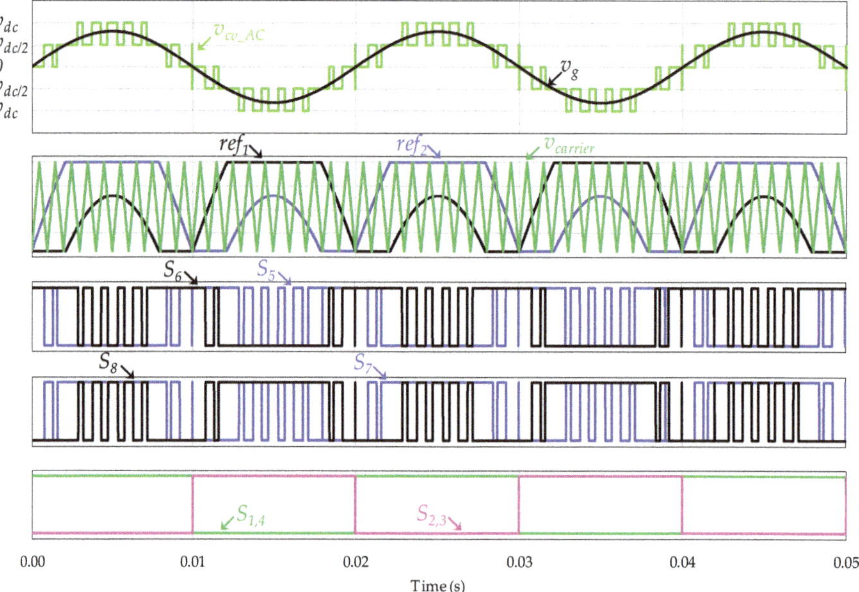

Figure 5. Modified pulse-width modulation (PWM) strategy for the grid-side converter during the operation as active rectifier or as grid-tied inverter.

2.2. Topology Description: Battery-Side Converter

The circuit topology of the battery-side converter implemented in the on-board EVBC is presented in Figure 2. The topology consists in four IGBTs connected to the split DC-link and a passive LC filter interfacing the batteries. Furthermore, this topology can produce three distinct voltage levels ($+v_{dc}$, $+v_{dc/2}$, 0) at the terminals of the converter (v_{cv_DC}). When the converter operates as a buck-type converter, the energy power flows of the DC-link to the batteries during the charging process. In this mode, the IGBTs (S_9, S_{10}) and the anti-parallel diodes of the IGBTs (S_{11}, S_{12}) are used. When only one of the IGBTs (S_9, S_{10}) is switched on, the inductors (L_3, L_4) and the batteries stores energy from the split DC-link (C_1 or C_2). When both IGBTs (S_9, S_{10}) are switched on, the inductors (L_3, L_4) and the batteries store energy from the DC-link (C_1, C_2). When both IGBTs (S_9, S_{10}) are switched off, the stored energy in the inductors (L_3, L_4) is released to the batteries. On the other hand, when the converter operates as boost-type converter, the energy power flows of the batteries to the DC-link, during the discharging process. During this mode, the IGBTs (S_{11}, S_{12}) and the anti-parallel diodes of the IGBTs (S_9, S_{10}) are used. When both IGBTs (S_{11}, S_{12}) are switched on, the inductors (L_3, L_4) stores energy from the batteries. When one of the IGBTs (S_{11}, S_{12}) is switched on, the split DC-link (C_1 or C_2) stores energy from the batteries and the inductors (L_3, L_4). Finally, when both IGBTs (S_{11}, S_{12}) are switched off, the DC-link (C_1, C_2) stores energy from the batteries and the inductors (L_3, L_4). Figures 6 and 7 show the operation stages for the battery-side converter during the operation as buck-type and boost-type, respectively.

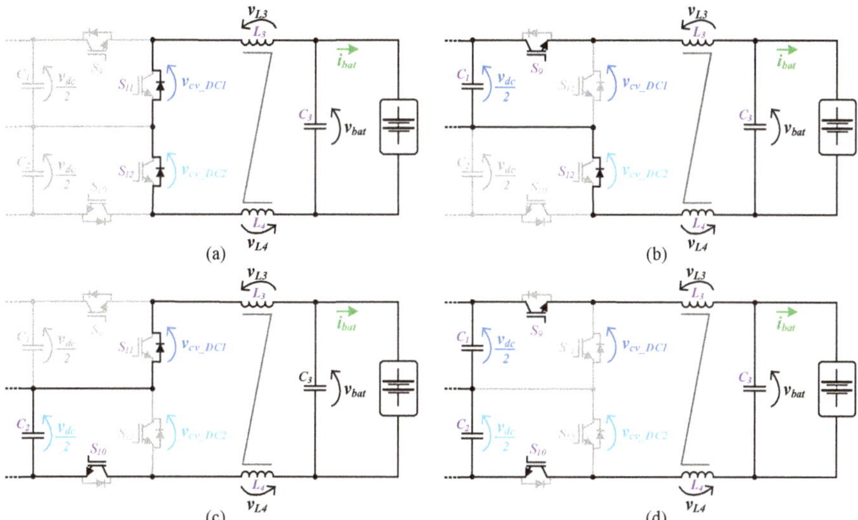

Figure 6. Operation stages for the battery-side converter during the operation as a buck-type converter: (a) The stored energy in L_3, L_4 is released to the batteries; (b) L_3, L_4 and the batteries stores energy from C_1; (c) L_3, L_4 and the batteries stores energy from C_2; (d) L_3, L_4 and the batteries store energy from C_1, C_2.

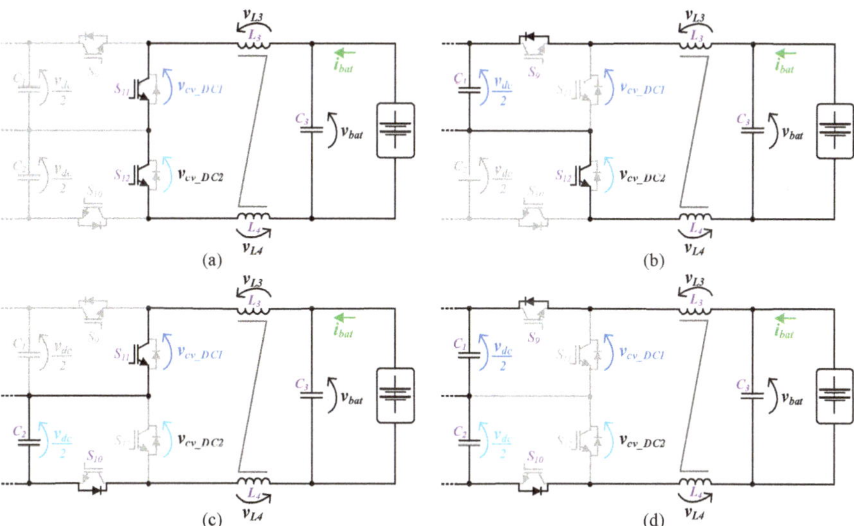

Figure 7. Operation stages for the battery-side converter during the operation as a boost-type converter: (a) L_3, L_4 stores energy from the batteries; (b) C_2 stores energy from the batteries and L_3, L_4; (c) C_1 stores energy from the batteries and L_3, L_4; (d) C_1, C_2 stores energy from the batteries and L_3, L_4.

Similar to the algorithm applied in the grid-side converter, in the battery-side converter was also applied a predictive current control technique and a modulation technique with a fixed switching frequency of 20 kHz, to control the current and voltage in the batteries during the charging or discharging processes. In the modulation technique, two 180° phase-shifted carriers were used, in order to reduce the ripple of the EV battery current and, hence, the frequency EV battery current is held twice of the switching frequency.

3. EV Battery Charger: Control Algorithms

This section presents the specifications and the methodology used for the control algorithms implementation, both for the grid-side and the battery-side converters. The control algorithm was designed for a digital platform, based on a Texas Instruments digital signal processor (DSP) F28335 (Texas Instruments, Inc., Dallas, TX, USA) and considering a sampling frequency of 40 kHz, obtained with a timer interruption.

3.1. Control Algorithm: Grid-Side Converter

Based on the voltages shown in Figure 2, Equation (2) can be established, where v_g represents the instantaneous value of the grid voltage, v_{L1} and v_{L2} the instantaneous value of the inductance voltage, and v_{cv_AC} is the instantaneous value (i.e., the voltage produced during each sampling period of the DSP) of the voltage produced by the converter:

$$v_g = v_{cv\_AC} + v_{L1} + v_{L2}.\tag{2}$$

It should be noted that, as represented in Figure 2, it was used a mutual coupling inductance. Therefore, replacing the inductance voltage by its intrinsic equation, and rewriting the equation as a function of the voltage produced by the converter, it is obtained:

$$v_{cv\_AC} = v_g - L_1 \frac{di_{ev}}{dt} - L_2 \frac{di_{ev}}{dt}.\tag{3}$$

Applying the progressive Euler method, illustrated in Equation (4), the derivative component of the current can be approximated by considering a very low Δt in order to obtain a good prediction of the system behavior:

$$\frac{di_{ev}(t)}{dt} = \frac{i_{ev}(t + \Delta t) - i_{ev}(t)}{\Delta t}.\tag{4}$$

Applying the Equation (4) in the Equation (3), and assuming a sampling frequency of $fs = 1/T_s$, results in the digital control Equation (5), where the term k represents the current sample and $[k + 1]$ represents the next sample:

$$v_{cv\_AC}[k] = v_g[k] - (L_1 + L_2)\frac{i_{ev}[k + 1] - i_{ev}[k]}{T_s}.\tag{5}$$

Since the law of predictive control consists of a closed-loop control and, if the reference current at time $[k + 1]$ is to be equal to the current produced by the converter at time $[k]$, the equation that translates the current control implemented can be defined by:

$$v_{cv\_AC}[k] = v_g[k] - (L_1 + L_2)\frac{i_{ev}^*[k] - i_{ev}[k]}{T_s}.\tag{6}$$

Since, the EVBC is proposed to operate with a sinusoidal current and unitary power factor in the grid-side converter, the instantaneous value of the power grid voltage is directly proportional to the EVBC current. However, aiming to prevent the inclusion of the harmonic distortion of the grid voltage into the current, it is used a phase-locked loop (PLL). Thus, instead of the grid voltage, it is used the output signal from the PLL, resulting in:

$$i_{ev} = G_{ev}v_{pll},\tag{7}$$

where v_{pll} is in phase with the power grid voltage and G_{ev} represents the equivalent conductance of the EVBC from the grid-side point of view, which can be defined according to the mean value of active power (P_{ev}) and the rms value of the power grid voltage (V_g):

$$G_{ev} = P_{ev}V_G^{-2}.$$ (8)

Applying the Equation (8) in the Equation (7), the reference of the EVBC current is obtained according to:

$$i_{ev}{}^* = P_{ev}V_G^{-2}v_{pll}.$$ (9)

The active power of the EVBC can be divided in two parts, namely the power to regulate the DC-link voltage and the power to regulate the batteries. Furthermore, because of the split DC-link, two proportional-integral (PI) are used to regulate the DC-link voltage independently in both capacitors (p_{dc1}, p_{dc2}). Therefore, during the G2V operation mode, the reference of the EVBC current can be defined as:

$$i_{ev}{}^* = (p_{dc1} + p_{dc1} + p_{bat})v_{pll}V_G^{-2}.$$ (10)

On the other hand, during the V2G operation mode, the reference of the EVBC current is established according to Equation (11), where $i_{bat}{}^*$ represent the reference of current to discharge the batteries.

$$i_{ev}{}^* = (p_{dc1} + p_{dc1} + i_{bat}{}^*v_{bat})v_{pll}V_G^{-2}.$$ (11)

3.2. Control Algorithm: Battery-Side Converter

During the process of charging the batteries, the battery-side converter operates as buck converter, controlling the charging current or the charging voltage for the batteries. In this way, based on the representations of the currents and voltages between the DC-link and the batteries (cf. Figure 2), it is possible to establish the Equation (12), where v_{cv_DC} represents the voltage produced by the converter (i.e., the sum between v_{cv_DC1} and v_{cv_DC2}), v_{L3} and v_{L4} represent the voltages in the inductances (L_3 and L_4), respectively, and v_{bat} represents the voltage in the batteries:

$$v_{cv\_DC} = v_{bat} + v_{L3} + v_{L4}.$$ (12)

Since the current in the inductance L_3 is the same as that in the inductance L_4, the Equation (12) can be rewritten, replacing the voltage in the inductance by its intrinsic equation:

$$v_{cv\_DC} = v_{bat} + (L_3 + L_4)\frac{di_{L3,L4}}{dt}.$$ (13)

Applying the progressive Euler method, the Equation (13) can be established in discrete time as:

$$v_{cv\_DC}[k] = v_{bat}[k] + (L_3 + L_4)(i_{L3,L4}[k+1] - i_{L3,L4}[k])T_s^{-1}.$$ (14)

Since it is desired that the reference current at time $[k + 1]$ should be equal to the current produced by the converter at time $[k]$, it is obtained:

$$v_{cv\_DC}[k] = v_{bat}[k] + (L_3 + L_4)(i_{L3,L4}{}^*[k] - i_{L3,L4}[k])T_s^{-1}.$$ (15)

When the same converter operates as buck converter, but controlling the charging voltage of the batteries, the reference voltage is established as:

$$v_{cv\_DC}[k] = v_{bat}{}^*[k] - (L_3 + L_4)(i_{L3,L4}[k] - i_{L3,L4}[k-1])T_s^{-1}.$$ (16)

The aforementioned equations were defined for the battery-side converter operating as buck converter, i.e., charging the batteries from the grid (G2V mode). On the other hand, a set of equations should also be defined for the battery-side converter operating as boost converter, i.e., discharging the batteries to the grid (V2G mode). Based on the representations of the currents and voltages in Figure 2, when the converter intends to discharge the batteries with a constant current, the following relation can be established:

$$v_{cv\_DC} = v_{bat} - v_{L3} - v_{L4}. \tag{17}$$

Applying the same aforementioned modeling reasoning, the discrete implementation of Equation (17) results in:

$$v_{cv\_DC}[k] = v_{bat}[k] - (L_3 + L_4)(-i_{L3,L4}{}^*[k] - i_{L3,L4}[k])T^{-1}. \tag{18}$$

From Equation (18), it is important to note that it is necessary to identify the positive direction of the current in the batteries (Figure 2), which is why the negative signal $-i_{L3,L4}{}^*[k + 1]$ is applied to the digital implementation of this equation. The obtained signal ($v_{cv\_DC}[k]$) is compared with two carriers in order to obtain the duty-cycle of the gate signals for S_9 and S_{10}, which is the same, but two carriers, delayed by 180 degrees, are employed. Applying this strategy, the frequency of the resultant ripple is the double of the switching frequency. The duty-cycle is determined according to the current control algorithm, e.g., in the previous equations is defined the a voltage that is compared with the carriers in order to obtain a current in the EV battery $i_{(L3,L4)}[k]$ equal to the reference current $i_{(L3,L4)}{}^*[k]$. Applying this strategy, the current ripple in the EV battery can be reduced when compared with a traditional buck or boost converter. In circumstances where the battery-side converter is responsible for controlling the voltage on the DC-link, the Equation (19) is implemented, where δ_{cv_DC} represents the duty-cycle that the converter must produce, $v_{dc}{}^*$ represents the reference voltage for the DC-link, and v_{bat} represents the battery voltage:

$$\delta_{cv\_DC}[k] = \frac{v_{dc}{}^*[k] - v_{bat}[k]}{v_{dc}{}^*[k]}. \tag{19}$$

4. EV Battery Charger: Experimental Validation

This section introduces the experimental validation, where is presented the hardware description, as well as the most relevant experimental results.

4.1. Description of the Developed Prototype

After the computer validation, a laboratory prototype of the on-board EVBC was implemented, which is mainly divided in two parts: the digital control platform and the power hardware. The digital control platform includes the DSP board (F28335), the current and voltage sensors (models LTSR15 NP and CYHVS025A from LEM (Geneva, Switzerland) and from ChenYang (Finsing, Germany), respectively), as well as the printed circuit boards of the signal conditioning, error detection, command, and gate drivers (developed with HCPL 3120 optocouplers from Avago). On the other hand, the power hardware includes both converters, constituted by discrete IGBTs (FGA25N120ANTD from Fairchild (Sunnyvale, CA, USA)) and by the DC-link capacitors (EETUQ2W561DA from Panasonic (Kadoma, Japan)). Since the main aim of the implementation was the development of a laboratory prototype as close as possible of the reality, namely in compactness and robustness, a three-dimensional modeling was carried out to determine the best method and solution of the component layout for its implementation. Figure 8 shows the internal and external view of the three-dimensional modeling of the EVBC. With this modeling, it was possible to implement the final laboratorial prototype, which is presented in Figure 9. However, it should be noted that the laboratory prototype was developed just aiming to validate the structure of the topology and to perform a laboratorial experimental validation. In fact, after the proof-of-concept in terms of topology and operation modes, some improvements are required focusing in the optimization of the switching devices (e.g., employing SiC devices) and in the optimization of the passive filters (e.g., employing a inductor-capacitor-inductor, LCL, filter) toward a pre-industrialized prototype.

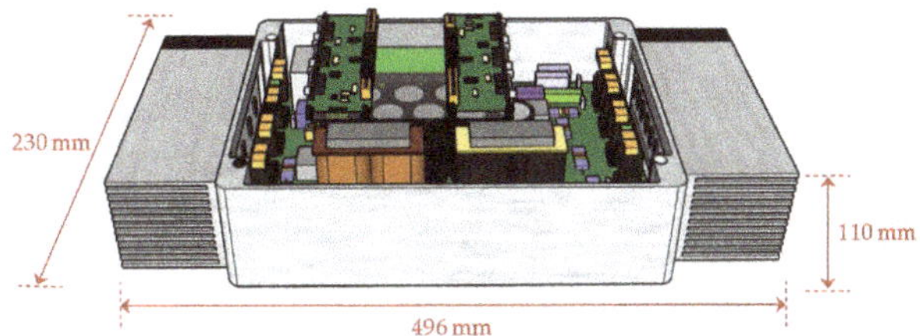

Figure 8. Three-dimensional modeling of the proposed on-board EVBC.

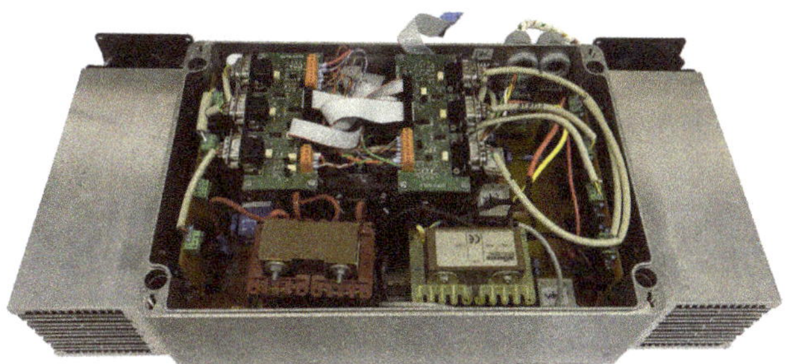

Figure 9. Laboratorial prototype of the developed EVBC.

The converters of the developed prototype were sized to meet the key requirements of the proposed on-board EVBC in terms of power quality, and the specifications given in Table 1. In this sense, the choice of components was mainly based on their operational characteristics, such as the maximum operating frequency and the rated current and voltage. In addition to the electrical aspects of the concerned application, mechanical and thermal aspects, such as dimensions, encapsulation and thermal dissipation of components, have also been taken into account, since the entire system is integrated in a metal box (Bernstein CA 380 (Porta Westfalica, Germany)), with dimensions of $330 \times 230 \times 110$ mm. As it can be seen, two heatsinks were used, one for each converter, which were fixed on each side of the metal box. In order to prevent electrical noise, it was decided to place the gate driver boards as close as possible to the power semiconductors, each one responsible for a power electronics converter. In the control system, electrical noise was also taken into account in order to optimize the signal-to-noise ratio, i.e., to minimize undesired signals superimposed on a measured signal. Based on the developed three-dimensional modeling and the implemented prototype, an estimation of the internal volume distribution associated to the different main parts of the on-board EVBC was established, which is presented in Figure 10.

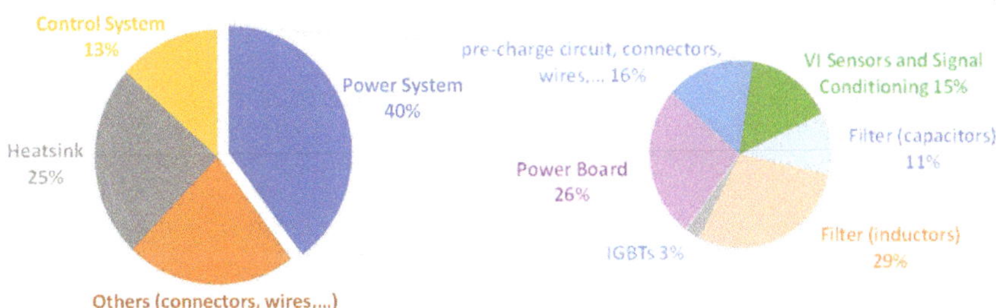

Figure 10. Internal volume distribution of the developed on-board EVBC.

4.2. Experimental Results

In the on-board EVBC laboratorial prototype described above, the G2V operation mode was initially validated, followed by the V2G. Besides the validation of both operations mode, the experimental results also demonstrate the validation of the PLL algorithm, modulation and current control strategy and DC-link voltage control. Figure 11 presents the general view of the laboratorial setup that was used during the experimental results, which were registered with a digital oscilloscope Yokogawa model DL708E (Yokogawa Electric, Tokyo, Japan), with a Fluke 435 power quality analyzer (Fluke Corporation, Everett, WA, USA), and with a FLIR i7 infrared thermal imaging camera (FLIR Systems, Wilsonville, OR, USA).

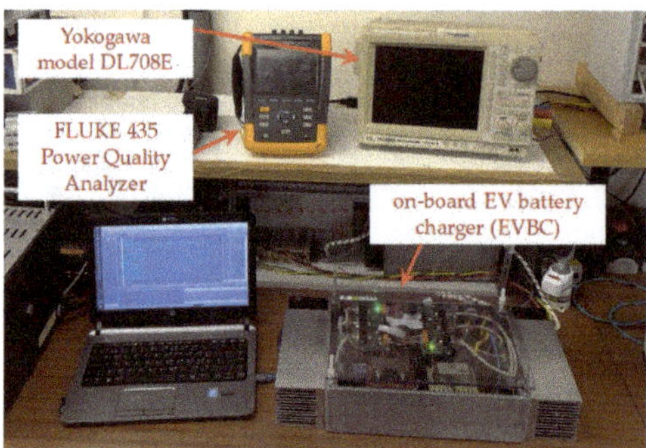

Figure 11. Laboratory setup used to obtain the experimental results.

4.2.1. Experimental Results: Grid-To-Vehicle (G2V) Operation

Figure 12 shows the experimental results during the initial stage of the EV charging process. First, during the time interval (1) the EVBC is not connected to the power grid. At time instant (2), the EVBC is connected to the power grid. In this stage, the DC-link start the pre-charge process through the anti-parallel diodes of the IGBTs (S_1, S_2, S_3, S_4), operating the grid-side converter as a traditional full-bridge rectifier. The pre-charge circuit contains a resistor to limit the typical peak currents of the capacitors. After the DC-link pre-charge process, at time instant (3), this resistor is bypassed so that the DC-link voltage remains close to the peak value of the grid voltage. At time instant (4), the split DC-link voltage, in each capacitor, is regulated to the operation voltage and only after stabilizing the DC-link voltage, at time instant (5), the current battery increases progressively (i.e., during the constant current algorithm), which is controlled by the battery-side converter operating as buck converter.

As clearly shown in Figure 13, the grid voltage is distorted due to the nonlinear electrical appliances linked in the electrical installation. However, during a steady-state operation, with the adopted current control strategy, the EVBC current is kept with a sinusoidal waveform and in phase with the grid voltage (i.e., operating with a unitary power factor, as shown in the zoom detail presented in this figure). Furthermore, the DC-link voltage in both capacitors presents an acceptable ripple for this type of application (i.e., about 10%). The average voltage value in each DC-link capacitor is controlled according to each reference, assuming a value which is greater than the maximum amplitude of the power grid voltage, and the ripple in the voltages of the capacitors has a frequency that is double the frequency of the power grid. In this operation mode the measured efficiency was 92%.

The experimental results shown in Figure 14 were attained to verify the switching states of the grid-side converter according the Figure 3. The IGBTs S_1, S_2, S_3, and S_4, as well as the IGBTs S_5 and S_6, have a fixed switching frequency of 50 Hz and 20 kHz, respectively. This figure also shows the digital signal obtained from the PLL algorithm, which is sinusoidal and in phase with grid voltage. This signal was visualized in the oscilloscope using an external digital-to-analog converter (DAC), model TLV5610 from Texas Instruments.

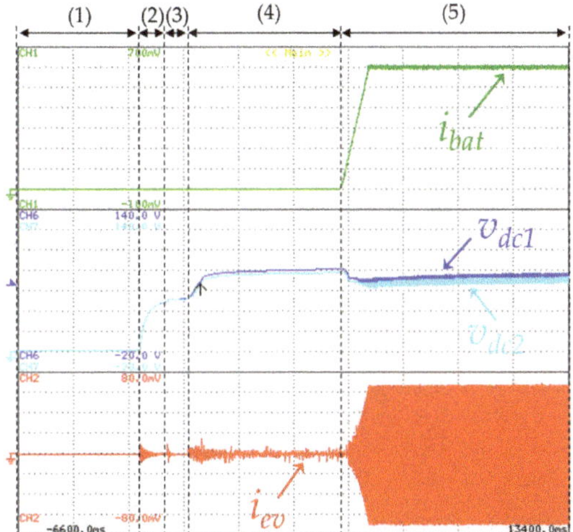

Figure 12. Experimental results during the initial stage of the electric vehicle (EV) battery charging process: EV battery current (i_{bat}: 1 A/div); DC-link voltages of both capacitors (v_{dc1}: 20 V/div, v_{dc2}: 20 V/div); EVBC current (i_{ev}: 2 A/div).

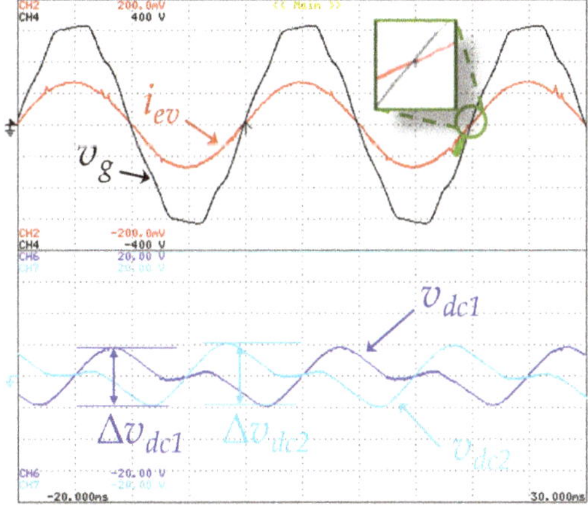

Figure 13. Experimental results in grid-to-vehicle (G2V) mode showing the EVBC current (i_{ev}: 5 A/div), the grid voltage (v_g: 100 V/div), a detail of zero crossing between the current and the voltage (i_{ev}, v_g), and the DC-link voltages ripple in both capacitors (Δv_{dc1}: 5 V/div, Δv_{dc2}: 5 V/div).

The PWM technique developed for the battery-side converter consists in two carriers with a 180° phase-shifting. With this phase-shifting strategy, one of the gate signals is applied to IGBT S_9 while the other is applied to the IGBT S_{10}, to reduce the ripple of the EV battery current. Thus, as it can be seen in Figure 15, the frequency of the current ripple in the inductor L_3 (i_{L3}) is 40 kHz, which is twice of the switching frequency. This experimental result shows the relation of the current in the inductor and the gate-emitter voltage, v_{ge_S9} and v_{ge_S10}, of the IGTBs S_9 and S_{10}. During this result, the registered value of the ripple current in the inductor was 0.16 A.

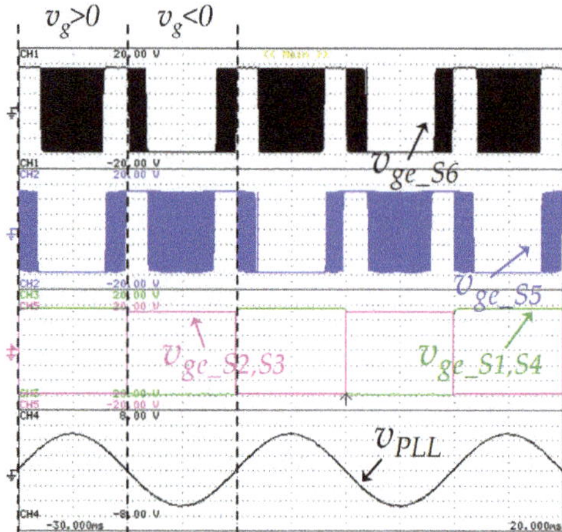

Figure 14. Experimental results during the G2V operation mode: Gate-emitter voltage of the grid-side insulated-gate bipolar transistors (IGBTs) (S_1, S_2, S_3, S_4, S_5, S_6: 5 V/div); Output digital signal of the phase-locked loop (PLL) (v_{PLL}: 150 V/div).

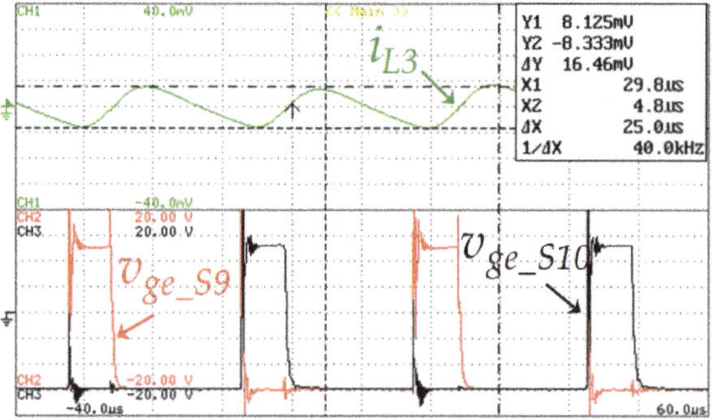

Figure 15. Experimental results showing the current in the inductor L_3 (i_{L3}: 0.1 A/div) and the gate-emitter voltage of the IGBTs S_9 and S_{10} (v_{ge_s9}: 5 V/div and v_{ge_10}: 5 V/div) during a time interval of 100 μs.

For further details on the relationship between the grid voltage and the EVBC grid-side current, it was used the *x-y* mode of the oscilloscope, as shown in Figure 16. Channel 4 is used in the *x*-axis and the channel 2 is used in the *y*-axis, representing the grid voltage and EVBC grid-side current, respectively. Thus, for the grid-side, this result shows the EVBC current in function of the grid voltage. It is relevant to highlight that this variation is not linear due to the harmonic distortion present in the grid voltage.

The experimental result shown in Figure 17 was obtained in *x-y* mode in order to identify the DC-link voltage regulation and to clearly identify the five distinct voltage levels ($+v_{dc}$, $+v_{dc/2}$, 0, $-v_{dc/2}$, $-v_{dc}$) produced by the grid-side converter. Thus, the DC-link voltage ripple (Δv_{dc1}, Δv_{dc2}) and the voltage levels (v_{cv_AC}), used in the *y*-axis, are a function of the grid voltage (v_g), used in the *x*-axis. In order to keep the DC-link regulated and balanced, during the positive half-cycle of the grid voltage, the voltage of the capacitor C_1 is regulated, and during the negative half-cycle of the grid voltage, the voltage of the capacitor C_2 is regulated.

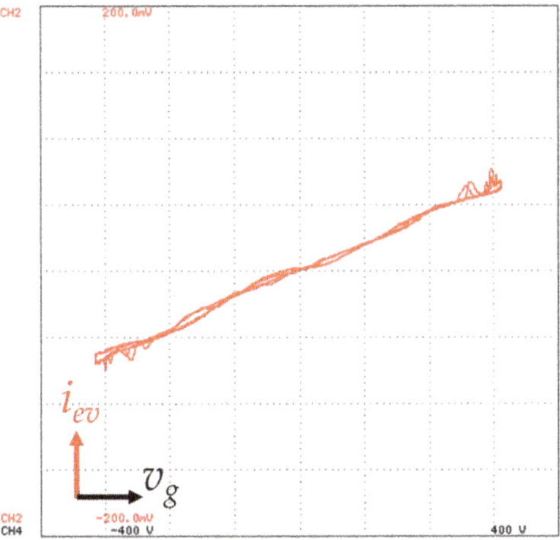

Figure 16. Experimental results in *x-y* mode showing the EVBC current (i_{ev}: 5 A/div) in function of the power grid voltage (v_g: 100 V/div).

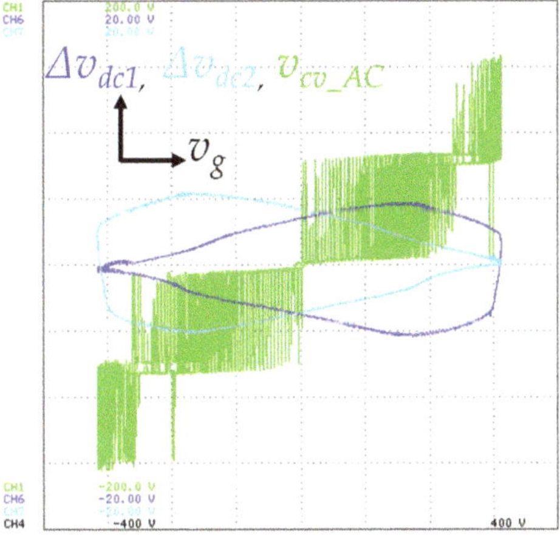

Figure 17. Experimental results in *x-y* mode showing the DC-link voltage ripple (Δv_{dc1}: 5 V/div, Δv_{dc2}: 5 V/div) (*y*-axis) and the voltage levels produced by the grid-side converter (v_{cv_AC}: 50 V/div) (*y*-axis), both in function of the power grid voltage (v_g: 100 V/div) (*x*-axis).

Using the Fluke power quality analyzer, in Figure 18a,b shows the harmonic spectrum of the power grid voltage and the EVBC current, measured total harmonic distortion (THD%) of 3.5% and 2.8%, respectively. In power electronics systems, the thermal characteristic is a factor that directly affects the performance of it. So, to analyse the thermal conditions of the EVBC, the experimental results of temperature measurements during the G2V operation mode are presented in Figure 19. Figure 19a shows the overall thermal distribution of the implemented EVBC, Figure 19b shows the measured temperature of IGBT S_9 (switched at a fixed frequency of 20 kHz), where was registered a temperature value of 47.8 °C, and Figure 19c shows the measured temperature of the IGBT S_{11} (only the antiparallel diode is used in this context), where was registered a temperature value of 36.8 °C.

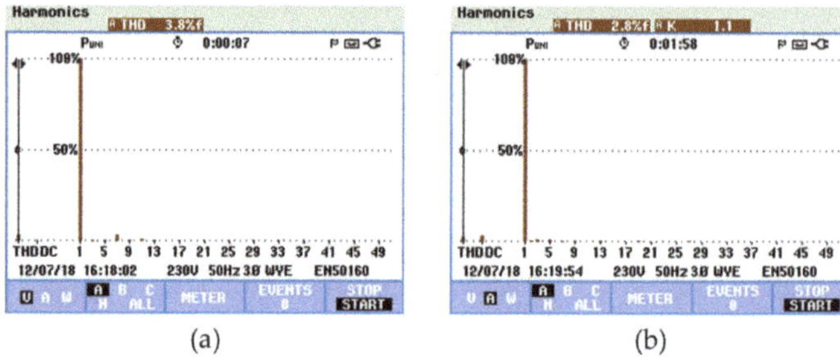

(a)   (b)

Figure 18. Experimental results in G2V mode of the total harmonic distortion and spectral analysis: (a) power grid voltage (v_g); and (b) EVBC current (i_{ev}).

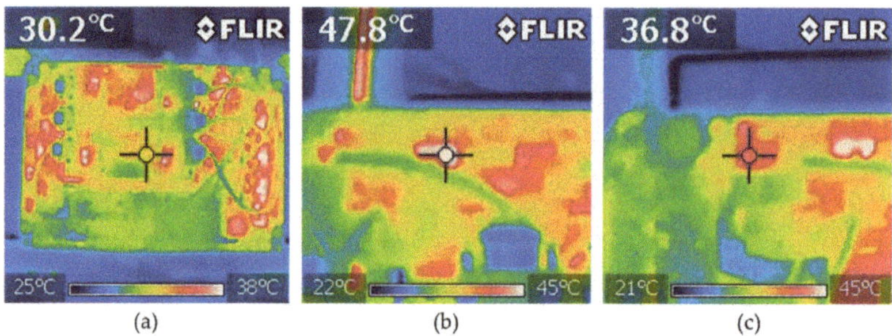

(a)   (b)   (c)

Figure 19. Experimental results of the temperature measurements during G2V operation mode: (a) overall thermal distribution of the developed EVBC prototype; (b) temperature at the IGBT S_9; and (c) temperature at the IGBT S_{11}.

4.2.2. Experimental Results: Vehicle-To-Grid (V2G) Operation

The developed EVBC was also validated during the V2G operation mode. Furthermore, once the IGBTs S_7 and S_8 of the grid-side converter has a fixed switching frequency of 20 kHz, to validate the modulation technique applied in these IGBTs, Figure 20 presents the reference signal adapted and the gate-emitter voltage of the respective IGBT. This voltage is a resulting signal of the comparison between the carrier signal and the reference signal. In this result, the reference signal was acquired using an external DAC.

Similar to the G2V operation mode, Figure 21 shows the switching states of the grid-side converter according to Figure 3 during the V2G operation mode. In this result, the IGBTs S_1, S_2, S_3, and S_4, as well as the IGBTs S_7 and S_8, have a fixed switching frequency of 50 Hz and 20 kHz, respectively. In this operation mode, the IGBTs (S_5, S_6) are always switched off, reason why they are not represented in this figure.

Figure 22 shows the V2G operation mode during a steady operation of the EVBC grid-side current (i_{ev}), the grid voltage (v_g), the voltage levels assumed by the grid-side converter (v_{cv_AC}), and the DC-link voltage of both capacitors (v_{dc1}, v_{dc2}). The EVBC grid-side current is sinusoidal, but in phase opposition with the grid voltage, meaning that the power follows from the batteries to the grid. Furthermore, the five distinct voltage levels ($+v_{dc}$, $+v_{dc/2}$, 0, $-v_{dc/2}$, $-vdc$) produced by the grid-side converter can be seen in this figure.

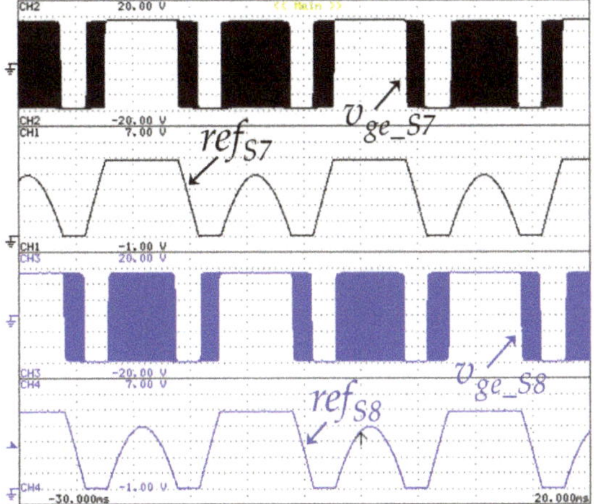

Figure 20. Experimental results showing the modulation technique applied to the IGBTs S_7 and S_8, namely, the reference signals adopted for the modulation (ref_{s7}: 1 V/div, ref_{s8}: 1 V/div) and the voltage gate-emitter of the respective IGBT (v_{ge_s7}: 5 V/div and v_{ge_s8}: 5 V/div).

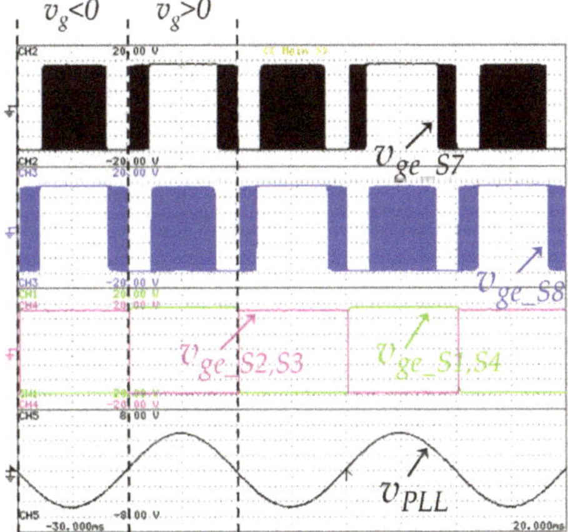

Figure 21. Experimental results during vehicle-to-grid (V2G) operation mode: Gate-emitter voltage of the grid-side IGBTs (S_1, S_2, S_3, S_4, S_7, S_8: 5 V/div), and output digital signal of the PLL (v_{PLL}: 150 V/div).

For further details, Figure 23 presents an experimental result, during a time interval of 50 ms, of the EVBC grid-side current (i_{ev}), the grid voltage (v_g), and the DC-link voltage ripple in both capacitors (Δv_{dc1}, Δv_{dc2}). With the detail of the current zero-crossing, it is possible to state that during these results the EVBC operates with unitary power factor. As aforementioned, during the positive half-cycle of the power grid voltage, the voltage of the capacitor C_1 is regulated and during the negative half-cycle of the power gird voltage, the voltage of the capacitor C_2 is regulated, which are controlled by the grid-side converter. Moreover, as it can be seen in this figure, the DC-link voltage has a voltage ripple of 3%.

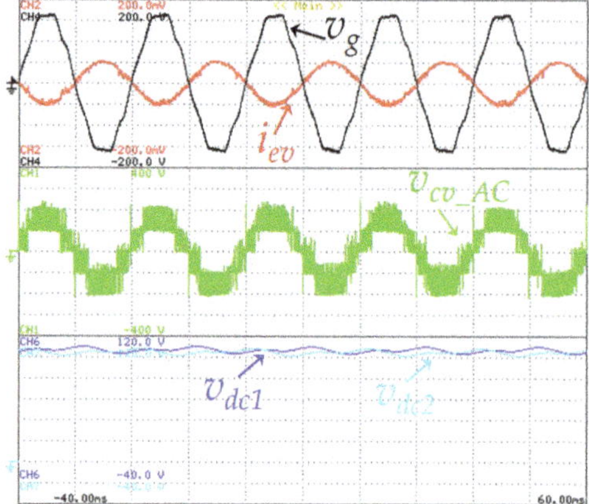

Figure 22. Experimental results in V2G operation mode showing the EVBC current (i_{ev}: 5 A/div), the power grid voltage (v_g: 50 V/div), the voltage levels produced by the grid-side converter (v_{cv_AC}: 100 V/div) and the DC-link voltage of both capacitors (v_{dc1}: 20 V/div, v_{dc2}: 20 V/div).

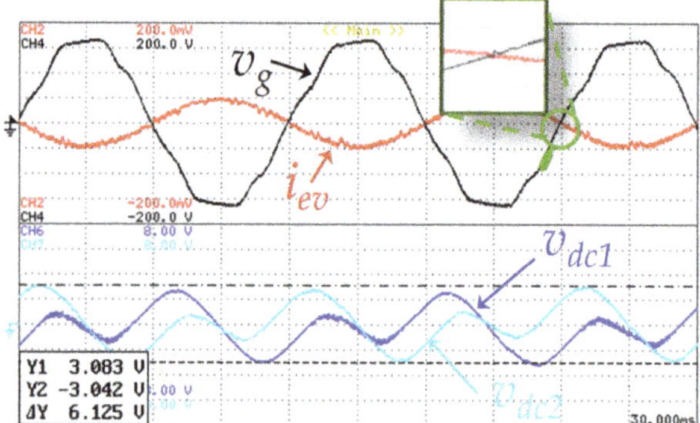

Figure 23. Experimental results in V2G operation mode showing the EVBC current (i_{ev}: 5 A/div), the grid voltage (v_g: 50 V/div), a detail of zero crossing between the current and the voltage (i_{ev}, v_g), and the DC-link voltages ripple in both capacitors (Δv_{dc1}: 2 V/div, Δv_{dc2}: 2 V/div).

Regarding the battery-side converter, the same modulation technique implemented in the G2V operation mode was used, namely the application of two 180° phase-shifted carrier signal. This strategy was adopted to reduce the ripple amplitude of the batteries current. In this sense, Figure 24 shows the current ripple in the inductor L_3 according to the gate-emitter voltages, v_{ge_S11} and v_{ge_S12}, of the IGBTs S_{11}, S_{12}. It is important to note that during this operation mode, the IGBTs S_9 and S_{10} are always off, reason why they are not shown in the figure. As it can be seen, the measured current ripple in the inductor L_3 was 0.13 A for a frequency of 40 kHz, which is twice of the switching frequency. According to this result, when the IGBT S_9 or the IGBT S_{10} is on, the inductor stores energy and during the state transition of one the IGBTs, the inductor releases this energy. Using the power quality analyzer, in Figure 25a,b, the harmonic spectrum of the power grid voltage and the EVBC current is shown, measured THD% of 4.2% and 3.5%, respectively. These figures were obtained employing a Fluke 435 power quality analyzer, and the high value of THD in the power grid voltage is caused by distorted voltage drop in the line impedance, which is produced by distorted current consumed by several nonlinear loads connected to the electrical installation.

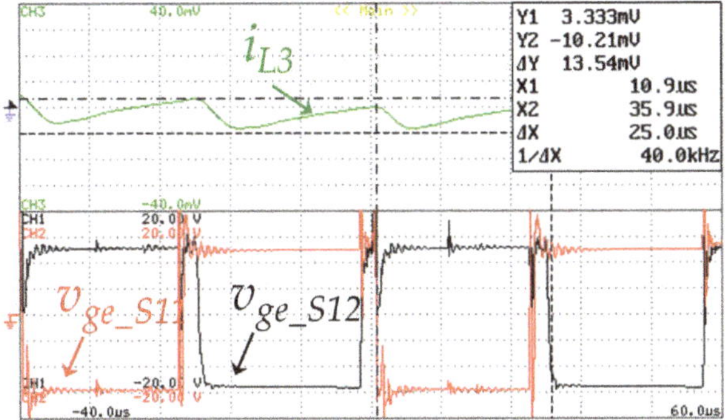

Figure 24. Experimental results showing the current in the inductor L_3 (i_{L3}: 0.1 A/div) and the gate-emitter voltage of the IGBTs S_{11} and S_{12} (v_{ge_s11}: 5 V/div and v_{ge_12}: 5 V/div) during a time interval of 100 μs.

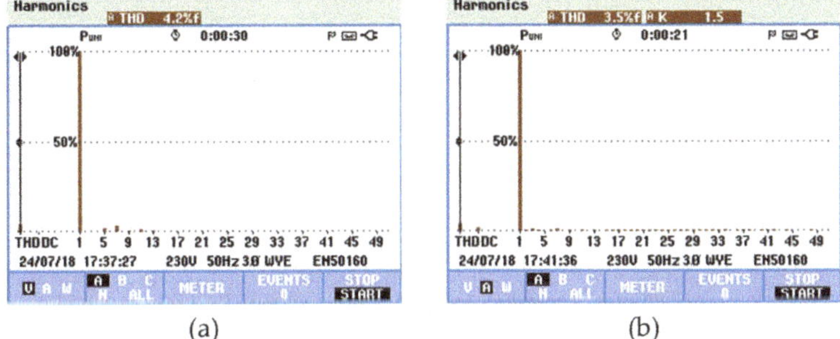

(a)   (b)

Figure 25. Experimental results in V2G operation mode of the total harmonic distortion and spectral analysis: (**a**) Power grid voltage (v_g); and (**b**) EVBC current (i_{ev}).

5. Conclusions

A novel on-board bidirectional EV battery charger (EVBC) was presented. It is constituted by a grid-side converter capable to operate with five voltage levels, and by a battery-side converter capable to operate with three voltage levels. The distinct voltage levels for both converters are obtained using a split DC-link. In order to ensure power quality features, the proposed EVBC operates with grid-side current controlled to improve power factor, and to preserve the battery lifetime the EVBC operates with battery-side controlled current or voltage. Throughout the paper is described the proposed hardware topology, the discrete-time predictive control algorithms used for the grid-side converter and for the battery-side converter, the developed full-scale laboratorial prototype of the EVBC, and the foremost experimental results considering operating modes for smart grids. The obtained results allow validating the key contributions of the paper, mainly, in terms of the bidirectional operation of the novel EVBC based on a multilevel topology. As the EVBC is controlled targeting the EV incorporation into smart grids, the grid-to-vehicle (G2V) and vehicle-to-grid (V2G) operation modes are discussed and evaluated.

Author Contributions: All authors contributed equally to the conceptualization and writing of the paper.

Funding: This work has been supported by COMPETE: POCI-01-0145-FEDER-007043 and FCT—Fundação para a Ciência e Tecnologia within the Project Scope: UID/CEC/00319/2013. This work is financed by the ERDF—European Regional Development Fund through the Operational Programme for Competitiveness and Internationalisation—COMPETE 2020 Programme, and by National Funds through the Portuguese funding agency, FCT—Fundação para a Ciência e a Tecnologia, within project SAICTPAC/0004/2015—POCI—01-0145-FEDER-016434. This work is part of the FCT project 0302836 NORTE-01-0145-FEDER-030283.

Conflicts of Interest: The authors declare no conflict of interest.

References

1. Idaho National Laboratory. *Charging and Driving Behavior of Nissan Leaf Drivers in the EV Project with Access to Workplace Charging*; EV Project; Idaho National Laboratory: Idaho Falls, ID, USA, 2014; pp. 1–4.
2. Matthé, R.; Eberle, U. The Voltec System—Energy Storage and Electric Propulsion. In *Lithium-Ion Batteries—Advances and Applications*; Elsevier: Amsterdam, The Netherlands, 2014; pp. 151–176.
3. Robledo, C.B.; Oldenbroek, V.; Abbruzzese, F.; van Wijk, A.J.M. Integrating a hydrogen fuel cell electric vehicle with vehicle-to-grid technology, photovoltaic power and a residential building. *Appl. Energy* **2018**, *215*, 615–629. [CrossRef]
4. Boulanger, A.G.; Chu, A.C.; Maxx, S.; Waltz, D.L. Vehicle Electrification: Status and Issues. *Proc. IEEE* **2011**, *99*, 1116–1138. [CrossRef]
5. Rajashekara, K. Present Status and Future Trends in Electric Vehicle Propulsion Technologies. *IEEE J. Emerg. Sel. Top. Power Electron.* **2013**, *1*, 3–10. [CrossRef]
6. Milberg, J.; Schlenker, A. Plug into the Future. *IEEE Power Energy Mag.* **2011**, *9*, 56–65. [CrossRef]
7. Monteiro, V.; Ferreira, J.C.; Melendez, A.A.N.; Couto, C.; Afonso, J.L. Experimental Validation of a Novel Architecture Based on a Dual-Stage Converter for Off-Board Fast Battery Chargers of Electric Vehicles. *IEEE Trans. Veh. Technol.* **2018**, *67*, 1000–1011. [CrossRef]
8. Basu, M.; Gaughan, K.; Coyle, E. Harmonic distortion caused by EV battery chargers in the distribution systems network and its remedy. In Proceedings of the International Conference UPEC—Universities Power Engineering Conference, Bristol, UK, 6–8 September 2004; pp. 869–873.
9. Monteiro, V.; Pinto, J.G.; Afonso, J.L. Operation Modes for the Electric Vehicle in Smart Grids and Smart Homes: Present and Proposed Modes. *IEEE Trans. Veh. Technol.* **2016**, *65*, 1007–1020. [CrossRef]
10. Kisacikoglu, M.C.; Ozpineci, B.; Tolbert, L.M. Examination of a PHEV Bidirectional Charger System for V2G Reactive Power Compensation. In Proceedings of the IEEE APEC Applied Power Electronics Conference and Exposition, Palm Springs, CA, USA, 21–25 February 2010; pp. 458–465.
11. Monteiro, V.; Ferreira, J.C.; Meléndez, A.A.N.; Afonso, J.L. Electric Vehicles On-Board Battery Charger for the Future Smart Grids. In *Technological Innovation for the Internet of Things*, 1st ed.; Camarinha-Matos, L.M., Tomic, S., Graça, P., Eds.; Springer: Berlin, Germany, 2013; Chapter 38; pp. 351–358.
12. Multin, M.; Allerding, F.; Schmeck, H. Integration of Electric Vehicles in Smart Homes—An ICT-based Solution for V2G Scenarios. In Proceedings of the IEEE ISGT PES Innovative Smart Grid Technologies, Washington, DC, USA, 16–20 January 2012; pp. 1–8.
13. Wong, N.; Kazerani, M. A Review of Bidirectional On-Board Charger Topologies for Plug-In Vehicles. In Proceedings of the IEEE CCECE Canadian Conference on Electrical and Computer Engineering, Montreal, QC, Canada, 29 April–2 May 2012; pp. 1–6.
14. Mei, J.; Xiao, B.; Shen, K.; Tolbert, L.M.; Zheng, J.Y. Modular Multilevel Inverter with new Modulation Method and its Application to Photovoltaic Grid-Connected Generator. *IEEE Trans. Power Electron.* **2013**, *28*, 5063–5073. [CrossRef]
15. Tolbert, L.M.; Peng, F.Z.; Habetler, T.G. Multilevel Converters for Large Electric Drives. *IEEE Trans. Ind. Appl.* **1999**, *35*, 36–44. [CrossRef]
16. Wang, H.; Kou, L.; Liu, Y.-F.; Sen, P.C. A New Six-Switch Five-Level Active Neutral Point Clamped Inverter for PV Applications. *IEEE Trans. Power Electron.* **2017**, *32*, 6700–6715. [CrossRef]
17. Monteiro, V.; Meléndez, A.A.N.; Afonso, J.L. Novel Single-Phase Five-Level VIENNA-Type Rectifier with Model Predictive Current Control. In Proceedings of the IEEE IECON Industrial Electronics Conference, Beijing, China, 29 October–1 November 2017; pp. 6413–6418.
18. Grbovic, P.; Lidozzi, A.; Solero, L.; Crescimbini, F. Five-Level Unidirectional T-Rectifier for High Speed Gen-Set Applications. *IEEE Trans. Ind. Appl.* **2016**, *52*, 1642–1651. [CrossRef]
19. Monteiro, V.; Ferreira, J.C.; Meléndez, A.A.N.; Afonso, J.L. Model Predictive Control Applied to an Improved Five-Level Bidirectional Converter. *IEEE Trans. Ind. Electron.* **2016**, *63*, 5879–5890. [CrossRef]
20. Monteiro, V.; Meléndez, A.A.N.; Ferreira, J.C.; Couto, C.; Afonso, J.L. Experimental Validation of a Proposed Single-Phase Five-Level Active Rectifier Operating with Model Predictive Current Control. In Proceedings of the IEEE IECON Industrial Electronics Conference, Yokohama, Japan, 9–12 November 2015; pp. 3939–3944.

21. Mokhberdoran, A.; Ajami, A. Symmetric and Asymmetric Design and Implementation of New Cascaded Multilevel Inverter Topology. *IEEE Trans. Power Electron.* **2014**, *29*, 6712–6724. [CrossRef]
22. Elsheikh, M.G.; Ahmed, M.E.; Abdelkarem, E.; Orabi, M. Single-Phase Five-Level Inverter with Less Number of Power Elements. In Proceedings of the IEEE INTELEC International Telecommunications Energy Conference, Orlando, FL, USA, 5–9 February 2012; pp. 1–8.
23. Loukriz, A.H.; Dudley, S.; Quinlan, T.; Walker, S.D. Experimental Realization of a Single-Phase Five Level Inverter for PV Applications. In Proceedings of the IEEE COMPEL Control and Modeling for Power Electronics, Trondheim, Norway, 27–30 June 2016; pp. 1–6.
24. Wang, K.; Xu, L.; Zheng, Z.; Li, Y. Capacitor Voltage Balancing of a Five-Level ANPC Converter Using Phase-Shifted PWM. *IEEE Trans. Power Electron.* **2015**, *30*, 1147–1156. [CrossRef]
25. Monteiro, V.; Pinto, J.G.; Sousa, T.J.C.; Meléndez, A.A.N.; Afonso, J.L. A Novel Single-Phase Five-Level Active Rectifier for On-Board EV Battery Chargers. In Proceedings of the IEEE ISIE International Symposium on Industrial Electronics, Edinburgh, UK, 19–21 June 2017; Volume 4, pp. 582–587.
26. Rosas-Caro, J.C.; Ramírez, J.M.; García-Vite, P.M. Novel DC-DC Multilevel Boost Converter. In Proceedings of the IEEE Power Electronics Specialists Conference, Rhodes, Greece, 15–19 June 2008; pp. 2146–2151.
27. Zhang, F.; Peng, F.Z.; Qian, Z. Study of the Multilevel Converters in DC-DC Applications. In Proceedings of the IEEE Annual Power Electronics Specialists Conference, Aachen, Germany, 20–25 June 2004.
28. Lee, P.; Lee, Y.; Cheng, D.K.W. Steady-state analysis of an interleaved boost converter with coupled inductors. *IEEE Trans. Ind. Electron.* **2000**, *47*, 787–795. [CrossRef]
29. Jiang, W.; Fahimi, B. Phase-Shift Controlled Multilevel Bidirectional DC/DC Converter: A Novel Solution to Battery Charge Equalization in Fuel Cell Vehicle. In Proceedings of the IEEE Vehicle Power and Propulsion Conference, Arlington, TX, USA, 9–12 September 2007; pp. 587–590.

Article

A Novel Phase Current Reconstruction Method for a Three-Level Neutral Point Clamped Inverter (NPCI) with a Neutral Shunt Resistor

Yungdeug Son [1] and Jangmok Kim [2,*]

[1] Department of Mechanical Facility Control Engineering, Korea University of Technology and Education; Cheonan, Chungnam 31253, Korea; ydson@koreatech.ac.kr

[2] Department of Electrical and Computer Engineering, Pusan National University, Busan 46241, Korea

* Correspondence: jmok@pusan.ac.kr; Tel.: +82-51-510-2366

Received: 1 September 2018; Accepted: 25 September 2018; Published: 1 October 2018

Abstract: This paper presents three phase current reconstruction methods for a three-level neutral point clamped inverter (NPCI) by measuring the voltage of a shunt resistor placed in the neutral point of the inverter. In order to accurately acquire the phase currents from the shunt resister, the dwell time of the active voltage vectors need to exceed the minimum time. On the other hand, if the time of active voltage is shorter than the minimum time, the current measurement becomes impossible. In this paper, unmeasurable regions for current are classified into three areas. Area 1 is a region in which both phase currents can be measure. Therefore, it is not necessary to restore the current. In Area 2, only one phase current can be measured. Thus, an estimation or restoration method is needed to measure another phase current. In this paper, the current estimation method using an electrical model of the motor is proposed. Area 3 is the region in which both phase currents can not be measured. In this case, it is necessary to move the voltage vector to the current measurable area by injecting the voltage. In this paper, Area 3 is divided into 36 sectors to inject optimal voltage. The proposed methods have the advantages of high current measurement accuracy and low THD (total harmonic distortion). The effectiveness of the proposed methods are verified through experimental results.

Keywords: alternating current (AC) motor drive; current estimation; current reconstruction method; current unmeasurable areas; total harmonic distortion (THD); single shunt resistor; space vector pulse width modulation (SVPWM); shift method; minimum voltage injection (MVI) method; three-level neutral point clamped inverter (NPCI)

---

1. Introduction

Two-level inverters are used in most home appliances, such as washing machines, refrigerators, and air conditioners, due to their simple structure, and high reliability and performance. However, in order to overcome the limitations of the efficiency and harmonics of the two-level inverter, three-level inverters have been recently investigated. The three-level neutral point clamped inverter (NPCI) has a structure characteristic of having a neutral point in the direct current (DC) stage, and thus has excellent electro magnetic interference (EMI) and electro magnetic compatibility (EMC) characteristics, due to a low voltage variation rate when switching [1–4]. Unlike a two-level inverter, each arm of a three-level NPCI consists of four switches and two clamping diodes, as shown in Figure 1. When the DC-link voltage is V_{dc}, the voltage of each capacitor is $V_{dc}/2$. The neutral point is connected to each phase output node by the clamp diodes and switches. Due to this structural feature, the three-level NPCI can output either $V_{dc}/2$ or $-V_{dc}/2$ by turning on the two switches located in the upper or lower side of a phase. When the switches of S_{x3} and S_{x4} are turned on, the node voltage of the x phase is 0 due to the connection with the neutral point through the diodes and switches, where x represents A, B, and C.

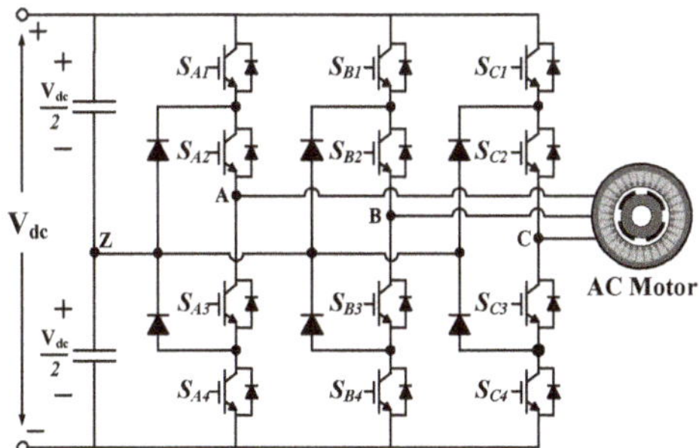

Figure 1. Three-level neutral point clamped (NPC) inverter.

For controlling the AC motor, the controller of the inverter requires the values of the three phase currents, which can be acquired through current sensors or a shunt resistor. A phase current sensing inverter (PCSI), as shown in Figure 2a, is a typical three-phase voltage source inverter with two phase current sensors. It requires at least two current sensors and sensing circuits, which raise the cost of the appliances [5]. For this reason, a DC link shunt resistor R_{shunt} can be used to measure the phase current, as shown in Figure 2b,c. A multi-shunt inverter (MSI) obtains the phase current from the shunt resistor located in the neutral and bottom of the DC link [6]. Measurement of the currents through the shunt resistors is possible when the current flows into the shunt resistor. Since the shunt resistors are located between the neutral and bottom of the DC link, it is possible to measure the phase current only if the active voltage vectors are combined with "O" or "N" switching states. However, although the active vectors are combined with "O" or "N" switching states, the phase current cannot be measured due to the short dwell time of the active vectors. The portion in the space vector diagram where the dwell time of the active vectors is not large enough to measure the current is called the current-unmeasurable area (CUA) [7–24]. A shunt resistor at the neutral point of the three-level NPCI can be used to measure the phase current, as shown in Figure 2c. This is similar to MSI, but a phase current is only obtainable when applying the state of "O" in the switching combination of the effective voltage vector. NPCI is effective in terms of volume and cost as compared to PCSI and MSI. It also has one current sensing circuit, which reduces the ripples caused by current sensor offset and scaling errors [6,7]. However, it has a limited time to sample the phase current in the shunt resistor over one period compared to the PCSI and MSI, so that the unmeasurable areas are widened in the output voltage hexagon. In order to overcome the limitations, some researchers have been interested in NPCI topology for phase current reconstruction.

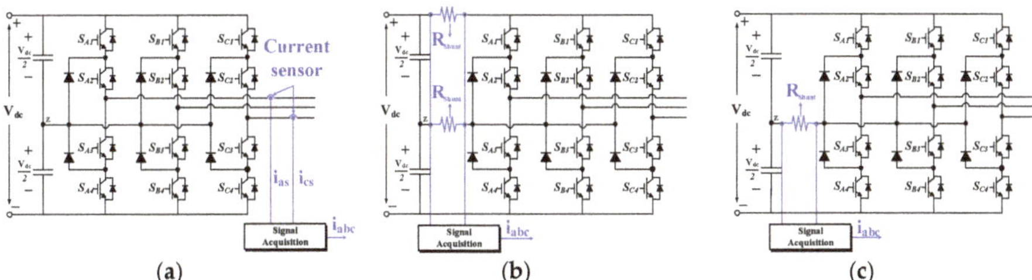

(a) (b) (c)

Figure 2. Measurement methods of phase current. (a) Conventional phase current sensing inverter (PCSI); (b) multi-shunt inverter (MSI); (c) neutral point clamped inverter (NPCI).

Previous research has established the effective voltage time by shifting the pulse width modulation (PWM) or injecting the voltage to restore the current [6–24]. However, these methods cause a high total harmonic distortion (THD) by injecting relatively large voltages. In [6], the minimum voltage injection (MVI) method minimizes voltage distortion and operating noise through THD reduction, but this method does not completely reconstruct the phase current at very high modulation index (MI). In addition, the PWM shifting method of [7] also generates harmonics due to asymmetric voltage modulation.

In this paper, the phase current unmeasurable region is classified into three areas, and the current reconstruction methods are proposed according to each area. First, two phase currents can be measured in Area 1, and the normal operation is executed in this area. On the other hand, only 1 phase current can be acquired in Area 2. In this case, the other phase current can be estimated by combining the q-axis current reference obtained from the speed controller and the electrical model of the motor [10]. Lastly, Area 3 is defined as an area where no current can be measured. For measuring the current in this area, the optimal voltage injection method is proposed [12]. To realize this, the hexagon of SVPWM is divided into 36 sectors, and the optimum injection voltage according to the sector is calculated. In addition, the current accuracy and THD are compared with the conventional method [6] in various MI conditions. The proposed method is verified through experimental results.

2. Acquiring Phase Current from Neutral Shunt Resistor

The operation of each three-level NPC inverter phase leg can be represented by a combination of the three switching states "P", "O", and "N". According to these switching states, the inverter has 27 possible combinations of switching states consisting of 24 effective voltage vectors and 3 zero voltage vectors. Because the NPCI has the shunt resistor at the neutral point, the phase current can be acquired only when the effective voltage vector includes the "O" switching state [6]. For example, the A phase current can be measured when the effective vector "O", "N", and "N" is applied to the inverter, as shown in Figure 3.

The 27 switching states of the neutral point clamped (NPC) inverter and the respective measurable phase currents are listed in Table 1. In addition, Figure 4 shows the switching vector in the spatial coordinates of the hexagon. Figure 4 and Table 1 show that the current cannot be measured through the neutral-point shunt resistor during the zero vector (V_0) and space vectors represented by V_1 to V_6.

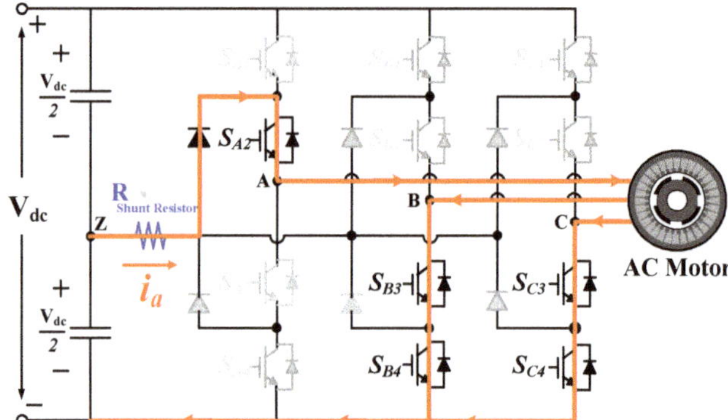

Figure 3. Current path of neutral shunt resistor when effective vector "O", "N", and "N" is applied to the inverter.

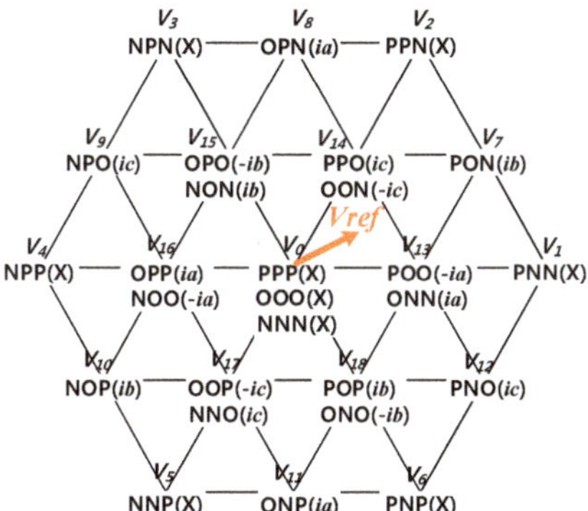

Figure 4. Three-level neutral shunt inverter effective-voltage vector hexagon.

Table 1. Switching states and acquiring the phase current from the shunt resistor.

Space Vector	Switching State		Acquiring Current from the Shunt Resistor		Vector Classification
V_0	[PPP], [OOO], [NNN]		X		Zero vector
	P-type	N-type	P-type	N-type	
V_{13}	[POO]	[ONN]	$-i_a$	i_a	
V_{14}	[PPO]	[OON]	i_c	$-i_c$	
V_{15}	[OPO]	[NON]	$-i_b$	i_b	Effective vector
V_{16}	[OPP]	[NOO]	i_a	$-i_a$	(Small vector)
V_{17}	[OOP]	[NNO]	$-i_c$	i_c	
V_{18}	[POP]	[ONO]	i_b	$-i_b$	
V_7, V_8	[PON], [OPN]		i_b, i_a		Effective vector
V_9, V_{10}	[NPO], [NOP]		i_c, i_b		(Medium vector)
V_{11}, V_{12}	[ONP], [PNO]		i_a, i_c		
V_1, V_2	[PNN], [PPN]		X, X		Effective vector
V_3, V_4	[NPN], [NPP]		X, X		(Large vector)
V_5, V_6	[NNP], [PNP]		X, X		

In one period of the three-level SVPWM, the sampling point and measurable phase current are expressed as shown in Figure 5. In this case, the "a" phase current can be measured in the "ONN" switching state and the "c" phase current can be measured in the "OON" switching state. The other phase current is calculated using Equation (1):

$$i_a + i_b + i_c = 0 \tag{1}$$

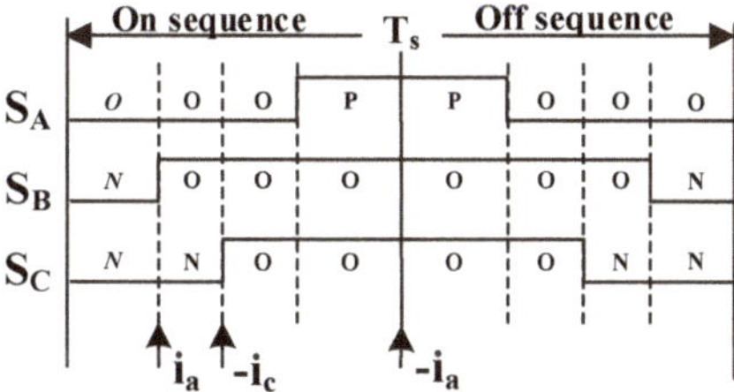

Figure 5. One period of the SVPWM method and sampling point.

3. Current Unmeasurable Areas

The sampling time for measuring the accurate current should have minimum delay from the point of switching time. This is to avoid the current ripple component in the current damping process by switching, as shown in Figure 6a. The minimum time T_{min} is determined using Equation (2) [9]:

$$T_{min} = T_{dead} + T_{settling} + T_{ad} \qquad (2)$$

where T_{dead} is the dead time to avoid arm-short of the inverter, $T_{settling}$ is the settling time of the neutral-point current, and T_{ad} is the sample and hold time of the A/D converter. Thus, in order to acquire the phase current properly, the switching time should be greater than T_{min}. Figure 6b shows switching state when the voltage modulation index is changed from Figure 5. In the case of "ONN", since the switching time T_a in the "O" state is larger than T_{min}, accurate "a" phase current can be obtained. However, in the case of "OON", since the switching time T_c of the "O" state is shorter than T_{min}, it is impossible to obtain current on the "c" phase in this state. The areas where the effective voltage dwell time is less than T_{min} are defined as CUAs (current unmeasurable areas). The CUAs in sector 1 of the three level SVPWM hexagon are shown in Figure 7a. In Figure 7a, Area 1 is a region where all phase currents are measurable, Area 2 is a region where only one phase current is measurable, and Area 3 is defined as a current unmeasurable region. The CUAs in all areas of the SVPWM are shown in Figure 7b.

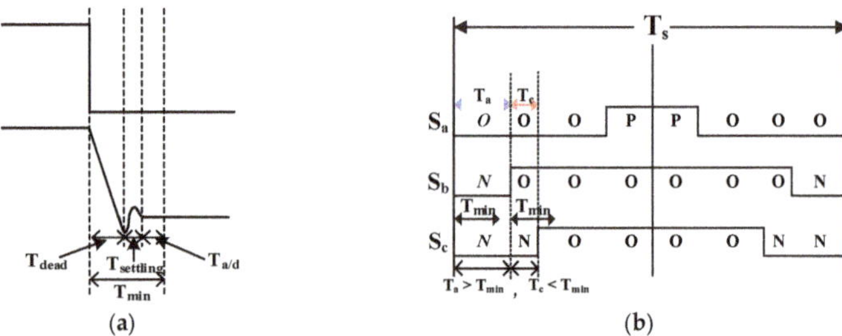

Figure 6. T_{min} and T_s of the SVPWM method. (**a**) Minimum time (T_{min}); (**b**) one PWM period (T_s).

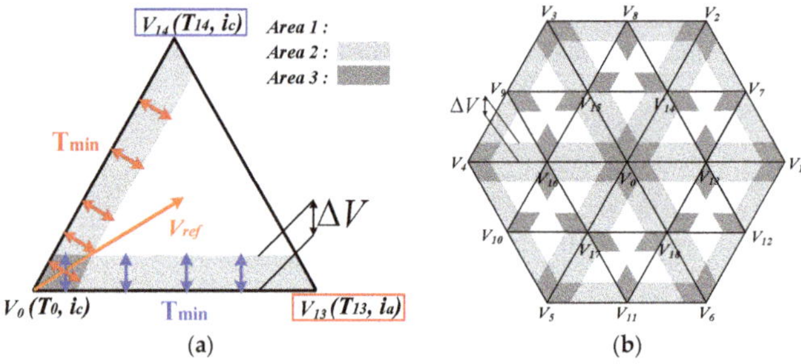

Figure 7. Current-unmeasurable areas (CUAs). (**a**) Sector 1; (**b**) reference-voltage vector hexagon.

In addition, the amplitude of the CUAs ΔV is obtained by Equation (3) [10].

$$T_{min} : \frac{T_s}{2} = \Delta V : \frac{V_{dc}}{\sqrt{3}}$$

$$\Delta V = \frac{2T_{min}}{\sqrt{3}T_s} V_{dc} \qquad (3)$$

where V_{dc} is the voltage of the DC link capacitor.

Sector 1 of the output voltage hexagon is divided into four regions composed of effective voltage vectors for the reference voltage vector V_{ref} duration, as shown in Figure 8. When the reference voltage vector is located as shown in Figure 8, the effective voltage vector and the duration time are expressed as follows:

$$V_{13}T_a + V_7T_b + V_{14}T_c = V_{ref}T_s$$

$$T_a + T_b + T_c = T_s \tag{4}$$

where T_a, T_b, and T_c are the duration time of V_{13}, V_7, and V_{14}, respectively.

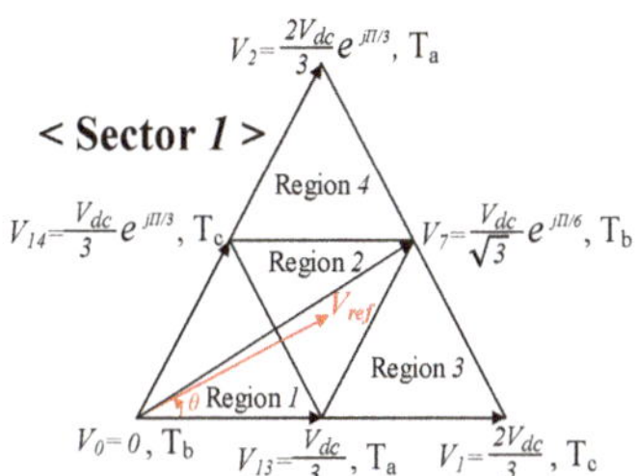

Figure 8. Sector 1 divided into four regions.

From Equation (4), the duration time of the effective voltage vectors T_a, T_b, and T_c in each region can be obtained through the equations given in Table 2. When T_a, T_b, and T_c are shorter than T_{min}, phase current cannot be obtained from the shunt resistor accurately. Therefore, it is a CUA.

Table 2. Dwell times of the voltage vector according to the regions.

Region	T_a	T_b	T_c
1	$T_s(2\sqrt{3}\frac{V_{ref}}{V_{dc}}\sin(\frac{\pi}{3}-\theta))$	$T_s(1-2\sqrt{3}\frac{V_{ref}}{V_{dc}}\sin(\frac{\pi}{3}+\theta))$	$T_s(2\sqrt{3}\frac{V_{ref}}{V_{dc}}\sin\theta)$
2	$T_s(1-2\sqrt{3}\frac{V_{ref}}{V_{dc}}\sin\theta)$	$T_s(2\sqrt{3}\frac{V_{ref}}{V_{dc}}\sin(\frac{\pi}{3}+\theta)-1)$	$T_s(1-2\sqrt{3}\frac{V_{ref}}{V_{dc}}\sin(\frac{\pi}{3}-\theta))$
3	$T_s(2-2\sqrt{3}\frac{V_{ref}}{V_{dc}}\sin(\frac{\pi}{3}+\theta))$	$T_s(2\sqrt{3}\frac{V_{ref}}{V_{dc}}\sin\theta)$	$T_s(2\sqrt{3}\frac{V_{ref}}{V_{dc}}\sin(\frac{\pi}{3}-\theta)-1)$
4	$T_s(2\sqrt{3}\frac{V_{ref}}{V_{dc}}\sin\theta-1)$	$T_s(2\sqrt{3}\frac{V_{ref}}{V_{dc}}\sin(\frac{\pi}{3}-\theta))$	$T_s(2-2\sqrt{3}\frac{V_{ref}}{V_{dc}}\sin(\frac{\pi}{3}+\theta))$

4. Conventional Method of Phase Current Reconstruction

4.1. Modified Voltage Modulation Method

The method in [7] with the alternative switching pattern is different from the classical SVPWM. The zero vector is replaced with a pair of effective voltage vectors in order to increase the duration time of the effective voltage vector, which does not ensure T_{min}. As a result, only one phase current measurable region (Area 2) can be compensated. If the reference voltage vector V_{ref} is in Area 2, as shown in Figure 9a, phase C current is unmeasurable and the zero vector [OOO] is replaced with a pair of effective voltage vectors V_{14} [OON] and V_{17} [NNO], as shown in Figure 10. Therefore, only the current of one phase can be reconstructed during one PWM period due to the variation of the switching time. However, if the reference voltage vector V_{ref} is in Area 3, as shown in Figure 9b, it needs two switching cycles. Therefore, the THD of phase current is high, because a pair of vectors, which are located at opposite positions to each other, are used to make the reference voltage.

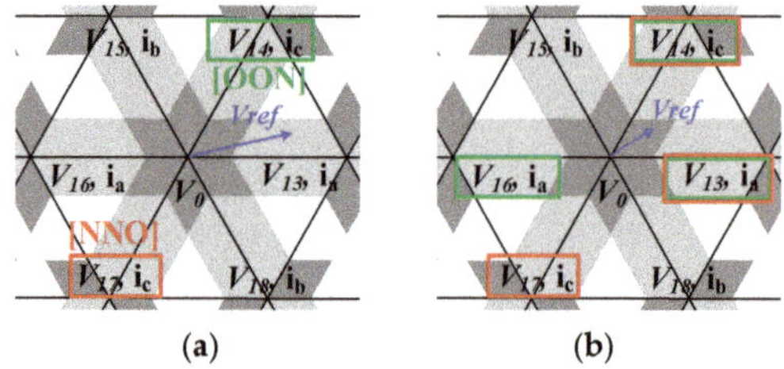

(a)   **(b)**

Figure 9. Duration of the reference voltage vector in CUAs. (**a**) Area 2; (**b**) Area 3.

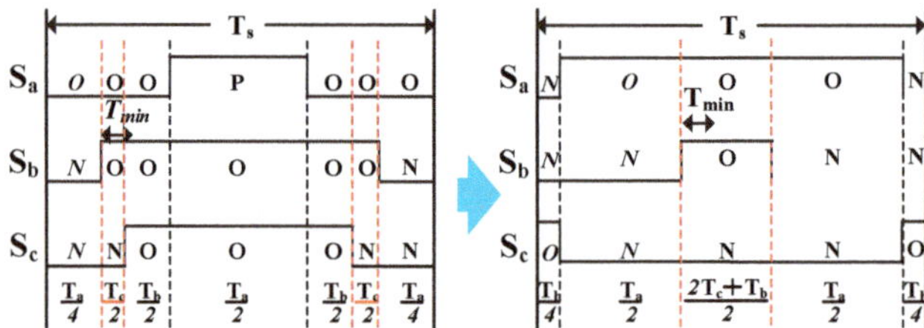

Figure 10. Switching sequences in conventional SVPWM and modified voltage modulation method. (Area 2).

4.2. Minimum Voltage Injection (MVI) Method

In this conventional method [6], the minimum voltage injection method is applied to measure the phase current, as shown in Figure 11. Figure 11a shows the reference voltage vector V_{ref} in Area 3, where two phase currents cannot be obtained. In this case, a constant voltage is added to the reference voltage vector to reconstruct the phase current. The reference voltage moved to measurable Area 1 is defined as V_m. The compensation voltage vector V_c is applied by subtracting the constant voltage which has the same magnitude as the added value to the command voltage vector. This compensation method is shown for one period of the switching pattern in Figure 11b. In this case, both "ONN" and "OON" are less than the minimum time T_{min}. So in a half period of modulation, the voltage V_m for the reconstruction is injected, and in the other half period, the compensation voltage V_c is injected to cancel the effect of the injected voltage.

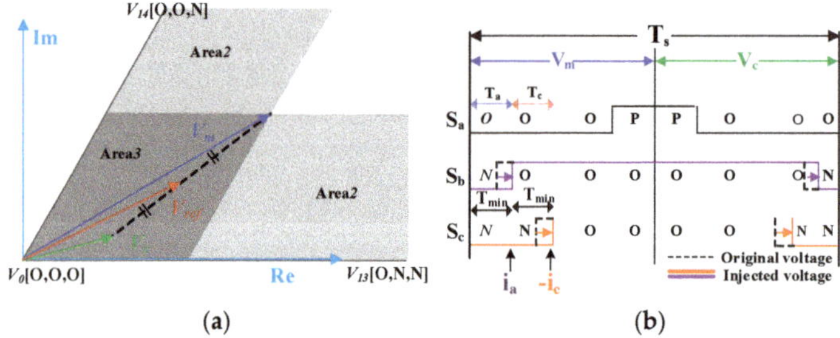

(a)   **(b)**

Figure 11. Minimum voltage injection method. (**a**) Voltage injection method in Area 3; (**b**) PWM switching patterns.

However, it is difficult to reconstruct the phase current using the MVI method in the high MI region shown in Figure 12. At this time, the compensation voltage V_c exceeds the linear modulation area. Therefore, it is impossible to reconstruct the phase current near the outermost edge of the hexagon.

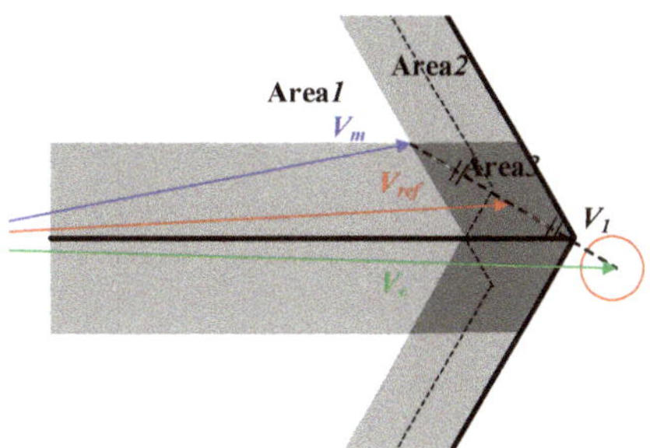

Figure 12. Voltage injection method in Area 3 beside a vertex of the hexagon.

5. Proposed Method of Phase Current Reconstruction

5.1. Based Method for Current Reconstruction in Area 2

In order to control the constant speed and constant torque of an AC motor, a current controller is essential. In general, a proportional integral (PI) controller is used on the synchronous reference frame of d–q axis [10]. Figure 13 shows the block diagram of the synchronous PI controller, and the electrical model of the motor system. Where R_a, L_a, K_p, and K_i mean the stator resistance, stator inductance, proportional gain, and integral gain, respectively. The proportional and integral gains of the PI controller are calculated using Equation (5):

$$K_p = L_a \omega_{cc}, \quad K_i = R_a \omega_{cc} \tag{5}$$

where L_a is the stator inductance, R_a is the stator resistance, and ω_{cc} is the bandwidth of the PI regulator. Then, the closed-loop transfer function $G_c(s)$ of the block diagram is given by Equation (6):

$$G_c(s) = \frac{i_{dq}^e(s)}{i_{dq}^{e*}(s)} = \frac{G_0(s)}{1 + G_0(s)} = \frac{\frac{\omega_{cc}}{s}}{1 + \frac{\omega_{cc}}{s}} = \frac{\omega_{cc}}{1 + \omega_{cc}} \tag{6}$$

According to Equations (5) and (6), the combination of the PI controller and model of the motor is equivalent to a first low-pass filter whose cutoff frequency is ω_{cc}. In this case, the real d–q axis current can be estimated using the current reference and low pass filter, as shown in Figures 13 and 14. Finally, the estimated three phase currents ($\hat{i}_{abc}$) can be obtained from reverse transformation of the estimated d–q axis currents ($\hat{i}_{dq}{}^e$). This estimated current is used to replace the unmeasurable current when the voltage command lies inside Area 2, where only one phase current is acquired. This current estimation method does not need PWM shift or voltage injection for current reconstruction.

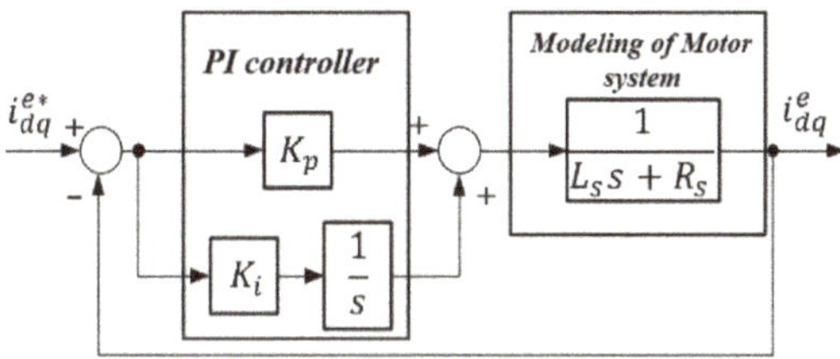

Figure 13. Block diagram of the PI controller and electrical model of the motor system.

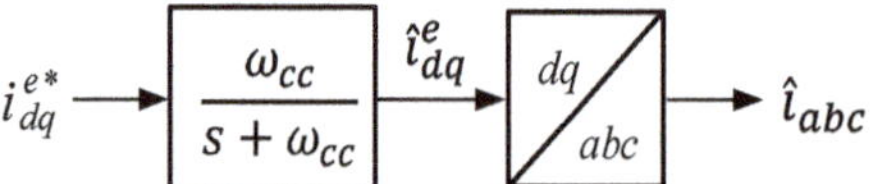

Figure 14. Block diagram of acquiring estimated current.

5.2. Proposed Method for Current Reconstruction in Area 3

In order to calculate the optimum injection voltage according to the switching sector of SVPWM, the conventional sector is classified into 36 switching sectors in this paper. These sectors can be represented by 18 straight lines, as shown in Figure 15. These straight lines can be obtained by using the two points of the hexagon. As a result, the 36 switching sectors can be defined by three straight lines. If the reference voltage vector V_{ref} is located as shown in Figure 15, straight Lines 5, 9, and 13 must be satisfied to discriminate the switching sector.

$$\text{Line 5}: V_{qs}^s \leq -\sqrt{3}\left(V_{ds}^s - \frac{2}{3}V_{dc}\right), \quad \text{Line 9}: V_{qs}^s \leq -\sqrt{3}\left(V_{ds}^s - \frac{2}{3}V_{dc}\right), \quad \text{Line 13}: V_{qs}^s \geq 0 \qquad (7)$$

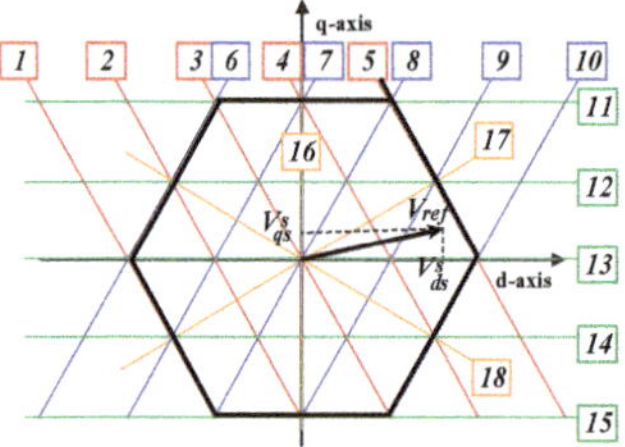

Figure 15. Lines (1–18) for the classification of 36 switching sectors.

For simple injection voltage calculation, the reference voltage vector $V_{ref}(V_{ds}^s, V_{qs}^s)$ that rotates the output voltage hexagon with electrical angle θ_e, is transformed as shown in Figure 16. V_{ref} is transformed to the shifted reference voltage vector $V_{dqs}^t(V_{ds}^t, V_{qs}^t)$ in sector 0 by Equation (8).

$$\begin{pmatrix} V_{ds}^t \\ V_{qs}^t \end{pmatrix} = \begin{pmatrix} \cos\theta_n & -\sin\theta_n \\ \sin\theta_n & \cos\theta_n \end{pmatrix} \begin{pmatrix} V_{ds}^s \\ V_{qs}^s \end{pmatrix} \qquad (8)$$

where $\theta_n = n \cdot \pi/3$.

The range of the shifted sector 0 is from $-\pi/6$ to $\pi/6$. As a result, the injected voltage is only calculated in the shifted sector 0.

As shown in Figure 17a, the shifted sector is symmetrical with respect to the d-axis. Therefore, when calculating the injection voltage, only the positive part needs to be calculated. In the negative region, the negative sign can be added to the magnitude of the q component.

$V_{dqs}^i\left(V_{ds}^i, V_{qs}^i\right)$ are the vector components for moving V_{dqs}^t to the measurable region. After calculating the injection voltage, V_{dqs}^i is added to V_{dqs}^t, and reverse-transformation is executed. This voltage vector is defined as the measurement voltage vector V_m^s. V_m^s is induced during the first half period of the modulation, and the compensation voltage V_c^s is induced during the other half period. The relation between V_{ref}, V_m^s, and V_c^s is given by Equation (9).

$$V_{ref} = \frac{1}{2}\left(V_m^s + V_c^s\right) \rightarrow V_c^s = 2\left(V_{ref} - V_m^s\right) \qquad (9)$$

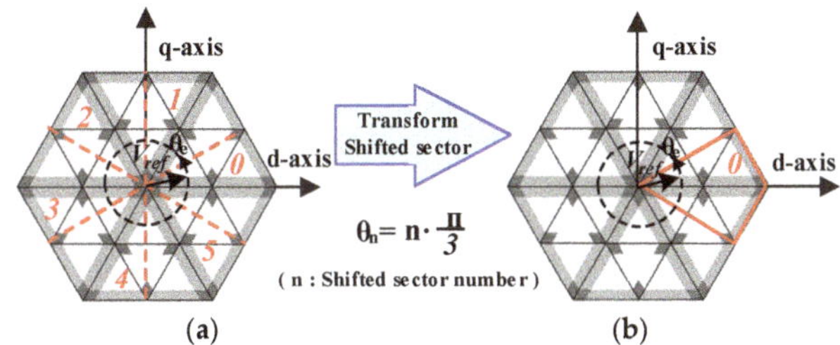

Figure 16. Transform shifted sector. (**a**) Shifted sectors (0–5); (**b**) shifted sector (0).

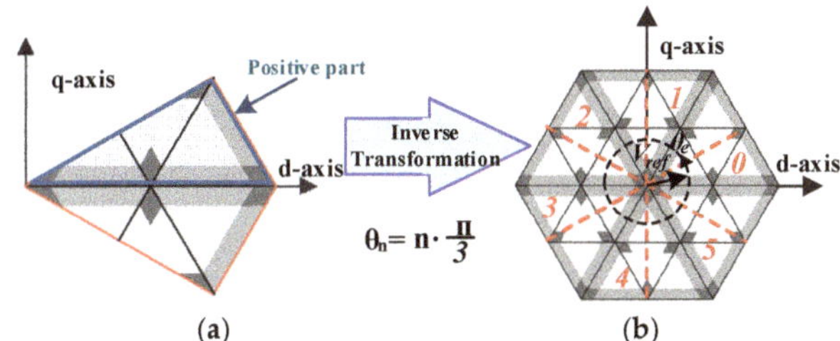

Figure 17. Inverse transformation shifted sector. (**a**) Shifted sector (0); (**b**) shifted sectors (0–5).

In order to recover the phase current by injecting the optimal voltage, the shifted sector 0 for the positive *q*-axis is divided into different parts, as shown in Figure 18a, according to the MI. As shown in Figure 18b, each part is more finely divided to restore the phase current by injecting the optimal voltage. Part.1 is further divided into Part.1_1 and Part.1_2, and Part.2 is divided into Part.2_1, Part.2_2, Part.2_3, and Part.2_4. Part.3 and Part.4 are also divided into two parts and four parts, respectively.

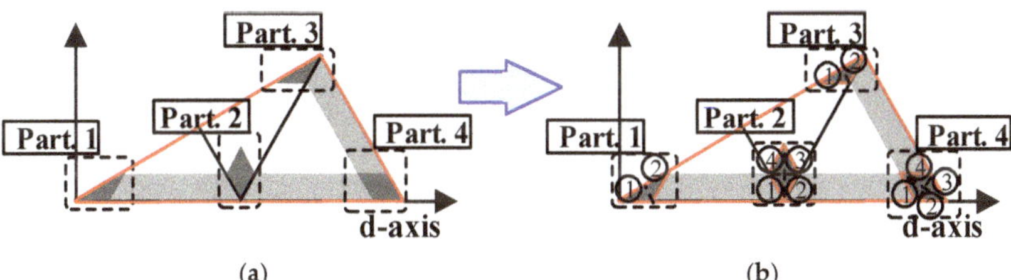

Figure 18. Detailed view of the parts of the positive shifted sector 0. (**a**) Part1–Part4; (**b**) segmentation of Parts.

When the reference voltage vector V_{ref} is located in Part.1_1, as shown in Figure 19, the optimal voltage is injected, and the measurement voltage vector V_m^s is pushed to the measurement point. At this time, the measurement point is on the boarder of Area 2. The magnitude of the injected voltage $V_{dqs}^i(V_{ds}^i, V_{qs}^i)$ can be calculated by using both the measurement point and V_{dqs}^t, as shown in Equation (10).

Part.1_1

$$V_{ds}^i = \left(2\Delta V / \sqrt{3}\right) - V_{ds}^t , \qquad V_{qs}^i = -V_{qs}^t \tag{10}$$

When the reference voltage vector is located in Part.1_2, as shown in Figure 20, the measurement voltage vector is moved to the measurement line by the optimal voltage injection. The measurement line is defined by Equation (11), with two points $(2\Delta V / \sqrt{3}, 0)$ and $(\sqrt{3}\Delta V, \Delta V)$, as the boarder of Area 2. The magnitude of the injected voltage is obtained by Equation (12), which gives the minimum distance from the point to the measurement line through the Pythagorean theorem, as shown in Equation (13). The optimal voltage injection method is used to classify each part of Area 3 in more detail, and then bring V_m^s to the closest measurement line or measurement point for reconstructing the phase currents. At this time, the measurement line or measurement point is at the boarder of Area 1 or Area 2. Each part of Area 3 has a different type of optimal voltage injection, as shown in Equations (10) and (13)–(23), and in Figure 21.

$$y = \sqrt{3}\left(x - \sqrt{3}\Delta V\right) + \Delta V \tag{11}$$

$$d = \frac{1}{2}\left|\sqrt{3}V_{ds}^t - V_{qs}^t - 2\Delta V\right| \tag{12}$$

Part.1_2

$$V_{ds}^i = \frac{\sqrt{3}}{2}d = \sqrt{3} \times 0.25\left(\sqrt{3}V_{ds}^t - V_{qs}^t - 2\Delta V\right),$$
$$V_{qs}^i = -\frac{1}{2}d = -0.25\left(\sqrt{3}V_{ds}^t - V_{qs}^t - 2\Delta V\right) \tag{13}$$

Part.2_1

$$V_{ds}^i = -\sqrt{3} \times 0.25\left(\sqrt{3}V_{ds}^t + V_{qs}^t - \frac{V_{dc}}{\sqrt{3}}\right), \quad V_{qs}^i = -\sqrt{3} \times 0.25\left(\sqrt{3}V_{ds}^t + V_{qs}^t - \frac{V_{dc}}{\sqrt{3}}\right) \tag{14}$$

Part.2_2

$$V_{ds}^i = \sqrt{3} \times 0.25\left(\sqrt{3}V_{ds}^t - V_{qs}^t - \frac{V_{dc}}{\sqrt{3}}\right), \quad V_{qs}^i = -0.25\left(\sqrt{3}V_{ds}^t - V_{qs}^t - \frac{V_{dc}}{\sqrt{3}}\right) \tag{15}$$

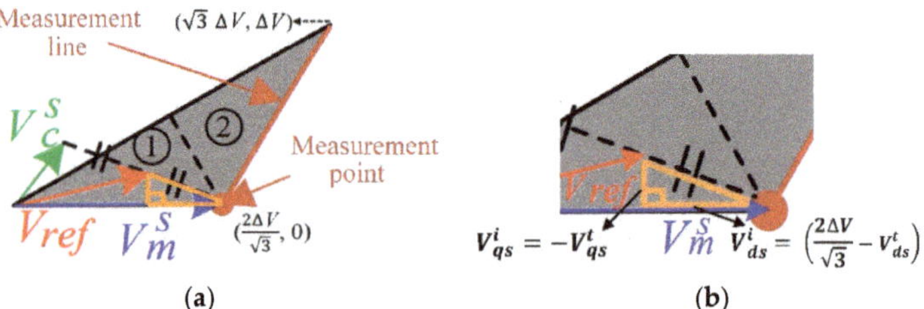

Figure 19. Optimal voltage injection method in Part.1_1. (**a**) Reference voltage vector on Part.1_1; (**b**) detailed view of the left one.

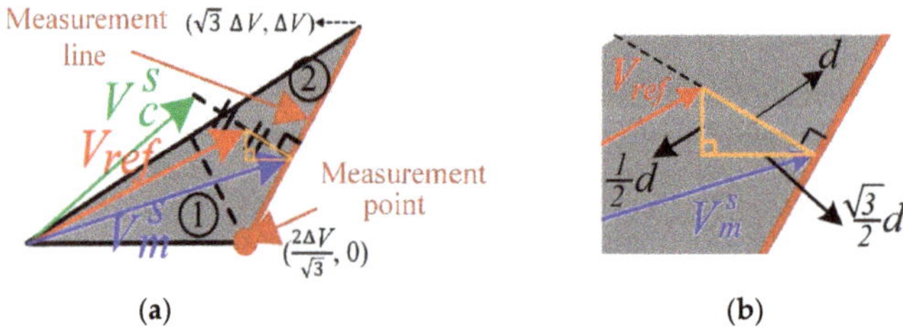

Figure 20. Optimal voltage injection method in Part.1_2. (**a**) Reference voltage vector on Part.1_2; (**b**) detailed view of the left one.

Part.2_3

$$V_{ds}^i = \sqrt{3} \times 0.25\left(\sqrt{3}V_{ds}^t + V_{qs}^t - \frac{V_{dc}}{\sqrt{3}} - 2\Delta V\right) , \quad V_{qs}^i = 0.25\left(\sqrt{3}V_{ds}^t + V_{qs}^t - \frac{V_{dc}}{\sqrt{3}} - 2\Delta V\right) \tag{16}$$

Part.2_4

$$V_{ds}^i = -\sqrt{3} \times 0.25\left(\sqrt{3}V_{ds}^t - V_{qs}^t - \frac{V_{dc}}{\sqrt{3}} + 2\Delta V\right) , \quad V_{qs}^i = 0.25\left(\sqrt{3}V_{ds}^t - V_{qs}^t - \frac{V_{dc}}{\sqrt{3}} + 2\Delta V\right) \tag{17}$$

Part.3_1

$$V_{ds}^i = 0 , \quad V_{qs}^i = \left(\frac{\sqrt{3}V_{dc}}{6} - \Delta V\right) - V_{qs}^t \tag{18}$$

Part.3_2

$$V_{ds}^i = \sqrt{3}\cdot 0.25\left(\sqrt{3}V_{ds}^t - V_{qs}^t - \frac{V_{dc}}{\sqrt{3}}\right) , \quad V_{qs}^i = -0.25\left(\sqrt{3}V_{ds}^t - V_{qs}^t - \frac{V_{dc}}{\sqrt{3}}\right) \tag{19}$$

Part.4_1

$$V_{ds}^i = -\sqrt{3}\cdot 0.25\left(\sqrt{3}V_{ds}^t + V_{qs}^t - \frac{2V_{dc}}{\sqrt{3}} + 2\Delta V\right),$$
$$V_{qs}^t = -0.25\left(\sqrt{3}V_{ds}^t + V_{qs}^t - \frac{2V_{dc}}{\sqrt{3}} + 2\Delta V\right) \tag{20}$$

Part.4_2

$$V_{ds}^i = \left(\frac{2V_{dc}}{3} - \frac{2\Delta V}{\sqrt{3}}\right) - V_{ds}^t , \quad V_{qs}^i = -V_{qs}^t \tag{21}$$

Part.4_3

$$V_{ds}^i = \left(\frac{2V_{dc}}{3} - \frac{\Delta V}{\sqrt{3}}\right) - V_{ds}^t, \quad V_{qs}^i = \Delta V - V_{qs}^t \tag{22}$$

Part.4_4

$$V_{ds}^i = 0, \quad V_{qs}^i = \Delta V - V_{qs}^t \tag{23}$$

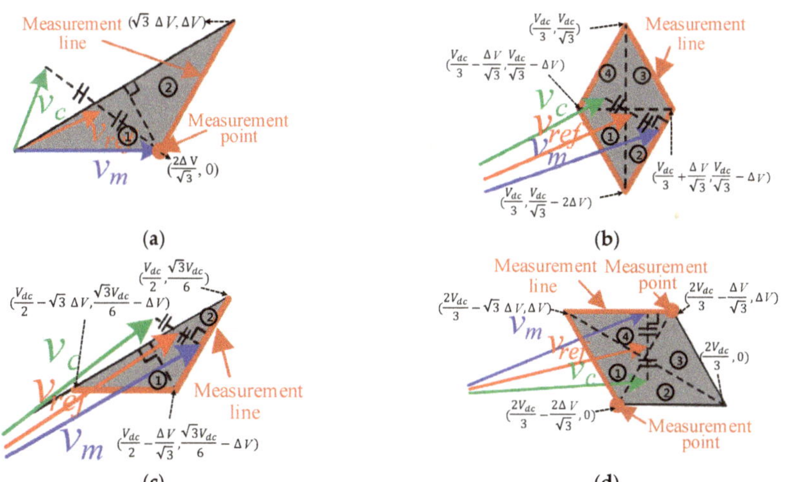

Figure 21. Optimal voltage injection method in Area 3 for each part. (**a**) Optimal voltage injection method in Part.1; (**b**) optimal voltage injection method in Part.2; (**c**) optimal voltage injection method in Part.3; (**d**) optimal voltage injection method in Part.4.

5.3. Comparison of Conventional and Proposed Method

Figure 22 shows the comparison between the proposed method and MVI method for a MI of 0.1 and 0.97. The proposed method restores the phase current by injecting a small amount of voltage when MI = 0.1 and MI = 0.97. However, the conventional method injected a large amount of voltage compared with the proposed method when MI = 0.1. In addition, the compensated voltage vector of the conventional method lies outside of the output voltage hexagon when the MI is 0.97, which makes the reconstruction of the phase current impossible.

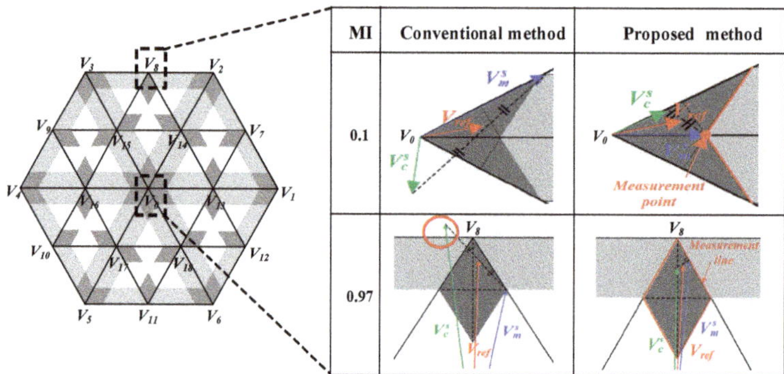

Figure 22. Comparison of conventional method and proposed method at MI = 0.1 and MI = 0.97.

As shown in Table 3, when the proposed method is applied, it can be seen that the area of the non-measurable region in the output voltage hexagon is reduced by about 93.7 % in comparison with the case in the MVI method.

Table 3. Unmeasurable area comparison.

3-Level NPCI One Shunt Unmeasurable Area	Conventional Method	Proposed Method
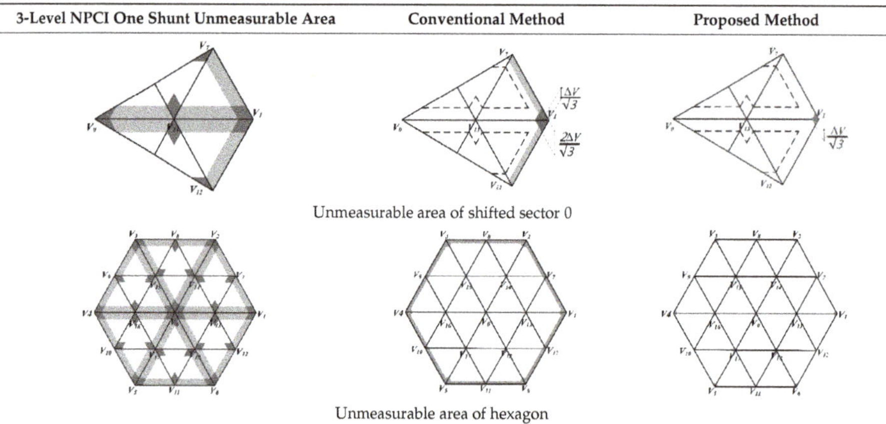

Unmeasurable area of shifted sector 0

Unmeasurable area of hexagon

6. Experimental Results

Figure 23 describes the system configuration of the experimental setup for the verification of the proposed current reconstruction method. The system was composed of the three-level NPC inverter, including the shunt resistor, connected to the neutral point, the control system based on the digital signal processor (DSP, TMS320C28346), and the resistive-inductive load. The upper and lower DC capacitors were connected with the 30 V power source.

A resistor of 10 Ω and inductor of 5 mH were used as the load. The resistance of the shunt resistor was 0.2 Ω and its power capacity was 3 W. In addition, the minimum time T_{min} needed to acquire the precise phase current from the shunt resistor was set to be 4.5 μs. The reconstructed current from the shunt resistor using the proposed algorithm was compared with an actual current measured using two current sensors in order to verify its precision.

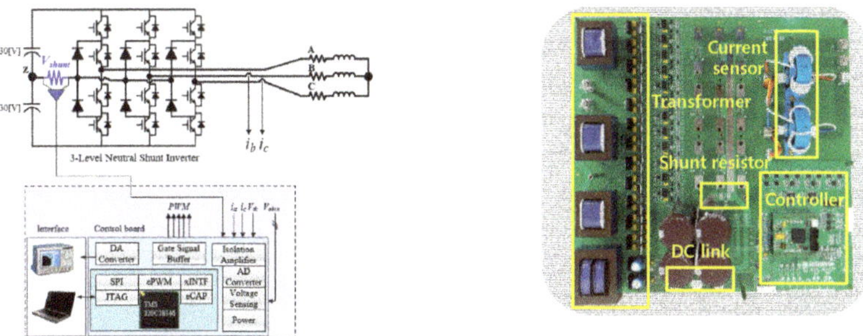

Figure 23. Configuration of hardware system and control board.

In the experiment, the inverter system was operated by the V/F open-loop control algorithm. The system was controlled with a fixed operating frequency of 12 Hz and variable output voltage to change the MI. The experimental results were classified in 6 cases to compare with the conventional MVI method. Figure 24a,b shows the experimental results with the condition of MI = 0.1 (V_{ref}= 3.56 V). In this case, the inverter was operated in Area 3 where the two currents cannot be measured from the shunt resistor. As presented in Figure 24b, the injected d–q voltage for the current reconstruction is smaller than the voltage in Figure 24a. The experimental results of Figure 24c,d was measured when the inverter was operated in Area 1 and 2 with MI of 0.4 (V_{ref}= 13.86 V). In Area 1, it is possible to obtain the two phase currents from the shunt resistor, and only one current can be acquired in Area 2. As shown in Figure 24c, the conventional method injects the q-axis voltage to reconstruct the phase current. In contrast, the proposed method can reconstruct the phase current without the voltage injection, as presented in Figure 24d, because the phase current is reconstructed using the estimated current. In Figure 24e,f, V_{ref} is passed through Area 1 and 3 when the MI is 0.6 (V_{ref}= 20.78 V). As shown in Figure 24f, a smaller voltage is injected to reconstruct the phase current in the proposed method in comparison with the conventional method. The waveforms presented in Figure 25e,f was measured when the MI was 0.97 (V_{ref}= 33.6 V). In this case, the trajectory of the voltage reference is near the inscribed circle of the space vector hexagon. In the case of the conventional method, the phase current cannot be reconstructed even if the voltage injection method is used, because the compensated voltage reference exceeds the hexagon. On the other hand, the proposed method can reconstruct the phase current, as presented in Figure 25f, because the voltage reference can be compensated on the boundary of the hexagon.

The accuracy of the phase currents reconstructed using the conventional and proposed methods were calculated to show the superiority of the proposed algorithm. The equation for the accuracy of the phase current is expressed in Equation (24).

$$Accurcy\ (\%) = \left(1 - \frac{Variance\ of\ [i_{sensor} - i_{reconstructed}]}{RMS\ of\ i_{sensor}}\right) \times 100 \tag{24}$$

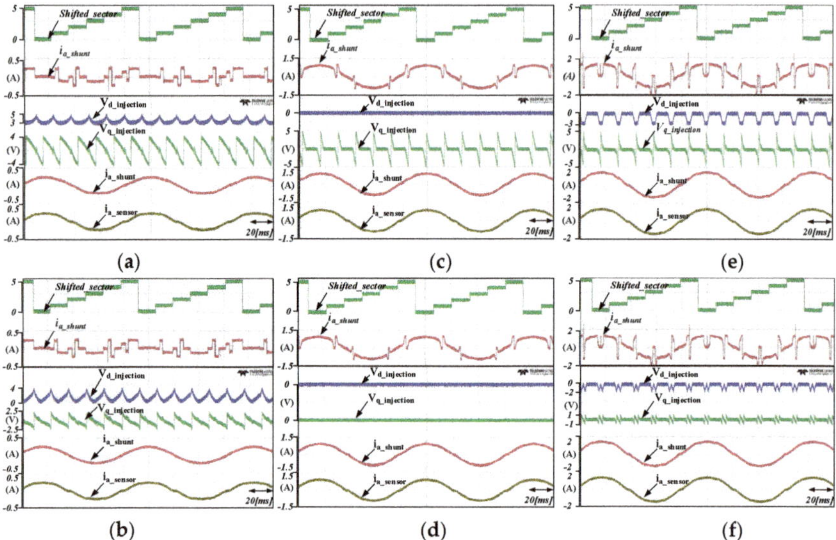

Figure 24. Comparison of the conventional and proposed methods at MI = 0.1, 0.4, and 0.6. (**a**) Conventiona method at MI = 0.1; (**b**) proposed method at MI = 0.1; (**c**) conventional method at MI = 0.4; (**d**) proposed method at MI = 0.4; (**e**) conventional method at MI = 0.6; (**f**) proposed method at MI = 0.6.

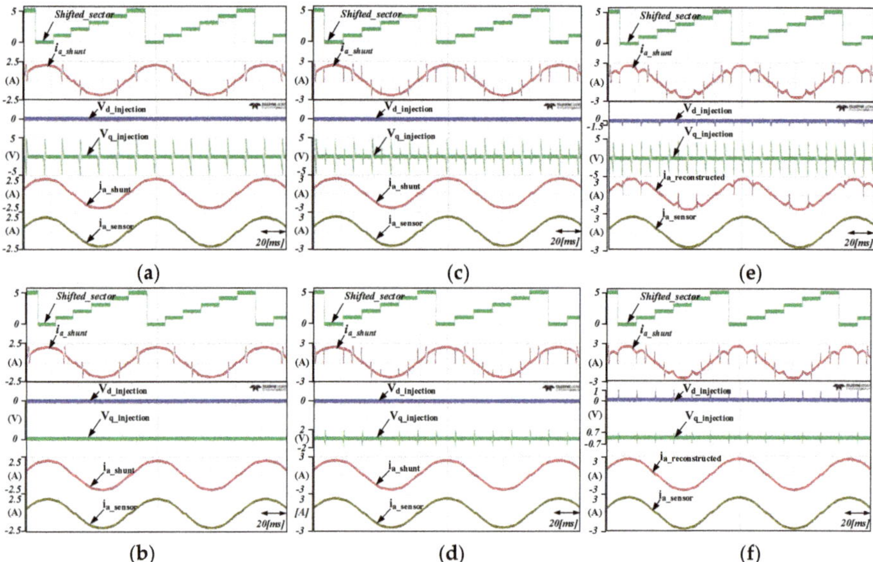

Figure 25. Comparison of the conventional and proposed methods at MI = 0.8, 0.9, and 0.97. (**a**) Conventional method at MI = 0.8; (**b**) proposed method at MI = 0.8; (**c**) conventional method at MI = 0.9; (**d**) proposed method at MI = 0.9; (**e**) conventional method at MI = 0.97; (**f**) proposed method at MI = 0.97.

As the difference between the current measured by the sensor and the reconstructed current is increased, the variance of the calculated results increases. This means the accuracy of the reconstruction is low. The difference in accuracy between the conventional method and the proposed method is compared according to the MI, and shown in Figure 26a. The accuracy of the proposed method is higher than the conventional method in the whole range of the MI.

The comparison results for the THD of the phase current between the conventional method and the proposed method are presented in Figure 26b.

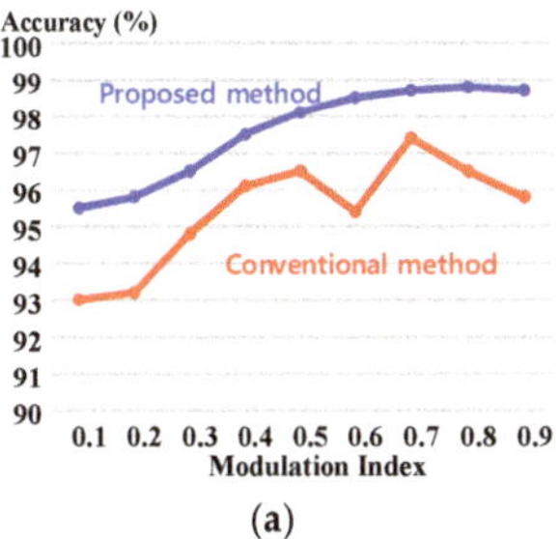

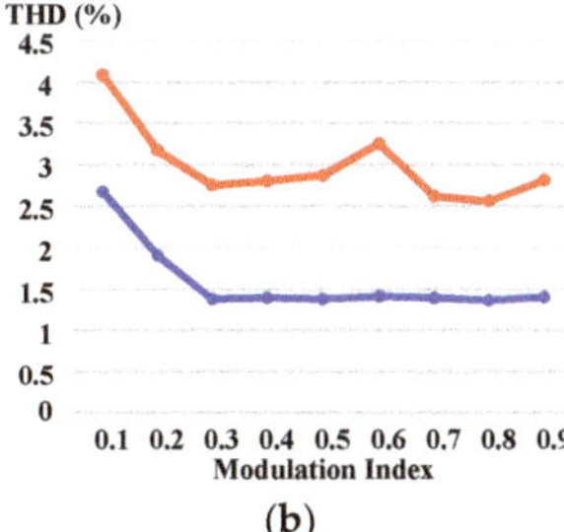

Figure 26. Comparison of the conventional and proposed methods according to the variation in the MI. (**a**) Phase current Accuracy; (**b**) total harmonic distortion (THD) of phase currents.

The THD result for the proposed method is lower than the result for the conventional method in the whole range of the MI, because the THD of the phase current is proportional to the injected voltage. On average the THD of the proposed method is improved by 44.5 % compared with that of the conventional method.

7. Conclusions

This paper proposed a phase current reconstruction method using a neutral shunt resistor, which lowers both the cost and volume of a three-level inverter system.

First, the paper proposes the method for dividing the switching sectors, the rotated sectors, and the region. Each region is separated by the equation of a straight line.

In addition, in Area 2 where there is only one measurable phase current, the remaining phase current is reconstructed by the estimation method. The electrical model of the motor and PI current controller are used to estimate the current.

Finally, the optimal voltage injection method is used in Area 3, where no current is acquired through the shunt resistor (CUAs). Area 3 is divided and defined as parts in more detail. When the reference is located on Area 3 of the hexagon, the phase currents are reconstructed by moving the reference vector to the border of Area 1 or Area 2.

The proposed method utilized a small magnitude injection voltage compared to the conventional method, and hence results in better performance in terms of THD and accuracy of the estimated currents. The validity of the proposed algorithm is proven by the experiments.

Author Contributions: Y.S. contributed to the research activity and to manuscript writing. J.K. provided technical feedback and did supervision of the overall work.

Funding: This research received no external funding.

Acknowledgments: This research was supported by the Basic Science Research Program through the National Research Foundation of Korea (NRF) funded by the Ministry of Education (NRF-2018R1D1A1B07048954).

Conflicts of Interest: The authors declare no conflict of interest.

References

1. De, S.; Banerjee, D.; Gopakumar, K.; Ramchand, R.; Patel, C. Multilevel inverters for low-power application. *IET Power Electron.* **2010**, *4*, 382–392. [CrossRef]
2. Lipo, T.A.; Manjrekar, M.D. Hybrid Topology for Multilevel Power Conversion. U.S. Patent 06,005,788, 21 December 1999.
3. Lai, R.; Wang, F.; Burgos, R.; Pei, Y.; Boroyevich, D.; Wang, B.; Lipo, T.A.; Immanuel, V.D.; Karimi, K.J. A systematic topology evaluation methodology for high-density three-phase PWM AC-AC converters. *IEEE Trans. Power Electron.* **2008**, *23*, 2665–2680.
4. Heldwein, M.L.; Kolar, J.W. Impact of EMC filters on the power density of modern three-phase PWM converters. *IEEE Trans. Power Electron.* **2009**, *24*, 1577–1588. [CrossRef]
5. Jarzebowicz, L. Error Analysis of Calculating Average d-q Current Components using Regular Sampling and Park Transformation in FOC Drives. In Proceedings of the 2014 International Conference and Exposition on Electrical and Power Engineering (EPE 2014), Iasi, Romania, 16–18 October 2014; pp. 901–905.
6. Shin, H.; Ha, J.I. Phase Current Reconstructions from DC-Link Currents in Three-Phase Three-Level PWM Inverters. *IEEE Trans. Power Electron.* **2014**, *29*, 582–593. [CrossRef]
7. Li, X.; Dusmez, S.; Akin, B.; Rajashekara, K. A New SVPWM for the Phase Current Reconstruction of Three-Phase Three-level T-type Converters. *IEEE Trans. Power Electron.* **2015**, *62*, 2627–2637.
8. Blaabjerg, F.; Pedersen, J.K. An ideal PWM-VSI inverter using only one current sensor in the DC-link. In Proceedings of the 1994 Fifth International Conference on Power Electronics and Variable-Speed Drives, London, UK, 26–28 October 1994; pp. 458–464.
9. Lee, W.C.; Hyun, D.S.; Lee, T.K. A Novel Control Method for Three-Phase PWM Rectifiers Using a Single Current Sensor. *IEEE Trans. Power Electron.* **2000**, *15*, 861–870.
10. Yeom, H.B.; Ku, H.K.; Kim, J.M. Current reconstruction method for PMSM drive system with a DC link shunt resistor. In Proceedings of the 2016 IEEE Energy Conversion Congress and Exposition (ECCE), Milwaukee, WI, USA, 18–22 September 2016; pp. 18–22.
11. Ha, J.I. Voltage Injection Method for Three-Phase Current Reconstruction in PWM Inverters Using a Single Sensor. *IEEE Trans. Power Electron.* **2009**, *24*, 767–775.
12. You, J.J.; Jung, J.H.; Park, C.H.; Kim, J.M. Phase current reconstruction of three-level Neutral-Point-Clamped (NPC) inverter with a neutral shunt resistor. In Proceedings of the 2017 IEEE Applied Power Electronics Conference and Exposition (APEC), Tampa, FL, USA, 26–30 March 2017; pp. 2598–2604.
13. Gu, Y.; Ni, F.; Yang, D.; Liu, H. Switching-State Phase Shift Method for Three-Phase-Current Reconstruction with a Single DC-Link Current Sensor. *IEEE Trans. Ind. Electron.* **2011**, *58*, 5186–5194.
14. Lin, Y.K.; Lai, Y.S. PWM Technique to Extend Current Reconstruction Range and Reduce Common-Mode Voltage for Three-Phase Inverter using DC-link Current Sensor Only. In Proceedings of the 2011 IEEE Energy Conversion Congress and Exposition, 12–16 September 2011; pp. 1978–1985.
15. Lai, Y.S.; Lin, Y.K.; Chen, C.W. New Hybrid Pulse width Modulation Technique to Reduce Current Distortion and Extend Current Reconstruction Range for a Three-Phase Inverter Using Only DC-link Sensor. *IEEE Trans. Power Electron.* **2013**, *28*, 1331–1337. [CrossRef]
16. Lu, H.; Cheng, X.; Qu, W.; Sheng, S.; Li, Y.; Wang, Z. A Three-Phase Current Reconstruction Technique Using Single DC Current Sensor Based on TSPWM. *IEEE Trans. Power Electron.* **2014**, *29*, 1542–1550.
17. Ying, L.; Ertugrul, N. An Observer-Based Three-Phase Current Reconstruction using DC Link Measurement in PMAC Motors. In Proceedings of the 2006 CES/IEEE 5th International Power Electronics and Motion Control Conference, Shanghai, China, 14–16 August 2006; pp. 1–5.
18. Saritha, B.; Janakiraman, P.A. Sinusoidal Three-Phase Current Reconstruction and Control Using a DC-Link Current Sensor and a Curve-Fitting Observer. *IEEE Trans. Ind. Electron.* **2007**, *54*, 2657–2664. [CrossRef]
19. Kim, K.; Yeom, H.; Ku, H.; Kim, J. Current reconstruction method with single DC-link. In Proceedings of the 2014 IEEE Energy Conversion Congress and Exposition, 14–18 September 2014; pp. 250–256.
20. Tomigashi, Y.; Hida, H.; Ueyama, K. Voltage vector correction based on a novel coordinate transformation for motor current detection using a single shunt resistor. In Proceedings of the 2009 13th European Conference on Power Electronics and Applications, 8–10 September 2009; pp. 1–8.

21. Kim, H.; Jahns, T.M. Integration of the Measurement Vector insertion Method (MVIM) with Discontinuous PWM for Enhanced Single Current Sensor Operation. In Proceedings of the 2006 IEEE Industry Applications Conference Forty-First IAS Annual Meeting, 8–12 October 2006; Volume 5, pp. 2459–2465.
22. Carpaneto, M.; Marchesoni, M.; Parodi, G. A Sensorless PMSM Drive Operating in the Field Weakening Region Using Only One Current Sensor. In Proceedings of the 2010 IEEE International Symposium on Industrial Electronics (ISIE), 4–7 July 2010; pp. 1199–1204.
23. Bing, Z.; Du, X.; Sun, J. Control of three-phase PWM rectifiers using a single DC current sensor. *IEEE Trans. Power Electron.* **2011**, *26*, 1800–1808. [CrossRef]
24. Sun, K.; Wei, Q.; Huang, L.; Matsuse, K. An overmodulation method for PWM-inverter-fed IPMSM drive with single current sensor. *IEEE Trans. Ind. Electron.* **2010**, *57*, 3395–3404. [CrossRef]

 *energies* 

Article

A Novel Three-Level Voltage Source Converter for AC–DC–AC Conversion

Zongbin Ye [1,2,*]**, Anni Chen** [1,2]**, Shiqi Mao** [1,2]**, Tingting Wang** [1,2]**, Dongsheng Yu** [1,2] **and Xianming Deng** [1,2]

[1] Jiangsu Province Laboratory of Mining Electric and Automation, China University of Mining and Technology, Xuzhou 221008, China; anne1993chen@163.com (A.C.); 17851147002@163.com (S.M.); wangtt_1994@163.com (T.W.); dongsiee@163.com (D.Y.); xmdengcumt@126.com (X.D.)

[2] School of Electrical and Power Engineering, China University of Mining and Technology, Xuzhou 221008, China

* Correspondence: yezongbin@163.com or 5120@cumt.edu.cn; Tel./Fax: +86-516-8359-2000

Received: 26 March 2018; Accepted: 27 April 2018; Published: 4 May 2018

Abstract: This paper presents a novel three-level voltage source converter for AC–DC–AC conversion. The proposed converter based on H-bridge structure is studied in detail. The control method with traditional double-closed-loop control strategy and voltage balancing algorithm is applied to the rectifier side. Correspondingly, a simplified modulation algorithm is applied to the inverter side, and the voltage balancing of inverter side is realized through the optimal selection of switching combination. Then, the application of the proposed topology is assessed in general and ideal operation conditions. Furthermore, the proposed topology with a variable voltage variable frequency (VVVF) is verified in experimental conditions. The performance of the proposed converter and control strategy is evaluated by experimental and simulation results.

Keywords: three-level converter; simplified PWM strategy; redundant switching combination; voltage balance control

---

1. Introduction

With the development of the multilevel converter (MC), it has become a cost-effective solution of medium-voltage AC drives [1]. Due to its merits compared with a conventional two-level voltage source converter—such as lower voltage stress on switches, improved output waveforms, reduced common mode voltage, and high voltage capability—MC has been applied to more emerging fields [2–4]. The areas of applications include renewable energy generation, electric vehicle traction [5], high-power energy storage system [6], micro-grids [7], high-voltage ac or dc transmission [8–10], and some newly-developing fields.

In general, there are two conventional types of AC–DC–AC multilevel converters in view of whether it has common dc-links. The diode-clamped MC (DCMC) [11] and fly-capacitor MC (FCMC) [12] are widespread adopted structures with common dc-links, which can operate in four quadrants and be supplied by single rectifier. Besides, there are some other topologies, such as five-level active neutral-point-clamped MC (5L-ANPC) [13], modular MC (MMC) [9,10], and some newly-developed MC [14–16]. However, these kinds of MCs, except MMC, are hard to extend towards higher output voltage levels and power grades because of the complicated structures. The other drawback of these types is the poor ability to deal with some special systems which have different voltage grades, e.g., connection of two grids with different voltage grades [16–18]. Separated dc-links are the features of the other types of MCs, including cascaded H-bridge MC (CHBMC) [17], five-level H-bridge NPC (5L-HNPC) [18], and some hybrid and asymmetrical cascaded H-bridge MCs with different sub-modules [19] or dc-link voltages [20]. It has the advantage of flexible extending of the output levels and power rating. However, the bulky and expensive phase-shifting transformers for isolated dc sources make it hard to increase the power density. A back-to-back CHB converter without any isolating device [21] can avoid these problems. However, short-circuits caused by the hardware topology are difficult to solve and the proposed topology cannot be expanded to a three phase system.

In this paper, a new three-level voltage source converter for AC–DC–AC conversion is proposed. It can be used in three-phase system and more easily to extend to higher voltage level than a back-to-back NPC converter. Compared to the back-to-back CHB converter proposed in [21], a half H-bridge cell used in the new topology provides more redundant vectors and makes it overcome the short-circuit problem, which simplifies the control method. In addition, the proposed topology utilizes fewer switches at the cost of increasing the number of dc-link capacitors, the separated dc links will decrease the total dc voltage of the system, which is beneficial for the insulation design in many fields [22].

The rest of the paper is organized as follows. In Section 2, the circuit configuration, characteristics and working principles of the proposed topology are studied in detail. The overall control strategy and pulse-width modulation strategy considering the voltage balance control is given in Section 3. In Section 4, two operation conditions are analyzed, and the simulation and experimental results demonstrate the effectiveness of the proposed control strategies. Section 5 concludes the paper.

2. Circuit Configuration of Proposed Three-Level Voltage Source Converter

2.1. Circuit Configuration

The proposed three-level converter is presented in Figure 1. It includes two basic submodules, power unit I and power unit II. Port 2 of power unit II in the three-phase topology is connected together, forming the neutral point N of the converter.

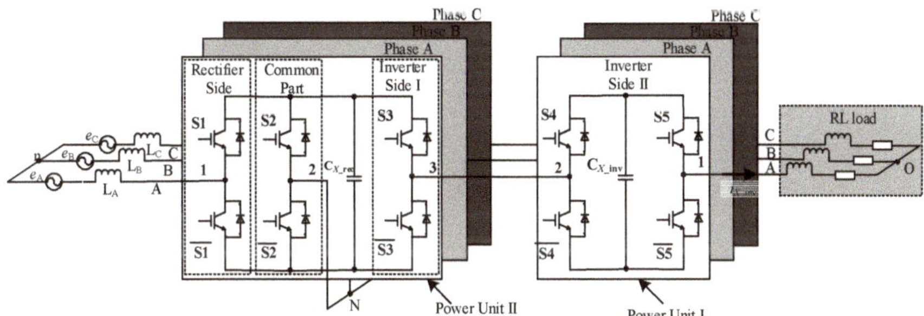

Figure 1. Circuit configuration of the proposed three-level voltage source converter.

For convenience, the rectifier side, common part, inverter side I, and inverter side II can be defined as shown in Figure 1. The rectifier side connects in series with three-phase inductors and the grid through three phase electrical terminals A, B, and C. The three electrical terminals A, B, and C of inverter side connect with the three-phase load.

2.2. Working Principle of Rectifier Side

All the dc-link voltage values are assumed to be equal to U_{dc}. Obviously, the output voltage levels relative to neutral point N are determined by S1/$\overline{S1}$ and S2/$\overline{S2}$. u_{X_rec} (X = A, B, C) is defined as the output voltage of rectifier side, which can be obtained as Equation (1).

$$u_{X\_rec} = (S1 - S2)U_{dc} \tag{1}$$

2.3. Working Principle of Inverter Side I

Combined with the common parts shown in Figure 1, inverter side I can produce three level voltages similar to u_{X_rec}. u_{X_inv} referenced to N is obtained as Equation (2).

$$u_{X\_inv} = (S3 - S2)U_{dc} \tag{2}$$

2.4. Master–Slave Control Principle

Combining with common part, there will be no problem obviously when rectifier side or inverter side I works independently. Due to the special structure, the operation principle of each side cannot be analyzed independently when they work together. In other words, there is a coupling relationship between two sides. Since any side can be chosen as the master control side, the rectifier side is chosen as an example. Hence the switching command of S2 is decided by rectifier side. Output voltage levels of u_{X_inv} will be limited in some switching combinations. For example, if rectifier side is P, S2 should be 0. However, if the inverter side I needs to be N, S2 should be 1. Consequently, a contradiction appears.

In order to reduce the coupling relationship, a submodule power unit I is added on the right of power unit II, which is defined as inverter side II, as shown in Figure 1. According to the switching states, the switching commands of S3S4S5 can be decided after switching commands of S1S2 are generated as shown in Table 1. The output voltage of inverter side, u_{X_inv} can be rewritten as Equation (3).

$$u_{X\_inv} = (S3 - S2)U_{dc} + (S5 - S4)U_{dc} \tag{3}$$

Table 1. Switching states of rectifier side and inverter side.

Rectifier State	S1 S2	Inverter State	S3 S4 S5
P	S1S2 = 10	P	S3 = 1, S4 = S5 or S3 = 0, S4 = 0, S5 = 1
		O	S3 = 0, S4 = S5 or S3 = 1, S4 = 1, S5 = 0
		N	S3 = 0, S4 = 1, S5 = 0
O	S1 = S2	P	S3 = S2, S4 = 0, S5 = 1 or S2 = 0, S3 = 1, S4 = S5
		O	S2 = S3, S4 = S5 or S2 = 1, S3 = 0,S4 = 0, S5 = 1 or S2 = 0, S3 = 1, S4 = 1, S5 = 0
		N	S2 = S3, S4 = 1, S5 = 0 or S2 = 1, S3 = 0, S4 = S5
N	S1S2 = 01	P	S3 = 1, S4 = 0, S5 = 1
		O	S3 = 1, S4 = S5 or S3 = 0, S4 = 0, S5 = 1
		N	S3 = 1, S4 = 1, S5 = 0 or S3 = 0, S4 = S5

2.5. Comparison with Classic Multilevel Topologies

For better understanding of the proposed technology, it is necessary to make a comparison over classic multilevel converter topologies. In order to achieve four-quadrant AC–DC–AC conversion, NPC, FC, and CHB are arranged in a back-to-back (B2B) scheme [23]. As a matter of convenience, the proposed topology is abbreviated as CMC. The state-of-the-art 4.5 kV, 450 A and 3.3 kV, 450 A IGBTs are applied in aforementioned three level and five level topologies, respectively, with the output line-to-line voltage V_{ll_rms} = 3 kV and power rating of 600 kW. It is assumed that the voltage rating of each clamping diode and flying capacitor is equal to the main switch device voltage rating. A comprehensive list of the requested components number of each converter topology is shown in Table 2 [24,25]. Obviously, the counts of active devices of these types are equal except the CMC, which needs two extra switches in each phase. A total of 36 diodes are requested in a five level B2B NPC converter, and the count will increase dramatically with the number of levels. Capacitors contain dc-link capacitors and flying capacitors, so the number of capacitors—as an example—for 5L B2B FC topology is 4 + 18. These large numbers of capacitors increase size and cost of the converter and reduce the reliability. Through the total component amount, CHB topology is extremely advantageous in quantity in the Table, but it must be equipped with a transformer and PWM rectifier for four-quadrant applications. In the rest of the topologies, CMC topology, without clamping diodes and capacitors, has a lower number of components than other topologies with the improvement of voltage level.

Table 2. Comparison of different topologies (V_{ll_rms} = 3 kV, I_{ph_rms} = 115.5 A, P = 600 kW).

Level	3L				5L			
Topology	NPC	FC	CHB	CMC	NPC	FC	CHB	CMC
Rated device voltage (IGBT)	4.5 kV	4.5 kV	4.5 kV	4.5 kV	3.3 kV	3.3 kV	3.3 kV	3.3 kV
Rated device current (IGBT)	450 A	450 A	450 A	450 A	450 A	450 A	450 A	450 A
IGBTs	24	24	24	30	48	48	48	54
Diodes	12	---	---	---	36	---	---	---
Capacitors	2 + 0	2 + 6	3 + 0	6 + 0	4 + 0	4 + 18	6 + 0	12 + 0
Total Components	38	32	23	36	88	70	54	66

To compare with the B2B 3L-NPC, the switching losses for both topologies are calculated and normalized according to the method proposed in [26] and the datasheet of IGBT. The result is shown in Figure 2 where the modulation index of the inverter side ranges from 0.5 to 1.15 and the power factor of the load ranges from 0.7 to 1.

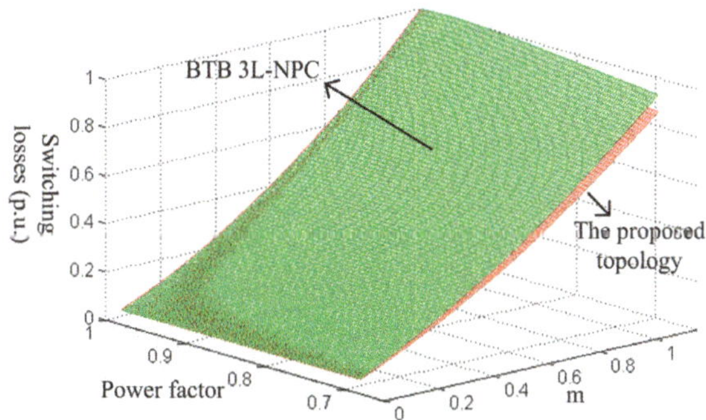

Figure 2. Switching losses comparison.

3. Control Method of the Proposed Three-Level Voltage Source Converter

3.1. Control Method

Since this work focuses on testing the proposed three-level converter topology, a common control method should be used. So dual close loop control structure in d–q synchronous reference frame is adopted in rectifier side [20]. The voltage loop contains a conventional proportional-integral (PI) controller to regulate the average value of capacitor voltage of C_{X_rec}, $U_{dc\_ave\_rec}$ to reference value U_{dcref_rec} (=U_{dc}). The reference current of the q-axis (i^*_q) is set to a certain value to adjust input power factor of the whole converter. Then, the inner current loops generate the reference voltage of rectifier side, $u^*_{X\_rec}$. Subsequently, the zero sequence voltage u_{z_rec} generated by the voltage balancing algorithm is injected to $u^*_{X\_rec}$ to control voltage values of C_{X_rec}. Then a simplified modulation algorithm in [27] is adopted to calculate the duration time of switching states, P/O/N, in the rectifier side and inverter side.

Due to the coupling relationship, proper switching commands of S2/S3/S4/S5 should be chosen to achieve voltage balancing of the capacitors C_{X_inv}. The optimal selection of switching combination (OSSC) is introduced later to generate the converter switching commands of S1~S5. The whole control block diagram of the proposed three-level converter is shown in Figure 3.

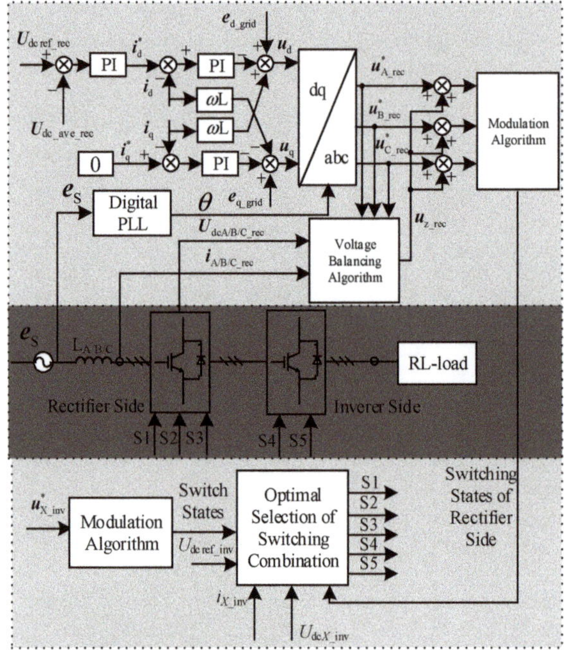

Figure 3. Control block diagram of the proposed multilevel converter.

3.2. Modulation Algorithm

A simplified PWM strategy [27] which is easier and more flexible to realize different targets was used as modulation algorithm. Taking inverter side as an example and assuming that $U_{\text{dcref\_inv}} = U_{\text{dc}}$, $u_{X\_\text{inv}}(t)$ consists of U_{dc} and 0 when the reference voltage $u^{*}_{X\_\text{inv}} > 0$; otherwise, $u_{X\_\text{inv}}(t)$ consists of $-U_{\text{dc}}$ and 0. This divides the space vector diagram into six sectors, as denoted by S in Figure 4.

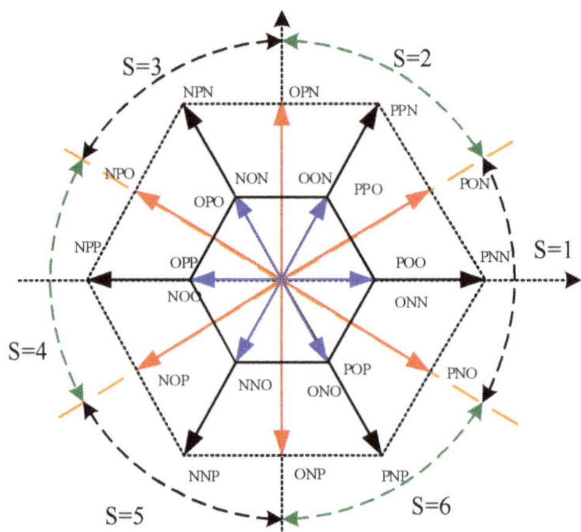

Figure 4. Sectors for the proposed three-level converter with the simplified PWM.

When S = 1, the voltage-second balancing principle can be represented by Equation (4), where u_z represents the equivalent zero-sequence voltage. The general solutions of (4) can be obtained as Equation (5).

$$\begin{cases} u^*_{A\_inv} \cdot T_s = \int_0^{T_s} u_{AN}(t)dt + \int_0^{T_s} u_z(t)dt \\ u^*_{B\_inv} \cdot T_s = \int_0^{T_s} u_{BN}(t)dt + \int_0^{T_s} u_z(t)dt \\ u^*_{C\_inv} \cdot T_s = \int_0^{T_s} u_{CN}(t)dt + \int_0^{T_s} u_z(t)dt \end{cases} \tag{4}$$

$$\begin{cases} T_{A\_inv} = (u^*_{A\_inv} - u_z) \cdot T_s/U_{dcA\_inv} \\ T_{B\_inv} = T_s + (u^*_{B\_inv} - u_z) \cdot T_s/U_{dcB\_inv} \\ T_{C\_inv} = T_s + (u^*_{C\_inv} - u_z) \cdot T_s/U_{dcC\_inv} \end{cases} \tag{5}$$

T_{X_inv} stands for the duration time of switching state P when $(u^*_{X\_inv} - u_z > 0)$ otherwise stands for the duration time of O.

3.3. Voltage Balancing Algorithm of Rectifier Side

There is only one capacitor in each phase. It only needs to consider the voltage balancing of C_{X_rec} between three-phase. Assuming that u_{z_rec} is the zero sequence voltage injected into $u^*_{X\_rec}$, which is used to realize the targets of voltage balancing of C_{X_rec}. The voltage-second balancing principle can be represented by Equation (6).

$$\begin{cases} T_{A\_rec} = (u^*_{A\_rec} - u_{z\_rec}) \cdot T_s/U_{dcA\_rec} \\ T_{B\_rec} = T_s + (u^*_{B\_rec} - u_{z\_rec}) \cdot T_s/U_{dcB\_rec} \\ T_{C\_rec} = T_s + (u^*_{C\_rec} - u_{z\_rec}) \cdot T_s/U_{dcC\_rec} \end{cases} \tag{6}$$

If U_{dcX_rec} is imbalanced, u_{z_rec} should be calculated to adjust the reference voltage $u^*_{X\_rec}$. As an example, if voltage values of C_{X_rec} satisfy $U_{dcA\_rec} > U_{dcB\_rec} > U_{dcC\_rec}$, it means that the magnitude of charge change within T_s should be $Q_A < Q_B < Q_C$. u_{z_rec} can be changed to adjust Q_X. Calculation of u_{z_rec} is as follows:

1. Q_X, $u^*_{X\_rec}$, and i_{X_rec} are sorted according to U_{dcX_rec}. In order to realize the voltage balancing, Q_X should satisfy Equation (7).

$$Q_{max} < Q_{mid} < Q_{min}, \tag{7}$$

Q_X is defined as Equation (8).

$$Q_X = i_{X\_rec} \cdot \frac{u^*_{X\_rec} - u_{z\_rec}}{U_{dcX\_rec}} \cdot T_s, \tag{8}$$

If $i_{X\_rec} > 0$ and $(u^*_{X\_rec} - u_{z\_rec}) > 0$, u_{X_rec} consists of P/O. The current paths of S1S2 are shown in Figure 5. Obviously, $Q_X > 0$ and C_{X_rec} is charged in this case. C_{X_rec} is discharged within T_s when $i_{X\_rec} < 0$ and $(u^*_{X\_rec} - u_{z\_rec}) > 0$.

2. Substituting (8) into (7) gives (9).

$$i_{max\_rec} \cdot \frac{u^*_{max\_rec} - u_{z\_rec}}{U_{dcmax\_rec}} \cdot T_s < i_{mid\_rec} \cdot \frac{u^*_{mid\_rec} - u_{z\_rec}}{U_{dcmid\_rec}} \cdot T_s < i_{min\_rec} \cdot \frac{u^*_{min\_rec} - u_{z\_rec}}{U_{dcmin\_rec}} \cdot T_s, \tag{9}$$

$$\begin{cases} a1 = i_{max\_rec} \cdot U_{dcmid\_rec} - i_{mid\_rec} \cdot U_{dcmax\_rec} \\ b1 = i_{max\_rec} \cdot u^*_{mid\_rec} \cdot U_{dcmid\_rec} - i_{mid\_rec} \cdot u^*_{mid\_rec} \cdot U_{dcmax\_rec} \\ a2 = i_{mid\_rec} \cdot U_{dcmin\_rec} - i_{min\_rec} \cdot U_{dcmid\_rec} \\ b2 = i_{mid\_rec} \cdot u^*_{mid\_rec} \cdot U_{dcmin\_rec} - i_{min\_rec} \cdot u^*_{min\_rec} \cdot U_{dcmid\_rec} \end{cases} \tag{10}$$

$$u_{temp1} = b1/a1, \text{ and } u_{temp2} = b2/a2.$$

3. The range of u_{z_rec} can be obtained from Equation (9), and u_{z_rec} can take any value within the range. However, it should satisfy Equation (11) to acquire a linear modulation.

$$\begin{cases} -U_{dcmax\_rec} \leq u^*_{max\_rec} - u_{z\_rec} \leq U_{dcmax\_rec} \\ -U_{dcmid\_rec} \leq u^*_{mid\_rec} - u_{z\_rec} \leq U_{dcmid\_rec} \quad, \\ -U_{dcmin\_rec} \leq u^*_{min\_rec} - u_{z\_rec} \leq U_{dcmin\_rec} \end{cases} \tag{11}$$

4. Calculating the limit value of u_{z_rec}: the corresponding limitations of the injected zero-sequence voltages are given in (12).

$$\begin{cases} u_{zmax} = \max(u^*_{max\_rec} - U_{dcmax\_rec}, u^*_{mid\_rec} - U_{dcmid\_rec}, \\ \quad u^*_{min\_rec} - U_{dcmin\_rec}) \\ u_{zmin} = \min(u^*_{max\_rec} + U_{dcmax\_rec}, u^*_{mid\_rec} + U_{dcmid\_rec}, \quad, \\ \quad u^*_{min\_rec} + U_{dcmin\_rec}) \end{cases} \tag{12}$$

Finally, u_{z_rec} can be obtained to realize the targets of voltage balancing as shown in Table 3. The voltage balancing algorithm is shown in Figure 6 in detail.

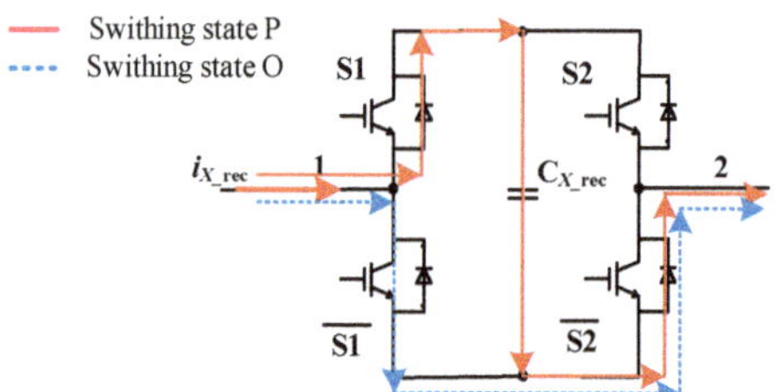

Figure 5. Voltage balancing algorithm of C_{X_rec}.

Table 3. Value of u_{z_rec}.

a1	a2	u_{temp1}, u_{temp2}	u_{z_rec}
	>0	$u_{temp1} > u_{temp2}$	u_{temp1}
		$u_{temp1} \leq u_{temp2}$	u_{temp2}
>0	<0	$u_{temp1} > u_{temp2}$	0
		$u_{temp1} \leq u_{temp2}$	$(u_{temp1} + u_{temp2})/2$
	>0	$u_{temp1} > u_{temp2}$	$(u_{temp1} + u_{temp2})/2$
<0		$u_{temp1} \leq u_{temp2}$	0
	<0	$u_{temp1} > u_{temp2}$	u_{temp2}
		$u_{temp1} \leq u_{temp2}$	u_{temp1}

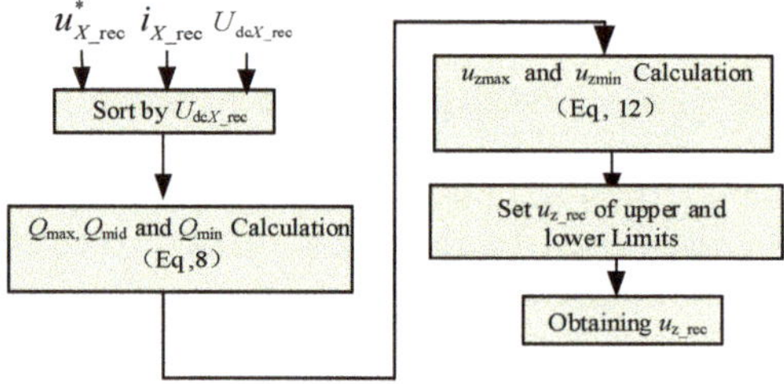

Figure 6. Current paths of S1S2 when $i_{X\_rec} > 0$ and $(u^*_{X\_rec} - u_{z\_rec}) > 0$.

3.4. Voltage Balancing Method of Inverter Side

Maintaining voltage balancing of the flying-capacitors in the inverter side is the main aim of this section. As introduced before, the coupling relationship shown in Table 1 can provide considerable number of redundant switching combinations. These combinations can provide a charging or discharging current paths for each flying-capacitors. The voltage balance control can be realized by selecting a proper combination. The optimal selection of switching combination can be generated as follows.

3.4.1. Effect of the Switching States on the Capacitors Voltages

According to Equation (3), the switching states of the inverter side P/O/N can be generated by inverter side I or II. However, only the switching states produced by inverter side II (S4S5) have an effect on the capacitors voltages $U_{\text{dc}X\_\text{inv}}$. Which inverter side is selected to generate the required switching states is decided by the inverter state, the direction of i_{X_inv}, and switching commands of S2 as listed in Table 4.

For example, when the inverter state is P, $i_{X\_\text{inv}} > 0$, and S2 = 0, the switching state can be generated as marked in the Table 4. The discharging and keeping paths of capacitor C_{X_inv} have been shown in Figure 7, respectively.

Table 4. Switching states of rectifier side and inverter side.

Inverter State	i_{X_inv}	S2	Inverter Side I	Inverter Side II	Switch Combinations	Charge State
P	>0	1	O	P	S3S4S5 = 101	D
		0	O	P	S3S4S5 = 001	D
			P	O	S3 = 1, S4 = S5	K
	≤0	1	O	P	S3S4S5 = 101	C
		0	P	O	S3 = 1, S4 = S5	K
			O	P	S3S4S5 = 001	C
O	>0	1	N	P	S3S4S5 = 001	D
			O	O	S3 = 1, S4 = S5	K
		0	O	O	S3 = 0, S4 = S5	K
			P	N	S3S4S5 = 110	C
	≤0	1	O	O	S3 = 1, S4 = S5	K
			N	P	S3S4S5 = 001	C
		0	P	N	S3S4S5 = 110	D
			O	O	S3 = 0, S4 = S5	K
N	>0	1	N	O	S3 = 0, S4 = S5	K
			O	N	S3S4S5 = 110	C
		0	O	N	S3S4S5 = 010	C
		1	O	N	S3S4S5 = 110	D
			N	O	S3 = 0, S4 = S5	K
		0	O	N	S3S4S5 = 010	D

C: Charging; D: Discharge; K: Keeping.

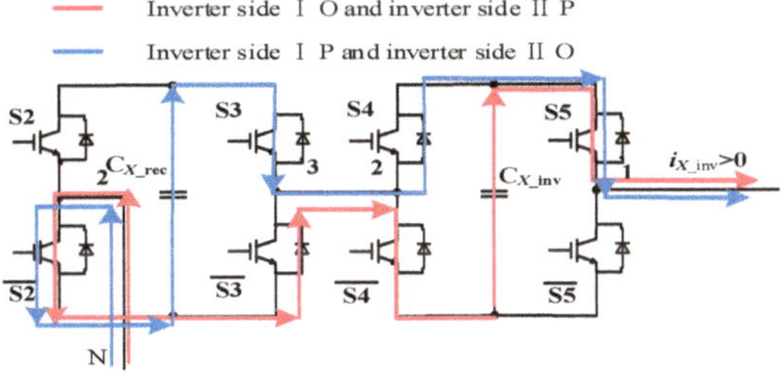

Figure 7. The discharging and keeping paths of capacitor C_{X_inv}.

3.4.2. Optimal Selection of Switching Combination (OSSC)

To balance the voltage of C_{X_inv}, OSSC is set to select proper switching combinations after the previous step (1). Before selecting the switching combination, the duration of switching state (T_{X_rec}/T_{X_inv}) is calculated through the simplified modulation algorithm in [27], thus the inverter state and rectifier state are determined. The switching commands of S2 should be a certain state 0(1) if the rectifier side is P(N). While it cannot be decided when rectifier side is O. Based on the actual situation, i_{X_inv} can be measured. To analyze the working principle of OSSC, the two examples are listed.

$$(u^*_{A\_rec} - u_{z\_rec}) < 0, (u^*_{A\_inv} - u_z) > \begin{cases} \text{Condition I}: & i_{X\_inv} > 0 \\ \text{Condition II}: & i_{X\_inv} > 0 \end{cases}, \qquad (13)$$
$$U_{dcA\_inv} > U_{dcref\_inv}$$

Condition I: U_{dcA_inv} should be decreased with a proper switching combination. Referring to Table 4, when the switching state of the rectifier and inverter sides are N (S2 = 1) and O, respectively, there are two switching combinations to choose from the Table 4. It is obvious that the combination S2S3S4S5 = 1001 is the optimal one to decrease the voltage deviation in condition I. In this way, the combination of switch can be selected out at different switching states as shown in Figure 8a.

Condition II: Due to $i_{X\_inv} < 0$, the P state should be generated by the inverter side I as much as possible. Similarly, the combination of switching can be acquired referring the Table 4. When the calculation result of duration satisfied the inequality $T_{A\_rec} < T_{A\_inv}$, the situation that the switching state of rectifier and inverter side are N (S2 = 1) and P will exist as shaded areas depicted in Figure 8b. In this situation, no discharge switching combination can be found except a charge combination in Table 4. Therefore, the deviation of C_{X_inv} is uncontrollable. Those situations, defined as 'uncontrollable switching combination' (USC), restrict the operation range of the converter.

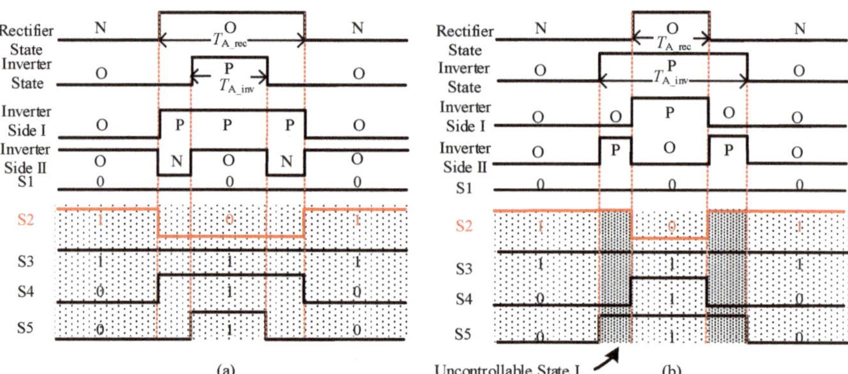

Figure 8. Converter state and switching commands. (**a**) $(u^*_{A\_rec} - u_{z\_rec}) < 0$, $(u^*_{A\_inv} - u_z) > 0$, and $U_{dcA\_inv} > U_{dcref\_inv}$; $i_{X\_inv} > 0$, $T_{A\_rec} > T_{A\_inv}$; (**b**) $(u^*_{A\_rec} - u_{z\_rec}) < 0$, $(u^*_{A\_inv} - u_z) > 0$, and $U_{dcA\_inv} > U_{dcref\_inv}$; $i_{X\_inv} < 0$, $T_{A\_rec} < T_{A\_inv}$.

3.5. Calculation of Duration Time of Each Arm

Based on the above analysis, the optimal selection of switching combination can be acquired. Then the duration time of S1~S5 in the proposed three-level converter can be calculated easily in each case as shown in Table 5. It should be noted that the high or low of S2 should be transformed as shown in Figure 8b. Then the trigger signals of each switch can be generated easily according to Table 5 in the proposed three-level converter.

Table 5. Switching states of rectifier side and inverter side.

$u^*_{A\_rec} - u_{z\_rec}$	$u^*_{A\_inv} - u_z$	U_{dcA_inv}	i_{A_inv}	t_{S1}	t_{S2}	t_{S3}	t_{S4}	t_{S5}
>0	>0	$U_{dcA\_inv} > U_{dcref\_inv}$	>0	T_{A_rec}	0	0	0	T_{A_inv}
			≤0	T_{A_rec}	0	T_s	T_s	T_{A_inv}
		$U_{dcA\_inv} \le U_{dcref\_inv}$	>0					
			≤0	T_{A_rec}	0	0	0	T_{A_inv}
	≤0	$U_{dcA\_inv} > U_{dcref\_inv}$	>0	T_s	T_{A_rec}	0	T_{A_rec}	T_{A_inv}
			≤0	T_{A_rec}	0	T_{A_inv}	T_s	0
		$U_{dcA\_inv} \le U_{dcref\_inv}$	>0					
			≤0	T_s	T_{A_rec}	0	T_{A_rec}	T_{A_inv}
≤0	>0	$U_{dcA\_inv} > U_{dcref\_inv}$	>0	T_{A_rec}	T_s	T_{A_inv}	0	T_s
		$U_{dcA\_inv} \le U_{dcref\_inv}$	≤0 / >0	0	T_{A_rec}	T_s	T_{A_rec}	T_{A_inv}
			≤0	T_{A_rec}	T_s	T_{A_inv}	0	T_s
	≤0	$U_{dcA\_inv} > U_{dcref\_inv}$	>0	T_{A_rec}	T_s	0	0	T_{A_inv}
			≤0	T_{A_rec}	T_s	T_s	T_s	T_{A_inv}
		$U_{dcA\_inv} \le U_{dcref\_inv}$	>0					
			≤0	T_{A_rec}	T_s	0	0	T_{A_inv}

$t_{S1\sim S5}$ is the duration time of each switch, S1~S5.

4. Simulation and Experimental Analysis

4.1. Operation of the Proposed Three-Level Voltage Source Converter

4.1.1. Ideal Operation Condition

The ideal operation condition of the proposed converter is that the sign of output voltages are synchronized to the input voltages, if Equation (14) is satisfied

$$\mathrm{Sgn}(u^*_{X\_rec} - u_{z\_rec}) = \mathrm{Sgn}(u^*_{X\_inv} - u_z), \tag{14}$$

there will be no uncontrollable cases based on the above analyses in this operation condition. That is, the voltage deviation of C_{X_inv} will be kept under control completely. Although this condition can balance the capacitor voltages well, the use of this structure is restricted in some applications such as power electronic transformers and AC regulators.

4.1.2. General Operation Condition

In this condition, there is no connection between $\mathrm{Sgn}(u^*_{X\_rec} - u_{z\_rec})$ and $\mathrm{Sgn}(u^*_{X\_inv} - u_z)$, the reference voltage of inverter side

$(u^*_{X\_inv} - u_z)$ can operate at the frequency and magnitude different with $(u^*_{X\_rec} - u_{z\_rec})$. Figure 9a has been drawn to illustrate the extreme case when $U_{dcX\_inv} < U_{dcref\_inv}$, $\mathrm{Sgn}(u^*_{X\_rec} - u_{z\_rec}) = -\mathrm{Sgn}(u^*_{X\_inv} - u_z)$. Based on Table 5, voltage deviation of C_{A_inv} is enlarged in most areas. However, the shadow areas can be removed under the condition that the modulation index of the rectifier side and inverter side satisfy Equation (15). Then, voltage deviation can be controlled in this extreme case.

$$m_{inv} \le 1 - m_{rec} = 1 - \frac{\mathrm{magnitude}(u^*_{X\_rec} - u_{z\_rec})}{U_{dc}}, \tag{15}$$

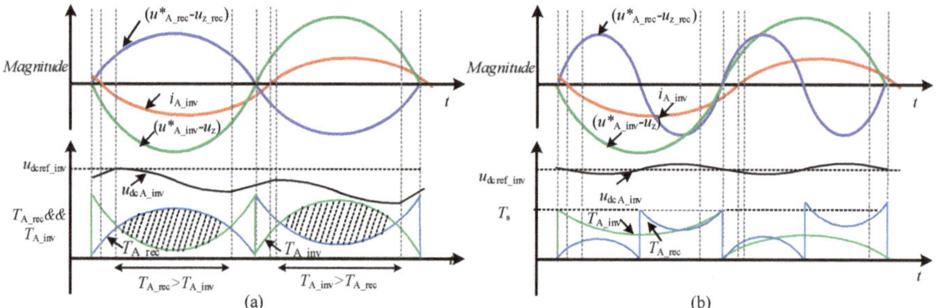

Figure 9. Analysis of voltage deviation with C_{A_inv}. (**a**) When $\mathrm{Sgn}\,|\,(u^*_{A\_rec} - u_{z\_rec})\,| = -\mathrm{Sgn}\,|\,(u^*_{A\_inv} - u_z)\,|$; (**b**) when $m_{inv} < 1 - m_{rec}$.

Including this special case, the uncontrollable states can be eliminated absolutely when $\mathrm{Sgn}(u^*_{X\_rec} - u_{z\_rec}) \neq \mathrm{Sgn}(u^*_{X\_inv} - u_z)$ and $m_{inv} \leq 1 - m_{rec}$ as shown in Figure 9b. Although the time of uncontrollable state can be quantified as shown in Figure 9b when $m_{inv} \geq 1 - m_{rec}$, the voltage deviation of C_{X_inv} still cannot be improved without efficient measures. Hence, the magnitude of output voltage will be limited. DC voltage deviation and low-frequency fluctuation will exist in the whole system.

4.2. Experimental Results

A low power prototype has been developed in lab conditions to verify the performance of the proposed three-level converter, as depicted in Figure 10. The three-level converter was built by using power IGBTs (TOSHIBA, Tokyo, Japan). The control method was implemented in a 150-MIPS float-point 32-bit TMS320F28335 board, and XC3S500E-4PQ208C of XILINX Company (San Jose, CA, USA) has been used to generate switching commands. The experimental parameter settings are shown in Table 6. In order to observe necessary signals, two scopes were used to monitor the signals after DA conversion. U_{AB_inv} was measured by voltage probes directly.

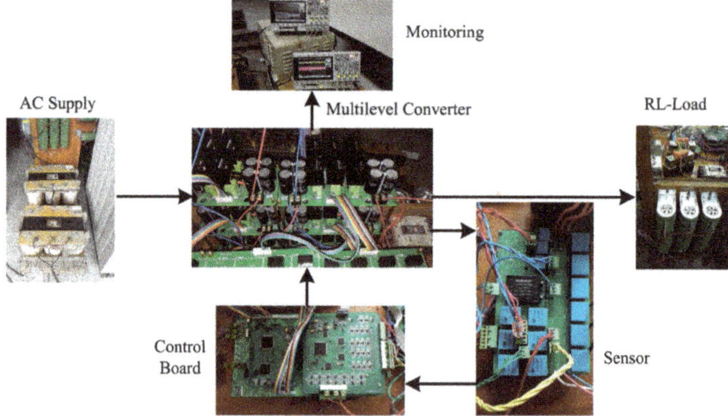

Figure 10. Experimental setup for the proposed multilevel converter.

Table 6. Parameter settings for simulation and experiment.

Parameter	Value
Source voltage, e_{X_rec}	55 V
DC-link voltage	100 V
DC-link capacitor	1200 µF
Filter-inductive	2.2 mH
Resistive-inductive load	20 Ω, 2.2 mH
Switching frequency	5 kHz

The experimental results obtained in Figure 11 show the voltage–current waveforms of the rectifier side and inverter side at different modulation indexes m_{inv} and switching frequency f during the whole working process. Figure 11a,c shows that the three-phase current i_{X_rec} rectifier side and i_{X_inv} inverter side increase with the increase of modulation index and frequency. In Figure 11c, the waveforms of line-to-line voltage u_{AB_inv} have three-levels when f = 20 Hz, m_{inv} = 0.4 and f = 30 Hz, m_{inv} = 0.6, while it changes to five-levels when f = 40 Hz, m_{inv} = 0.8 and f = 50 Hz, m_{inv} = 0.9. U_{dcX_rec} and U_{dcX_inv} are shown in Figure 11b,d are the waveforms of three-phase capacity of C_{X_rec} and C_{X_inv}. It can be seen that U_{dcX_rec} and U_{dcX_inv} do not change with the modulation index and frequency after the system is working. Capacitor voltages can be balanced well, and better performance of the proposed multilevel converter is verified in this process. Figure 12 shows the performance of the converter in transient-state condition with the modulation index m_{inv} changing from 0.4 to 0.6 and output frequency f changing from 20 Hz to 30 Hz. Figure 12a,b shows the input voltage–current waveforms and voltage waveforms of C_{X_rec}. Figure 12c show the waveforms of line-to-line voltage u_{AB_inv} and three-phase currents i_{X_inv}. The capacitor voltages of C_{X_inv}, U_{dcX_rec} are shown in Figure 12d.

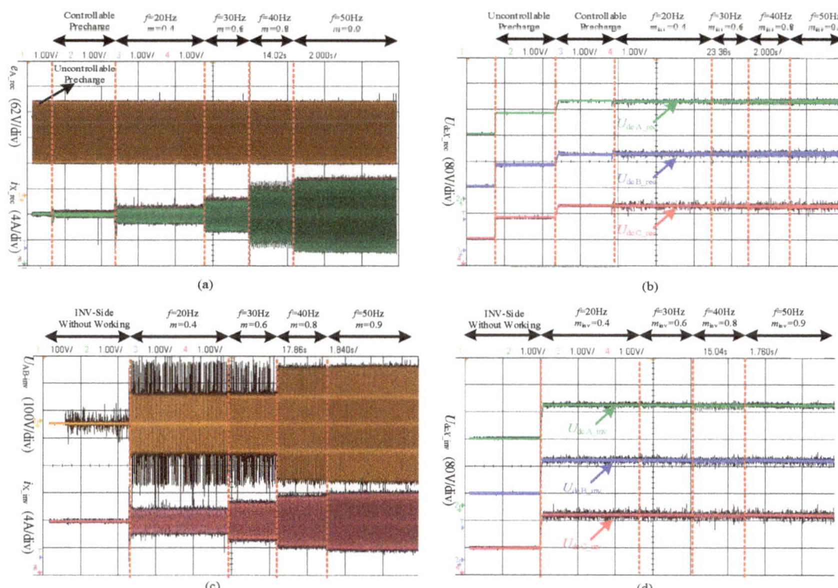

Figure 11. Experimental results of the whole working-process in transient-state condition; (**a**) Input voltage–current waveforms, e_{A_rec} and i_{X_rec}; (**b**) voltages of C_{X_rec}; (**c**) output voltage–current waveforms, u_{AB_inv} and i_{X_inv}; (**d**) voltages of C_{X_inv}.

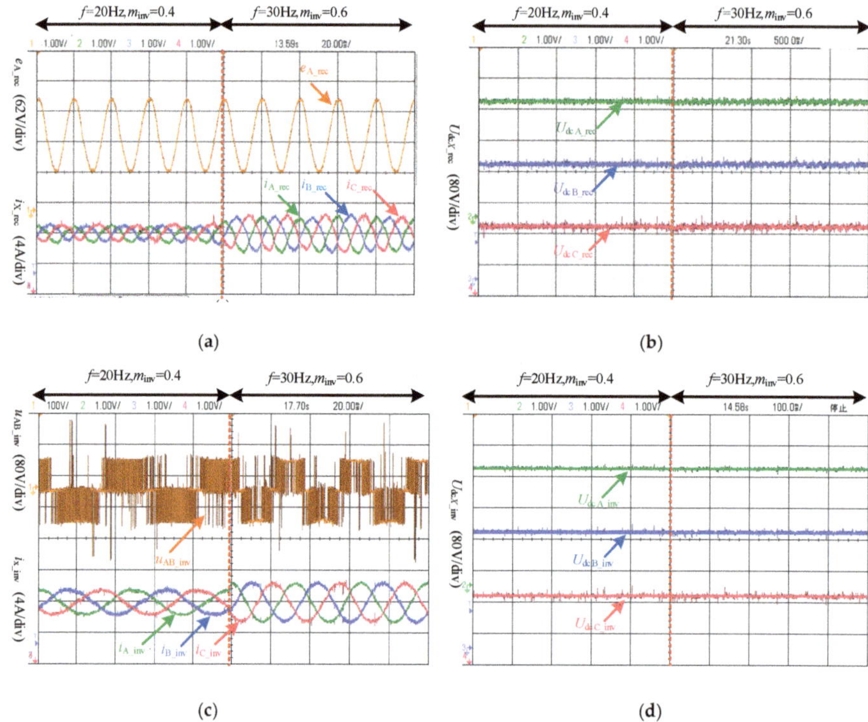

Figure 12. Experimental results of the whole working process in transient-state conditions; (**a**) Input voltage–current waveforms, e_{A_rec} and i_{X_rec}; (**b**) voltages of C_{X_rec}; (**c**) output voltage–current waveforms, u_{AB_inv} and i_{X_inv}; (**d**) voltages of C_{X_inv}.

As can be seen from Figure 11a, when the output frequency f and modulation index m_{inv} are 20 Hz and 0.4, the peak value of three-phases on the rectifier side current i_{X_rec} has low-frequency fluctuations, and the sine effect is not ideal; when switching to $f = 20$Hz and $m_{inv} = 0.4$, the three-phase current i_{X_rec} stabilizes rapidly after about 25 ms, the sine is good, and the amplitude is basically the same. In the process of switching, the entire control system can achieve a balanced three-phase current and unity power factor control, and show good robust performance. U_{dcX_rec} and U_{dcX_inv} shown in Figure 11b,d have almost no change when the frequency and modulation index switching. They are constantly maintained at a fixed value, showing strong anti-interference performance. As shown in Figure 11c, after switching, the inverter side line-to-line voltage u_{AB_inv} and three-phase currents i_{X_inv} are rapidly stabilized, and the three-phase current change trend remains the same.

Obviously, the inverter side of the converter performs well in this case. Figure 13 shows the experimental results in transient-state conditions with the modulation index set at 0.8 and 0.9 and the output frequency f set from 40 Hz to 50 Hz. Figure 13a,c shows the same results as Figure 12 and will not be repeated here. According to Figure 13b,d, voltages of rectifier side and inverter side are maintained at their given values. At the same time, it becomes more stable after switching. Thus, the effectiveness of the proposed three-level converter to capacitor voltage equalization control is verified.

4.3. Simulation Results

Figure 14 shows the curves of the voltage weight total harmonic distortion WTHD with different modulation indexes m_{inv} and switching frequency f_{switch} based on MATLAB/Simulink.

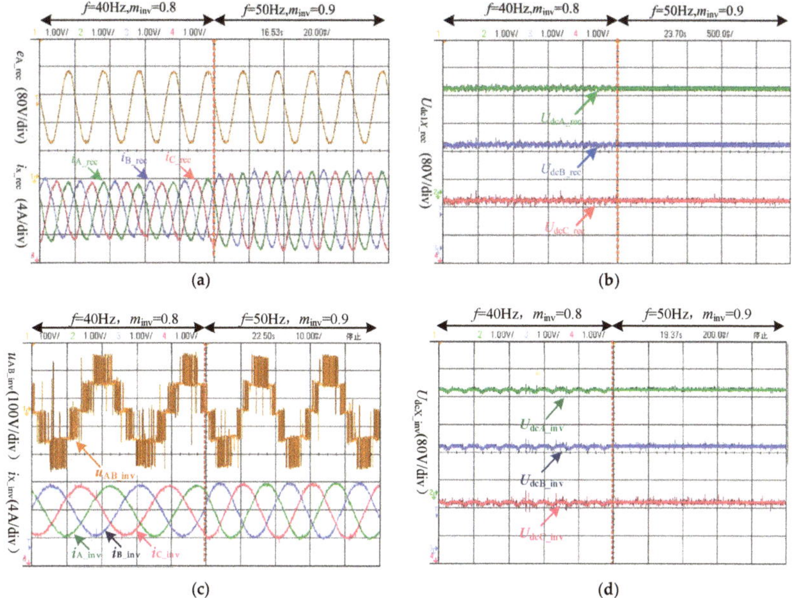

Figure 13. Experimental results of the multilevel converter in transient-state conditions; m_{inv} changes from 0.8 to 0.9; (**a**) Input voltage-current waveforms, e_{A_rec} and i_{X_rec}; (**b**) voltages of C_{X_rec}; (**c**) output voltage–current waveforms, u_{AB_inv} and i_{X_inv}; (**d**) voltages of C_{X_inv}.

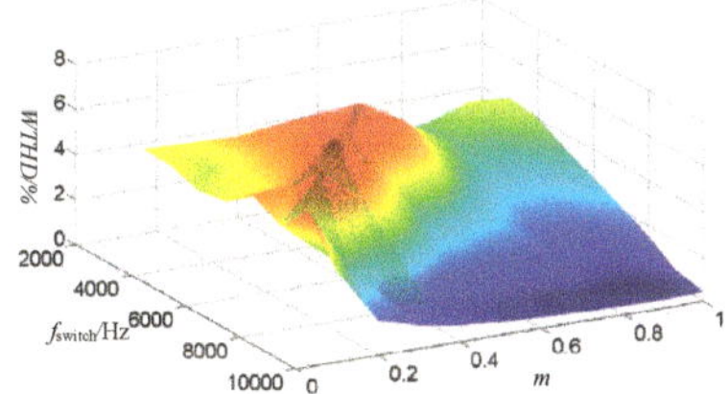

Figure 14. Harmonic characteristic results of WTHD curves of u_{AB_inv} with f_{switch} and m_{inv}.

WTHD is defined in Equation (16), where V_1 and V_n mean the fundamental and n order harmonic components in line-to-line voltage respectively. As shown in Figure 14, WTHD of u_{AB_inv} increases with the decrease of switching frequency f_{switch} and modulation index m_{inv}. It shows better performance when $m_{inv} > 0.4$, while WTHD becomes taller when $m_{inv} < 0.4$ in some areas. In general, the performance of the proposed converter can operate well.

$$\text{WTHD} = \sqrt{\sum_{n=2}^{\infty} \frac{V_n^2}{n^2}} / V_1 \ , \tag{16}$$

4.4. Simulation Analysis of 5/3 Level Voltage Source Converter

This new topology can be expanded asymmetrically, which means the rectifier side and inverter side can work with different nominal voltages. It is possible to the proposed topology to connect the asynchronous multi-scale power network. On the basis of the proposed three level voltage source converter, the voltage level in rectifier side has been expended to five level. The circuit configuration of 5/3 level converter has shown in Figure 15. Due to the similar structure, the control methods of 5/3 level voltage source converter are as same as the aforementioned methods of the three-level converter.

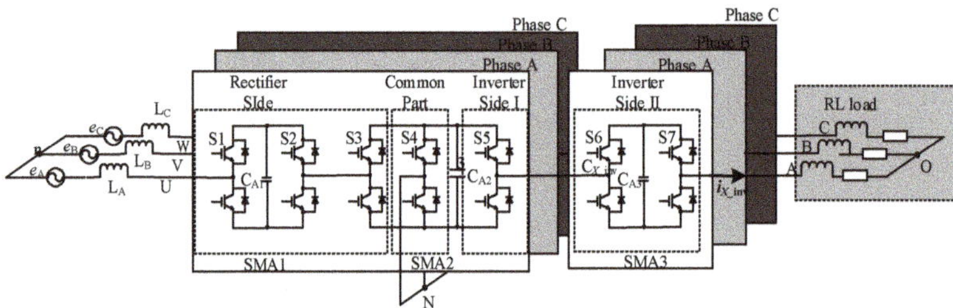

Figure 15. Circuit configuration of 5/3 level voltage source converter.

The simulation result is shown in Figure 16. The whole working process shown in Figure 16a, is divided into three sections: uncontrollable precharge, controllable precharge, and inverter side working. In the uncontrollable precharge section, uncontrollable full wave rectification is achieved only by diodes with anti-parallel device. Then the rectifier side starts in Power Unit II, and the voltages of modules SMX1 and SMX2 are selected as 50 V and 100 V, respectively. Figure 16b,c shows line-to-line voltages and three-phase currents of the rectifier side under the modulation index set at 1. Obviously, the voltage reaches nine levels and the currents are undistorted sinusoidal waveforms.

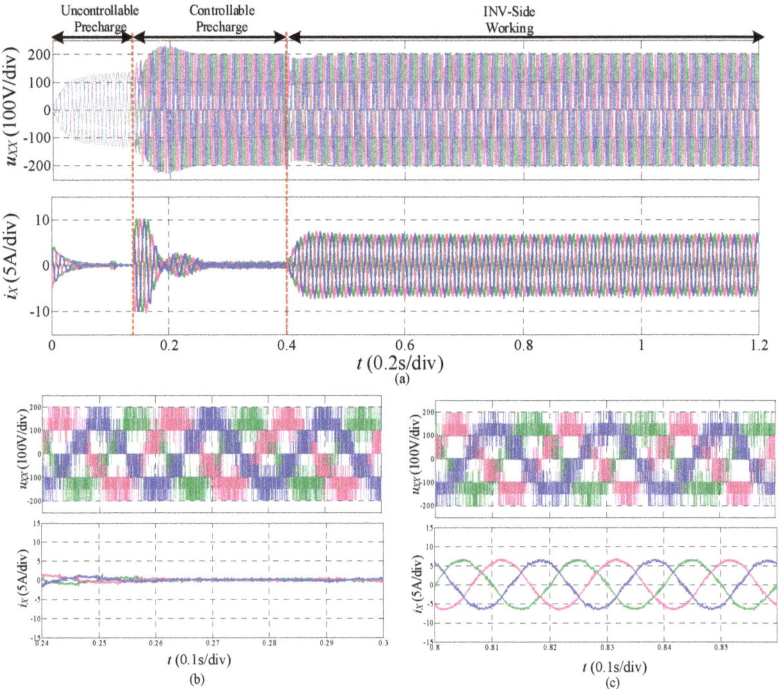

Figure 16. Simulation results of the whole working process; (**a**) Output voltage–current waveforms, u_{xx} and i_X; (**b**) Output voltage–current waveforms, u_{xx} and i_X ; time from 0.24s to 0.3s;(**c**) Output voltage–current waveforms, u_{xx} and i_X; time from 0.8 s to 0.86 s.

5. Conclusions

In order to balance the voltage of flying-capacitors, a novel three-level voltage source converter for AC–DC–AC conversion was proposed in this paper. The circuit configuration and work principle of the proposed three-level voltage source converter were studied in detail. The dual double-closed-loop control strategy and voltage balancing algorithm, especially the method of inverter capacitors with OSSC, were introduced to elaborate the control method of a three-level converter. Then, two operation conditions were analyzed to assess the operating characteristics of the proposed converter. Finally, the balanced control capabilities of this new topology to the three-phase suspension capacitor voltage of the rectifier side and inverter side was verified by simulations and experiments.

Author Contributions: Conceptualization, Z.Y. and X.D.; Methodology, A.C.; Software, S.M.; Validation, A.C., T.W. and S.M.; Formal Analysis, D.Y.; Investigation, T.W.; Resources, Z.Y.; Data Curation, D.Y.; Writing-Original Draft Preparation, A.C.; Writing-Review & Editing, T.W.; Visualization, S.M.; Supervision, S.M.; Project Administration, Z.Y.; Funding Acquisition, X.D.

Funding: This work has been partially supported by the Fundamental Research Funds for the Central Universities under Award 2015XKMS030 and the National Natural Science Foundation of China (U1610113).

References

1. Nguyen, L.V.; Tran, H.D.; Johnson, T.T. Virtual prototyping for distributed control of a fault-tolerant modular multilevel inverter for photovoltaics. *IEEE Trans. Energy Convers.* **2014**, *29*, 841–850. [CrossRef]
2. Alishah, R.S.; Hosseini, S.H.; Babaei, E.; Sabahi, M. A New General Multilevel Converter Topology Based on Cascaded Connection of Submultilevel Units With Reduced Switching Components, DC Sources, and Blocked Voltage by Switches. *IEEE Trans. Ind. Electron.* **2016**, *63*, 7157–7164. [CrossRef]
3. Buccella, C.; Cecati, C.; Cimoroni, M.G.; Razi, K. Analytical method for pattern generation in five-level cascaded H-bridge inverter using selective harmonic elimination. *IEEE Trans. Ind. Electron.* **2014**, *61*, 5811–5819. [CrossRef]
4. Zamiri, E.; Vosoughi, N.; Hosseini, S.H.; Barzegarkhoo, R.; Sabahi, M. A New Cascaded Switched-Capacitor Multilevel Inverter Based on Improved Series–Parallel Conversion with Less Number of Components. *IEEE Trans. Ind. Electron.* **2016**, *63*, 3582–3594. [CrossRef]
5. Tolbert, L.M.; Fang, Z.P.; Thomas, G.H. Multilevel Converters for Large Electric Drives. *IEEE Trans. Ind. Appl.* **1999**, *35*, 36–44. [CrossRef]
6. Teymour, H.R.; Sutanto, D.; Muttaqi, K.M. Solar PV and Battery Storage Integration using a New Configuration of a Three-Level NPC Inverter With Advanced Control Strategy. *IEEE Trans. Energy Convers.* **2014**, *29*, 354–365.
7. Pouresmaeil, E.; Montesinos-Miracle, D.; Gomis-Bellmunt, O. Control scheme of three-level NPC inverter for integration of renewable energy resources into AC grid. *IEEE Syst. J.* **2012**, *6*, 242–253. [CrossRef]
8. Peng, F.Z.; Lai, J.S.; McKeever, J.W. A multilevel voltage-source inverter with separate DC sources for static var generation. *IEEE Trans. Ind. Appl.* **1996**, *32*, 1130–1138. [CrossRef]
9. Pou, J.; Ceballos, S.; Knstantinou, G. Circulating current injection methods based on instantaneous information for the modular multilevel converter. *IEEE Trans. Ind. Electron.* **2015**, *62*, 777–788. [CrossRef]
10. Meshram, P.M.; Borghate, V.B. A simplified nearest level control (NLC) voltage balancing method for modular multilevel converter (MMC). *IEEE Trans. Power Electron.* **2015**, *30*, 450–462. [CrossRef]
11. Hatti, N.; Hasegawa, K.; Akagi, H. A 6.6-kV transformerless motor drive using a five-level diode-clamped PWM inverter for energy savings of pumps and blowers. *IEEE Trans. Power Electron.* **2009**, *24*, 796–803. [CrossRef]
12. Ghias, A.M.Y.M.; Pou, J.; Agelidis, V.G.; Ciobotaru, M. Optimal switching transition-based voltage balancing method for flying capacitor multilevel converters. *IEEE Trans. Power Electron.* **2015**, *30*, 1804–1817. [CrossRef]
13. Oikonomou, N.; Gutscher, C.; Karamanakos, P.; Kieferndorfand, F.D.; Geyer, T. Model predictive pulse pattern control for the five-level active neutral-point-clamped inverter. *IEEE Trans. Ind. Appl.* **2013**, *49*, 2583–2592. [CrossRef]

14. Roshankumar, P.; Rajeevan, P.P.; Mathew, K.; Leon, J.I.; Franquelo, L.G. A five-level inverter topology with single-DC supply by cascading a flying capacitor inverter and an H-bridge. *IEEE Trans. Power Electron.* **2012**, *27*, 3505–3512. [CrossRef]

15. Ooi, G.H.P.; Maswood, A.L.; Lim, Z. Five-level multiple-pole PWM AC-AC converters with reduced components count. *IEEE Trans. Ind. Electron.* **2015**, *62*, 4739–4748. [CrossRef]

16. Narimani, M.; Wu, B.; Cheng, Z.; Zargari, N.R. A new nested neutral point-clamped (NNPC) converter for medium-voltage (MV) power conversion. *IEEE Trans. Power Electron.* **2014**, *29*, 6375–6382. [CrossRef]

17. Cortes, P.; Wilson, A.; Kouro, S.; Rodriguez, J.; Abu-Rub, H. Model predictive control of multilevel cascaded H-bridge inverters. *IEEE Trans. Ind. Electron.* **2010**, *57*, 2691–2699. [CrossRef]

18. Kouro, S.; Malinowski, M.; Gopakumar, K.; Pou, J.; Franquelo, L.G.; Wu, B.; Rodriguez, J.; Perez, M.A.; Leon, J.I. Recent Advances and Industrial Applications of Multilevel Converters. *IEEE Trans. Ind. Electron.* **2010**, *57*, 2553–2580. [CrossRef]

19. Veenstra, M.; Rufer, A. Hybrid Cascaded Control of a Hybrid Asymmetric Multilevel Inverter for Competitive Medium-Voltage Industrial Drives. *IEEE Trans. Ind. Appl.* **2005**, *41*, 655–664. [CrossRef]

20. Venugopal, J.; Subarnan, G.M. Hybrid cascaded MLI topology using ternary voltage progression technique with multicarrier strategy. *J. Electr. Eng. Technol.* **2015**, *10*, 1610–1620. [CrossRef]

21. Miranbeigi, M.; Iman-Eini, H.; Asoodar, M. A new switching strategy for transformer-less back-to-back cascaded H-bridge multilevel converter. *IET Power Electron.* **2014**, *7*, 1868–1877. [CrossRef]

22. Kongnun, W.; Sangswang, A.; Chotigo, S. Design of a cascaded H-bridge converter for insulation testing. In Proceedings of the IEEE International Symposium on Industrial Electronics, Seoul, Korea, 5–8 July 2009; pp. 859–863.

23. Blahnik, V.; Peroutka, Z.; Talla, J. Control of cascaded H-bridge active rectifier providing active voltage balancing. In Proceedings of the 40th Annual Conference of the IEEE Industrial Electronics Society, IECON 2014, Dallas, TX, USA, October–1 November 2015; pp. 4589–4594.

24. Islam, M.R.; Guo, Y.; Zhu, J.G. Performance and cost Comparison of NPC, FC and SCHB Multilevel Converter Topologies for High-voltage Applications. In Proceedings of the International Conference on Electrical Machines and Systems, Beijing, China, 20–23 August 2011; pp. 1–6.

25. Fazel, S.S.; Bernet, S.; Krug, D.; Jalili, K. Design and Comparison of 4-kv Neutral-point-clamped, Flying-capacitor, and Series-connected H-bridge Multilevel Converters. *IEEE Tran. Ind. Appl.* **2007**, *43*, 1032–1040. [CrossRef]

26. Minambres-Marcos, V.; Guerrero-Martinez, M.A.; Romero-Cadaval, E.; Milanes-Montero, M.I. Generic Losses Model for Traditional Inverters and Neutral Point Clamped Inverters. *Electron. Electr. Eng.* **2014**, *20*, 84–88. [CrossRef]

27. Ye, Z.B.; Xu, Y.M.; Li, F.; Deng, X.M.; Zhang, Y.Z. Simplified PWM Strategy for Neutral-Point-Clamped (NPC) Three-Level Converter. *J. Power Electron.* **2014**, *14*, 519–530. [CrossRef]

Article

A Reverse Model Predictive Control Strategy for a Modular Multilevel Converter

Weide Guan, Shoudao Huang, Derong Luo * and Fei Rong

College of Electrical and Information Engineering, Hunan University, Changsha 410082, China;
guanweide@hnu.edu.com (W.G.); hsd1962@hnu.edu.cn (S.H.); rongfei@hnu.edu.cn (F.R.)
* Correspondence: hdldr@hnu.edu.cn; Tel.: +86-731-8882-2461

Received: 19 December 2018; Accepted: 15 January 2019; Published: 18 January 2019

Abstract: In recent years, modular multilevel converters (MMCs) have developed rapidly, and are widely used in medium and high voltage applications. Model predictive control (MPC) has attracted wide attention recently, and its advantages include straightforward implementation, fast dynamic response, simple system design, and easy handling of multiple objectives. The main technical challenge of the conventional MPC for MMC is the reduction of computational complexity of the cost function without the reduction of control performance of the system. Some modified MPC scan decrease the computational complexity by evaluating the number of on-state sub-modules (SMs) rather than the number of switching states. However, the computational complexity is still too high for an MMC with a huge number of SMs. A reverse MPC (R-MPC) strategy for MMC was proposed in this paper to further reduce the computational burden by calculating the number of inserted SMs directly, based on the reverse prediction of arm voltages. Thus, the computational burden was independent of the number of SMs in the arm. The control performance of the proposed R-MPC strategy was validated by Matlab/Simulink software and a down-scaled experimental prototype.

Keywords: model predictive control (MPC); computational burden; reverse prediction; modular multilevel converter (MMC)

---

1. Introduction

Multilevel converters have been widely used in medium and high voltage applications in recent years [1–3]. Among various multilevel converters, the modular multilevel converter (MMC) has become more popular because of its scalability, modularity, and redundancy. An MMC can be used in many high-power applications; for example, high-voltage direct current (HVDC) systems, static synchronous compensators, grid-connected systems, and medium/high voltage motor drive systems [4–6].

There are some technical challenges for the control of an MMC, such as the balance of sub-module (SM) capacitor voltages, the suppression of circulating currents, and the tracking of output currents. Many control and modulation methods have been proposed to address these issues. Among them, model predictive control (MPC) is an interesting control scheme for the MMC. Its advantages include straightforward implementation, fast dynamic response, and suitability for dealing with multiple objectives [7–9].

Reference [10] first applied the conventional MPC scheme to MMC, in which the output currents, SM capacitor voltages, and circulating currents were controlled together by the evaluation for all the possible switching states in a cost function. In Reference [11], the proportional integral (PI) control method was experimentally compared with the conventional MPC scheme for MMC, and the conclusion was that the MPC scheme had better control performance than PI method, either for steady-state or dynamic performance. In Reference [12,13], similar methods based on conventional MPCs were used to control the MMC by evaluating for all the possible switching states of SMs.

However, it is difficult to implement a conventional MPC scheme in practical applications due to the huge computation complexity because the number of all the possible switching states of SMs for an MMC with N SMs (N is the number of SMs) in each arm is as large as C_{2N}^{N} (e.g., MMC used for HVDC usually has more than 100 SMs, thus the number of all possible control options is more than 9×10^{58}).

In the literature, several modified strategies have been developed to reduce the computational burden [14–22]. Reference [14] proposed a modified method to reduce the subset of control options, which can reduce the computational burden of MPC to a certain extent. For an MMC with eight SMs per bridge, the number of switching states can be reduced to 361. In Reference [15], an integrated MPC combined with the classical energy balancing approach was proposed to reduce the number of switching states to $(N + 1)^3$. In Reference [16], by combining with the conventional sorting algorithm, an indirect MPC was proposed to reduce the calculation burden, in which the computation complexity was determined by the SM number instead of the switching states of SMs. The number of switching states can be reduced to $(N + 1)^2$, thus the computation complexity of indirect MPC is significantly reduced. In Reference [17], a grouping-sorting-optimized MPC strategy was proposed, in which the SMs in each arm were divided into several groups, and the computational load is determined by the number of groups and SMs of each group. The number of control options can be reduced to $2X + M + 3$, where M is the number of groups, and X is the number of SMs in each group. Reference [18] proposed a fast MPC method, in which the number of control options could be significantly reduced to two or three by limiting the change of output voltage level in each control cycle within two or three levels near the previous output voltage level. However, the cost was the reduction of dynamic performance because the variation of the output voltage was limited in each control cycle. A dual-stage MPC scheme for MMC was proposed in Reference [19], in which the control objectives were achieved by a two-stage prediction algorithm. Compared with fast MPC method or indirect MPC method, the dual-stage MPC scheme had better dynamic performance, but the computational burden was increased. In Reference [20], a modulated MPC method combined with the sorting algorithm for MMC was proposed, in which the SMs were selected by evaluating the output voltage level and inserted at regular intervals to obtain a fixed switching frequency. The number of switching states could be reduced to $N + 1$. In Reference [21], aiming at the control objectives of output currents and circulating currents, and based on evaluating the output voltage levels, the overall computation complexity of indirect MPC was reduced to $N + 4$. A similar approach was also adopted in Reference [22]. However, for the MMC used in HVDC (usually with hundreds of SMs), the computational burden of indirect MPC was still too large.

In this paper, a reverse MPC (R-MPC) strategy for the MMC was proposed to further reduce computation complexity. Based on predicted output voltage of MMC, the number of control options was further reduced by calculating the number of on-state SMs directly and decoupling the SM capacitor voltage control. The control of capacitor voltage balance task was carried out in an external control loop. Thus, the computational complexity was independent of the SM number of MMC. This strategy could be used for the MMC with hundreds of SMs.

The rest of this paper is arranged as follows. Section 2 presents the topology, basic operation, and mathematical model of system. In Section 3, the details of the conventional MPC, modified MPC, and proposed R-MPC are explained. The control performance of the proposed R-MPC strategy is validated by Matlab/Simulink software and a down-scaled experimental prototype in Sections 4 and 5, respectively. In Section 6, the conclusions are drawn.

2. Mathematical Model

The topology of the MMC and the single-phase equivalent circuit are shown in Figure 1. The MMC was comprised of three phase legs, and each leg contained an upper arm and a lower arm, which were represented by the subscript "p" and "n", respectively. Each arm included an arm inductor L_0 and N SMs. The arm inductor limited the di/dt of the circulating currents caused by instantaneous voltage differences within the arms. The SMs usually adopted half-bridge structure, which contained two switches (T_1 and T_2) and one capacitor (C_{SM}). The SMs generally worked in two switching states, namely, on-state and off-state. On-state: The SM output voltage was the capacitor voltage u_{cij} (i = p, n, j = 1, 2, ... , N) when T_1 was turned ON and T_2 was OFF. Off-state: The output voltage of the SM is zero when T_1 was turned OFF and T_2 was ON. The switching states of SMs could be written as follows:

$$S_{ij} = \begin{cases} 1, & T_1 \text{ is ON, } T_2 \text{ is OFF} \\ 0, & T_1 \text{ is ON, } T_2 \text{ is OFF} \end{cases} \tag{1}$$

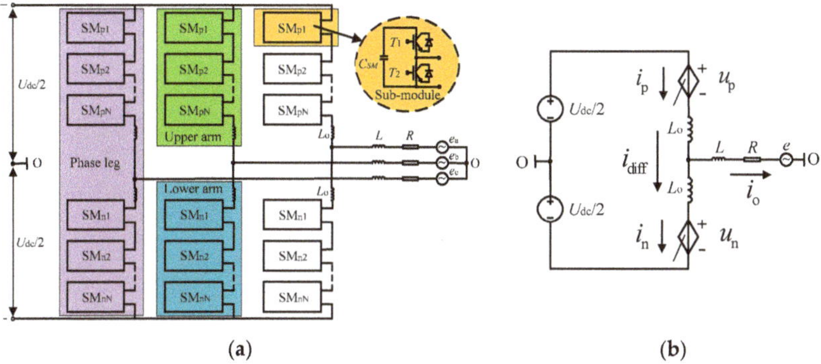

(a)   (b)

Figure 1. The illustrative diagram of a modular multilevel converter (MMC) (a) The topology and (b) the single-phase equivalent circuit.

The capacitor voltage dynamic equation for each SM of the MMC is expressed as follows:

$$\frac{du_{cij}}{dt} = i_{cij}/C_{SM} \tag{2}$$

where i_{cij} is the capacitor current, which can be obtained from the switching state S_{ij} and the arm current i_i as follows:

$$i_{cij} = S_{ij}i_i \tag{3}$$

Based on Figure 1, the arm current i_p and i_n can be written as follows:

$$\begin{cases} i_p = \frac{1}{2}i_o + i_{diff} \\ i_n = -\frac{1}{2}i_o + i_{diff} \end{cases} \tag{4}$$

where i_{diff} is the circulating current, i_o is the output current, and they can be calculated by:

$$\begin{cases} i_o = i_p - i_n \\ i_{diff} = \frac{1}{2}(i_p + i_n) \end{cases} \tag{5}$$

Similarly, the arm voltage equations are as follows:

$$\begin{cases} \dfrac{U_{dc}}{2} = u_p + L_o\dfrac{di_p}{dt} + Ri_o + L\dfrac{di_o}{dt} + e \\ \dfrac{U_{dc}}{2} = u_n + L_o\dfrac{di_n}{dt} - Ri_o - L\dfrac{di_o}{dt} - e \end{cases} \tag{6}$$

where u_p is the voltage of upper arm, u_n is the voltage of lower arm, U_{dc} is the voltage of dc link, e is the voltage of grid, L_o represents the arm inductance, and R and L represent the equivalent resistance and inductance of the load circuit.

Based on Equations (4)–(6), the dynamic equation of the MMC can be obtained as:

$$
\begin{cases}
\dfrac{di_o}{dt} = \dfrac{u_n - u_p - 2Ri_o - 2e}{L_o + 2L} \\[2ex]
\dfrac{di_{\text{diff}}}{dt} = \dfrac{U_{dc} - (u_n + u_p)}{2L_o}
\end{cases}
\tag{7}
$$

Equation (7) indicates that the output current and the circulating current can be regulated directly by the voltage difference and voltage summation between the lower arm voltage u_n and upper arm voltage u_p, respectively.

3. MPC Strategy

3.1. Conventional MPC

In general, there are three control targets for conventional MPC used for MMCs. The first control target is the balancing of SM capacitor voltages. The second control target is the tracking of output current correctly, including magnitude, frequency, and phase angle. The third control target is the suppressing of circulating current, removing its AC component and only keeping its DC component.

The discrete-time model of MPC can be obtained by the following forward Euler approximation equation:

$$
\frac{dx}{dt} = \frac{x(k+1) - x(k)}{T_s}
\tag{8}
$$

where x is the variable of control objectives, $x(k+1)$, $x(k)$ are the variable values at time $k+1$ and k, respectively, and T_s is the sampling period.

The discrete-time dynamic models of the SM capacitor voltages, output currents, and circulating currents can be obtained as follows:

$$
u_{cij}(k+1) = u_{cij}(k) + \frac{S_{ij} i_i(k) T_s}{C_{SM}}
\tag{9}
$$

$$
i_o(k+1) = \left(1 - \frac{2RT_s}{L_o + 2L}\right) i_o(k) + \frac{[u_n(k) - u_p(k) - 2e(k)]T_s}{L_o + 2L}
\tag{10}
$$

$$
i_{\text{diff}}(k+1) = i_{\text{diff}}(k) + \frac{[U_{dc} - u_n(k) - u_p(k)]T_s}{2L_o}
\tag{11}
$$

For MMC, the balancing of SM capacitor voltages, tracking of output current, and suppressing of circulating current should be achieved simultaneously. Because these control variables (SM capacitor voltages, output currents, circulating currents) interact with each other, they can be included in a multivariable cost function with weighting factors. More details of the MPC cost function can be found in [10–12]. The cost function can be defined as follows:

$$
g = \lambda_1 \left| i_o^*(k+1) - i_o^P(k+1) \right| + \lambda_2 \left| i_{\text{diff}}^*(k+1) - i_{\text{diff}}^P(k+1) \right| + \lambda_3 \left| U_{dc}/N - u_{cij}^P(k+1) \right|
\tag{12}
$$

where λ_1, λ_2, and λ_3 are the function weighing factors, $i_o^*(k+1)$, $i_{\text{diff}}^*(k+1)$, U_{dc}/N are the reference values of control objectives, and $i_o^P(k+1)$, $i_{\text{diff}}^P(k+1)$, and $u_{cij}^P(k+1)$ are the next-step predicted values of control objectives, respectively.

The block diagram of conventional MPC is shown in Figure 2a. Within each sampling period of MPC strategy, the cost function was evaluated one step ahead, and the switching state that minimized the cost function was chosen and used to control the converter at next switching cycle by evaluating for all the possible switching states of the converter.

Nevertheless, the number of all possible switching states for an MMC with N SMs in each bridge is C_{2N}^N. For example, an MMC used for HVDC usually has more than 100 SMs per arm, thus the number of all possible switching states in one phase is more than 9×10^{58}. Therefore, it is challenging to implement the conventional MPC by the existing digital processors in practical applications due to large computation load.

3.2. Modified MPC

Some modified MPC strategies have been developed to reduce the computational burden. In combination with decoupling the SM capacitor voltage control and carrying it out in an independent control loop, indirect MPC [16,21] (shown in Figure 2b) can be used to reduce the computation complexity by evaluating the number of inserted SMs instead of the all possible switching states. However, the computational burden is still too high when the MMC has huge number of SMs.

In addition, the fast MPC [18] method can significantly reduce the computational burden to two or three by limiting the change of output voltage level in every control cycle within two or three levels near the previous output voltage level. However, the cost would be the reduction of dynamic performance because the variation of the output voltage was limited in each control cycle.

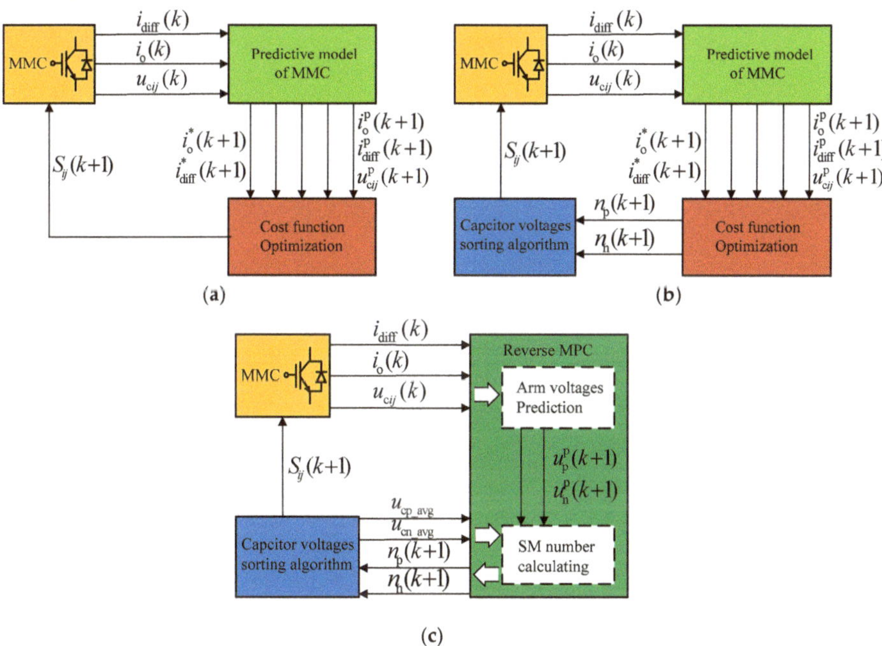

Figure 2. Control schemes of a model predictive control (MPC). (**a**) A conventional MPC, (**b**) an indirect MPC, and (**c**) the proposed inverse MPC.

3.3. Reverse MPC

Figure 2c shows the reverse MPC (R-MPC) strategy, which was proposed to further reduce the computational complexity of MPC by decoupling the SM capacitor voltage control and carrying it out in an independent control loop. Assuming the on-state SM numbers of the upper arm and lower arm at next step are n_{p} and n_{n}, the SM capacitor voltages should be resorted in either ascending or descending order, according to the direction of the corresponding arm current. n_i $(i = \mathrm{p}, \mathrm{n})$ SMs with the lowest voltages are chosen to be inserted to the arm when the arm current is positive, and the other SMs are bypassed. On the contrary, n_i $(i = \mathrm{p}, \mathrm{n})$ SMs with the highest voltages are selected to be inserted to the arm when the arm current is negative, and the other SMs are bypassed.

The balancing of SM capacitor voltage is one of key issues in MMC. The modified MPC generally uses the conventional sorting method (e.g., bubble sorting) to balance the SM capacitor voltages [16,18,20,22], in which all of the SM capacitor voltages have to be sorted in ascending or descending order to determine the inserted SMs with the highest or lowest voltages. However, the sorting algorithm itself is not a trivial task when MMC has huge number of SMs. Therefore, many improved sorting methods were presented to reduce the computational complexity, such as grouping sorting method [17], fundamental-frequency sorting method [23], limited sorting method [24], etc. In general, as long as several SMs with the highest or lowest voltages are selected, the voltage balance task can be completed within an allowable range of voltages ripples, while the rest SMs need not be sorted to reduce the computational burden. In addition, using the parallel computing method, FPGA (Field Programmable Gate Array) can be used to complete the SM capacitor voltage balancing task more quickly, which does not occupy CPU resources [9,16].

After the sorting task, assuming all of the capacitor voltages are balanced well, then the number of on-state SMs for upper arm and lower arm at next-step can be calculated if the next-step arm voltage can be predicted.

From Equation (7), the arm voltages can be written as follows:

$$\begin{cases} u_\mathrm{p} = \frac{U_\mathrm{dc}}{2} - L_0\frac{di_\mathrm{diff}}{dt} - (\frac{L_0}{2}+L)\frac{di_\mathrm{o}}{dt} - Ri_\mathrm{o} - e \\ u_\mathrm{n} = \frac{U_\mathrm{dc}}{2} - L_0\frac{di_\mathrm{diff}}{dt} + (\frac{L_0}{2}+L)\frac{di_\mathrm{o}}{dt} + Ri_\mathrm{o} + e \end{cases} \tag{13}$$

By using the backward Euler approximation,

$$\frac{dx}{dt} = \frac{x(k) - x(k-1)}{T_\mathrm{s}} \tag{14}$$

The discrete-time dynamic models of the arm voltages of MMC can be written as follows:

$$\begin{cases} u_\mathrm{p}(k) = \frac{U_\mathrm{dc}}{2} - \frac{L_0}{T_\mathrm{s}}[i_\mathrm{diff}(k) - i_\mathrm{diff}(k-1)] - (\frac{L_0/2+L}{T_\mathrm{s}} + R)i_\mathrm{o}(k) + \frac{L_0/2+L}{T_\mathrm{s}}i_\mathrm{o}(k-1) - e(k) \\ u_\mathrm{n}(k) = \frac{U_\mathrm{dc}}{2} - \frac{L_0}{T_\mathrm{s}}[i_\mathrm{diff}(k) - i_\mathrm{diff}(k-1)] + (\frac{L_0/2+L}{T_\mathrm{s}} + R)i_\mathrm{o}(k) - \frac{L_0/2+L}{T_\mathrm{s}}i_\mathrm{o}(k-1) + e(k) \end{cases} \tag{15}$$

Assuming the next-step reference values of control objectives can be tracked without error, thus, the next-step predicted arm voltages $u_\mathrm{p}^\mathrm{P}(k+1)$ and $u_\mathrm{n}^\mathrm{P}(k+1)$ can be obtained by shifting forward (Equation (15)), as follows:

$$\begin{cases} u_\mathrm{p}^\mathrm{P}(k+1) = \frac{U_\mathrm{dc}}{2} - \frac{L_0}{T_\mathrm{s}}[i_\mathrm{diff}^*(k+1) - i_\mathrm{diff}(k)] - (\frac{L_0/2+L}{T_\mathrm{s}} + R)i_\mathrm{o}^*(k+1) + \frac{L_0/2+L}{T_\mathrm{s}}i_\mathrm{o}(k) - e^*(k+1) \\ u_\mathrm{n}^\mathrm{P}(k+1) = \frac{U_\mathrm{dc}}{2} - \frac{L_0}{T_\mathrm{s}}[i_\mathrm{diff}^*(k+1) - i_\mathrm{diff}(k)] + (\frac{L_0/2+L}{T_\mathrm{s}} + R)i_\mathrm{o}^*(k+1) - \frac{L_0/2+L}{T_\mathrm{s}}i_\mathrm{o}(k) + e^*(k+1) \end{cases} \tag{16}$$

where $i_\mathrm{diff}^*(k+1)$ is the next-step reference value of circulating current, $i_\mathrm{o}^*(k+1)$ is the next-step reference value of output current, and $e^*(k+1)$ is the next-step reference values of grid voltage.

For the circulating current, the main frequency component is DC, its next-step reference value $i_\mathrm{diff}^*(k+1)$ can be replaced by the previous reference value $i_\mathrm{diff}^*(k)$. But for output current and grid voltage, in order to improve the control accuracy, the next-step reference value $i_\mathrm{o}^*(k+1)$, $e^*(k)$ can be predicted by the formula of the Lagrange extrapolation [25], as follows:

$$i_\mathrm{o}^*(k+1) = 3i_\mathrm{o}^*(k) - 3i_\mathrm{o}^*(k-1) + i_\mathrm{o}^*(k-2) \tag{17}$$

$$e^*(k+1) = 3e^*(k) - 3e^*(k-1) + e^*(k-2) \tag{18}$$

At last, the number of next-step on-state SMs can be calculated as follows:

$$\begin{cases} n_\mathrm{p}^\mathrm{P}(k+1) = \mathrm{round}(\frac{u_\mathrm{p}^\mathrm{P}(k+1)}{u_\mathrm{cp\_avg}^\mathrm{P}(k)}) \\ n_\mathrm{n}^\mathrm{P}(k+1) = \mathrm{round}(\frac{u_\mathrm{n}^\mathrm{P}(k+1)}{u_\mathrm{cn\_avg}^\mathrm{P}(k)}) \end{cases} \tag{19}$$

where $u_{\text{cp\_avg}}(k)$ and $u_{\text{cn\_avg}}(k)$ are the average arm voltages of the MMC (upper arm and lower arm).

The number of control options of the proposed R-MPC strategy was further reduced to one by calculating the number of on-state SMs directly based on the reverse prediction of arm voltages. Thus, the computational burden was independent of the number of SMs in the arm. This strategy was especially suitable for the MMC which has huge number of SMs. The number of control options for different MPC strategies are listed in Table 1. As can be seen from this table, fast MPC and proposed R-MPC have the least computational complexity, especially for the MMC with hundreds of SMs.

It should be noted that the computational burden of R-MPC is independent of the number of SMs, but the computational burden of sorting algorithm is still related to the number of SMs. However, the sorting algorithm can be improved by other means [9,16,17,23,24] to reduce its computational load.

Table 1. Number of control options of different MPC strategies.

	Number of SM (N)	**4**	**10**	**50**	**100**	**200**
	Conventional MPC [10–13]	70	1.8×10^5	1.0×10^{29}	9.1×10^{58}	1.0×10^{119}
	Integrated MPC [15]	125	1331	1.3×10^5	1.0×10^6	8.1×10^6
	Indirect MPC-I [16]	25	121	2601	1.0×10^4	4.0×10^4
Number of	Indirect MPC-II [21]	8	14	54	104	204
control options	Modulated MPC [20]	5	11	51	101	201
	Indirect MPC-III [22]	5	11	51	101	201
	Fast MPC [18]	2~3	2~3	2~3	2~3	2~3
	Proposed MPC (R-MPC)	1	1	1	1	1

4. Simulation Results

As shown in Figure 1, a three-phase MMC system was investigated in MATLAB/Simulink softwareto validate the control performance of the proposed R-MPC strategy. Table 2 lists the parameters of the MMC.

Table 2. Parameters of MMC system.

Parameter	Simulation	Experiment
Rated power (kVA)	5000	2
Rated line voltage (V)	10,000	200
DC bus voltages (V)	20,000	400
SMs per arm	32	8
SM capacitance (μF)	4700	1000
SM capacitor voltage (V)	625	50
Arm buffer inductance (mH)	2.8	2.8
Load inductance (mH)	1	1
Load resistance (Ω)	0.01	1.6
Output frequency (Hz)	50	50
Sampling period (μs)	100	100

4.1. Steady-State Operation

Figure 3 shows the steady-state operation results of a fast MPC [18], an indirect MPC [21], and the proposed R-MPC strategy, respectively. Three MPC strategies have almost the same steady state performance. All the capacitor voltages are well balanced, which are around 3% of their rated values. All the circulating currents are also well-suppressed, with peak-peak ripple values of 28 A, 20 A and 26 A, respectively. The total harmonic distortions (THD) of output currents are only 2.13%, 2.43%, and 2.02%, respectively. The results show that the three strategies all had good steady-state performance. But for an MMC with hundreds of SMs, fast MPC and R-MPC have more advantages because of less computational burden.

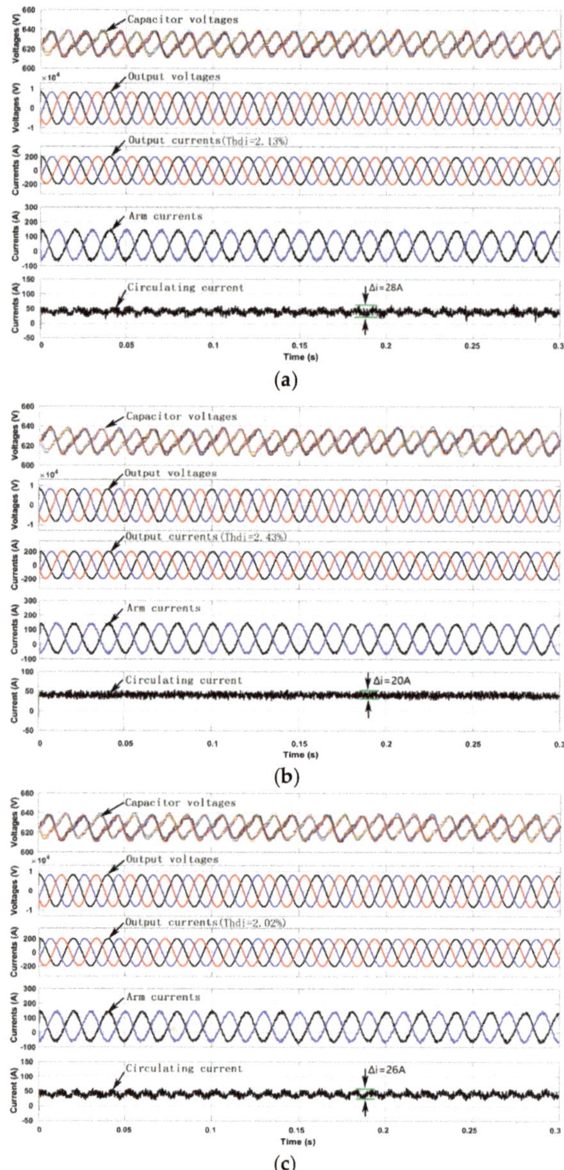

Figure 3. Simulation results of steady-state operation: (**a**) fast MPC, (**b**) indirect MPC, (**c**) proposed R-MPC. Subplots: from top to bottom, capacitor voltages of phase a, three-phase output voltages, three-phase output currents, arm currents of phase a, and circulating current of phase a.

Figure 4 shows the harmonic spectrums of the output voltages and currents with proposed R-MPC. The THDs of output voltages and currents were only 1.88% and 2.02%, respectively. The results showed that the low-order harmonics were very small and within acceptable range. Thus, the noises generated by prediction did not affect the quality of the generated voltages/currents.

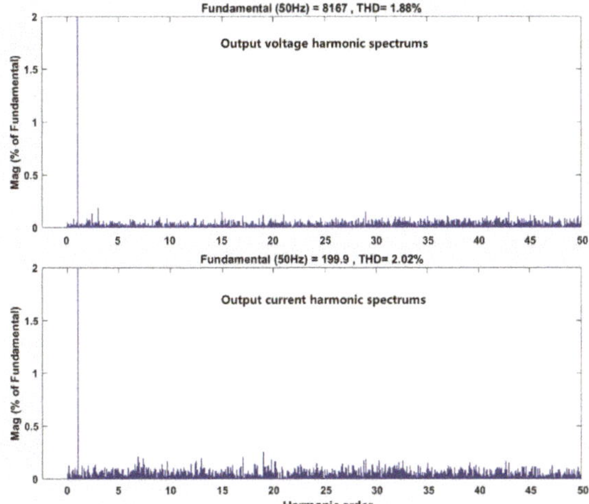

Figure 4. Harmonic spectrums of the output voltages and currents with proposed reverse model predictive control (R-MPC).

Figure 5 shows the comparison of output SM states between the traditional nearest level modulation (NLM) method and the R-MPC strategy proposed in this paper. The NLM method just obtains the nearest level state from the reference voltage, according to the tracking of output currents, however, R-MPC strategy considers not only the tracking of output current but also the suppressing of circulating current and the system delay compensation. Thus, the two methods generated different SM states and switching sequences.

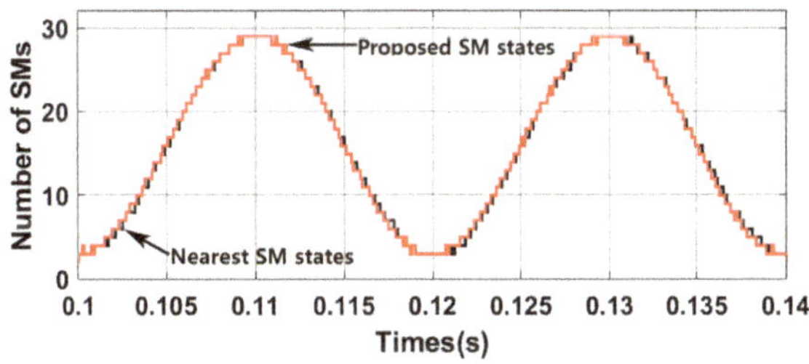

Figure 5. Comparison of output sub-module (SM) states between the traditional nearest level modulation (NLM) method and the R-MPC strategy.

4.2. Dynamic Operation

Figure 6 shows the step response results of fast MPC [18], an indirect MPC [21], and the proposed R-MPC strategy, respectively. Before $t = 0.1$ s, the MMC system reached a stable state, after which the magnitudes of the reference output currents experienced two step changes, stepped up from 100 A to 200 A at $t = 0.1$ s and stepped down from 200 A to 100 A at $t = 0.2$ s. In Figure 6b,c the capacitor voltages were well balanced after the step changes, the output currents of the indirect MPC and the proposed R-MPC could rapidly track their references, and the circulating currents were still well-suppressed. However, as seen in Figure 6a the output currents of the fast MPC took more time to reach the references because the change of output voltage level in each control cycle was limited within two or three levels near the previous output voltage level.

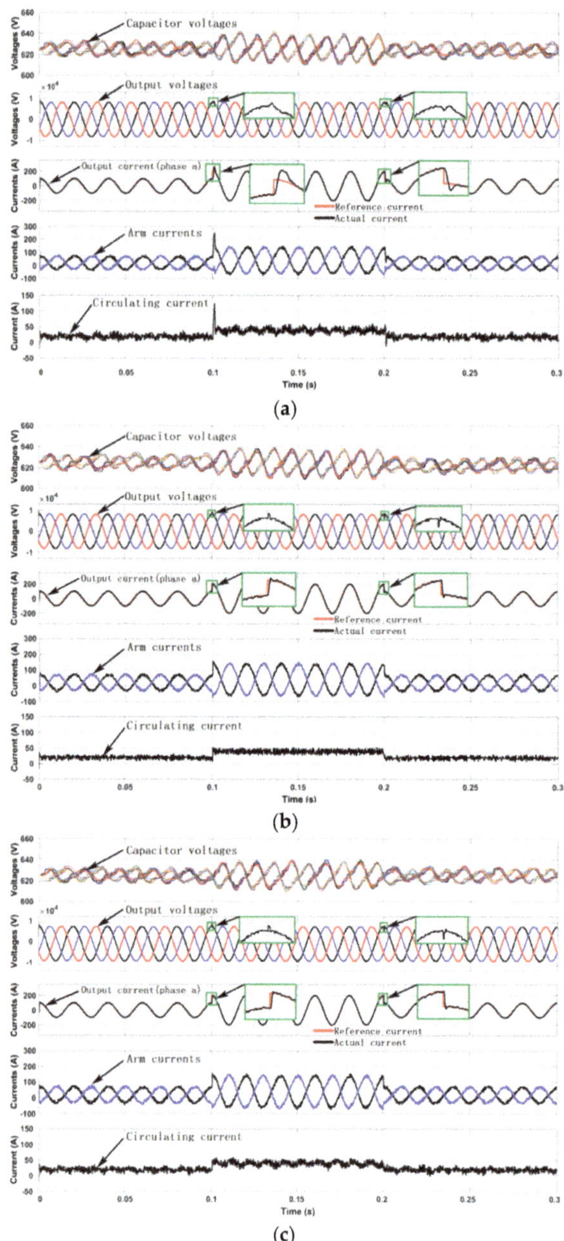

Figure 6. Simulation results of dynamic operation: (**a**) a fast MPC, (**b**) an indirect MPC, and (**c**) the proposed R-MPC. Subplots: from top to bottom, capacitor voltages of phase a, three-phase output voltages, output current of phase a, arm currents of phase a, and circulating current of phase a.

The results showed that the indirect MPC and the proposed R-MPC both had better dynamic-state performance than the fast MPC. However, for the MMC with hundreds of SMs, the R-MPC had more advantages due to less computational burden.

5. Experimental Results

Figure 7 shows a down-scaled experimental prototype of the MMC (2 kVA), setup to verify the control performance of the proposed R-MPC.A DSP/FPGA-based central control board was chosen to complete the control scheme. The DSP (TI TMS320F28335) was used for mathematical calculations and the MPC algorithm. The FPGA (Altera EP2C8Q208C8) was used for Pulse-width modulation (PWM) generation, the sorting algorithm of capacitor voltages, and fault protection. Table 2 lists the parameters of the experimental prototype.

It should be noted that a larger load resistance than that of simulation was chosen because the experimental prototype could only operate in the passive inversion state (limited by the experimental conditions), but the simulation model operated in the active inversion state, which simulated the working characteristics of HVDC. In addition, also limited by the experimental conditions, the SM capacitance was also smaller than that of simulation, which would increase the capacitor voltage ripple slightly. Despite the above differences in load and capacitor parameters, it would not affect the performance comparison and analysis of different control strategies.

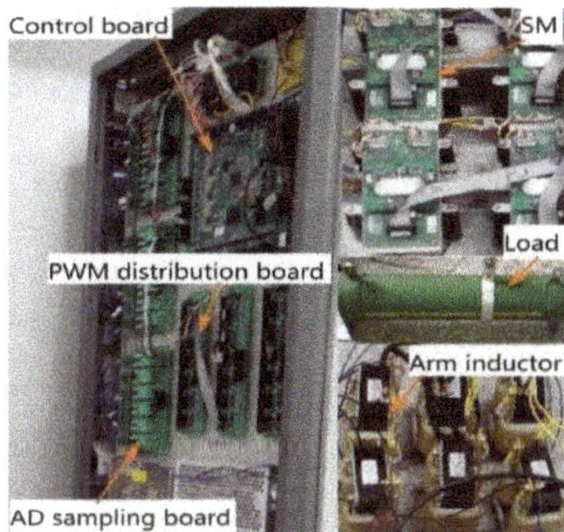

Figure 7. Experimental prototype.

5.1. Steady-State Operation

Figure 8 gives the steady-state experimental results of fast MPC [18], an indirect MPC [21], and the proposed R-MPC strategy, respectively. As seen in Figure 8, three MPC strategies had almost the same steady state performance, where the total harmonic distortions (THD) of output currents were 5.1%, 5.4%, and 5.0%, respectively. In addition, the capacitor voltages of the MMC were well-balanced and the circulating currents were also well-suppressed. The results showed that the three strategies all had good steady-state performance. But, for MMC with hundreds of SMs, fast MPC and R-MPC had more advantages because of less computational burden.

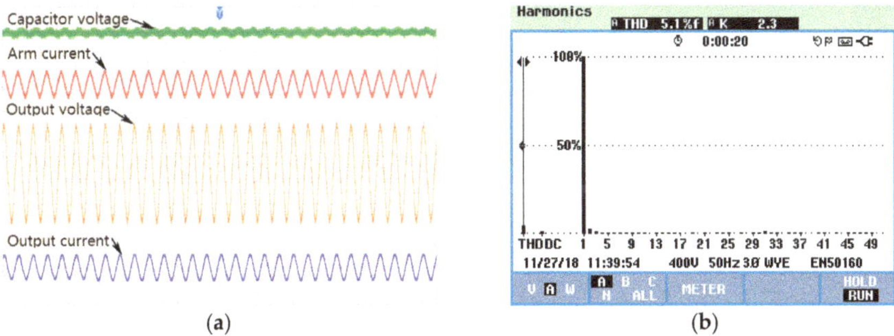

Figure 8. *Cont.*

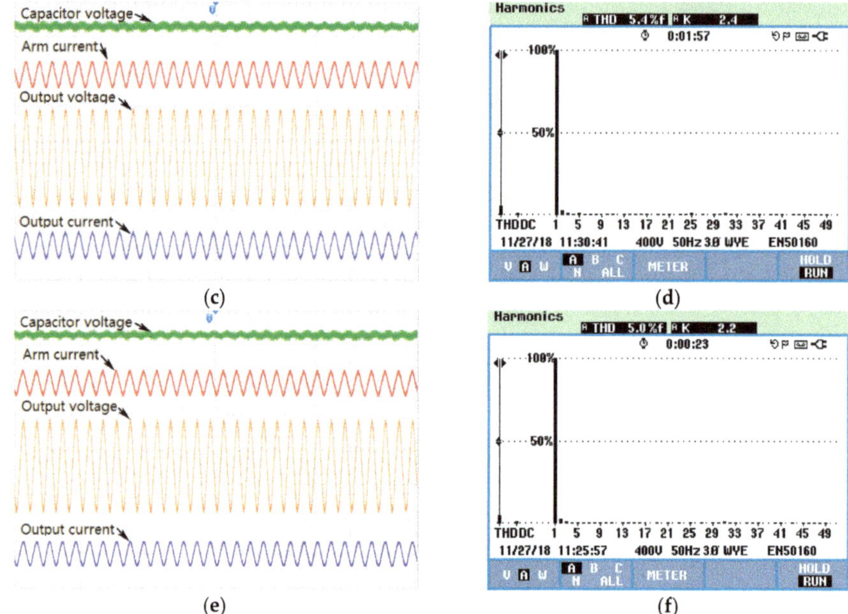

Figure 8. Experimental results of steady-state operation: (**a**,**b**) a fast MPC, (**c**,**d**) an indirect MPC, (**e**,**f**) the proposed R-MPC. Scopes: from top to bottom, first capacitor voltage of phase a (25 V/div), arm current of phase a (5 A/div), output voltage of phase a (100 V/div), and output current of phase a (10 A/div). Time scale: 50 ms/div.

5.2. Dynamic Operation

Figure 9 shows the step response results of fast MPC [18], an indirect MPC [21], and the proposed R-MPC strategy, respectively. Before t = 1.2 s, the MMC system reached a stable state, after which the magnitudes of the reference output currents experienced one step change, stepped up from 2 A to 4 A. From Figure 9b,c it is easy to see that the capacitor voltages could be balanced well after step changes, the circulating currents were well-suppressed, and the output currents could be rapidly tracked to their references, for both indirect MPC and proposed R-MPC strategy. However, as seen in Figure 9a, the output currents of the fast MPC took more time to reach the references because the change of output voltage level in each control cycle was limited within two or three levels near the previous output voltage level.

The results show that the indirect MPC and proposed R-MPC both have better dynamic-state performance than the fast MPC. However, for MMC with hundreds of SMs, the R-MPC have more advantages due to less computational burden.

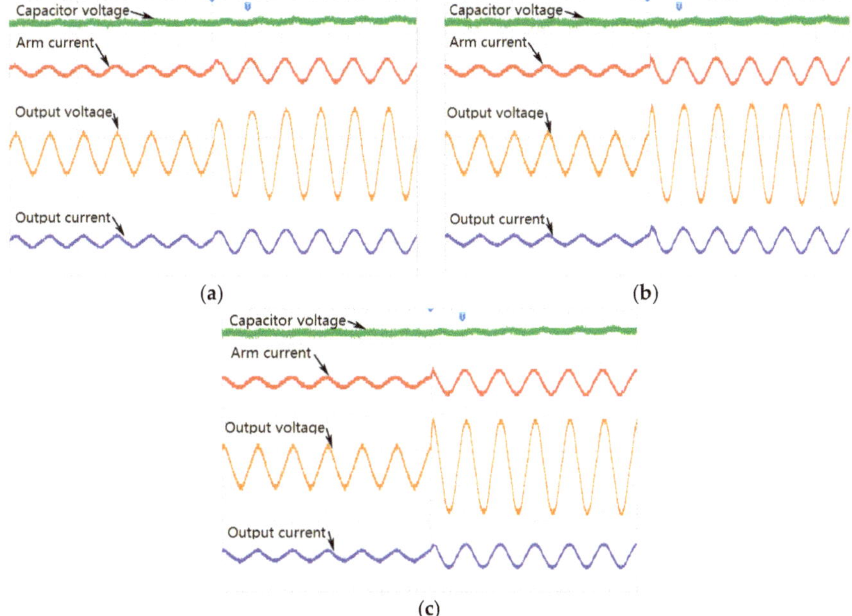

Figure 9. Experimental results of dynamic operation: (**a**) a fast MPC, (**b**) an indirect MPC, (**c**) the proposed R-MPC. Scopes: from top to bottom, first capacitor voltage of phase a (25 V/div), arm current of phase a (5 A/div), output voltage of phase a (100 V/div), and output current of phase a (10 A/div). Time scale: 20 ms/div.

6. Conclusions

In order to reduce the computational complexity of MPC for MMC, this paper proposed a R-MPC strategy. Compared with the fast MPC and indirect MPC, the number of control options to be calculated of proposed R-MPC was greatly reduced to one by calculating the number of on-state SMs directly based on the reverse prediction of arm voltages. The simulation and experimental results of steady-state operation showed that the fast MPC, an indirect MPC, and the proposed R-MPC all had good steady-state performance, but fast MPC and the proposed R-MPC had more advantages than indirect MPC because of less computational burden. However, the simulation and experimental results of dynamic operation showed that the proposed R-MPC had better dynamic-state performance than the fast MPC. Therefore, the proposed R-MPC strategy would be especially suitable for an MMC with hundreds of SMs because of less computational burden and good performance of both steady-state and dynamic operation.

Author Contributions: W.G. and D.L. put forward the idea and designed the proposed strategy; W.G. and F.R. completed the simulation and experiment; S.H. managed the project; and W.G. wrote the paper. All authors gave advice for the manuscript.

Funding: This research was funded in part by the State Key Program of National Natural Science of China (Grant number 51737004) and in part by the Natural Science Foundation of Hunan Province, China (Grant number 2018JJ2045).

Conflicts of Interest: The authors declare no conflict of interest.

References

1. Dekka, A.; Wu, B.; Fuentes, L.R.; Perez, M.; Zargari, R.N. Evolution of Topologies, Modeling, Control Schemes, and Applications of Modular Multilevel Converters. *IEEE J. Emerg. Sel. Top. Power Electron.* **2017**, *5*, 1631–1656. [CrossRef]
2. Grandi, G.; Loncarski, J.; Dordevic, O. Analysis and Comparison of Peak-to-Peak Current Ripple in Two-Level and Multilevel PWM Inverters. *IEEE Trans. Ind. Electron.* **2015**, *62*, 2721–2730. [CrossRef]
3. Zhetessov, A.; Ruderman, A. Time-domain optimization of current THD for a single-phase three-level inverter modulation. In Proceedings of the 19th International Conference on Electrical Drives and Power Electronics (EDPE), Dubrovnik, Croatia, 4–6 October 2017; pp. 82–87. [CrossRef]
4. He, L.; Zhang, K.; Xiong, J.; Fan, S. A Repetitive Control Scheme for Harmonic Suppression of Circulating Current in Modular Multilevel Converters. *IEEE Trans. Power Electron.* **2015**, *30*, 471–481. [CrossRef]
5. Zhang, M.; Huang, L.; Yao, W.; Lu, Z. Circulating Harmonic Current Elimination of a CPS-PWM-Based Modular Multilevel Converter with a Plug-In Repetitive Controller. *IEEE Trans. Power Electron.* **2014**, *29*, 2083–2097. [CrossRef]
6. Zeng, R.; Xu, L.; Yao, L.; Finney, J.S.; Wang, Y. Hybrid HVDC for Integrating Wind Farms with Special Consideration on Commutation Failure. *IEEE Trans. Power Deliv.* **2016**, *31*, 789–797. [CrossRef]
7. Dekka, A.; Wu, B.; Yaramasu, V.; Fuentes, L.R.; Zargari, R.N. Model Predictive Control of High-Power Modular Multilevel Converters–An Overview. *IEEE J. Emerg. Sel. Top. Power Electron.* **2018**, (Early Access). [CrossRef]
8. Leuer, M.; Lönneker, M.; Böcker, J. Model predictive control strategy for multi-phase thyristor matrix converters—Advantages, problems and solutions. In Proceedings of the 18th European Conference on Power Electronics and Applications (EPE'16 ECCE Europe), Karlsruhe, Germany, 6–8 September 2016; pp. 1–10. [CrossRef]
9. Guo, P.; He, Z.; Yue, Y.; Xu, Q.; Huang, X.; Chen, Y.; Luo, A. A Novel Two-Stage Model Predictive Control for Modular Multilevel Converter with Reduced Computation. *IEEE Trans. Ind. Electron.* **2019**, *66*, 2410–2422. [CrossRef]
10. Qin, J.; Saeedifard, M. Predictive Control of a Modular Multilevel Converter for a Back-to-Back HVDC System. *IEEE Trans. Power Deliv.* **2012**, *27*, 1538–1547. [CrossRef]
11. Böcker, J.; Freudenberg, B.; The, A.; Dieckerhoff, S. Experimental Comparison of Model Predictive Control and Cascaded Control of the Modular Multilevel Converter. *IEEE Trans. Power Electron.* **2015**, *30*, 422–430. [CrossRef]
12. Dekka, A.; Wu, B.; Yaramasu, V.; Zargari, R.N. Model Predictive Control With Common-Mode Voltage Injection for Modular Multilevel Converter. *IEEE Trans. Power Electron.* **2017**, *32*, 1767–1778. [CrossRef]
13. Zhou, D.; Yang, S.; Tang, Y. A Voltage-Based Open-Circuit Fault Detection and Isolation Approach for Modular Multilevel Converters with Model-Predictive Control. *IEEE Trans. Power Electron.* **2018**, *33*, 9866–9874. [CrossRef]
14. Perez, A.M.; Rodriguez, J.; Fuentes, J.E.; Kammerer, F. Predictive Control of AC–AC Modular Multilevel Converters. *IEEE Trans. Ind. Electron.* **2012**, *59*, 2832–2839. [CrossRef]
15. Dekka, A.; Wu, B.; Yaramasu, V.; Zargari, R.N. Integrated model predictive control with reduced switching frequency for modular multilevel converters. *IET Electr. Power Appl.* **2017**, *11*, 857–863. [CrossRef]
16. Vatani, M.; Bahrani, B.; Saeedifard, M.; Hovd, M. Indirect Finite Control Set Model Predictive Control of Modular Multilevel Converters. *IEEE Trans. Smart Grid* **2015**, *6*, 1520–1529. [CrossRef]
17. Liu, P.; Wang, Y.; Cong, W.; Lei, W. Grouping-Sorting-Optimized Model Predictive Control for Modular Multilevel Converter with Reduced Computational Load. *IEEE Trans. Power Electron.* **2016**, *31*, 1896–1907. [CrossRef]
18. Gong, Z.; Dai, P.; Yuan, X.; Wu, X.; Guo, G. Design and Experimental Evaluation of Fast Model Predictive Control for Modular Multilevel Converters. *IEEE Trans. Ind. Electron.* **2016**, *63*, 3845–3856. [CrossRef]
19. Dekka, A.; Wu, B.; Yaramasu, V.; Zargari, R.N. Dual-Stage Model Predictive Control with Improved Harmonic Performance for Modular Multilevel Converter. *IEEE Trans. Ind. Electron.* **2016**, *63*, 6010–6019. [CrossRef]
20. Mahmoudi, H.; Aleenejad, M.; Ahmadi, R. Modulated Model Predictive Control of Modular Multilevel Converters in VSC-HVDC Systems. *IEEE Trans. Power Deliv.* **2018**, *33*, 2115–2124. [CrossRef]

21. Moon, J.; Gwon, J.; Park, J.; Kang, D.; Kim, J. Model Predictive Control with a Reduced Number of Considered States in a Modular Multilevel Converter for HVDC System. *IEEE Trans. Power Deliv.* **2015**, *30*, 608–617. [CrossRef]

22. Wang, Y.; Cong, W.; Li, M.; Li, N.; Cao, M.; Lei, W. Model predictive control of modular multilevel converter with reduced computational load. In Proceedings of the IEEE Applied Power Electronics Conference and Exposition–APEC, Fort Worth, TX, USA, 16–20 March 2014; pp. 1776–1779. [CrossRef]

23. Qin, J.; Saeedifard, M. Reduced Switching-Frequency Voltage-Balancing Strategies for Modular Multilevel HVDC Converters. *IEEE Trans. Power Deliv.* **2013**, *28*, 2403–2410. [CrossRef]

24. Tu, Q.; Xu, Z.; Xu, L. Reduced Switching-Frequency Modulation and Circulating Current Suppression for Modular Multilevel Converters. *IEEE Trans. Power Deliv.* **2011**, *26*, 2009–2017. [CrossRef]

25. Cortes, P.; Rodriguez, J.; Silva, C.; Flores, A. Delay Compensation in Model Predictive Current Control of a Three-Phase Inverter. *IEEE Trans. Ind. Electron.* **2012**, *59*, 1323–1325. [CrossRef]

Article

A Simplified Model Predictive Control for T-Type Inverter with Output LC Filter

Van-Quang-Binh Ngo [1,2], Minh-Khai Nguyen [3], Tan-Tai Tran [1], Young-Cheol Lim [1] and Joon-Ho Choi [1,*]

[1]   Department of Electrical Engineering, Chonnam National University, Gwangju 500-757, Korea; qbinhtam@gmail.com (V.-Q.-B.N.); trantantaikdd@gmail.com (T.-T.T.); yclim@chonnam.ac.kr (Y.-C.L.)
[2]   Department of Physics, College of Education, Hue University, Thua Thien Hue 530000, Vietnam
[3]   Department of Electrical Engineering, Chosun University, Gwangju 61452, Korea; nmkhai00@gmail.com
*   Correspondence: joono@chonnam.ac.kr

Received: 14 November 2018; Accepted: 19 December 2018; Published: 22 December 2018

Abstract: In this paper, a model predictive control scheme for the T-type inverter with an output LC filter is presented. A simplified dynamics model is proposed to reduce the number of the measurement and control variables, resulting in a decrease in the cost and complexity of the system. Furthermore, the main contribution of the paper is the approach to evaluate the cost function. By employing the selection of sector information distribution in the reference inverter voltage and capacitor voltage balancing, the execution time of the proposed algorithm is significantly reduced by 36% compared with conventional model predictive control without too much impact on control performance. Simulation and experimental results are studied and compared with conventional finite control set model predictive control to validate the effectiveness of the proposed method.

Keywords: finite control set model predictive control; T-type inverter; computational cost; LC filter; DC-link capacitor voltage balancing

---

1. Introduction

Recently, the multilevel converter has been widely applied to various applications such as renewable energy system, flexible AC transmission systems and electric drives thanks to its benefits: increase the power capacity of the converter and improve the quality of the system [1–3]. In particular, compared with the neutral-point-clamped (NPC) type, the T-type inverter topology has the advantage in terms of the efficiency for medium switching frequency [4–6]. Thus, the T-type inverter is considered to be an alternative solution for multilevel inverter. Like the NPC converter, the unbalance of neutral-point potential is the drawback of this topology which causes the distortion of the output voltage and current. However, several approaches have been introduced to solve this problem [7–10].

A linear controller with proportional-integral (PI) is typically applied to control the converter because of its simplicity and stability [11,12]. However, this approach has a low dynamic response and requires a complex modulation technique for balancing the DC-link capacitor voltage. Recently, direct power control [13] which uses a switching look-up table for determining the switching state has been introduced to improve the performance. Nonetheless, it requires a high sampling frequency to achieve an acceptable steady-state and high dynamic performances. To deal with this disadvantage, several control approaches have been proposed such as using direct power control with space vector modulation [14], fuzzy control [15], sliding mode control [16], and predictive control [17–19].

In recent years, a finite control set model predictive control (FCS-MPC) is considered as an attractive alternative control strategy for power converters due to its simple structure, facilitating implementation, and fast dynamic response [20–25]. Furthermore, compared with classical control, the FCS-MPC provides the advantages such as easy inclusion of nonlinearities and constraints in the controller. However, at each sampling period, the prediction of control variables is 27, corresponding to the three-level T-type inverter, leading to producing a high computational cost. In [26], a simplified FCS-MPC for three-level voltage source converter is introduced. In order to reduce the computational time, this approach used the two-level switching state group for prediction and optimization. Another approach is presented in [27], which employs equivalent transformations in the cost function for the optimization loop. Another approach is proposed in [28,29] based on using a modified sphere decoding algorithm for multilevel converters. In [30], a sector distribution and non-zero voltage vectors are exploited with the aim to reduce the computational burden for two-level converters. Nonetheless, the main disadvantage of this method is the nonexistence of zero voltage leading to an increase of the total harmonic distortion (THD) in the load current. In [31], the control approach is suggested based on the candidate region that minimizes the sub-cost function to reduce the execution time. The presented technique in [32] combined the conventional FCS-MPC, a look-up table, and steady-state evaluation to reduce the computational burden. However, this algorithm can have a large amount of computational cost like the conventional FCS-MPC in the worst case.

With a three-level T-type inverter, control variables are predicted by using the predictive model and measured variables such as DC-link capacitor voltage, output voltage, filter current, and output current. In order to reduce the cost and complexity of this system, a simplified dynamics model is presented in this paper. Moreover, the highlight of this research is the significant computational cost reduction without decreasing the quality of control by preselecting the required inverter output voltage. The balance of DC-link capacitor voltage is guaranteed by determining the suitable small voltage vectors resulting in the elimination of the weighting factor in the cost function. As a consequence, the amount of predictive state for loop optimization is reduced from 27 to 6 compared with the conventional FCS-MPC method. This means that it is easy to implement the proposed algorithm in a real-time system with a low-cost processor and to extend with a long prediction horizon for improving the control performance. The simulation and experimental results validate the effectiveness of the proposed control strategy.

The rest of this paper is organized as follows: a reduced model predictive control for the three-level T-type inverter is presented in Section 2. Next, the proposed algorithm is explained in Section 3 for reducing the computational cost. In Section 4, a comparative study of the conventional FCS-MPC and the proposed method is examined. Finally, the conclusions are given in Section 5.

2. Model Predictive Control for a Three-Level T-Type Inverter

2.1. Topology

A simple topology of the three-level T-type inverter (3L-T-type) is shown in Figure 1. The basic principle of this configuration can be expressed by three switching states [P], [N] and [O] which correspond to three inverter output voltages $+U_{dc}/2$, $-U_{dc}/2$ and 0. Consequently, 27 possible switching configurations are considered for a 3L-T-type inverter. Table 1 presents the summary of the operating principle for 3L-T-type.

Table 1. Operating status of inverter leg $x \in \{a, b, c\}$.

State	Switch				Inverter Output Voltage
S_x	S_{1x}	S_{2x}	S_{3x}	S_{4x}	u_{xZ}
P	1	1	0	0	$U_{dc}/2$
O	0	1	1	0	0
N	0	0	1	1	$-U_{dc}/2$

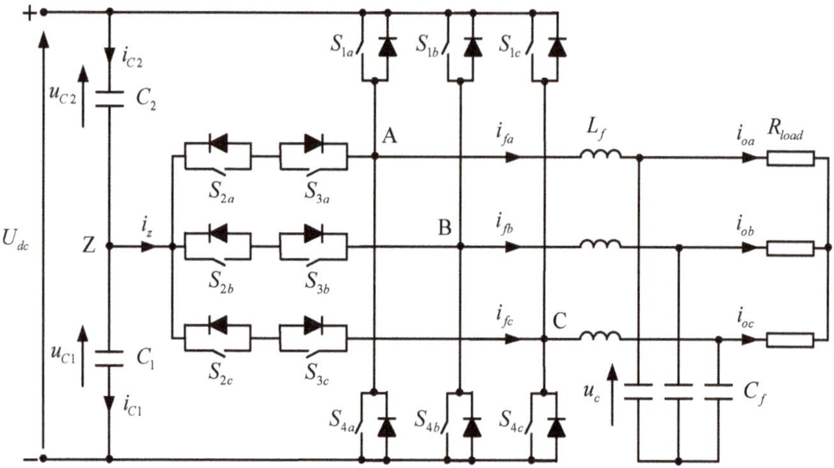

Figure 1. Three-level T-type inverter topology.

2.2. Mathematical Modeling of the System

The inverter output voltage produced by the 3L-T-type inverter is given by:

$$u_{inv} = \frac{2}{3}\left(u_{AZ} + ku_{BZ} + k^2 u_{CZ}\right),\tag{1}$$

where u_{AZ}, u_{BZ}, and u_{CZ} are the output phase voltages; $k = e^{j2\pi/3} = -\frac{1}{2} + j\frac{\sqrt{3}}{2}$.

The phase voltage u_{xZ} is calculated in terms of DC-link voltage U_{dc} and switching state S_x as [9,22]:

$$u_{xZ} = S_x \frac{U_{dc}}{2},\tag{2}$$

where S_x represents the switching status and has three possible values: $\{-1, 0, 1\}$ with the index $x \in \{a, b, c\}$.

The dynamic behavior of LC filter can be described by the following:

$$L_f \frac{di_f}{dt} = u_{inv} - u_c,\tag{3}$$

$$C_f \frac{du_c}{dt} = i_f - i_o,$$

where u_{inv} and u_c are the inverter and output capacitor voltage vectors; i_f and i_o are the filter and output load current vectors and L_f, C_f are the filter inductance and capacitance.

The control variables u_c and i_f are measured while u_{inv} is obtained from Equations (1) and (2). In general, i_o is measured or estimated by using an observer, leading to an increase in the cost and complexity of the system. In this paper, to achieve a simple model, we assume that the output load current is derived from output capacitor voltage. Thus, Equation (3) is rewritten as:

$$\frac{du_c}{dt} = \frac{1}{C_f}\left(i_f - \frac{u_c}{R_{Load}}\right),\tag{4}$$

$$\frac{di_f}{dt} = \frac{1}{L_f}\left(u_{inv} - u_c\right),$$

where R_{load} is the load resistance.

In order to reduce the number of the control variable, the neutral-point voltage is taken into account in the model instead of two capacitor voltages (u_{C1}, u_{C2}). The neutral-point voltage (u_z) can be expressed based on the assumption that the DC-link voltage is kept constant and $C_1 = C_2 = C$ as follows:

$$\frac{du_z}{dt} = \frac{d\left(u_{C1} - u_{C2}\right)}{dt} = -\frac{1}{C}i_z = -\frac{1}{C}\left((1 - |S_a|)i_{fa} + (1 - |S_b|)i_{fb} + (1 - |S_c|)i_{fc}\right). \tag{5}$$

Consequently, we have a representation of the dynamics model based on Equations (4) and (5) as:

$$\frac{du_c}{dt} = \frac{1}{C_f}\left(i_f - \frac{u_c}{R_{Load}}\right), \tag{6}$$

$$\frac{di_f}{dt} = \frac{1}{L_f}\left(u_{inv} - u_c\right),$$

$$\frac{du_z}{dt} = -\frac{1}{C}\left((1 - |S_a|)i_{fa} + (1 - |S_b|)i_{fb} + (1 - |S_c|)i_{fc}\right).$$

3. Model Predictive Control with Selection Sector Distribution

The main goal of the proposed control scheme is to minimize the error between the predicted output voltage and its reference value and to maintain capacitor voltage balancing. Furthermore, additional terms can be taken into account in the objective function such as switching frequency, current limitation, but this is not the main focus of this research and will not be developed here. As a result, the cost function for 3L-T-type inverter is expressed as [21–23]:

$$g = \left|u_{c\alpha}^*(k+1) - u_{c\alpha}^p(k+1)\right| + \left|u_{c\beta}^*(k+1) - u_{c\beta}^p(k+1)\right| + \lambda_{uz}\left|u_z^p(k+1)\right|, \tag{7}$$

where $u_{c\alpha}^*(k+1)$, $u_{c\beta}^*(k+1)$ and $u_{c\alpha}^p(k+1)$, $u_{c\beta}^p(k+1)$ indicate the real and imaginary components of the reference and predicted output capacitor voltages at instant $k+1$, respectively. λ_{uz} is the weighting factors of the capacitor voltage balancing.

To achieve the discrete-time model, the first-order Euler approximation is used as:

$$\frac{dx}{dt} = \frac{x(k) - x(k-1)}{T_s}, \tag{8}$$

where T_s is the sampling time.

By approximating Equation (6) with Equation (8), the discrete-time representation of output capacitor voltage can be obtained as:

$$u_c(k) = \frac{T_s R_{load}}{C_f R_{load} + T_s}i_f(k) + \frac{C_f R_{load}}{C_f R_{load} + T_s}u_c(k-1). \tag{9}$$

By shifting the output voltage in Equation (9) into one future sample, we have the predicted output voltage at instant $k+1$:

$$u_c^p(k+1) = \frac{T_s R_{load}}{C_f R_{load} + T_s}i_f^p(k+1) + \frac{C_f R_{load}}{C_f R_{load} + T_s}u_c(k). \tag{10}$$

The discrete-time form for the filter current is given by using the forward Euler approximation as:

$$i_f^p(k+1) = i_f(k) + \frac{T_s}{L_f}\left(u_{inv}(k) - u_c(k)\right). \tag{11}$$

Similarly, the discrete-time of neutral-point potential is expressed by:

$$u_z^p(k+1) = u_z(k) - \frac{T_s}{C}\left((1-|S_a|)\,i_{fa}(k) + (1-|S_b|)\,i_{fb}(k) + (1-|S_c|)\,i_{fc}(k)\right). \tag{12}$$

Substituting Equation (11) into Equation (10), the predicted output voltage is rewritten as:

$$u_c^p(k+1) = \frac{T_s R_{load}}{C_f R_{load} + T_s}\,i_f(k) + \frac{T_s^2 R_{load}}{L_f\left(C_f R_{load} + T_s\right)}\,u_{inv}(k) + \frac{R_{load}}{C_f R_{load} + T_s}\left(C_f - \frac{T_s^2}{L_f}\right)u_c(k). \tag{13}$$

A control input is a sequential switch state $S_p = \begin{bmatrix} S_{pa} & S_{pb} & S_{pc} \end{bmatrix}^T$, symbolized as a set of p vector $S_p \in \{1,...,27\}$. Furthermore, the switching inputs a finite set: $S_{px} \in \{-1,0,1\}$ with the index $x \in \{a,b,c\}$. As a result, the optimal switching input S_{opt} is achieved as the result of Equation (14):

$$\begin{aligned} S_{opt} = \arg\{\min g\}, \quad p = 1,...,27 \\ \text{subject to (7), (12) and (13).} \end{aligned} \tag{14}$$

The space voltage vector of 3L-T-type inverter can be classified into four groups: zero vectors (from u_{25} to u_{27}), small vectors (from u_{13} to u_{24}), medium vectors (u_2, u_4, u_6, u_8, u_{10}, and u_{12}) and large vectors (u_1, u_3, u_5, u_7 and u_9), wherein the small vectors are divided into two types: positive state (P) and negative state (N) such as u_{14} and u_{13}, respectively. The neutral-point voltage is increased with the positive state and decreased with the negative state, respectively [33]. The zero, medium and large vectors do not affect the neutral-point voltage deviation. In the conventional FCS-MPC, the capacitor voltage balancing can be solved by adjusting the weighting factor in the cost function. However, it is not easy to obtain the optimal weighting factor value leading to affecting the THD of the load current. In this study, the capacitor voltages are balanced by selecting the suitable small vectors that depend on the predicted neutral-point voltage. Therefore, the proposed method is simple due to no requirement of the weighting factor for balancing capacitor voltages in the cost function.

For the 3L-T-type inverter, 27 switching states are considered to evaluate the cost function. Long prediction horizon can improve the control performance. However, the computational cost is increased exponentially corresponding to the prediction horizon. Therefore, it leads to a large computational cost which makes it difficult to implement the algorithm in common digital signal processing. In this paper, the selection of sector distribution is employed with the aim to solve this problem. The main idea of the proposed method is to determine the position of inverter reference voltage which is obtained from the predictive model. In this case, the required inverter voltage $u_{inv}^*(k)$ is achieved based on Equation (13) by replacing the predicted output voltage $u_c^p(k+1)$ with its reference. Then, the location of the reference voltage $u_{inv}^*(k)$ is determined by its components $u_{inv\alpha}^*$ and $u_{inv\beta}^*$. In the proposed method, we divide the space vector of the 3L T-type inverter into six sectors as illustrated in Figure 2. For example, when the reference voltage $u_{inv}^*(k)$ is in sector I, there are only 10 voltage vectors which are selected for the evaluation of the cost function. As previously discussed, the neutral-point voltage is predicted based on the previous optimal switching states and filter currents by using Equation (5). In order to achieve the balance of capacitor voltages, two cases are considered: the first one corresponds to $u_z \le 0$ and the second one to $u_z > 0$. The positive small vectors (u_{14}, u_{15}) and negative small vectors (u_{13}, u_{16}) are considered with the condition $u_z \le 0$ and $u_z > 0$, respectively. Zero vectors can reduce from 3 to 1 due to the same value and without the effect of voltage imbalance. In this case, the feasible voltage vectors are u_1, u_2, u_3, u_{14}, u_{15}, u_{25} for $u_z \le 0$, whereas they are u_1, u_2, u_3, u_{13}, u_{16}, u_{25} for $u_z > 0$, respectively. Table 2 illustrates the available inverter voltage vectors for a 3L-type inverter after obtaining the appropriate sector. Thus, the prediction of the control variable for cost function loop optimization is decreased from 27 to 6 with the proposed method. As a result, compared with the conventional FCS-MPC method, the computational cost is appreciably reduced by about 77% in the proposed algorithm. It is obvious that this advantage is more attractive to real-time implementation with low-cost digital hardware and long prediction horizon. The overall control strategy of the proposed method is shown in Figure 3. Then, the optimal switching state is applied to the inverter by minimizing this cost function:

$$g_{mdf} = \left| u_{c\alpha}^*(k+1) - u_{c\alpha}^p(k+1) \right| + \left| u_{c\beta}^*(k+1) - u_{c\beta}^p(k+1) \right|, \tag{15}$$

$$S_{opt} = \arg\{\min g_{mdf}\}, \quad p = 1,...,6.$$

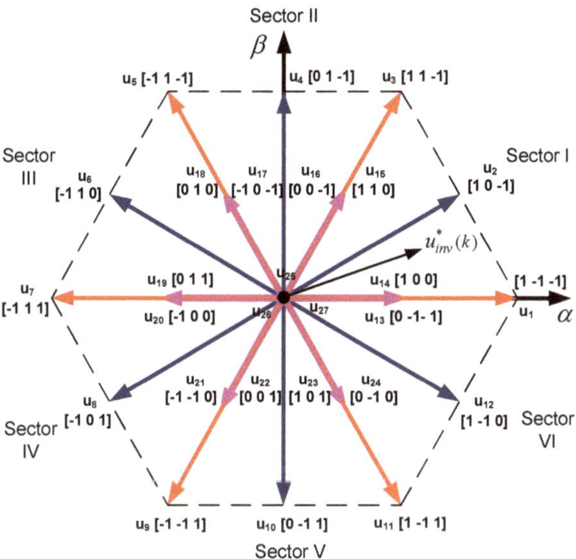

Figure 2. Voltage vectors distribution of T-type inverter.

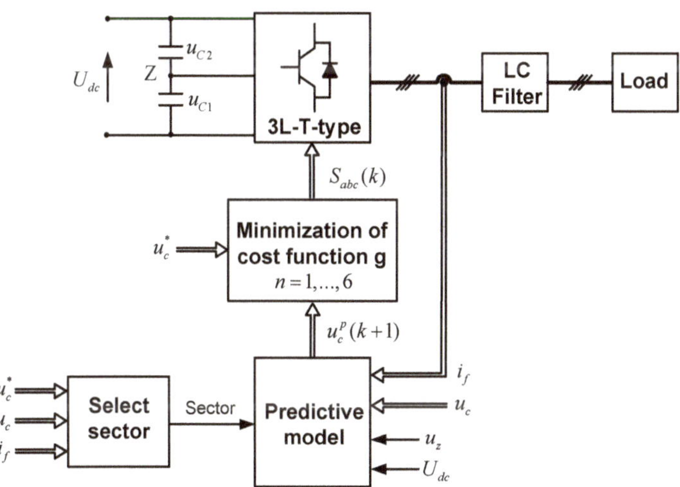

Figure 3. Block diagram of the proposed control strategy.

Table 2. Feasible voltage vectors for each sector.

Sector	Feasible Voltage Vectors	
	$u_z \leq 0$	$u_z > 0$
I	$u_1, u_2, u_3, u_{14}, u_{15}, u_{25}$	$u_1, u_2, u_3, u_{13}, u_{16}, u_{25}$
II	$u_3, u_4, u_5, u_{15}, u_{18}, u_{25}$	$u_3, u_4, u_5, u_{16}, u_{17}, u_{25}$
III	$u_5, u_6, u_7, u_{18}, u_{19}, u_{25}$	$u_5, u_6, u_7, u_{17}, u_{20}, u_{25}$
IV	$u_7, u_8, u_9, u_{19}, u_{22}, u_{25}$	$u_7, u_8, u_9, u_{20}, u_{21}, u_{25}$
V	$u_9, u_{10}, u_{11}, u_{22}, u_{23}, u_{25}$	$u_9, u_{10}, u_{11}, u_{21}, u_{24}, u_{25}$
VI	$u_{11}, u_{12}, u_1, u_{23}, u_{14}, u_{25}$	$u_{11}, u_{12}, u_1, u_{24}, u_{13}, u_{25}$

Finally, the proposed control algorithm is described in Figure 4.

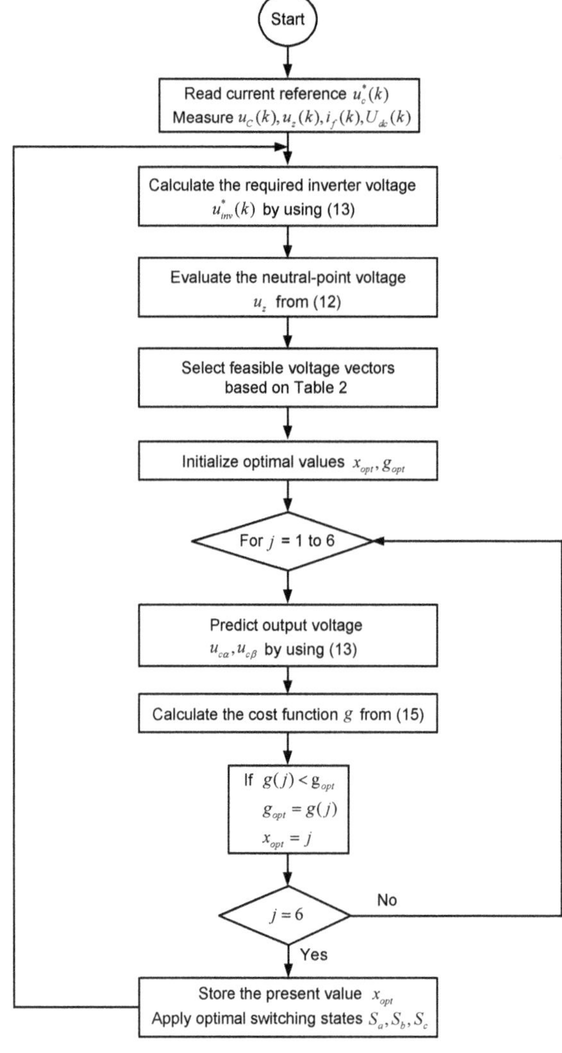

Figure 4. Flowchart of the proposed control strategy.

4. Simulation and Experimental Results

4.1. Simulation Results

Simulation analyses were performed in a Matlab/Simulink environment with version 2015a to verify the control performance of the proposed strategy for the T-type inverter as illustrated in Figure 5. The SimPowerSystems toolbox was used to create the 3L-T-type inverter with output LC filter. The Matlab Function block is employed to easily implement the control algorithm in the simulation environment. The parameters of the system are listed in Table 3.

Table 3. System parameters.

Parameter	Value	Description
U_{dc}	600 [V]	DC-link voltage
C	1000 [μF]	DC-link capacitance
L_f	3 [mH]	Filter inductance
C_f	40 [μF]	Filter capacitance
R_{load}	20 [Ω]	Load resistance
f_s	20 [kHz]	Sampling frequency
f	50 [Hz]	Frequency of the grid

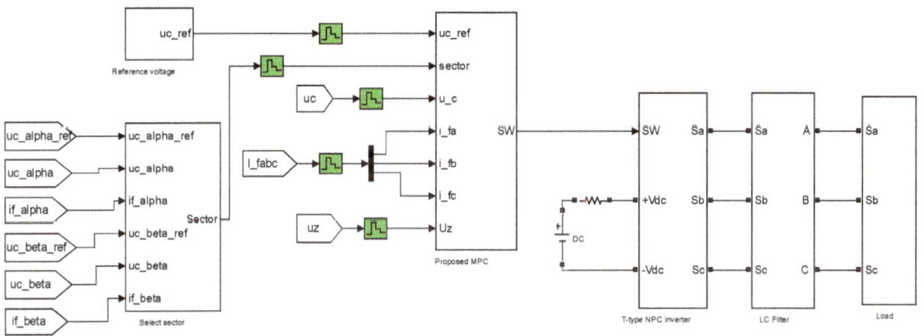

Figure 5. Block diagram of the proposed strategy in Matlab/Simulink.

Figure 6 shows the steady-state of the proposed method with the output voltage at 155 V. As depicted in Figure 6, the proposed method obtains the sinusoidal output voltage and the balance of DC-link capacitor voltage. The characteristic of the variable switching frequency is illustrated in Figure 6c. This can increase the THD of the load current, but this does not affect the control performance too much. The THD of the load current can receive further improvements by using alternative methods. However, this is not the main focus of this paper and will not be developed here.

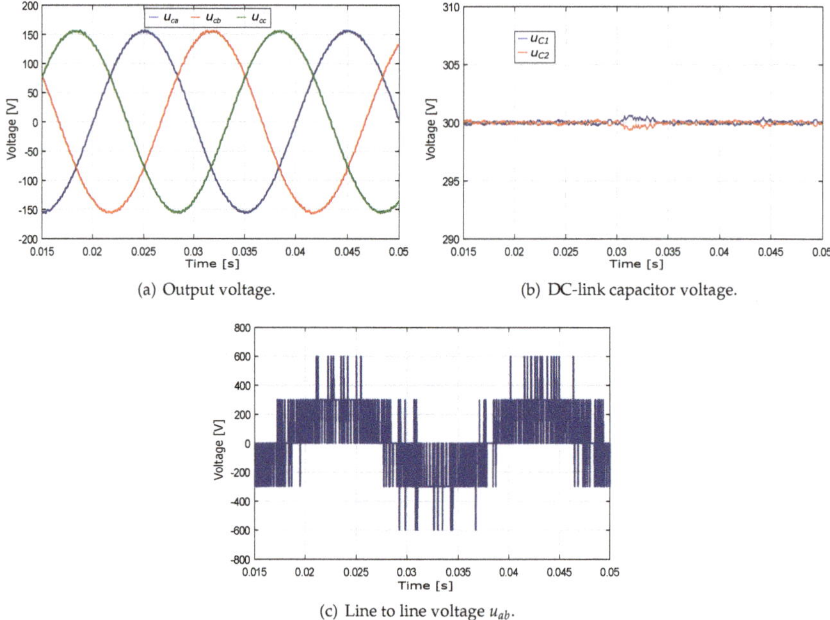

(a) Output voltage.

(b) DC-link capacitor voltage.

(c) Line to line voltage u_{ab}.

Figure 6. Steady-state of the proposed method at the output voltage of 155 V.

In order to show the efficiency of the control strategy, a comparison between the proposed method and the conventional FCS-MPC [23] were carried out under different operating conditions and the same parameters. The amplitude of the reference voltage changed from 155 to 311 V in the first scenario and stepped from 311 to 155 V at $t = 0.03$ s in the second scenario as illustrated in Figures 7 and 8. The corresponding dynamic current response is shown in Figures 7b and 8b. As can be seen, it is clear that the proposed method achieves sinusoidal current with the different reference amplitude. In addition, one important issue associated with the T-type inverter is the balance of DC-link capacitor voltage. Figures 7c and 8c indicate that the voltage of the DC-link capacitor is balanced despite the change in reference. The maximum absolute error of this voltage at steady-state are about 1 and 3 V for output voltage of 155 and 311 V, respectively. Figure 9 demonstrates single phase output voltage of the proposed and conventional FCS-MPC methods. The simulation results indicate the ability of the proposed method to accurately track and accomplish the steady-state with a fast dynamic response.

With the aim to evaluate the steady-state performance, the harmonic spectra of load current for the conventional FCS-MPC and proposed methods are also examined in Figure 10a,b. These figures show that the THD of the load current is increased slightly from 0.45% to 0.58% with the proposed method. The comparison of two control methods is summarized in Table 4. Although the THD of the load current is not perfect, we nevertheless believe that the slight increase does not affect the control performance too much. Specifically, the computation time of the proposed algorithm is greatly reduced compared with the conventional FCS-MPC as shown in Figure 11a. In fact, the minimum, average and maximum computation times of the proposed algorithm are 3, 6 and 9 µs in a 2.0 GHz, i5 4310 CPU, while their corresponding values are 4, 10 and 16 µs with conventional FCS-MPC. The performance of FCS-MPC method is influenced by the sampling time which is improved by choosing the smaller value. To investigate the effect of sampling time on the quality of the current, two controllers are employed with different sampling times. Figure 11b shows that the quality of load current is the best with sampling time 40 µs and the worst with 100 µs. However, there is a limitation of sampling time due to the requirement of execution time such as computation time and measurement of the signal.

Therefore, this method exhibits a valuable alternative to reduce the sampling time and extend with a long prediction horizon, which improves the control performance.

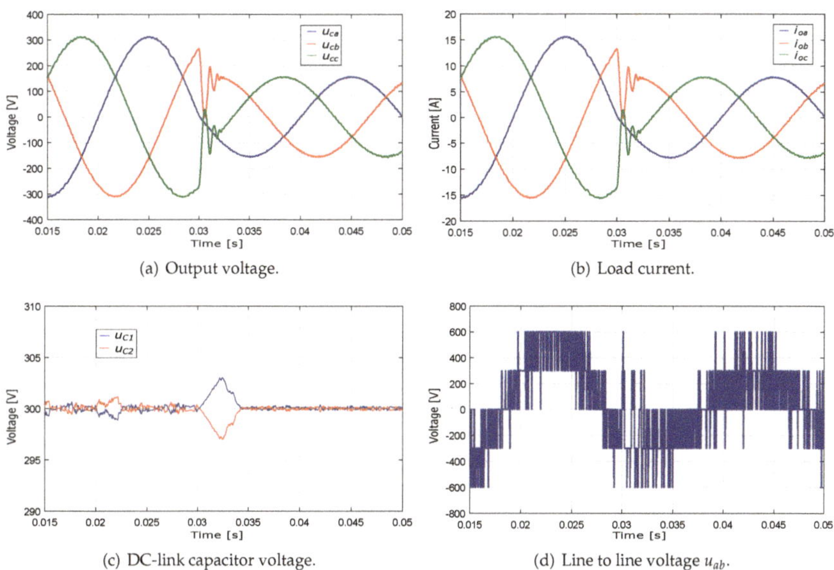

(a) Output voltage.

(b) Load current.

(c) DC-link capacitor voltage.

(d) Line to line voltage u_{ab}.

Figure 7. The dynamic response of the proposed method for step change from 311 to 155 V.

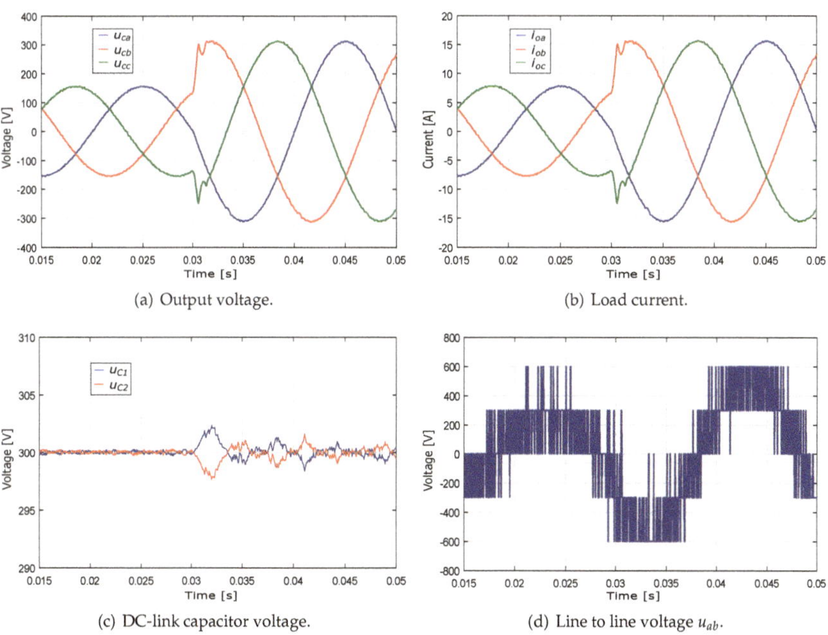

(a) Output voltage.

(b) Load current.

(c) DC-link capacitor voltage.

(d) Line to line voltage u_{ab}.

Figure 8. The dynamic response of the proposed method for step change from 155 to 311 V.

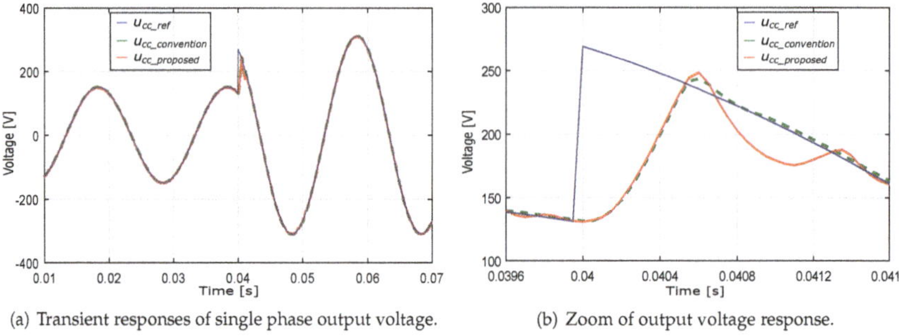

(a) Transient responses of single phase output voltage.

(b) Zoom of output voltage response.

Figure 9. The dynamic response of the output voltage for the conventional FCS-MPC and proposed methods.

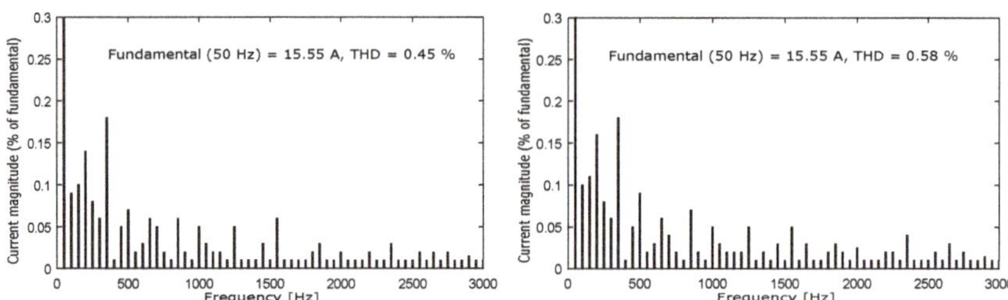

(a) Current harmonic spectrum of conventional FCS-MPC. (b) Current harmonic spectrum of the proposed method.

Figure 10. The harmonic spectrum of the load current for the conventional FCS-MPC and proposed methods.

Table 4. Comparison of transient performance for two controllers.

Reference Step	$u_c^* = 155 \rightarrow 311$ (V)	
	Conventional FCS-MPC	Proposed Method
State of loop optimization	27	6
Rise time (ms)	0.5	0.5
Settling time (ms)	0.7	1.3
THD of current (%)	0.45	0.58

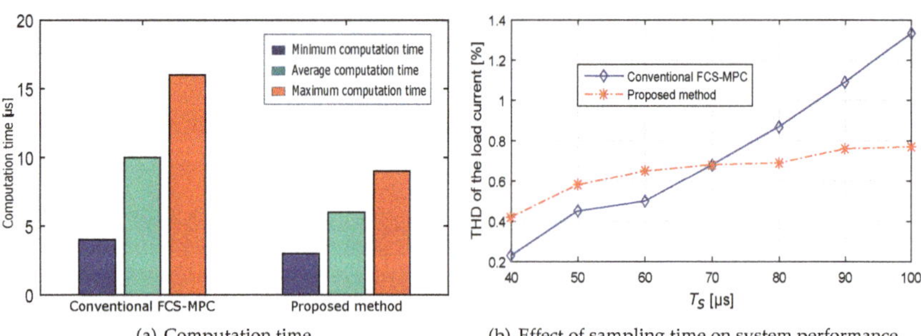

(a) Computation time.

(b) Effect of sampling time on system performance.

Figure 11. Comparison of two control methods.

The behavior of the system is also examined under time varying load step as illustrated in Figure 12. At the initial state, the system operates at no load condition; then, the load is set to 20 Ω at

$t = 0.02$ s. According to Figure 12, this change does not impact on the quality of the output voltages. A resistive-inductive load is imposed for the same test as shown in Figure 13. The load resistance and inductance are set to 40 Ω and 10 mH, respectively. It can be seen from Figure 13 that no deterioration of output voltage is observed in this case.

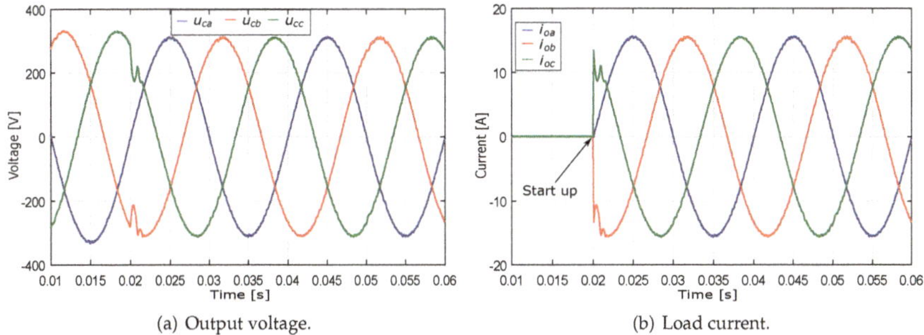

(a) Output voltage.

(b) Load current.

Figure 12. Output voltage and current with the resistive load step at $t = 0.02$ s.

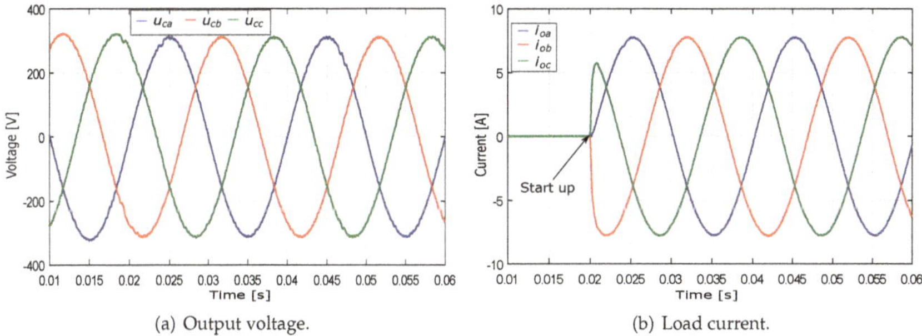

(a) Output voltage.

(b) Load current.

Figure 13. Output voltage and current with the resistive–inductive load step at $t = 0.02$ s.

To investigate the influence of frequency variations, a step change in the voltage from 60 Hz to 50 Hz at $t = 0.03$ s with $R_{load} = 10\ \Omega$ is examined in this study. Figure 14 indicates that the proposed method can achieve a reasonable reference tracking despite the sudden change in the frequency.

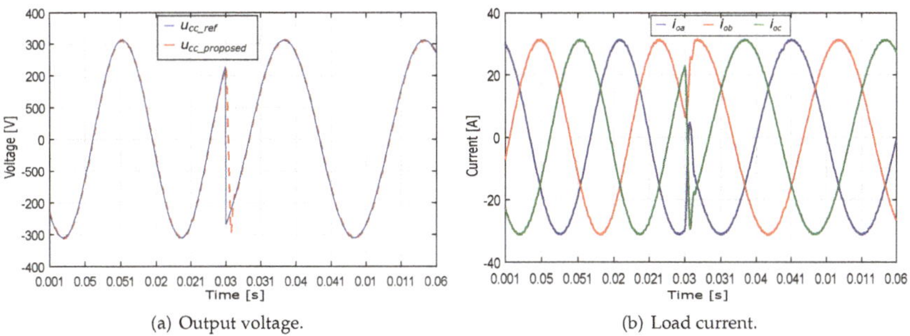

(a) Output voltage.

(b) Load current.

Figure 14. The output voltage and current responses under dynamic change in frequency.

A nonlinear load test is also performed in this study with a diode rectifier and resistive-inductive load ($R = 60\,\Omega$, $L_{nl} = 10$ mH) as shown in Figure 15c. Figure 15 illustrates that the output voltages give a small distortion, but it still acquires sinusoidal in spite of the high distorted load currents.

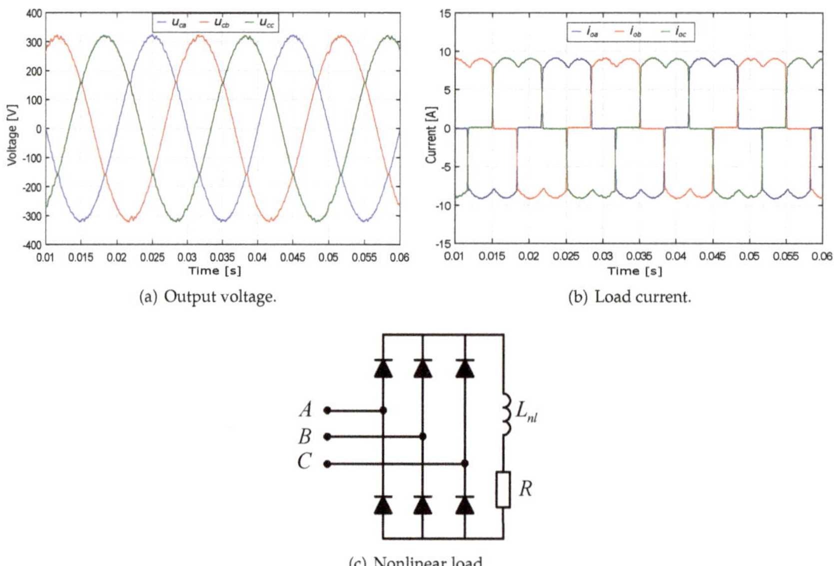

(a) Output voltage.

(b) Load current.

(c) Nonlinear load.

Figure 15. Output voltage and current with nonlinear load.

To confirm the robustness of the controller against parameter variations, we have considered a change of parameters with two cases. In the first case, the filter inductance and capacitance have been decreased to 40% of their real values as illustrated in Figure 16a. On the other hand, the load resistance has been increased to 50% of its value as shown in Figure 16b. It can be observed that the proposed method is continued to obtain sinusoidal current with small deviations. The load current increases from 0.58% to 1.5%, but it still meets within the limit required of the IEEE 519 standard.

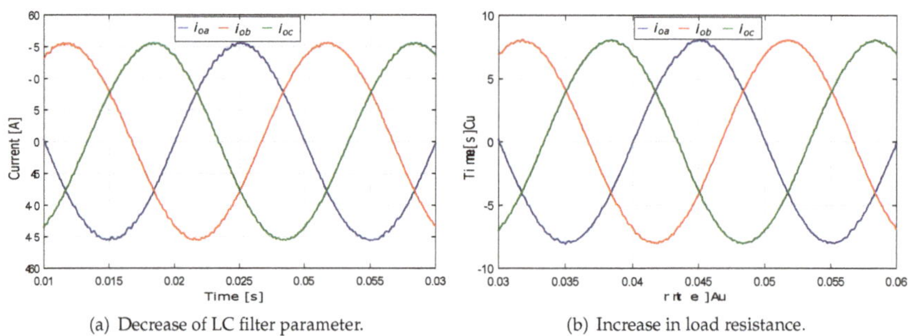

(a) Decrease of LC filter parameter.

(b) Increase in load resistance.

Figure 16. Output current with parameter variations.

4.2. Experimental Results

In order to validate the effectiveness of the proposed control strategy, a laboratory prototype with small power was constructed as shown in Figure 17. A digital signal processor TMS320F28335 [34] was employed to implement the control method. The algorithm was programmed using S-function builder block in the Matlab/Simulink with embedded coder tools [35]. Twelves modules FGH40T120SMD for IGBT were applied in the three-phase inverter. Furthermore, two capacitors B43305A9108M 1000 µF-400 V were used for DC-link voltage. The parameters of the LC filter were maintained at 3 mH and 40 µF. The LV 25-P and LA 25-P sensors were used to measure the output voltage, filter current and capacitor voltages. The digital signal processing (DSP) generates the signals for 12 switches of 3L-T-type inverter via general-purpose input/output (GPIO) outputs.

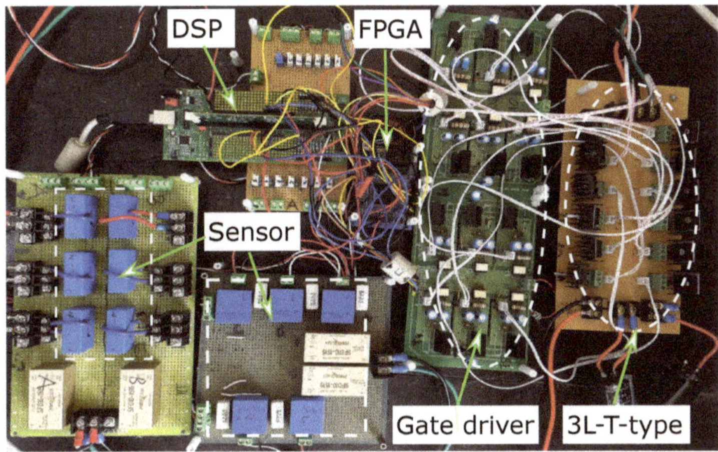

Figure 17. Experimental test bench in the laboratory.

The DC input voltage is set at 180 V while the load resistance is kept at 30 Ω. The reference of the peak phase output voltage is stepped from 90 to 60 V corresponding to the change in output current from 3 to 2 A. Figure 18 indicates that the proposed method has a fast dynamic response and a good balance of DC-link capacitor voltage. As illustrated in Figure 19, the steady-state of three-phase sinusoidal load current confirms the control performance of the proposed method. Furthermore, the execution time of the proposed and conventional FCS-MPC methods are 41 and 64 µs, respectively, as shown in Figure 20. This highlights that the execution time is effectively reduced 36% by the proposed method. Therefore, the sampling time of the conventional method is increased compared with the proposed method resulting in a decrease in the quality of control performance. The load current of the conventional FCS-MPC is depicted in Figure 21. In this case, the THD of the load current of the proposed method is reduced from 1.6% to 1.0% compared with the conventional method as illustrated in Figure 22. Thus, the better performance of the proposed algorithm can be obtained with the low-cost processor.

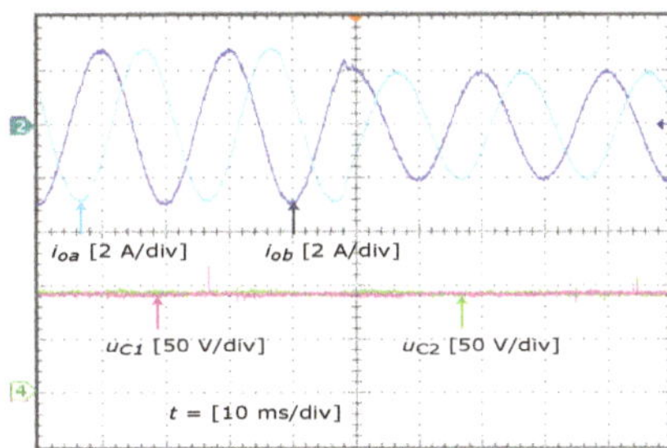

Figure 18. Experimental results for step change in the output voltage.

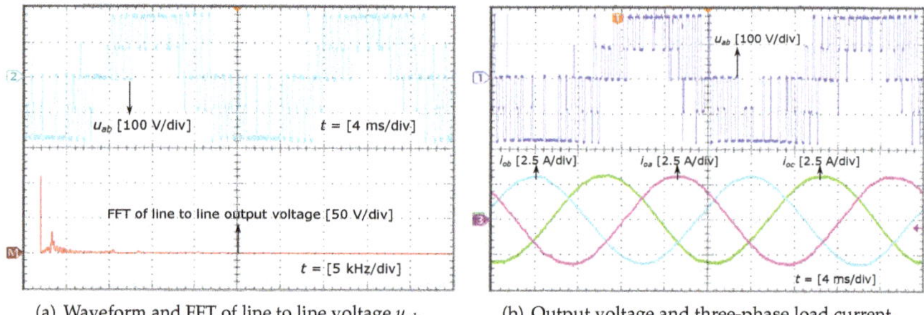

(a) Waveform and FFT of line to line voltage u_{ab}.   (b) Output voltage and three-phase load current.

Figure 19. Experimental results of the proposed method with $I_f = 3$ A.

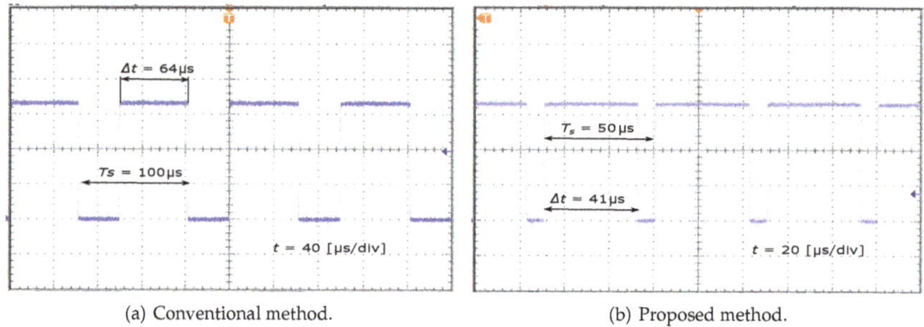

(a) Conventional method.   (b) Proposed method.

Figure 20. Execution time of the conventional and proposed methods.

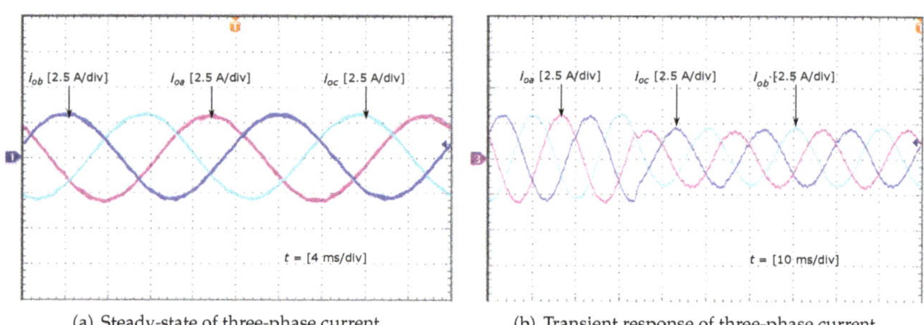

(a) Steady-state of three-phase current.   (b) Transient response of three-phase current.

Figure 21. Experimental results of the conventional FCS-MPC.

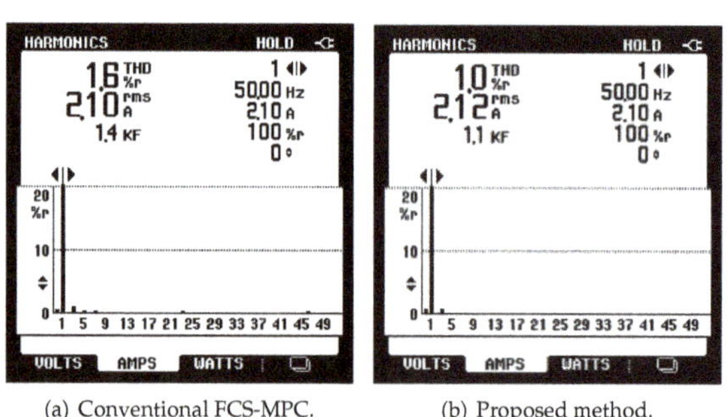

(a) Conventional FCS-MPC.   (b) Proposed method.

Figure 22. THD of load current.

5. Conclusions

This paper presents a simplified model predictive control method for a three-level T-type inverter. A reduced dynamics model is proposed to decrease the cost and the complexity of the system. Moreover, the execution time is greatly reduced compared with the conventional FCS-MPC by applying the preselection of reference inverter voltage and capacitor voltage balancing, allowing an easy real-time implementation. In order to show the effectiveness of the control strategy, a comparative study of the proposed method and conventional FCS-MPC is performed. Simulation and experimental results prove the feasibility of the proposed approach.

Author Contributions: V.-Q.-B.N. established the major part of this paper which includes modeling, simulation investigation, and original draft preparation. M.-K.N. contributed in review and editing. T.-T.T. contributed to validating in the real system. Y.-C.L. provided resources and supervision. J.-H.C. provided resources and funding acquisition.

Acknowledgments: This research was supported by Korea Electric Power Corporation (Grant number: R18XA04).

Conflicts of Interest: The authors declare no conflict of interest

References

1. Blaabjerg, F.; Liserre, M.; Ma, K. Power Electronics Converters for Wind Turbine Systems. *IEEE Trans. Ind. Appl.* **2012**, *48*, 708–719. [CrossRef]
2. Portillo, R.C.; Prats, M.M.; Leon, J.I.; Sanchez, J.A.; Carrasco, J.M.; Galvan, E.; Franquelo, L.G. Modeling Strategy for Back-to-Back Three-Level Converters Applied to High-Power Wind Turbines. *IEEE Trans. Ind. Electron.* **2006**, *53*, 1483–1491. [CrossRef]
3. Kouro, S.; Malinowski, M.; Gopakumar, K.; Pou, J.; Franquelo, L.G.; Wu, B.; Rodriguez, J.; Perez, M.A.; Leon, J.I. Recent Advances and Industrial Applications of Multilevel Converters. *IEEE Trans. Ind. Electron.* **2010**, *57*, 2553–2580. [CrossRef]
4. Schweizer, M.; Kolar, J.W. Design and Implementation of a Highly Efficient Three-Level T-Type Converter for Low-Voltage Applications. *IEEE Trans. Power Electron.* **2013**, *28*, 899–907. [CrossRef]
5. Lee, K.; Shin, H.; Choi, J. Comparative analysis of power losses for 3-Level NPC and T-type inverter modules. In Proceedings of the 2015 IEEE International Telecommunications Energy Conference (INTELEC), Osaka, Japan, 18–22 October 2015; pp. 1–6.
6. Qin, C.; Zhang, C.; Chen, A.; Xing, X.; Zhang, G. A Space Vector Modulation Scheme of the Quasi-Z-Source Three-Level T-Type Inverter for Common-Mode Voltage Reduction. *IEEE Trans. Ind. Electron.* **2018**, *65*, 8340–8350. [CrossRef]
7. Vahedi, H.; Labbe, P.A.; Al-Haddad, K. Balancing three-level NPC inverter DC bus using closed-loop SVM Real time implementation and investigation. *IET Power Electron.* **2016**, *9*, 2076–2084. [CrossRef]
8. Xiang, C.; Shu, C.; Han, D.; Mao, B.; Wu, X.; Yu, T. Improved Virtual Space Vector Modulation for Three-Level Neutral-Point-Clamped Converter With Feedback of Neutral-Point Voltage. *IEEE Trans. Power Electron.* **2018**, *33*, 5452–5464. [CrossRef]
9. Ngo, B.Q.V.; Ayerbe, P.R.; Olaru, S. Model Predictive Control with Two-step horizon for Three-level Neutral-Point Clamped Inverter. In Proceedings of the 20th International Conference on Process Control, Strbske Pleso, Slovakia, 9–12 June 2015; pp. 215–220.
10. In, H.C.; Kim, S.M.; Lee, K.B. Design and Control of Small DC-Link Capacitor-Based Three-Level Inverter with Neutral-Point Voltage Balancing. *Energies* **2018**, *11*, 1435. [CrossRef]
11. Wilamowski, B.M.; Irwin, J.D. *Power Electronics and Motor Drives*; Taylor and Francis Group: Boca Raton, FL, USA, 2011.
12. Zorig, A.; Belkheiri, M.; Barkat, S. Control of three-level T-type inverter based grid connected PV system. In Proceedings of the 13th International Multi-Conference on Systems, Leipzig, Germany, 21–24 March 2016; pp. 66–71.
13. Malinowski, M.; Kazmierkowski, M.P.; Hansen, S.; Blaabjerg, F.; Marques, G.D. Virtual-flux-based direct power control of three-phase PWM rectifiers. *IEEE Trans. Ind. Appl.* **2001**, *37*, 1019–1027. [CrossRef]

14. Malinowski, M.; Jasinski, M.; Kazmierkowski, M.P. Simple direct power control of three-phase PWM rectifier using space-vector modulation (DPC-SVM). *IEEE Trans. Ind. Electron.* **2004**, *51*, 447–454. [CrossRef]

15. Farahani, H.F.; Rashidi, F. Direct power control of a grid-connected photovoltaic system using a fuzzy-logic based controller. *Simul.-Trans. Soc. Model. Simul. Int.* **2017**, *93*, 213–223. [CrossRef]

16. Sebaaly, F.; Vahedi, H.; Kanaan, H.Y.; Moubayed, N.; Al-Haddad, K. Design and Implementation of Space Vector Modulation-Based Sliding Mode Control for Grid-Connected 3L-NPC Inverter. *IEEE Trans. Ind. Electron.* **2016**, *63*, 7854–7863. [CrossRef]

17. Donoso, F.; Mora, A.; Cárdenas, R.; Angulo, A.; Sáez, D.; Rivera, M. Finite-Set Model-Predictive Control Strategies for a 3L-NPC Inverter Operating With Fixed Switching Frequency. *IEEE Trans. Ind. Electron.* **2018**, *65*, 3954–3965. [CrossRef]

18. Vazquez, S.; Mohan, C.; Franquelo, L.G.; Marquez, A.; Leon, J.I. A Generalized Predictive control for T-type power inverters with output LC filter. In Proceedings of the 9th International Conference on Compatibility and Power Electronics (CPE), Costa da Caparica, Portugal, 24–26 June 2015; pp. 20–24.

19. Nauman, M.; Hasan, A. Efficient Implicit Model-Predictive Control of a Three-Phase Inverter With an Output LC Filter. *IEEE Trans. Power Electron.* **2016**, *31*, 6075–6078. [CrossRef]

20. Xing, X.; Chen, A.; Zhang, Z.; Chen, J.; Zhang, C. Model predictive control method to reduce common-mode voltage and balance the neutral-point voltage in three-level T-type inverter. In Proceedings of the IEEE Applied Power Electronics Conference and Exposition, Long Beach, CA, USA, 20–24 March 2016; pp. 3453–3458.

21. Kouro, S.; Cortes, P.; Vargas, R.; Ammann, U.; Rodriguez, J. Model Predictive Control:A Simple and Powerful Method to Control Power Converters. *IEEE Trans. Ind. Electron.* **2009**, *56*, 1826–1838. [CrossRef]

22. Rodriguez, J.; Kazmierkowski, M.P.; Espinoza, J.R.; Zanchetta, P.; Abu-Rub, H.; Young, H.A.; Rojas, C.A. State of the Art of Finite Control Set Model Predictive Control in Power Electronics. *IEEE Trans. Ind. Inf.* **2013**, *9*, 1003–1016. [CrossRef]

23. Rodriguez, J.; Cortes, P. *Predictive Control of Power Converters and Electrical Drives*; John Wiley: New York, NY, USA, 2012.

24. Ngo, B.Q.V.; Rodriguez-Ayerbe, P.; Olaru, S.; Niculescu, S.I. Model predictive power control based on virtual flux for grid connected three-level neutral-point clamped inverter. In Proceedings of the 18th European Conference on Power Electronics and Applications, Karlsruhe, Germany, 5–9 September 2016; pp. 1–10.

25. Singh, V.K.; Tripathi, R.N.; Hanamoto, T. HIL Co-Simulation of Finite Set-Model Predictive Control Using FPGA for a Three-Phase VSI System. *Energies* **2018**, *11*, 909. [CrossRef]

26. Shi, T.; Zhang, C.; Geng, Q.; Xia, C. Improved model predictive control of three-level voltage source converter. *Electr. Power Compon. Syst.* **2014**, *42*, 1029–1138. [CrossRef]

27. Xia, C.; Liu, T.; Shi, T.; Song, Z. A Simplified Finite-Control-Set Model-Predictive Control for Power Converters. *IEEE Trans. Ind. Inf.* **2014**, *10*, 991–1002.

28. Geyer, T.; Quevedo, D.E. Performance of Multistep Finite Control Set Model Predictive Control for Power Electronics. *IEEE Trans. Power Electron.* **2015**, *30*, 1633–1644. [CrossRef]

29. Aguilera, R.P.; Baidya, R.; Acuna, P.; Vazquez, S.; Mouton, T.; Agelidis, V.G. Model predictive control of cascaded H-bridge inverters based on a fast-optimization algorithm. In Proceedings of the 41st Annual Conference of the IEEE Industrial Electronics Society, Yokohama, Japan, 9–12 November 2015; pp. 4003–4008.

30. Liu, X.; Wang, D.; Peng, Z. Improved finite-control-set model predictive control for active front-end rectifiers with simplified computational approach and on-line parameter identification. *ISA Trans.* **2017**, *69*, 51–64. [CrossRef] [PubMed]

31. Zhang, Z.; Hackl, C.M.; Kennel, R. Computationally Efficient DMPC for Three-Level NPC Back-to-Back Converters in Wind Turbine Systems With PMSG. *IEEE Trans. Power Electron.* **2017**, *32*, 8018–8034. [CrossRef]

32. Taheri, A.; Zhalebaghi, M.H. A new model predictive control algorithm by reducing the computing time of cost function minimization for NPC inverter in three-phase power grids. *ISA Trans.* **2017**, *71*, 391–402. [CrossRef] [PubMed]

33. Wu, B.; Narimani, M. *High-Power Converters and AC Drives*; Wiley-IEEE Press: New York, NY, USA, 2017.

34. TMS320F28335 controlCARD. Available online: http://www.ti.com/tool/tmdscncd28335 (accessed on 28 October 2018).
35. Elrajoubi, A.; Ang, S.S.; Abushaiba, A. TMS320F28335 DSP programming using MATLAB Simulink embedded coder: Techniques and advancements. In Proceedings of the 2017 IEEE 18th Workshop on Control and Modeling for Power Electronics (COMPEL), Stanford, CA, USA, 9–12 July 2017; pp. 1–7.

Article

An Averaged-Value Model of an Asymmetrical Hybrid Multi-Level Rectifier

Salvatore Foti [1],*, Giacomo Scelba [2], Antonio Testa [1] and Angelo Sciacca [3]

[1]   DI, University of Messina, 98122 Messina, Italy; atesta@unime.it
[2]   DIEEI, University of Catania, 95131 Catania, Italy; giacomo.scelba@dieei.unict.it
[3]   ST Microelectronics, I-95121 Catania, Italy; angelo.sciacca@st.com
*   Correspondence: sfoti@unime.it; Tel.: +39-3206487898

Received: 30 December 2018; Accepted: 6 February 2019; Published: 13 February 2019

Abstract: The development and the validation of an averaged-value mathematical model of an asymmetrical hybrid multi-level rectifier is presented in this work. Such a rectifier is composed of a three-level T-type unidirectional rectifier and of a two-level inverter connected to an open-end winding electrical machine. The T-type rectifier, which supplies the load, operates at quite a low switching frequency in order to minimize inverter power losses. The two-level inverter is instead driven by a standard sinusoidal pulse width modulation (SPWM) technique to suitably shape the input current. The two-level inverter also plays a key role in actively balancing the voltage across the DC bus capacitors of the T-type rectifier, making unnecessary additional circuits. Such an asymmetrical structure achieves a higher efficiency compared to conventional PWM multilevel rectifiers, even considering extra power losses due to the auxiliary inverter. In spite of its advantageous features, the asymmetrical hybrid multi-level rectifier topology is a quite complex system, which requires suitable mathematical tools for control and optimization purposes. This paper intends to be a step in this direction by deriving an averaged-value mathematical model of the whole system, which is validated through comparison with other modeling approaches and experimental results. The paper is mainly focused on applications in the field of electrical power generation; however, the converter structure can be also exploited in a variety of grid-connected applications by replacing the generator with a transformer featuring an open-end secondary winding arrangement.

Keywords: electrical drives; energy saving; multilevel power converters; permanent magnet synchronous generator; open-end winding configuration; voltage balancing; power factor

---

1. Introduction

Multi-level converters have proved in the last decades to be a viable alternative to conventional topologies in medium-voltage, high-power, industrial applications, but today, their field of applications is rapidly spreading toward low-power and low-voltage ranges. Main advantages of multi-level converters are basically those of an improved harmonic content of AC voltages and currents and of a reduction of power switch voltage ratings [1,2], the main drawback being a greater complexity. Open-winding (OW) configurations, consisting of an AC machine fed by two power converters [3–6], can be deemed as a special kind of multi-level converter [7]. Different configurations, control schemes, and modulation techniques dealing with OW systems have been discussed in the literature [8–10]. Some OW configurations embedding multi-level converters have also been recently developed [11–13]. Among them, a high efficiency asymmetrical hybrid multilevel inverter for motor drives has been presented and analyzed in [12] featuring a particular asymmetrical structure where two different kinds of converters are connected at the two sides of an OW AC machine with different functions. Specifically, a main multilevel converter supplies the load, and an auxiliary two-level inverter acts as an active power filter. Such an approach has also been used in [13] to realize an asymmetrical hybrid unidirectional T-type rectifier (AHUTR) for gen-set applications, tailored around an open-end winding permanent magnet synchronous generator (PMSG), as shown in Figure 1. According to the AHUTR topology, the open-end winding PMSG on one side supplies the electrical load through the main converter, a T-type rectifier (TTR), also commonly known as a Vienna rectifier, and on the other side, it is connected to an auxiliary two-level inverter (TLI). The main converter processes the whole power delivered to the load, and thus, it is operated at the fundamental frequency in order to minimize the switching power losses. The TLI is instead driven by a high switching frequency PWM technique to suitably shape the phase currents. Therefore, a stable output DC voltage and almost sinusoidal input currents are obtained, achieving a higher efficiency than comparable conventional PWM rectifiers [12]. The AHUTR structure is also of general applicability, being exploitable in grid-connected applications by replacing the generator with a transformer featuring an open-end secondary winding, as shown in Figure 1, but it is more complex than conventional rectifiers, requiring suitable mathematical tools for control and optimization purposes. The aim of this work is thus to provide an essential tool for the design of the control system of an AHUTR by developing an averaged-value model (AVM) of the system. In general, averaged-value techniques approximate the model of a switching converter to a continuous system by considering the values taken by the variables along a switching period as constant. They are useful when designing and testing control algorithms, as well as to develop efficiency optimization techniques, because a high frequency dynamic analysis is not required, differently than power circuits and filters design. Specifically, an AHUTR AVM has been developed with the aim to support the design of effective solutions to maximize system efficiency, to provide a stable DC output voltage, to cancel low-order undesired harmonics from the phase currents, to equalize the Vienna rectifier DC bus capacitor voltages, and to control the TLI DC bus voltage. Furthermore, the developed model is valuable in tuning voltage and current regulators.

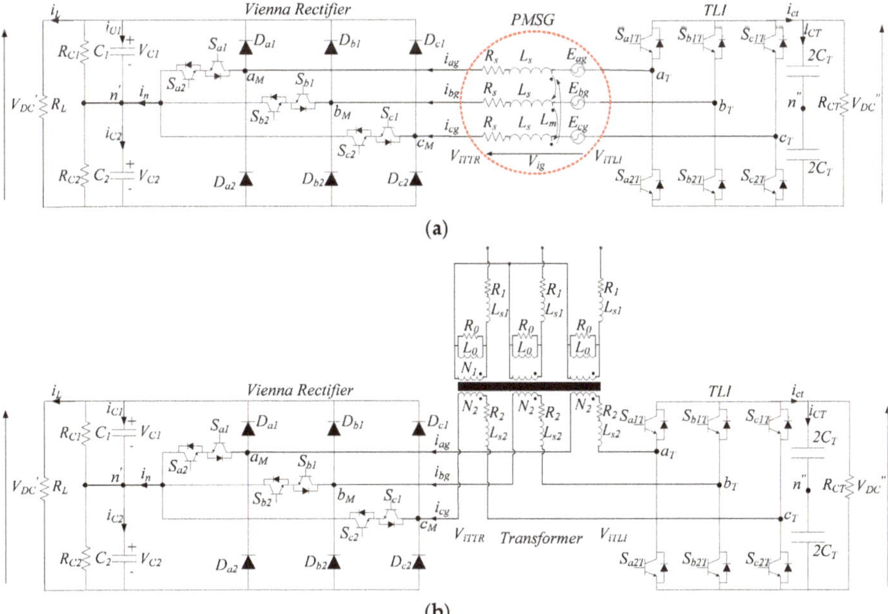

Figure 1. AHUTR for electrical power generation (**a**) and grid-connected (**b**) applications.

2. Asymmetrical Hybrid Unidirectional T-Type Rectifier

According to Figure 1, an AHUTR supplies the load through a Vienna rectifier switching at fundamental frequency. In electricity generation applications, this rectifier is connected to one end of an open winding electrical generator, very often a PMSG. For grid-connected applications, the electrical generator is replaced by a transformer with an open-end secondary winding. While remarkably reducing the switching power losses, low switching frequency operations would, however, produce highly distorted phase currents. This is prevented by an active power filter based on a conventional TLI, which is connected to the other end of the electrical machine winding. Such an inverter features a lower DC bus voltage compared to the Vienna rectifier and exploits a floating capacitor to reduce the complexity of the system and to prevent the occurrence of zero sequence currents [11–13]. The efficiency of the Vienna rectifier can be increased by using low on-state voltage drop power devices, thus optimizing the design of this converter for low conduction power losses. On the other hand, the design of the TLI can be optimized for high switching frequency operation, by using fast power devices with lower voltage ratings. A key feature of the AHUTR topology is that the voltages of the two Vienna rectifier DC bus capacitors can be independently regulated through the TLI, thus making unnecessary additional power converters or special PWM strategies.

In the AHUTR topology, three bidirectional switches S_{ij}, ($i = a$, b, c and $j = 1$, 2) are connected between the midpoint n' of the Vienna rectifier and the rectifier poles [14]. The generic i-phase voltage V_{iTTR} between the rectifier input terminal i_M and the mid-point n'' of the Vienna rectifier DC bus is given by

$$V_{iTTR} = \frac{l_i' - 1}{2} V_{DC}', \ l_i' = 0, 1, 2 \tag{1}$$

where V_{DC}' is the DC bus voltage. Hence, three different levels can be taken by the Vienna rectifier input voltage, namely: $-V_{DC}'/2$, $V_{DC}'/2$, and 0, according to the rectifier i-pole state l_i'.

On the TLI side, the voltage between the TLI i-phase output terminal i_T and the mid-point n' of the TLI DC bus is given by:

$$V_{iTLI} = \frac{2l_i'' - 1}{2} V_{DC}'', \ l_i'' = 0, 1 \tag{2}$$

providing two voltage levels, namely, $-V_{DC}''/2$ and $V_{DC}''/2$, according to the inverter i-pole state l_i''.

The voltage across a phase winding is given by

$$V_{ig} = V_{iTTR} - V_{iTLI} - V_{n'n''} = \frac{l_i'' - 1}{2} V_{DC}' - \frac{2l_i'' - 1}{2} V_{DC}'' - V_{n'n''} \tag{3}$$

where V_{DC}' and V_{DC}'' are the DC bus voltages of the Vienna rectifier and the TLI, respectively, and $V_{n'n''}$ is the voltage between the mid points n' and n'' of the two DC buses, which can be expressed as

$$V_{n'n''} = \frac{1}{3}(V_{aTTR} + V_{bTTR} + V_{cTTR}) - \frac{1}{3}(V_{aTLI} + V_{bTLI} + V_{cTLI}). \tag{4}$$

According to (2) and (3), the OW structure of Figure 1, featuring twelve power switches, is equivalent to a six-level neutral point clamped (NPC) or flying capacitor (FC) converter, which would, however, encompass thirty power switches [12]. As shown in Figure 2, the AHUTR requires a complex control system to suitably coordinate the operations of the two converters in order to regulate the DC output voltage, to cancel low-order harmonics from phase currents, to equalize the Vienna rectifier DC bus capacitor voltages, and to control the TLI DC bus voltage [14,15].

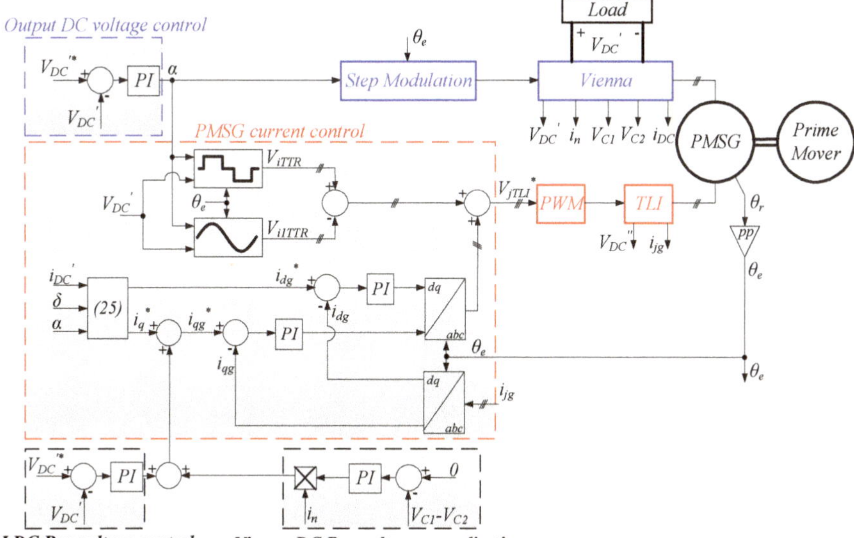

Figure 2. Block diagram of the control system of the AHUTR for electrical power generation applications.

3. Averaged-value Model of the System

The averaged-value mathematical model of the system includes three sub-models: of the electrical machine, of the Vienna rectifier, and of the TLI.

3.1. Open-Winding PMSG Model

It is assumed that the stator windings produce sinusoidal magnetomotive forces; moreover, effects of the saturation of the magnetic core are neglected. Under these assumptions, the surface-mounted PMSG model in an orthogonal qd reference frame synchronous to the rotor flux is given by the following sets of Equations:

$$V_{qs} = R_s i_{qs} + \frac{d}{dt}\lambda_{qs} + \omega_{re}\lambda_{ds}$$
$$V_{ds} = R_s i_{ds} + \frac{d}{dt}\lambda_{ds} - \omega_{re}\lambda_{qs} \tag{5}$$

$$\lambda_{qs} = L_s i_{qs}$$
$$\lambda_{ds} = L_s i_{ds} + \lambda_{pm} \tag{6}$$

$$T_e = \frac{3}{2}pp(\lambda_{ds}i_{qs} - \lambda_{qs}i_{ds})$$
$$T_e - T_L = J\frac{d}{dt}\omega_r + F\omega_r \tag{7}$$

where i_{qs}, i_{ds}, V_{qs}, V_{ds}, λ_{qs}, and λ_{ds} are the components of stator current, voltage, and flux in the qd axis; L_s is the stator inductance; λ_{pm} is the linkage flux of permanent magnets; T_e is the electromagnetic torque; J is the total mechanical inertia; F is the rotor friction; $\omega_{re} = pp\omega_r$ is the rotor speed; and pp is the amount of pole pairs. The rotational terms $\omega_{re}\lambda_{ds}$ and $\omega_{re}\lambda_{qs}$ account for the qd axis back-emf E_q and E_d, respectively.

The averaged-value PMSG phase voltage V_{ig} is obtained as the difference between the fundamental harmonic $\overline{V}_{iTTR}$ of the Vienna rectifier input voltage and the fundamental harmonic $\overline{V}_{iTLI}$ of the TLI output voltage. The voltage $V_{n'n''}$ between the mid points of the two DC buses can be neglected for averaged-value analysis, since it only includes high frequency harmonics [13].

PMSG phase voltages can be expressed in a qd synchronous reference frame to the back-EMF vector as a function of qd components of voltages $\overline{V}_{iTTR}$ and $\overline{V}_{iTLI}$ by:

$$\begin{vmatrix} \overline{V}_{qTTR} \\ \overline{V}_{dTTR} \end{vmatrix} = \frac{2}{3} \begin{vmatrix} \cos(\omega_{re}t) & \cos(\omega_{re}t - \frac{2}{3}\pi) & \cos(\omega_{re}t + \frac{2}{3}\pi) \\ \sin(\omega_{re}t) & \sin(\omega_{re}t - \frac{2}{3}\pi) & \sin(\omega_{re}t + \frac{2}{3}\pi) \\ \frac{1}{2} & \frac{1}{2} & \frac{1}{2} \end{vmatrix} \begin{vmatrix} \overline{V}_{aTTR} \\ \overline{V}_{bTTR} \\ \overline{V}_{cTTR} \end{vmatrix} \tag{8}$$

$$\begin{vmatrix} \overline{V}_{qTLI} \\ \overline{V}_{dTLI} \end{vmatrix} = \frac{2}{3} \times \begin{vmatrix} \cos(\omega_{re}t) & \cos(\omega_{re}t - \frac{2}{3}\pi) & \cos(\omega_{re}t + \frac{2}{3}\pi) \\ \sin(\omega_{re}t) & \sin(\omega_{re}t - \frac{2}{3}\pi) & \sin(\omega_{re}t + \frac{2}{3}\pi) \\ \frac{1}{2} & \frac{1}{2} & \frac{1}{2} \end{vmatrix} \times \begin{vmatrix} \overline{V}_{aTLI} \\ \overline{V}_{bTLI} \\ \overline{V}_{cTLI} \end{vmatrix} \tag{9}$$

$$\overline{V}_{qg} = \overline{V}_{qTTR} - \overline{V}_{qTLI}, \ \overline{V}_{dg} = \overline{V}_{dTTR} - \overline{V}_{dTLI} \tag{10}$$

A block scheme of the PMSG model is shown in Figure 3.

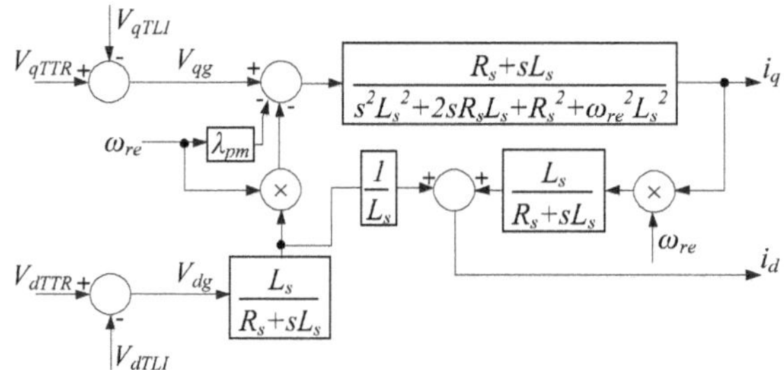

Figure 3. Block scheme of the permanent magnet synchronous generator (PMSG) model.

Similarly, a three-phase open secondary winding transformer (OSWT) can be modeled in an orthogonal qd reference frame synchronous to the primary voltage vector according to:

$$\begin{aligned} V_{q1} &= R_1 i_{q1} + \tfrac{d}{dt}\lambda_{q1} + \omega_e \lambda_{d1} \\ V_{d1} &= R_1 i_{d1} + \tfrac{d}{dt}\lambda_{d1} - \omega_e \lambda_{q1} \\ V_{q2} &= R_2 i_{q2} + \tfrac{d}{dt}\lambda_{q2} + \omega_e \lambda_{d2} \\ V_{d2} &= R_2 i_{d2} + \tfrac{d}{dt}\lambda_{d2} - \omega_e \lambda_{q2} \end{aligned} \tag{11}$$

$$\begin{aligned} \lambda_{q1} &= L_{s1} i_{q1} + L_m i_{q2} \\ \lambda_{d1} &= L_{s1} i_{d1} + L_m i_{d2} \\ \lambda_{q2} &= L_{s2} i_{q2} + L_m i_{q1} \\ \lambda_{d2} &= L_{s2} i_{d2} + L_m i_{d1} \\ L_{s1} &= L_{l1} + L_m \\ L_{s2} &= L_{l2} + L_m \end{aligned} \tag{12}$$

where i_{q1}, i_{d1}, i_{q2}, and i_{d2} are the q- and d-axis components of the primary and secondary winding currents, while V_{q1}, V_{d1}, V_{q2}, V_{d2} and λ_{q1}, λ_{d1}, λ_{q2}, λ_{d2}, are the q- and d-axis components of the primary and secondary winding voltages and fluxes. L_{s1} and L_{s2} are the self-inductances and L_m is the magnetization inductance. The angular frequency of the grid voltage is indicated as ω_e. The secondary windings are connected to the TTR and TLI, and thus, the phase winding voltages are given by:

$$\overline{V}_{q2} = \overline{V}_{qTTR} - \overline{V}_{qTLI}, \quad \overline{V}_{d2} = \overline{V}_{dTTR} - \overline{V}_{dTLI} \tag{13}$$

3.2. Vienna Rectifier Model

The Vienna rectifier switches at the fundamental frequency, according to Table 1, where θ_e is the angular displacement of the fundamental harmonic of the winding phase voltage and α is the switching angle of S_{ij}, ($i = a, b, c$ and $j = 1, 2$).

Table 1. Vienna rectifier switching table.

Phase *a*	$0 < \theta_e < \alpha$ $\pi - \alpha < \theta_e < \pi + \alpha$ $2\pi - \alpha < \theta_e < 2\pi$	if $i_{ag} > 0 \Rightarrow S_{a1}$ ON S_{a2} OFF if $i_{ag} < 0 \Rightarrow S_{a1}$ OFF S_{a2} ON
Phase *b*	$2/3\pi < \theta_e < \alpha + 2/3\pi$ $5/3\pi - \alpha < \theta_e < 5/3\pi + \alpha$ $2/3\pi - \alpha < \theta_e < 2/3\pi$	if $i_{bg} > 0 \Rightarrow S_{b1}$ ON S_{b2} OFF if $i_{bg} < 0 \Rightarrow S_{b1}$ OFF S_{b2} ON
Phase *c*	$4/3\pi < \theta_e < \alpha + 4/3\pi$ $1/3\pi - \alpha < \theta_e < 1/3\pi + \alpha$ $4/3\pi - \alpha < \theta_e < 4/3\pi$	if $i_{cg} > 0 \Rightarrow S_{c1}$ ON S_{c2} OFF if $i_{cg} < 0 \Rightarrow S_{c1}$ OFF S_{c2} ON

Assuming the output voltage V_{DC}' is constant, actual values of Vienna rectifier input phase voltages V_{iTTR} are thus given by:

$$V_{iTTR} = S_{ij} \frac{l_{ij} - 1}{2} V_{DC}', \quad \begin{array}{l} -\alpha < \varphi_{TTR} < \alpha \\ l_{ij} = 0, 1, 2. \end{array} \tag{14}$$

To avoid improper operations leading to extra power losses and voltage distortion, the angular displacement φ_{TTR} between the fundamental harmonics of voltage V_{iTTR} and current must be set lower than $|\alpha|$. Dealing with an electrical power generation application, a vector diagram of AC variables is shown in Figure 4a, where φ is the phase displacement between the PMSG back-EMF $\overline{E}_g$ and the current $\overline{I}$. δ represents the angle between the voltage $\overline{V}_{TTR}$ and the q axis, and is set to allow a reactive power flow between the Vienna rectifier and PMSG, associated to the inductive elements of the electrical machine.

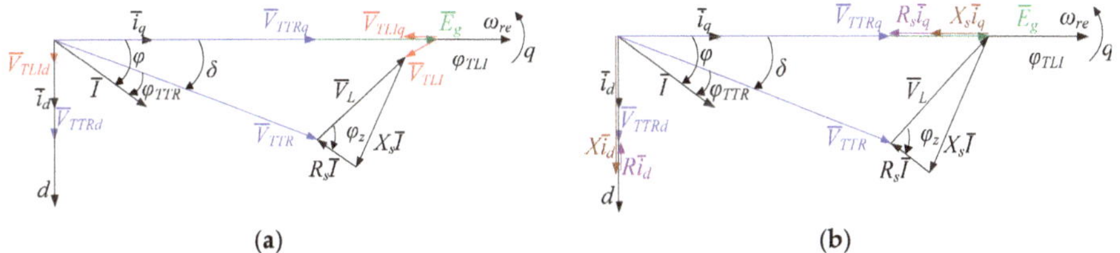

Figure 4. Vector diagram of AC variables: (**a**) considering V_{TLI}, (**b**) neglecting V_{TLI}.

Neglecting, for simplicity, the voltage V_{TLI} generated by the auxiliary inverter, which is an independent variable and whose amplitude is significantly lower than V_{i1TTR}, the amplitude of the fundamental harmonic of the TTR input voltage V_{i1TTR} is obtained as a function of the switching angle α and DC bus voltage V_{DC}' as follows:

$$|\overline{V}_{i1TTR}| = \frac{2}{\pi} V_{DC}' \cos(\alpha), \quad m_{TTR} = \frac{|V_{i1TTR}|}{V_{DC}'} \tag{15}$$

where m_{TTR} is the modulation index of the Vienna rectifier. According to the vector diagram of Figure 4b, qd components of the voltage can be written as:

$$\left\{\begin{array}{l} \overline{V}_{qTTR} = |\overline{V}_{i1TTR}|\cos(\delta) \\ \overline{V}_{dTTR} = |\overline{V}_{i1TTR}|\sin(\delta) \end{array}\right.' \left\{\begin{array}{l} \overline{i}_q = |\overline{I}|\cos(\varphi) \\ \overline{i}_d = |\overline{I}|\sin(\varphi) \end{array}\right.' \left\{\begin{array}{l} \overline{E}_{qg} = |\overline{E}_g| \\ \overline{E}_{dg} = 0 \end{array}\right.' \left\{\begin{array}{l} X\overline{i}_q = -|X_s\overline{I}|\sin(\varphi) \\ X\overline{i}_d = +|X_s\overline{I}|\cos(\varphi) \end{array}\right.' \left\{\begin{array}{l} R\overline{i}_q = -|R\overline{I}|\cos(\varphi) \\ R\overline{i}_d = -|R\overline{I}|\sin(\varphi) \end{array}\right. \tag{16}$$

while the active and reactive powers are given by:

$$\left\{\begin{array}{l} P_{TTR} = \tfrac{3}{2}|\overline{V}_{i1TTR}\overline{I}|\cos(\delta - \varphi) \\ Q_{TTR} = \tfrac{3}{2}|\overline{V}_{i1TTR}\overline{I}|\sin(\varphi - \delta) \end{array}\right.' \left\{\begin{array}{l} P_R = -\tfrac{3}{2}R|\overline{I}|^2 \\ Q_X = -\tfrac{3}{2}X|\overline{I}|^2 \end{array}\right.' \left\{\begin{array}{l} P_g = \tfrac{3}{2}|\overline{E}_{qg}\overline{I}|\cos(\varphi) \\ Q_g = \tfrac{3}{2}|\overline{E}_{qg}\overline{I}|\sin(\varphi) \end{array}\right. \tag{17}$$

where P_{TTR} and Q_{TTR} are the active and reactive power, respectively, processed by the Vienna rectifier, P_R and Q_X are the active power wasted in the stator resistance R and the reactive power due to the PMSG synchronous reactance X_s, respectively, while P_g and Q_g are the active and reactive power delivered by the PMSG, respectively.

Neglecting the rectifier power losses, the AC power generated by the PMSG is equal to the sum of the power dissipated in the DC bus capacitor resistances R_{C1} and R_{C2} and the power delivered to the load R_L. In the Laplace domain, V_{DC}' and the capacitor voltages V_{C1} and V_{C2} are thus given by

$$\left\{\begin{array}{l} V_{DC}'(s) = \sqrt{R_L\left(P_{AC}(s) - \dfrac{V_{C_1}^2(s)}{R_{C1}} - \dfrac{V_{C_2}^2(s)}{R_{C2}}\right)} \\[2ex] V_{C1}(s) = V_{DC}'(s)\dfrac{sR_{C1}(1+R_{C2}C_2)}{R_{C1}+R_{C1}+sR_{C1}R_{C2}(C_1+C_2)} \\[2ex] V_{C2}(s) = V_{DC}'(s) - V_{C1}(s) \\[1ex] P_{AC}(s) = \tfrac{3}{2}\left(V_{qTTR}(s)i_q(s) + V_{dTTR}(s)i_d(s)\right) \end{array}\right. \tag{18}$$

where i_n is mainly given by the difference between the currents flowing through the two DC bus capacitors and it can be also computed as the sum of the currents flowing through the three branches of the Vienna rectifier:

$$i_n = S_{aj}i_{ag} + S_{bj}i_{bg} + S_{cj}i_{cg} \tag{19}$$

The averaged-value of i_n during a switching period T is given by

$$\overline{i}_n = \frac{1}{T}\left(T_{ONaj}i_{ag} + T_{ONbj}i_{bg} + T_{ONcj}i_{cg}\right) = \left(d_{aj}i_{ag} + d_{bj}i_{bg} + d_{cj}i_{cg}\right) \tag{20}$$

where $d_{ij} = T_{ONij}/T$ are the duty cycles of the bidirectional switches S_{ij}, according to Table 2. Figure 5 shows some simulations dealing with balanced and unbalanced DC bus voltages operations, while a block diagram of the Vienna rectifier mathematical model is shown in Figure 6.

Table 2. d_{aj}, d_{bj} and d_{cj}.

Sector I	Sector II	Sector III	Sector VI	Sector V	Sector IV
$V_{a1TTR} > 0$	$V_{a1TTR} > 0$	$V_{a1TTR} < 0$	$V_{a1TTR} < 0$	$V_{a1TTR} < 0$	$V_{a1TTR} > 0$
$V_{b1TTR} < 0$	$V_{b1TTR} > 0$	$V_{b1TTR} > 0$	$V_{b1TTR} > 0$	$V_{b1TTR} < 0$	$V_{b1TTR} < 0$
$V_{c1TTR} < 0$	$V_{c1TTR} < 0$	$V_{c1TTR} < 0$	$V_{c1TTR} > 0$	$V_{c1TTR} > 0$	$V_{c1TTR} > 0$
$d_{aj} = \dfrac{V_{a1TTR}}{V_{DC}'}$	$d_{aj} = -\dfrac{V_{a1TTR}}{V_{DC}'}$	$d_{aj} = -\dfrac{V_{a1TTR}}{V_{DC}'}$	$d_{aj} = -\dfrac{V_{a1TTR}}{V_{DC}'}$	$d_{aj} = \dfrac{V_{a1TTR}}{V_{DC}'}$	$d_{aj} = \dfrac{V_{a1TTR}}{V_{DC}'}$
$d_{bj} = \dfrac{V_{b1TTR}}{V_{DC}'}$	$d_{bj} = \dfrac{V_{b1TTR}}{V_{DC}'}$	$d_{bj} = -\dfrac{V_{b1TTR}}{V_{DC}'}$	$d_{bj} = -\dfrac{V_{b1TTR}}{V_{DC}'}$	$d_{bj} = -\dfrac{V_{b1TTR}}{V_{DC}'}$	$d_{bj} = -\dfrac{V_{b1TTR}}{V_{DC}'}$
$d_{cj} = -\dfrac{V_{c1TTR}}{V_{DC}'}$	$d_{cj} = -\dfrac{V_{c1TTR}}{V_{DC}'}$	$d_{cj} = \dfrac{V_{c1TTR}}{V_{DC}'}$	$d_{cj} = \dfrac{V_{c1TTR}}{V_{DC}'}$	$d_{cj} = \dfrac{V_{c1TTR}}{V_{DC}'}$	$d_{cj} = -\dfrac{V_{c1TTR}}{V_{DC}'}$

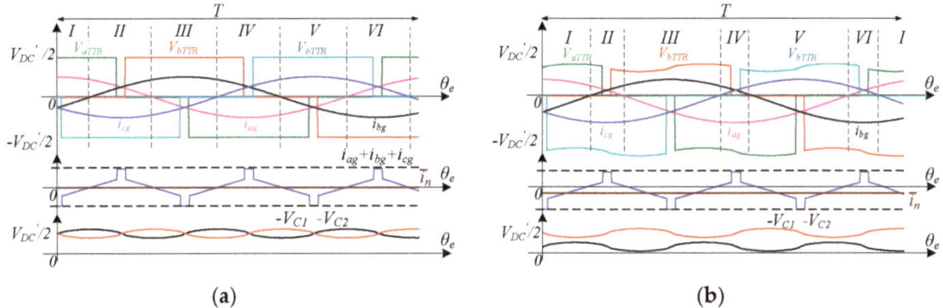

Figure 5. Averaged-value i_n, i_{abcg}, V_{c1}, V_{c2}, and V_{iTTR}. (**a**) Balanced DC bus voltages, and (**b**) unbalanced DC bus voltages.

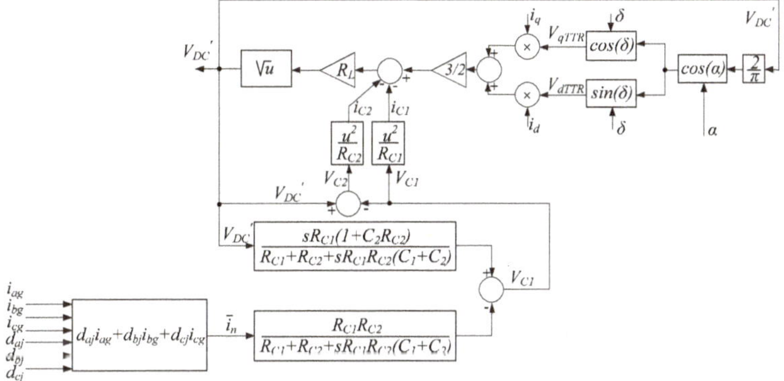

Figure 6. Block diagram of the Vienna rectifier model.

A non-null average i_n leads to unbalanced DC bus voltages [16–18]; moreover, the mean value of fundamental voltages V_{a1TTR} becomes negative if $V_{C1} < V_{C2}$ or positive if $V_{C2} < V_{C1}$. This is included in the TTR model by adding the term $\Delta V_{DC}{}' = V_{C1} - V_{C2}$:

$$\begin{cases} V_{a1TTR} = |V_{a1TTR}| \sin(\theta_e) + \Delta V_{DC}{}' \\ V_{b1TTR} = |V_{b1TTR}| \sin(\theta_e - \tfrac{2}{3}\pi) + \Delta V_{DC}{}' \\ V_{c1TTR} = |V_{c1TTR}| \sin(\theta_e + \tfrac{2}{3}\pi) + \Delta V_{DC}{}' \end{cases} \tag{21}$$

According to Table 2, by replacing (21) into (20), $\bar{i}_n$ is given by

$$\bar{i}_n = \begin{cases} m_{TTR}0.5I\left[-\cos(\varphi_{TTR}) - 2\cos(2\theta_e - \tfrac{4\pi}{3} - \varphi_{TTR})\right] - 2I\tfrac{\Delta V_{DC}{}'}{V_{DC}}\sin(\theta_e - \tfrac{2\pi}{3} - \varphi_{TTR}), & 0 < \theta_e < \tfrac{\pi}{3} \\ m_{TTR}0.5I[\cos(\varphi_{TTR}) + 2\cos(2\theta_e - \varphi_{TTR})] - 2I\tfrac{\Delta V_{DC}{}'}{V_{DC}}\sin(\theta_e - \varphi_{TTR}), & \tfrac{\pi}{3} < \theta_e < \tfrac{2\pi}{3} \\ m_{TTR}0.5I\left[-\cos(\varphi_{TTR}) - 2\cos(2\theta_e - \tfrac{2\pi}{3} - \varphi_{TTR})\right] - 2I\tfrac{\Delta V_{DC}{}'}{V_{DC}}\sin(\theta_e + \tfrac{2\pi}{3} - \varphi_{TTR}), & \tfrac{2\pi}{3} < \theta_e < \pi \\ m_{TTR}0.5I\left[\cos(\varphi_{TTR}) + 2\cos(2\theta_e - \tfrac{4\pi}{3} - \varphi_{TTR})\right] + 2I\tfrac{\Delta V_{DC}{}'}{V_{DC}}\sin(\theta_e - \tfrac{2\pi}{3} - \varphi_{TTR}), & \pi < \theta_e < \tfrac{4\pi}{3} \\ m_{TTR}0.5I[-\cos(\varphi_{TTR}) - 2\cos(2\theta_e - \varphi_{TTR})] - 2I\tfrac{\Delta V_{DC}{}'}{V_{DC}}\sin(\theta_e - \varphi_{TTR}), & \tfrac{4\pi}{3} < \theta_e < \tfrac{5\pi}{3} \\ m_{TTR}0.5I\left[\cos(\varphi_{TTR}) + 2\cos(2\theta_e - \tfrac{2\pi}{3} - \varphi_{TTR})\right] + 2I\tfrac{\Delta V_{DC}{}'}{V_{DC}}\sin(\theta_e + \tfrac{2\pi}{3} - \varphi_{TTR}), & \tfrac{5\pi}{3} < \theta_e < \pi \end{cases} \tag{22}$$

3.3. TLI Model

A key task of the TLI present in the AHUTR topology is to compensate all low-order voltage harmonics generated by the step-modulated Vienna rectifier [12]. For this reason, the TLI reference phase voltage is equal to the difference between the AC side input voltage V_{iTTR} and its fundamental component V_{i1TTR}, as shown in Figure 7.

$$V_{iTLI}^{*} = V_{iTTR} - V_{i1TTR} \tag{23}$$

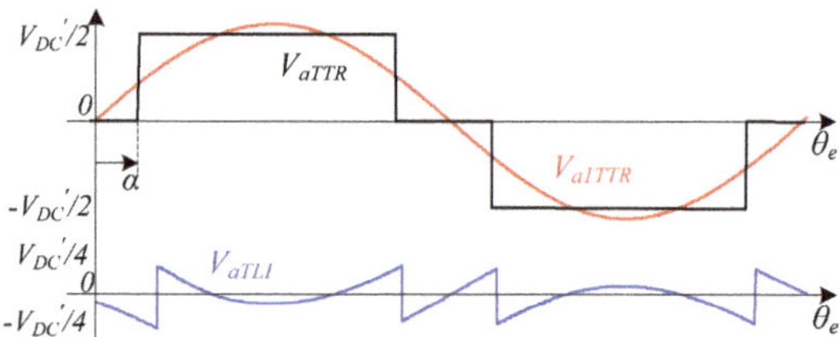

Figure 7. Two-level inverter (TLI) reference voltage for active power filtering.

Phase voltages V_{iTTR} encompass some zero sequence components, such as the 3rd, 9th, 27th, and 81st, that will not result in corresponding currents in the PMSG because the considered open-end winding topology is composed by two isolated converters. Hence, these harmonics can be neglected in the TLI reference voltages V_{iTLI}^{*}. This leads to a reduction of TLI DC bus voltage and, accordingly, to a positive impact on TLI losses. TLI reference voltages V_{abcTLI}^{*} are thus given by

$$\begin{cases} V_{aTLI}^{*}(n,\theta_e) = \sum\limits_{n=5,7,11,13} b_{an} \times \sin(n\theta_e - \varphi_n) \\ V_{bTLI}^{*}(n,\theta_e) = \sum\limits_{n=5,7,11,13} b_{bn} \times \sin(n\theta_e - \varphi_n - \frac{2\pi}{3}) \\ V_{cTLI}^{*}(n,\theta_e) = \sum\limits_{n=5,7,11,13} b_{cn} \times \sin(n\theta_e - \varphi_n + \frac{2\pi}{3}) \end{cases} \tag{24}$$

Figure 8 shows the V_{aTLI}^{*} waveform when considering a different set of zero sequence components. For each case, the minimum V_{DC}''/V_{DC}' requirement has been computed as shown in Figure 9, while current and voltage THDs are provided in Figure 10. At medium-high values of the modulation index m_{TTR}, a proper suppression of the effects of the low-order voltage harmonics produced by the Vienna rectifier is simply achieved by compensating the 5th and 7th harmonics. However, at low m_{TTR}, additional harmonics must be considered to keep the THDs suitably low.

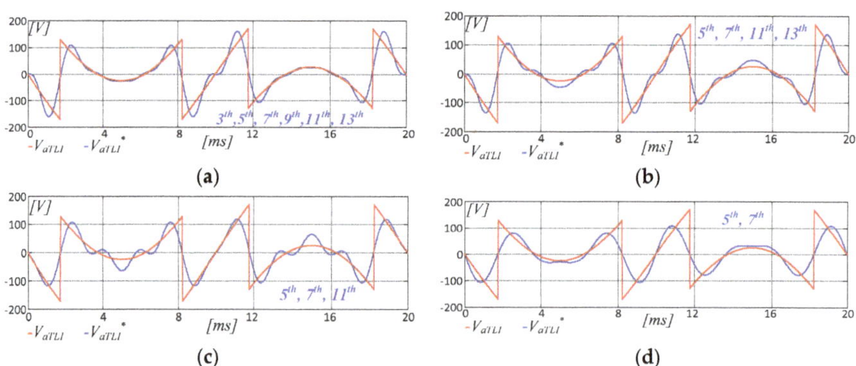

Figure 8. TLI reference voltage approximation. (**a**) 3rd, 5th, 7th, 9th, 11th, 13th. (**b**) 5th, 7th, 11th, 13th. (**c**) 5th, 7th, 11th. (**d**) 5th, 7th.

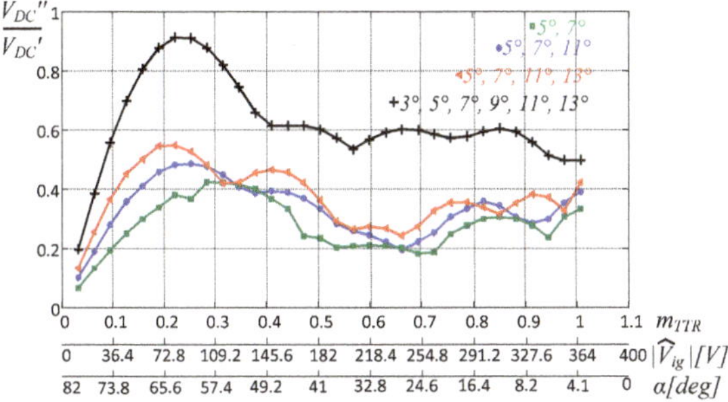

Figure 9. Minimum V_{DC}''/V_{DC}' requirement vs. peak amplitude of PMSG phase voltage, m_{TTR}, and α.

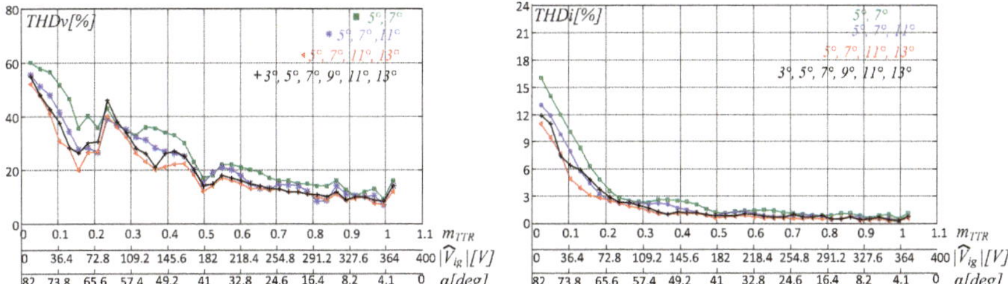

Figure 10. $THDv$ and $THDi$ vs. the peak amplitude of PMSG phase voltage.

As shown in Figure 2, a closed loop input current control system is added to the predictive filter in order to cope with unmodeled non-linearities and improve the input current waveform as well as the system dynamic response. By equaling the active power generated by the PMSG to the output DC power, the reference q-axis current i_q^* is computed from actual values of α, δ and the output DC current i_{DC} as:

$$i_q^* = \frac{\pi i_{DC}}{3\cos(\alpha)\cos(\delta)}, \quad i_d^* = 0 \tag{25}$$

The d-axis reference current i_d^* can be simply set to zero or suitably determined in case of interior permanent magnet structures in order to operate the PMSG according to a maximum power per ampere strategy.

Another key function of the TLI is to balance the voltage across the DC bus capacitors of the Vienna rectifier, making unnecessary additional circuits. As shown in Figure 2, this goal is obtained by acting on the q-axis component of the TLI reference current in order to control the amplitude of i_n, which is given by the difference between the currents flowing through the two DC bus capacitors.

The DC side of the TLI is modeled by balancing the AC and DC side power, neglecting the power switches losses (Equation (26)). The TLI DC-link includes the resistance R_{CT} representing the floating capacitor losses, while V_{qTLI} and V_{dTLI} are the voltage components of TLI V_{jTLI} in the qd axis, as shown in Figure 11.

$$\begin{cases} P_{AC} = P_{DC2} = \frac{3}{2}\left(V_{qTLI}i_q + V_{dTLI}i_d\right) \\ P_{DC2} = V_{DC}''i_{DC}'' + \dfrac{V_{DC}''^2}{R_{CT}} = V_{DC}''C_{TS}V_{DC}'' + \dfrac{V_{DC}''^2}{R_{CT}} = \frac{3}{2}\left(V_{qTLI}i_q + V_{dTLI}i_d\right) \end{cases} \tag{26}$$

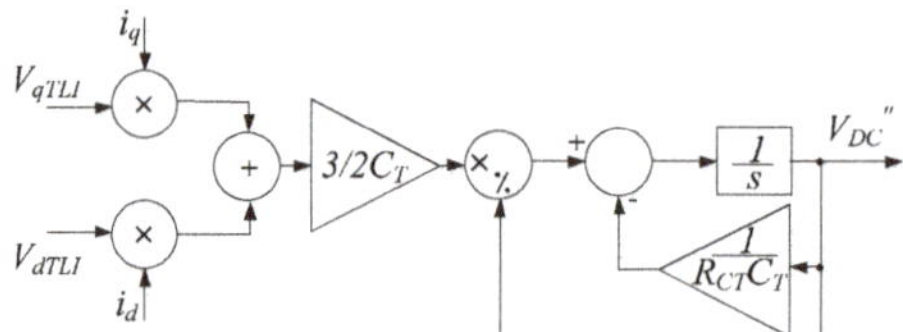

Figure 11. Block diagram of TLI model.

4. Model Validation

An electric power generation application has been considered for validating the value-averaged model. Specifically, the proposed model represented with the blocks scheme of Figure 12 has been implemented in a Simulink environment and compared to a detailed model of the system developed in the same environment exploiting the SimPower System Toolbox, which is a circuit-based modeling platform widely used for the simulation of power electronic converters, electromechanical systems, and their control systems. The last model includes both converter topologies. The control scheme used on both models is shown in Figure 2, including low-order harmonic compensation and DC bus capacitor voltages balancing [14]. Simulation settings are summarized in Table 3, where $k_{P\alpha}$ and $k_{I\alpha}$ are the proportional and integral gains of the output DC voltage controller, while k_{Piq}, k_{Iiq}, k_{Pid}, and k_{Iid} are the proportional and integral gains of qd PMSG current regulators; k_{Pin} and k_{Iin} are the proportional and integral gains of the Vienna rectifier DC bus voltage equalization system; and k_{PTLI} and k_{ITLI} are the proportional and integral gains of the TLI DC Bus voltage controller. Figures 13 and 14 show simulation results obtained with the SimPower System model and the averaged-value model, showing a purposely generated Vienna rectifier DC bus voltage unbalance with the balance system not activated. Specifically, capacitor voltages V_{C1} and V_{C2}, which at the beginning are equal because R_{C1} and R_{C2} are both set to 1000 Ω, diverge after $t = 3$ s because R_{C2} is changed to 600Ω in order to generate the voltage unbalance. The i_n current is zero when capacitor voltages are balanced and greater than zero after $t = 3$ s, while DC bus voltages V_{DC}' and V_{DC}'' do not vary. A zoomed-in view of the balanced and unbalanced steady-state operations of Figures 13 and 14 are shown in Figures 15 and 16, confirming a good matching between the results obtained with the two models. Figures 15d and 16d show the instantaneous Vienna rectifier power losses P_{TTR}, TLI power losses P_{TLI}, and PMSG power losses P_{Lg} during balanced DC bus capacitors. A one-time variation of the references of the output voltage and the TLI DC bus voltage is considered in Figures 17 and 18, while a load variation is shown in Figures 19 and 20. The results achieved with the two models are very close, but using the averaged-value model, the simulation time is roughly one third. In particular, all simulations have been accomplished on an Intel®CoreTM i7 CPU with 2.60 GHz and 16 GB RAM running a 64-bit Windows 10 operating system. Simulation results shown in Figures 13–20 required three minutes computing time using the SimPower System model with a 10^{-6} s time step. A 10^{-5} s time step can be used with the averaged-value model, because high frequency voltage and current harmonics are neglected, leading to only ten seconds to accomplish the same simulation.

Table 3. System parameters.

PMSG		Vienna		TLI	Control Gains
Power Rating	3 kW	IGBT Ratings	600 V, 20 A	200 V, 10 A	$K_{P\alpha} = 0.1$, $K_{I\alpha} = 10$
Rated Voltage	575 V_{DC}	DC-Link Voltage	200 V	100 V	$K_{Piqg} = 80$, $K_{Iiqg} = 1000$
Rated Current	6.5 A	DC Bus Capacitors	470 µF (C_1, C_2)	470 µF (C_T)	$K_{Pidg} = 80$, $K_{Iidg} = 1000$
Phase Inductance	20 mH	Load	50 Ω	//	$K_{Pin} = 0.2$, $K_{Iin} = 2$
Stator Resistance	4.3 Ω	Capacitors Resistance	1000 Ω (R_{C1}, R_{C2})	600 Ω (R_{CT})	$K_{PTLI} = 2$, $K_{ITLI} = 30$
PM Flux	0.57 Wb	Switching Frequency	50 Hz	40 kHz	

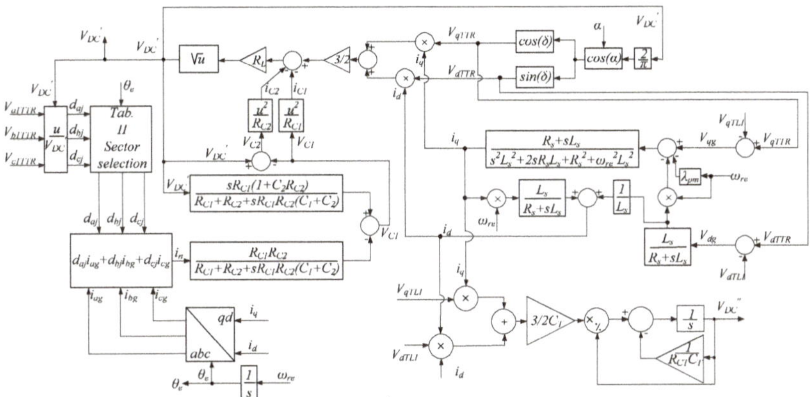

Figure 12. Block diagram of the developed averaged-value model.

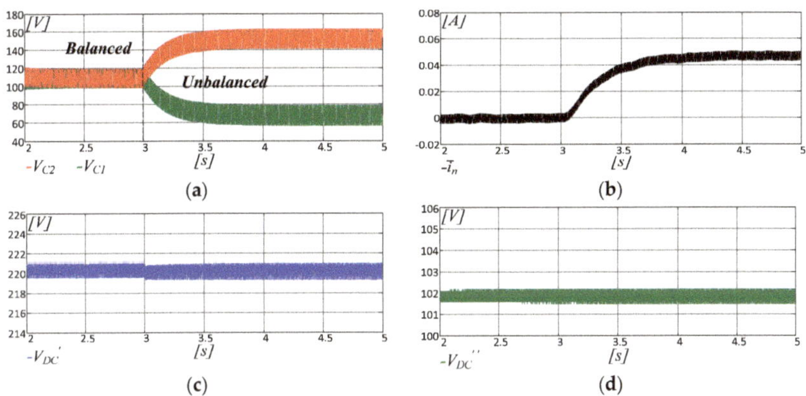

Figure 13. SimPower System model. (**a**) DC bus capacitor voltages V_{C1} and V_{C2}. (**b**) $i_n = i_{C1} - i_{C2}$. (**c**) output voltage V_{DC}'. (**d**) TLI DC bus voltage V_{DC}''.

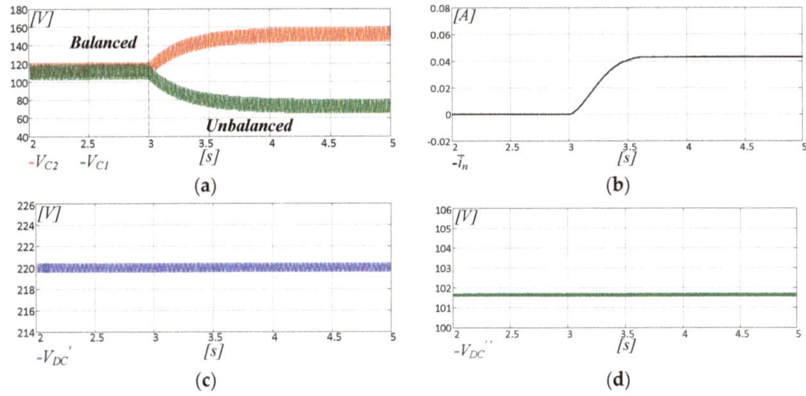

Figure 14. Averaged-value model. (**a**) DC bus capacitor voltages V_{C1} and V_{C2}. (**b**) $i_n = i_{C1} - i_{C2}$. (**c**) output voltage V_{DC}'. (**d**) TLI DC bus voltage V_{DC}''.

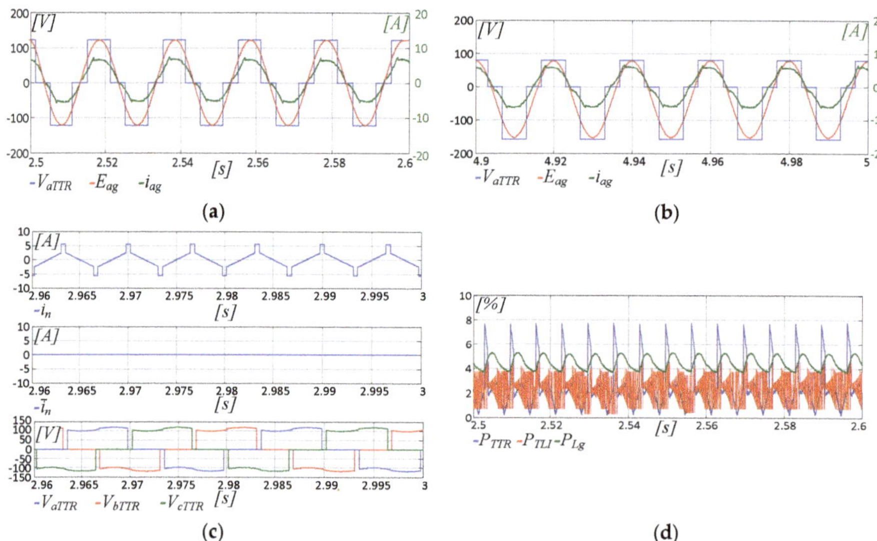

Figure 15. SimPower System model. (**a**) Balanced DC bus capacitor voltage operations, V_{aTTR}, E_{ag}, and i_{ag}. (**b**) Unbalanced DC bus capacitor voltage operations. (**c**) Current i_n, average value $\bar{i}_n$, and AC input Vienna voltages V_{iTTR}. (**d**) TRR power losses P_{TTR}, TLI power losses P_{TLI}, and PMSG power losses P_{Lg}.

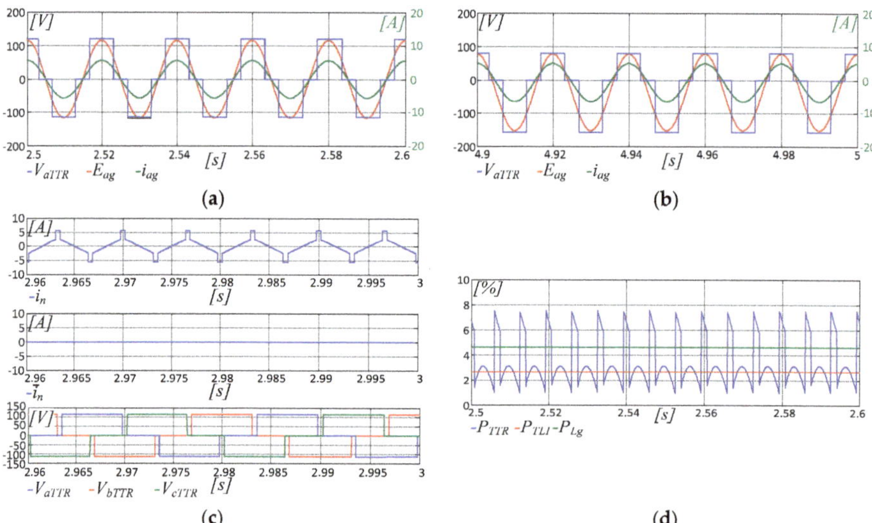

Figure 16. Averaged-value model. (**a**) Balanced DC bus capacitor voltage operations, V_{aTTR}, E_{ag}, and i_{ag}. (**b**) Unbalanced DC bus capacitor voltage operations. (**c**) Current i_n, average value $\bar{i}_n$, and AC input Vienna voltages V_{iTTR}. (**d**) TRR power losses P_{TTR}, TLI power losses P_{TLI}, and PMSG power losses P_{Lg}.

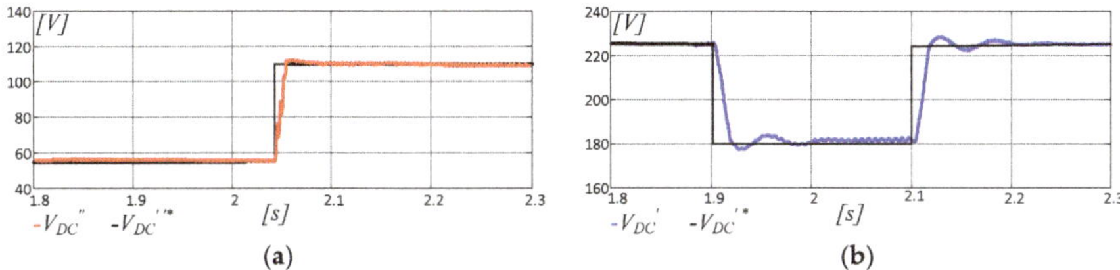

Figure 17. SimPower System model. One-time variation of $V_{DC}{}'$ and $V_{DC}{}''$ references. (**a**) TLI DC bus voltage $V_{DC}{}''$. (**b**) Output voltage $V_{DC}{}'$.

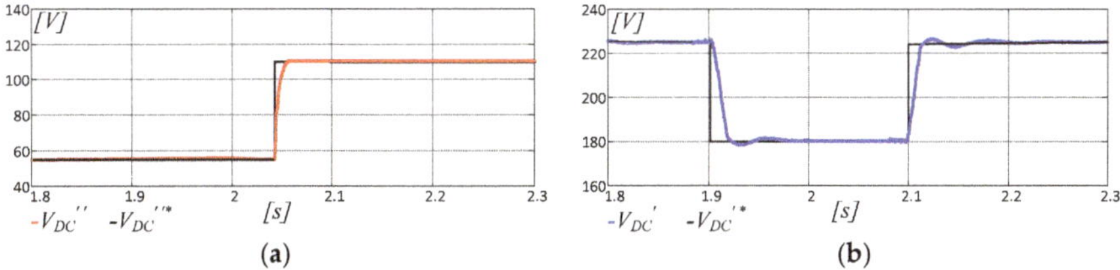

Figure 18. Averaged-value model. One-time variation of $V_{DC}{}'$ and $V_{DC}{}''$ references. (**a**) TLI DC bus voltage $V_{DC}{}''$. (**b**) Output voltage $V_{DC}{}'$.

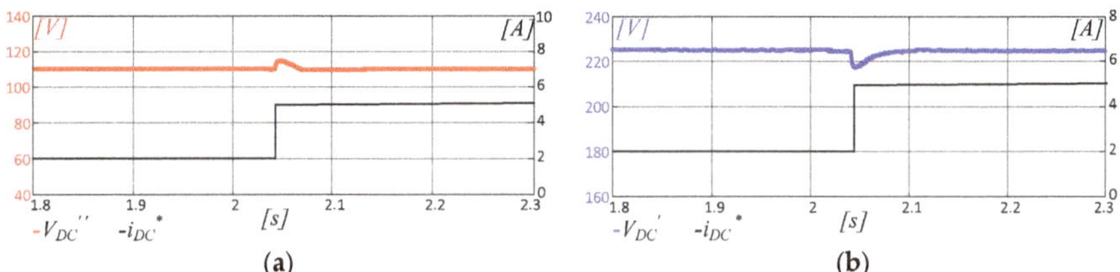

Figure 19. SimPower System model. Load variation $i_{DC}{}^*$. (**a**) TLI DC bus voltage $V_{DC}{}''$. (**b**) Output voltage $V_{DC}{}'$.

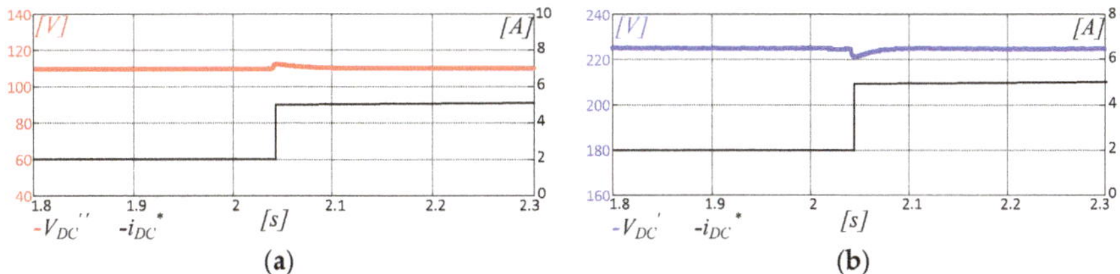

Figure 20. Averaged-value model. Load variation $i_{DC}{}^*$. (**a**) TLI DC bus voltage $V_{DC}{}''$. (**b**) Output voltage $V_{DC}{}'$.

Figure 21 displays the maximum errors between the quantities carried out by the two models, confirming a good accuracy of the proposed average model in a wide range of load conditions.

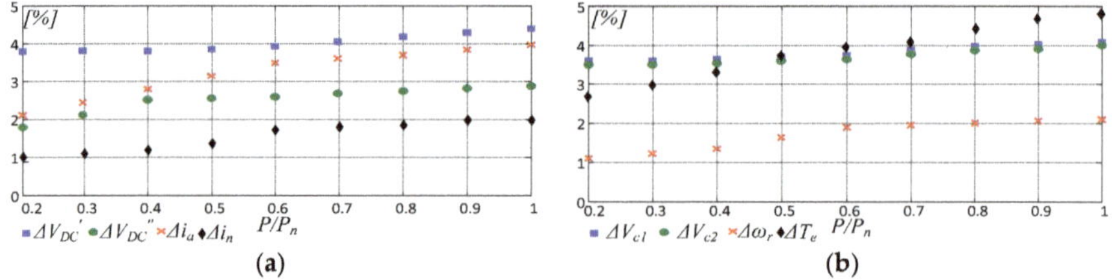

Figure 21. Percentage error between SymPower System and averaged model vs. the power expressed in per unit P/P_n. (**a**) Errors of V_{DC}', V_{DC}'', i_a, and i_n. (**b**) Errors of V_{c1}, V_{c2}, ω_r and T_e. Note: $\omega_r = 200$ rad/s, $V_{DC}' = 400$ V, and $V_{DC}'' = 100$ V.

5. Experimental Assessment

The accuracy of the AHUTR analytical model has also been verified comparing the results from the model with those from an experimental test rig consisting of 1kW AHUTR supplying an open-end-winding PMSG, mechanically coupled to a 2.6 kW PM synchronous motor drive used as a prime mover. Technical specifications of the PMSG are given in Table 4. This AHUTR supplied DC loads at 400V through the Vienna rectifier equipped with insulated gate bipolar transistors (IGBTs) whose technical data are listed in Table 5. The TLI was realized with low-voltage power metal-oxide-semiconductor field-effect transistor (MOSFETs) and operated at 40 kHz, $V_{DC}'' = 100$ V. Technical data of the power MOSFETs are reported in Table 6. The TLI floating capacitor and both capacitors C_1, C_2 were equal to 470µF. The DC load was modified using a variable power resistor. A single dSPACE DS1103 control board was used to control the Vienna rectifier and the TLI, while a 2048 ppr encoder was used to measure the rotor position θ_r of the PMSG. The experimental setup is shown in Figure 22. The currents and voltages were measured by using a dedicated sensing board equipped with the current transducer LEM LA 55-P and voltage transducer LEM LV 25-P.

Table 4. PMSG technical data.

P_n (kW)	L_s (mH)	V_s (V)	R_s (Ω)	I_s (A)	λ_{PM} (Wb)	ω_r (krpm)	Pole Pairs
1	20	565	4.8	6.5	1.53	2	3

Table 5. Technical specifications of STGW30NC60KD IGBT.

V_{ce} (V)	$V_{ce}(on)$ (V)	i_{RMS} (A)	t_{rise} (ns)	t_{fall} (ns)
600	2.1	30	27	160

Table 6. Technical specifications of IRFB5615PBF MOSFET.

V_{DS} (V)	R_{DS} (m)	I_D (A)	t_{rise} (ns)	t_{fall} (ns)
150	32	35	8.9	17.2

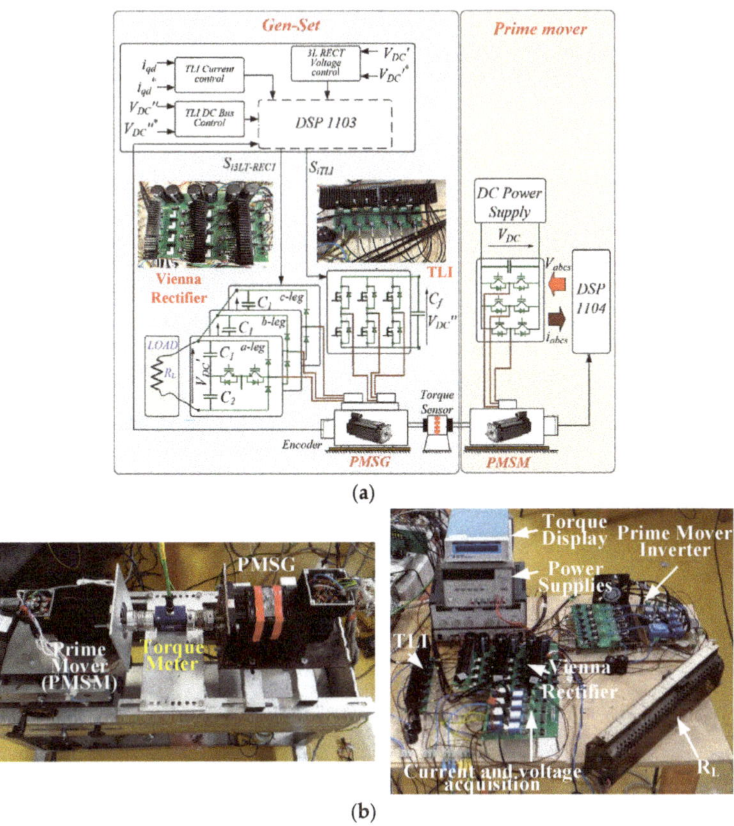

(a)

(b)

Figure 22. Experimental test bench. (**a**) Block scheme. (**b**) Experimental setup.

The experimental results shown in Figure 23 were obtained by imposing a transient voltage to V_{DC}' from 400 V to 320 V by keeping a constant resistor value $R_L = 80\ \Omega$ and with the PMSG spinning at $\omega_r = 200$ rad/s. Note a satisfying accuracy in the mechanical and electrical quantities estimated by the model. The voltage V_{DC}'' was properly modified by the control algorithm in order to keep the optimal ratio between the DC bus voltages V_{DC}' and V_{DC}''. A different test is displayed in Figure 24 in which a speed transient was forced by acting on the prime mover. More specifically, the rotational speed ω_r was changed from 200 rad/s to 260 rad/s while the resistive load was still kept constant. Even in this case, the model accurately predicted the behavior of the drive, both at steady-state and transient. The DC bus voltages were both affected by the speed variation, but the feedback control loops restored the reference values. A step load variation was imposed in the test of Figure 25, where the DC load was purposely doubled by switching from $T_L = 2$ Nm to $T_L = 4$ Nm. In this case, a more remarkable difference was observed between the model and the experimental results. Finally, the effectiveness of the model to predict the balancing of the voltages across the DC bus capacitors is shown in Figure 26. Initially, the balancing algorithm described in the previous sections was inactive, and thus, the voltages at the terminals of C_1 and C_2 were significantly different. At the instant t^*, the voltage-balancing approach was activated, nullifying $V_{C1} - V_{C2}$. The results of Figure 26 confirm the capability of the model to accurately simulate even this critical issue of the AHUTR. Maximum percentage errors between the outputs of the SimPower System and the averaged-value model are summarized in

Table 7, where the quantities with the suffix Δ are the errors in estimating $V_{DC}{}'$, ω_r, T_e, $V_{DC}{}''$, i_n, V_{c1}, and V_{c2}.

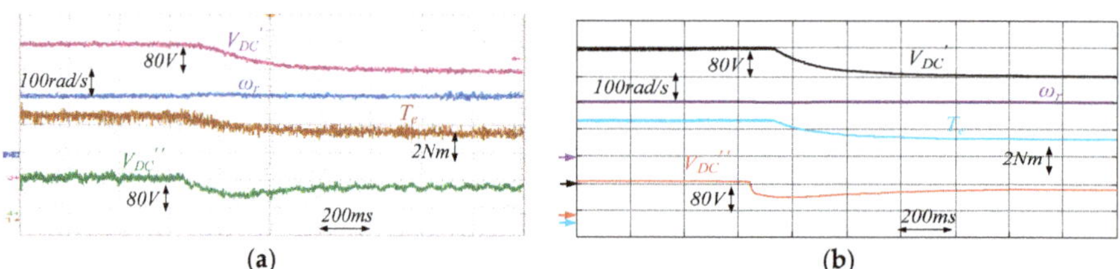

Figure 23. Voltage transient of $V_{DC}{}'$ from 400 V to 320 V under a constant resistor value $R_L = 80\ \Omega$. Output voltage $V_{DC}{}'$, TLI DC bus voltage $V_{DC}{}''$, rotor speed ω_r, electromagnetic torque T_e. (**a**) Experimental results. (**b**) Simulation results.

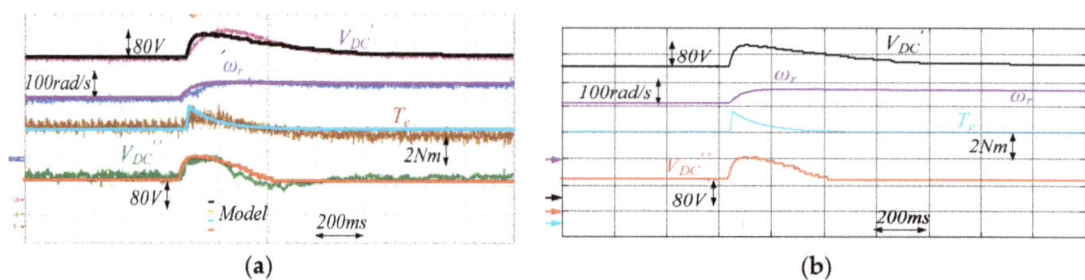

Figure 24. Speed transient from $\omega_r = 200$ rad/s to $\omega_r = 260$ rad/s under a constant resistor value $R_L = 80\ \Omega$ and $V_{DC}{}' = 400$ V, $V_{DC}{}'' = 100$ V. Output voltage $V_{DC}{}'$, TLI DC bus voltage $V_{DC}{}''$, rotor speed ω_r, electromagnetic torque T_e. (**a**) Experimental results. (**b**) Simulation results.

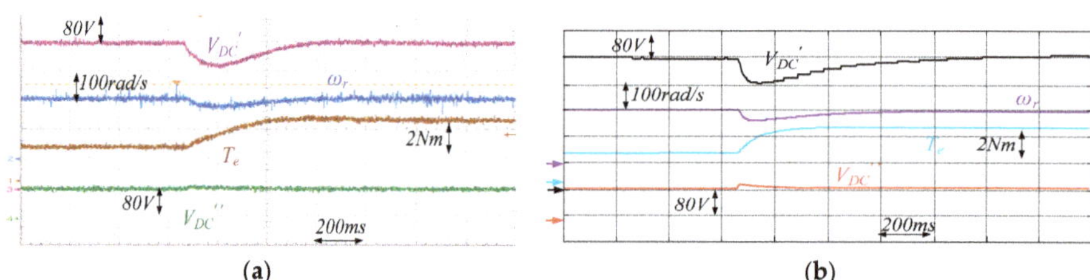

Figure 25. Load transient from $T_L = 2$ Nm to $T_L = 4$ Nm at $\omega_r = 200$ rad/s and $V_{DC}{}' = 400$ V, $V_{DC}{}'' = 100$ V. Output voltage $V_{DC}{}'$, TLI DC bus voltage $V_{DC}{}''$, rotor speed ω_r, electromagnetic torque T_e. (**a**) Experimental results. (**b**) Simulation results.

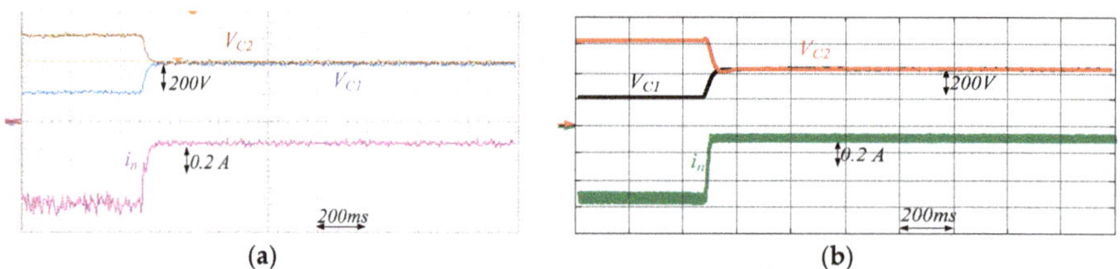

Figure 26. DC bus voltage balancing: $T_L = 4$ Nm, $\omega_r = 200$ rad/s and $V_{DC}{}' = 400$ V, $V_{DC}{}'' = 100$ V. V_{c1}, V_{c2} and i_n. (**a**) Experimental results. (**b**) Simulation results.

Table 7. Errors between experimental results and those obtained with the averaged-value model.

Test	$\Delta V_{DC}{'}$ (%)	$\Delta\omega_r$ (%)	ΔT_e (%)	$\Delta V_{DC}{''}$ (%)	$\Delta i_n{'}$ (%)	ΔV_{C1} (%)	ΔV_{C2} (%)
Figure 22	2.2	1.9	3.7	4.7	/	/	/
Figure 23	2.1	2.6	5	7.7	/	/	/
Figure 24	4.5	4.9	6	5	/	/	/
Figure 25	/	/	/	/	6.6	10	4

6. Conclusions

The asymmetrical hybrid unidirectional T-type rectifier is more efficient than a conventional PWM rectifier, mainly due to line frequency operation of the main converter, a T-type rectifier. However, it features a more complex structure composed of three main components, namely a TTR, a TLI, and an open winding electric machine, all interacting. The development of an accurate averaged-value mathematical model of the AHUTR topology aimed to optimally design control and management algorithms has been faced in the paper. Simulations and experimental results show that the proposed model is able to reproduce the static and dynamic behavior of the AHUTR with good accuracy. Furthermore, the obtained mathematical representation made a fast analysis of the system during TTR DC bus voltage unbalance operations possible. This has been exploited to design an active balancing system acting on the TLI side—a task which would be time-consuming with circuit-oriented simulator models. Furthermore, the averaged-value model has been used to define the entire AHTUR control and management system tasked to deal with efficiency maximization, input power factor control, TTR DC bus capacitor voltage balance, and the control of TLI floating DC bus voltage.

Author Contributions: This work was carried out in collaboration between all authors. S.F., G.S. and A.T. designed the study, wrote the manuscript and analyzed simulations and experimental results. A.S. undertook all of the experimental measurements.

Conflicts of Interest: The authors declare no conflict of interest.

References

1. Kaarthik, R.S.; Gopakumar, K.; Mathey, J.; Underland, T. Medium Voltage Drive for Induction Machine with Multilevel Dodecagon Voltage Space Vectors with Symmetric Triangles. *IEEE Trans. Ind. Electron.* **2015**, *62*, 79–87. [CrossRef]
2. Kouro, S.; Malinowski, M.; Gopakumar, K.; Pou, J.; Franquelo, L.G.; Wu, B.; Rodriguez, J.; Perez, M.A.; Leon, J.I. Recent Advances and Industrial Applications of Multilevel Converters. *IEEE Trans. Ind. Electron.* **2010**, *57*, 2553–2580. [CrossRef]
3. De Caro, S.; Foti, S.; Scimone, T.; Testa, A.; Cacciato, M.; Scarcella, G.; Scelba, G. THD and efficiency improvement in multi-level inverters through an open end winding configuration. In Proceedings of the IEEE 2016 Energy Conversion Congress and Expo, Milwaukee, WI, USA, 18–22 September 2016; pp. 1–7.
4. Edpuganti, A.; Rathore, A. A Survey of Low Switching Frequency Modulation Techniques for Medium-Voltage Multilevel Converters. *IEEE Trans. Ind. Appl.* **2015**, *51*, 4212–4228. [CrossRef]
5. Mondal, G.; Sivakumar, K.; Ramchand, R.; Gopakumar, K.; Levi, E. A Dual Seven-Level Inverter Supply for an Open-End Winding Induction Motor Drive. *IEEE Trans. Ind. Electron.* **2009**, *56*, 1665–1673. [CrossRef]
6. Wang, Y.; Panda, D.; Lipo, T.A.; Pan, D. Open-Winding Power Conversion Systems Fed by Half-Controlled Converters. *IEEE Trans. on Power Electron.* **2013**, *28*, 2427–2436. [CrossRef]
7. Mohapatra, K.K.; Gopakumar, K.; Somasekhar, V.T.; Umanand, L. A Harmonic Elimination and Suppression Scheme for an Open-End Winding Induction Motor Drive. *IEEE Trans. Ind. Electron.* **2003**, *50*, 1187–1198. [CrossRef]
8. Kawabata, Y.; Nasu, M.; Nomoto, T.; Ejiogu, E.C.; Kawabata, T. High-efficiency and low acoustic noise drive system using open-winding AC motor and two space-vector-modulated inverters. *IEEE Trans. Ind. Electron.* **2002**, *49*, 783–789. [CrossRef]
9. Sivakumar, K.; Das, A.; Ramchand, R.; Patel, C.; Gopakumar, K. A hybrid multilevel inverter topology for an open-end winding induction-motor drive using two-level inverters in series with a capacitor-fed H- bridge cell. *IEEE Trans. Ind. Electron.* **2010**, *57*, 3707–3714. [CrossRef]

10. Somasekhar, V.T.; Gopakumar, K.; Bajiu, M.R.; Mohapatra, K.K.; Umanand, L. A multilevel inverter system for an inductor motor with open-end windings. *IEEE Trans. Ind. Electron.* **2015**, *52*, 824–836. [CrossRef]

11. Edpuganti, A.; Rathore, A.K. Optimal Pulsewidth Modulation for Common-Mode Voltage Elimination Scheme of Medium-Voltage Modular Multilevel Converter-Fed Open-End Stator Winding Induction Motor Drives. *IEEE Trans. Ind. Electron.* **2017**, *64*, 848–856. [CrossRef]

12. Foti, S.; Testa, A.; Scelba, G.; De Caro, S.; Cacciato, M.; Scarcella, G.; Scimone, T. An Open-End Winding Motor Approach to Mitigate the Phase Voltage Distortion on Multilevel Inverters. *IEEE Trans. Power Electron.* **2018**, *33*, 2404–2416. [CrossRef]

13. Foti, S.; Testa, A.; Scelba, G.; Sabatini, V.; Lidozzi, A.; Solero, L. A symmetrical hybrid unidirectional T-type rectifier for high-speed gen-set applications. In Proceedings of the IEEE Energy Conversion Congress and Expo, Cincinnati, OH, USA, 1–5 October 2017; pp. 4887–4893.

14. Foti, S.; De Caro, S.; Scelba, G.; Scimone, T.; Testa, A.; Cacciato, M.; Scarcella, G. An Optimal Current Control Strategy for Asymmetrical Hybrid Multilevel Inverters. *IEEE Trans. Ind. Appl.* **2018**, *54*, 4425–4436. [CrossRef]

15. Mengoni, M.; Amerise, A.; Zarri, L.; Tani, A.; Serra, G.; Casadei, D. Control Scheme for Open-Ended Induction Motor Drives with a Floating Capacitor Bridge Over a Wide Speed Range. *Trans. Ind. Applications* **2017**, *53*, 4504–4514. [CrossRef]

16. Pan, Z.; Peng, F.Z.; Corzine, K.A.; Stefanovic, V.R.; Leuthen, J.M.; Gataric, S. Voltage balancing control of diode-clamped multilevel rectifier/inverter systems. *IEEE Trans. Ind. Applications* **2005**, *41*, 1698–1706. [CrossRef]

17. Yan, G.; Duan, S.; Zhao, S.; Li, G.; Wu, W.; Li, H. Research on the mechanism of neutral-point voltage fluctuation and capacitor voltage balancing control strategy of three-phase three-level T-type inverter. *J. Electr. Eng. Technol.* **2017**, *12*, 2227–2236.

18. Zhang, Y.; Li, J.; Li, X.; Cao, Y.; Sumner, M.; Xia, C. A Method for the Suppression of Fluctuations in the Neutral-Point Potential of a Three-Level NPC Inverter with a Capacitor-Voltage Loop. *IEEE Trans. Power Electron.* **2017**, *32*, 825–836. [CrossRef]

Article

An Improved Imperialist Competitive Algorithm to Solve the Selected Harmonic Elimination Pulse-Width Modulation in Multilevel Converters

Zheng Gong [1], Qi Cui [1], Xi Zheng [1], Peng Dai [1,* and Rongwu Zhu [2]

[1] Jiangsu Province Laboratory of Mining Electric and Automation, China University of Mining and Technology, Xuzhou 221008, China; gzcumt@163.com (Z.G.); kuangdarencuiqi@sina.com (Q.C.); 15262017302@163.com (X.Z.)
[2] Chair of Power Electronics, Chritian-Albrechts- University of Kiel, 24118 Kiel, Germany; rzh@tf.uni-kiel.de
*   Correspondence: 13329285666@189.cn; Tel.: +86-150-0521-5026

Received: 18 October 2018; Accepted: 5 November 2018; Published: 8 November 2018

Abstract: The traditional intelligent algorithms for the selected harmonic elimination pulse-width modulation (SHEPWM) of multilevel converters provide low convergent rate and low accuracy of solutions when solving quarter-wave symmetry nonlinear equations. To obviate this problem and obtain a better modulating performance, an improved imperialist competition algorithm is proposed. The proposed algorithm enhances the global search ability by using moving imperialists. Also, a novel type of particles, named independent countries, are proposed to help the algorithm jump out of the local optimum. These independent countries change their positions using swarm intelligence. Compared with the existing particle swarm algorithm and genetic algorithm, the proposed algorithm has significant advantages by improving the accuracy of solutions and the rate of convergence. Finally, the correctness and effectiveness of the proposed algorithm are verified and evaluated by simulation and experimental results.

Keywords: multilevel converter; selected harmonic elimination; genetic algorithm; imperialist competitive algorithm

---

1. Introduction

Multilevel converters have been widely applied in high-voltage and high-power applications because of their advantages of effectively improving the quality of output voltage waveform, large output capacity, and high inverting efficiency [1,2]. They have been employed for many industrial applications, such as electrical motor drives [3], energy storage systems [4], and renewable power generators [5]. Moreover, they are also considered as active power filters [6,7] to satisfy the urgent grid-friendly requirements. There are several methods, such as sinusoidal pulse-width modulation (SPWM), space vector pulse-width modulation (SVPWM), and selective harmonic elimination pulse-width modulation (SHEPWM), that can be applied to multilevel converters. As shown in Table 1, compared with the other two methods, the most attractive advantage of the SHEPWM is that the low-order harmonics can be controlled. In addition, SHEPWM also keeps the advantages of a wide modulation index and a high utilization of DC voltage [4]. It was also verified that SHEPWM could be implemented with the objective of minimizing total harmonic distortion (THD) [8].

The key challenge for the SHEPWM technique is to solve nonlinear equations containing trigonometric functions to obtain the right switching angles. A lot of contributions have been made to address this issue in recent years. In a study [9], the Walsh functions are employed to transform nonlinear equations into a series of linear algebraic equations, which can be easily calculated online, providing various sets of solutions. Nevertheless, the transitions between the Walsh series and the

Fourier series requires an effective algorithm, and the high-accuracy requirement may cause an extra heavy computation [10]. In reference [11], high-order equations are transferred to simple trigonometric polynomials, and the calculation mass is greatly reduced by employing the resultant theory. However, the increase of the switching angles enlarges the polynomial order and makes it very difficult to be implemented; also, if there are several DC sources, the expression of the resultant polynomials becomes more complicated [12]. The Homotopy algorithm is adopted to solve the equations in other studies [13,14]. It features a rapid convergence and can be extended to the high-level converters without any extra analytical calculations. Moreover, only one set of solutions can be obtained by this algorithm. In references [15,16], the Newton–Raphson algorithm is applied to solve nonlinear equations. This algorithm keeps a high accuracy and a low computation burden, but its implementation strongly relies on the accuracy of the initial values. If the predictions of these initial values are not around the right solution, the solving process would plunge into local optima. Meanwhile, there is still only one sets of solutions that can be determined in each solving instant. A novel Groebner–Bases-based method is presented in reference [17], which transforms the nonlinear equations into an equivalent canonical system and improves the accuracy of the solutions. However, the transformation becomes very complex when the order of the equations is high, making this method more complicated than the previous methods.

Table 1. Comparison of sinusoidal pulse-width modulation (SPWM), space vector pulse-width modulation (SVPWM), and selected harmonic elimination pulse-width modulation (SHEPWM).

Modulation Type	SPWM	SVPWM	SHEPWM
DC voltage utilization	0~0.866	0~1	0~1.12
Switching frequency	Medium	High	Low
Complexity of strategy	Low	High	High
Implementation approach	Online	Online	Offline

On the basis of the review above, it is seen that the aforementioned solving methods for the SHEPWM are commonly complex and not available for all sets of solutions. Actually, the problem of solving the SHEPWM can be reformulated into a constraint optimization problem, and many intelligent algorithms can also be adopted to solve this problem [18–22]. They not only can find all solutions, but also can be expediently applied to the multilevel converters with equal or non-equal DC voltage sources.

The genetic algorithm (GA) has been used in SHEPWM for many years. It is an algorithm which was inspired by natural biological evolution. This approach employs selection, crossover, and mutation operators and starts by randomly generating the individuals of a population. The individuals represent the properties of a solution and can be mutated and crossover to evolve toward better solutions. As we can see, the GA is simple and easy to understand but, as the required switching angles increase, the distortion of the output voltage gets worse. A new GA algorithm in reference [18] divided a population into several independent populations whose individuals can migrate from one to another. This improved algorithm, named multi-population genetic algorithm (MGA), can settle the problem of precociousness efficiently, but the accuracy of solutions still demands improvement. Particle swarm optimization (PSO), a common algorithm, simulates the predation behaviour of bird flocks and has been proved very efficient in precision and convergent rate compared with GA. Each particle of PSO has personal and global best positions which guide it towards to the optimal solution. The PSO in reference [19] is used to solve a problem that reduces the computation. In reference [20], a constriction factor was introduced, and a method to obtain proper constriction factor and acceleration coefficients was described. The advanced algorithm improves the precociousness problem but still lacks global search ability.

The imperialist competitive algorithm (ICA) is inspired by imperialistic competition and includes two primary categories: imperialists and colonies. The major steps are colonies assimilation and empires' competition. The research presented in references [21,22] has proved the ascendency of the

ICA in rate and speed of convergence. The contribution described in reference [23] is the first attempt to apply the ICA to solve the SHEPWM of power converters. It has been proved that this conventional ICA algorithm can solve nonlinear equations when the dimension of the variable is low. However, the convergent rate decreases, and the accuracy of solutions of the conventional ICA becomes worse with the increasing dimension of multilevel converters. Hence, the solving execution time will be longer, and the modulation accuracy will not be guaranteed.

To improve the ICA-based SHEPWM for multilevel converters, a novel kind of solving method based on the particles named the independent countries is proposed in this paper. The optimized design is conducted from two starting points: one is to increase the diversity and movements of the imperialists, the other is to enhance the ability to jump out of the local optimal. This optimized design is expected to obtain high convergent rates and high accuracy solutions when the converter' dimension increases. In addition, with taking the neutral-point-clamped H-Bridge (NPC/H-Bridge) five-level converter [24] as a typical multilevel converter for the application, sufficient comparative analysis and simulation research regarding various kind of solving methods for the multilevel SHEPWM are carried out to present the characteristics and advantages of the proposed method. Finally, experimental research is also conducted with a downscaled NPC/H-Bridge five-level converter prototype to validate its practical applicability.

The rest of this paper is divided as follows. Section 2 introduces the topology of the NPC/H-Bridge five-level converter and the basic principle for solving the SHEPWM. Section 3 presents the novel intelligent algorithm and compares it to three existing algorithms in detail. Simulation results and best solutions regarding the condition of two switching angles per quarter wave are described and evaluated in Section 4. The experimental results are shown in Section 5, and Section 6 concludes this paper.

2. NPC/H-Bridge Five-Level Converter and SHEPWM Switching Strategy

2.1. NPC/H-Bridge Five-Level Converter

Figure 1 shows the topology of the NPC/H-Bridge five-level converter. As shown, compared to the conventional neutral-point-clamped (NPC) [25], flying capacitor (FC) [26] topologies, this topology needs fewer clamping diodes or capacitors, which reduces the cost and improves the stability of system. Besides, compared to the modular multilevel converters (MMC) [27] cascaded H-Bridge (CHB) [28] topology, the number of DC-side power supplies can be reduced, and then the large number of capacitors or the bulky multiple isolation transformers can be reduced. For the NPC/H-Bridge five-level converter, each phase unit consists of two single-phase three-level NPC-type bridge arms, two DC-side capacitors, and one DC voltage source. In each NPC-type bridge arm, there are four power switching devices (along with the forward diodes) and two clamped power diodes.

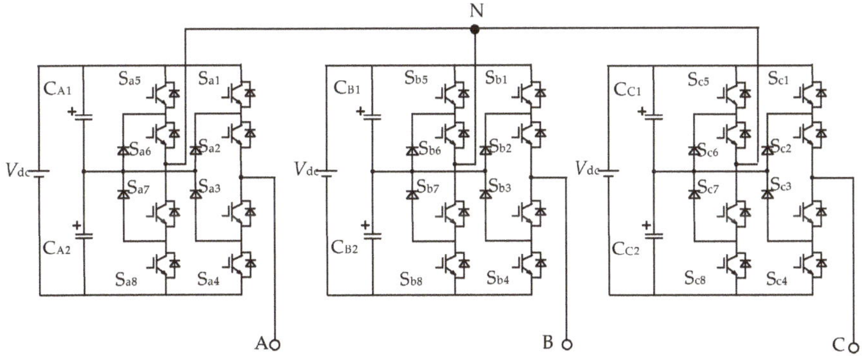

Figure 1. Three-phase NPC/H bridge five-level inverter topology.

Taking phase A as an example, it can be seen that the switching devices Sa1–Sa4 cascade the right bridge, while the Sa5–Sa8 cascade the left bridge. Each bridge can generate three voltages, i.e., $+1/2V_{dc}$, 0, and $-1/2V_{dc}$ by different switching angles, so the output voltages of phase A include $+V_{dc}$, $+1/2V_{dc}$, 0, $-1/2V_{dc}$, $-V_{dc}$. Table 2 shows the switching states of the NPC/H bridge five-level inverter. S_{x1}-S_{x8} represent the switching states of IGBTs. V_{XN} is the value of output voltage. In this paper, the switching states 1, 2, 5, 8, and 9 are chose to synthesize the output waveforms.

Table 2. Switching states of the NPC/H bridge five-level inverter.

Number	S_{x1}, S_{x2}, S_{x3}, S_{x4}, S_{x5}, S_{x6}, S_{x7}, S_{x8}	V_{XN}
	(x = a, b, c)	(X = A, B, C)
1	11000011	V_{dc}
2	11000110	$V_{dc}/2$
3	11000110	
4	11001100	
5	01100110	0
6	00110011	
7	01101100	$-V_{dc}/2$
8	00110110	
9	00111100	$-V_{dc}$

2.2. Basic Principle for Solving the SHEPWM

Figure 2 shows the output voltage waveforms at the high and low modulation shown in Figure 1. Every quarter-wave of a waveform includes two switching angles. These angles are symmetric to $\pi/2$ in a half period and conform to the so-called quarter-wave symmetry technique [29]. Compared to the half-wave symmetry and asymmetry techniques [15,30], the quarter-wave symmetry technique can simplify nonlinear equations and reduce the computation burden.

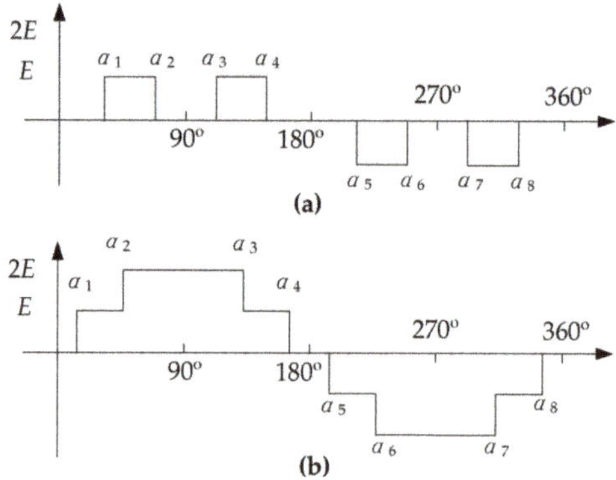

Figure 2. Pulse width modulation (PWM) waveforms definition of two switching angles according to the quarter-wave SHEPWM techniques. (**a**) Waveform at low modulation width; (**b**) waveform at high modulation width.

Equation (1) shows the Fourier series expansion of the full cycle.

$$V(t) = a_0 + \sum_{n=1}^{\infty} (a_n \cos n\omega t + b_n \sin n\omega t) \tag{1}$$

where $n = 1, 2, 3 \ldots$

Because of the characters of quarter-wave symmetry, the DC component a_0 and the cosine component a_n are equal to zero. The sine component is zero when n is even. Equation (1) is simplified as follow:

$$\begin{cases} a_n = 0, \ n = 0,1,2,3\dots \\ b_n = \begin{cases} 0, \ n \text{ is even} \\ \frac{4E}{n\pi} \sum\limits_{k=1}^{N} (-1)^{k+1} \cos n\alpha_k \end{cases} \end{cases} \tag{2}$$

Because all triple harmonics are removed in line voltage, only the $6k \pm 1$th harmonic should be eliminated. Equation (3) is the simplified Equation (2):

$$\begin{cases} \frac{2}{\pi} \sum\limits_{k=1}^{N} p_k \cos \alpha_k = M \\ \sum\limits_{k=1}^{N} p_k \cos n\alpha_k = 0 \quad n = 5,7,11,\dots,6k \pm 1 \end{cases} \tag{3}$$

where p_k indicates the rising or falling edge of the output voltage, which is given by Equation (4):

$$p_k = \begin{cases} 1 & \text{rising edge} \\ -1 & \text{falling edge} \end{cases} \tag{4}$$

According to the waveform of the five-level converter, the polynomial equations can be simplified to Equation (5):

$$\begin{cases} \cos \alpha_1 + \cos \alpha_2 - M\pi/2 = \varepsilon_1 \\ \cos 5\alpha_1 + \cos 5\alpha_2 = \varepsilon_2 \end{cases} \tag{5}$$

$$f(\alpha) = \varepsilon_1{}^2 + \varepsilon_2{}^2 \tag{6}$$

Equation (6) is the cost function of the intelligent algorithm, which is built on Equation (3). Different switching angles can be calculated when the modulation index is changed.

3. The Proposed Improved Imperialist Competitive Method for Multilevel SHEPWM

3.1. Design of the Improved Imperialist Competitive Algorithm

ICA establishes a mathematical model based on imperialistic competition and belongs to the random optimization search methods. ICA generates initial countries (which equal to particles of PSO or chromosomes of GA) within the search space. A number of countries are selected as imperialists according to their fitness, and the remaining countries are randomly allocated to the imperialists as colonies. An imperialist and its colonies constitute an empire. Then, each imperialist assimilates its relative colonies, and the empires compete with each other until only one empire remains or the terminal condition is achieved. The main steps of ICA are as follows:

(1) Generating initial countries. The algorithm randomly generates N_{pop} initial countries within the search space. For a N_{var} dimensional problem, a country is defined according to Equation (7). Then, the first N_{imp} powerless countries are selected as imperialists, according to Equations (8)–(10). Colonies are assigned to each imperialist, according to Equation (11).

$$\text{country} = [d_1, d_2, d_3, \dots, d_{N_{\text{var}}}] \tag{7}$$

$$\text{cost} = f(\text{country}) = f(d_1, d_2, d_3, \dots, d_{N_{\text{var}}}) \tag{8}$$

$$C_n = c_n - \max_i \{c_i\} \tag{9}$$

$$p_n = \left| \frac{C_n}{\sum\limits_{i=1}^{N_{imp}} C_i} \right| \tag{10}$$

$$N.C_n = round\{ p_n \times N_{col} \} \tag{11}$$

where c_n is the cost of the nth imperialist, and C_n and p_n are its normalized value and power. $N.C_n$ in Equation (11) represents the number of colonies of every imperialist.

(2) Assimilating colonies. Each imperialist improves its colonies by changing their positions. All colonies move toward the imperialist according to Equations (12) and (13):

$$Vector_n = imp^{i-1} - col_n^{i-1} \tag{12}$$

$$col_n^i = col_n^{i-1} + r_4 * \text{rand}(1, N_{var}). * Vector \tag{13}$$

col_n^i and col_n^{i-1} mean the current and previous positions of the nth colony, and r_4 is the assimilation coefficient. During this process, if the cost of any colony is less than the cost of the relative imperialist, the imperialist position is updated.

(3) Imperialistic competition. The most powerful empire takes possession of the worst colony of the weakest empire by two procedures.

(a) Calculating the total costs of empires.

$$T.C_n = f(imp_n) + \xi \times \frac{\sum\limits_{i=m}^{N.C._n} f(col_m)}{N.C_n} \tag{14}$$

In Equation (14), $T.C_n$ means the total cost of the nth empire, and ξ is a positive number less than 1.

(b) The empire which has the largest $T.C_n$ is regarded as the weakest empire, and its worst colony is captured by other empires, according to Equations (15) and (16):

$$N.T.C_n = T.C_n - \max_i \{ T.C_i \} \tag{15}$$

$$R_n = \left| \frac{N.T.C_n}{\sum\limits_{i=1}^{N_{imp}} N.T.C_n} \right| \tag{16}$$

where $T.C_n$ and $N.T.C_n$ indicate the total cost and normalized total cost, respectively. R_n is the probability of the nth empire capturing the worst colony.

(4) Eliminating the worst empire. When an empire loses all its colonies, it is deleted.

(5) Reset colonies' positions. When the total cost of the best empire does not change for a long time, reset all colonies' positions of each empire stochastically.

In this paper, not only the colony countries change positions, but also the imperialists move their positions to improve their performance. Also, a type of particles called independent countries is presented to improve the diversity of ICA, which change positions using swarm intelligence. This new hybrid algorithm consists of both PSO and ICA and, thus, it is called PSOICA algorithm. The implementation steps of the proposed PSOICA algorithm are as follows.

(a) Select the first ($N_{imp} + 1 \sim N_{imp} + N_{ind}$) powerless countries as independent countries according to Equations (7)–(9). N_{ind} is the number of independent countries.

(b) Move imperialists and independent countries in line with the PSO procedures. In this step, imperialists and independent countries may gain better positions. This step enhances the global search ability of the algorithm.

$$imp_n^i = imp_n^{i-1} + V_{imp}^i \tag{17}$$

$$V_{imp}^i = c_1 * r_1 * (best_{imp} - imp_n^{i-1}) \tag{18}$$

Move the imperialists on the basis of Equations (17) and (18); imp_n^i and V_{imp}^i indicate the current position and velocity of the nth imperialist, imp_n^{i-1} means the previous position of the nth imperialist, c_1 is a positive number less than 1, and r_1 is a random number between 0 and 1. If the cost of the current position is less than that of the previous position, update the imperialist, otherwise, keep the previous position.

$$V_n^i = \omega * V_n^{i-1} + c_2 * r_2 * (P_l^{i-1} - ind_n^{i-1}) + c_3 * r_3 * (P_g^{i-1} - ind_n^{i-1}) \tag{19}$$

$$ind_n^i = ind_n^{i-1} + V_n^i \tag{20}$$

Equations (19) and (20) show the movement of independent countries, where P_l and P_g are the personal and global best positions of the nth independent countries, V_n^i is the current velocity, ind_n^i is the current position of the nth independent country, ω is the inertia weight, c_2 and c_3 are acceleration constants, and r_2 and r_3 are random numbers between 0 and 1. If the cost of P_g is better for the imperialist, the imperialist moves to P_g, and vice versa if it is worse.

Figure 3 shows the moving steps of the ICA and the proposed PSOICA.

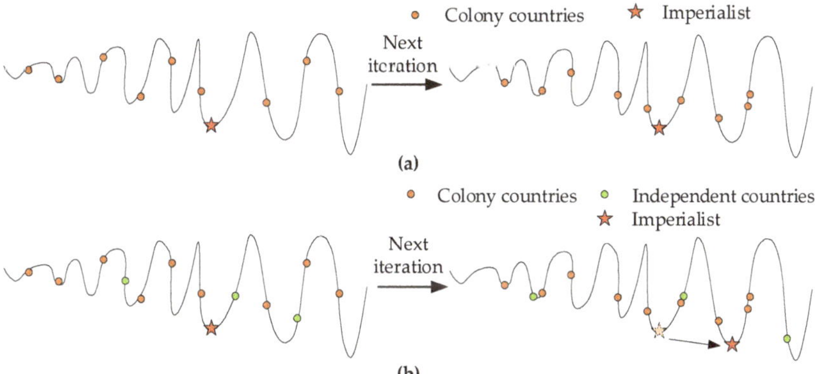

Figure 3. (a) Moving steps of the imperialist competitive algorithm (ICA); (b) moving steps of the particle swarm optimization (PSO) ICA (PSOICA).

Figure 3a shows the moving steps of the conventional ICA [20,22] when all the colony countries are assimilated by their imperialist, and the imperialist never moves. Figure 3b shows the moving steps of the proposed PSOICA, with all the colony countries moving towards their imperialist. Meanwhile, the imperialist changes its position and then it may move to a better position. The independent countries move to their next positions on the basis of the swarm intelligence, and they may get new better positions or stay in the former positions.

Figure 4 shows the flow chart of the proposed PSOICA.

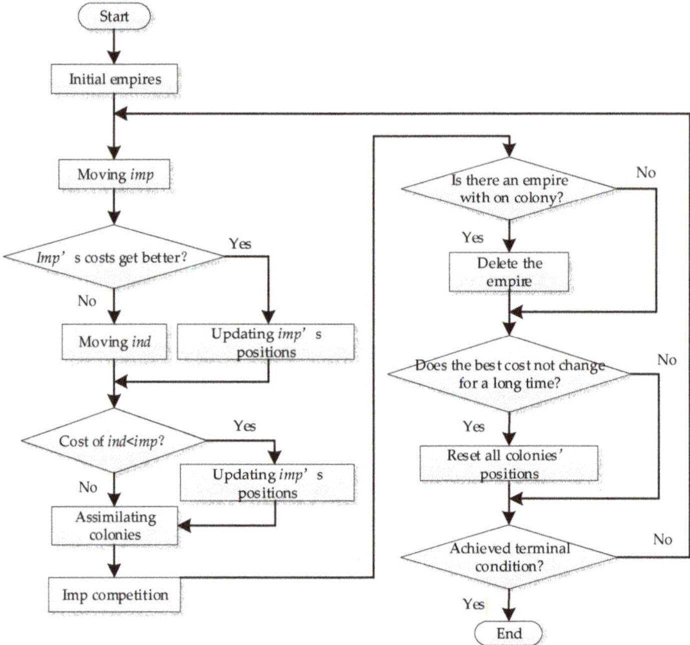

Figure 4. Flow chart of the proposed PSOICA.

3.2. Comparisons with the Existing Methods

To make the comparisons, the MGA [18], PSO [20], ICA [23], and the proposed PSOICA are considered to calculate their cost function. The parameters of three algorithms are set as follows: for MGA, the populations re 10, and each of them has 40 chromosomes; the initial crossover rates are random numbers between 0.7 and 0.9, and the mutation rates are random numbers between 0.2 and 0.35; for PSO, the particles are 400, both the acceleration coefficients are 1.49445, and the inertia weight increases along with iteration from 0.21 to 0.7; for ICA, there are 400 initial countries and 20 imperialists, and c_1, c_2, c_3 are equal to 0.95, 0.5, 0.5, ω is 0.7298, and the assimilation coefficient r_4 is 2.5; for the proposed PSOICA, there are 400 initial countries, 20 imperialists, 20 independent countries, and c_1, c_2, c_3, ω, and r_4 are the same as for ICA.

In Table 3, the efficiency and feasibility of the proposed PSOICA for the optimization of SHEPWM are compared with those of three other algorithms.

Table 3. Comparing the performance of the PSOICA to those of the multi-population genetic algorithm (MGA), PSO, and ICA.

m	Switching Angles		THD%	$f(\alpha)$			
				MGA	PSO	ICA	PSOICA
1.1	6.715	42.72	18.34	2.95×10^{-11}	4.93×10^{-32}	4.93×10^{-32}	4.93×10^{-32}
1	16.33	52.33	19.27	9.80×10^{-11}	4.93×10^{-32}	4.93×10^{-32}	4.93×10^{-32}
0.9	23.99	59.99	26.58	1.02×10^{-11}	4.93×10^{-32}	4.93×10^{-32}	2.77×10^{-32}
0.8	30.65	66.65	35.21	5.06×10^{-17}	0	0	0
0.7	36.7	72.68	43.95	1.08×10^{-8}	0	0	0
0.6	42.3	78.3	52.64	8.53×10^{-10}	1.233×10^{-32}	1.23×10^{-32}	1.11×10^{-31}
0.5	47.61	83.61	60.7	3.62×10^{-11}	4.93×10^{-32}	4.93×10^{-32}	1.97×10^{-31}
0.4	52.71	88.71	65.51	1.64×10^{-11}	4.93×10^{-32}	4.93×10^{-32}	4.93×10^{-32}
0.3	57.69	86.31	88.04	1.01×10^{-6}	3.081×10^{-33}	3.08×10^{-33}	0
0.2	62.5	81.5	128.72	5.04×10^{-8}	1.233×10^{-32}	1.23×10^{-32}	1.77×10^{-30}
0.1	67.26	76.74	205.63	5.12×10^{-8}	7.704×10^{-34}	7.70×10^{-34}	7.70×10^{-34}

It should be noticed that all four algorithms converge successfully, and all the THD under the same modulation index unify, which means that all the algorithms can solve the nonlinear equations efficiently when only two switching angles are included. The cost values show that the PSO, ICA, and the proposed PSOICA are better than the MGA in local search. The cost of the MGA is about 1×10^{-10}, and those of the PSO, ICA, and PSOICA are about 1×10^{-30}.

At low modulation index, if more selected harmonics of output voltage have to be eliminated, the switching angles per quarter wave should be increased. Table 4 shows the results of the four algorithms when we compare the best costs of five computations. Each quarter wave includes four angles.

Table 4. Switching angles and costs at 0.2 modulation index.

Algorithms	$\alpha1$	$\alpha2$	$\alpha3$	$\alpha4$	$f(\alpha)$
MGA	51.219	57.315	73.357	84.342	0.001912
PSO	50.893	57.74	72.439	85.149	1.76×10^{-31}
ICA	50.894	57.74	72.439	85.148	2.38×10^{-10}
PSOICA	50.893	57.74	72.439	85.149	1.68×10^{-30}

As can be seen, the switching angles of the MGA deteriorated, and the cost are bigger than 1×10^{-5}. Figure 5a shows the simulation results of the MGA. The output voltage distortion could not be ignored. In contrast, the costs of the PSO, ICA, and the proposed PSOICA are much smaller than 1×10^{-5}. However, with the increasing switching angles, the accuracy of the solution could worsen. Comparing Table 3 with Table 4, the accuracy of the solutions calculated by the ICA or the MGA decreases rapidly with the increasing switching angles, while the costs of the PSO and the PSOICA have no significant changes. Hence, the advantages of the PSO and PSOICA algorithms in local searching ability are further confirmed. Figure 5b,c exhibits the phase voltages and corresponding spectra.

As shown in Figure 5, when the switching angles increase to 4, the MGA cannot eliminate the 5th harmonic completely, and the phase voltage distortion raises to 9.2%. In contrast to the MGA, the 5th, 7th, and 11th harmonics are eliminated by the PSO, ICA, and proposed PSOICA. At the same time, the phase voltages also meet the desired values.

The results show that the dimension of the variables has a large influence on the precision of solutions. In Figure 6, the phase voltage distortion rates and the THDs of the four algorithms are plotted with respect to the modulation degree in the situation of four switching angles per quarter wave.

As we can see, the phase voltage distortion rates (PVDR) of the MGA even reach 40%, which is quite higher than the those of PSO, conventional ICA, and proposed PSOICA, whose PVDR ar on the average, 1%. That means that the MGA performs worse in high variable optimization problems. In contrast, the PSO, conventional ICA, and proposed PSOICA can output the desired voltage effectively and eliminate the selective harmonics. However, there is a big difference between these three algorithms. They have different THD of phase voltages. For PSO and ICA, the THDs decrease with modulation increase but, in some situations, the THDs get higher. For the proposed PSOICA, THD is an obvious attenuation curve. This is because in low modulation index, the nonlinear equations with the four switching angles have three different solutions, and each of them has different THDs. The PSOICA finds the best optimal solution that has the lowest THD each time. However, the ICA and PSO frequently result in suboptimal solutions, with higher THDs due to the searching limitation.

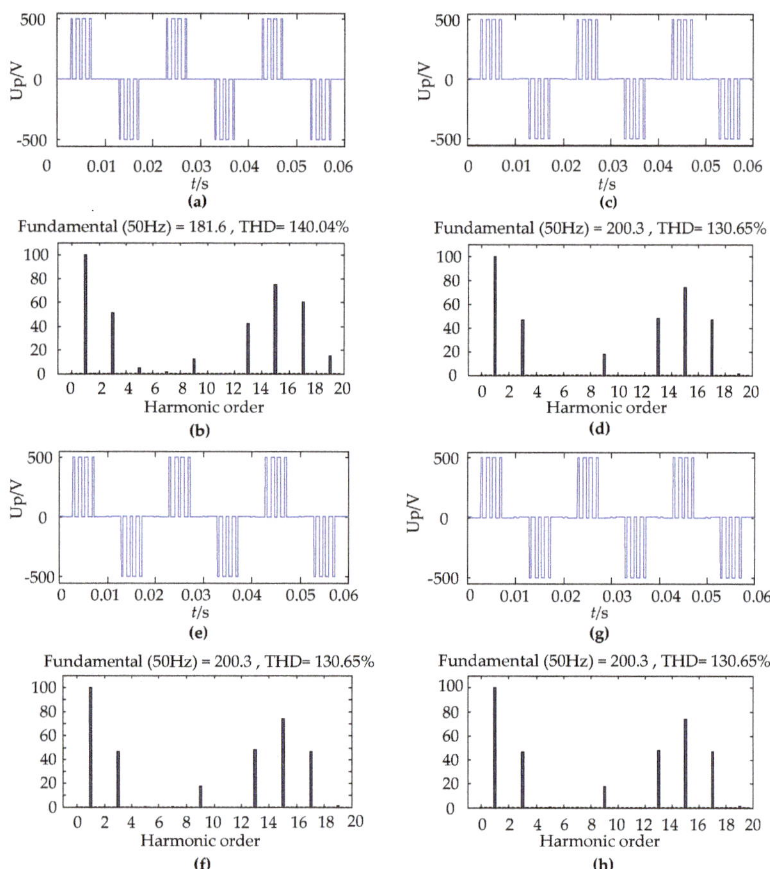

Figure 5. Waveforms of phase voltage and spectrum analysis of four algorithms: (**a,b**) MGA; (**c,d**) PSO; (**e,f**) ICA; (**g,h**) the proposed PSOICA. THD: total harmonic distortion.

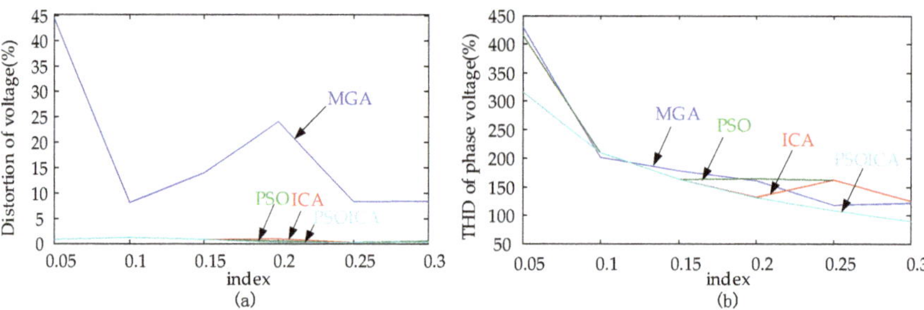

Figure 6. Characteristic analysis: (**a**) voltage distortion rate; (**b**) THDs.

Therefore, we come to the conclusion that, when the dimension of the cost function increases, the local search ability of the MGA decreases, and the PSO, ICA, and proposed PSOICA meet the precision requirements. However, the results of the PSOICA include different solutions which always contain the optimal solution. That means that the PSOICA may present the best global searching ability among the four studied algorithms.

Table 5 is built to compare the global searching ability which is based on the data of 100 runnings for every algorithm.

Table 5. Comparing the results of four switching angles.

	Convergent Rate	One Solution	Two Solutions	Three Solutions
MGA	96%	60%	35%	1%
PSO	91%	91%	0	0
ICA	95%	32%	45%	18%
PSOICA	100%	5%	41%	54%

Compared to the proposed PSOICA, the conventional ICA has a lower convergent rate and a higher possibility of getting one solution. It shows that the proposed PSOICA enhances the global searching ability of the ICA. The PSOICA has a 54% chance of getting all three sets of solutions. The MGA barely obtains three solutions, and the PSO only has one solution in each calculation instant. Moreover, the convergent rate of PSOICA is also better than those of the MGA and PSO, which demonstrates the superiority of the PSOICA in global search.

According to the comprehensive comparative analysis above, it can be seen that the proposed PSOICA has advantages over the MGA, PSO, and ICA, regarding both global and local searching abilities.

4. Simulation Analysis

To identify the correctness of solutions, we selected two different solutions, which are [20.3232, 56.3232] at 0.95 modulation index, and [62.4933, 81.5067] at 0.2 modulation index, and used Matlab/Simulink for simulation and harmonic analysis. The simulation conditions consisted of a fundamental output frequency of 50 Hz and a DC voltage of 1000 V.

In Figure 7, the phase voltage waveforms conform to the waveforms at low modulation shown in Figure 1a, and the amplitude matches the desired values exactly. A spectrum analysis shows that the selected 5th harmonic in phase voltage as well as the 5th and triple harmonics in line voltage are completely eliminated.

In Figure 8, the phase voltage waveforms are five-level, as shown in Figure 1b. The output voltages match the desired values exactly, and the selected 5th harmonic in phase voltage as well as the 5th and triple harmonics in line voltage are equal to zero. This verifies the correctness of switching angles and the effectiveness of the SHEPWM strategy.

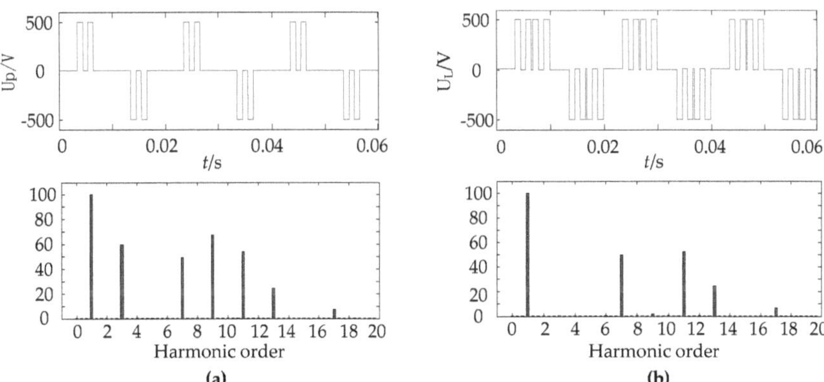

Figure 7. Simulation results with a modulation index of 0.2: (**a**) waveforms of phase voltage and its spectrum analysis; (**b**) waveforms of line voltage and its spectrum analysis.

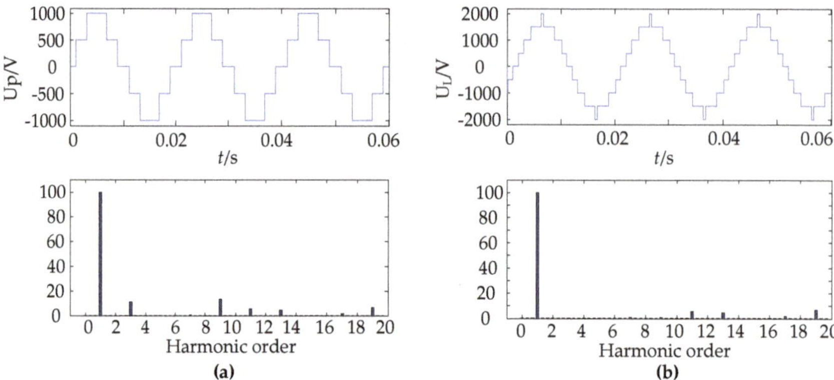

Figure 8. Simulation and analysis with a 0.95 modulation index: (**a**) waveforms of phase voltage and its spectrum analysis; (**b**) waveforms of line voltage and its spectrum analysis.

Figure 9a displays the trends of two switching angles, and second solutions are found in the modulation index range of 0–0.605. The fact that the two solutions change regularly and work out at all modulation indexes is convenient for building a look-up table. Both these solutions meet various conditions of the five-level SHEPWM strategy and have different THDs. Figure 9b shows the THDs of the two solutions. For the first solutions, low and high waveforms correspond to the modulation index ranges 0–0.375 and 0.375–1.12, respectively. The second solutions merely work for waveforms of low modulation index, between 0 and 0.605. Taking THD into consideration, for the situation of two switching angles per quarter wave, we can conclude that, in the ranges of 0–0.551 and 0.605–1.12, the first solutions are better, whereas, in the remaining range, the second solutions are more suitable.

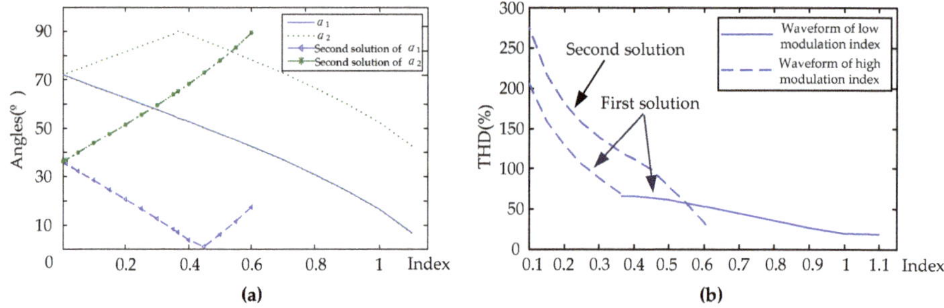

Figure 9. Trends of two switching angles and second solutions with the varying of the modulation index: (**a**) switching angles of the PSOICA; (**b**) THDs of the two different solutions.

5. Experimental Results

A downscaled three-phase NPC/H-Bridge five-level converter experimental prototype, shown in Figure 10, is set up to validate the proposed PSOICA. The power devices of the experimental prototype are insulated gate bipolar transistors (IGBT) modules (Infineon-BSM50GB120DLC). Meanwhile, a high-speed digital control platform based on the digital signal processor (DSP) (TI-TMS320F28335) and field programmable gate array (FPGA) (Xilinx-XC3S500E) is also adopted to implement the proposed PSOICA. The experimental conditions consist of the fundamental output frequency of 50 Hz and the DC voltage of 100 V.

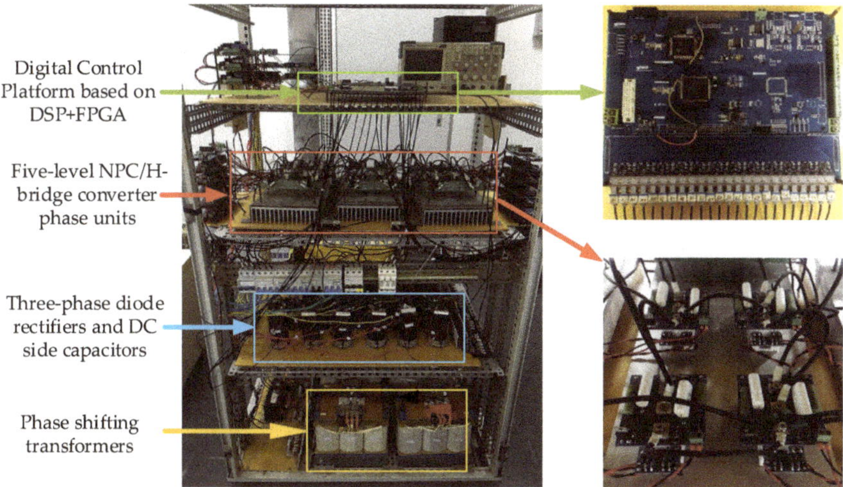

Figure 10. Three-phase neutral-point-clamped (NPC)/H-Bridge five-level converter experimental prototype.

Figure 11a shows the phase voltage and line voltage and their spectrum analysis under 0.2 modulation index. The waveforms are the same as those of the simulation results, and the selected 5th harmonic and triple harmonics are totally eliminated. Figure 11b exhibits the phase voltage and line voltage and their spectrum analysis under 0.95 modulation index. The phase voltage is five-level and the line voltage is nine-level, which are similar to the simulation results. Moreover, the selected harmonic and triple harmonics in line voltage are also removed. Hence, the analysis and simulation results are validated by these experimental results.

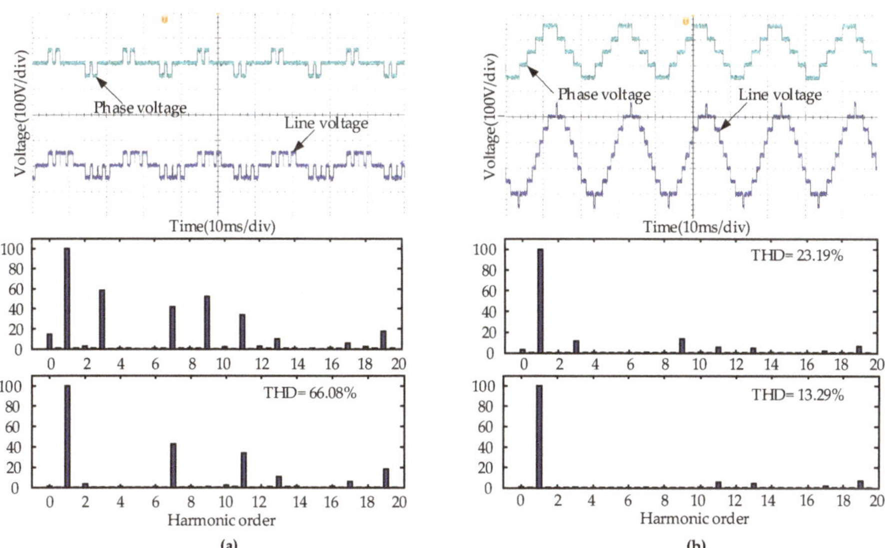

Figure 11. Experimental analysis of the waveforms: (**a**) waveforms of phase voltage and line voltage and their spectrum analysis under a modulation index of 0.2; (**b**) waveforms of phase voltage and line voltage and their spectrum analysis under a modulation index of 0.9.

As seen, a minor difference can be observed regarding the harmonic distribution characteristics between the simulation results (see Figures 7 and 8) and the experimental results. We think that this is mainly the result of two factors: one is the model difference between the simulation model and the actual experimental prototype, the other is the dead-time effect. As known, the IGBT model in the Matlab/Simulink is relatively ideal and can be regarded as a switch without the turn-on/turn-off time, so the dead-time is not set in the simulation research in this paper. However, for the experimental setup, the actual IGBT modules cannot switch as fast as the simulation models. Hence, for the experimental research described above, the dead-time was set to 4 μs in order to prevent a short circuit in each bridge arm. Considering that the SHEPWM is strongly associated with the switching time of the IGBTs, the above factors may cause the aforementioned minor difference.

6. Conclusions

In this paper, an improved ICA algorithm named PSOICA has been proposed for solving the multilevel SHEPWM. Two starting points were followed for designing the PSOICA: one was to increase the diversity and movements of the imperialists, the other was to enhance the ability to jump out of the local optimal. Hence, a novel type of particles, named independent countries, were brought into the conventional ICA to solve the nonlinear equations for the multilevel SHEPWM, especially for the situations with two and four switching angles per quarter wave. Compared to the existing MGA, PSO algorithm, and conventional ICA, the proposed PSOICA shows better performances both in global and local searching abilities, which verify the superiority of the proposed PSOICA. At the same time, the switching angles at different modulation indexes were calculated more accurately by the proposed PSOICA, which could help to obtain a more accurate modulating performance for various kinds of multilevel converters. Moreover, since a higher convergent rate can be obtained by the proposed algorithm, the execution time for solving the multilevel SHEPWM can also be reduced. Thus, the proposed PSOICA could provide a more appropriate solving approach for the SHEPWM with a high number of voltage levels.

Author Contributions: The individual contribution of each co-author with regards to the reported research and writing of the paper is as follows. Z.G. and P.D. conceived the idea, Q.C. and X.Z. performed data analysis and the simulation/experimental results, R.Z. analyzed the results and all authors wrote the paper. All authors have read and approved the manuscript.

Funding: This research was funded by the Fundamental Research Funds for the Central Universities (China University of Mining and Technology), grant number 2018QNB02.

Conflicts of Interest: The authors declare no conflict of interest.

References

1. Son, Y.; Kim, J. A novel phase current reconstruction method for a three-level neutral point clamped inverter (NPCI) with a neutral shunt resistor. *Energies* **2018**, *11*, 2616. [CrossRef]
2. Gong, Z.; Wu, X.; Dai, P.; Zhu, R. Modulated model predictive control for MMC-based active front-end rectifiers under unbalanced grid conditions. *IEEE Trans. Ind. Electron.* **2019**, *66*, 2398–2409. [CrossRef]
3. Rathore, R.; Holtz, H.; Boller, T. Generalized optimal pulse-width modulation of multilevel inverters for low-switching-frequency control of medium-voltage high-power industrial AC drives. *IEEE Trans. Ind. Electron.* **2013**, *60*, 4215–4224. [CrossRef]
4. Wang, Q.S.; Cheng, M.; Jiang, Y.L.; Zuo, W.J.; Buja, G. A simple active and reactive power control for applications of single-phase electric springs. *IEEE Trans. Ind. Electron.* **2018**, *65*, 6291–6300. [CrossRef]
5. Pouresmaeil, E.; Mehrasa, M.; Shokridehaki, M.A.; Rodrigues, E.M.G. Integration of renewable energy for the harmonic current and reactive power compensation. In Proceedings of the 2015 IEEE 5th International Conference on Power Engineering, Energy and Electrical Drives (POWERENG), Riga, Latvia, 11–13 May 2015; pp. 32–36.
6. Mehrasa, M.; Pouresmaeil, E.; Akorede, M.F. Multilevel converter control approach of active power filter for harmonics elimination on electric grids. *Energy* **2015**, *84*, 722–731. [CrossRef]

7.   Mehrasa, M.; Pouresmaeil, E.; Zabihi, S.; Rodrigues, E.M.G. A control strategy for the stable operation of the shunt active power filters in power grids. *Energy* **2016**, *96*, 325–334. [CrossRef]

8.   Srndovic, M.; Zhetessov, A.; Alizadeh, T.; Familiant, Y.L.; Grandi, G.; Ruderman, A. Simultaneous Selective Harmonic Elimination and THD Minimization for a Single-Phase Multilevel Inverter with Staircase Modulation. *IEEE Trans. Ind. Appl.* **2018**, *54*, 1532–1541. [CrossRef]

9.   Zheng, C.H.; Zhang, B.; Qiu, D.Y. Solving switching angles for the inverter's selected harmonic elimination technique with walsh function. In Proceedings of the 2005 International Conference on Electrical Machines and Systems, Nanjing, China, 27–29 September 2005; pp. 1366–1370.

10. Dahidah, M.S.A.; Konstantinou, G.; Agelidis, V.G. A review of multilevel selective harmonic elimination PWM: Formulations, solving algorithms, implementation and applications. *IEEE Trans. Power Electron.* **2015**, *30*, 4091–4106. [CrossRef]

11. Chiasson, J.N.; Tolbert, L.M.; McKenzie, K.J.; Du, Z. Elimination of harmonics in a multilevel converter using the theory of symmetric polynomials and resultants. *IEEE Trans. Control Syst. Technol.* **2005**, *13*, 216–223. [CrossRef]

12. Jacob, T.; Suresh, L.P. A review paper on the elimination of harmonics in multilevel inverters using bioinspired algorithms. In Proceedings of the 2016 International Conference on Circuit, Power and Computing Technologies (ICCPCT), Nagercoil, India, 18–19 March 2016; pp. 1–8.

13. Guan, E.; Song, P.G.; Ye, M.Y.; Wu, B. Selective harmonic elimination techniques for multilevel cascaded H-bridge inverters. In Proceedings of the International Conference on Power Electronics and Drives Systems, Kuala Lumpur, Malaysia, 28 November–1 December 2005; pp. 1441–1446.

14. Aghdam, M.G.H.; Fathi, S.H.; Gharehpetian, G.B. Elimination of harmonics in a multi-level inverter with unequal DC sources using the homotopy algorithm. *IEEE Int. Symp. Ind. Electron.* **2007**, 578–583. [CrossRef]

15. Fei, W.M.; Du, X.L.; Wu, B. A Generalized half-wave symmetry SHE-PWM formulation for multilevel voltage inverters. *IEEE Trans. Ind. Electron.* **2010**, *57*, 3030–3038. [CrossRef]

16. Kumar, A.; Chatterjee, D.; Dsagupta, A. Harmonic mitigation of cascaded multilevel inverter with non-equal DC sources using hybrid newton raphson method. In Proceedings of the 2017 4th International Conference on Power, Control & Embedded Systems (ICPCES), Allahabad, India, 9–11 March 2017; pp. 1–5.

17. Yang, K.H.; Zhang, Q.; Yu, W.S.; Yuan, H.X. Selective harmonic elimination with groebner bases and symmetric polynomials. *IEEE Trans. Power Electron.* **2016**, *31*, 2742–2752. [CrossRef]

18. Ye, M.Y.; Zhou, Q.Q.; Hong, C.; Song, L. Multiple population genetic algorithm based on multi-band SHEPWM control technology for multi-level inverter. *Trans. China Electrontech. Soc.* **2015**, *30*, 111–119. [CrossRef]

19. Kumar, N.V.; Chinnaiyan, V.K.; Pradish, M.; Divekar, M.S. Enhanced power quality of MLI using PSO based selective harmonic elimination. In Proceedings of the 2015 International Conference on Green Computing and Internet of Things (ICGCIoT), Noida, India, 8–10 October 2015; pp. 42–47.

20. Kouzou, A.; Mahmoudi, M.O.; Boucherit, M.S. Application of SHE-PWM for seven-level inverter output voltage enhancement based on particle swarm optimization. In Proceedings of the 2010 7th International Multi- Conference on Systems, Signals and Devices, Amman, Jordan, 27–30 June 2010; pp. 1–6.

21. Atashpaz-Gargari, E.; Lucas, C. Imperialist competitive algorithm: An algorithm for optimization inspired by imperialistic competition. In Proceedings of the 2007 IEEE Congress on Evolutionary Computation, Singapore, 25–28 September 2007; pp. 4661–4667.

22. Mitras, B.A.; Sultan, J.A. A novel hybrid imperialist competitive algorithm for global optimization. *Aust. J. Basic Appl. Sci.* **2013**, *7*, 330–341.

23. Etesami, M.H.; Farokhnia, N.; Fathi, S.H. A method based on imperialist competitive algorithm (ICA), aiming to mitigate harmonics in multilevel inverters. In Proceedings of the 2011 2nd Power Electronics, Drive Systems and Technologies Conference, Tehran, Iran, 16–17 February 2011; pp. 32–37.

24. Cheng, Z.; Wu, B. A Novel Switching Sequence Design for five-level NPC/H-bridge inverters with improved output voltage spectrum and minimized device switching frequency. *IEEE Trans. Power Electron.* **2007**, *22*, 2138–2145. [CrossRef]

25. Chang, H.; Wei, R.; Ge, Q.; Wang, X.; Zhu, H. Comparison of SVPWM for five level NPC/H-Bridge inverter and traditional five level NPC inverter based on line-voltage coordinate system. In Proceedings of the 2015 18th International Conference on Electrical Machines and Systems (ICEMS), Pattaya, Thailand, 25–28 October 2015; pp. 574–577.

26. Taleb, R.; Helaimi, M.; Benyoucef, D.; Boudjema, Z. A comparative analysis of multicarrier SPWM strategies for five-level flying capacitor inverter. In Proceedings of the 2016 8th International Conference on Modelling, Identification and Control (ICMIC), Algiers, Algeria, 15–17 November 2016; pp. 608–611.
27. Mehrasa, M.; Pouresmaeil, E.; Akorede, M.F.; Zabihi, S.; Catalao, J.P.S. Function-based modulation control for modular multilevel converters under varying loading and parameters conditions. *IET Gener. Transm. Distrib.* **2017**, *11*, 3222–3230. [CrossRef]
28. Ayyappa, P.N.V.S.; Srinivas, C.; Singh, T.R.S. A novel way to deal with harmonic elimination in multi-level CHB inverter using without filtering technique. In Proceedings of the 2017 International conference of Electronics, Communication and Aerospace Technology (ICECA), Coimbatore, India, 20–22 April 2017; pp. 62–67.
29. Fei, W.; Wu, B.; Wu, Q. A novel SHE-PWM method for five level voltage inverters with quarter-wave symmetry. In Proceedings of the 2009 Canadian Conference on Electrical and Computer Engineering, St. John's, NL, Canada, 3–6 May 2009; pp. 1034–1038.
30. Dahidah, M.S.A.; Agelidis, V.G.; Rao, M.V. On abolishing symmetry requirements in the formulation of a five-level selective harmonic elimination pulse-width modulation technique. *IEEE Trans. Power Electron.* **2006**, *21*, 1833–1837. [CrossRef]

Article

An Input-Parallel-Output-Series Switched-Capacitor Three-level Boost Converter with a Three-Loop Control Strategy

Jianfei Chen [1,*], Caisheng Wang [2] and Jian Li [1]

[1] State Key Laboratory of Power Transmission Equipment & System Security and New Technology, Chongqing University, Chongqing 400044, China; lijian@cqu.edu.cn
[2] Department of Electrical and Computer Engineering, Wayne State University, Detroit, MI 48202, USA; cwang@wayne.edu
* Correspondence: cjf6221@gmail.com; Tel: +86-188-8387-6496

Received: 2 September 2018; Accepted: 28 September 2018; Published: 2 October 2018

Abstract: There has been increasing interest for industry applications, such as solar power generation, fuel cell systems, and dc microgrids, in step-up dc-dc converters with reduced number of components, low component stress, small input ripples and high step-up ratios. In this paper, an input-parallel-output-series three-level boost (IPOS-SC-TLB) converter is proposed. In addition to achieving the required performance, the input and output terminals can share the same ground and an automatic current balance function is also achieved in the IPOS-SC-TLB converter. Besides, a capacitor voltage imbalance mechanism was revealed and a three-loop control strategy composed of output voltage loop, input current loop and voltage-balance loop was proposed to address the voltage imbalance issue. Finally both simulation and experiment studies have been conducted to verify the effectiveness of the IPOS-SC-TLB converter and the three-loop control strategy.

Keywords: three-level boost; automatic current balance; three-loop; voltage imbalance

1. Introduction

Multilevel step-up dc-dc converters are widely employed in wind farms [1–6], solar power generation systems [7–11], fuel cell systems [12–15], high-power charging stations for electric cars [16,17], and dc microgrids [18–20]. In these systems, a multilevel step-up dc-dc converter helps regulate a varying low-level input voltage to a stable high-level voltage, which usually serves as the dc link voltage of a grid-connected inverter. It is desirable to achieve both low voltage stress and low current stress across components to reduce power losses and save cost. Besides, input current ripple is another important issue that should be considered for these systems, especially for fuel cell or battery storage systems. As multilevel conversion techniques have evolved, many multilevel step-up dc-dc converters have been proposed. In terms of non-isolated multilevel step-up dc-dc converters, the three-level boost converter was firstly proposed and then adopted to combine with a three-level diode-clamped inverter to achieve medium voltage and high power [2,6]. The corresponding four-level boost converter was subsequently proposed to output higher voltage level and higher power [4]. Owing to the interleaved scheme, small input current ripple and low component stress could be easily realized in these multilevel boost converters. However, the input terminal and the output terminal in the two converters do not share the same ground, which can bring in severe EMI problem [7]. One flying-capacitor-based three-level boost converter was proposed to address this problem and good effect has been achieved [12]. However, all these multilevel boost converters face the same inherent limitation, i.e., the voltage gain is limited to be $1/(1-d)$, where d is the duty ratio. Unfortunately, practical considerations limit its output voltage to approximately four times its input voltage. To supply a high output voltage, it must operate at extremely high duty-cycle whereas extreme duty-cycles impose inefficient small off times. Small off times will cause severe diode reverse-recovery currents, increasing electromagnetic interference (EMI) levels [9].

Another flying-capacitor-based three-level boost converter with intrinsic voltage doubler was proposed in [21,22]. In addition to the advantages of topology described in [8], the two input inductor currents of this converter could be self-balanced due to the flying-capacitor. Moreover, the voltage gain of this converter is $2/(1-d)$ instead of $1/(1-d)$. However, the voltage stresses across the output diode and the output capacitor equals to the output voltage, which is a disadvantage. On the other hand, the voltage stresses across the output diodes could be decreased by half of the output voltage in the converter with two symmetrical flying-capacitors [23]. Nevertheless, one more diode and one more capacitor are necessary, and voltage stress across the output capacitor is still very high. These shortcomings are also presented in the topologies proposed in [24,25]. In general, a list of split capacitors connected in series is a good solution to reducing voltage stress across the output capacitor. One solution is the application of a diode-capacitor voltage multiplier on a classical non-isolated boost converter, which also presents a high voltage gain and self-balanced function for capacitor voltages [26]. However, a large input current ripple and a high current stress across the single switch exist inevitably as no interleaved scheme is adopted in these converters. Modular multilevel dc-dc converter is a good choice for high voltage applications, such as high voltage direct current (HVDC) and high voltage drive areas [27–31]. However, it is not a good choice for medium-voltage applications. Reference [32] proposed a modular multilevel dc-dc converter composed of multiple buck-boost converter modules, which is suitable for medium-voltage and high-power applications. However, the output voltage of the lower module multiplying by $d/(1-d)$ serves as the input voltage of the upper module, to achieve a high voltage gain. The voltage gain is smaller than $2/(1-d)$ and all switches do not share the same ground. Recently, a switched-capacitor technique has begun to be employed in medium-voltage and high-power dc-dc converters with good performance [33–36].

To address the abovementioned issues and to achieve a reduced number of components, low component stress, small input ripples and high step-up ratio, an input-parallel-output-series switched-capacitor three-level boost (IPOS-SC-TLB) converter is proposed in this paper. In addition to achieving the required performance target, the proposed IPOS-SC-TLB converter also has automatic current balancing capability. Compared with the existing three-level boost converters, the proposed converter has the advantages of high voltage gain at full duty cycle range, small component stress, a reduced number of components, common ground for the input and output terminals, and automatic current balancing. All capacitors, diodes, switches only endure half of the output voltage, enabling components with less voltage rating used in the proposed IPOS-SC-TLB converter. Common ground for the input and output terminals not only save power supplies for designing drivers, also helps reduce EMI. Automatic current balancing capability avoid additional current-balance control strategy that is required in a multi-converter system. To address the voltage imbalance issue, a three-loop control strategy composed of an output voltage loop, an input current loop and a voltage-balance loop is developed for the IPOS-SC-TLB converter.

The remainder of the paper is organized as follows: Section 2 introduces the topology derivation and operating principle of the proposed IPOS-SC-TLB converter. Performance analysis is subsequently presented in Section 3 and the three-loop control strategy is given in Section 4. Both simulation and experimental verifications have been done in Section 5 and finally the conclusions of the paper are drawn in Section 6.

2. IPOS-SC-TLB Converter

2.1. Topology Derivation

Until now, interleaved techniques adopted in multilevel dc-dc converters can be divided into two different types: serial-interleaved (SI) techniques (Figure 1a) and parallel-interleaved (PI) techniques

(Figure 1b). As it can be seen, the total components of the two topologies are equal except the numbers of inductors and capacitors. One inductor is necessary in the SI structure while $(n-1)$ inductors are employed in the PI structure. The SI structure needs $(n-1)$ capacitors, while one capacitor is necessary in the PI structure.

A comparative analysis between the two techniques are presented in Table 1. On the one hand, $(n-1)$ voltage levels U_0, U_1, ... U_{n-1}, are achieved due to the $(n-1)$ split capacitors in the SI structure while only one output voltage level U_{n-1} is achieved in the PI structure. On the other hand, the total input current flow through $(n-1)$ split inductors in the PI structure while through only one inductor in the SI structure. As a result, the SI structure has output voltage divider function and voltage-balance control strategy is necessary to realize voltage balance. The PI structure has input current shunt function and current-balance control strategy is necessary to balance all split inductor currents. All the drive circuits of the switches must be isolated in the SI topology, i.e., $(n-1)$ isolated drive sources are necessary in the SI structure. However, this drawback does not exist in the PI structure because all the switches share the same ground.

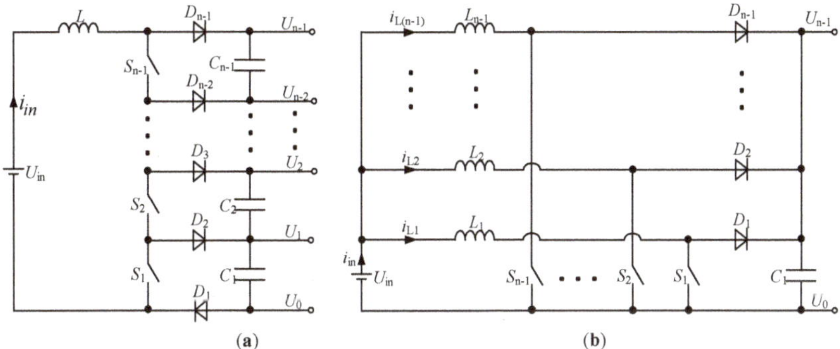

Figure 1. Two interleaved structures: (**a**) SI; (**b**) PI.

Table 1. Comparative analysis between SI and PI topologies.

Structure	Voltage Level	Function	Control
SI	$U_0, U_1, \ldots U_{n-1}$	Divide voltage	Voltage-balance
PI	U_{n-1}	Shunt current	Current-balance

The conventional three-level boost converter is based on the SI structure in Figure 2a. To distinguish it from other topologies in this paper, the converter in Figure 2a is called SI-TLB. The converter in Figure 2b is named as PIB as it is based on the PI topology. The input terminal and the output terminal of SI-TLB do not share the same ground, which easily results in electromagnetic interference (EMI) problems. Moreover, the voltage stresses across all the components in a PIB converter are high since no multilevel technique is employed.

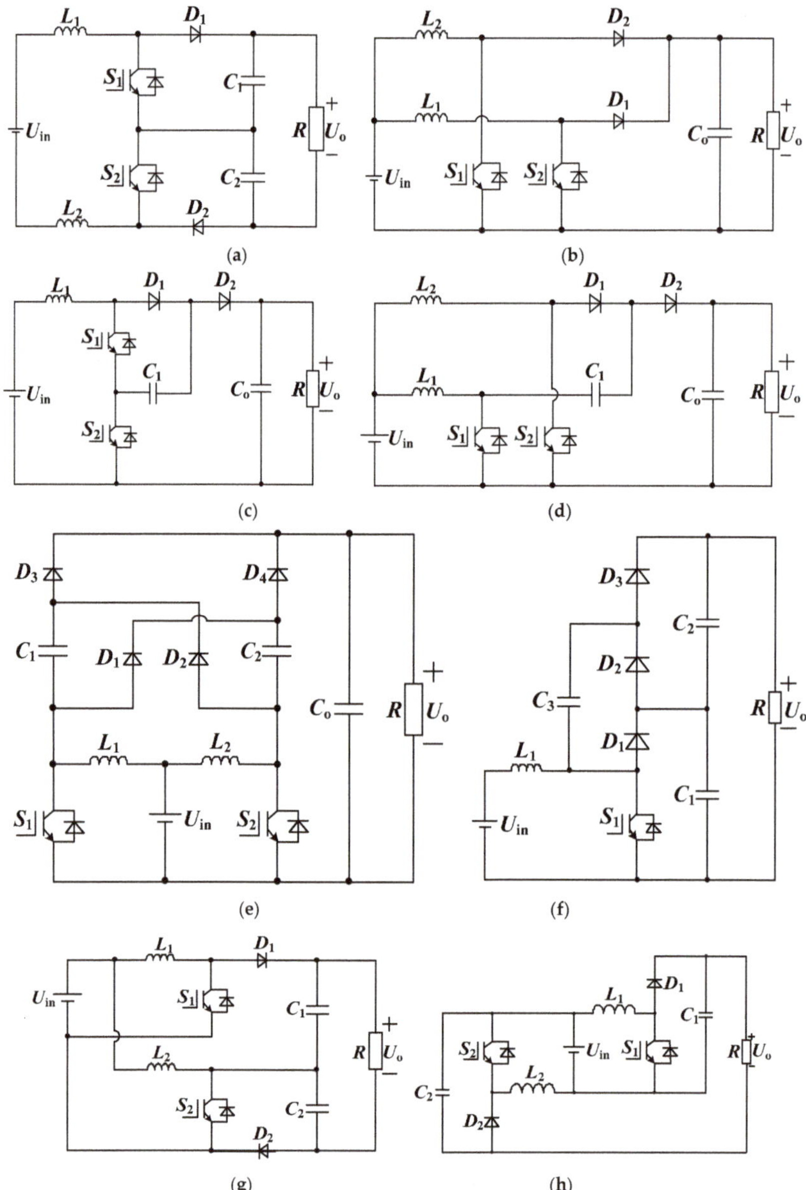

Figure 2. Development of TLB converters: (**a**) SI-TLB; (**b**) PIB; (**c**) SI-FC-TLB; (**d**) PI-FC-TLB; (**e**) PI-SFC-TLB; (**f**) SC-TLB; (**g**) IPOS-TLB1; (**h**) IPOS-TLB2.

To avoid these shortcomings mentioned above, flying-capacitor technique has been introduced into SI-TLB and PIB converters to develop new three-level boost converters. The converter called SI-FC-TLB in Figure 2c is derived by employing one flying-capacitor. The input terminal and the output terminal share the same ground and all switches and diodes are clamped at the half of the output voltage by controlling the voltage of the flying-capacitor C_f to be half of the output voltage [21,22].

However, it can be seen that SI-TLB and SI-FC-TLB both have a limited voltage gain due to the SI structure. Thus, the converter called PI-FC-TLB in Figure 2d was proposed based on PI structure and one flying-capacitor [21,22]. The voltage gain of PI-FC-TLB is as two times as that of SI-TLB and SI-FC-TLB. But the voltage stress across the output diode is high, equal to the output voltage. Also, another converter PI-SFC-TLB based on the PI structure and two symmetrical flying-capacitors in Figure 2e was proposed to reduce the voltage stresses across the output diodes [23]. However, the voltage stress across the output capacitor is still equal to the output voltage and many capacitors and diodes are necessary. As analyzed above, flying-capacitor technique introduced into multilevel boost converters based on SI structure could help solve the problem that input and output terminals do not share the same ground while flying-capacitor technique introduced into multilevel boost converters based on PI structure could help enhance voltage gains. However, there is a common disadvantage among SI-FC-TLB, PI-FC-TLB, and PI-SFC-TLB converters that the output terminal is constructed by only one output capacitor, which not only bears a high voltage stress, but also does not help reduce the voltage stress across output diodes and output capacitors. Additionally, the voltage-balance control is not easy to realize as one or more flying-capacitor voltages should be control independently. Even though the output capacitor can be replaced by two split capacitors in series in the output terminal, the two split capacitors could not be self-balanced and could not be controlled by voltage-balance control strategy either. On the other hand, a three-level boost converter based on switched-capacitor network is proposed in [26]. For simplification, the converter is name as SC-TLB, which not only has two split capacitors at the output terminal, but also has self-balancing function for capacitor voltages. As a result, there is no need to employ any voltage-balance control strategies to solve the voltage imbalance issue. However, as analyzed in Section 1, there is a big disadvantage that SC-TLB has high input current and high input current ripple since no interleaved structures are employed. As a result, high power losses are inevitable in the SC-TLB converter.

There are also two input-parallel-output-series (IPOS) boost converters shown in Figure 2g (called by IPOS-TLB1) and Figure 2h (called by IPOS-TLB2) from references [24,25]. Like the ISOS-TLB converter, the input terminal and the output terminal do not share the same ground and the voltage stress across diode D_1 is equal to the output voltage in the IPOS-TLB1 converter. Although the topology IPOS-TLB2 is simple, its voltage gain is smaller than the other topologies and the input terminal and the output terminal do not share the same ground, either.

According to the comparative analysis mentioned above, the SI structure is suitable for high input voltage and high output voltage applications while the PI structure is suitable for high input current and high output current applications. As shown in Figure 1, the input terminal, output terminal and all switches share the same ground in the PI topology. The flying-capacitor technique helps enhance the voltage gains of the converters based on the PI structure. Besides, the switched-capacitor technique, which could be deemed as an extension of flying-capacitor technique, not only increases the voltage gain, but also brings a self-balancing function for capacitor voltages. On the whole, there are three techniques could be employed in multilevel dc/dc converters, i.e., interleaved technique, flying-capacitor technique, and switched-capacitor technique. Until now, only one or two of the three techniques were employed in a single power converter.

This paper proposes an IPOS-SC-TLB converter in Figure 3 and presents a detailed analysis of the converter. IPOS-SC-TLB combines the parallel-interleaved technique, flying-capacitor technique, and switched-capacitor technique together. In Figure 3, L_1, L_2, S_1, S_2 formulate the PI structure, while L_1, S_1, D_1, C_1 form Boost I and L_2, S_2, D_3, C_2 form Boost II. Besides, C_f, D_2 and S_2, C_1 formulate a switched-capacitor network, which makes the two input terminals in parallel and the output terminals in series for Boost I and Boost II.

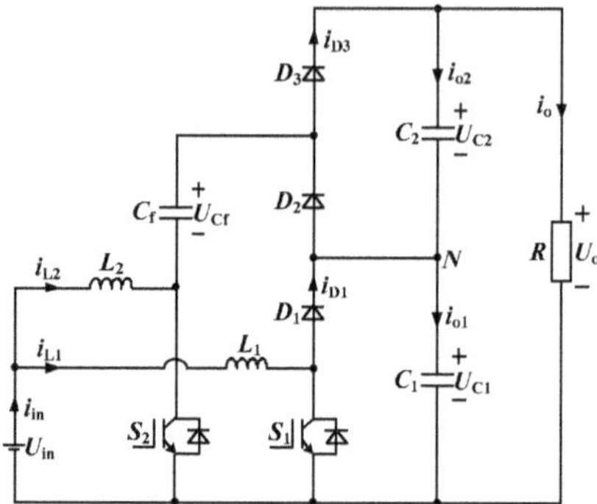

Figure 3. The proposed IPOS-SC-TLB converter.

2.2. Operating Principle

In the IPOS-SC-TLB converter, all the inductors, capacitors, switches, diodes have the same respective parameters, i.e.,

$$L_1 = L_2 = L \tag{1}$$

$$C_1 = C_2 = C \tag{2}$$

Considering the voltage drops of IGBT and diode, and the equivalent series resistor of inductor, the equivalent circuits of the IPOS-SC-TLB converter are presented in Figure 4. In the interleaved scheme, the operating stages of IPOS-SC-TLB could be divided into two modes according to duty cycle: when d is greater than 0.5 and when d is smaller than 0.5.

(1)　When the duty cycle d is greater than 0.5, the IPOS-SC-TLB operates at the periodic stages of I, II, I, and III.

Stage I: Both switches S_1, S_2 are turned on and the diode D_2 is forward biased as the capacitor voltage U_{C1} is still a little bigger than the capacitor voltage U_{Cf} after the stage III. During Stage I, both the two inductors L_1, L_2 are charged by the input source U_{in}. Thus, there are:

$$L_1 \frac{di_{L1}}{dt} = L_2 \frac{di_{L2}}{dt} = U_{in} - I_{L1} r_L - U_S \tag{3}$$

$$U_{C1} = U_{Cf} + U_S + U_D \tag{4}$$

As the two capacitor voltages U_{C1}, U_{Cf} are charged in parallel, the voltage differences between U_{C1} and U_{Cf} are small but cannot be ignored. As a result, the current flowing through D_2 caused by the small voltage difference is labelled as I_1. The current flowing through S_1 is equal to i_{L1} while the current flowing through S_2 is the sum of i_{L2} and I_1.

Stage II: When the switch S_1 is turned on and the switch S_2 is turned off, the diode D_3 is forward. The inductor L_1 is still charged by the input source U_{in}, which also supplies energy to the load together with the inductor L_2 and the flying-capacitor C_f. Thus, there are:

$$L_1 \frac{di_{L1}}{dt} = U_{in} - I_{L1} r_L - U_S \tag{5}$$

$$L_2 \frac{di_{L2}}{dt} = U_{in} + U_{Cf} - U_D - U_{C2} - U_{C1} \tag{6}$$

According to (4) and (6), there is:

$$L_2 \frac{di_{L2}}{dt} = U_{in} - U_{C2} - U_S - 2U_D \tag{7}$$

During Stage II, the capacitor voltage U_{Cf} decreases while the capacitor voltage U_{C1} increases. As a result, the voltage difference between U_{C1} and U_{Cf} becomes bigger and bigger and finally reaches its maximum at the end of Stage II. The current flowing through the switch S_1 is still equal to i_{L1} while no currents flows through the switch S_2 and the diode D_2 during this stage.

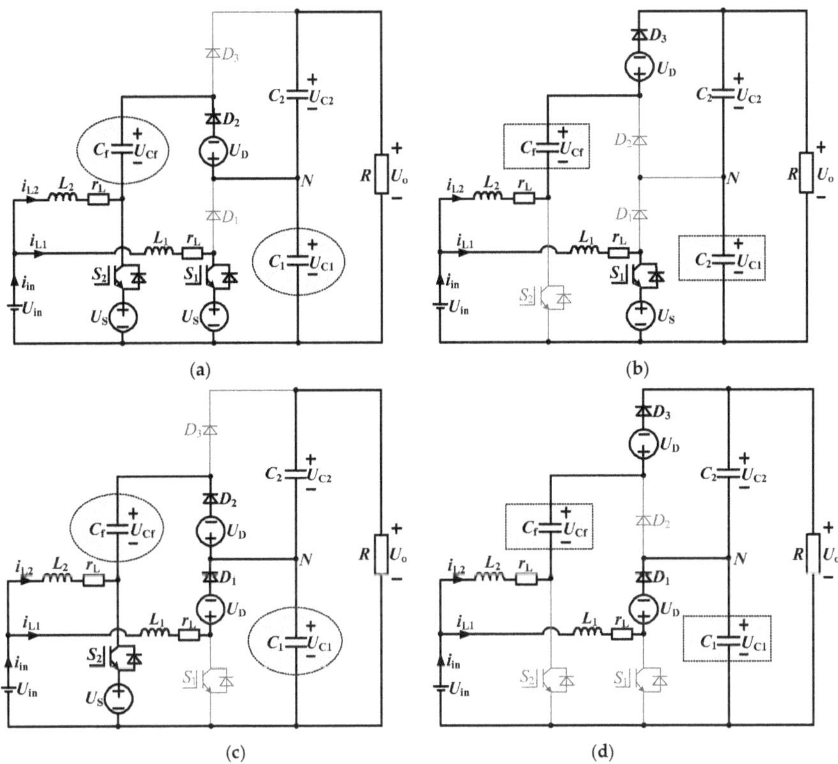

Figure 4. Equivalent circuits of the IPOS-SC-TLB converter: (**a**) stage I; (**b**) stage II; (**c**) stage III; (**d**) stage IV.

Stage I: The converter repeats Stage I and the same output results could be achieved like (3) and (4). However, as analyzed in Stage II, the voltage difference between U_{C1} and U_{Cf} reaches its maximum value, the current flowing through the diode D_2 reaches its maximum value, labelled as I_2. The current flowing through S_1 turns to be the same as i_{L1} again while the current flowing through S_2 is the sum of i_{L2} and I_2.

Stage III: When the switch S_1 is turned off while the switch S_2 is turned on, the diodes D_1, D_2 are both forward. The inductor L_2 is charged by the input source U_{in}, which also supplies energy to C_1 and C_f together with the inductor L_1. Thus, there are:

$$L_1 \frac{di_{L1}}{dt} = U_{in} - I_{L1}r_L - U_D - U_{C1} \tag{8}$$

$$L_2 \frac{di_{L2}}{dt} = U_{in} - I_{L1}r_L - U_S \tag{9}$$

During Stage III, the two capacitors C_1, C_f are connected in parallel and thus the voltage difference between them is small, which results in a small current flowing through the diode D_2. It has been mentioned in the first Stage I, labelled as I_1. As a consequence, the current flowing through S_1 turns to be zero while the current flowing through S_2 is the sum of i_{L2} and I_1.

(2)   When the duty cycle d is smaller than 0.5, the IPOS-SC-TLB converter operates at the periodic stages of IV, II, IV, and III.

Stage IV: Both switches S_1, S_2 are turned off while D_1 and D_3 are on forward biased:

$$L_1 \frac{di_{L1}}{dt} = U_{in} - I_{L1}r_L - U_D - U_{C1} \tag{10}$$

$$L_2 \frac{di_{L2}}{dt} = U_{in} + U_{Cf} - U_D - U_{C2} - U_{C1} = U_{in} - U_{C2} - U_S - 2U_D \tag{11}$$

As the first Stage IV begins after Stage III, the capacitor voltage U_{Cf} decreases while the capacitor voltage U_{C1} increases. During this stage, no currents pass through the two switches S_1, S_2 and the diode D_2 as they are all switched off.

Stage II: The same results could be achieved like (5)–(7) and the voltage difference between U_{C1} and U_{Cf} continues increasing during this stage. The current flowing through the switch S_1 is still equal to i_{L1} while no currents flows through the switch S_2 and the diode D_2 during this stage.

Stage IV: The converter enters into another Stage IV, where the voltage difference between U_{C1} and U_{Cf} continues increasing and reach its maximum value at the end of the stage. And no currents no currents pass through the two switches S_1, S_2 and the diode D_2.

Stage III: At the beginning of the stage III, the current flowing through the diode D_2 reaches its maximum value I_2. But later becomes a smaller value I_1 as the two capacitors C_1, C_f are charged in parallel. Thus, the current flowing through the switch S_1 is zero while the current flowing through the switch S_2 is the sum of i_{L2} and I_2 and then the sum of i_{L2} and I_1 during this stage. For any duty cycle d, two equations can be attained based on Voltage-Second Balance Principle during one switching period:

$$dT_s(U_{in} - I_{L1}r_L - U_S) + (1 - d)T_s(U_{in} - I_{L1}r_L - U_D - U_{C1}) = 0 \tag{12}$$

$$dT_s(U_{in} - I_{L2}r_L - U_S) + (1 - d)T_s(U_{in} - U_{C2} - U_S - 2U_D) = 0 \tag{13}$$

During the switching period, the output voltage of the converter is always described by:

$$U_o = U_{C1} + U_{C2} \tag{14}$$

According to (28), there is:

$$I_{L1} = I_{L2} = \frac{U_o}{R(1 - d)} \tag{15}$$

Therefore, the voltage gain G and the capacitor voltages could be derived by:

$$G = \frac{U_o}{U_{in}} = \frac{2 + \frac{(2d-3)U_D - U_S}{U_{in}}}{1 - d + \frac{r_L(1+d)}{R(1-d)}} \tag{16}$$

The capacitor voltages are calculated by:

$$\begin{cases} U_{C1} = \frac{U_{in} - I_{L1}r_L - U_D}{1 - d} \\ U_{C2} = \frac{U_{in} - dI_{L2}r_L - U_S + (2d-2)U_D}{1 - d} \\ U_{Cf} = \frac{U_{in} - I_{L1}r_L - (1-d)U_S - (2-d)U_D}{1 - d} \end{cases} \tag{17}$$

When the parasitic parameters are ignored, there are:

$$\begin{cases} G = \frac{2}{1-d} \\ U_{C1} = U_{C2} = U_{Cf} = \frac{1}{2}U_o \end{cases} \tag{18}$$

3. Performance Analysis

3.1. Component Stress

According to the analysis mentioned above, the key voltage waveforms of the IPOS-SC-TLB converter are presented in Figure 5. The voltage stresses across all switches, diodes and capacitors are half of the output voltage:

$$U_{S1} = U_{S2} = U_{D1} = U_{D2} = U_{D3} = U_{C1} = U_{C2} = U_{Cf} = \frac{1}{2}U_o \tag{19}$$

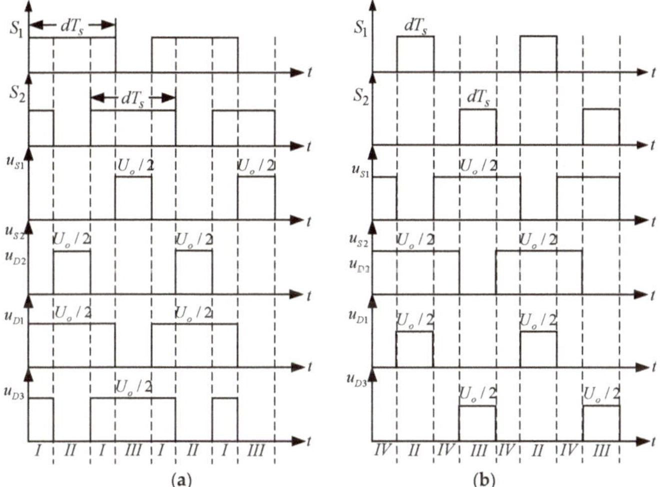

Figure 5. Key voltage waveforms: (**a**) $d > 0.5$; (**b**) $d \leq 0.5$.

The current waveforms of the IPOS-SC-TLB converter are presented in Figure 6. From Figure 6, whatever the duty cycle d is, the average current across S_1, S_2 can be obtained as follows:

$$\begin{cases} I_{S1} = dI_{L1} \\ I_{S2} = I_{L2} \end{cases} \tag{20}$$

The average currents across D_1, D_2, D_3 identical with value equal to the average output current are determined as follows:

$$I_{D1} = I_{D2} = I_{D3} = I_o = \frac{U_o}{R} \tag{21}$$

When the duty cycle d is over 0.5, the operating period of Stage II in Figure 4b can be expressed by $(1 - d)T_s$ during one switching period. During Stage II, the capacitor C_2 is charged with the current expressed by:

$$i_{C2\_charged} = I_{L2} - \frac{U_o}{R} \tag{22}$$

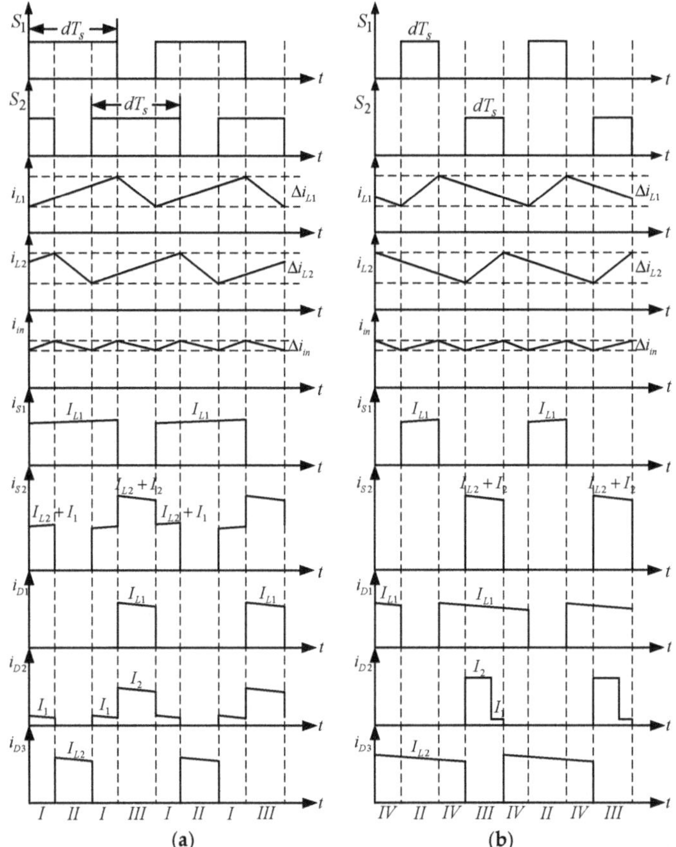

Figure 6. Key current waveforms: (a) $d > 0.5$; (b) $d \leq 0.5$.

During the remained operating period dT_s, C_2 is discharged with the current expressed by:

$$i_{C2\_discharged} = -\frac{U_o}{R} \tag{23}$$

According to Ampere-Second Balance Principle, there is:

$$(1-d)T_s\left(I_{L2} - \frac{U_o}{R}\right) + dT_s\left(-\frac{U_o}{R}\right) = 0 \tag{24}$$

We can obtain the average current of inductor L_2 by simplifying (24) as below:

$$I_{L2} = \frac{U_o}{R(1-d)} \tag{25}$$

When the duty cycle is smaller than 0.5, the same formula as (25) can be obtained. It should be noted that the average current of inductor L_2 could be also derived as below. During one whole switching period, the average charging current flowing through C_f is the same as the average current flowing through D_2. So the increased charges of C_f during one switching period is $I_{D2}*T_s$. In addition, when d is over 0.5, the flying-capacitor C_f is only discharged during Stage II and the average discharging current flowing through C_f is I_{L2} with the discharging time $(1-d)T_s$. When d is smaller than 0.5, the flying-capacitor C_f is discharged during Stage II and Stage IV with the average discharging current I_{L2} and the total discharged time $(1 - d)T_s$. It can be seen that the decreased charges of C_f during one switching period is $I_{L2}*(1 - d)T_s$ no matter what the duty cycle d is. Therefore, by applying Ampere-Second Balance Principle on C_f, we have:

$$I_{D2} * T_s = I_{L2} * (1 - d)T_s \tag{26}$$

According to (26), the same formula as (25) can be achieved. On the other hand, the average current of L_1 can be easily obtained as below:

$$I_{L1} = \frac{I_{D1}}{1 - d} = \frac{U_o}{R(1 - d)} \tag{27}$$

According to (20)–(27), the average currents across all switches and diodes are:

$$\begin{cases} I_{L1} = I_{L2} = \frac{U_o}{(1-d)R} \\ I_{S1} = \frac{dU_o}{(1-d)R} \\ I_{S2} = \frac{U_o}{(1-d)R} \\ I_{D1} = I_{D2} = I_{D3} = \frac{U_o}{R} \end{cases} \tag{28}$$

3.2. Switched-Capacitor Network

For two typical boost converters, their input terminals cannot be simply connected in parallel and while their output terminals are connected in series simultaneously. The flying-capacitor C_f and the diode D_2 in the proposed IPOS-SC-TLB converter are used to realize the input-parallel output-series topology. Because a switched-capacitor network is constructed and it helps support the output voltage of the Boost I for the Boost II. As shown in Figure 4, the flying-capacitor C_f is clamped with the capacitor C_1 during Stage I and Stage III, labelled as the oval areas, i.e., the two capacitor voltages are identical. During Stage II and Stage IV, the flying-capacitor C_f serves as the voltage support for the Boost II. So, it could be thought of as that the output capacitor C_2 is charged by the input source because the capacitor voltage U_{Cf} offsets the capacitor voltage U_{C1}, which are labelled as the rectangular areas. Furthermore, it can be seen from (26) that the flying-capacitor C_f could automatically balance the average currents of the two inductors L_1 and L_2. Thus, the IPOS-SC-TLB converter does not need any current-balance circuit or current-balance control strategy that is required in the conventional parallel-interleaved dc/dc converters.

3.3. Ripple Analysis

In the switched-capacitor network, the flying-capacitor C_f could be served as an energy buffer. According to (26) and (27), the increased or decreased charges on C_f is U_o*T_s/R, which could be described by another way of $C_f*\Delta u_{Cf}$, where Δu_{Cf} represents the voltage ripple of C_f. Finally, the voltage ripple of C_f is derived by:

$$\Delta u_{Cf} = \frac{U_o}{RC_f f_s} \tag{29}$$

Besides, it is easy to attain the voltage ripples of C_1 and C_2:

$$\Delta u_{C1} = \Delta u_{C2} = \frac{U_o d}{RC f_s} \tag{30}$$

Additionally, the current ripples of L_1 and L_2 could be obtained by:

$$\Delta i_{L1} = \Delta i_{L2} = \frac{U_{in}}{L}\frac{d}{f_s} \tag{31}$$

The input current ripple can be calculated by:

$$\Delta i_{in} = \begin{cases} \frac{U_{in}}{L}\frac{2d-1}{f_s}d & d > 0.5 \\ \frac{U_{in}}{L}\frac{d(1-2d)}{f_s(1-d)}d & d \leq 0.5 \end{cases} \tag{32}$$

3.4. Inrush Current Suppression

In practical application, IGBT and diode usually have some voltage drops and capacitors has equivalent serial resistors. Thus, it is inevitable to see some voltage differences between C_1 and C_f, which can be described by:

$$\Delta U = U_{C1} - U_{Cf} = U_S + U_D \tag{33}$$

U_D and U_S are assumed to be the voltage drop of one diode and the voltage drop of one IGBT. Figure 7 shows the equivalent circuit of the switched-capacitor network when S_2 turns on.

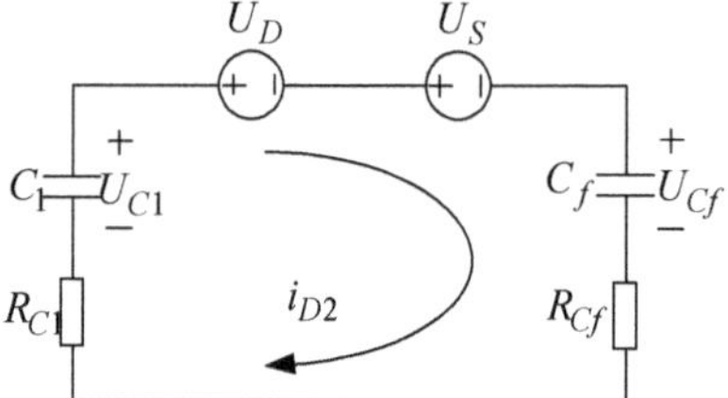

Figure 7. Equivalent circuit of the switched-capacitor network.

It can be seen that the output capacitor C_1 is connected with the flying-capacitor C_f in parallel. R_{C1}, R_{Cf} means the equivalent serial resistors of C_1 and C_f, respectively. The current i_{D2} flowing through the diode D_2 could be calculated by:

$$i_{D2} = \frac{\Delta U}{R_{C1} + R_{Cf}} \tag{34}$$

In (34), the equivalent serial resistors R_{C1}, R_{Cf} are usually very small, which are in the range of milliohms. As a result, although ΔU is small, it may bring in very high inrush current i_{D2} flowing the switched-capacitor network when S_2 is turned on and D_2 is forward instantaneously. Moreover, this will result in more conduction losses across the switch S_2 and the diode D_2.

From (10), one way to suppress i_{D2} is to reduce the voltage difference ΔU is by using wide bandgap semiconductors, such as SiC or GaN components that have smaller voltage drops compared with Si-based components. However, ΔU cannot be reduced to zero and this may still bring in a certain inrush current. Another method is to increase the impedance of the switched-capacitor network. Placing a serial resistor with high resistance could increase the impedance but extra power losses are produced. As shown in Figure 8, this paper proposes to put a small stray inductor L_s together with D_2. In this way, the loop impedance is increased by $2\pi f_s L_s$ and then the inrush current i_{D2} is reduced to:

$$i_{D2} = \frac{\Delta U}{2\pi f_s L_s + R_{C1} + R_{Cf}} \tag{35}$$

3.5. Comparative Analysis

Comparative analyses of SC-TLB, SI-TLB, SI-FC-TLB, PI-FC-TLB and the proposed IPOS-SC-TLB are presented in Table 2. L, S, D, and C represent the quantities of inductors, switches, diodes and capacitors, respectively. DS means the quantity of driver supplies and G means the voltage gains. Besides, U_{VPS}, U_{VPD}, and U_{VPC} respectively represent the voltage stresses across switches, diodes, capacitors; and I_{VPS1}, I_{VPS2}, and I_{VPD} represent the average current across switches S_1, S_2 and diodes, respectively. "Self-balance" means the input inductor currents could be self-balanced and "same ground" means the input terminal and the output terminal share the same ground. In addition, the voltage gain comparison curves are presented in Figure 9.

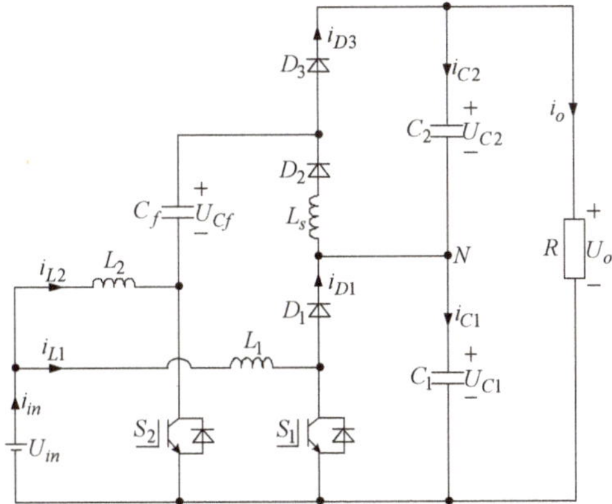

Figure 8. IPOS-SC-TLB with a small stray inductor L_s.

Among these seven TLB converters, the common performance parameters are the voltage stress across switches and the input current ripple. The TLB converters based on SI structure need two isolated drive power supplies and have a low voltage gain, while those TLB converters based on PI structure need only one power supply and show a higher voltage gain. The SI-FC-TLB and PI-FC-TLB are very similar except for different interleaved structures. From these two converters, it could be seen that the voltage stresses across the output diodes are low in the SI structure while high in the PI structure; and the average current stresses across switches are high in the SI structure while low in the PI structure. The smaller average current stress across switches should be attributed to the PI structure. Among the five converters, the quantity of components are not the most in the proposed IPOS-SC-TLB, and high voltage gain, small voltage stress and small current stress are achieved. Moreover, voltage-balance control could be easily achieved with the input terminal and the output terminal sharing the same ground. In other words, the proposed IPOS-SC-TLB converter integrates nearly all the merits of the other four TLB converters. However, there is also a disadvantage that the imbalance current between the two power switches S_1, S_2. As analyzed in Equation (28), the average current of S_2 is U_o/R higher than the average current of S_1.

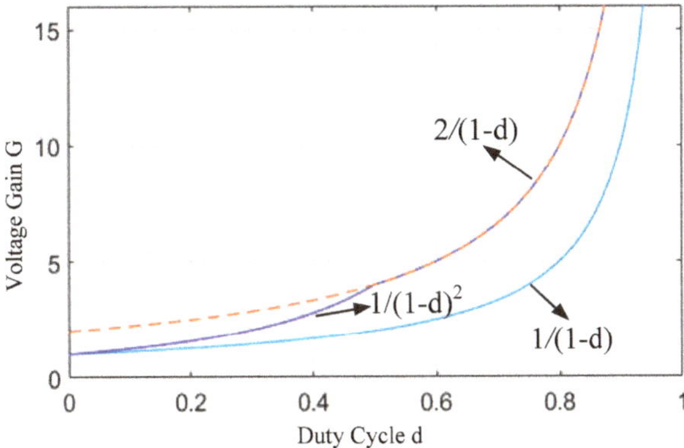

Figure 9. Voltage gain comparison.

Table 2. Comparative analysis among SC-TLB, SI-TLB, SI-FC-TLB, PI-FC-TLB, SC-TLB and the proposed IPOS-SC-TLB.

Topology	L	S	D	C	DS	G $(d \leq 0.5)$	G $(d > 0.5)$	U_{VPS}	U_{VPD}	U_{VPC}	I_{VPS1}	I_{VPS2}	I_{VPD}	Current Ripple	Self-Balance	Same Ground
SC-TLB	1	1	3	3	1	$2/(1-d)$		$0.5U_o$	$0.5U_o$	$0.5U_o$	dI_{in}	-	I_o	Large	Yes	Yes
SI-TLB	2	2	2	2	2	$1/(1-d)$		$0.5U_o$	$0.5U_o$	$0.5U_o$	dI_{in}	dI_{in}	I_o	Small	No	No
SI-FC-TLB	2	2	2	2	2	$1/(1-d)$		$0.5U_o$	$0.5U_o$	$0.5U_o$	dI_{in}	dI_{in}	I_o	Small	No	Yes
PI-FC-TLB	2	2	2	2	1	$1/(1-d)^2$	$2/(1-d)$	$0.5U_o$	U_o	U_o	$0.5dI_{in}$	$0.5dI_{in}$	I_o	Small	Yes	Yes
PI-SFC-TLB	2	2	4	3	1	$2/(1-d)$		$0.5U_o$	U_o	U_o	$(1+d)I_{in}/4$	$(1+d)I_{in}/4$	$0.5I_o$	Small	Yes	Yes
IPOS-TLB1	2	2	2	2	1	$1/(1-d)^2$	$2/(1-d)$	$0.5U_o$	U_o	$0.5U_o$	$0.5dI_{in}$	$0.5dI_{in}$	I_o	Small	No	Yes
IPOS-TLB2	2	2	2	2	2	$(1+d)/(1-d)$		$U_o/(1+d)$	$U_o/(1+d)$	$U_o/(1+d)$	$0.5dI_{in}$	$0.5dI_{in}$	I_o	Small	No	No
Proposed	2	2	3	3	1	$2/(1-d)$		$0.5U_o$	$0.5U_o$	$0.5U_o$	$0.5dI_{in}$	$0.5I_{in}$	I_o	Small	Yes	Yes

4. Three-Loop Control Strategy

4.1. Voltage Imbalance Mechanism

η_1, η_2 are labelled as the conversion efficiencies of Boost I and Boost II, respectively. Thus, there is:

$$\begin{cases} \eta_1 U_{in} I_{L1} = U_{C1} I_{D1} \\ \eta_2 U_{in} I_{L2} = U_{C2} I_{D3} \end{cases} \tag{36}$$

As the output terminals of Boost I and Boost II are connected in series, the two Boost modules have the same output current. As the average currents across C_1 and C_2 are both equal to zero during one switching period, there is:

$$I_{D1} = I_{D3} \tag{37}$$

In addition, the power losses of D_2 and S_2 produced in the switched-capacitor network are small but could not be ignored. But the power losses should be attributed to the Boost II as D_2 and S_2 help formulate the Boost II. As a result, there is:

$$\eta_1 > \eta_2 \tag{38}$$

Based on (36)–(38), we have:

$$U_{C1} > U_{C2} \tag{39}$$

As analyzed above, the two split inductor currents could be self-balanced, but the two output capacitor voltages could not be self-balanced. Considering the voltage drops of IGBT and diode, the voltage difference between C_1 and C_2 could be described by the sum of the voltage drop of one IGBT and the voltage drop of one diode. Besides, the parasitic resistances of L_1 and L_2 are labelled as r_L and the parasitic resistance of C_f is labelled as r_{Cf}. As the average currents across L_1 and L_2 are high, the voltage drops of parasitic resistances are large and could not be ignored. Under this condition, the two output capacitor voltages could be rewritten as:

$$U_{C1} = \frac{1}{1-d}(U_{in} - I_{L1}r_L) - U_D \tag{40}$$

$$U_{C2} = \frac{1}{1-d}[U_{in} - I_{L2}(r_L + r_{Cf})] - \Delta U - U_D \tag{41}$$

The two split inductors are designed to attain the same parameters. Owing to the automatic balanced inductor currents, the voltage difference between C_1 and C_2 could be described by:

$$\Delta U = U_{C1} - U_{C2} = \frac{I_{L2}r_{Cf}}{1-d} + \Delta U \tag{42}$$

Considering (15)–(42) is further simplified as:

$$\Delta U = U_{C1} - U_{C2} = \frac{I_{L2}r_{Cf} + \Delta U_{SC}}{1-d} = \frac{I_{in}r_{Cf}}{2(1-d)} + U_S + U_D \tag{43}$$

It can be seen from (43) that the voltage imbalance issue is related to the output characteristic and the parasitic parameters, including the average input current I_{in}, the duty cycle d, the equivalent series resistance r_{Cf} of the flying-capacitor, and the voltage drops of IGBTs and diodes. The capacitances of the two output capacitors have no effect on the voltage imbalance issue, which is quite different from the conventional three-level boost converter shown in Figure 2a.

4.2. Three-Loop Control Strategy

To address the voltage imbalance issue and to achieve stale operation of the IPOS-SC-TLB converter, a three-loop control strategy including an output voltage loop, an input current loop and a voltage-balance loop is proposed in this section. The voltage loop and the current loop respectively controls the output voltage and the input inductor currents, while the voltage-balance loop helps alleviate the voltage imbalance issue. However, the voltage loop and the voltage-balance loop will influence each other if no decoupling scheme is employed. To decouple the output voltage loop and the voltage-balance loop, the derivation analysis has been done as follows.

Duty cycles d_1, d_2 in (44) are both composed of the common duty cycle d and the voltage-balance duty cycles Δd_1, Δd_2. Also, I_{L1}, I_{L2} in (45) are both composed of the average inductor current I_L and the voltage-balance inductor current ΔI_{L1}, ΔI_{L2}:

$$\begin{cases} d_1 = d + \Delta d_1 \\ d_2 = d + \Delta d_2 \end{cases} \tag{44}$$

$$\begin{cases} I_{L1} = I_L + \Delta I_{L1} \\ I_{L2} = I_L + \Delta I_{L2} \end{cases} \tag{45}$$

In the IPOS-SC-TLB converter, the relationship of the input inductor currents and the output current could be described by:

$$\begin{cases} (1 - d_1)I_{L1} = I_o \\ (1 - d_2)I_{L2} = I_o \end{cases} \tag{46}$$

By substituting (44) and (45) into (46), there is:

$$\begin{cases} (1 - d)\Delta I_{L1} - \Delta d_1 I_L = \Delta I_o \\ (1 - d)\Delta I_{L2} - \Delta d_2 I_L = \Delta I_o \end{cases} \tag{47}$$

When the output voltage is not disturbed, the output current variation ΔI_o is equal to zero. Thus, (47) could be simplified by:

$$\begin{cases} \Delta d_1 = \frac{1-d}{I_L}\Delta I_{L1} \\ \Delta d_2 = \frac{1-d}{I_L}\Delta I_{L2} \end{cases} \tag{48}$$

When the IPOS-SC-TLB converter works at stable steady state, ΔI_{L1} and ΔI_{L2} indirectly reflect the values of Δd_1, Δd_2. According to (38), it is not difficult to deduce the following formula:

$$\Delta I_{L1} + \Delta I_{L2} = \frac{I_L}{1-d}(\Delta d_1 + \Delta d_2) \tag{49}$$

In the three-loop control strategy, to decouple the voltage loop and the voltage-balance loop, the sum of Δd_1 and Δd_2 should be equal to zero. Thus, according to (49), there is:

$$\Delta I_{L1} + \Delta I_{L2} = 0 \tag{50}$$

Then, the reference inductor currents of Boost I and Boost II could be concluded as follows:

$$\begin{cases} I_{L1}{}^* = I_L + \Delta I_{L1} = I_L - \Delta I_L \\ I_{L2}{}^* = I_L + \Delta I_{L2} == I_L + \Delta I_L \end{cases} \tag{51}$$

According to (51), the three-loop control strategy is presented in Figure 10. The regulators of the output voltage loop and the voltage-balance loop adopt proportional-integral controller while the regulator of the current loop adopts proportional controller. The controllers can be designed based on a small-signal linearized model of the dc/dc converter, which can be developed according to the classic average modeling method for power converters [35,36]. The inner current control loop is designed to respond faster than the outer voltage control loop so that the two control loops can be designed independently. As a result, when dealing with the inner loop, we take the outer loop as a constant input. On the other hand, the inner current control loop can be approximated as a simple lag block when we proceed with the voltage loop. The voltage-balance loop has the slowest response. When designing the controller for the voltage-balance loop, the voltage and current control loop can be considered being in steady state already. Classic Bode-plot and root-locus proportional-integral controller design procedures [36] can be used to obtain the parameters for the controllers. Nevertheless, it should be noted that due to the nonlinearity of power devices, the designed controller parameters need to be further tuned for the actual circuit. Besides, the carrier signals C_{a1}, C_{a2} are with phase-shifted 180 degrees to realize interleaved scheme for the switches S_1 and S_2.

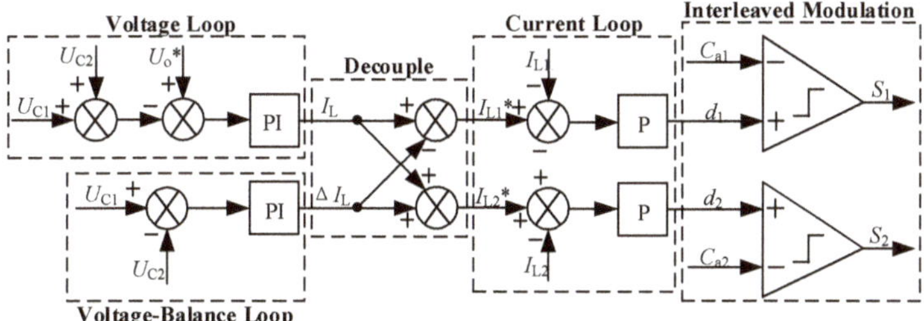

Figure 10. The three-loop control strategy.

The two sampled capacitor voltages U_{C1}, U_{C2} are added together and then compared with the output voltage reference $U_o{}^*$ to output the average inductor current I_L through the voltage loop regulator. ΔI_L is achieved through the voltage-balance loop regulator by comparing U_{C1} and U_{C2}. ΔI_{L2} equals to $-\Delta I_{L1}$ according to (50). The inductor current references $I_{L1}{}^*$, $I_{L2}{}^*$ in (51) could be achieved through the decoupled scheme. Then, $I_{L1}{}^*$ and $I_{L2}{}^*$ compares with I_{L1} and I_{L2}, and pass though the two current loop regulators to output the duty cycles of Boost I and Boost II as follows:

$$\begin{cases} d_1 = d + \Delta d_1 = d - \Delta d \\ d_2 = d + \Delta d_2 = d + \Delta d \end{cases} \tag{52}$$

When U_{C1} is bigger than U_{C2}, the voltage-balance process is: ΔI_L becomes positive, which makes $I_{L1}{}^*$ decrease and $I_{L2}{}^*$ increase. As a result, d_1 decreases while d_2 increases, i.e., the turn-on time of S_1 decreases while that of S_2 increases. Thus, U_{C1} decreases while U_{C2} increases. Finally, U_{C1} equals to U_{C2} after several switching periods. When U_{C1} is smaller than U_{C2}, U_{C1} and U_{C2} could be also balanced according to a similar voltage-balance process.

5. Simulation and Experimental Verification

5.1. Simulation Verification

To verify the correctness and feasibility of the IPOS-SC-TLB converter, a simulation model adopting the proposed three-loop control strategy with 400 W output power has been implemented. The detailed simulation and experimental parameters are presented in Table 3.

Table 3. Simulation & Experimental Parameters.

Components	Parameters
Input voltage U_{in}	48 V–120 V
Output voltage U_o	400 V
Switching frequency f_s	25 kHz
Output power P_o	400 W
Switches S_1, S_2	G80N60, 2.4 V voltage drop
Diodes D_1, D_2	DSEP30-06B, 2.0 V voltage drop
Inductors L_1, L_2	915 μH, 895 μH, 0.1 ohm equivalent series resistance
Capacitors C_1, C_2, C_f	470 μF, 0.28 ohm equivalent series resistance
Driver	A3120

The input voltage varies between 48 V and 120 V, and the output voltage is controlled to be stable at 400 V. The switching frequency of the converter is set as 25 kHz. Two inductors are both chosen as about 900 μH with 0.1 ohm equivalent series resistance. Three capacitors are all set as 470 μF with 0.28 ohm equivalent series resistance. Each of the two IGBT switches has a voltage drop of 2.4 V and each of the three diodes has a voltage drop of 2.0 V. Figure 11 shows the simulation results when the input voltage is 48 V and Figure 12 shows the simulation results when the input voltage is 120 V. It can be seen that the IPOS-SC-TLB converter can output a stable dc voltage of 400 V under both the two different input voltages. The voltage difference between U_{C1} and U_{C2} is about 20 V under the input voltage 48 V and 10 V under the input voltage 120 V without voltage-balance control. However, once the voltage-balance control loop is added, U_{C1} and U_{C2} are balanced with the same voltage 200 V. Besides, in the whole experimental process, a small voltage difference between U_{C1} and U_{Cf} is about 4.4 V, which is the sum of the voltage drop of one IGBT and the voltage drop of one diode.

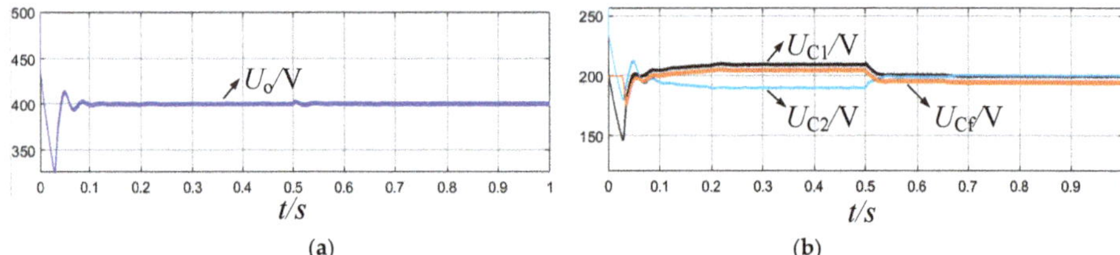

Figure 11. Simulated voltage waveforms when U_{in} is 48 V: (a) U_o; (b) U_{C1}, U_{C2}, and U_{Cf}.

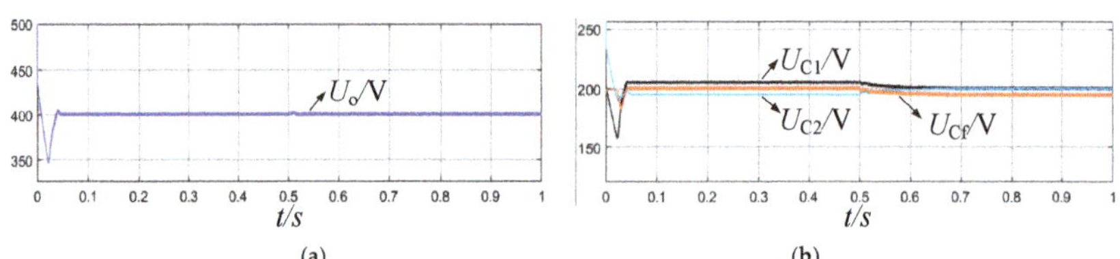

Figure 12. Simulated voltage waveforms when U_{in} is 120 V: (a) U_o; (b) U_{C1}, U_{C2}, and U_{Cf}.

More importantly, Figure 13 shows the two split inductor current waveforms of the converter without voltage-balance control and with voltage-balance control when the input voltage is 48 V. Under the condition without voltage-balance control, the average values of the two inductor currents are equal while a little different under the condition with voltage-balance control. Because the duty cycle d_1 and the duty cycle d_2 are the same under the condition without voltage-balance control but d_1 is a little smaller than d_2 under the condition with voltage-balance control. Besides, the input current ripple is smaller than the inductor current ripples, and input current ripple frequency is 50 kHz, which is two times the switching frequency 25 kHz.

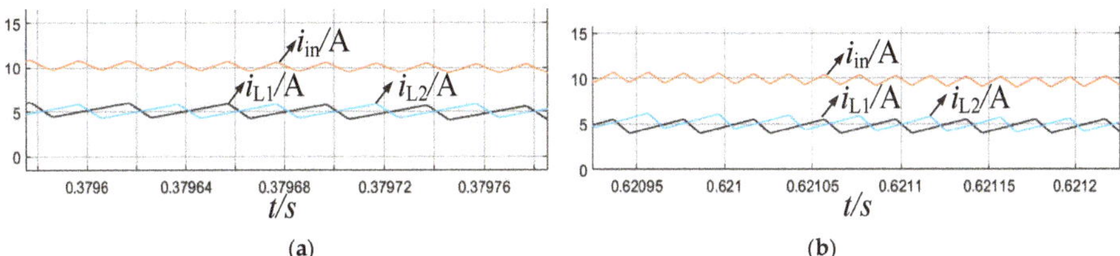

Figure 13. Simulated current waveforms when U_{in} is 48 V: (**a**) without voltage-balance control; (**b**) with voltage-balance control.

The simulated voltage waveforms and current waveforms are presented in Figure 14. The voltage stress across all power devices are half of the output voltage. The average current across every diode is 1 A, which is the same as the output current. Additionally, the average current across the switch S_2 is bigger than S_1 because the current across the diode D_2 added on the current of S_2. These results prove correctness of the theoretically derived results shown in (28). On the whole, the simulation results basically verify the effectiveness of the IPOS-SC-TLB converter and the proposed three-loop control strategy.

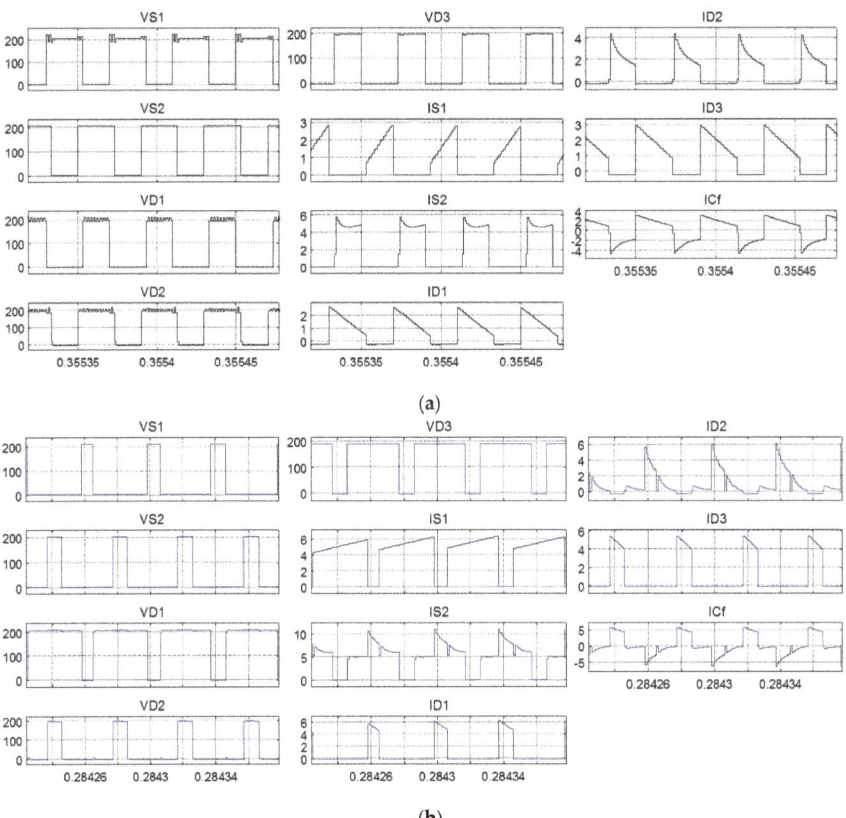

Figure 14. Simulated voltage and current waveforms: (**a**) U_{in} = 48 V; (**b**) U_{in} = 120 V.

5.2. Experimental Verification

To verify the converter and three-loop control strategy further, an experimental prototype with the same parameters shown in simulation model is built and it is given in Figure 15. It should be noted that the inductors L_1, L_2 are respectively designed to be 915 µH, 895 µH with some deviations in fact. The two switches are both selected as IGBT G80N60, which have a voltage drop of 2.4 V and the three diodes are selected as DSEP30-06B, which have a voltage drop of 2.0 V. The control loop of the converter was implemented based on Dspace 1103.

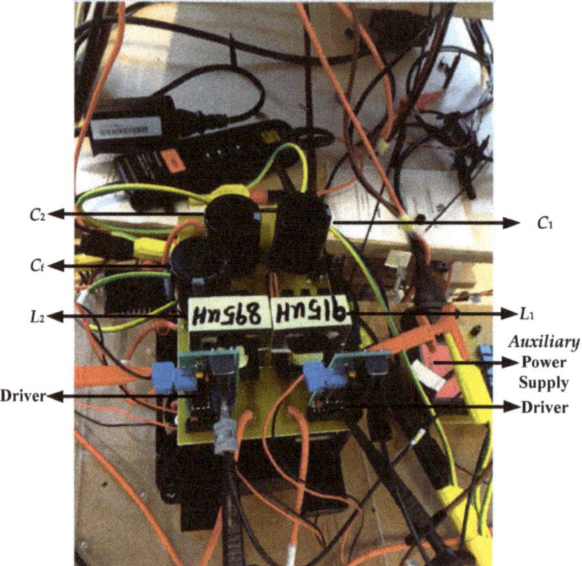

Figure 15. The experimental prototype.

The input current and capacitor voltages of the IPOS-SC-TLB converter under different input voltages are presented in Figure 16, and the corresponding capacitor voltages are presented in Figure 17. The inductor currents across L_1, L_2 and the drive signals of S_1, S_2 are presented in Figure 18. It can be seen that the output voltage is stable at 400 V and the three capacitor voltages are stable with 200 V under different input voltages. The input current is continuous with a small current ripple and the input current ripple frequency is 50 kHz, which is two times the switching frequency 25 kHz. Moreover, it is easy to observe that as the duty cycle approaches 0.50, the input current ripple becomes almost zero, which verifies (32).

To show voltage stresses across all the switches and diodes, the terminal voltage waveforms of S_1, S_2, D_1, D_2, and D_3 are presented in Figures 19–21. It should be noted that u_{S1}, u_{S2} are defined to describe the voltage difference between the drain terminal and the source terminal of S_1 and S_2, respectively. u_{D1}, u_{D2}, u_{D3} are the voltage differences between the cathode and the anode of D_1, D_2 and D_3. It can be seen that all the voltage stresses of the switches and diodes are 200 V, which is half of the output voltage 400 V. It matches with (12). In addition, the current I_{S2} is the sum of I_{S1} and I_{D2}, which matches with (20) and (21). For example, when the input voltage is 48 V, I_{S1}, I_{S2}, I_{D2} are 3.74 A, 4.78 A and 1.13 A, respectively. It is not difficult to know that the switching state of D_1 is complementary to that of D_2, and the switching state of D_1 is 180 degrees shifted from that of D_3. The switching state of S_1 is also 180 degrees shifted from that of S_2. All of these results can verify the correctness of the operating principle of the IPOS-SC-TLB converter.

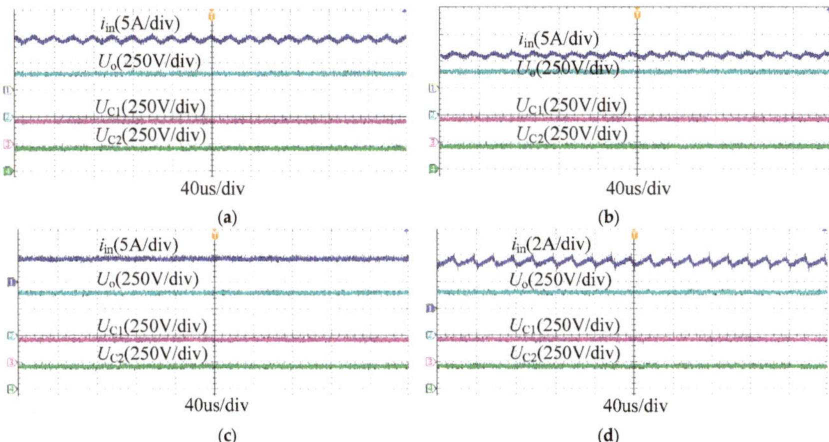

Figure 16. Input current and capacitor voltage waveforms: (**a**) U_{in} = 48 V; (**b**) U_{in} = 72 V; (**c**) U_{in} = 100 V; (**d**) U_{in} = 120 V.

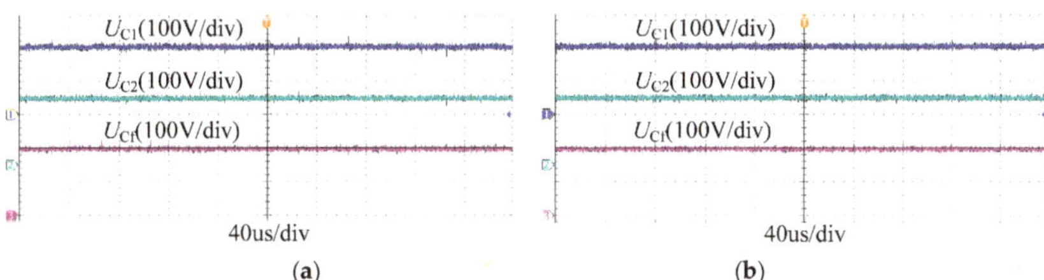

Figure 17. Capacitor voltage waveforms: (**a**) U_{in} = 48 V; (**b**) U_{in} = 120 V.

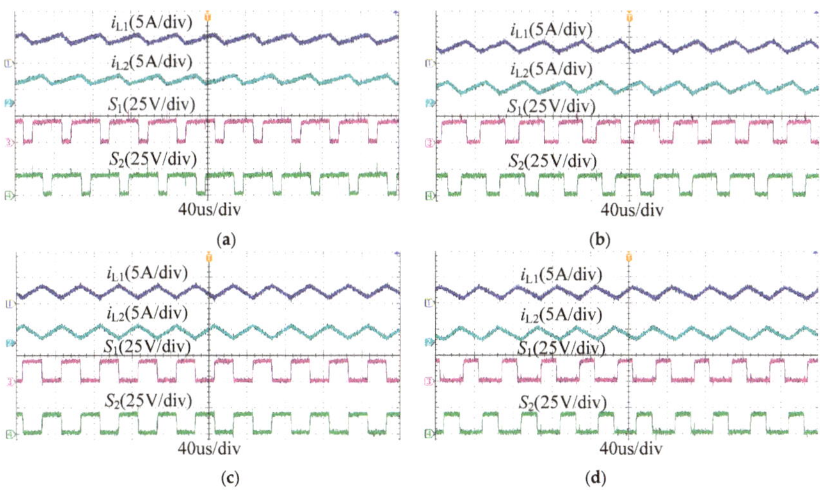

Figure 18. Inductor current and drive signal waveforms: (**a**) U_{in} = 48 V; (**b**) U_{in} = 72 V; (**c**) U_{in} = 100 V; (**d**) U_{in} = 120 V.

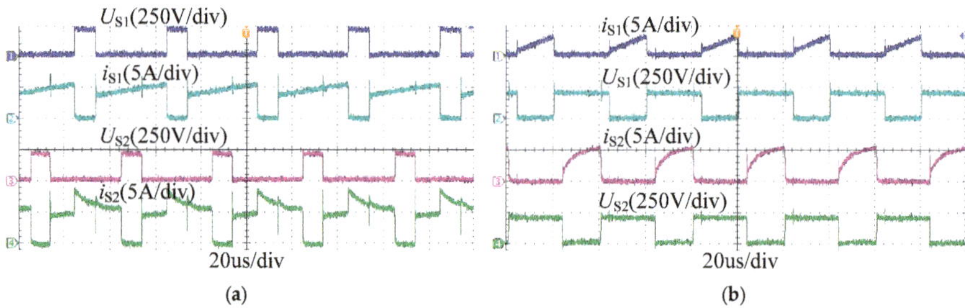

Figure 19. Tested voltage and current waveforms of S_1, S_2: (a) U_{in} = 48 V; (b) U_{in} = 120 V.

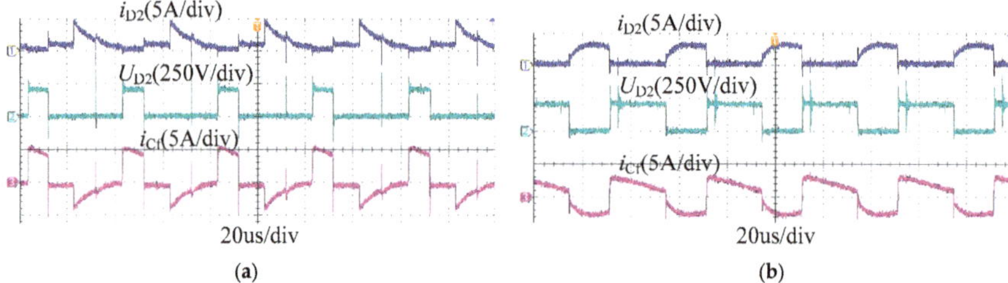

Figure 20. Tested voltage and current waveforms of D_2 and current waveform of C_f: (a) U_{in} = 48 V; (b) U_{in} = 120 V.

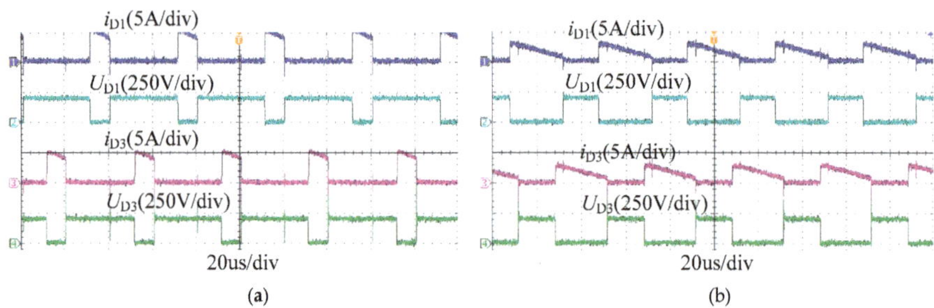

Figure 21. Tested voltage and current waveforms of D_1, D_3: (a) U_{in} = 48 V; (b) U_{in} = 120 V.

More importantly, the voltage-balance experimental waveforms of the IPOS-SC-TLB converter are presented in Figures 22 and 23. The experimental results indicate that when the voltage-balance loop is not added, there is about 13.0 V voltage difference between U_{C1} and U_{C2}. For example, when the input voltage is 48 V, the tested duty cycle is around 0.83. According to (43), the voltage difference between U_{C1} and U_{C2} is 11.26 V under the input voltage of 48 V. The tested voltage difference of 13.00 V basically matches the theoretical value 11.26 V with some voltage error. When the voltage-balance loop is added, the voltage difference becomes nearly zero.

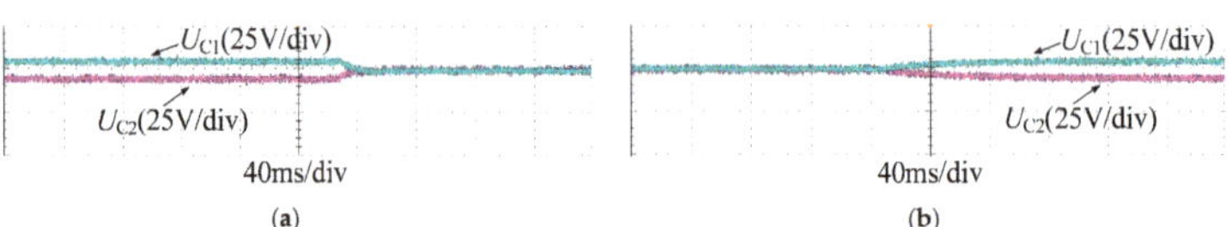

Figure 22. Voltage balance process of C_1, C_2 when U_{in} is 48 V: (a) from no voltage-balance control to voltage-balance control; (b) from voltage-balance control to no voltage-balance control.

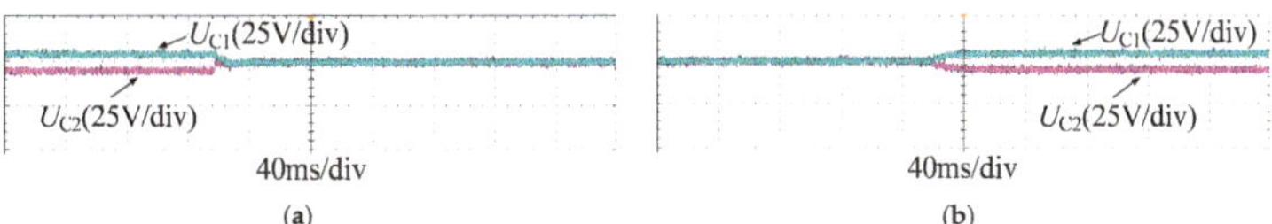

Figure 23. Voltage balance process of C_1, C_2 when U_{in} is 120 V: (**a**) from no voltage-balance control to voltage-balance control; (**b**) from voltage-balance control to no voltage-balance control.

Figure 24 shows the theoretical voltage gain and the experimental voltage gain versus duty cycle when U_{in} is 48 V. It can be seen that the theoretical voltage gain and the experimental voltage gain have the same increasing trend though some deviations exist. The experimental voltage gain basically matches the theoretical voltage gain when the duty cycle varies between 0.2 and 0.5. However, when the duty cycle is over 0.50, the experimental theoretical voltage gain is less than the theoretical voltage gain, and their difference increases with the duty-cycle increasing. This phenomenon may be due to the non-linearity of power electronic components and the fact the true values of parasitic parameters are hard to obtain.

The conversion efficiency curves versus output power for the IPOS-SC-TLB converter under different input voltages are given in Figure 25. The minimum efficiency and the maximum efficiency are 92.08% and 94.20%, respectively, at an input voltage of 48 V; 95.13% and 96.55% at the input voltage of 72 V; 96.08% and 97.32% at the input voltage 100 V; 96.62% and 98.57% at an input voltage of 120 V. It can be seen that the proposed converter is not efficient in low voltage levels, such as 48 V in the experiment. To make it efficient, the converter should be implemented with optimized design, including component selection, coupling inductor design and applying soft switching technique. For component selection, wide bandgap device (SiC, GaN) with much smaller parasitic parameters should be a good solution, which could not only reduce conduction and switching losses, but also enhance the switching frequency to reduce passive components' size and parasitic parameters as well. Coupling design for the two inductors L_1 and L_2 will help reduce size and improve efficiency of the converter. Soft switching technique applied on this converter will help enhance conversion efficiency.

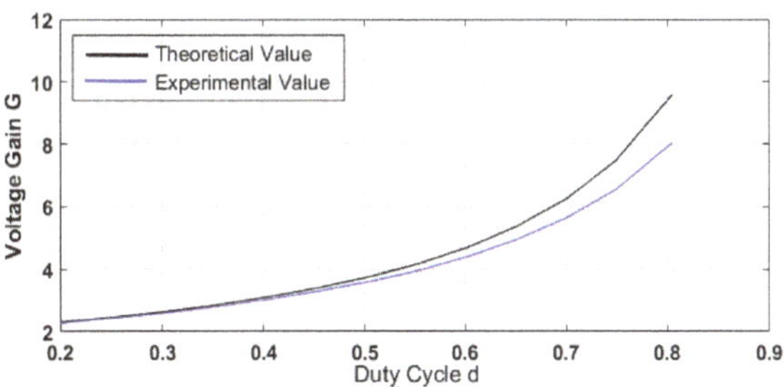

Figure 24. The theoretical voltage gain and the experimental voltage gain versus duty cycle when U_{in} is 48 V.

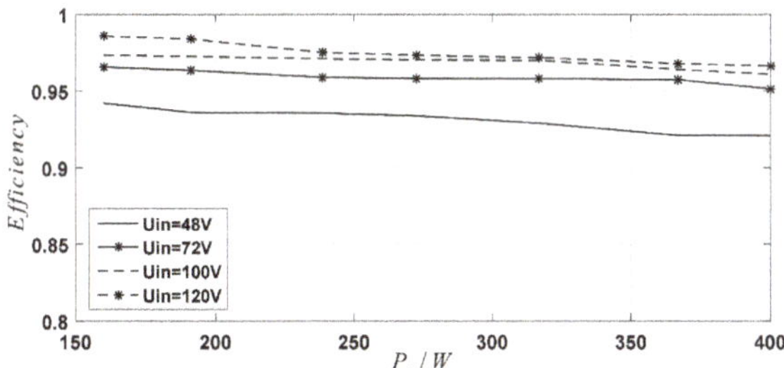

Figure 25. Efficiency curves under different input voltages.

Based on all the experimental results, the theoretical analysis of the IPOS-SC-TLB converter is correct and the three-loop control strategy is feasible. The effectiveness of the proposed IPOS-SC-TLB converter has been verified.

6. Conclusions

This paper presents an input-parallel-output-series three-level Boost converter, which can step up the input voltage to a high voltage level, as well as attaining low voltage stress, low current stress and small input current ripple. Another advantage of the proposed topology is the automatic current balancing function. There is also a disadvantage that the imbalance current between the two power switches S_1, S_2. The average current of S_2 is U_o/R higher than the average current of S_1.

Author Contributions: J.C. proposed the topology and control strategy, and built the simulation model and experimental prototype. C.W. and J.L. helped do comparative analysis and wrote the paper.

Funding: This research was funded by China Postdoctoral Science Foundation, grant number 2017M612908.

Conflicts of Interest: The authors declare no conflict of interest.

Nomenclature

List of Abbreviations

IPOS-SC-TLB	input-parallel-output-series switched-capacitor three-level boost
HVDC	high voltage direct current
SI	serial-interleaved
PI	parallel-interleaved
SI-TLB	serial-interleaved three-level boost
PIB	parallel-interleaved boost
SI-FC-TLB	serial-interleaved flying-capacitor three-level boost
PI-FC-TLB	parallel-interleaved flying-capacitor three-level boost
PI-SFC-TLB	parallel-interleaved symmetric flying-capacitor three-level boost
SC-TLB	switched-capacitor three-level boost
IPOS-TLB1	input-parallel-output-series three-level boost 1
IPOS-TLB2	input-parallel-output-series three-level boost 2

List of Symbols

U_{in}	average input voltage
U_o, I_o	average output voltage and average output current
$U_o{}^*$	output voltage reference
U_{C1}, U_{C2}, U_{Cf}	average voltages of capacitors C_1, C_2, C_f
ΔU	voltage difference between C_1 and C_f,
i_{L1}, i_{L2}	currents of inductors L_1, L_2
I_{L1}, I_{L2}	average currents of inductors L_1, L_2
$I_{L1}{}^*, I_{L2}{}^*$	reference currents of inductors L_1, L_2
$\Delta i_{L1}, \Delta i_{L2}$	current ripples of inductors L_1, L_2
Δi_{in}	input current ripple
I_{D1}, I_{D2}, I_{D3}	average currents of diodes D_1, D_2, D_3
I_{S1}, I_{S2}	average currents of switches S_1, S_2
d	duty cycle
d_1	duty cycle of boost 1
d_2	duty cycle of boost 2
Δd	duty cycle difference

References

1. Yaramasu, V.; Wu, B.; Sen, P.C.; Kouro, S.; Narimani, M. High-power Wind Energy Conversion Systems: State-of-the-art and Emerging Technologies. *Proc. IEEE* **2015**, *103*, 740–788. [CrossRef]
2. Yaramasu, V.; Wu, B. Predictive Control of a Three-Level Boost Converter and an NPC Inverter for High-Power PMSG-Based Medium Voltage Wind Energy Conversion Systems *IEEE Trans. Power Electron.* **2014**, *29*, 5308–5322. [CrossRef]
3. Parastar, A.; Seok, J.-K. High Power Step-Up Modular Resonant DC/DC Converter for Offshore Wind Energy Systems. *IEEE Energy Convers. Congr. Expos. (ECCE)* **2014**, 3341–3348. [CrossRef]
4. Yaramasu, V.; Wu, B.; Rivera, M.; Rodriguez, J. A New Power Conversion System for Megawatt PMSG Wind Turbines Using Four-Level Converters and a Simple Control Scheme Based on Two-Step Model Predictive Strategy-Part I: Modeling and Theoretical Analysis. *IEEE J. Emerg. Sel. Top. Power Electron.* **2014**, *2*, 3–13. [CrossRef]
5. Paez, J.D.; Frey, D.; Maneiro, J.; Bacha, S.; Dworakowski, P. Overview of DC-DC Converters Dedicated to HVDC Grids. *IEEE Trans. Power Deliv.* **2018**. [CrossRef]
6. Xia, C.; Gu, X.; Shi, T.; Yan, Y. Neutral-Point Potential Balancing of Three-Level Inverters in Direct-Driven Wind Energy Conversion System. *IEEE Trans. Energy Convers.* **2011**, *26*, 18–29. [CrossRef]
7. Chen, J.; Wang, C.; Li, J.; Jiang, C.; Duan, C. A Nonisolated Three-Level Bidirectional DC-DC Converter. In Proceedings of the 2018 IEEE Applied Power Electronics Conference and Exposition (APEC), San Antonio, TX, USA, 4–8 March 2018; pp. 1566–1570.
8. Hou, S.; Chen, J.; Sun, T.; Bi, X. Multi-input Step-Up Converters Based on the Switched-Diode-Capacitor Voltage Accumulator. *IEEE Trans. Power Electron.* **2016**, *31*, 381–393. [CrossRef]
9. Zhou, H.; Zhou, J.; Hu, B.; Tong, C. A New Interleaved Three-level Boost Converter and Neutral-point Potential Balancing. In Proceedings of the 2013 2nd International Symposium on Instrumentation and Measurement, Sensor Network and Automation (IMSNA), Toronto, ON, Canada, 23–24 December 2013.
10. Zhang, Y.; Sun, J.-T.; Wang, Y.-F. Hybrid Boost Three-Level DC-DC Converter With High Voltage Gain for Photovoltaic Generation Systems. *IEEE Trans. Power Electron.* **2013**, *28*, 3659–3664. [CrossRef]
11. Andrade, A.M.S.S.; Martins, M.L.D.S. Quadratic-Boost with Stacked Zeta Converter for High Voltage Gain Applications. *IEEE J. Emerg. Sel. Top. Power Electron.* **2017**, *5*, 1787–1796. [CrossRef]
12. Jin, K.; Yang, M.; Ruan, X.; Xu, M. Three-Level Bidirectional Converter for Fuel-Cell/Battery Hybrid Power System. *IEEE Trans. Ind. Electron.* **2010**, *57*, 1976–1986. [CrossRef]
13. Rahul, J.R.; Kirubakaran, A.; Vijayakumar, D. A New Multilevel DC-DC Boost Converter for Fuel Cell based Power System. In Proceedings of the 2012 IEEE Students' Conference on Electrical, Electronics and Computer Science (SCEECS), Bhopal, India, 1–2 March 2012.
14. Hegazy, O.; Mierlo, J.V.; Lataire, P. Analysis, Modeling, and Implementation of A Multidevice Interleaved DC/DC Converter for Fuel Cell Hybrid Electric Vehicles *IEEE Trans. Power Electron.* **2012**, *27*, 4445–4458. [CrossRef]
15. Ni, L.; Patterson, D.J.; Hudgins, J.L. High Power Current Sensorless Bidirectional 16-Phase Interleaved DC-DC Converter f or Hybrid Vehicle Application. *IEEE Trans. Power Electron.* **2012**, *27*, 1141–1151. [CrossRef]
16. Rivera, S.; Wu, B.; Kouro, S.; Yaramasu, V.; Wang, J. Electric Vehicle Charging Station Using A Neutral-point Clamped Converter with Bipolar DC Bus. *IEEE Trans. Ind. Electron.* **2015**, *62*, 1999–2009. [CrossRef]
17. Tan, L.; Wu, B.; Yaramasu, V.; Rivera, S.; Guo, X. Effective Voltage Balance Control for Bipolar-DC-Bus Fed EV Charging Station with Three-Level DC-DC Fast Charger. *IEEE Trans. Ind. Electron.* **2016**, *63*, 4031–4041. [CrossRef]
18. Adam, G.P.; Gowaid, I.A.; Finney, S.J.; Holliday, D.; Williams, B.W. Review of DC-DC Converters for Multi-terminal HVDC Transmission Networks. *IET Power Electron.* **2016**, *9*, 281–296. [CrossRef]
19. Chen, J.; Hou, S.; Deng, F.; Chen, Z. An Interleaved Five-Level Boost Converter with Voltage-Balance Control. *J. Power Electron.* **2016**, *16*, 1735–1742. [CrossRef]
20. Kakigano, H.; Miura, Y.; Ise, T. Distribution Voltage Control for DC Microgrids Using Fuzzy Control and Gain-Scheduling Technique. *IEEE Trans. Power Electron.* **2013**, *28*, 2246–2258. [CrossRef]
21. Jang, Y.; Jovanovic, M.M. Interleaved Boost Converter with Intrinsic Voltage-Doubler Characteristic for Universal-Line PFC Front End. *IEEE Trans. Power Electron.* **2007**, *22*, 1394–1401. [CrossRef]

22. Zhou, L.-W.; Zhu, B.-X.; Luo, Q.-M.; Chen, S. Interleaved Non-isolated High Step-up DC-DC Converter Based on the Diode-capacitor Multiplier. *IET Power Electron.* **2014**, *7*, 390–397. [CrossRef]
23. Franco, L.C.; Pfitscher, L.L.; Gules, R. A New High Static Gain Nonisolated DC-DC Converter. In Proceedings of the 34th Annual Power Electronics Specialist Conference IEEE, Acapulco, Mexico, 15–19 June 2003.
24. Kajangpan, K.; Neammenee, B. High Gain Double Interleave Technique with Maximum Peak Power Tracking for Wind Turbine Converter. In Proceedings of the 2009 6th International Conference on Electrical Engineering/Electronics, Computer, Telecommunications and Information Technology Pattaya, Chonburi, Thailand, 6–9 May 2009.
25. Forest, F.; Huselstein, J.J.; Martiré, T.; Flumian, D.; Meynard, T.A.; Abdelli, Y.; Lienhardt, A.M. A Nonreversible 10-kW High Step-Up Converter Using a Multicell Boost Topology. *IEEE Trans. Power Electron.* **2018**, *33*, 151–160. [CrossRef]
26. Rosas-Caro, J.C.; Ramirez, J.M.; Peng, F.Z.; Valderrabano, A. A DC-DC Multilevel Boost Converter. *IET Power Electron.* **2010**, *3*, 129–137. [CrossRef]
27. Parastar, A.; Kang, Y.C.; Seok, J.-K. Multilevel Modular DC-DC Power Converter for High-Voltage DC-Connected Offshore Wind Energy Applications. *IEEE Trans. Ind. Electron.* **2015**, *62*, 2879–2890. [CrossRef]
28. Zhang, X.; Green, T.C. The Modular Multilevel Converter for High Step-Up Ratio DC-DC Conversion. *IEEE Trans. Ind. Electron.* **2015**, *62*, 4925–4936. [CrossRef]
29. Ferreira, J.A. The Multilevel Modular DC Converter. *IEEE Trans. Power Electron.* **2013**, *28*, 4460–4465. [CrossRef]
30. Zhang, X.; Green, T.C. The New Family of High Step Ratio Modular Multilevel DC-DC Converters. In Proceedings of the 2015 IEEE Applied Power Electronics Conference and Exposition (APEC), Charlotte, NC, USA, 15–19 March 2015.
31. Hagar, A.A.; Lehn, P.W. Comparative Evaluation of a New Family of Transformerless Modular DC-DC Converters for High-Power Applications. *IEEE Trans. Power Deliv.* **2014**, *29*, 444–452. [CrossRef]
32. Filsoof, K.; Lehn, P.W. A Bidirectional Modular Multilevel DC-DC Converter of Triangular Topology. *IEEE Trans. Power Electron.* **2015**, *30*, 54–64. [CrossRef]
33. Athab, H.; Yazdani, A.; Wu, B. A Transformerless DC-DC Converter with Large Voltage Ratio for MV DC Grids. *IEEE Trans. Power Deliv.* **2014**, *29*, 1877–1885. [CrossRef]
34. Chen, W.; Huang, A.Q.; Li, C.; Wang, G.; Gu, W. Analysis and Comparison of Medium Voltage High Power DC-DC Converters for Offshore Wind Energy Systems. *IEEE Trans. Power Electron.* **2013**, *28*, 2014–2023. [CrossRef]
35. Mohan, T.; Undeland, M.; Robbins, W.P. *Power Electronics—Converters, Applications, and Design*; John Wiley & Sons Inc.: Hoboken, NJ, USA, 2003.
36. Erickson, R.W.; Maksimovic, D. *Fundamentals of Power Electronics*, 2nd ed.; Kluwer Academic Publisher: Norwell, MA, USA, 2001.

Article

Capacitors Voltage Switching Ripple in Three-Phase Three-Level Neutral Point Clamped Inverters with Self-Balancing Carrier-Based Modulation

Manel Hammami, Gabriele Rizzoli *, Riccardo Mandrioli and Gabriele Grandi

Department of Electrical, Electronic, and Information Engineering, University of Bologna, 40136 Bologna, Italy; manel.hammami2@unibo.it (M.H.); riccardo.mandrioli4@unibo.it (R.M.); gabriele.grandi@unibo.it (G.G.)
* Correspondence: gabriele.rizzoli@unibo.it; Tel.: +39-051-20-9-3585

Received: 23 October 2018; Accepted: 19 November 2018; Published: 22 November 2018

Abstract: This paper provides a comprehensive analysis of the capacitors voltage switching ripple for three-phase three-level neutral point clamped (NPC) inverter topologies. The voltage ripple amplitudes of the two dc-link capacitors are theoretically estimated as a function of both amplitude and phase angle of output current and the inverter modulation index. In particular, peak-to-peak distribution and maximum amplitudes of the capacitor voltage switching ripple over the fundamental period are obtained. A comparison is made considering different carrier-based pulse-width modulations in the case of almost all sinusoidal load currents, representing either grid connection or passive load with a negligible current ripple. Based on the voltage switching ripple requirements of capacitors, a simple and effective original equation for a preliminary sizing of the capacitors has been proposed. Numerical simulations and experimental tests have been carried out in order to verify the analytical developments.

Keywords: voltage ripple; voltage source inverter; three-phase inverter; DC-link capacitor design

---

1. Introduction

In industrial applications, the most used switching inverter is the two-level converter. Due to mass production, it is a relatively cheap and reliable configuration. Furthermore, as the number of semiconductor devices is low, they can be simply controlled by different types of pulse-width modulation (PWM) techniques. However, the main drawback of the two-level converter is the high harmonic content of the output voltage, which makes the use of bulky output filter necessary, increasing the cost of the system and the losses. The harmonic content can be reduced simply by increasing the PWM switching frequency, which leads to an increase of the switching losses. The use of power filters and high switching frequency has to be balanced to achieve reasonable converter costs with acceptable efficiency.

During the last decades, the drawbacks of the two-level converter motivated researchers to develop new converter topologies. A very promising inverter family, called multilevel inverters (MLIs), offer better quality output voltage waveforms with a reduced harmonic content comparing to the conventional two-level inverter topologies, increase the overall voltage and power rating of the converter, and generally mitigate the electromagnetic interferences.

The penetration of multilevel inverters has been steadily increasing due to their widespread usage in manufacturing, transport, energy, high-power drives and other industry applications. Several MLI topologies have been introduced and extensively studied in the literature. Most currently used multilevel topologies can be grouped: cascaded H-bridge (CHB), neutral point clamped (NPC), and flying capacitors (FC).

Multilevel inverters were initiated by the invention of the so-called NPC inverter in 1981 by Nabae et al. [1]. The NPC inverter uses a single DC bus subdivided into a number of voltage levels by a series string of capacitors. The voltages across the individual switches are clamped by diodes at the voltage level of only one capacitor of the series DC-link string. The major difficulty associated with control of the diode-clamped inverter is the balancing of the capacitor voltages [2,3]. To achieve that, many modulation techniques have been developed [4–8]. In particular, carrier-based modulation and space vector modulation (SVM) strategies, similar to those employed for conventional two-level three-phase inverters, can be readily modified and extended to fulfill the multilevel NPC requirements.

An essential part of MLI design is the selection of DC-link capacitors. The capacitors are a sensitive element of the inverter and a common source of failures. So far, regarding two-level converter systems, some papers have investigated the minimum DC-link capacitance and proposed methods for its size reduction [9,10]. Recently, based on the DC-link voltage analysis and considering both low- and/or high-frequency DC voltage components, simple and effective guidelines for designing the DC-link capacitor have been presented in [11] for single-phase H-bridge inverter, and in [12] for three-phase three-level flying capacitor inverter. Methods used to derive expressions for RMS value and harmonic spectrum of the capacitor current in two-level inverters, are extended to the three-level inverters in [13]. An analytical expression for calculating RMS current through the DC-link capacitor in a three-level NPC inverter is given in [14]. The analysis of the DC-link capacitor current in three-level NPC and CHB inverters and a new numerical approach for calculating the RMS value of the capacitor current is proposed in [15].

Evaluation of the low-frequency neutral-point voltage oscillations in the three-level NPC inverter using space vector modulation (SVM) techniques has been analyzed in [16]. A novel modulation strategy for the NPC inverter is proposed in [17] to overcome one of the main problems of this converter, which is the low-frequency voltage oscillation that appears in the neutral point. The proposed modulation strategy can completely remove this oscillation for all the operating conditions and for any kind of loads, even unbalanced and nonlinear loads.

Nowadays, NPC inverters are the most widely used three-level inverter topologies in the industrial applications, considering both the conventional and the T-type configurations (Figure 1). In order to choose the most suitable topology and hereby increase power efficiency, the comparison between of conventional NPC and T-type inverters has been investigated and reported in the literature [18], also considering the loss evaluation. An efficiency comparison of both over-mentioned topologies is presented in [19] and the result shows that the T-type inverter is generally more efficient at lower switching frequencies.

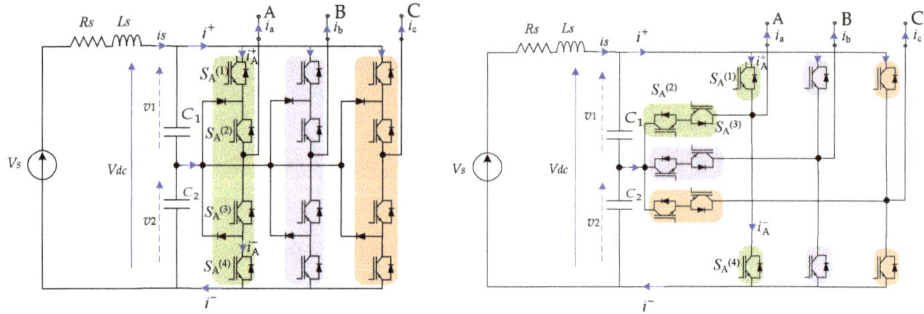

Figure 1. Circuit schemes of the three-phase three-level neutral point clamped (NPC) inverter: conventional type (left side) and T-type (right side).

The analysis of inverter DC-link input current and voltage is essential for sizing and designing the DC-link capacitor since it directly impacts the price, the lifetime and the failure rate of a converter system. A DC-link capacitor has to deal with the harmonics of the inverter input current and to avoid the high DC-link voltage ripple appearance. In general, the low-frequency voltage ripple component is more important for the required capacitance design since it has a higher value comparing to the switching frequency voltage ripple component. Although correctly sizing the DC-link capacitances is important to control the magnitude of the low-frequency voltage ripple, it is not the only thing that must be considered.

The calculation of low-frequency input current and input voltage ripples on DC-link capacitors have been analyzed and presented in the literature by other authors [15,16]. The evaluation of voltage switching ripple for the NPC converters has not been reported yet. The analysis of the voltage switching ripple in DC-link capacitors of three-level three-phase NPC inverters, applicable to both conventional and T-type configurations (Figure 1), is presented in this paper, with reference to different carrier-based PWM techniques. The peak-to-peak capacitor voltage switching ripple amplitudes are analytically determined as a function of modulation index and output phase angle. Based on the limitation on the peak-to-peak capacitor voltage switching ripple amplitudes, simple and practical expressions for sizing the capacitors have been proposed.

The paper is organized as follows. Section 2 introduces the system configuration and the modulation principles. Section 3 presents the analysis of low-frequency and switching frequency input current. Section 4 presents the analysis of the capacitor voltage switching ripple. Section 5 defines the guidelines for a preliminary design of the DC-link capacitors. In Section 6 simulation and experimental verifications have been reported, and Section 7 presents the conclusion.

2. System Configuration and Modulation Principles

2.1. System Configuration

The circuit scheme of a three-level three-phase neutral point clamped inverter (NPC) is shown in Figure 1, for both conventional and T-type configurations. It consists of a DC voltage source (V_s) with series RL impedance, representing either a simplified model of a real DC source or a DC filter (series reactor). Each leg of the inverter is composed of four power switches (for leg A: $S_A{}^{(1)}$ to $S_A{}^{(4)}$, the same for legs B and C). Two capacitors C_1 and C_2 are connected to the neutral point of the inverter and serve as a voltage divider. For proper operation of the NPC inverter, the neutral point must be kept at one half of DC-link voltage by using a proper modulation strategy able to achieve voltage balancing between the capacitors.

2.2. Modulation Principles

In case of balanced modulation and within the linear modulation range, the output voltages normalized by V_{dc} and averaged over the switching period ($T_{sw} = 1/f_{sw}$) correspond to the modulating signals [20]:

$$\begin{cases} u_A = m\,\sin(\vartheta) + C_m = u_A^* + C_m \\ u_B = m\,\sin(\vartheta - \frac{2\pi}{3}) + C_m = u_B^* + C_m \\ u_C = m\,\sin(\vartheta - \frac{4\pi}{3}) + C_m = u_C^* + C_m \end{cases} . \tag{1}$$

where $\vartheta = \omega t$, ω is the fundamental angular frequency ($\omega = 2\pi f$), f is the fundamental frequency, m is the inverter modulation index, u_i^* are the normalized reference output voltages of each phase (i = A, B or C) and C_m represents the injected common mode signal.

The voltages across the two capacitors C_1 and C_2 can be spontaneously regulated to half of the DC-link voltage by the use of proper modulation technique with self-balancing capability.

Correspondingly, the following averaged switching functions for the upper and lower switches $\overline{S}_i^{(1)}$ and $\overline{S}_i^{(4)}$ can be written (averaging is denoted by overline):

$$\begin{cases} \overline{S}_i^{(1)} = u_i + |u_i| \\ \overline{S}_i^{(4)} = -u_i + |u_i| \end{cases}. \tag{2}$$

With reference to the modulation (1) and considering phase A, the averaged switching functions of the upper and the lower switches, Equation (2), become:

$$\begin{cases} \overline{S}_A^{(1)} = m\,\sin(\vartheta) + C_m + |m\,\sin(\vartheta) + C_m| \\ \overline{S}_A^{(4)} = -m\,\sin(\vartheta) + C_m + |m\,\sin(\vartheta) + C_m| \end{cases}. \tag{3}$$

Due to the modulation symmetry among the three phases, the switching functions for the phases B and C are readily obtained considering the phase displacement of the normalized output voltages given in Equation (1).

In case of sinusoidal PWM (SPWM), the injected common-mode signal is zero:

$$C_m = 0. \tag{4}$$

However, in case of centered PWM (CPWM), the injected common-mode signal is:

$$C_m = -\frac{1}{2}(\max(u_A^*, u_B^*, u_C^*) + \min(u_A^*, u_B^*, u_C^*)). \tag{5}$$

In this last case, C_m can be rewritten as:

$$C_m = \frac{1}{2}m \begin{cases} \sin\vartheta,\ -\frac{\pi}{6} \le \vartheta \le \frac{\pi}{6},\ \frac{5\pi}{6} \le \vartheta \le \frac{7\pi}{6} \\ \sin\left(\vartheta + \frac{2\pi}{3}\right),\ \frac{\pi}{6} \le \vartheta \le \frac{\pi}{2},\ \frac{7\pi}{6} \le \vartheta \le \frac{3\pi}{2} \\ \sin\left(\vartheta - \frac{2\pi}{3}\right),\ \frac{\pi}{2} \le \vartheta \le \frac{5\pi}{6},\ \frac{3\pi}{2} \le \vartheta \le \frac{11\pi}{6} \end{cases}. \tag{6}$$

A straightforward method to implement carrier-based optimized centered PWM (OCPWM) for three-phase three-level inverters has been proposed in [20]. The procedure is based on applying the traditional min/max centering separately to pivot voltages and residual two-level voltages. Pivot voltages are determined by a simple polarity combination of reference voltages. The resulting common-mode voltage that has to be injected in reference voltages is determined in few simple steps as described in the following equations:

$$C_m = -\frac{1}{2}\left(\max(u_A^p, u_B^p, u_C^p) + \min(u_A^p, u_B^p, u_C^p)\right) - \frac{1}{2}\left(\max(u_A^{2L}, u_B^{2L}, u_C^{2L}) + \min(u_A^{2L}, u_B^{2L}, u_C^{2L})\right) \tag{7}$$

being

$$\begin{cases} u_A^p = \frac{1}{4}\left[sign(u_A^*) - \frac{1}{3}\left(sign(u_A^*) + sign(u_B^*) + sign(u_C^*)\right)\right] \\ u_B^p = \frac{1}{4}\left[sign(u_B^*) - \frac{1}{3}\left(sign(u_A^*) + sign(u_B^*) + sign(u_C^*)\right)\right] \\ u_C^p = \frac{1}{4}\left[sign(u_C^*) - \frac{1}{3}\left(sign(u_A^*) + sign(u_B^*) + sign(u_C^*)\right)\right] \end{cases} \tag{8}$$

$$\begin{cases} u_A^{2L} = u_A^* - u_A^p \\ u_B^{2L} = u_B^* - u_B^p \\ u_C^{2L} = u_C^* - u_C^p \end{cases} \tag{9}$$

Figure 2 shows the three carrier-based modulations considered in this paper.

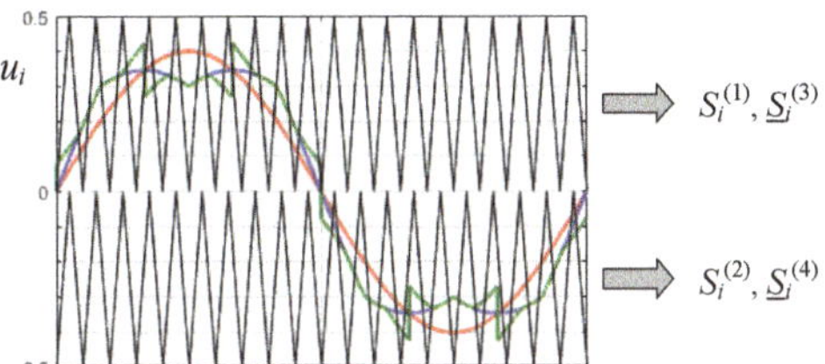

Figure 2. Carrier-based PWM modulation logic for each leg (*i*) of NPC inverters: SPWM (red trace), CPWM (blue trace), and OCPWM (green trace) in case of *m* = 0.4. Underline denotes complementary.

3. Input Current Analysis

Input Current Components

With reference to Figure 1, the instantaneous input currents $i^+(t)$ and $i^-(t)$ are composed of the averaged value over the switching period, $\bar{i}^+$ and $\bar{i}^-$, and the switching frequency component, Δi^+ and Δi^-. Similarly, $\bar{i}^+$ and $\bar{i}^-$ consist of DC component over the fundamental period, I_{dc}, and the alternating low-frequency component, $\tilde{i}^+$ and $\tilde{i}^-$, leading to:

$$\begin{cases} i^+(t) = \bar{i}^+ + \Delta i^+ = I_{dc} + \tilde{i}^+ + \Delta i^+ \\ i^-(t) = \bar{i}^- + \Delta i^- = I_{dc} + \tilde{i}^- + \Delta i^- \end{cases} \tag{10}$$

In case of balanced load and neglecting the output current ripple, the corresponding three-phase output currents can be written as:

$$\begin{cases} i_a = I_{ac} \sin(\vartheta - \varphi) \\ i_b = I_{ac} \sin\left(\vartheta - \frac{2\pi}{3} - \varphi\right) \\ i_c = I_{ac} \sin\left(\vartheta - \frac{4\pi}{3} - \varphi\right) \end{cases} \tag{11}$$

where I_{ac} is the output current amplitude and φ is the power phase angle.

The averaged component (over the switching period) of each leg input current can be determined by multiplying the switching function of upper or lower switch of each phase by the corresponding output current. In the case of leg A it leads to:

$$\begin{cases} \bar{i}_A^+ = \bar{S}_A^{(1)} i_a \\ \bar{i}_A^- = -\bar{S}_A^{(4)} i_a \end{cases} \tag{12}$$

By introducing Equations (3) and (11) in Equation (12), the averaged currents of the upper and lower switch of the leg A become:

$$\begin{cases} \bar{i}_A^+ = I_{ac} \sin(\vartheta - \varphi)[m \sin(\vartheta) + C_m + |m \sin(\vartheta) + C_m|] \\ \bar{i}_A^- = I_{ac} \sin(\vartheta - \varphi)[-m \sin\vartheta - C_m + |m \sin(\vartheta) + C_m|] \end{cases} \tag{13}$$

The total input currents $\bar{i}^+$ and $\bar{i}^-$ can be calculated as the sum of the three leg currents as:

$$\begin{cases} \bar{i}^+ = \bar{S}_A^{(1)} i_a + \bar{S}_B^{(1)} i_b + \bar{S}_C^{(1)} i_c \\ \bar{i}^- = -\bar{S}_A^{(4)} i_a - \bar{S}_B^{(4)} i_b - \bar{S}_C^{(4)} i_c \end{cases} \tag{14}$$

Due to the three-phase symmetry of modulation and output currents within the fundamental period T, both input currents have a periodicity of $T/3$. As a consequence, the analysis can be restricted to an angle range of $2\pi/3$, and all the input harmonics are multiple of 3. i.e., apart from the DC component, the lowest harmonic order component is the 3rd.

With reference to the phase angle range $2\pi/3 \le \vartheta \le 4\pi/3$, two sub-ranges $\pi/3$ can be identified. Considering Equations (2), (11) and (14), the input currents $\bar{i}^+$ and $\bar{i}^-$ can be expressed as:

$$\bar{i}^+ = \begin{cases} \overline{S}_A^{(1)}\, i_a + \overline{S}_B^{(1)}\, i_b, & \frac{2\pi}{3} \le \vartheta \le \pi \\ \overline{S}_B^{(1)}\, i_b, & \pi \le \vartheta \le \frac{4\pi}{3} \end{cases} \tag{15}$$

$$\bar{i}^- = \begin{cases} -\overline{S}_C^{(4)}\, i_c, & \frac{2\pi}{3} \le \vartheta \le \pi \\ -\overline{S}_A^{(4)}\, i_a - \overline{S}_C^{(4)}\, i_c, & \pi \le \vartheta \le \frac{4\pi}{3} \end{cases} \tag{16}$$

Introducing Equations (1), (2), and (11) in Equations (15) and (16), and setting $C_m = 0$ as in case of SPWM, leads to:

$$\bar{i}^+ = \begin{cases} mI_{ac}\left[2\cos(\varphi) + \cos\left(2\vartheta - \frac{2\pi}{3} - \varphi\right)\right], & \frac{2\pi}{3} \le \vartheta \le \pi \\ mI_{ac}\left[\cos(\varphi) + \cos\left(2\vartheta - \frac{\pi}{3} - \varphi\right)\right], & \pi \le \vartheta \le \frac{4\pi}{3} \end{cases} \tag{17}$$

$$\bar{i}^- = \begin{cases} mI_{ac}\left[\cos(\varphi) - \cos\left(2\vartheta - \frac{2\pi}{3} - \varphi\right)\right], & \frac{2\pi}{3} \le \vartheta \le \pi \\ mI_{ac}\left[2\cos(\varphi) - \cos\left(2\vartheta - \frac{\pi}{3} - \varphi\right)\right], & \pi \le \vartheta \le \frac{4\pi}{3} \end{cases} \tag{18}$$

Similarly, replacing Equation (6) in Equation (1) in case of CPWM, input currents $\bar{i}^+$ and $\bar{i}^-$ can be expressed as

$$\bar{i}^+ = \begin{cases} \frac{mI_{ac}}{2}\left[\sqrt{3}\sin(2\vartheta - \varphi) + \sin\left(\frac{\pi}{6} + \varphi\right) + 4\cos(\varphi)\right], & \frac{2\pi}{3} \le \vartheta \le \frac{5\pi}{6} \\ \frac{mI_{ac}}{2}\left[\sqrt{3}\sin\left(2\vartheta - \frac{\pi}{3} - \varphi\right) - \sin\left(\varphi - \frac{\pi}{6}\right) + 4\cos(\varphi)\right], & \frac{5\pi}{6} \le \vartheta \le \pi \\ \frac{mI_{ac}}{2}\sqrt{3}\left[\cos\left(2\vartheta - \frac{\pi}{6} - \varphi\right) + \cos\left(\varphi + \frac{\pi}{6}\right)\right], & \pi \le \vartheta \le \frac{7\pi}{6} \\ \frac{mI_{ac}}{2}\sqrt{3}\left[\sin(2\vartheta - \varphi) + \sin\left(\varphi + \frac{\pi}{3}\right)\right], & \frac{7\pi}{6} \le \vartheta \le \frac{4\pi}{3} \end{cases} \tag{19}$$

$$\bar{i}^- = \begin{cases} \frac{mI_{ac}}{2}\left[-\sqrt{3}\sin(2\vartheta - \varphi) - \sin\left(\frac{\pi}{6} + \varphi\right) + 2\cos(\varphi)\right], & \frac{2\pi}{3} \le \vartheta \le \frac{5\pi}{6} \\ \frac{mI_{ac}}{2}\left[-\sqrt{3}\sin\left(2\vartheta - \frac{\pi}{3} - \varphi\right) + \sin\left(\varphi - \frac{\pi}{6}\right) + 2\cos(\varphi)\right], & \frac{5\pi}{6} \le \vartheta \le \pi \\ \frac{mI_{ac}}{2}\left[-\sqrt{3}\cos\left(2\vartheta - \frac{\pi}{6} - \varphi\right) - \sqrt{3}\cos\left(\varphi + \frac{\pi}{6}\right) + 6\cos(\varphi)\right], & \pi \le \vartheta \le \frac{7\pi}{6} \\ \frac{mI_{ac}}{2}\left[-\sqrt{3}\sin(2\vartheta - \varphi) - \sqrt{3}\sin\left(\varphi + \frac{\pi}{3}\right) + 6\cos(\varphi)\right], & \frac{7\pi}{6} \le \vartheta \le \frac{4\pi}{3} \end{cases} \tag{20}$$

Although the above analytical calculations are based on SPWM and CPWM, the analysis could be readily extended to other PWM techniques, such as OCPWM, leading to more complex expressions not presented in this paper.

Consequently, the low-frequency input current components are readily determined in case of SPWM as:

$$I_{dc} = \frac{3}{2}mI_{ac}\cos(\varphi) \tag{21}$$

$$\tilde{i}^+ = \begin{cases} \frac{1}{2}mI_{ac}\left[\cos(\varphi) + 2\cos\left(2\vartheta - \frac{2\pi}{3} - \varphi\right)\right], & \frac{2\pi}{3} \le \vartheta \le \pi \\ -\frac{1}{2}mI_{ac}\left[\cos(\varphi) - 2\cos\left(2\vartheta - \frac{\pi}{3} - \varphi\right)\right], & \pi \le \vartheta \le \frac{4\pi}{3} \end{cases} \tag{22}$$

$$\tilde{i}^- = \begin{cases} -\frac{1}{2}mI_{ac}\left[\cos(\varphi) + 2\cos\left(2\vartheta - \frac{2\pi}{3} - \varphi\right)\right], & \frac{2\pi}{3} \le \vartheta \le \pi \\ \frac{1}{2}mI_{ac}\left[\cos(\varphi) - 2\cos\left(2\vartheta - \frac{\pi}{3} - \varphi\right)\right], & \pi \le \vartheta \le \frac{4\pi}{3} \end{cases} \tag{23}$$

In the case of CPWM, the low-frequency input currents are expressed as:

$$
\tilde{i}^+ = \begin{cases}
\frac{mI_{ac}}{2}\left[\sqrt{3}\sin(2\vartheta - \varphi) + \sin\left(\frac{\pi}{6} + \varphi\right) + \cos(\varphi)\right], & \frac{2\pi}{3} \leq \vartheta \leq \frac{5\pi}{6} \\
\frac{mI_{ac}}{2}\left[\sqrt{3}\sin(2\vartheta - \frac{\pi}{3} - \varphi) - \sin\left(\varphi - \frac{\pi}{6}\right) + \cos(\varphi)\right], & \frac{5\pi}{6} \leq \vartheta \leq \pi \\
\frac{mI_{ac}}{2}\sqrt{3}\left[\cos(2\vartheta - \frac{\pi}{6} - \varphi) + \cos\left(\varphi + \frac{\pi}{6}\right) - 3\cos(\varphi)\right], & \pi \leq \vartheta \leq \frac{7\pi}{6} \\
\frac{mI_{ac}}{2}\sqrt{3}\left[\sin(2\vartheta - \varphi) + \sin\left(\varphi + \frac{\pi}{3}\right) - 3\cos(\varphi)\right], & \frac{7\pi}{6} \leq \vartheta \leq \frac{4\pi}{3}
\end{cases}
\tag{24}
$$

$$
\tilde{i}^- = \begin{cases}
-\frac{mI_{ac}}{2}\left[\sqrt{3}\sin(2\vartheta - \varphi) + \sin\left(\frac{\pi}{6} + \varphi\right) + \cos(\varphi)\right], & \frac{2\pi}{3} \leq \vartheta \leq \frac{5\pi}{6} \\
-\frac{mI_{ac}}{2}\left[\sqrt{3}\sin(2\vartheta - \frac{\pi}{3} - \varphi) - \sin\left(\varphi - \frac{\pi}{6}\right) + \cos(\varphi)\right], & \frac{5\pi}{6} \leq \vartheta \leq \pi \\
-\frac{mI_{ac}}{2}\left[\sqrt{3}\cos(2\vartheta - \frac{\pi}{6} - \varphi) + \sqrt{3}\cos\left(\varphi + \frac{\pi}{6}\right) - 3\cos(\varphi)\right], & \pi \leq \vartheta \leq \frac{7\pi}{6} \\
-\frac{mI_{ac}}{2}\left[\sqrt{3}\sin(2\vartheta - \varphi) + \sqrt{3}\sin\left(\varphi + \frac{\pi}{3}\right) - 3\cos(\varphi)\right], & \frac{7\pi}{6} \leq \vartheta \leq \frac{4\pi}{3}
\end{cases}
\tag{25}
$$

4. Input Voltage Analysis

4.1. Input Voltage Components

Based on the analysis of the inverter input current ripple components, the instantaneous voltages v_1 and v_2 across DC-link capacitors can be written as:

$$
\begin{cases}
v_1 = V_1 + \tilde{v}_1 + \Delta v_1 \\
v_2 = V_2 + \tilde{v}_2 + \Delta v_2
\end{cases}
\tag{26}
$$

where V_1 and V_2 are the DC components averaged over the fundamental period, $\tilde{v}_1$ and $\tilde{v}_2$ are the alternating low-frequency ripple components, and Δv_1 and Δv_2 are the switching frequency ripple components of the voltages across capacitors C_1 and C_2, respectively. Being the low-frequency ripple components widely studied in literature, the analysis is focused on the switching ripple component.

4.2. Peak-to-Peak Voltage Switching Ripple Evaluation

In order to calculate the voltage switching ripple of DC-link capacitors, the amount of the switching frequency component of currents Δi_1 and Δi_2 circulating through the DC-link capacitor C_1 and C_2 should be determined. Assuming that the DC source impedance at the switching frequency is much higher than the capacitor's reactance, the whole current component Δi^+ and Δi^- are circulating through the capacitors C_1 and C_2, i.e., $\Delta i_1 = \Delta i^+$ and $\Delta i_2 = \Delta i^-$. In this case, the corresponding DC voltage variations (peak-to-peak) over the sub-periods $[0\text{--}\Delta t_1]$ and $[0\text{--}\Delta t_2]$ can be expressed as

$$
\Delta V_1 = \frac{1}{C_1}\int_0^{\Delta t_1} \Delta i^+ \, dt, \quad \Delta V_2 = \frac{1}{C_2}\int_0^{\Delta t_2} \Delta i^- \, dt.
\tag{27}
$$

being Δt_1 and Δt_2 specific switching time intervals. The instantaneous input current is considered constant within each considered time interval.

Due to the periodicity of the input currents i^+ and i^-, the evaluation of peak-to-peak voltage ripple is limited to the phase angle range $2\pi/3 \leq \vartheta \leq 4\pi/3$. Analyzing the voltage ripple, two cases have been identified: the first case considering zero phase angle ($\varphi = 0°$) and the second considering $\varphi = 60°$. The peak-to-peak voltage switching ripple of capacitors has been analytically calculated as:

In case of $\varphi = 0°$:

$$
\Delta V_1 = \frac{T_{sw}}{C_1}\begin{cases}
\left(1 - \overline{s}_A^{(1)}\right)\tilde{i}^+, & \frac{2\pi}{3} \leq \vartheta \leq \frac{5\pi}{6} \\
\left(1 - \overline{s}_B^{(1)}\right)\tilde{i}^+, & \frac{5\pi}{6} \leq \vartheta \leq \frac{4\pi}{3}
\end{cases}
\tag{28}
$$

$$\Delta V_2 = \frac{T_{sw}}{C_2} \begin{cases} \left(1 - \overline{S}_{C}^{(4)}\right)\overline{i}^{-}, & \frac{2\pi}{3} \leq \vartheta \leq \frac{7\pi}{6} \\ \left(1 - \overline{S}_{A}^{(4)}\right)\overline{i}^{-}, & \frac{7\pi}{6} \leq \vartheta \leq \frac{4\pi}{3} \end{cases} \tag{29}$$

In case of $\varphi = 60°$:

$$\Delta V_1 = \frac{T_{sw}}{C_1} \begin{cases} \left(1 - \overline{S}_{A}^{(1)}\right)\overline{i}^{+}, & \frac{2\pi}{3} \leq \vartheta \leq \frac{5\pi}{6} \\ \left(i_a + i_b - \overline{i}^{+}\right)\overline{S}_{A}^{(1)}, & \frac{5\pi}{6} \leq \vartheta \leq \pi \\ \left(1 - \overline{S}_{B}^{(1)}\right)\overline{i}^{+}, & \frac{5\pi}{6} \leq \vartheta \leq \frac{4\pi}{3} \end{cases} \tag{30}$$

$$\Delta V_2 = \frac{T_{sw}}{C_2} \begin{cases} \left(1 - \overline{S}_{C}^{(4)}\right)\overline{i}^{-}, & \frac{2\pi}{3} \leq \vartheta \leq \frac{7\pi}{6} \\ \left(-i_c - i_a - \overline{i}^{-}\right)\overline{S}_{C}^{(4)}, & \frac{7\pi}{6} \leq \vartheta \leq \frac{4\pi}{3} \end{cases} \tag{31}$$

being $S_{i}^{(1)}$ and $S_{i}^{(4)}$ the switching function for the upper and lower switches of phase i (being i = A, B or C), given by Equation (2). The total averaged input currents $\overline{i}^{+}$ and $\overline{i}^{-}$ can be calculated by Equations (17) and (18) in case of sinusoidal PWM, and by Equations (19) and (20) in case of centered PWM. In case of optimized centered PWM the input currents can be calculated introducing Equations (2), (7) and (11) in Equations (15) and (16), leading to more complex developments.

Equations (28)–(31) suggest normalization for ΔV_1 and ΔV_2, as follow:

$$\Delta V_1 = \frac{I_{ac}}{f_{sw}C_1}\Delta U_1, \; \Delta V_2 = \frac{I_{ac}}{f_{sw}C_2}\Delta U_2. \tag{32}$$

being ΔU_1 and ΔU_2 the normalized peak-to-peak voltage switching ripple amplitude of capacitors.

Figure 3 shows the distribution of the normalized peak-to-peak voltage switching ripple calculated based on Equations (28), (29), (30) and (31) over the period [0, 120°]. Two modulation indices have been selected, m = 0.3 and m = 0.5 and two output phase angles are considered, φ = 0 and φ = 60°.

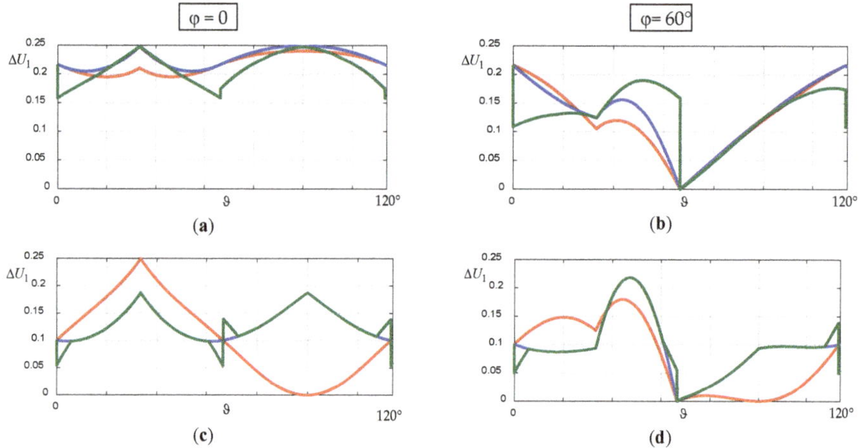

Figure 3. Normalized peak-to-peak voltage ripple amplitude across the capacitors over the period [0, 120°] for two modulation indices m = 0.3 (**a,b**) and m = 0.5 (**c,d**) and output phase angles φ = 0 (**a,c**) (left) and 60° (**b,d**) (right) in case of SPWM (red), CPWM (blue) and OCPWM (green).

Figure 3 presents the normalized peak-to-peak voltage ripple amplitude across the upper capacitor since the normalized peak-to-peak voltage ripple amplitude across the lower capacitor has exactly the same profile with a phase shift corresponding to 60° (1/2 of the considered period).

According to Figure 3, it can be seen a wide excursion of the normalized peak-to-peak voltage ripple amplitude, generally ranging between 0 (min) and 0.25 (max). Despite there are evident differences in the voltage switching ripple envelope profile among the three modulation strategies, it cannot be identified a modulation clearly better than the others from this point of view. This consideration is also supported by the diagrams presented in Figure 4, representing the maximum of the normalized peak-to-peak ripple amplitude, calculated numerically, over the whole modulation index range, for $\varphi = 0°, 30°, 60°$, and $90°$, considering SPWM, CPWM, and OCPWM. Again, these three modulation techniques give similar and comparable results also with reference to the maximum of the peak-to-peak voltage switching ripple. For all the cases, the absolute maximum of normalized peak-to-peak voltage ripple amplitude can be assumed as 0.25.

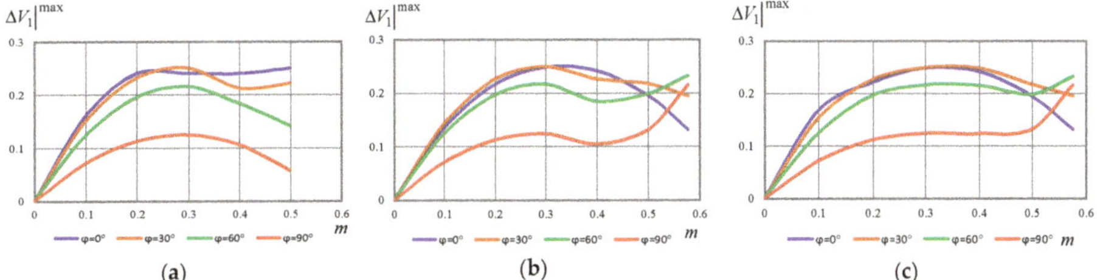

Figure 4. Maximum peak-to-peak value of the normalized voltage switching ripple vs. modulation index m in case of (**a**) sinusoidal pulse-width modulation (SPWM), (**b**) centered PWM (CPWM) and (**c**) optimized centered PWM (OCPWM) for different output phase angles.

5. Dc-link Preliminary Capacitor Design

Based on the analysis of capacitor voltage ripple components, simple and effective guidelines for a preliminary design of the capacitors C_1 and C_2 are proposed in this section. In particular, the capacitances can be calculated taking into account requirements or restrictions referred to the switching frequency and/or low-frequency voltage ripple components.

Despite the design of DC-link capacitors in the three-phase three-level inverter has been widely addressed in literature considering always the low-frequency voltage ripple, a guideline for the design of the DC-link capacitors for this multilevel inverter configuration based on the switching voltage ripple has not been developed yet.

In general, the capacitor voltage switching ripple amplitude could have additional specific restrictions to limit switching noise, electromagnetic interferences, and voltage stress on the DC-link.

In this paper, the selection of the DC-link capacitors C_1 and C_2 can be performed on the basis of the maximum amplitude of the peak-to-peak voltage switching ripple at the switching frequency (that is in the order of kHz).

According to Figure 4, it can be noted that the maximum amplitude of the peak-to-peak ripple is determined as:

$$\Delta U_1^{\max} = \Delta U_2^{\max} = \frac{1}{4} \tag{33}$$

In this case, the capacitances can be readily calculated on the basis of Equations (32) and (33):

$$C_1 \geq \frac{1}{4}\frac{I_{ac}}{f_{sw}\Delta V_1^{\max}}, C_2 \geq \frac{1}{4}\frac{I_{ac}}{f_{sw}\Delta V_2^{\max}} \tag{34}$$

6. Results

In order to verify proposed theoretical developments valid for both the considered NPC multilevel inverter configurations (Figure 1), numerical simulations and corresponding experimental tests are carried out. Inverter is controlled by carrier-based multilevel PWM (Figure 2) with reference to the three considered modulation techniques (SPWM, CPWM, and OCPWM). The switching frequency is set 2.5 kHz to better emphasize the switching ripple components, and two different output phase angles (0 and 60°) are considered, as for the specific analytical developments. The main circuit parameters are summarized in Table 1 for both simulations and experiments.

Table 1. Simulation/experiment circuit parameters.

Label	Description	Parameters
V_s	DC voltage source	100 V
R_s	DC source resistance	5 Ω
L_s	DC source inductance	10.15 mH
C_1, C_2	DC-link capacitors	1.12 mF
f, f_{sw}	fundamental and switching frequencies	50 Hz, 2.5kHz

6.1. Simulation Results

Simulation results are carried out by implementing the power circuit scheme and the PWM techniques by Matlab/Simulink, considering SPWM, CPWM, and OCPWM.

The first simulation tests (Figures 5 and 6) are concerning the input inverter currents. In this case, unity sinusoidal output current (I_{ac} = 1A) are considered to easily verify the analytical developments presented in Section 3, with specific reference to the considered output phase angles $\varphi = 0°$ and $\varphi = 60°$.

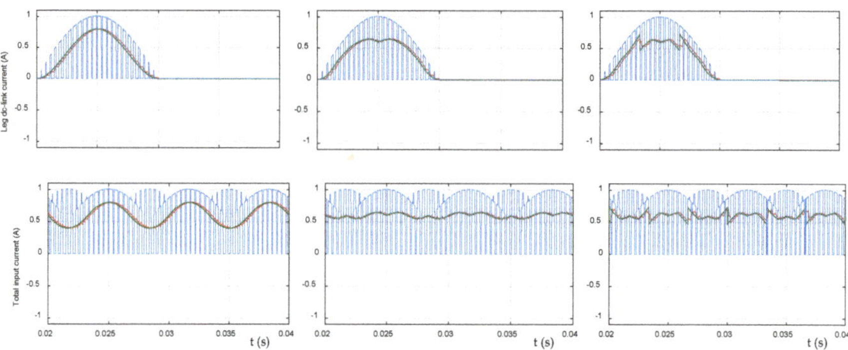

Figure 5. One leg dc-link current (**top**) and total input current (**bottom**): instantaneous value (blue trace), its averaged value over the switching period (red), and calculated value (green) in case of SPWM, CPWM and OCPWM (from right to left) for $m = 0.4$, I_{ac} = 1A and $\varphi = 0°$.

The top traces in Figures 5 and 6 show the simulation results comparing the instantaneous input current of leg A i_A^+ (blue trace) with its averaged value over the switching period $\bar{i}_A^+$ (red trace), and the corresponding low-frequency current component calculated by Equation (13) (green trace) in case of $m = 0.4$ and for two cases of output phase angles $\varphi = 0°$ (Figure 5) and $\varphi = 60°$ (Figure 6). The bottom traces in Figures 5 and 6 show the total input current, with emphasis to instantaneous value i^+ (blue trace) and its averaged value over the switching period $\bar{i}^+$ (red trace). The low-frequency current component (green trace) is determined analytically in both cases of SPWM and CPWM by Equations (22) and (24), respectively, and is calculated by replacing Equations (2), (7), and (11) in Equation (15), in the case of OCPWM.

The numerical results generally show a perfect matching with the theoretical values. The small delay between averaged and theoretical currents ($T_{sw}/2$) is due to the averaging process itself.

The second group of simulation tests is concerning the switching voltage ripple across the two DC-link capacitors. In this case, the load circuit model is corresponding to the real experimental setup, made with the purpose to easily adapt the output phase angles to the considered cases ($\varphi = 0$ and $\varphi = 60°$), according to Figure 7. The corresponding circuit parameters are given in Table 2.

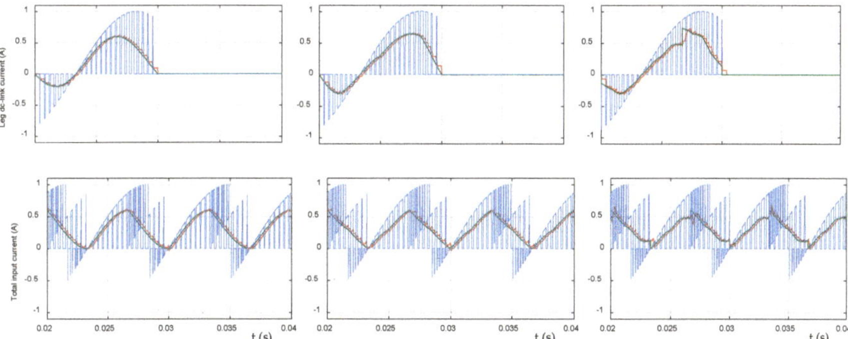

Figure 6. One leg dc-link current (**top**) and total input current (**bottom**): instantaneous value (blue trace), its averaged value over the switching period (red), and calculated value (green) in case of SPWM, CPWM and OCPWM (from right to left) for $m = 0.4$, $I_{ac} = 1$A and $\varphi = 60°$.

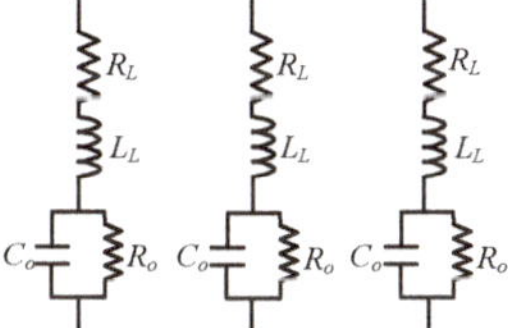

Figure 7. Three-phase load circuit.

Table 2. Load parameters.

Load	$\varphi = 0$	$\varphi = 60°$
R_L	3.16 Ω	3.16 Ω
L_L	20.1 mH	20.1 mH
R_o	20 Ω	0
C_o	58 μF	0

The voltage switching ripple is determined by filtering away the low-frequency components form the instantaneous capacitor voltages.

Figure 8 presents the instantaneous voltage switching ripple across the capacitors (blue traces) in case of sinusoidal PWM together with the theoretical envelopes $\pm\Delta V_1/2$ (upper capacitor) and $\pm\Delta V_2/2$ (lower capacitor), analytically evaluated by (28) and (29) (red traces) for two cases of modulation index $m = 0.3$ (top) and 0.5 (bottom), considering the output phase angle $\varphi = 0$. The same quantities are presented in Figure 9 with reference to output phase angle $\varphi = 60°$. In this case, envelopes $\pm\Delta V_1/2$ and $\pm\Delta V_2/2$ are analytically evaluated by (30) and (31).

Similarly, Figures 10 and 11 present the instantaneous voltage switching ripple across the capacitors (blue traces) in case of centered PWM together with the theoretical envelopes $\pm\Delta V_1/2$ (upper capacitor) and $\pm\Delta V_2/2$ (lower capacitor) (red traces) for two cases of modulation index $m = 0.3$ (top) and 0.5 (bottom), considering the output phase angles $\varphi = 0$ (Figure 10) and $\varphi = 60°$ (Figure 11).

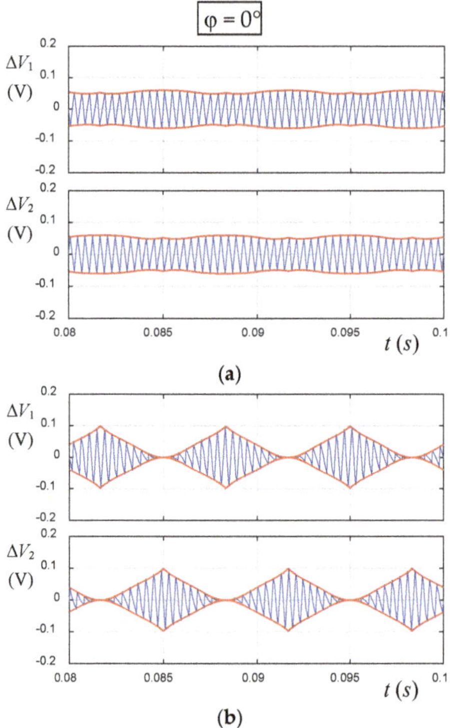

Figure 8. Capacitor voltage switching ripple (SPWM): simulation results (blue trace) and calculated peak- to-peak envelope (red traces) over a period for $m = 0.3$ (**a**) and $m = 0.5$ (**b**) in case of $\varphi = 0°$.

As mentioned in Section 4, the analytical developments could be readily extended to other more sophisticated modulation strategies, such as optimized centered PWM, but leading to more complex and less meaningful expressions. For this reason, the envelopes of voltage switching ripple have not explicitly obtained in case of OCPWM, just numerically derived introducing Equations (2), (7), (11), (15), and (16) in the basic Equations (28)–(31). Similarly, to previous cases, Figures 12 and 13 present the instantaneous voltage switching ripple across the capacitors (blue traces) in case of optimized centered PWM together with the envelopes $\pm\Delta V_1/2$ (upper capacitor) and $\pm\Delta V_2/2$ (lower capacitor) (red traces) for two cases of modulation index $m = 0.3$ (top) and 0.5 (bottom), considering the output phase angles $\varphi = 0$ (Figure 12) and $\varphi = 60°$ (Figure 13).

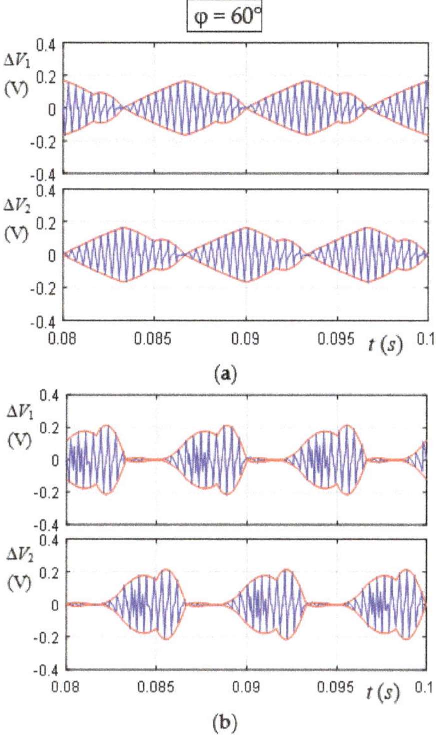

Figure 9. Capacitor voltage switching ripple (SPWM): simulation results (blue trace) and calculated peak- to-peak envelope (red traces) over a period for $m = 0.3$ (**a**) and $m = 0.5$ (**b**) in case of $\varphi = 60°$.

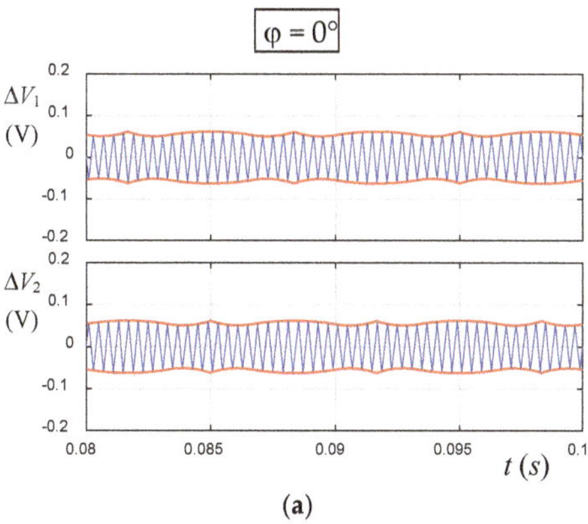

Figure 10. *Cont.*

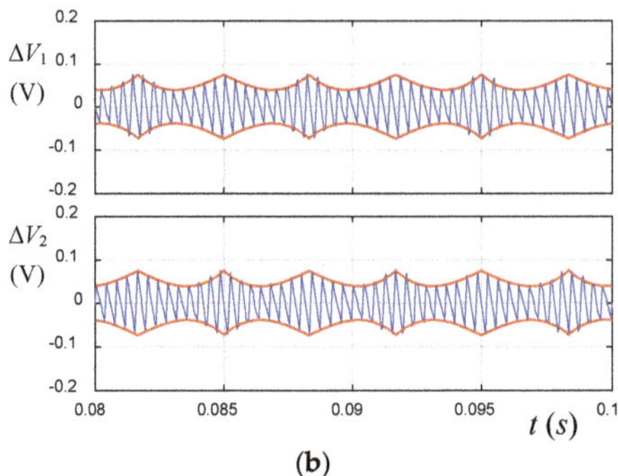

(b)

Figure 10. Capacitor voltage switching ripple (CPWM): simulation results (blue trace) and calculated peak- to-peak envelope (red traces) over a period for $m = 0.3$ (a) and $m = 0.5$ (b) in case of $\varphi = 0°$.

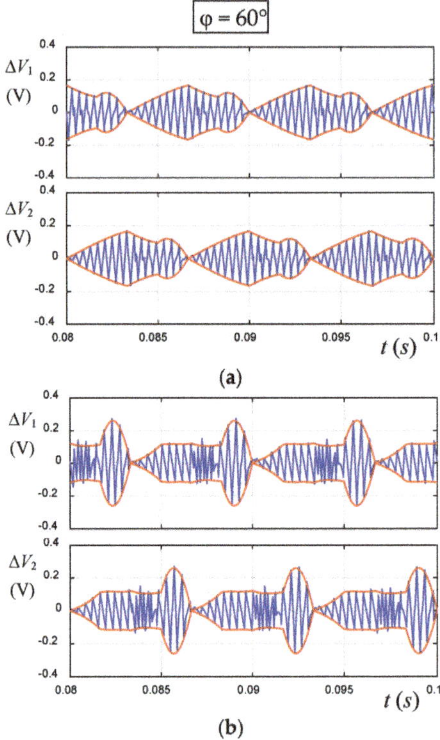

(a)

(b)

Figure 11. Capacitor voltage switching ripple (CPWM): simulation results (blue trace) and calculated peak- to-peak envelope (red traces) over a period for $m = 0.3$ (a) and $m = 0.5$ (b) in case of $\varphi = 60°$.

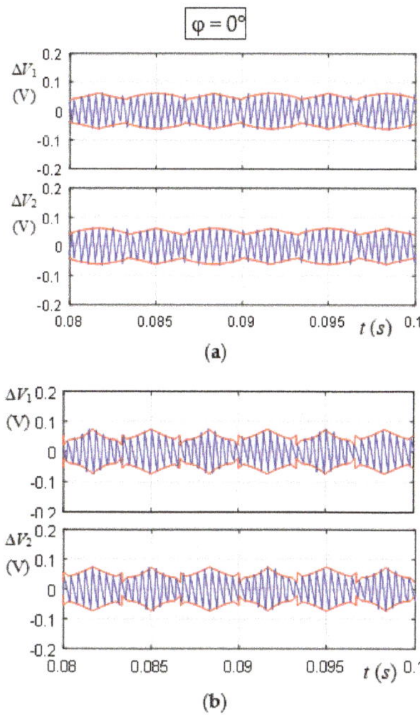

(a)

(b)

Figure 12. Capacitor voltage switching ripple (OCPWM): simulation results (blue trace) and calculated peak- to-peak envelope (red traces) over a period for *m* = 0.3 (**a**) and *m* = 0.5 (**b**) in case of φ = 0°.

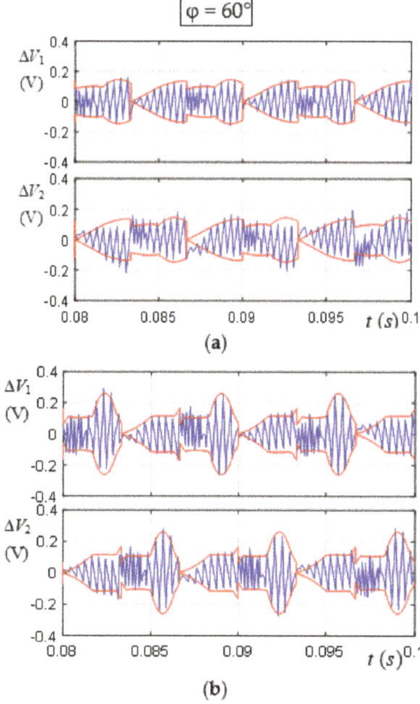

(a)

(b)

Figure 13. Capacitor voltage switching ripple (OCPWM): simulation results (blue trace) and calculated peak- to-peak envelope (red traces) over a period for *m* = 0.3 (**a**) and *m* = 0.5 (**b**) in case of φ = 60°.

6.2. Experimental Results

A picture view of the whole experimental setup is shown in Figure 14. It consists of a three-phase T-type NPC inverter implemented by 12 discrete Silicon Carbide (SiC) power MOSFETs (CREE C2M0080120D) rated for 1200 V and 36 A. The three considered PWM techniques and the calculations to analytically determine the envelopes of voltage switching ripple across the two dc-link capacitors are implemented by a TMS320F28335 floating point DSP control board. Code Composer Studio (CCS) is adopted for programming the DSP board, with the possibility of real-time adjustment of modulation parameters by computer interface. The main circuit parameters are given in Tables 1 and 2, i.e., the same used for the second group of simulations.

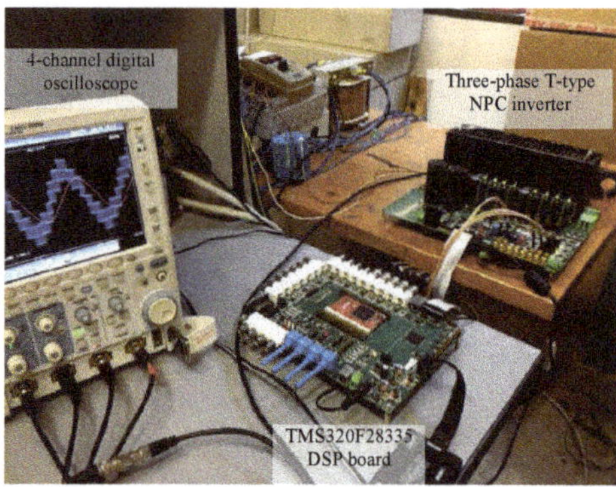

Figure 14. Picture view of the experimental setup.

Experimental results are shown by Yokogawa DLM 2024 oscilloscope screenshots. Figure 15 presents an example of load voltage and current (blue and red traces, respectively) obtained by the laboratory setup in case of sinusoidal PWM and unity power factor (m = 0.5).

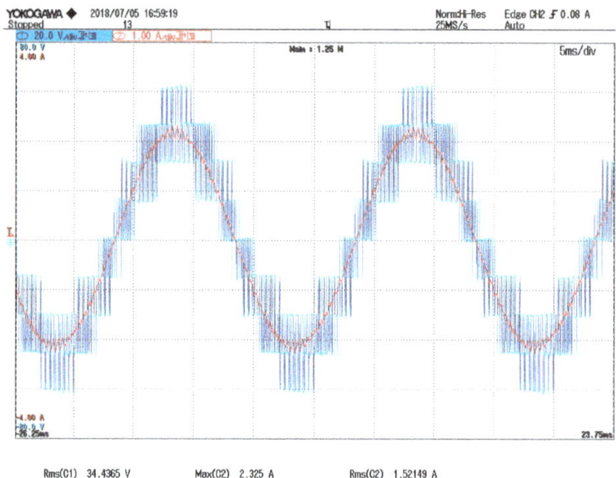

Figure 15. Example of load voltage (blue) and current (red) for SPWM (m = 0.5, φ = 0°).

Figures 16 and 17 present the results with reference to sinusoidal PWM and centered PWM techniques, respectively. Two values of modulation index: m = 0.3 and 0.5 (from top to bottom) and two values of the output phase angles φ = 0 and φ = 60° are considered (left and right column, respectively). In all screenshots, upper traces present the capacitor voltage and its averaged counterpart (blue and green traces, respectively) and the bottom traces present the measured capacitor voltage switching ripple and the calculated peak-to-peak envelopes provided by the DSP board and displayed using DAC block with a proper voltage scaling (red and green traces, respectively).

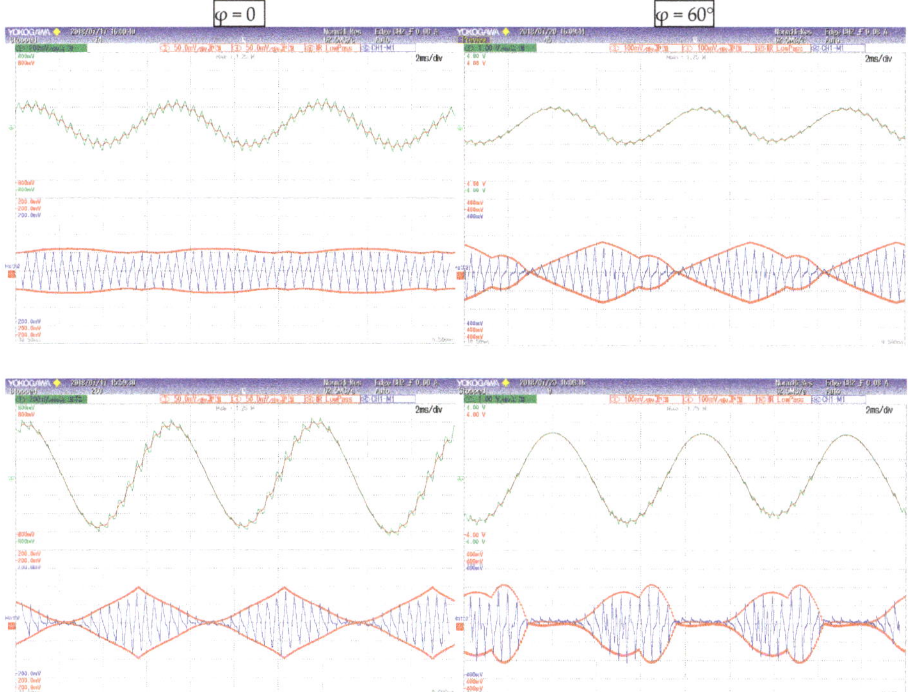

Figure 16. Experimental results for φ = 0° (**left**) and φ = 60° (**right**). Upper half: capacitor voltage and its averaged value. Lower half: calculated peak-to-peak envelope and measured capacitor voltage switching ripple with different modulation indexes: m = 0.3 and 0.5 (from top to bottom) in case of SPWM.

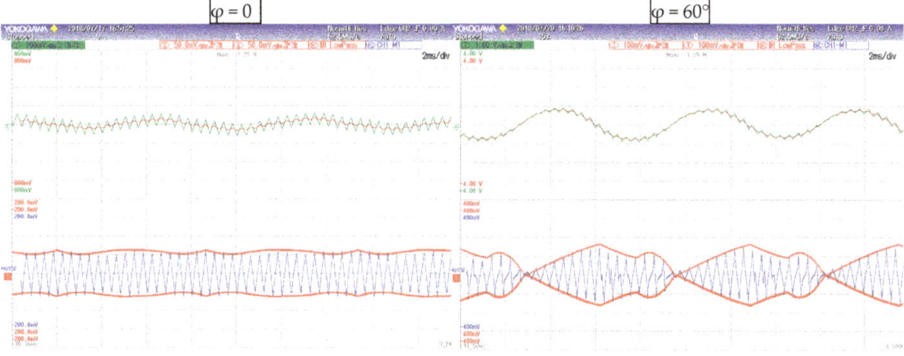

Figure 17. *Cont.*

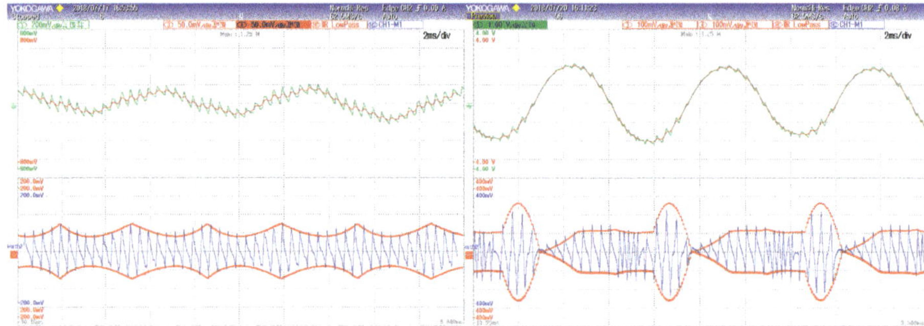

Figure 17. Experimental results for $\varphi = 0°$ (**left**) and $\varphi = 60°$ (**right**). Upper half: capacitor voltage and its averaged value. Lower half: calculated peak-to-peak envelope and measured capacitor voltage switching ripple with different modulation indexes: $m = 0.3$ and 0.5 (from top to bottom) in case of CPWM.

The capacitor voltage switching ripple has been obtained experimentally by simply using the "ac coupling" built-in function of the oscilloscope together with the built-in low-pass filter, on the basis of the instantaneous capacitor voltage, according to Equation (26).

Simulation and experimental results have a good matching for all the considered cases, as proved by comparing Figures 8 and 9 with Figure 16 in case of SPWM and by comparing Figures 10 and 11 with Figure 17 in case of CPWM. Note that the same scale is adopted in corresponding simulation and experimental di agrams to facilitate the comparison.

7. Conclusions

This paper deals with input current analysis and determination of capacitors voltage switching ripple in three-phase three-level neutral point clamped inverters. Reference is made to the basic modulation strategies, namely sinusoidal PWM, centered PWM, and optimized centered PWM, but the proposed method be easily extended to other modulation techniques. The switching frequency current and voltage ripple components have been analytically determined for the two dc-link capacitors. In particular, the peak-to-peak voltage ripple amplitudes have been calculated as a function of the inverter modulation index and the output current amplitude. Simple and effective guidelines for a preliminary design the dc-link capacitors of the NPC configuration have been also introduced. Developments have been carried out in case of general output power factor, representing either grid connections, motor, or passive loads.

The mathematical developments have been verified, both numerically and experimentally, for different values of modulation indices and specifically for two output phase angles $\varphi = 0$ (corresponding to most of the grid-connected applications) and $\varphi = 60°$. A very satisfactory matching between analytical, numerical, and experimental results has been achieved, proving the validity of the proposed approach.

Author Contributions: M.H. developed the theoretical analysis and the simulation results, also providing for the manuscript arrangement in cooperation with R.M. G.R. performed the experimental test and revised the manuscript, supported by M.H. and R.M. G.G. generally supervised and finalized the work.

Funding: This research received no external funding.

Conflicts of Interest: The authors declare no conflict of interest.

References

1. Nabae, A.; Takahashi, I.; Akagi, H. A New Neutral-Point-Clamped PWM Inverter. *IEEE Trans. Ind. Appl.* **1981**, *17*, 518–523. [CrossRef]
2. Giri, S.K.; Mukherjee, S.; Kundu, S.; Banerjee, S.; Chakraborty, C. A Carrier-Based PWM Scheme for Neutral Point Voltage Balancing in Three-Level Inverter Extending to Full Power Factor Range. *IEEE J. Emerg. Sel. Top. Power Electron.* **2017**, *64*, 1873–1883. [CrossRef]
3. Liu, P.; Duan, S.; Yao, C.; Chen, C. A Double Modulation Wave CBPWM Strategy Providing Neutral-Point Voltage Oscillation Elimination and CMV Reduction for Three-Level NPC Inverters. *IEEE Trans. Ind. Electron.* **2018**, *65*, 16–26. [CrossRef]
4. Dordevic, O.; Jones, M.; Levi, E. A Comparison of Carrier-Based and Space Vector PWM Techniques for Three-Level Five-Phase Voltage Source Inverters. *IEEE Trans. Ind. Inform.* **2013**, *9*, 609–618. [CrossRef]
5. Petrella, R.; Pevere, A. Interleaved Carrier-Based Modulations for Reducing Low-Frequency Neutral Point Voltage Ripple in the Three-Phase Neutral Point Clamped Inverter. In Proceedings of the 2014 IEEE Applied Power Electronics Conference and Exposition APEC, Fort Worth, TX, USA, 16–20 March 2014; pp. 2386–2393.
6. Xia, C.L.; Zhang, G.Z.; Yan, Y.; Gu, X.; Shi, T.N.; He, X.N. Discontinuous Space Vector PWM Strategy of Neutral-Point-Clamped Three-Level Inverters for Output Current Ripple Reduction. *IEEE Trans. Power Electron.* **2017**, *32*, 5109–5121. [CrossRef]
7. López, I.; Ceballos, S.; Pou, J.; Zaragoza, J.; Andreu, J.; Ibarra, E.; Konstantinou, G. Generalized PWM-Based Method for Multiphase Neutral-Point-Clamped Converters with Capacitor Voltage Balance Capability. *IEEE Trans. Power Electron.* **2017**, *32*, 4878–4890. [CrossRef]
8. Pevere, A.; Petrella, R. Discontinuous Hybrid Modulation Technique for Three-Phase Three-Level Neutral Point Clamped Inverters. In Proceedings of the 2013 IEEE Energy Conversion Congress and Exposition ECCE, Denver, CO, USA, 15–19 September 2013; pp. 3992–3999.
9. Bianchi, N.; Di Pre, M. Active Power Filter Control Using Neural Network Technologies. *IEE Proc. Electr. Power Appl.* **2003**, *150*, 139–145.
10. Vujacic, M.; Hammami, M.; Srndovic, M.; Grandi, G. Evaluation of DC Voltage Ripple InSingle-Phase H-Bridge PWM Inverters. In Proceedings of the IEEE 26th International Symposium on Industrial Electronics, Scotland, UK, 19–21 June 2017; pp. 711–716.
11. Vujacic, M.; Hammami, M.; Srndovic, M.; Grandi, G. Theoretical and Experimental Investigation of Switching Ripple in the DC-Link Voltage of Single-Phase H-Bridge PWM Inverters. *Energies* **2017**, *10*, 1189. [CrossRef]
12. Hammami, M.; Vujacic, M.; Grandi, G. Dc-Link Current and Voltage Ripple Harmonics in Three-Phase Three-Level Flying Capacitor Inverters with Sinusoidal Carrier-Based PWM. In Proceedings of the IEEE International Conference Industrial Technology, Lyon, France, 20–22 February 2018.
13. Liang, Z.; Li, M.; Dong, Z.; Tian, S.; Li, K. Analytical Closed-Form Expressions of DC Current Ripple for Three-Level Neutral Point Clamped Inverters with Space-Vector Pulse-Width Modulation. *IET Power Electron.* **2016**, *9*, 930–937.
14. Gopalakrishnan, K.S.; Das, S. Analytical Expression for RMS DC Link Capacitor Current in a Three-Level Inverter. In Proceedings of the Centenary Conference Electrical Engineering, Indian Institute of Science, Bangalore, India, 15–17 December 2011.
15. Orfanoudakis, G.I.; Sharkh, S.M.; Yuratich, M.A. Analysis of Dc-Link Capacitor Current in Three-Level Neutral Point Clamped and Cascaded H-Bridge Inverters. *IET Power Electron.* **2013**, *6*, 1376–1389. [CrossRef]
16. Pou, J.; Pindado, R.; Boroyevich, D.; Rodríguez, P. Evaluation of the Low-Frequency Neutral-Point Voltage Oscillations in the Three-Level Inverter. *IEEE Trans. Ind. Electron.* **2005**, *52*, 1582–1588. [CrossRef]
17. Pou, J.; Zaragoza, J.; Rodríguez, P.; Ceballos, S.; Sala, V.M.; Burgos, R.P.; Boroyevich, D. Fast-Processing Modulation Strategy for the Neutral-Point-Champed Converter with Total Elimination of Low-Frequency Voltage Oscillations in the Neutral Point. *IEEE Trans. Ind. Electron.* **2007**, *54*, 2288–2294. [CrossRef]
18. Fujii, K.; Kikuchi, T.; Koubayashi, H.; Yoda, K. 1-MW Advanced T-Type NPC Converters for Solar Power Generation System. In Proceedings of the 2013 15th European Conference on Power Electronics and Applications (EPE), Lille, France, 2–6 September 2013.

19. Zhang, Z.; Anthon, A.; Andersen, M.A.E. Comprehensive Loss Evaluation of Neutral-Point-Clamped (NPC) and T-Type Three-Level Inverters Based on a Circuit Level Decoupling Modulation. In Proceedings of the 2014 International Power Electronics and Application Conference and Exposition PEAC, Shanghai, China, 5–8 November 2014.

20. Grandi, G.; Loncarski, J. Simplified Implementation of Optimised Carrier-Based PWM in Three-Level Inverters. *Electron. Lett.* **2014**, *50*, 631–633. [CrossRef]

Article

Design and Controller-In-Loop Simulations of a Low Cost Two-Stage PV-Simulator

Sridhar Vavilapalli [1], Umashankar Subramaniam [1], Sanjeevikumar Padmanaban [2,*] and Frede Blaabjerg [3]

[1]   Department of Energy and Power Electronics, School of Electrical Engineering, VIT University, Vellore 632014, India; sridhar.spark@gmail.com (S.V.); shankarums@gmail.com (U.S.)
[2]   Department of Energy Technology, Aalborg University, 6700 Esbjerg, Denmark
[3]   Center of Reliable Power Electronics (CORPE), Department of Energy Technology, Aalborg University, 9100 Aalborg, Denmark; fbl@et.aau.dk
*   Correspondence: san@et.aau.dk; Tel.: +45-209-751-79

Received: 2 September 2018; Accepted: 3 October 2018; Published: 16 October 2018

Abstract: A PV-Simulator is a DC power source in which the current-voltage (I-V) characteristics of different PV arrays can be programmed. With a PV-simulator, the operation of the solar power conditioning systems can be validated at a laboratory level itself before actual field trials. In this work, design, operation and controls for a two-stage programmable PV-simulator required for the testing of solar power conditioning systems are presented. The proposed PV-simulator consists of a three-level T-type active front-end converter in the first stage and a buck-chopper-based DC-DC converter in the second stage. An active front-end rectifier using a three-level T-type IGBT-based converter is used at the input stage to help in operating the system at unity power factor. A DC-DC converter at the output stage of the simulator is regulated to obtain the I-V characteristics of the programmed PV-Array. Hardware-In-Loop simulations are carried out to validate the proposed system and the associated controls implemented in the controller. As a case study, this PV-simulator is programmed with electrical parameters of a selected PV-array and the characteristics obtained from the PV-simulator are compared with the actual PV-array characteristics. The dynamic response of the system for sudden changes in the load and sudden changes in irradiance values are studied.

Keywords: buck-chopper; PV-simulator; T-type converter; real time simulator

---

1. Introduction

The operation and efficiency of a solar power conditioning system at different operating points can be tested using a variable DC source in the laboratory, but to validate the ability to track the maximum power point (MPP) in the power conditioning system, it is necessary to test the system with an actual PV array, but with an actual PV array, it is difficult to test the system at predefined operating points due to varying climatic changes. The space and cost required for the installation of an actual PV array are also more. With an actual PV array on site, it is necessary to alter the series/parallel combination of PV modules for testing different rated PV power conditioning systems. Hence a PV simulator is required to validate the solar power conditioning system at the laboratory level itself before any actual site trials [1,2]. A PV-simulator system is a DC power source in which the current-voltage (I-V) characteristics of different PV arrays can be programmed so that the operation of the solar power conditioning system can be validated. With a PV-simulator, it is possible to test the power conditioning systems for different voltage current combinations within the rated values of the PV-simulator.

An LLC resonant DC-DC converter-based PV-simulator discussed in [3,4] consists of a current driven centre tapped transformer which provides galvanic isolation between input and output circuits. A PV-simulator based on an interleaved buck converter is presented in [5,6]. A buck-boost chopper-based PV-simulator with double current mode control is proposed in [7] and a buck chopper-based PV-simulator with a two stage LC filter is discussed in [8]. A PV-simulator based on a two-quadrant DC-DC converter is proposed in [9]. A buck-chopper-based DC-DC converter with a single stage L-C filter is presented in [10–12]. The above discussed configurations and the components required for each configuration are summarized in Table 1 and it is observed that the buck-chopper-based PV-simulator is the most economical compared to other configurations due to its lesser number of magnetic components and power switches.

Table 1. Parameters of Selected PV-array to Be Programmed.

SL. No.	Reference	Configuration	Remarks
1	[3,4]	An LLC resonant DC-DC converter-based PV-simulator	This configuration requires a centre-tapped transformer and two inductors, two IGBT/MOSFET switches, etc. Cost and size of the system increases for higher rated systems due to the higher number of magnetic components.
3	[5,6]	PV-simulator based on an interleaved buck converter	This configuration requires three inductors and six IGBT/MOSFET switches. The cost of the system is comparatively lesser than [3,4].
3	[7]	Buck-boost chopper-based PV-simulator	Cost of this system is comparatively lesser than [3–6] as this configuration requires two inductors and four IGBT/MOSFET switches.
2	[8]	Buck chopper-based PV-simulator with two stage LC filter	Two inductors are required at the output side of the chopper. The cost of the system is comparatively lesser than [3–7], but an additional L-C filter stage on the output side is costlier than a buck-chopper-based PV-simulator with a single L-C stage.
2	[9]	Two quadrant DC-DC converter-based PV-simulator with two stage LC filter	This configuration requires only one inductor on the output stage and two IGBT/MOSFET switches. Hence the cost of this configuration is less that that of the configurations presented in [3–8].
6	[10–12]	Buck chopper-based PV-simulator with two stage LC filter	Buck-chopper-based PV-simulator is cheaper than the systems presented in [3–9] since this configuration requires only one inductor on the output stage and only one IGBT/MOSFET switch.

In the works discussed in [4–12], a DC source is considered as input, hence the AC to DC conversion stage is not discussed. A PV-simulator manufactured by M/s Chroma with type number 62000H-S series, is suitable for the input sources such as single-phase 220 Vac and three-phase 440 Vac which are usually available at laboratories. This simulator is suitable for the testing of inverters up to the rating of 1000 V, 25 A. Multiple such PV simulators can be connected for testing of higher rated power conditioning systems. Since AC is the commonly available supply in laboratories, it is preferred to have a rectifier at the input stage of the PV simulator.

A PV-simulator with an AC input source presented in [13,14] consists of a single phase diode front end rectifier and a buck chopper-based DC-DC converter. Due to the uncontrolled single phase diode rectifier in the input section, the input THD and power factor are poor with the presented configuration. To make the system operate at unity power factor with a better THD, it is desirable to have an active front end rectifier in the input stage. A three level front end converter is preferable over a conventional two level inverter as the total power loss in a three-level converter is comparatively lesser than that of a two-level converter [15,16]. Also the voltage and current harmonics in a three-level converter are comparatively lesser than those of a two-level converter [17]. The dv/dt in a three-level converter is less than that of a two-level inverter; hence the voltage stresses on the devices are also minimized. Three-level front-end converters can be designed either with a diode clamped converter or a T-type converter. A T-type three-level converter is suitable for low voltage applications as the switching and conduction losses in T-type three-level converters are less compared to that of a diode clamped three level inverter [18]. Due to the above advantages a front-end converter based on a T-type configuration is selected in this work.

In this work, the design of a low cost PV-simulator with a single phase front end converter and a buck-chopper-based DC-DC converter is proposed. The block diagram of the proposed simulator is shown in Figure 1. Parameters such as open-circuit voltage, short-circuit current, MPP voltage, MPP current, power rating, series resistance, parallel resistance and temperatures coefficients of the selected PV array which are usually available from the product datasheet and can be programmed in the PV-simulator. The variable inputs such as irradiance on the PV array and the temperature of PV array can be varied from the user interface for testing the power conditioning system at various operating points. The maximum ratings of the PV array that can be programmed are restricted to the rated output voltage and currents of PV-simulator.

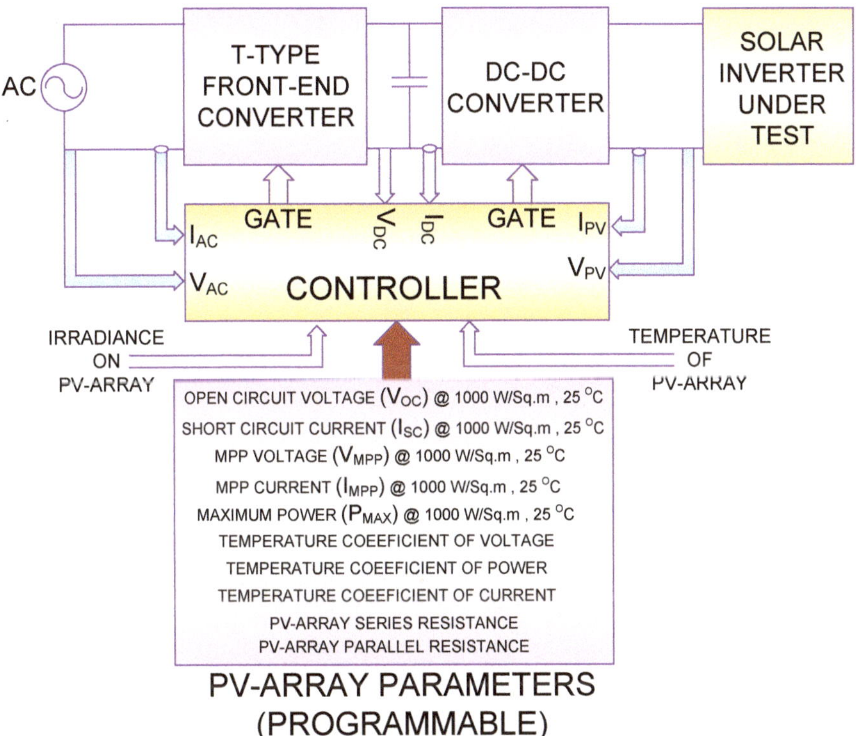

Figure 1. Block diagram of the proposed PV-Simulator.

Due to the advantages of a T-type configuration over a conventional two-level converter mentioned earlier, a front-end converter based on a T-type configuration is selected in the present work. The control philosophy for a single phase active front-end rectifier is presented in [19]. Closed loop voltage control is adapted to regulate the DC link voltage. Closed loop voltage control for a three-phase active front-end rectifier is discussed in [20]. Reference DC link voltage for FEC is always adjusted more than the open circuit voltage of the programmed PV array. An isolation transformer is used at the input side of the simulator. The isolation transformer enables the operation of multiple PV-simulators connected in parallel for testing of higher rated solar power conditioning systems. The input inductance required for the boosting operation of front-end rectifier can be incorporated in the transformer itself so that the component count, cost and weight can be minimized.

The second stage of power conversion consists of a buck-chopper-based DC-DC converter. The output voltage of the chopper ranges between the minimum PV voltage i.e., zero and the maximum PV voltage i.e., the open circuit voltage of the programmed PV array. Current control is adapted for the DC-DC converter to replicate the I-V characteristics of programmed PV array. Reference output current is obtained from the programmed PV parameters, while irradiance input, and operating temperature input are adjusted by the user and the instantaneous output voltage. In this work, the PV-simulator is designed to be programmed up to the ratings of the selected PV array shown in Table 2. A PV array comprised of multiple PV modules (Type number SPR-435NE-WHT-D of M/s Sun Power) connected in series-parallel combination is considered. The T-type front-end converter, buck-chopper-based DC-DC converter and its controls are discussed in detail. Real-time simulations are carried out to validate the system using an Opal-RT brand real-time simulator. I-V characteristics obtained from the PV-simulator are compared with the actual PV-array characteristics to validate the controls.

Table 2. Parameters of Selected PV-Array to Be Programmed.

SL. No.	Electrical Parameter	Value	Units
	PV Module Ratings		
1	Module Power (P_Mod)	435	W
2	Open Circuit Voltage (Voc_Mod)	85.6	V
3	Short Circuit Current (Isc_Mod)	6.43	A
	PV Array Ratings		
4	No of Series Modules in PV Array (Nse)	4	No's
5	No of Parallel Modules in PV Array (Np)	10	No's
6	PV Array Power (P_Mod × Nse × Np)	17.40	kW
7	Maximum DC Voltage (Nse × Voc_Mod)	342	V
8	Maximum Output Current (Np × Isc_Mod)	64.3	A

2. A Three-Level T-Type Front-End Rectifier

As the input to the PV-simulator is an AC source, a rectifier is to be used on the input stage. Instead of using an uncontrolled diode rectifier, an active front-end converter (FEC) is proposed in this system to obtain unity power factor on the input side. A three-level T-Type FEC is selected over a conventional two-level H-bridge-based FEC to obtain better input THD and low dv/dt. A transformer is also proposed on the input side for isolation purposes in the present work. The input inductance required for the boost operation can also be incorporated in the input transformer. The isolation transformer also enables the parallel operation of multiple PV-simulators during the testing of higher rated power conditioning systems. Since the open-circuit voltage of the PV-array selected is 342 V, the DC link voltage (Vdc) should always be more than the open circuit voltage i.e., 342 V. In the present system, Vdc selected is 500 V i.e., approximately 1.5 times the open-circuit voltage of the PV-array. Commercially available power conditioning systems are listed in [21] and it is observed that the maximum PV voltages are in the range of 1000 V to 1500 V DC. The proposed system can be extended to higher voltages by suitably selecting the turn ratio of the input transformer. The power rating of the FEC should be more than the maximum power rating of the PV-array. By considering the efficiency of the PV-simulator as 85%, the power rating of the FEC obtained is 20.5 kW.

The power circuit and the control block diagram for the proposed FEC are shown in Figure 2. The controller monitors the input AC voltage and using a phase locked loop (PLL), a unit signal which is in-phase with the input voltage is generated. The reference DC link voltage (Vdc_ref) is adjusted to 500 V in this system and actual DC link voltage (Vdc) is monitored and the error is applied to a PI controller. A feedforward control is used for improving the dynamic response of the system to sudden changes in DC current (Idc). The output of the feed-forward loop is multiplied with the unit signal to obtain the reference input current. The reference AC current is compared with the actual AC current and the error is applied to a PI controller to obtain the modulating signal for the rectifier. The dynamic response of the system can be studied by applying a step change in reference DC voltage (Vdc_Ref) and the load current (Idc) [22].

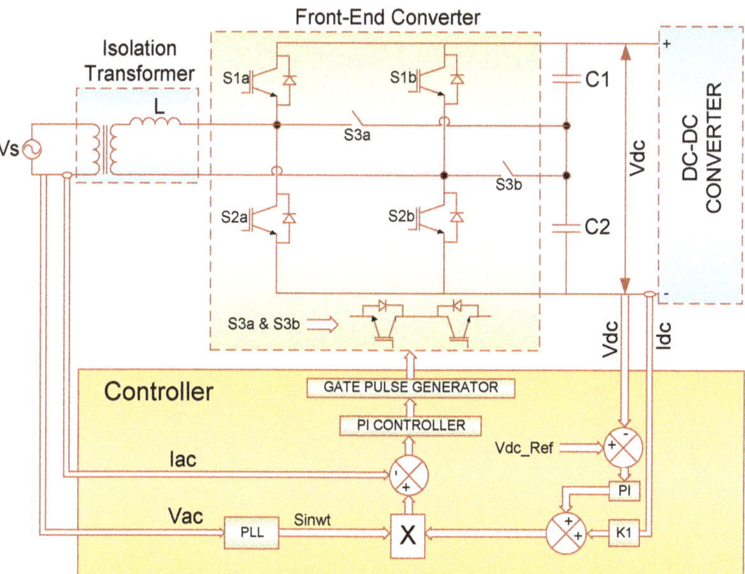

Figure 2. Control Block diagram of T-Type Front-End Converter.

Operation of FEC and the dynamic response of the system are validated through real-time simulations using an Opal RT real-time simulator. The electrical scheme for real-time simulation validation is shown in Figure 3. In this setup, the plant, comprised of an input supply, transformer, T-Type converter, DC-link capacitors, and the load are modeled in Matlab-Simulink with a sample time of 10 microseconds. The simulated model is compiled and loaded in the high speed processor of the real-time simulator. The control software for T-type front end converter is developed through a Matlab embedded coder and loaded in the controller card based on a Texas Instruments TMS320F2812 digital single processor. Details of the controller card used and the real-time simulator are shown in Table 3.

Start/Stop commands are given to the controller from external pushbuttons. The DC link voltage reference VDC_Ref signal is also given from user interface to the analog input channel of the controller card. Using the analog output channels of the real-time simulator, the voltage and current signals from the plant are given to the analog input channels of the controller card.

The controller card provides gate signals for the FEC through the digital input channels of the real-time simulator. A simulated 'Stack faulty signal' in the plant is given to the controller card through the digital output channel of the real-time simulator. On receiving the stack faulty signal from the plant, the controller blocks the gate pulses to the FEC and also provides an off command to the input AC breaker Q1. To provide step changes in the load, two contactors Q2, Q3 are simulated and the control signals for these load switches are provided by the controller through the digital input channels of the real-time simulator. The dynamic response of the FEC is observed for sudden changes in the DC link voltage reference (VDC_Ref) and for sudden changes in load.

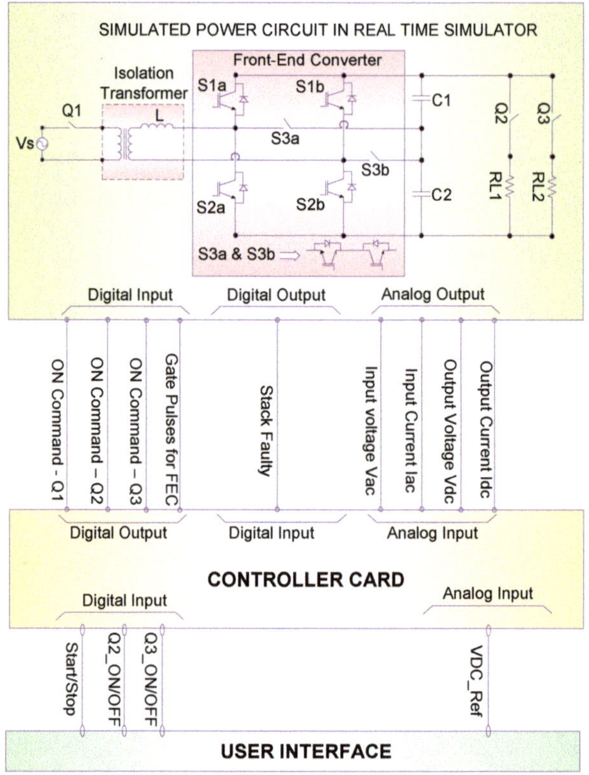

Figure 3. Electrical scheme for controller-In-Loop simulation validation of T-Type Front-End Converter.

Table 3. Details of the Hardware Used in the Controller-In-Loop Simulations.

SL. No.	Parameter	Description
	Real-Time Simulator	
1	Manufacturer	Opal-RT
2	Processor in Real-Time Simulator	Intel Xeon Quadcore 2.50 GHz
3	Operating System	QNX
4	IO Interface	FPGA Based
5	No of Analog Input Channels	16
6	No of Analog Output Channels	16
7	No of Digital Input Channels	16
8	No of Digital Output Channels	16
9	Analog Channels voltage Range	-10 to $+10$ V
10	Digital Channels Voltage range	0 V = Logic Low, 15 V = High
11	Front-End Modeling tool	Matlab-Simulink
	Controller Details	
12	DSP Used	TMS320F2812 DSP
13	No of Analog Input Channels	16
14	No of Digital Input Channels	16
15	No of Digital Output Channels	16
16	No of PWM Outputs	12
17	Analog Channels voltage Range	-10 to $+10$ V
18	Digital Channels Voltage range	0 V = Logic Low, 15 V = High

2.1. Dynamic Response of FEC for a Step Change in Vdc_Ref

A fixed resistive load is connected across the output DC terminals of FEC and a Vdc_ref is adjusted to 500 V. Since a T-type configuration is used, the PWM voltage at the rectifier AC terminals has three levels. The voltage step in the PWM voltage is equal to half of the value of the DC link voltage as shown in Figure 4a. Since the FEC maintains unity power factor, the input voltage and current are in-phase with each other, as shown in Figure 4b. A step change in the reference DC link voltage is applied and it is observed that the actual DC link voltage follows the reference DC voltage. Since the new Vdc_Ref is 600 volts, the voltage step in the three-level PWM voltage also varied and becomes 300 volts, i.e., half of the DC link voltage as shown in Figure 4a. Since a fixed resistive load is connected across DC terminals of the FEC, the load current also increases which results in an increase in the input side current. It is observed that the input current is also increased but still maintains the unity power factor as shown in Figure 4b.

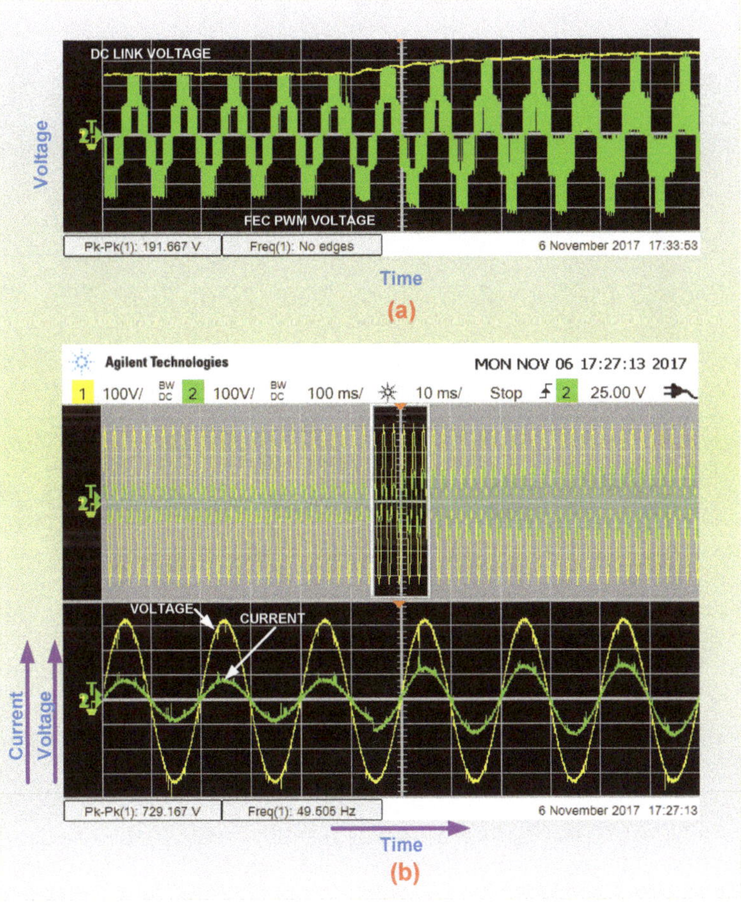

Figure 4. (**a**) Voltage and Current Waveforms of T-Type Front-End Converter (**b**) Input Voltage and Current Waveforms of Front-End Converter during Transition time for a step change in DC link voltage.

2.2. Dynamic Response of FEC for a Step Change in Idc

In this case, the reference DC link voltage is maintained at 600 volts and a step change in the load is applied. Since the Vdc ref is 600 volts, the DC link voltage is 600 volts and the voltage step in the three-level PWM voltage is 300 volts, i.e., half of the DC link voltage as shown in Figure 5a. FEC is operating at unity power factor, hence the input voltage and current are in-phase with each other as shown in Figure 5b.

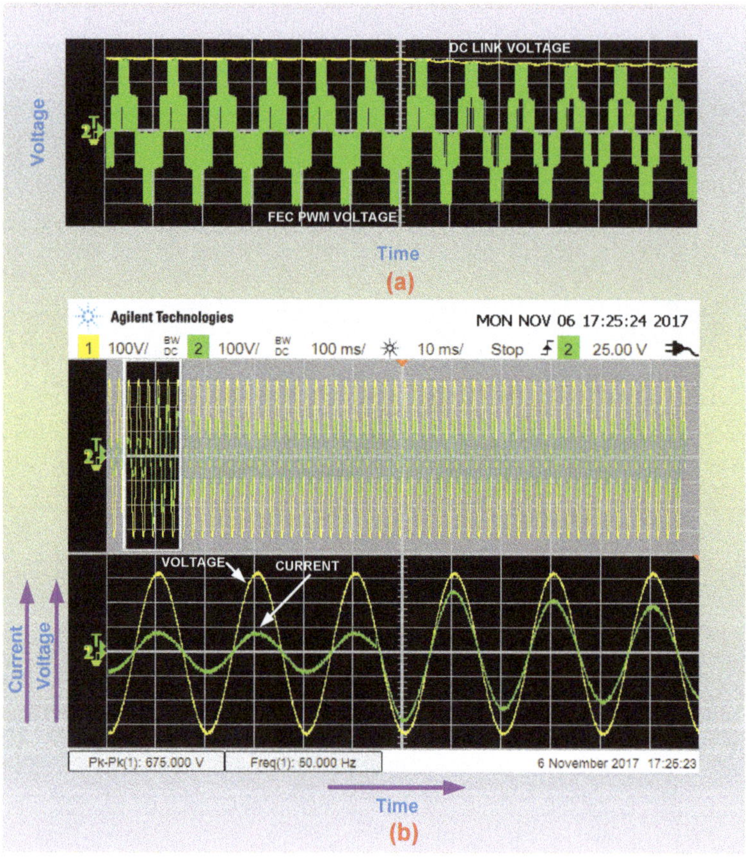

Figure 5. (**a**) Input Voltage and Current Waveforms of T-Type Front-End Converter (**b**) Input Voltage and Current Waveforms of T-Type Front-End Converter during Transition time for a step change in Load.

A step change in DC load is applied and it is observed that the actual DC link voltage drops by approximately 40 volts at that instant and increases gradually to the reference value, i.e., 600 volts, as shown in Figure 5a. Since the load is increased on the DC side, there in an increase in the input side current. It is observed that the input current is increased but the unity power factor is still maintained as shown in Figure 5b.

From the presented real-time simulation results, it is observed that the FEC controls programmed in the real controller are satisfactory as the system operation is as desired under different operating conditions. The dynamic response of the system and the associated controls are also good as the settling time and overshoots are less. For PV-simulator applications, the reference DC link voltage of the FEC is always adjusted to be more than the open circuit voltage of the programmed PV array. The FEC always operates at unity power factor and the input current depends on the DC-DC converter operation which is connected as load to the FEC. Detailed discussions on DC-DC converter and associated controls are presented in the next section.

3. Buck-Chopper-Based DC-DC Converter

The current-voltage (I-V) characteristics of a programmed PV array are realized through a DC-DC converter in the PV-simulator. A buck-chopper based DC-DC converter is used in this work and the closed loop current control is applied, to make the buck-chopper work like a current source similar to a PV array. Reference output current (I_{PV_Ref}) is compared with actual output current (I_{PV}) and the error signal is applied to PI controller to obtain the reference duty cycle for the chopper. A hysteresis controller can also be adapted for the current control through buck-chopper. The PWM block generates the gate pulse for the chopper. The detailed control block diagram is as shown in Figure 6.

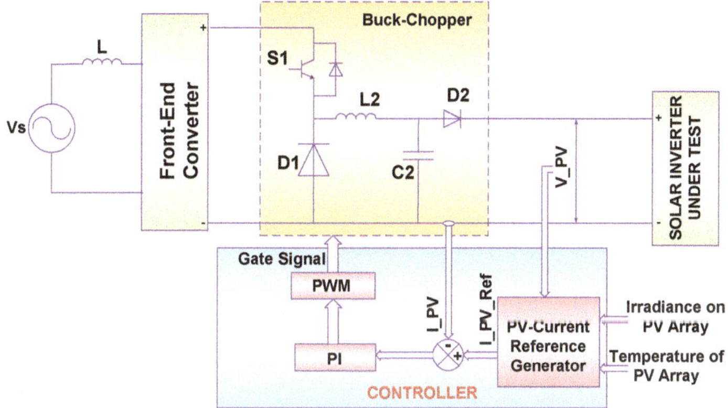

Figure 6. Control Block diagram of a DC-DC Converter.

In this work, the reference DC current is obtained from the PV current equation given in [11] as shown in Equation (1):

$$I_{PV\_Ref} = I_{ph} - (I_s*(e^{(V + I*Rse)/N*Vt-1)})) - ((V + I*R_{se})/R_p) \tag{1}$$

where I = PV current, V = PV voltage, I_{sc} = short circuit current and V_{oc} = Open circuit Voltage

$$\text{Photon current } I_{ph} = \text{Irradiance X } (I_{sc}/I_{ro}) \tag{2}$$

$$\text{Saturation current } Is = I_{sc}/(exp(V_{oc}/(N*V_t))-1) \tag{3}$$

where I_{ro} = measured irradiance (1000 W/m^2), R_{Se} = series resistance, R_P = parallel resistance and N is the quality factor

$$\text{Thermal voltage } Vt = k*T/q \tag{4}$$

where Boltzmann's constant (k) = 1.3806×10^{-23}, operating temperature (T) = 25 °C and the charge of an electron (q) = 1.602×10^{-19}.

Irradiance and temperature of PV array signals are adjustable by the user in this PV-simulator to test the solar power conditioning system under different operating conditions. To verify the operation of the PV-simulator, a PV array with the specifications listed in Table 4 and the proposed PV-simulator circuit are simulated for the same operating conditions and compared with each other.

Table 4. Parameters of Programmed PV-Array at Irradiance Inputs of 1000 W/Sq.m and 600 W/Sq.m.

SL. No.	Electrical Parameter	Value	Units
	PV Module Ratings		
1	Maximum power of module at 1000 W/Sq.m	435	W
2	Maximum power of module at 600 W/Sq.m	261	W
3	MPP Voltage (Vmpp) at 1000 W/sq.m	72.9	V
4	MPP Current (Impp) at 1000 W/sq.m	5.97	A
	PV Array Ratings		
5	No of Series Modules in PV Array (Nse)	4	No's
6	No of Parallel Modules in PV Array (Np)	10	No's
7	PV Array Power at 1000 W/sq.m	17.40	kW
8	MPP Current of Array (Impp × Np) at 1000 W/sq.m	59.7	A
9	Maximum PV Array Power at 600 W/sq.m	10	kW
10	MPP Current of Array at 600 W/sq.m	36	A

The PV array and the PV-simulator are simulated in two different cases. In case-1, load resistance is maintained constant and the irradiance value is varied in steps. In case-2, irradiance is kept constant and the load resistance is varied.

Case-1: A fixed resistive load is connected to the PV array and the PV-array output voltage, current and power are monitored by varying the irradiance. PV current is proportional to the irradiance and the PV voltage is the product of PV current and the load resistance. Similarly the PV-simulator is also connected to the same load and tested for the same conditions to compare the results with the results obtained by simulating a PV array. It is observed that the outputs of PV array and the PV-simulator match with each other as shown in Figure 7. Irradiance input to the simulator is reduced in steps of 200 W/sq.m at each 0.5 s time and it is observed that the reference current to the chopper is also reduced proportionally to the irradiance. The controller ensures the actual output current remains equal to the reference output current with the help of a closed-loop current controller.

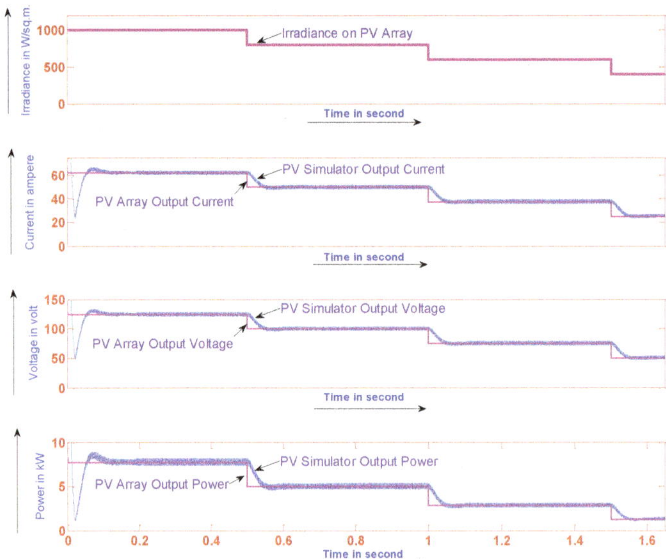

Figure 7. Output Voltage, Current and Powers from an Actual PV array and the PV-simulator for Different Irradiance Values.

Case-2: An irradiance of 600 W/Sq.m is applied as input to both the PV array and PV-simulator. Table 3 shows the parameters of the PV array at 600 W/sq.m which are derived based on the parameters in the datasheet of the PV module (Type number SPR-435NE-WHT-D of M/s Sun Power) selected. Independent variable loads are connected to PV array and the PV-simulator outputs. Loads on the PV-Array and the PV-simulator are increased in steps and monitored the output voltages, currents and powers. From the results shown in Figure 8, it is observed that the output values obtained from the PV-simulator are identical to the values obtained from the PV array.

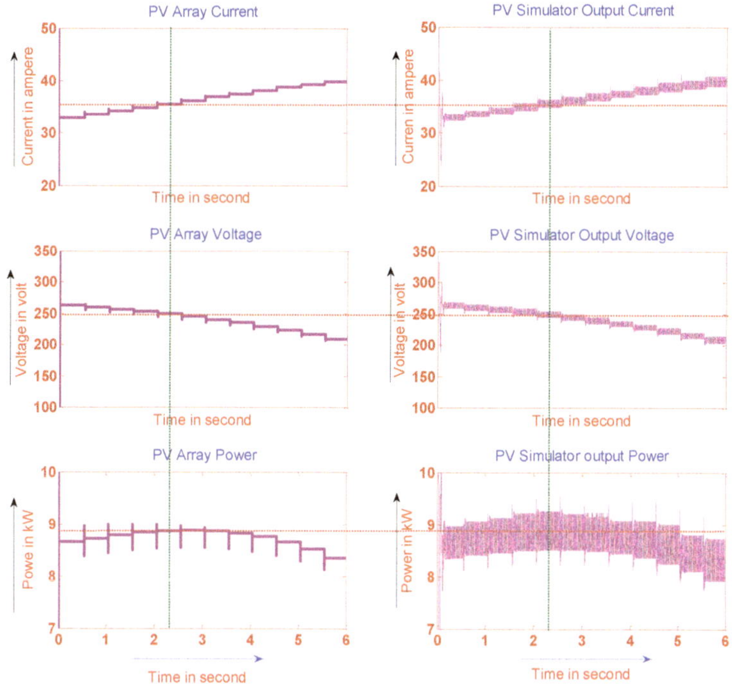

Figure 8. Output Voltage, Current and Powers from an Actual PV array and the PV-Simulator for at an Irradiance of 600 W/Sq.m.

From the simulation results shown in Figures 7 and 8, it is evident that the proposed PV-simulator can replicate the actual PV-array by programming the actual PV-array parameters, so that it can be used for testing solar power conditioning systems.

4. Real-Time Simulation Results

To validate the controls for the proposed PV-simulator, controller in loop simulations are carried out with the help of an Opal-RT real-time simulator similar to the method presented in Section 2. The details of the hardware used for controller-in-loop simulations are listed in Table 3 and the scheme for the controller in loop simulations is shown in Figure 9.

The plant, comprised by an input source, FEC, DC-DC converter, and the load are modeled in Matlab-Simulink with a sample time of 10 microseconds and loaded into the real-time simulator. The control software is developed through Matlab embedded coder and the loaded in the DSP-based controller card. Start/Stop commands are given to the controller from external pushbuttons. Irradiance input and the temperature of PV array signals generated with the help of potentiometers and given to the analog input channels of the controller card. The controller card also receives the analog signals required from the plant such as input voltage, input current, output voltage and output currents from the simulated plant through the analog out (AO) channels of the real-time simulator.

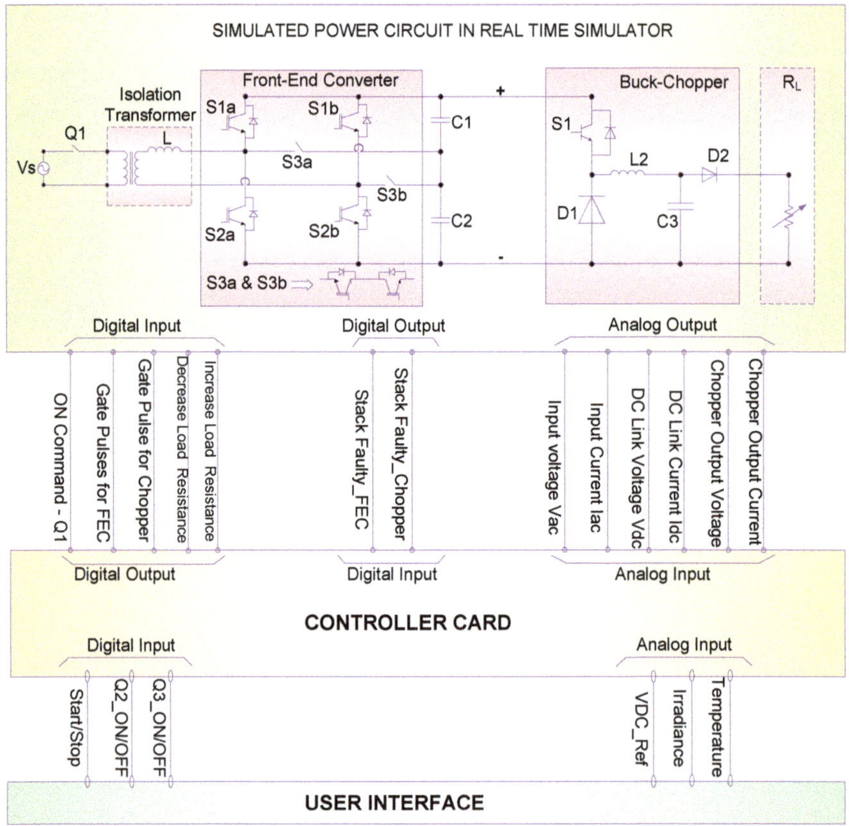

Figure 9. Controller-In-Loop Simulation setup for validation of proposed PV-Simulator Controls.

As discussed in previous sections, on receiving the actual analog signals from the plant, the controller generates the gate pulses to maintain the actual values equal to the reference values for both FEC and DC-DC converters.The controller provides the gGate pulses to the simulated FEC and DC-DC converter through the digital input channels of the real-time simulator. The real-time simulator converts the gate pulses received into data to trigger the IGBTs. All this process happens with real-time speed.

In this work, the PV array with the parameters shown in Table 4 are programmed in the PV-simulator and the load current is varied from short circuit current to zero current. The voltage and power of the PV array are monitored for different irradiance Inputs. From the presented results shown in Figure 10, it is observed that for different irradiance inputs, the MPP voltage is almost constant with a value of approximately 250 volts. The MPP voltages and powers obtained from the PV-simulator at different irradiance inputs are compared with the PV module datasheet (SPR-435NE-WHT-D) and it is found that the proposed PV-simulator can replicate the actual PV array characteristics.

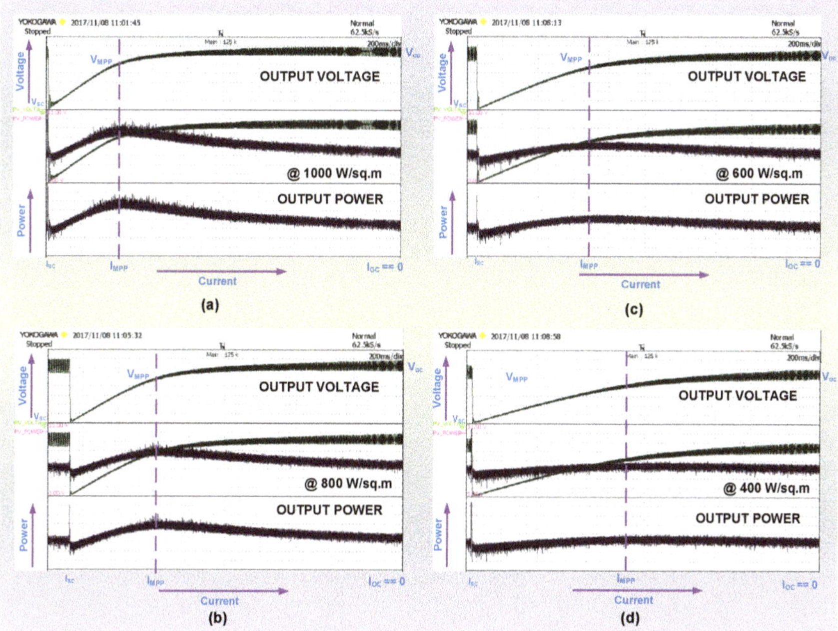

Figure 10. Output Voltage and Powers of PV-Simulator for Different Irradiance Inputs.

5. Future Scope

In the present work, operation of a PV-simulator is validated through controller in loop simulations by interfacing the real controller hardware with the simulated power circuit by using a real-time simulator. Through controller-in-loop simulations, tuning of control software and the power circuit parameters is carried out. As a future work, it is envisaged to build a PV-simulator based on the present work.

A dual-active bridge (DAB)-based DC-DC converter with high frequency transformer presented in [23] is also investigated along with the chopper-based DC-DC converter for the present application. As a future work, it is envisaged to study a DAB-based PV-simulator to reduce the size of the equipment.

6. Conclusions

The proposed PV-simulator configuration provides better input THD as the front-end converter of a three-level rectifier. Unity power factor operation is achieved with this configuration which results in an optimized power rating of the simulator. FEC also helps in boosting the DC link voltage without using a step-up transformer on the AC side which results in lesser cost. From the presented results it can be concluded that the dynamic response of the system is satisfactory for sudden changes in load, sudden changes in the irradiance inputs and sudden changes in the DC link voltage reference. Instead of using an isolation transformer on the input side, it is also feasible to use a high-frequency transformer in the DC-DC converter which can reduce the size of the simulator. With the proposed configuration, multiple numbers of PV simulators can be connected in parallel for testing higher rated power conditioning systems. Operation of two PV-simulators connected in parallel subjected to unequal irradiance inputs can be studied which should help in analyzing the system operation during partial shading conditions.

Author Contributions: S.V., S.P., U.S., had developed the original proposed research work and implemented with numerical simulation software and real time RTDS system for investigation and performance validation. F.B. contributed his expertise in the proposed subject of research and verification of the obtained results based on theoretical concepts and insight background. All authors involved to articulate the research work for its final depiction as research paper.

Conflicts of Interest: The authors declare no conflict of interest.

References

1. Jin, S.; Zhang, D.; Wang, C. UI-RI Hybrid Lookup Table Method with High Linearity and High-Speed Convergence Performance for FPGA-Based Space Solar Array Simulator. *IEEE Trans. Power Electron.* **2018**, *33*, 7178–7192. [CrossRef]
2. Zhou, Z.; Macaulay, J. An Emulated PV Source Based on an Unilluminated Solar Panel and DC Power Supply. *Energies* **2017**, *10*, 2075. [CrossRef]
3. Chang, C.; Chang, E.; Cheng, H. A High-Efficiency Solar Array Simulator Implemented by an LLC Resonant DC–DC Converter. *IEEE Trans. Power Electron.* **2013**, *28*, 3039–3046. [CrossRef]
4. Chang, C.H.; Cheng, C.A.; Cheng, H.L. Modeling and design of the LLC resonant converter used as a solar-array simulator. *IEEE J. Emerg. Sel. Top. Power Electron.* **2014**, *2*, 833–841. [CrossRef]
5. Mahmud, M.H.; Zhao, Y.; Wang, L. A high-bandwidth PV source simulator using a sliding mode controlled interleaved buck converter. In Proceedings of the IEEE International Conference IECON, Beijing, China, 29 October–1 November 2017; pp. 2326–2331.
6. Koran, A.; LaBella, T.; Lai, J.S. High efficiency photovoltaic source simulator with fast response time for solar power conditioning systems evaluation. *IEEE Trans. Power Electron* **2014**, *29*, 1285–1297. [CrossRef]
7. Zhang, W.; Kimball, J.W. DC–DC Converter Based Photovoltaic Simulator with a Double Current Mode controller. *IEEE Trans. Power Electron.* **2018**, *33*, 5860–5868. [CrossRef]
8. Koran, A.; Sano, K.; Kim, R.Y.; Lai, J.S. Design of a Photovoltaic Simulator with a Novel Reference Signal Generator and Two-Stage LC Output Filter. *IEEE Trans. Power Electron.* **2010**, *25*, 1331–1338. [CrossRef]
9. Cordeiro, A.; Foito, D.; Pires, V.F. A PV panel simulator based on a two quadrant DC/DC power converter with a sliding mode controller. In Proceedings of the IEEE International Conference on Renewable Energy Research and Applications (ICRERA), Padova, Italy, 22–25 November 2015; pp. 928–932.
10. Kawano, M.; Sato, T.; Taguchi, T.; Genzima, Y.; Teramoto, S. Simple and low cost PV simulator using diode characteristics for development of PV system. In Proceedings of the INTELEC 2015—IEEE International Conference on Telecommunications Energy, Osaka, Japan, 18–22 October 2015; pp. 1–5.
11. Özden, Ö.; Duru, Y.; Zengin, S.; Boztepe, M. Design and implementation of programmable PV simulator. In Proceedings of the IEEE International Conference on Fundamentals of Electrical Engineering (ISFEE), Bucharest, Romania, 30 June–2 July 2016; pp. 1–5.
12. Seo, Y.T.; Park, J.Y.; Choi, S.J. A rapid IV curve generation for PV model-based solar array simulators. In Proceedings of the IEEE International Conference on Energy Conversion Congress and Exposition (ECCE), Milwaukee, WI, USA, 18–22 September 2016; pp. 1–5.
13. Shinde, U.K.; Kadwane, S.G.; Keshri, R.K.; Gawande, S.P. Dual Mode Controller-Based Solar Photovoltaic Simulator for True PV Characteristics. *Can. J. Electr. Comput. Eng.* **2017**, *40*, 237–245.
14. Piao, Z.G.; Gong, S.J.; An, Y.H.; Cho, G.B. A study on the PV simulator using equivalent circuit model and look-up table hybrid method. In Proceedings of the IEEE International Conference on Electrical Machines and Systems (ICEMS), Busan, Korea, 26–29 October 2013; pp. 2128–2131.
15. Alemi, P.; Lee, D.C. Power loss comparison in two-and three-level PWM converters. In Proceedings of the IEEE International Conference on Power Electronics and ECCE Asia (ICPE & ECCE), Jeju, Korea, 30 May–3 June 2011; pp. 1452–1457.
16. Schweizer, M.; Friedli, T.; Kolar, J.W. Comparative evaluation of advanced three-phase three-level inverter/converter topologies against two-level systems. *IEEE Trans. Ind. Electron.* **2013**, *60*, 5515–5527. [CrossRef]
17. Masisi, L.; Choudhury, A.; Pillay, P.; Williamson, S. Performance comparison of a two-level and three-level inverter permanent magnet synchronous machine drives for HEV application. In Proceedings of the IEEE International Conference on Industrial Electronics Society (IECON), Vienna, Austria, 10–13 November 2013; pp. 7262–7266.

18. Shin, H.; Lee, K.; Choi, J.; Seo, S.; Lee, J. Power loss comparison with different PWM methods for 3L-NPC inverter and 3L-T type inverter. In Proceeding of the IEEE Power Electronics and Application Conference and Exposition (PEAC), Shanghai, China, 5–8 November 2014; pp. 1322–1327.

19. Labaki, R.; Kedjar, B.; Al-Haddad, K. Single-Phase Active Front End Converter with series compensation. In Proceedings of the IEEE International Symposium on Industrial Electronics, Montreal, QC, Canada, 9–13 July 2006; pp. 769–774.

20. Huerta, F.; Stynski, S.; Cobreces, S.; Malinowski, M.; Rodriguez, F.J. Novel Control of Three-Phase Active Front-End Converter with Compensation of Unknown Grid-Side Inductance. *IEEE Trans. Ind. Electron.* **2011**, *58*, 3275–3286. [CrossRef]

21. Sridhar, V.; Umashankar, S. A comprehensive review on CHB MLI based PV inverter and feasibility study of CHB MLI based PV-STATCOM. *Renew. Sustain. Energy Rev.* **2017**, *78*, 138–156. [CrossRef]

22. Hou, C.C.; Cheng, P.T. Experimental Verification of the Active Front-End Converters Dynamic Model and Control Designs. *IEEE Trans. Power Electron.* **2011**, *26*, 1112–1118. [CrossRef]

23. Sridhar, V.; Sanjeevikumar, P.; Umashankar, S.; Mihet-Popa, L. Power Balancing Control for Grid Energy Storage System in Photovoltaic Applications-Real Time Digital Simulation Implementation. *Energies J.* **2017**, *10*, 928.

Article

Experimental Evaluation of the Performance of a Three-Phase Five-Level Cascaded H-Bridge Inverter by Means FPGA-Based Control Board for Grid Connected Applications

Fabio Viola

Dipartimento Energia, ingegneria dell'Informazione e modelli Matematici, DEIM, University of Palermo, 90133 Palermo, Italy; fabio.viola@unipa.it; Tel.: +39-091-238-60253

Received: 22 October 2018; Accepted: 22 November 2018; Published: 26 November 2018

Abstract: Over the last decades, plants devoted to the generation of green energy significantly increased their number, together with the demand of same electrical energy, also stored in battery systems. This fact produced the growth of energy conversion systems with advanced performances with respect to the traditional ones. In this circumstance, multilevel converters play a significant role for their great advantages in performances, flexibility, fault-tolerability, employment of renewable energy sources and storage systems and finally yet importantly reduced filter requirements. In this context, this paper faces the performance of a cascaded H-bridge 5 level inverter in terms of harmonic distortion generated and injected into the grid. Through an accurate analysis that takes into account the pulse width modulation (PWM) multicarrier modulation techniques (phase disposition PD, phase opposition disposition POD, alternative phase opposition disposition APOD, phase shifted PS) and related reference signals (sinusoidal reference; third harmonic injection THI reference, switching frequency optimal SFO reference), a framework of distorting harmonics is presented by comparing twelve cases. The results obtained from the simulations are reproduced and validated in a prototype system of five level cascaded H-bridge multilevel inverter. A deep discussion of control and filtering system is provided to justify the choice of the best modulation technique to adopt.

Keywords: Cascaded H-bridge multilevel inverter (CHBMI); field-programmable gate array; total harmonic distortion (THD); modulation techniques

---

1. Introduction

The increased demand of green energy has led to the development of even more performing structures allowing the generation and storage of energy in DC form. The drawbacks are due to the increased number of harmonics introduced and filtered in the power grids.

The three main multilevel power inverter (MPI) structures proposed in technical literature, with their related benefits and disadvantages, according to [1–3] are diode clamped converter (DCC), capacitor-clamped inverter (CCI) and cascaded H-bridge (CHB) multilevel inverter.

The neutral point clamped converter (NPC) was the first multilevel structure proposed. The common DC-link is composed of four capacitors connected in series that split the voltage into four level. The middle point of the capacitors n is used as neutral point. The peculiar components, that differentiate this circuit from the others multilevel inverters, are the clamping diodes that allow to subdivided the DC voltage on the switches. Thus, the voltage across on the switches is limited to one capacitor voltage equal to $V_{dc}/(n_L - 1)$, where n_L is the number of level. By supposing that for each blocking diode its voltage value is identical to the voltage rating of active device, the number of diodes requested for each phase will be $(n_L - 1)\cdot(n_L - 2)$. This converter presents some operative limits as: (1) max number of levels is five, due both to the complexity of the circuit and both to the large number of components demanded; (2) uneven distribution of semiconductor power losses among the switches, which reduces the switching frequency and the output power; (3) unbalanced capacitor voltages which generate low frequency harmonics; (4) the system cannot involve a modular structure (non-modular topology structure).

Flying capacitor inverters (FCIs) or CCIs are an alternative to overcome some of the DCC disadvantages. The structure of CCIs have similarity to NPC inverter except the CCI uses several capacitors in the place of the clamping diodes. The main advantage of the CCI are the redundant states to obtain the voltage levels. In this way, it is possible an even distribution of semiconductor losses among the switches but it is necessary a dedicated control algorithm to balance the capacitors voltage.

The increase of voltage levels confines the proper charging and discharging mechanism of capacitors. By considering the economical aspect, the cost of the inverter follows the increase of the number of levels, but also the device becomes bulkier and its lifetime decreases due to the growing number of used capacitors. For a n_L-level converter, it is necessary $(n_L - 1)\cdot(n_L - 2)/2$ clamping capacitors per phase in addition to the $(n_L - 1)\cdot$main dc bus capacitors. Thus, the high number of capacitors limit the use to three or five levels. Moreover, lack of modularity and high quantity of capacitors for higher number of voltage levels reduces the reliability of this converter.

Figure 1 shows the topology structure of a single-phase five-level cascaded H-bridge multilevel inverter.

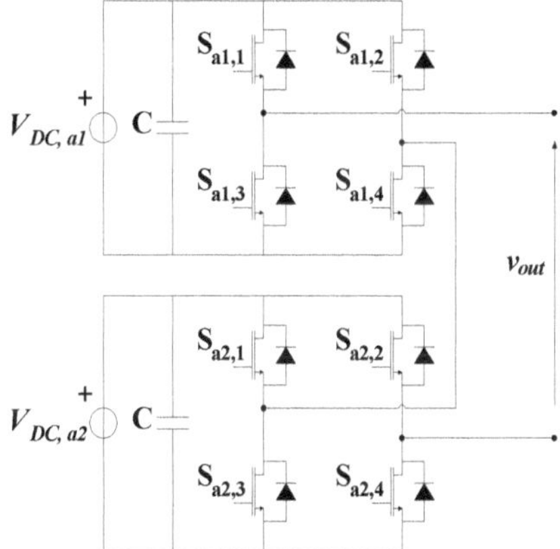

Figure 1. Topology structure of a single-phase five-level cascaded H-bridge (CHB) inverter.

This topology has a main advantage: the modular configuration, blocks can be added to reach voltage levels, control is easily performed, and maintenance, in case of fault, requires the disconnection of a block to keep the system working. Thus, each module can be either half- or full-bridge with separated DC source and can be controlled as a single-phase converter. This topology reaches in output medium voltage levels, by enforcing only common low-voltage components so there are not operative limits about max number of the voltage levels. Matching the number of capacitors and diodes between the cited topologies yields that CHB converter has the least number of components.

The phase voltage is synthesized by the addition of the voltages generated by the different modules. Thus, the voltage levels n_L depend of the number of modules connected in series for phase through the equation:

$$n_L = 2n_{HB} + 1 \tag{1}$$

where n_{HB} is the number of cells connected in series for phase.

Separated DC sources are an advantage in many applications but this feature leads to a more complex DC-voltage regulation loop.

Others topology structures of multilevel inverters and different classification methods were reported in literature [4]. In [5], an interesting classification into two comprehensive categories according to the applied DC source structure were discussed. A high number of topology structures were developed since separated DC sources are very diffused in renewable energy plants (PV, Wind farm, Fuel cell, etc.).

1.1. Overview of Pulse Width Modulation (PWM) Modulation Techniques

Pulse width modulation (PWM) techniques found large use in many industrial applications due to their easy implementation in the modern control systems and the high flexibility. Generally, PWM techniques used for multilevel inverters are an extension of modulation techniques for the traditional two-level voltage source inverters (VSI). A general classification of the modulation strategies for MPI presents two categories: "Fundamental switching frequency" and "High switching frequency".

Generally, the first category have been used in application where it is necessary to reduce the switching losses (i.e., high power electrical drive) while the second category have been used in applications where it is necessary to reduce the harmonic content on the output voltage (i.e., grid connected systems).

In literature [6–8] were reported many multicarrier modulation PWM methods, which differ for the reference signals and carrier signals. About the carrier signals, there are the "amplitude shifted" multicarrier PWM strategies and the "phase shifted" multicarrier PWM.

Amplitude shifted multicarrier PWM presents three alternative PWM strategies with the identical peak-to-peak amplitude and different phase relationships between the carriers, which are:

- Phase Disposition (PD) (Figure 2a), where all carriers are in phase;
- Phase Opposition Disposition (POD) (Figure 2b), where the carriers above the reference zero point have a difference of phase respect those below the zero point of π;
- Alternative Phase Opposition Disposition (APOD) (Figure 2c), where each carrier is phase shifted by π from its adjacent carriers.

The carrier number n_c of the level shifted multicarrier PWM in function of the number of the converter level n_L, is equal to:

$$n_c = n_L - 1 \tag{2}$$

These strategies lead to elimination of all carriers and related sideband harmonics up to the switching frequency.

Phase shifted multicarrier PWM strategy, shown in Figure 2d, is an extension of the unipolar PWM for traditional single-phase two-level inverter. For this technique, the modulation of the H-bridge inverters in each phase leg is modular. Thus, the reference waveforms for the two-phase legs inverter are phase shifted by π. The number of the carrier signals is equal to n_{HB} while the phase shifted optimum (PSO) to obtain harmonic cancellation, is achieved:

$$PSO = \frac{(i-1)\pi}{n_{HB}} \tag{3}$$

where i is the ith H-Bridge series connected per phase. For a five-level inverter two carrier signals with mutual phase shift equal to $\pi/2$, Figure 2b, are necessary. This scheme leads to elimination of all carriers and associated sideband harmonics up to the $2n_{HB}$ times of the switching frequency.

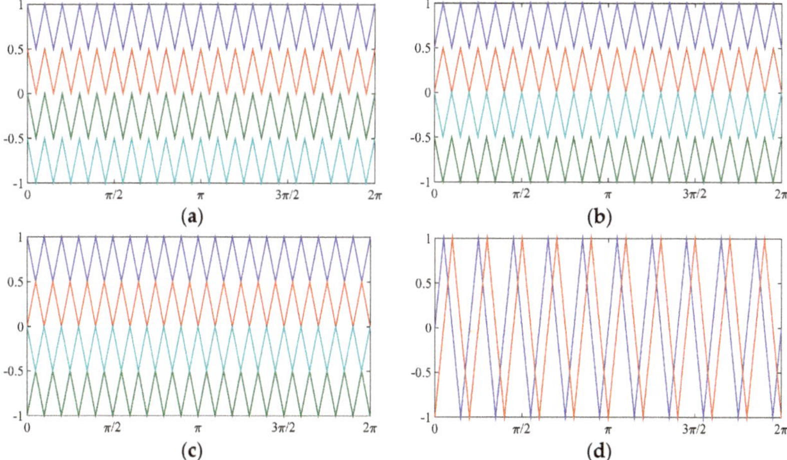

Figure 2. Multicarrier strategies for five-level converter: (**a**) Phase Disposition (PD); (**b**) Phase Opposition Disposition (POD); (**c**) Alternative Phase Opposition Disposition (APOD); (**d**) Phase Shifted (PS). For the first three modulation techniques, four carriers are required, for example in PD technique blue and red lines enable the voltage control of higher bridge, green and cyan the lower bridge. In PS each only two carriers are required since each phase leg has a modular control.

About the reference signals, there are three alternative:

1. Sinusoidal reference;
2. Third harmonic injection (THI);
3. Switching frequency optimal (SFO).

The THI allows overcoming the limit of the three-phase inverters about the reduction of the maximum peak fundamental line voltage of $\sqrt{3}V_{DC}/2$ (86.60% of V_{DC}). Modulation index can be increased by including a common mode third-harmonic term into the reference signal of each phase leg, as shown in Figure 3a (green curve).

This third-harmonic component does not effect on the fundamental line-to-line voltage because the common mode voltages cancel between the phase legs. According to [9], the optimum third-harmonic injection component must have a magnitude of 25% of the fundamental reference. In this way, it is possible to obtain an increasing of the modulation index up to 1.12 and a maximum value of the peak fundamental line-to-line voltage equal to 97% of V_{DC}.

Figure 3b shows the SFO signal (green curve) for a phase of the converter. As demonstrated in [9], SFO is a space vector equivalent reference voltage that can be used in PWM modulation to produce output voltages with the same average low-frequency content.

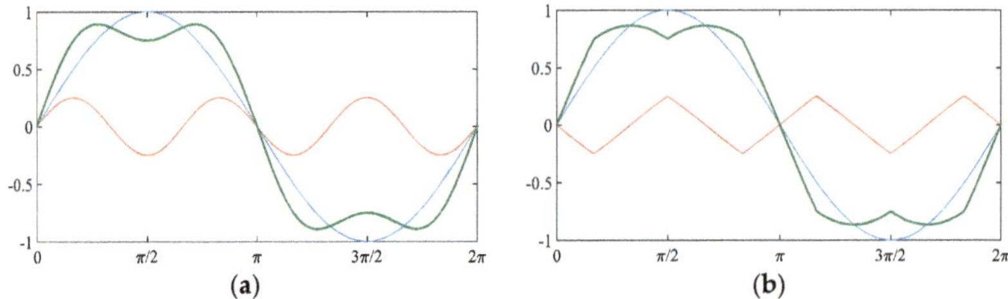

Figure 3. Reference signal for a phase of the converter: (**a**) third harmonic injection (THI); (**b**) switching frequency optimal (SFO). Blue waveform represents the fundamental, orange is the adjustment signal and blue is the modified reference.

Like the THI, a sinusoidal reference and the three-times-fundamental-frequency triangular reference, called "voltage offset" v_{offset}, compose SFO reference signal. The mathematical expression of the SFO signal for the three-phase system is:

$$
\begin{aligned}
v_a^*(t) &= v_a(t) - v_{offset} \\
v_b^*(t) &= v_b(t) - v_{offset} \\
v_c^*(t) &= v_c(t) - v_{offset}
\end{aligned}
\tag{4}
$$

where $v_a(t)$, $v_b(t)$ and $v_c(t)$ are the sinusoidal reference that can be expressed in function of the modulation index M as (5):

$$
\begin{aligned}
v_a(t) &= M\sin(\omega t) \\
v_b(t) &= M\sin\left(\omega t - \tfrac{2\pi}{3}\right) \\
v_c(t) &= M\sin\left(\omega t - \tfrac{4\pi}{3}\right)
\end{aligned}
\tag{5}
$$

The voltage offset v_{offset} can be expressed as (6):

$$
v_{offset} = \frac{\max(v_a, v_b, v_c) + \min(v_a, v_b, v_c)}{2}
\tag{6}
$$

The arrangement between carrier signals and modulating references produces twelve modulation techniques, graphically summarized in Figures 4–6.

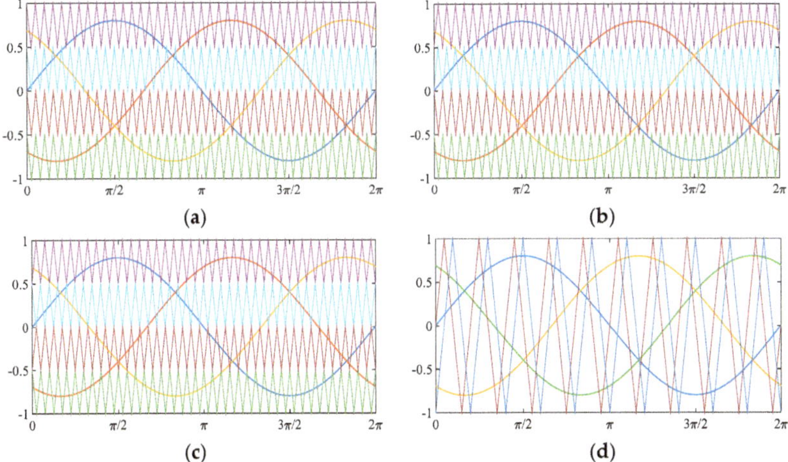

Figure 4. Proposed modulation techniques with sinusoidal reference: (**a**) Phase Disposition (PD); (**b**) Phase Opposition Disposition (POD); (**c**) Alternative Phase Opposition Disposition (APOD); and (**d**) Phase Shifted (PS). Blue, red and orange sinusoidal signals represent the reference signals; interferences with the triangular signals generate the modulation angles for the four switches.

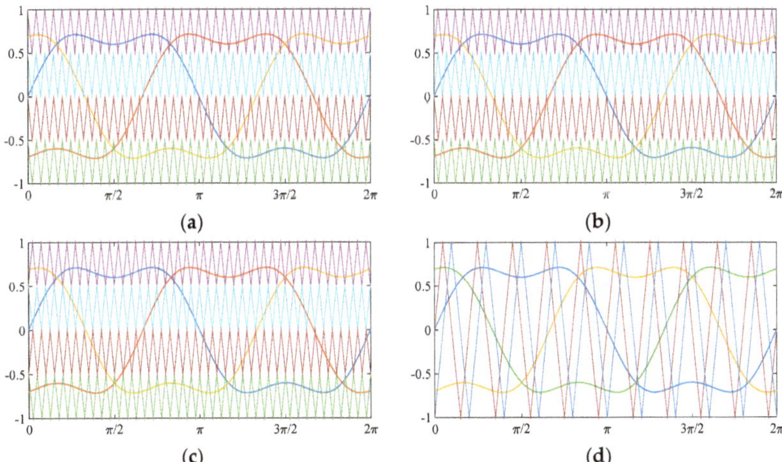

Figure 5. Proposed modulation techniques with THI reference: (**a**) Phase Disposition PD; (**b**) Phase Opposition Disposition POD; (**c**) Alternative Phase Opposition Disposition APOD; and (**d**) Phase Shifted PS. Blue, red and orange THI signals represent the reference signals; interferences with the triangular signals generate the modulation angles for the four switches.

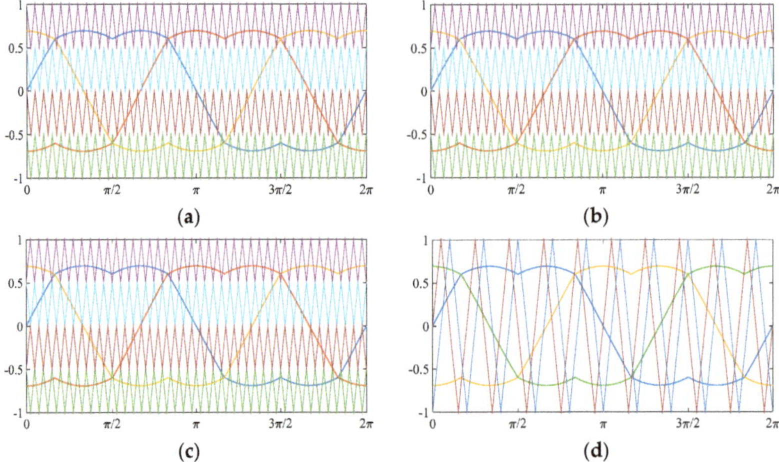

Figure 6. Proposed modulation techniques with SFO reference: (**a**) Phase Disposition (PD); (**b**) Phase Opposition Disposition (POD); (**c**) Alternative Phase Opposition Disposition (APOD); and (**d**) Phase Shifted (PS). Blue, red and orange SFO signals represent the reference signals; interferences with the triangular signals generate the modulation angles for the four switches.

An interesting deep discussion on the previous proposed techniques can be found in [10], in which some features of the proposed technique can be found without the control issue and filter design; preliminary simulations of the multicarrier PWM modulation techniques for a three-phase five-level cascaded H-bridge multilevel inverter (CHBMI) were also reported in [11,12]. In these works, the authors addressed that the modulation techniques with sinusoidal reference should present the lower values of the total harmonic distortion rate. A complementary study on the use of B-Spline functions as carrier signals replacing triangular waveforms, can be found in [13,14]. This study confirms that the traditional triangle waveforms as carrier signals are solutions that allow best performances.

1.2. Digital Control Boards for Power Converters

The fast technological growth of the electronic design automation (EDA) and the very large scale integration (VLSI) has significantly contributed to the development of programmable digital systems with high performances both in terms of execution time and compactness for the realization of control systems. In addition, the recent advances of software for the implementation, simulation and validation of digital systems, dedicated to the control of specific applications, has contributed to simplify and speed-up the overall design process of the digital controller, which represents the core of modern systems for the electrical energy conversion. Figure 7 shows a block diagram of a typical grid-connected system.

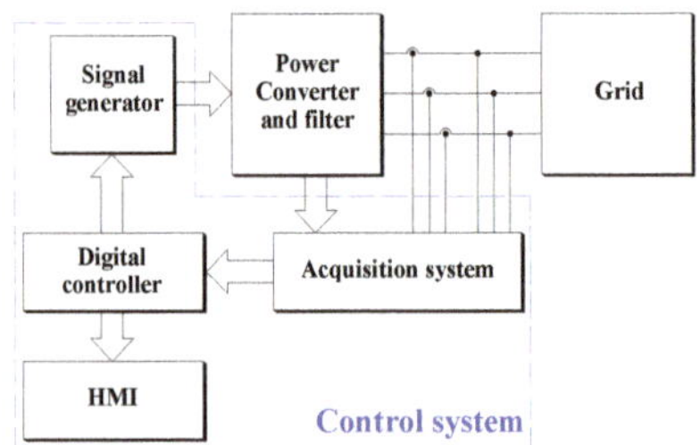

Figure 7. Block diagram of a typical grid-connected system. HMI: human–machine interface.

The block named Power Converter and filter requires a control system supervising the behavior (harmonic reduction, filter efficiency, etc.). Generally, the control system consists of four parts:

1. Acquisition system, which provides signal conditioning and digital acquisition of electrical measures (usually current, voltage, frequency) and other quantities (usually solar irradiation, temperature, etc.);
2. Digital controller, required for algorithms employment (filtering, identification, control, modulation of output signals and others);
3. Human–machine interface (HMI), suitable for setup phase as well as for monitoring functions;
4. Signal generator, allowing conversion of the digital signals to analog signals for the power components.

Digital controller represents the core of the control system, different are the digital controller available in the market. A first example of digital controller is the Microcontrollers or DSP (digital signal processor), allowing the implementation of the control algorithms through a purely software programming (with C or C++). The constructor defines the DSP hardware and it is composed by several peripherals, such as the RAM or the ROM. However, an already designed hardware structure reduces the flexibility in the use of the microcontroller. In fact, a complex issue is related to the time of computing of the control algorithm, due to the fact that all the operation needed for this computing are executed in a sequential routine, causing losses in terms of efficiency of the control system [15].

The FPGA (field programmable gate array) is another example of digital controller, composed by a matrix of configurable logic blocks (CLBs) with completely reprogrammable connections. This fact leads to a higher flexibility with respect to a DSP, allowing the realization of specific hardware structures in dependence of the nature of applications. In addition, this feature allows the realization of a system of logic operations developed in parallel, reducing the time of computing. Thus, by means of an FPGA, high-performance control systems can be realized, even comparable with equivalent controllers composed by analogical components, as reported in [16–18].

For instance, the main advantages provided by the adoption of an FPGA-based system for the current control in AC machine drives is presented in [19]. In particular, the FPGA system significantly reduces the execution time of the control algorithm, increasing, therefore, the performances of the related system.

1.3. Literature Survey

Research in the world of multilevel inverter is very wide. Different reviews can be found in the literature [20–23], regarding topologies, switching frequencies, employment of photovoltaic sources (PV), control and cost of inverters, depending on different factors such number of sources, switches, and connections.

Novel structures are continuously suggested [24,25] and ways to control them [26–28]. Efficiency, dimension, weight, and reliability influence the cost of manufacturing inverters. Nowadays, multilevel inverters reached in efficiency the value of 98% and the achievement of the next 1% increase is a hard challenge, and ever-more efficient and advanced modulation techniques are required, which are embracing two different ways: low switching frequency modulation techniques and high switching frequency PWM.

Two systems coexist. The main advantage of the employment of the low switching frequency modulation technique is the reduction of the switching losses to a minimum assessment and the confining the stress on the power components (less overshoot). Through one period of the fundamental reference, low switching frequencies techniques generally act one or two commutations of the switches, so generating a staircase waveform. However, the output voltage waveform has different low order harmonics, with amplitudes similar to the fundamental one, hard to be filtered.

Different are the works that face the issue of reducing the harmonics distortion rate (THD) [29–35]. Some of them exploit the problematic of extraction maximum power from PV modules [29] or implement innovative fast switching modulation, with or without considering power losses [30,31]. As reference, the authors of [32] used a single-phase multilevel inverter scheme, employing three series connected full-bridge stages and a single half-bridge inverter. The result was a reduction of harmonics, evaluated in terms of total harmonic distortion rate, about 9.85%. Spice models help researchers to consider coupled issues such as THD and storage elements [33]. Again, simulation allows the tracking of the performance of a multilevel inverter in partial shaded condition of PV panels [34]. Simulation seems to be the best way to predict how reduce the THD with grid connected systems [35]. The simulation produces a 15-level output voltage, the total harmonic distortion was about 8.12%, a very low level was reached, but similar multilevel inverters will be overpriced.

On the opposite side, higher frequency modulation presents harmonics with higher frequencies, which requires an economic filters, but high frequency switching brings also higher losses due circulating currents. An evaluation of harmonics content, useful to better define the output voltage waveforms, are reported [10–12], and will be objects of deeper discussions in this paper.

High power electrical drives applications need mainly the reduction of the switching losses and electromagnetic interferences (EMI). Losses concur to define efficiency of converter, so soft switching modulation techniques, employing selective harmonic elimination (SHE) technique and selective harmonic mitigation (SHM) technique, are frequently chosen. SHE method requires the choosing of a single h^{th} harmonic to be removed, so a set of non-linear sinusoidal formulas is solved by choosing the switching angles. The SHM techniques, instead, mitigates simultaneously different harmonics by correctly choosing the switching angles.

Whichever technique is employed, SHE or SHM, to resolve the group of transcendental equations and to discover the related switching angles, different approaches can be taken into account. Obviously, the simplest approach develops iterative methods such as Newton–Rhapson. As reference in [36], the Newton–Rhapson iterative method is employed for the assessment of switching angles for a seven level inverter. The total harmonic distortion of the staircase voltage output is equal to 11.8%.

The authors of [37] present an evaluation between different modulation techniques, applied for a five-level cascaded H-bridge multilevel inverter. The employed control scheme enforces three different pre-defined arrangements for the switching angles. By the uses of these schemes there is an achievement of a minimum THD around 17.07% for the waveform of voltage. The work presented in [38], employs a particularly fast optimal solution of harmonic elimination techniques, used inverter is a five-level multilevel one, and also non-equal DC sources feed the different levels. The solution of the problem is entrusted to a novel particle swarm optimization (PSO) algorithm, and THD achieved a minimum value of 5.44%.

Authors of [39] proposed an optimal SHM technique for a seven-level inverter scheme. The individual harmonic to be mitigated and the THD are subjected to satisfaction of three voltage harmonic standards, named EN50160 [40], CIGRE JWG C4.07 [41] and IEC61000-3-6 [42].

Paper [43] proposes again the PSO technique in presence of PV sources with different voltage levels, the non-linear transcendental equations were solved offline. THD was minimized by employing of pre-calculated switching angles. By the dual employment of the adaptive neuro fuzzy inference system (ANFIS) and also maximum power point tracker (MPPT) algorithm, PV DC sources were transformed in identical DC source. The resulting THD was around 3.7%, less than the ones recommended by IEEE-519 (5%).

Author of [44] again employed an adaptive neuro-fuzzy interference system in order to eliminate voltage harmonics. The proposed comparison shows a best performance of ANFIS referred to neuro fuzzy controller (NFC), in the case of studying a seventh level inverter with active filter. Again the option of an active filter, used in order to improve the performance of the control, can be find in [45].

In [46] the method known as "voltage cancelation" is used for single-phase H-bridge inverters, and in [47] was applied for a single-phase five-level CHBMI.

In this paper the structures used in [47] with a soft switching modulation is used with high switching PWM in order to achieve the same or a better THD.

1.4. Contributions and the Organization of Paper

The purpose of this article is to define in detail all the information required to implement the modeling, the operation and the control of a CHBMI.

By starting from general information on CHBMI a detailed report is presented. Informations for the reproduction of the results are not omitted, different tables report the used parameters, from the impedances used in the filtering systems for the various PWM employed techniques (twelve different cases) to the delay times used for driving the system.

Particular attention is devoted to the following aspects: LCL filter design, control design, harmonic content and validation of the proposed approach.

Section 2 is devoted to the simulation of a CHBMI for grid connected applications. The instantaneous model of the converter will be presented in Section 2.1 and after a CHBMI average model. The LCL filter design will be introduced in Section 2.2 and the controller design in Section 2.3. Finally, the simulated performances are evaluated in Section 2.4.

Section 3 is devoted to the experimental validation. The Test Bench will be described in Section 3.1. The control algorithm design will be presented in Section 3.2. The model validation will be discussed in Section 3.3 and finally the grid connected application will be introduced in Section 3.4.

Section 4 recalls and discusses the obtained results, and finally Section 5 concludes the paper.

2. Cascaded H-Bridge Inverter for Grid Connected Applications: Modelling and Control

The purpose of this section is to provide all the useful information to realize a virtual model of a CHBMI.

2.1. Mathematical Model of the System

Among the classic structures of multilevel inverters presented in literature, this work considers the three-phase, five-level Cascaded H-Bridge inverter topology. Figure 8 shows the topology structure of a three-phase five-level CHB inverter connected to the grid through a LCL filter and a transformer (used for boost voltage and security purpose).

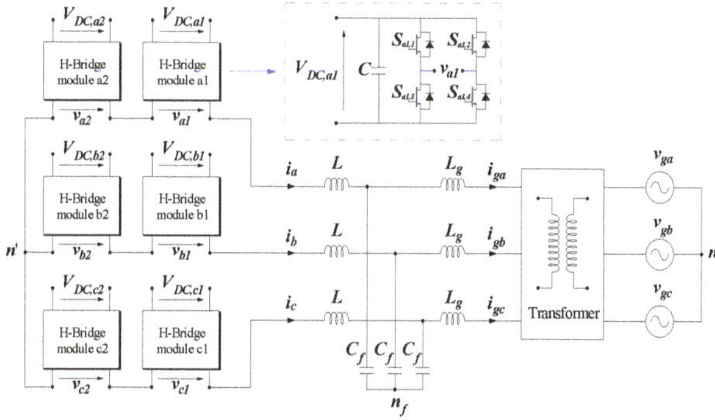

Figure 8. Topology structure of a three-phase five-level CHB inverter system.

Each phase of the CHB inverter consists of two cascaded H-bridges in series connected. Thus, the phase voltages $v_a(t)$, $v_b(t)$ and $v_c(t)$, referred on the n' point, is obtained by summing output voltage of the series connected H-Bridges (7):

$$
\begin{aligned}
v_a(t) &= v_{a1}(t) + v_{a2}(t) \\
v_b(t) &= v_{b1}(t) + v_{b2}(t) \\
v_c(t) &= v_{c1}(t) + v_{c2}(t)
\end{aligned}
\tag{7}
$$

Taking into account the switching state of the power components, the instantaneous model of the three-phase five-level CHBMI in both the AC and DC side can be totally described by the following equations:

$$
\begin{aligned}
v_a(t) &= V_{DC,a1} \cdot (S_{a1,1} - S_{a1,2}) + V_{DC,a2} \cdot (S_{a2,1} - S_{a2,2}) \\
v_b(t) &= V_{DC,b1} \cdot (S_{b1,1} - S_{b1,2}) + V_{DC,b2} \cdot (S_{b2,1} - S_{b2,2}) \\
v_c(t) &= V_{DC,c1} \cdot (S_{c1,1} - S_{c1,2}) + V_{DC,c2} \cdot (S_{c2,1} - S_{c2,2})
\end{aligned}
\tag{8}
$$

$$
\begin{aligned}
i_{DC,a1} &= i_a \cdot (S_{a1,1} - S_{a1,2}) & i_{DC,a2} &= i_a \cdot (S_{a2,1} - S_{a2,2}) \\
i_{DC,b1} &= i_b \cdot (S_{b1,1} - S_{b1,2}) & i_{DC,b2} &= i_b \cdot (S_{b2,1} - S_{b2,2}) \\
i_{DC,c1} &= i_c \cdot (S_{c1,1} - S_{c1,2}) & i_{DC,c2} &= i_c \cdot (S_{c2,1} - S_{c2,2})
\end{aligned}
\tag{9}
$$

$$
\begin{aligned}
C\frac{dV_{DC,a1}}{dt} &= i_{in,a1} - i_{DC,a1} & C\frac{dV_{DC,a2}}{dt} &= i_{in,a2} - i_{DC,a2} \\
C\frac{dV_{DC,b1}}{dt} &= i_{in,b1} - i_{DC,b1} & C\frac{dV_{DC,b2}}{dt} &= i_{in,b2} - i_{DC,b2} \\
C\frac{dV_{DC,c1}}{dt} &= i_{in,c1} - i_{DC,c1} & C\frac{dV_{DC,c2}}{dt} &= i_{in,c2} - i_{DC,c2}
\end{aligned}
\tag{10}
$$

where $S_{ji,k}$ ($j = a \ldots c$; $i,k = 1$ or 2) are the switching state in which "1" represents that the switch is ON and "0" represents that the switch is OFF.

Equations (8) and (9), can be simplified in (12) and (13) by considering the same DC voltage V_{DC} for each H-Bridges and by defining the switching functions $S_{ji} \in \{-1, 0, 1\}$ as:

$$
S_{ji} = S_{ji,k} - S_{ji,k+1}
\tag{11}
$$

Thus, Equations (14) and (15) can be rewritten as:

$$
\begin{aligned}
v_a(t) &= V_{DC} \cdot (S_{a1} + S_{a2}) \\
v_b(t) &= V_{DC} \cdot (S_{b1} + S_{b2}) \\
v_c(t) &= V_{DC} \cdot (S_{c1} + S_{c2})
\end{aligned}
\tag{12}
$$

$$
\begin{aligned}
i_{DC,a1} &= i_a \cdot S_{a1} & i_{DC,a2} &= i_a \cdot S_{a2} \\
i_{DC,b1} &= i_b \cdot S_{b1} & i_{DC,b2} &= i_b \cdot S_{b2} \\
i_{DC,c1} &= i_c \cdot S_{c1} & i_{DC,c2} &= i_c \cdot S_{c2}
\end{aligned}
\tag{13}
$$

Equations (12) and (13) represent the instantaneous model of the three-phase five-level CHBMI in terms of the phase voltage and DC current. The line-to-line voltage as described by Equation (14):

$$
\begin{aligned}
v_{ab}(t) &= V_{DC} \cdot (S_{a1} + S_{a2} - S_{b1} - S_{b2}) \\
v_{bc}(t) &= V_{DC} \cdot (S_{b1} + S_{b2} - S_{c1} - S_{c2}) \\
v_{ca}(t) &= V_{DC} \cdot (S_{c1} + S_{c2} - S_{a1} - S_{a2})
\end{aligned}
\tag{14}
$$

The model is complete with the equations to describe the LCL filter behavior and the transformer. Thus, the equivalent circuit have to be taken in to account where r_{TR} and L_{TR} ($r_{TR} = 25$ mΩ and $L_{TR} = 108.23$ µH) represent the short-circuit impedance reported on the low side of the transformer.

Finally, the following equations can be used for the rating the output currents $i_{\{a,b,c\}}$ of the converter (15), grid currents $i_{g\{a,b,c\}}$ (16) and the capacitor voltages $v^*_{\{a,b,c\}}$ (17), where r and r_g are the resistance of the inductance L and L_g of LCL filter.

$$
\begin{aligned}
L\frac{di_a}{dt} &= v_a - i_a r - v_a^* - v_{n_f n'} \\
L\frac{di_b}{dt} &= v_b - i_b r - v_b^* - v_{n_f n'} \\
L\frac{di_c}{dt} &= v_c - i_c r - v_c^* - v_{n_f n'}
\end{aligned}
\tag{15}
$$

$$\left(L_g + L_{TR}\right) \cdot \frac{di_{ga}}{dt} = v_a^* - i_{ga} \cdot \left(r_g + r_{TR}\right) - v_{ga} - v_{nn_f}$$
$$\left(L_g + L_{TR}\right) \cdot \frac{di_{gb}}{dt} = v_b^* - i_{gb} \cdot \left(r_g + r_{TR}\right) - v_{gb} - v_{nn_f} \tag{16}$$
$$\left(L_g + L_{TR}\right) \cdot \frac{di_{gc}}{dt} = v_c^* - i_{gc} \cdot \left(r_g + r_{TR}\right) - v_{gc} - v_{nn_f}$$

$$C_f \frac{dv_a^*}{dt} = i_a - i_{ga}$$
$$C_f \frac{dv_b^*}{dt} = i_b - i_{gb} \tag{17}$$
$$C_f \frac{dv_c^*}{dt} = i_c - i_{gc}$$

where $v_{n_f n'}$ is the voltage between the point n_f and n' and v_{nn_f} is the voltage between the point n and n_f that can be expressed by the Equation (18).

$$v_{n_f n'} = \tfrac{1}{3}\left[\left(v_a + v_b + v_c\right) - r\left(i_a + i_b + i_c\right) - L\left(\tfrac{di_a}{dt} + \tfrac{di_b}{dt} + \tfrac{di_c}{dt}\right)\right]$$
$$v_{nn_f} = \tfrac{1}{3}\left[\left(v_a^* + v_b^* + v_c^*\right) - r_g\left(i_{ga} + i_{gb} + i_{gc}\right) - L_g\left(\tfrac{di_{ga}}{dt} + \tfrac{di_{gb}}{dt} + \tfrac{di_{gc}}{dt}\right)\right] \tag{18}$$

In the case of the equilibration system, the voltage v_{nn_f} is equal to zero. In order to design the control system, it is necessary to develop an average model of the system. Generally, the average model takes into account the average values in a switching period T_{sw}. In this way, the average phase voltages and average currents can be expressed as (19) and (20):

$$\bar{v}_a(t) = \tfrac{1}{T_{sw}} \int_0^{T_{sw}} V_{DC} \cdot (S_{a1} + S_{a2}) dt$$
$$\bar{v}_b(t) = \tfrac{1}{T_{sw}} \int_0^{T_{sw}} V_{DC} \cdot (S_{b1} + S_{b2}) dt \tag{19}$$
$$\bar{v}_c(t) = \tfrac{1}{T_{sw}} \int_0^{T_{sw}} V_{DC} \cdot (S_{c1} + S_{c2}) dt$$

$$\bar{i}_{DC,a1} = \tfrac{1}{T_{sw}} \int_0^{T_{sw}} i_a \cdot S_{a1} dt \quad \bar{i}_{DC,a2} = \tfrac{1}{T_{sw}} \int_0^{T_{sw}} i_a \cdot S_{a2} dt$$
$$\bar{i}_{DC,b1} = \tfrac{1}{T_{sw}} \int_0^{T_{sw}} i_b \cdot S_{b1} dt \quad \bar{i}_{DC,b2} = \tfrac{1}{T_{sw}} \int_0^{T_{sw}} i_b \cdot S_{b2} dt \tag{20}$$
$$\bar{i}_{DC,c1} = \tfrac{1}{T_{sw}} \int_0^{T_{sw}} i_c \cdot S_{c1} dt \quad \bar{i}_{DC,c2} = \tfrac{1}{T_{sw}} \int_0^{T_{sw}} i_c \cdot S_{c2} dt$$

As demonstrated in [46], the equations of the average voltages and average currents in a switching period become (21) and (22):

$$\bar{v}_a(t) = V_{DC} \cdot (m_{a1} + m_{a2})$$
$$\bar{v}_b(t) = V_{DC} \cdot (m_{b1} + m_{b2}) \tag{21}$$
$$\bar{v}_c(t) = V_{DC} \cdot (m_{c1} + m_{c2})$$

$$\bar{i}_{DC,a1} = i_a \cdot m_{a1} \quad \bar{i}_{DC,a2} = i_a \cdot m_{a2}$$
$$\bar{i}_{DC,b1} = i_b \cdot m_{b1} \quad \bar{i}_{DC,b2} = i_b \cdot m_{b2} \tag{22}$$
$$\bar{i}_{DC,c1} = i_c \cdot m_{c1} \quad \bar{i}_{DC,c2} = i_c \cdot m_{c2}$$

where m_{ji} ($j = a \ldots c$; $i = 1$ or 2) is the modulation index of each H-Bridge module. Equations (21) and (22) represent the *average model of the converter* in a switching period.

2.2. LCL Filter Design

The power quality is an important aspect for grid-connected systems. According to [48,49], there are several limits related to the power injection on the grid that have to be respected, as: voltage unbalance (three-phase inverters should not exceed 3%), DC current injection ($I_{DC} < 0.5\%$ for IEEE 1574 and $I_{DC} < 1\%$ for IEC 61727) and current harmonics. The standard harmonic current limits, defined by IEEE 1574 and IEC 61727 at the point of common coupling (PCC), are summarized in Table 1. Therefore, the level fixed of the total harmonic distortion (THD%) is <5%.

Table 1. Current harmonic limits.

Harmonic Order, h	Limit in % of Rated Current
$h < 11$	4.0
$11 \leq h < 17$	2.0
$17 \leq h < 23$	1.5
$23 \leq h < 35$	0.6
$h \geq 35$	0.3

On the grid side, a LCL-filter to reduce high-order harmonics that can interfere with other equipment is typically adopted [50]. A step-by-step design procedure for an LCL filter has been proposed in [51,52]. The proposed method employs different factors as inputs such the power rating of the converter, the chosen line frequency and obviously the switching frequency.

According to [52,53], the converter side inductance L is defined in order to bound the current ripple produced by the converter. Grid side inductance L^{*}_{g} can be determined as a function of L, using the index r ($L^{*}_{g} = r \cdot L$). While, Capacitor value C_f can be determined as a percentage $x\%$ of the delivered reactive power under rated conditions.

Aim of this section is to design the LCL filter parameters with the step-by-step method for each modulation techniques taken into account. The step-by-step procedure was applied by considering a system with a line voltage of 50 V_{rms}, frequency 50 Hz and rated power of 600 W (parameters of the test bench in Laboratory of Electrical Applications-LEAP of the University of Palermo).

In Table 2 are reported the converter side inductance L values for each modulation techniques taken into account. It should be noted that the lower values of L have been obtained with phase disposition based modulation techniques. Moreover, also phase shifted modulation techniques based present interesting results.

Table 2. Grid side inductance values.

	PD	POD	APOD	PS
Sine	0.260 mH	0.882 mH	0.530 mH	0.371 mH
THI	0.222 mH	1.938 mH	0.584 mH	0.393 mH
SFO	0.260 mH	0.882 mH	0.530 mH	0.371 mH

Addition, in the design of the grid side inductance L^{*}_{g} has been taken into account the inductance of the transformer, thus, can be expressed as $L^{*}_{g} = L_g + L_{TR}$. Table 3 reports the current ripple attenuation depending of the r and x values.

Table 3. Current Ripple, r and x values.

	Sine			THI			SFO		
	r	$x\%$	i_g/i	r	$x\%$	i_g/i	r	$x\%$	i_g/i
PD	1.40	1.94%	10.09%	1.40	2.38%	10.29%	1.20	2.38%	10.03%
POD	0.40	1.94%	10.21%	0.10	3.25%	10.85%	0.40	1.94%	10.21%
APOD	0.90	1.50%	10.81%	0.60	1.94%	10.51%	0.90	1.50%	10.18%
PS	1.00	1.94%	10.21%	1.20	1.50%	10.65%	1.00	1.94%	10.21%

Thus, fixing different values of x and r, in particular $x\%$ less than 5% according to limit reported in [52], have been evaluated the current ripple for each modulation techniques taken into account. In this way, have been identified the values of r and $x\%$, summarized in Table 3, that allow to obtain a current ripple approximately then 10%. By using the values reported in Table 3, have been calculated the filter parameters and are summarized in Table 4.

Table 4. LCL filters parameters and frequency resonant.

	L (mH)	C_f (μF)	L^*_g (mH)	L_g (mH)
SPD	0.260	8.04	0.365	0.257
SPOD	0.882	8.04	0.352	0.244
SAPOD	0.530	6.21	0.477	0.369
SPS	0.371	8.04	0.371	0.263
THIPD	0.222	9.86	0.311	0.203
THIPOD	1.938	13.47	0.193	0.085
THIAPOD	0.584	8.04	0.350	0.242
THIPS	0.393	6.21	0.471	0.363
SFOPD	0.260	9.86	0.313	0.204
SFOPOD	0.882	8.04	0.352	0.244
SFOAPOD	0.530	6.21	0.424	0.316
SFOPS	0.371	8.04	0.371	0.263

An interesting consideration is about the PD based modulation techniques because the converter side inductance L and grid side inductance L_g present the lower values. Higher value of the converter side inductance were obtained with THIPOD modulation technique. This phenomenon is attributable at the higher number of the side band harmonics generated by POD carrier signals. In fact, also SPOD and SFOPOD present higher values of the converter side inductance compared others modulation techniques. Regarding the capacitor filter values C_f, it is interesting to note that have been obtained similar values among the modulation techniques taken into account.

As stated earlier, it is necessary to evaluate the limits on the parameter values, introduced in step-by-step method, to verify the filter effectiveness. In Table 5 are reported the limits on the LCL parameters values for each modulation techniques.

Table 5. Limits on the LCL parameter values.

	ΣL (p.u.)	x (%)	f_{res} (kHz)
SPD	0.025	1.94%	4.54
SPOD	0.050	1.94%	3.53
SAPOD	0.041	1.50%	4.02
SPS	0.030	1.94%	4.11
THIPD	0.021	2.38%	4.44
THIPOD	0.087	3.25%	3.26
THIAPOD	0.038	1.94%	3.79
THIPS	0.035	1.50%	4.35
SFOPD	0.023	2.38%	4.24
SFOPOD	0.050	1.94%	3.53
SFOAPOD	0.039	1.50%	4.15
SFOPS	0.030	1.94%	4.11

Each LCL configuration designed allows to respect the limits introduced in step-by-step method. In the next section, is reported the design of the control system for each modulation techniques taken into account.

2.3. Controller Design

The objective of the control strategy is to guarantee the synchronization with the main grid and regulate the power injection through the current control loop. Moreover, the current loop is accountable of the power quality issues and of protection for high values of current. Thus, low harmonic content on the current and dynamic response are the important properties of the control system.

According to [54–56], the control strategy is based on the synchronous reference frame, also called *dq* control. Figure 9 shows the schematic block diagram of the *dq* control.

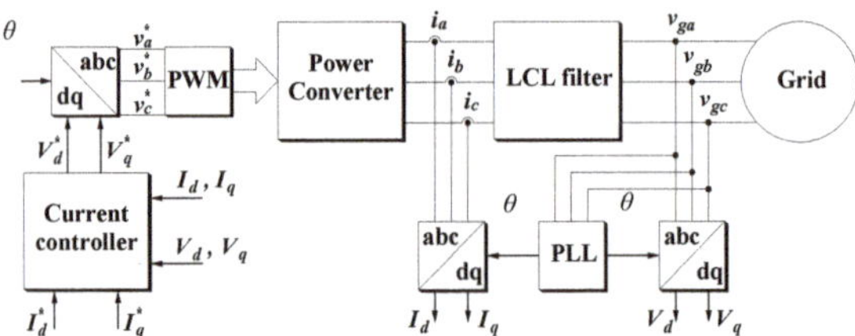

Figure 9. Synchronous reference frame control strategy.

The system needs the measurement of the current and voltage through the sensors. Figure 9 shows that the current sensors are on the converter side, since in more applications they are also employed to protect the power converter. Moreover, the LCL filter design and the control design are influenced by the position of the sensors [51].

Synchronous reference frame control is based on the Park's transformation to express both grid currents and voltages into a reference structure rotating synchronously with the grid frequency. In this way, the *dq* components of the voltage and current assume continuous trend and it is possible to use the PI regulator.

The phase-locked loop (PLL) technique [55], allows extracting the instantaneous phase angle of the grid voltage in order to synchronize the voltage waveforms.

For the control design, the instantaneous model of the system, Equations (15) and (16), can be simplified neglecting the filter capacitor C_f.

Using the Park's transformation, the Equations (15) and (16) can be rewritten as (23):

$$
\begin{aligned}
v_d(t) &= r_T i_d(t) + L_T \frac{di_d(t)}{dt} - \omega_g L_T i_q(t) + v_{gd}(t) \\
v_q(t) &= r_T i_q(t) + L_T \frac{di_q(t)}{dt} + \omega_g L_T i_d(t) + v_{gq}(t)
\end{aligned}
\tag{23}
$$

where r_T is the sum of the internal resistance of the inductors ($r_T = r + r_g + r_{TR}$) and the L_T is the sum of the inductance of the LCL filter and the short-circuit inductance reported on the low side of the transformer ($L_T = L + L_g + L_{TR}$).

In synchronous reference frame, it is possible to control the *dq* components independently thanks the decoupling of the two-channel control. In this way, through the *d* component it is possible to control the active power while through the *q* component it is possible to control the reactive power.

For the design of the proportional-integral (PI) regulators, the method used to tune the parameters of the PI is the "technical optimum" criterion where both plant and PI regulator have the same time constant in order to simplify the closed loop transfer function.

By according the PI integrator time constant T_I equal to the plant time constant $T_I = L_T/r_T$, with the aim to delete the slower plant pole and supposing a perfect pole-zero cancellation, the current closed-loop transfer function $W(s)$ become of the second order Equation (24).

$$
W(s) = \frac{\frac{k_p}{1.5 L_T T}}{s^2 + \frac{1}{1.5 T} s + \frac{k_p}{1.5 L_T T}}
\tag{24}
$$

where k_p is the proportional gain, T_I is the integral time constant and T is the sampling period.

By the analysis of Equation (24), the proportional gain k_p depends on the inductance of the LCL filter. In this way, it is possible to evaluate the PI regulators parameters for each modulation techniques choosing a damping coefficient equal to 0.707. In Table 6 are reported the PI regulator parameters for each modulation techniques taken into account.

Table 6. PI regulator parameters.

	Sine			THI			SFO		
	k_p	T_I (ms)	k_i	k_p	T_I (ms)	k_i	k_p	T_I (ms)	k_i
PD	2.09	15.65	133.73	1.78	13.36	133.73	1.91	14.35	133.73
POD	4.12	30.88	133.73	7.11	53.31	133.73	4.11	30.88	133.73
APOD	3.36	25.20	133.73	3.12	23.36	133.73	3.18	23.87	133.73
PS	2.48	18.58	133.73	2.88	21.63	133.73	2.48	18.58	133.73

In the next section, simulation results were reported to evaluate the performance of the system with each modulation techniques taken into account.

2.4. Performances Evaluation

Aim of this section is to evaluate the performance of the system through a simulation analysis for each modulation techniques taken into account. In particular, the main purpose is to investigate the effectiveness of the control strategy and the LCL filter in terms of the harmonic content in the currents and voltages in order to determinate the best solution for grid connected applications. The simulation have been carried out in Matlab/Simulink® (version 4.1.1, The MathWorks, Inc., Natick, MA, USA) environment with the same parameters used for each modulation techniques taken into account and reported in Table 7.

Table 7. Simulation parameters.

Electric parameter	Value
Grid line Voltage	50 V
Rated current	6 A
Grid frequency	50 Hz
DC Voltages	24 V
Switching frequency	10 kHz
Inductance and resistance of the transformer (low side reported)	108.23 µH 25 mΩ

LCL filter requirements and parameters of regulators were determined in the precedent sub-sections (2.2. LCL Filter Design and 2.3. Controller Design). In the follow are reported the results obtained in simulation analysis for each modulation techniques taken into account and have been compared the results among the modulation with the same carrier signals.

2.4.1. Phase Disposition

Generally, the main characteristics of the modulation techniques with PD as carrier signals is that the harmonic spectra of the phase voltage presents a predominant harmonic centered on the switching frequency and side band harmonics. Thus, the difference among SPD, THIPD, and SFOPD are in the amplitude on the harmonic centered on the switching frequency and side band harmonics as shown in Figure 10. However, THI and SFO introduce a third harmonic component on the phase voltage that disappear on the line voltage (three-wired systems).

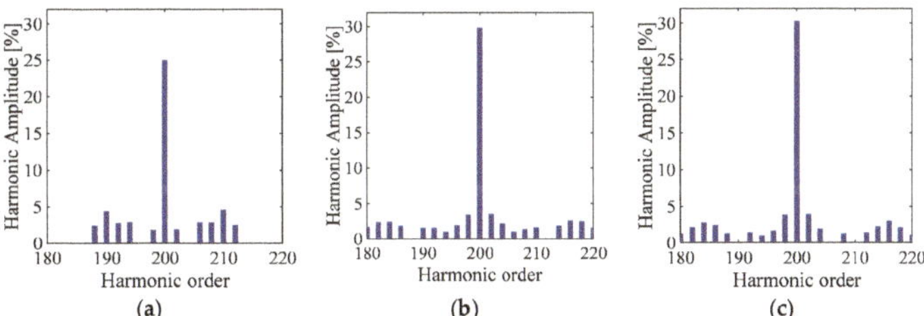

Figure 10. Phase voltage harmonic spectra centered around the switching frequency (10kHz) in percent respect to the fundamental amplitude of (**a**) SPD, (**b**) THIPD and (**c**) SFOPD.

This phenomenon explains the different values of THD% and different LCL filter requirements. Consequently, the performance in terms of the harmonic content on the current will be different. Figures 11–13 show the transitory of the converter side currents and grid side currents from zero to the rated current obtained with SPD, THIPD, and SFOPD, respectively. As mentioned above, are visible little differences in the currents trend among SPD, THIPD, and SFOPD, as emphasized by red and green zoom windows. In particular, these differences are present in terms of harmonic content around the switching frequency and the low order harmonics.

Overall, the three-grid side current have a total harmonic distortion—THD% less then 5% according to IEEE 1574 and IEC 61727.

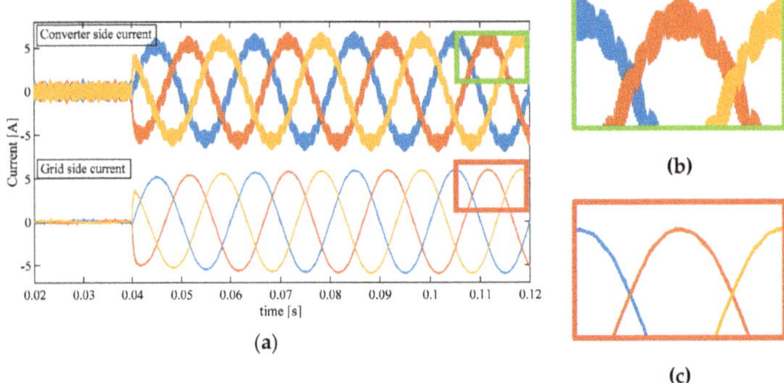

Figure 11. Converter side and grid side three-phase current with SPD. (**a**) Transient behavior in multiple cycles; (**b**) Ripple magnification for converter side current; (**c**) Ripple magnification for grid side current.

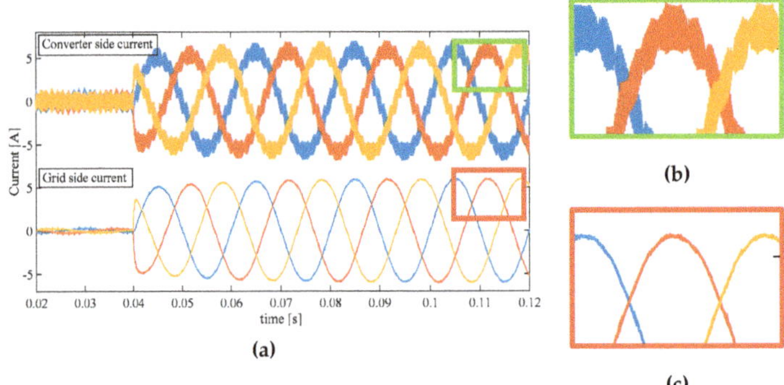

Figure 12. Converter side and grid side three-phase current with THIPD. (**a**) Transient behavior in multiple cycles; (**b**) Ripple magnification for converter side current; (**c**) Ripple magnification for grid side current.

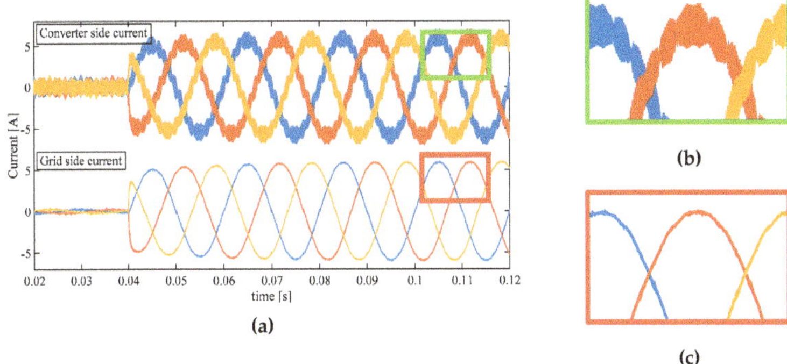

Figure 13. Converter side and grid side three-phase current with SFOPD. (**a**) Transient behavior in multiple cycles; (**b**) Ripple magnification for converter side current; (**c**) Ripple magnification for grid side current.

Figure 14 shows the low order harmonics (from third to fortieth harmonic) in grid side current I_{ga} obtained with SPD, THIPD, and SFOPD, respectively.

In the first all, it is interesting to note that the amplitude of the lower order harmonics are below of the standard harmonic current limits defined by IEEE 1574 and IEC 61727 at the PCC. However, is visible only a predominant fifth harmonic in low order spectra obtained with SPD while there are also the eleventh and thirteenth harmonics in the low order spectra obtained with THIPD and SFOPD.

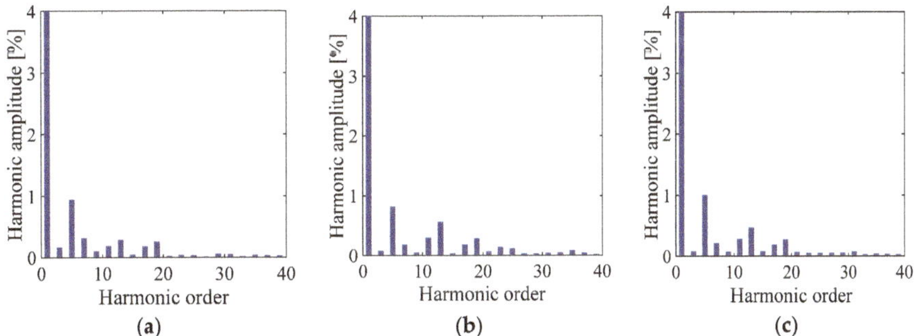

Figure 14. Low order harmonics on the grid side current I_{ga} for (**a**) SPD, (**b**) THIPD, and (**c**) SFOPD.

Figure 15 shows a comparison among harmonic spectra centered around the switching frequency of 10 kHz among line voltage V_{ab} (blue bars), converter side current I_a (red bars) and grid side current I_{ga} (yellow bars) obtained with SPD, THIPD, and SFOPD, respectively. The lower values of the grid side current harmonics, less of 0.3% referred to the fundamental amplitude, demonstrate the LCL filter effectiveness. It should be noted that in the spectra of the line voltage and currents appear only side band harmonics thanks to the three-wired connection.

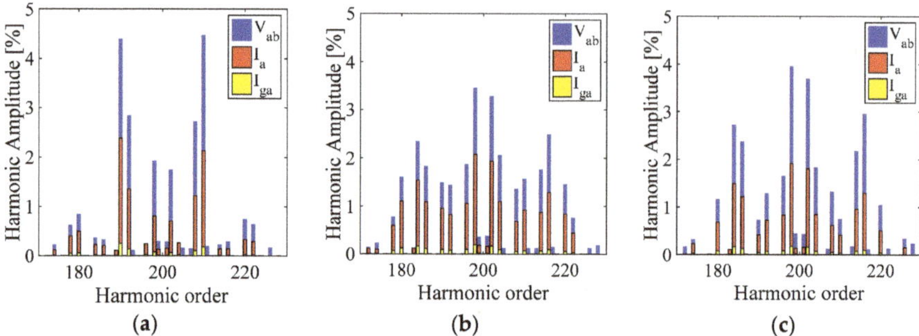

Figure 15. Comparison of line voltage harmonic spectra V_{ab}, converter side current I_a and grid side current I_{ga} centered around the switching frequency (10 kHz) in percent respect to the fundamental amplitude for (**a**) SPD, (**b**) THIPD, and (**c**) SFOPD.

In particular, harmonic spectra obtained with SPD modulation technique presents three-pair predominant of the side band harmonics while the harmonic spectra obtained with THIPD and SFOPD are different. The tool used to compare the harmonic content around the switching frequency among the SPD, THIPD, and SFOPD is the "Partial Total Harmonic Distortion" (*PTHD*%) defined as:

$$PTHD\% = \frac{\sqrt{\sum_{f_{SW}-n/2}^{f_{SW}+n/2} V_h^2}}{V_1} \cdot 100 \tag{25}$$

where f_{SW} is the switching frequency, n is the bandwidth centered around the switching frequency, h is the harmonic order and V_1 is the fundamental amplitude.

In Table 8 are summarized the *PTHD*% calculated for line voltage V_{ab}, converter side current I_a and grid side current I_{ga}.

Table 8. Partial Total Harmonic Distortion" (*PTHD*%) values obtained with SPD, THIPD, and SFOPD.

	V_{ab}	I_a	I_{ga}
SPD	7.94%	3.92%	0.38%
THIPD	7.86%	4.61%	0.43%
SFOPD	8.29%	4.08%	0.38%

It is interesting to note that the SPD and SFOPD present the lower values of the *PTHD*% as regard to the currents. While, the lower values of the *PTHD*% of the line voltage has been obtained with THIPD modulation technique. Obviously, both the modulation technique and the LCL filter concur to define the harmonic content.

2.4.2. Phase Opposition Disposition and Alternative Phase Opposition Disposition

In harmonic spectra obtained with modulation techniques based POD or APOD as carrier signals, the harmonic component at switching frequency does not appear but there are only side band.

Figure 16 shows the harmonic spectra of the phase voltage centered around the switching frequency obtained with SPOD, THIPOD, SFOPOD, SAPOD, THIAPOD, and SFOAPOD, respectively.

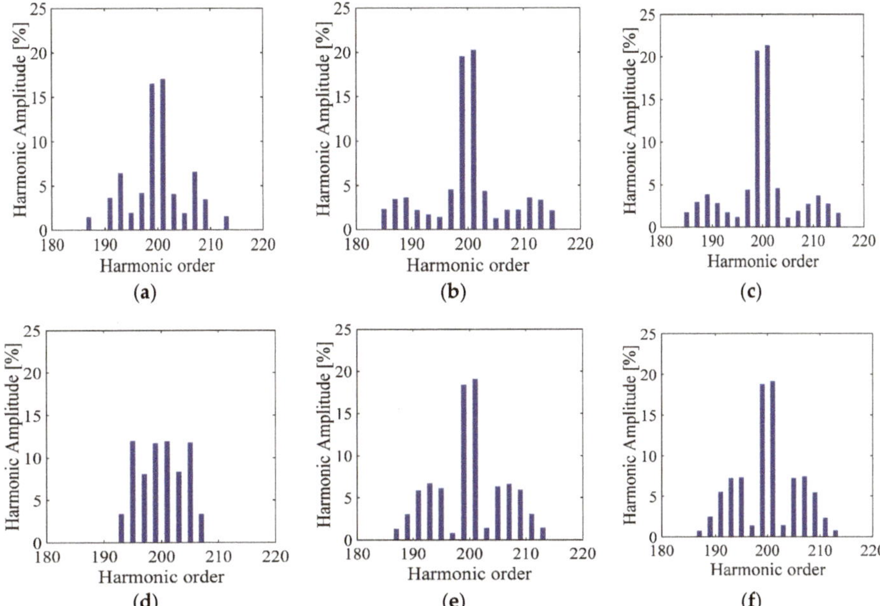

Figure 16. Phase voltage harmonic spectra centered around the switching frequency (10kHz) in percent respect to the fundamental amplitude of (**a**) SPOD, (**b**) THIPOD, (**c**) SFOPOD, (**d**) SAPOD, (**e**) THIAPOD, and (**f**) SFOAPOD.

As shown in Figure 16, the harmonic spectra are similar with a pair component predominant respect to the others while only the SAPOD (Figure 16d) presents little differences. Respect the modulation techniques PD or PS based, in the sub-section 2.4. LCL Filter Design the higher values of the filter requirements were obtained with POD and APOD.

Figures 17–22 show the transitory of the converter side currents and grid side currents from zero to the rated current obtained with SPOD, THIPOD, SFOPOD, SAPOD, THIAPOD, and SFOAPOD respectively.

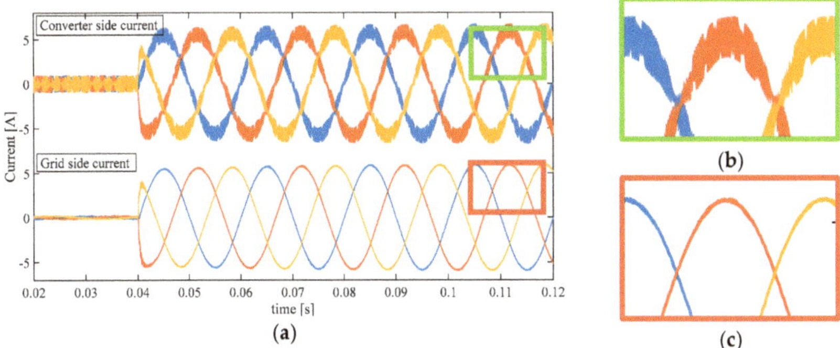

Figure 17. Converter side and grid side three-phase current with SPOD. (**a**) Transient behavior in multiple cycles; (**b**) Ripple magnification for converter side current; (**c**) Ripple magnification for grid side current.

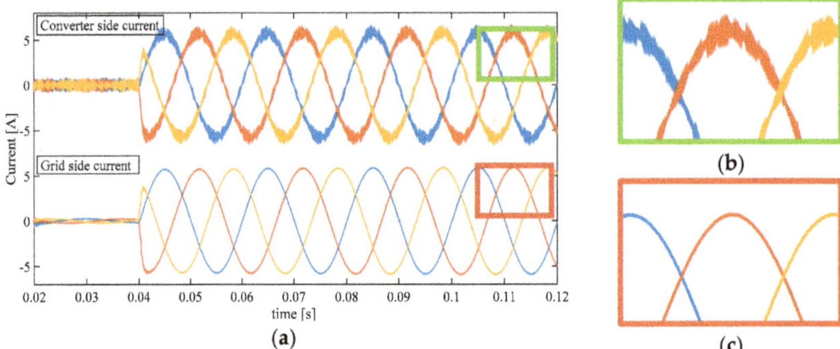

Figure 18. Converter side and grid side three-phase current with THIPOD. (**a**) Transient behavior in multiple cycles; (**b**) Ripple magnification for converter side current; (**c**) Ripple magnification for grid side current.

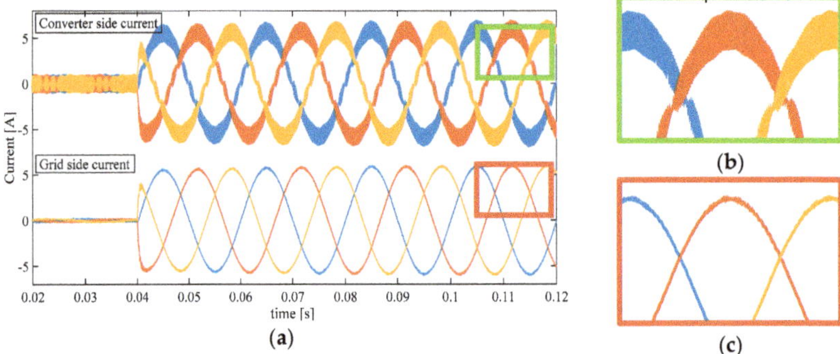

Figure 19. Converter side and grid side three-phase current with SFOPOD. (**a**) Transient behavior in multiple cycles; (**b**) Ripple magnification for converter side current; (**c**) Ripple magnification for grid side current.

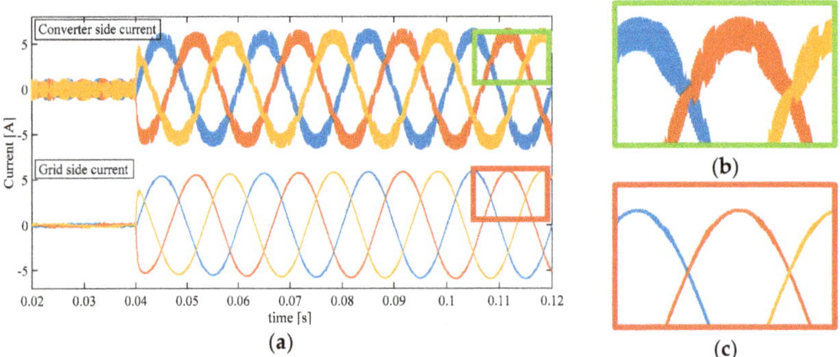

Figure 20. Converter side and grid side three-phase current with SAPOD. (**a**) Transient behavior in multiple cycles; (**b**) Ripple magnification for converter side current; (**c**) Ripple magnification for grid side current.

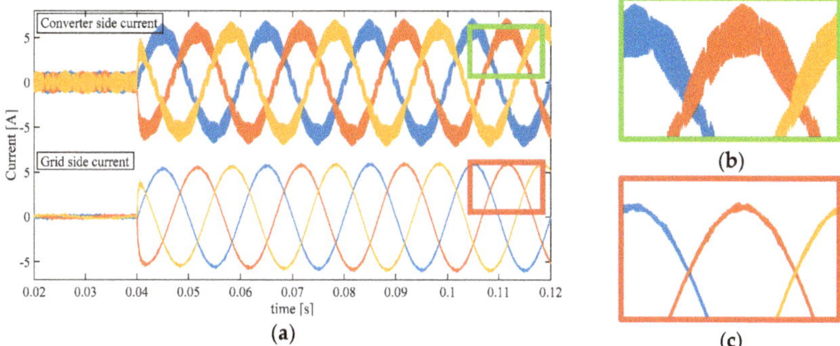

Figure 21. Converter side and grid side three-phase current with THIAPOD. (**a**) Transient behavior in multiple cycles; (**b**) Ripple magnification for converter side current; (**c**) Ripple magnification for grid side current.

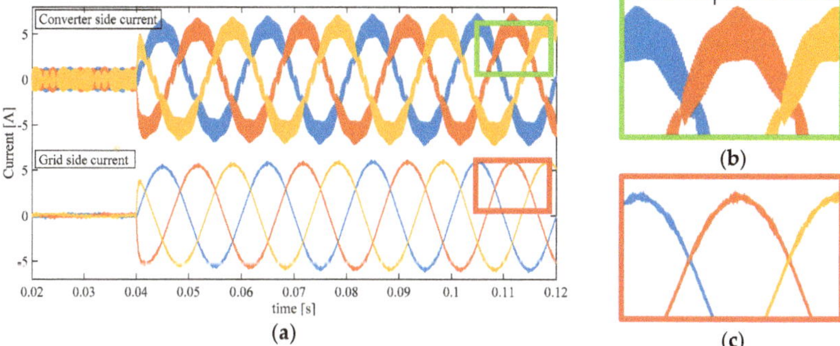

Figure 22. Converter side and grid side three-phase current with SFOAPOD. (**a**) Transient behavior in multiple cycles; (**b**) Ripple magnification for converter side current; (**c**) Ripple magnification for grid side current.

As before, the grid currents trend presents a THD% less then 5% according to IEEE 1574 and IEC 61727, but it is necessary to study the lower order harmonics and the ones around the switching frequency, in order to investigate in depth the performances. Figure 23 shows the low order harmonics (from third to fortieth harmonic) in grid side current I_{ga} obtained with SPOD, THIPOD, SFOPOD, SAPOD, THIAPOD, and SFOAPOD, respectively.

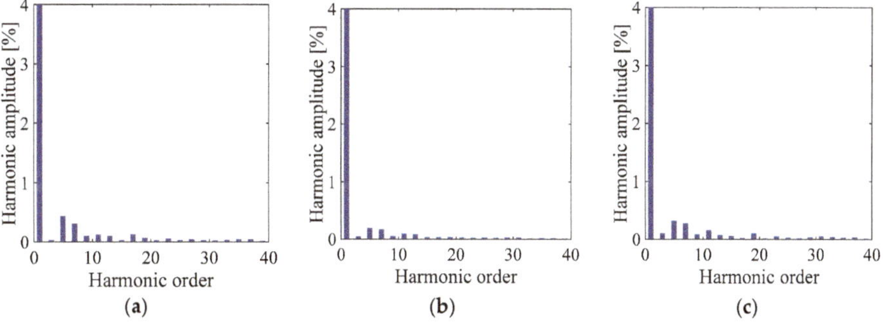

Figure 23. *Cont.*

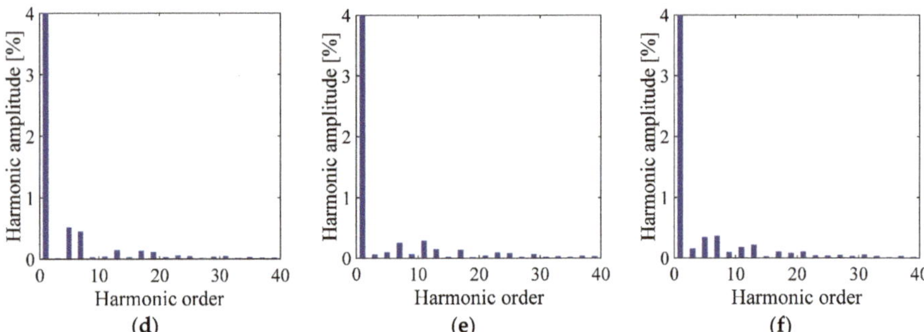

Figure 23. Low order harmonics on the grid side current *Iga* for (**a**) SPOD, (**b**) THIPOD, (**c**) SFOPD, (**d**) SAPOD, (**e**) THIAPOD, and (**f**) SFOAPOD.

It should be noted that the amplitude of the lower order harmonics are below of the standard harmonic current limits defined by IEEE 1574 and IEC 61727 at the PCC. Interesting results were obtained with THIPOD (Figure 23b) considering that are visible only fifth and seventh components with the lower amplitude compared to others modulation techniques POD and APOD based. Moreover, this is the best results also compared with modulation techniques PD based. This phenomenon is also explained by the higher values of the LCL filter parameters.

Figure 24 shows a comparison among harmonic spectra centered around the switching frequency of 10 kHz among line voltage V_{ab} (blue bars), converter side current I_a (red bars) and grid side current I_{ga} (yellow bars) obtained with SPOD, THIPOD, SFOPOD, SAPOD, THIAPOD, and SFOAPOD, respectively.

The lower values of the grid side current harmonics, less of 0.3% referred to the fundamental amplitude, demonstrate the LCL filter effectiveness. While, the predominant pair of the side band harmonic that appear in all harmonics spectra explains the higher values of the LCL filter parameters compared with modulation techniques PD or PS based. An interesting consideration is about the grid side current, the values are very lower respect to the modulation techniques PD based.

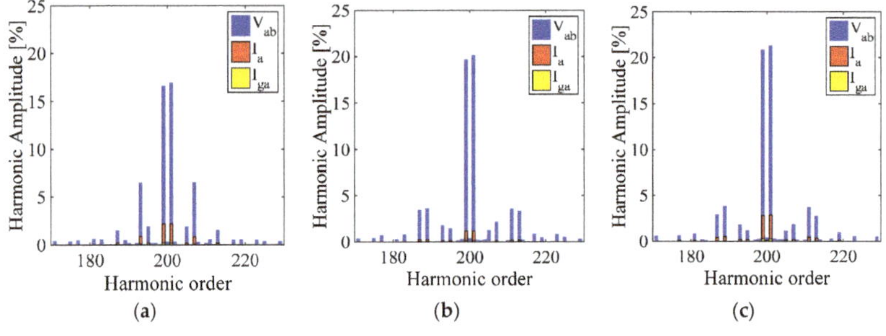

Figure 24. *Cont.*

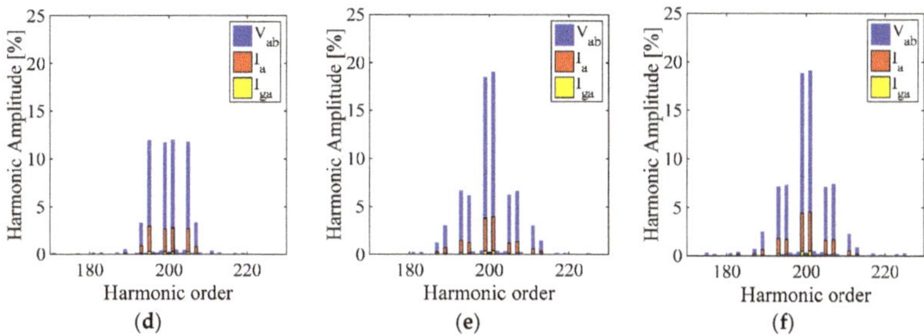

Figure 24. Comparison of line voltage harmonic spectra V_{ab}, converter side current I_a and grid side current I_{ga} centered around the switching frequency (10 kHz) in percent respect to the fundamental amplitude for (**a**) SPOD, (**b**) THIPOD, (**c**) SFOPOD, (**d**) SAPOD, (**e**) THIAPOD, and (**f**) SFOAPOD.

In Table 9 are reported the values of the PTHD% for line voltage V_{ab}, converter side current I_a and grid side current I_{ga}.

Table 9. *PTHD*% values obtained with SPOD, THIPOD, SFOPD, SAPOD, THIAPOD, and SFOAPOD.

	V_{ab}	I_a	I_{ga}
SPOD	25.62%	3.37%	0.33%
THIPOD	29.12%	1.72%	0.18%
SFOPOD	30.60%	4.12%	0.40%
SAPOD	24.17%	5.68%	0.53%
THIAPOD	29.78%	6.16%	0.61%
SFOAPOD	30.65%	7.21%	0.77%

As expected, the *PTHD*% relatively of the line voltages present higher values compared with modulation techniques PD based. While, the low values of *PTHD*% relatively of the currents are similar respect to the values calculated with modulation techniques PD based.

2.4.3. Phase Shifted Disposition

Last type of the carrier signals analyzed in this work is the Phase Shifted PS that allows shifting the harmonic to $2nHB$ times of the switching frequency. In the case of the five-level converter, nHB is equal to 2, so the harmonic content is shifting to four times the switching frequency (40 kHz). Moreover, the harmonics are centered around four times the switching and are present only side band harmonics like in modulation techniques POD and APOD based. Figure 25 shows the harmonic spectra, centered around four times of the switching frequency, obtained with SPS, THIPS, and SFOPS, respectively.

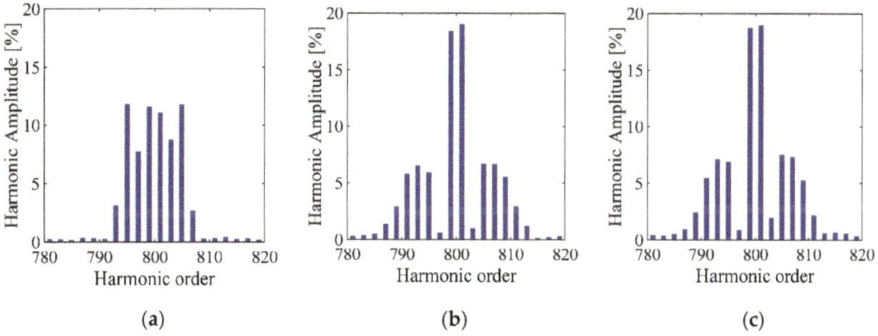

Figure 25. Phase voltage harmonic spectra centered around four times of the switching frequency (10 kHz) in percent respect to the fundamental amplitude of (**a**) SPS, (**b**) THIPS, and (**c**) SFOPS.

The harmonic spectra are similar respect to the harmonic spectra obtained with POD and APOD but the amplitude of the predominant harmonics are lower.

Figures 26–28 show the transitory of the converter side currents and grid side currents from zero to the rated current obtained with SPS, THIPS, and SFOPS, respectively.

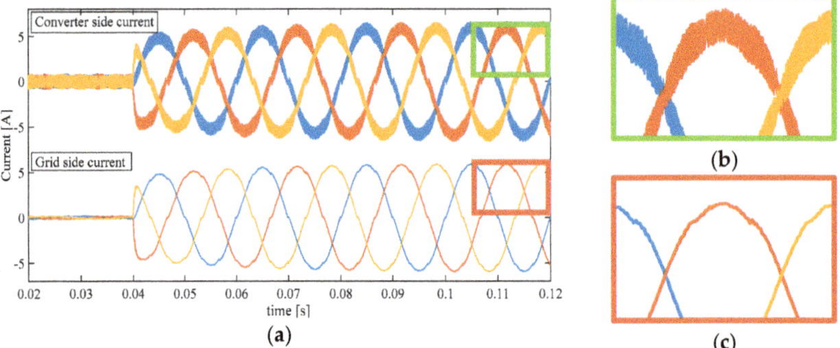

Figure 26. Converter side and grid side three-phase current with SPS. (**a**) Transient behavior in multiple cycles; (**b**) Ripple magnification for converter side current; (**c**) Ripple magnification for grid side current.

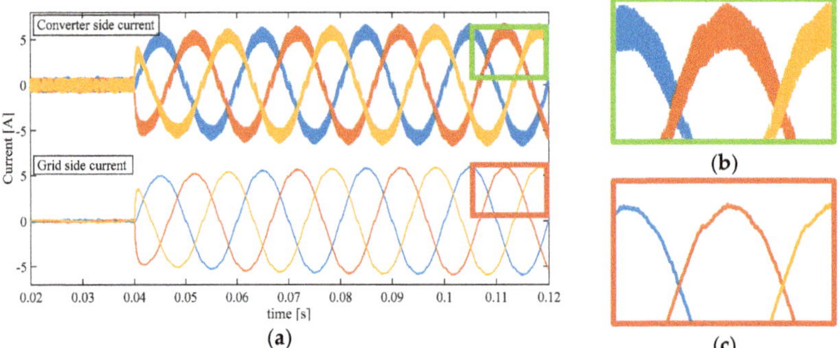

Figure 27. Converter side and grid side three-phase current with THIPS. (**a**) Transient behavior in multiple cycles; (**b**) Ripple magnification for converter side current; (**c**) Ripple magnification for grid side current.

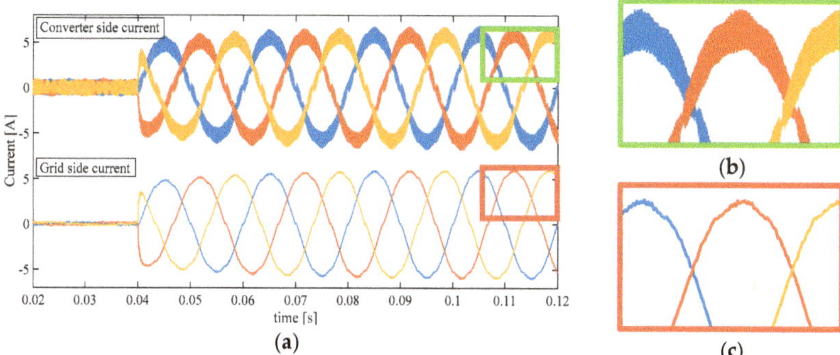

Figure 28. Converter side and grid side three-phase current with SFOPS. (**a**) Transient behavior in multiple cycles; (**b**) Ripple magnification for converter side current; (**c**) Ripple magnification for grid side current.

As previously noted, the grid currents trend presents a THD% less then 5% according to IEEE 1574 and IEC 61727 but observing the grid currents trend is evident the presence of low order harmonics.

Figure 29 shows the low order harmonics (from third to fortieth harmonic) in grid side current I_{ga} obtained with SPS, THIPS, and SFOPS, respectively.

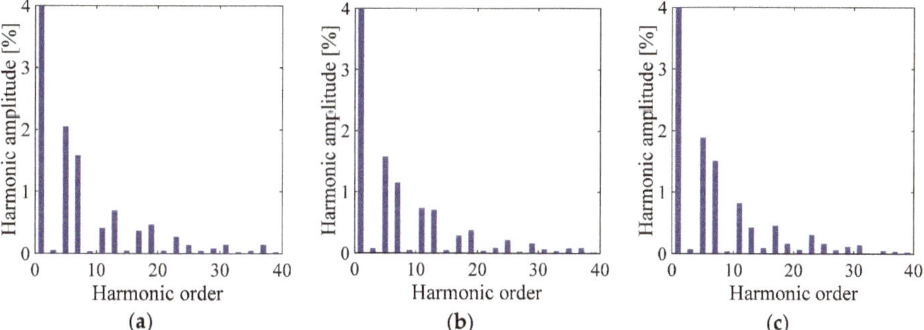

Figure 29. Low order harmonics on the grid side current I_{ga} for (**a**) SPS, (**b**) THIPS, and (**c**) SFOPS.

As presumed, by analyzing Figure 29, low order harmonics contents are present, in particular the fifth, seventh are predominant respect to the others, which anyway are under the harmonics current limits. Modulation techniques PS based have the higher values of the lower order harmonics compared with all modulation techniques previously described. Figure 30 shows a comparison among harmonic spectra centered around the switching frequency of 10 kHz among line voltage V_{ab} (blue bars), converter side current I_a (red bars) and grid side current I_{ga} (yellow bars) obtained with SPS, THIPS, and SFOPs, respectively.

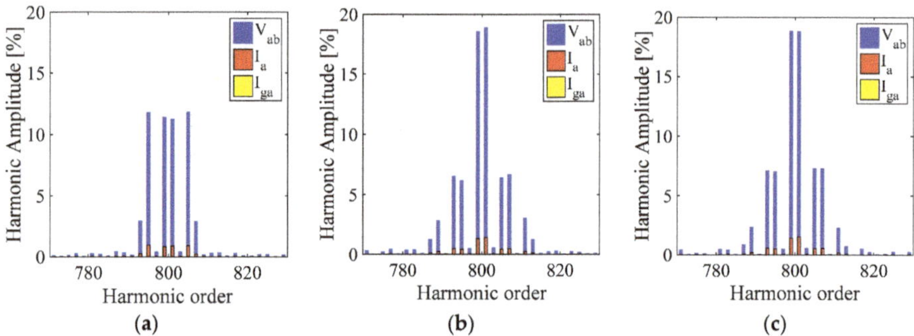

Figure 30. Comparison of line voltage harmonic spectra V_{ab}, converter side current I_a and grid side current I_{ga} centered around the switching frequency (10 kHz) in percent respect to the fundamental amplitude for (**a**) SPS, (**b**) THIPS, and (**c**) SFOPS.

Also for modulation techniques PS based, the grid side current harmonics are less than 0.3% referred to the fundamental amplitude demonstrating the LCL filter effectiveness. In terms of harmonic spectra of the line voltage, there are a pair of the predominant side band harmonics that are present also in the current spectra. In order to compare the performance were evaluated the *PTHD*% where the values are summarized in Table 10.

Table 10. *PTHD*% values obtained with SPS, THIPS, and SFOPS.

	V_{ab}	I_a	I_{ga}
SPS	23.51%	1.80%	0.0095%
THIPS	29.84%	2.16%	0.0117%
SFOPS	30.51%	2.35%	0.0124%

As shown in Table 10, the values of the *PTHD*% are lower compared with other values previously calculated. These are interesting results because on the grid side current are present only low order harmonics that are under the current limits (IEEE 1574 and IEC 61727).

In conclusion, modulation techniques PD based allow obtaining good results in terms of the harmonic content on the grid current with the lower values of the LCL filter parameters. In particular, SPD represent the best solution. Interesting results have been obtained also with modulation techniques PS based thanks to the feature that allows to shift the harmonic content respect to the switching frequency. In fact, have been obtained the lower values of the *PTHD*% about the grid side currents among all the modulation techniques taken into account. Moreover, modulation techniques PS based allow to control the power flow from the DC sources because each level can be controlled as a single-phase converter. This feature is important in very applications like PV systems for example. In the next section the experimental validation to confirm the simulation results were reported.

3. Experimental Validation

The purpose of this section is to provide all the useful information to describe the employed of a prototype CHBMI.

The first aim of the experimental validation is to validate the model of the system previously described and to confirm the effectiveness of the LCL filter design and the control strategy. In particular, the experimental tests have been executed with a prototype of a three-phase five-level multilevel converter with topology structure cascaded H-bridge inverter. Moreover, also the control board FPGA based is a prototype designed for power electronics applications. In this section were reported detailed descriptions of the test bench, control algorithm design and the experimental results. By the simulation analysis, reported in section "2.4 Performances Evaluation", the experimental validation was focused on SPD and SPS modulation techniques.

3.1. Test Bench

In order to carry out the experimental analysis, a three-phase, five-level multilevel inverter prototype with a CHB circuital structure were assembled.

The test bench is shown in Figure 31 and it is mainly composed by:

- a prototype of FPGA-based control board (produced by DigiPower s.r.l);
- Six prototypes of H-bridges (produced by DigiPower s.r.l);
- A Three-phase LCL filter especially designed (produced by SDESLAB and LEAP of the University of Palermo);
- Six DC sources with 24 V of rated voltage;
- Three-phase variac to grid interface;
- A Teledyne LeCroy WaveRunner 6Zi, scope, employed for the real-time acquisition of the waveforms and monitoring of the system.

Figure 31. A photograph of the test bench.

Figure 32a shows the prototype of the H-Bridge that is based on power Mosfet (International Rectifier—model IRFB4115PbF [57]) whose technical features are reported in Table 11. While, Figure 32b shows the FPGA-based control board where the FPGA is produced by Altera—model Cyclone III and the features reported in [58,59].

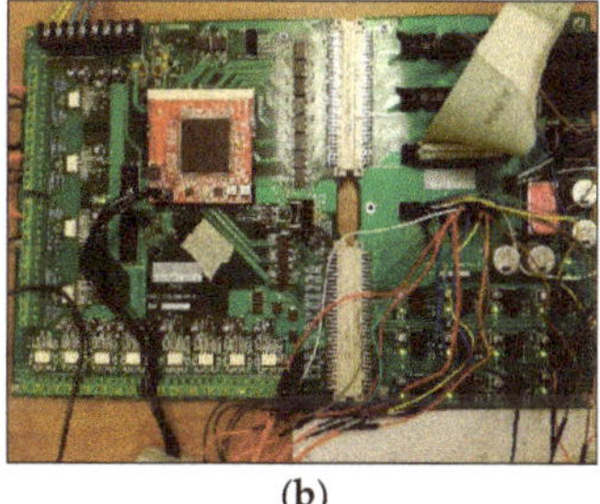

(a)   (b)

Figure 32. A photograph of the prototype (**a**) H-Brides and (**b**) field programmable gate array (FPGA) control board.

Table 11. Technical features of the IRFB4115PBF device [57].

Voltage V_{dss}	150 V
Resistance $R_{ds(on)}$	9.3 mΩ
Current Id (silicon limited)	104 A
Turn on delay $t_{D(on)}$	18 ns
Rise time t_R	73 ns
Turn off delay $T_{D(off)}$	41 ns
Fall time t_F	39 ns
Reversal recovery t_{RR}	86 ns

Figure 33 shows the three-phase LCL filter especially designed and assembled with commercial components at SDESLAB and LEAP laboratories of the University of Palermo. For the converter and grid side inductance have been used commercial inductance of 560 µH with rated current of 4A in parallel connected to obtain an inductance then 280 µH with rated current of 8A.

Figure 33. Tree-phase LCL filter.

While, for the capacitor filter have been used the ceramic capacitors of 4.7 µF with rated voltage of 100 V in parallel connected to obtain a capacitor of 9.4 µF.

3.2. Control Algorithm Design

The FPGA is commonly used in order to implement complex functions, such as arithmetic logic unit (ALU), memories, communication units and so on [16]. The main difference with other programmable systems used in industrial applications (µ-controller or DSP) is that through a software programming it is possible to describe an especially designed hardware for specific application. For this reason, the FPGA allows to obtain high flexibility and very fast execution time that allows using in very large of applications field. Actually, in power electronics application the complexity of the control algorithms, due also to the application type is increasing. For example, in grid connected applications there are very different control algorithm to control the power flow, power quality, synchronization with the grid, parallel control of the DC side and AC side and so on. Thus, are necessary programmable systems with fast execution time and the clock of the digital system should be adapted to the specific application. Aim of this section, is to investigate on the use of the FPGA for grid-connected application in order to validate the simulation analysis previously described and to optimize the control algorithm for the application under test. The control algorithm was implemented by means of an FPGA Altera Cyclone III EP3C40Q. The control software is Quartus II by Altera and the used programming language is the VHDL [60–62].

The structure of the control algorithm implemented can be explained by means the schematic block diagram shown in Figure 34.

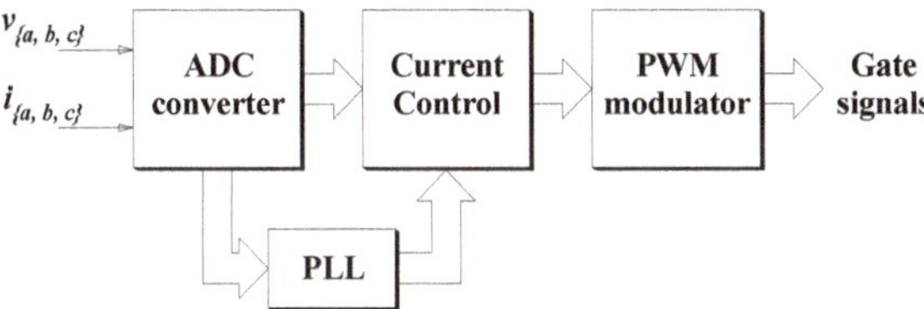

Figure 34. Schematic block diagram of the control algorithm.

The block named "ADC converter" represent the algorithm to manage the acquisition of the electrical quantities. In the prototype FPGA-Based control board is available an ADC converter with 16 channel (no simultaneously), 1 MSPS, 12 bit successive approximation ADC produced by Analog Devices model AD7490 16. The conversion process is managed by a clock signal reference with a frequency at 10MHz and conversion time (analog to digital signal) is equal to 2 µs. For the system under test, three voltage (v_a, v_b and v_c) and three current (i_a, i_b and i_c) are acquired; the conversion process is subdivided in two sub-conversion process relatively for the voltages and currents with a conversion time equal to 6 µs, respectively. In this way, the operation are executed in parallel, so when finished the first sub-conversion process relatively to the voltages, the mathematic elaborations for PLL with voltages samples start jointly with the second sub-conversion process relatively to the currents. After the conversion process, the mathematic elaboration with a resolution of 32 bit Floating Point (FP) single precision and a clock reference equal to 100 MHz starts. In Table 12 are summarized the execution time of the main mathematic operation.

Table 12. Execution time of the mathematic operation in FP 32bit.

Mathematic operation	Time
Conversion Integer to Floating (Integer 13 bit, FP 32bit)	60 ns
Sum or subtraction (FP 32 bit)	140 ns
Product (FP 32 bit)	110 ns

It should be noted that between two operations there is a delay time equal to 10 ns in order to stabilize and address the digital signals.

The equivalent schematic block diagram of the PLL to describe the implementation in VHDL is shown in Figure 35.

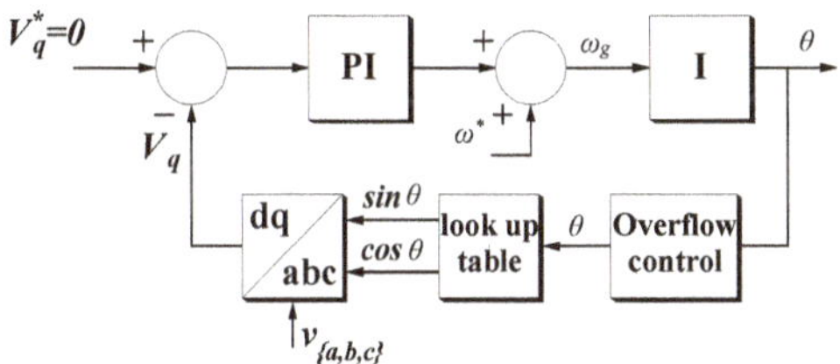

Figure 35. Equivalent schematic block diagram of the PLL. PI: proportional-integral; I: integral.

The PI regulator have been implemented in the discretion form as:

$$u(k) = k_p e(k) + \frac{k_p T}{T_i} e(k) + u_i(k-1) \tag{26}$$

where k indicate the k-sample, $u(k)$ is the output term of the PI regulator, $e(k)$ is the error, T is the sampling time, T_i is the time-integral and $u_i(k-1)$ is the $k-1$ integral output. In order to optimize the execution time to evaluate the $sin\theta$ and $cos\theta$, where θ is the instantaneous phase of the space vector voltage, a look-up table was used. Each value of the instantaneous phase was used as an address (from 0 to 6280) to determinate the values of the *sinθ* and *cosθ*. In the look-up table there are only a quarter of the sine waveform with a number of the sample equal to 1570 and it is possible to determinate the *sinθ* and *cosθ* values through a logic circuit. The block "Overflow Control" is necessary to limit the instantaneous phase value to 2π. In this way, the instantaneous phase of the space vector voltage θ assumes the sawtooth trend. In Table 13 are summarized the execution time of the main block of the PLL algorithm.

Table 13. Execution time main block of the PLL.

Operation	Time
ABC to DQ transformation (FP 32bit)	880 ns
PI regulator (FP 32 bit)	540 ns
Integral (FP 32 bit)	270 ns
Overflow control and look-up table (FP 32 bit)	450 ns

The total execution time algorithm from the end of the acquisition voltage to the instantaneous phase θ is equal to 3 μs. Figure 36 shows the experimental PLL effectiveness through a comparison of the grid voltage (red trend) and phase voltage of the converter (yellow trend) before carrying out the parallel with the grid.

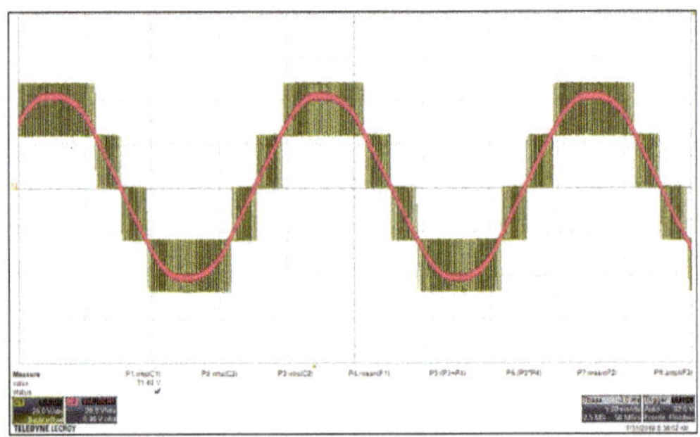

Figure 36. Experimental PLL effectiveness.

Figure 37 shows the execution time of the current control scheme where the total execution time is 3.13 μs.

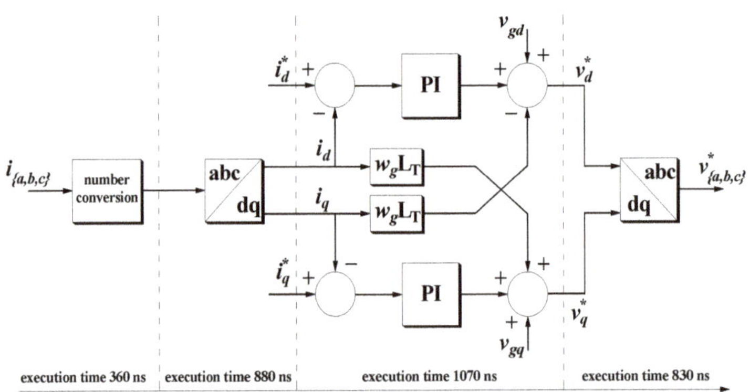

Figure 37. Execution time of the current control scheme.

The block "PWM generator" generates the gate signals for the converter. Generally, a carrier generator, comparator circuit and a logic circuit to generate the "dead time" compose the "PWM generator". Figure 38 shows the screenshot of the schematic block diagram of the PWM modulator implemented in Quartus II environment for SPD modulation technique.

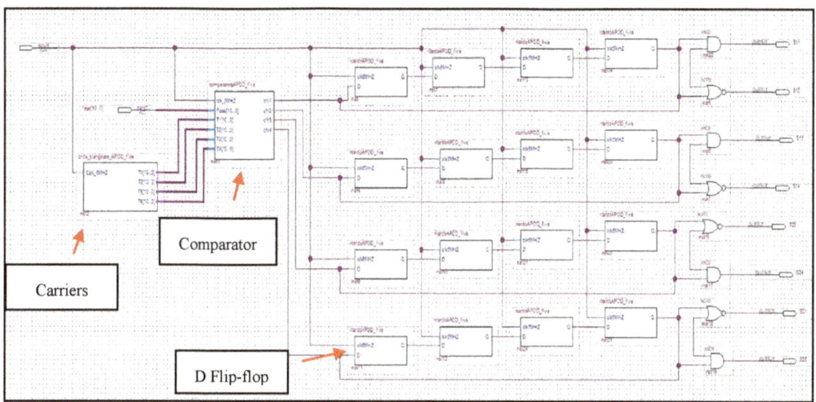

Figure 38. Screenshot of the schematic block diagram of the pulse width modulation (PWM) modulator implemented in Quartus II environment for SPD modulation technique.

Carrier waveform is generated by means a 13 bit up-down counter with a resolution of 1000 sample. The frequency of the clock reference is 10 MHz in order to obtain a frequency-switching equal to 10 kHz. The comparator circuit carried out the comparison between the modulating signal and carrier generator with a frequency equal to 10 MHz. The generation of the command signals of the components of the H-Bridge legs, as well as the obtainment of the dead-time for the protection of the series-connected components, is achieved through means of the logic circuit shown in Figure 39.

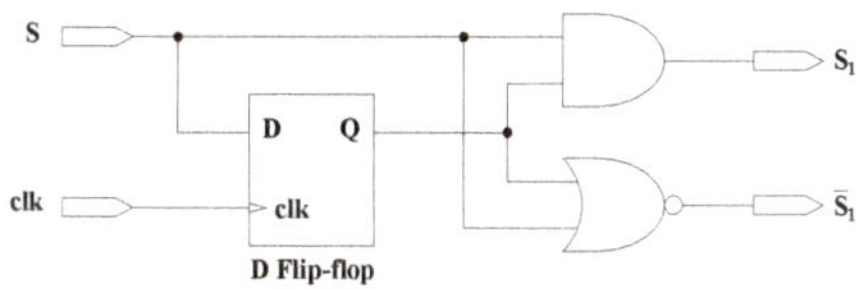

Figure 39. Logic circuit to generate the gate signals with dead time.

The delayed signal is obtained by using several cascaded-connected D flip-flops, whose number is dependent on the adopted clock signal. In order to obtain 400 ns of delay, four D flip-flops have been connected and managed with the 10 MHz clock signal. Figure 40 shows the simulation of the "PWM Generator" carried out in ModelSim environment relatively a phase of the converter.

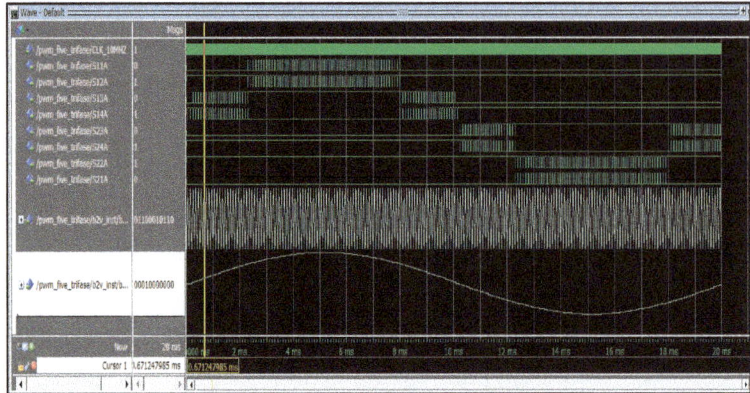

Figure 40. Simulation of the "PWM generator" in ModelSim environment.

From the technical features of the IRFB4115PBF reported in Table 12, the minimum dead time is equal to 100 ns, approximately. Thus, has been chosen for safe reason a dead time equal to 400 ns.

Figure 41 shows a screenshot of the experimental validation between gate signals of the same leg in order to establish the proper operation of the digital system.

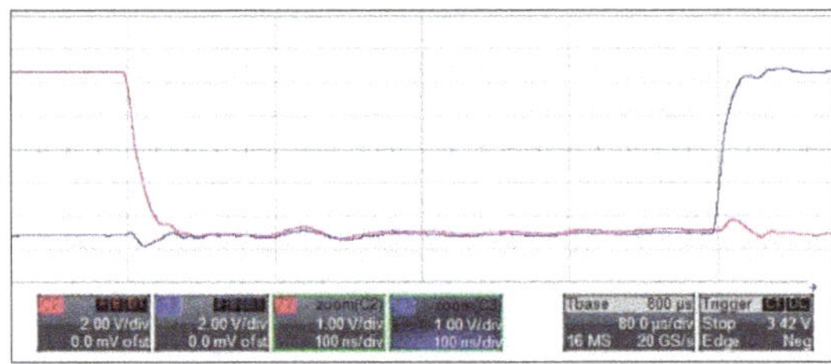

Figure 41. Experimental validation between gate signals of the same leg.

It should be noted that the dead time obtained is equal to 400 ns.

3.3. Model Validation

By the employment of the previously described test bench, the suggested techniques were experimentally implemented in order to validate the model of the system and to compare the simulation and experimental results.

The *Teledyne LeCroy WaveRunner 6Zi* acquisition system recorded the voltage waveforms. For the modulation PD, POD and APOD based techniques a sampling frequency of 50 MHz and a number of samples equal to 1 Ms were used; an observation window was choosen with a time interval equal to 20 ms.

The PS modulation techniques required an acquisition of 5 Ms of samples, equivalent to a sampling frequency of 250 MHz. The used tool to compare the simulation results and experimental results is the THD%, as reported in (27) [63]:

$$\text{THD\%} = \sqrt{\frac{V_{rms}^2 - V_{rms,1}^2}{V_{rms,1}^2}} \cdot 100 \tag{27}$$

where V_{rms} is the root mean square (rms) value of the phase voltage and $V_{rms,1}$ defines the rms value of the fundamental harmonic.

Figures 42–44 show the comparison between the simulated (blue bars) and the experimental (yellow bars) THD% values obtained with Sinusoidal (Figure 42), THI (Figure 43) and SFO (Figure 44) as reference signals for each modulation techniques taken into account with the designed filter discussed in Section 2.2.

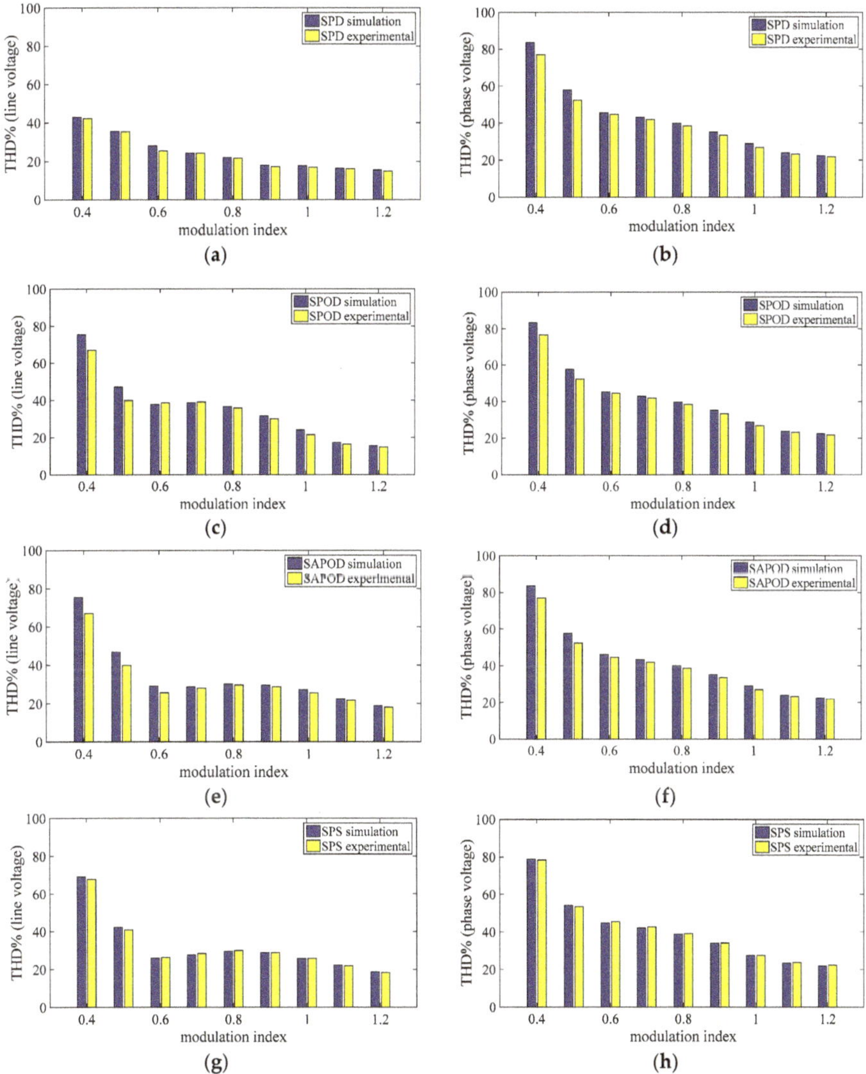

Figure 42. Comparison between the simulated (blue) and the experimental (yellow) THD% values: (**a**) SPD line voltage, (**b**) SPD phase voltage, (**c**) SPOD line voltage, (**d**) SPOD phase voltage, (**e**) SAPOD line voltage, (**f**) SAPOD phase voltage, (**g**) SPS line voltage, and (**h**) SPS phase voltage.

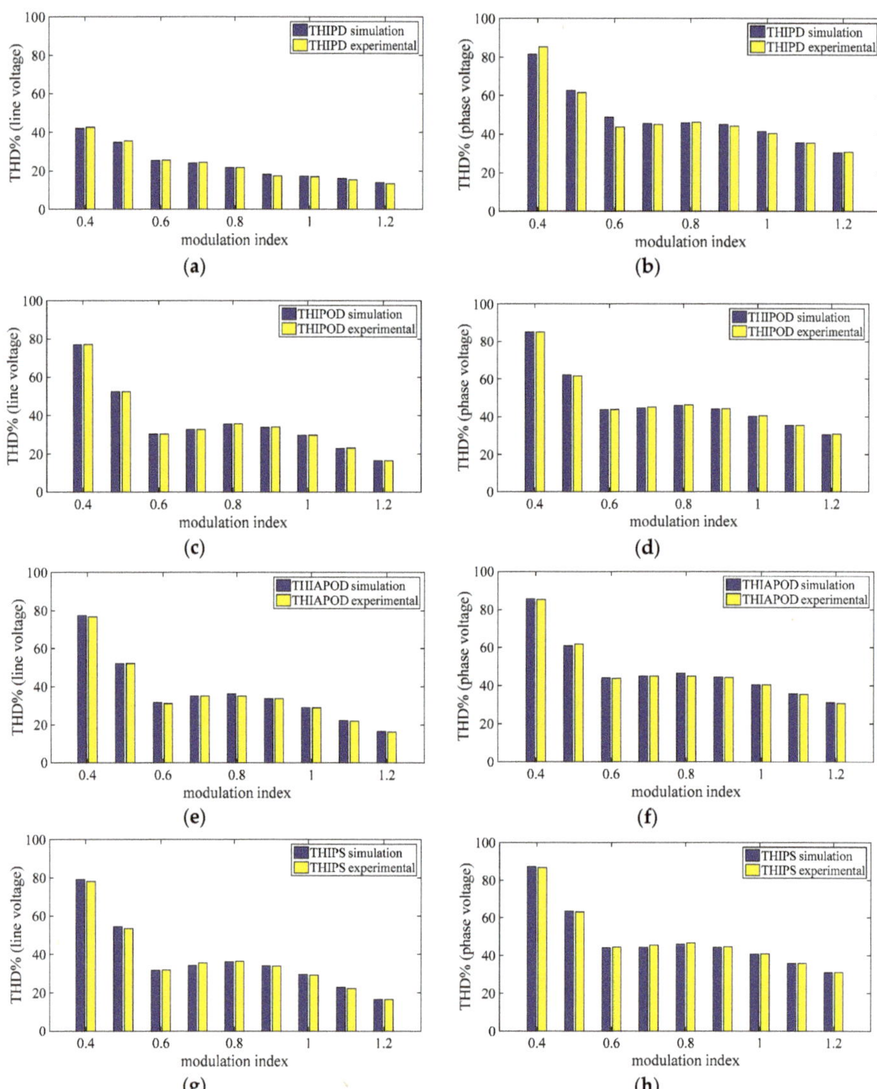

Figure 43. Comparison between the simulated (blue) and the experimental (yellow) THD% values: (**a**) THIPD line voltage, (**b**) THIPD phase voltage, (**c**) THIPOD line voltage, (**d**) THIPOD phase voltage, (**e**) THIAPOD line voltage, (**f**) THIAPOD phase voltage, (**g**) THIPS line voltage, and (**h**) THIPS phase voltage.

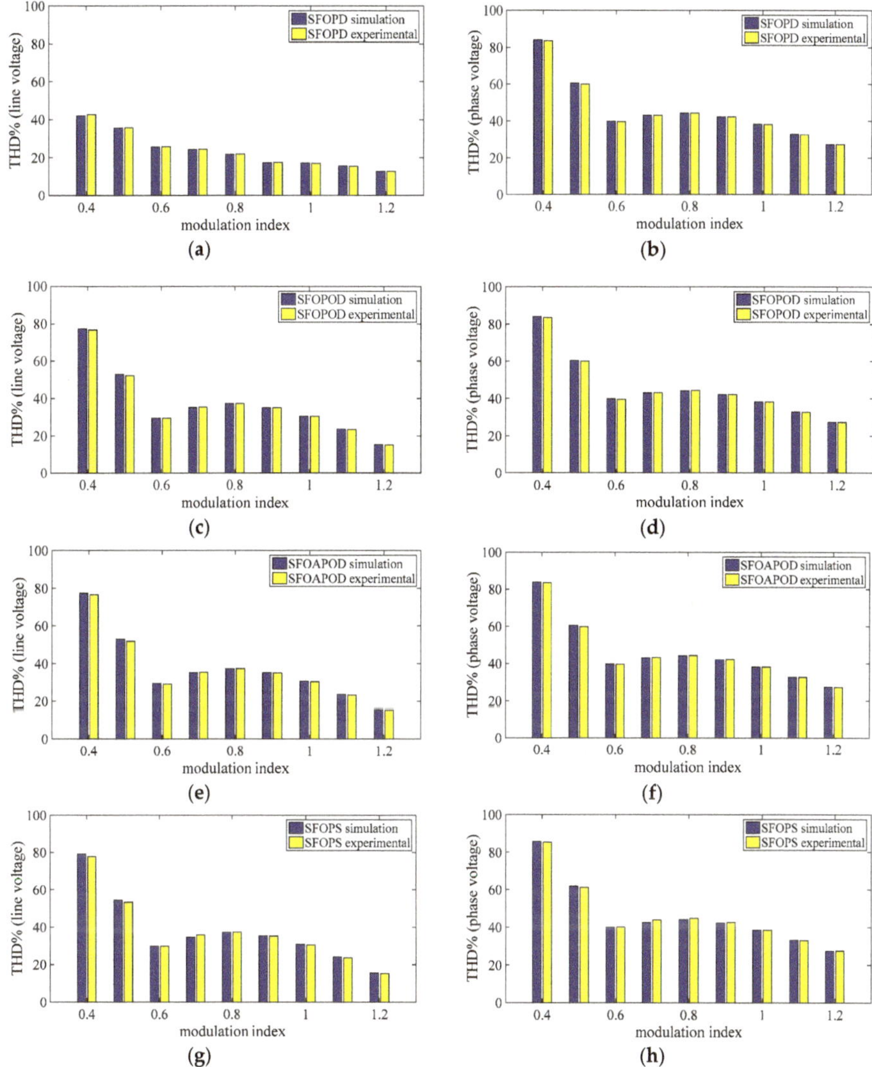

Figure 44. Comparison between the simulated (blue) and the experimental (yellow) THD% values: (**a**) SFOPD line voltage, (**b**) SFOPD phase voltage, (**c**) SFOPOD line voltage, (**d**) SFOPOD phase voltage, (**e**) SFOAPOD line voltage, (**f**) SFOAPOD phase voltage, (**g**) SFOPS line voltage, and (**h**) SFOPS phase voltage.

It should be noted that the simulated and experimental THD% presents similar values. For this reason, it is possible to establish the effectiveness of the model implemented.

Interesting comparison among the experimental THD% values for each reference signals taken into account, is shown in Figure 45. Modulation technique with PD as carrier signals and sinusoidal reference SPD seems to be the best solution in terms of the harmonic content. Moreover, also the SPS is a good solution for grid-connected applications.

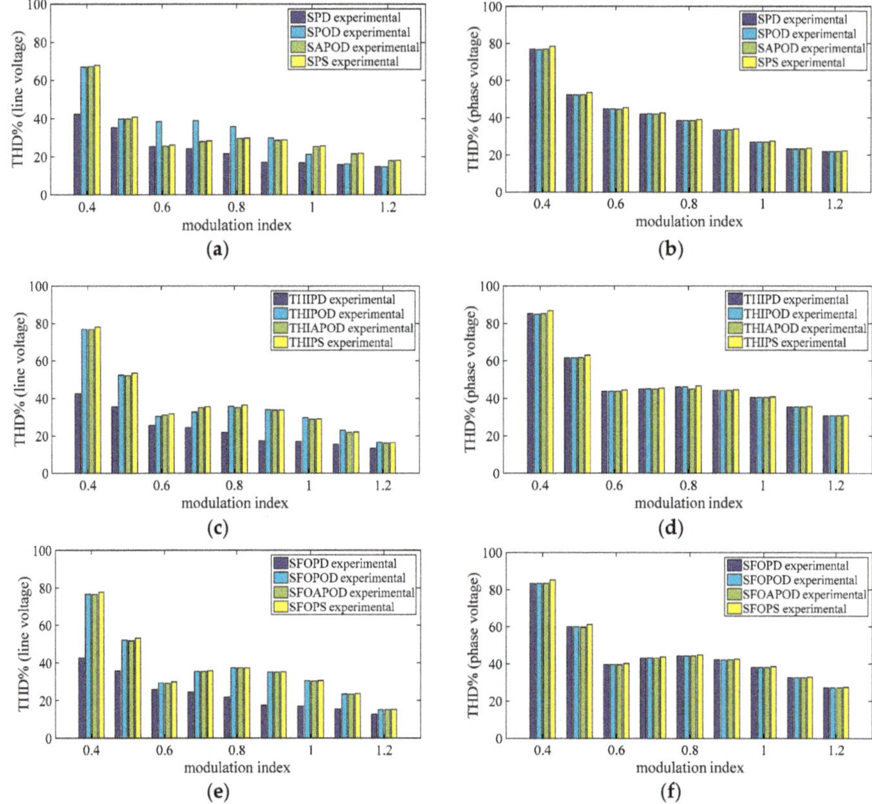

Figure 45. Comparison between the experimental THD% results: (**a**) Sinusoidal line voltage, (**b**) Sinusoidal phase voltage, (**c**) THI line voltage, (**d**) THI phase voltage, (**e**) SFO line voltage, and (**f**) SFO phase voltage.

In conclusion, modulation technique with PD as carrier signals and sinusoidal reference SPD present interesting results. Moreover, also the PS carrier signal is a good solution due to high order harmonic components respect other carrier signals.

In the next section, experimental validation of the grid-connected application is reported. The experimental validation considers only the SPD and SPS modulation techniques.

3.4. Grid Connected Application

Aim of this subsection is to validate the simulation results, reported in section "2.1 Performances evaluation", in which the best performances were obtained with SPD and SPS modulation techniques. In particular, the purpose is to validate by means experimental tests the effectiveness of the control strategy and the LCL filter. Thus, the experimental tests were carried out only with SPD and SPS modulation techniques.

3.4.1. Phase Disposition

Figure 46 shows the measured grid phase voltages and grid currents of the phase a and b obtained with SPD modulation technique at the rated power. It is interesting to note that the phase angle between voltage and current of the same phase is equal to zero. This result demonstrates the effectiveness of the control strategy because, as explained in the section "2.3 Controller Design", through the d component it is possible to control the active power while through the q component it is possible to control the reactive power. Thus, fixing q component of the current equal to zero and d component of the current equal to rated value (6A) it is possible to inject only active power on the grid as shown in Figure 46.

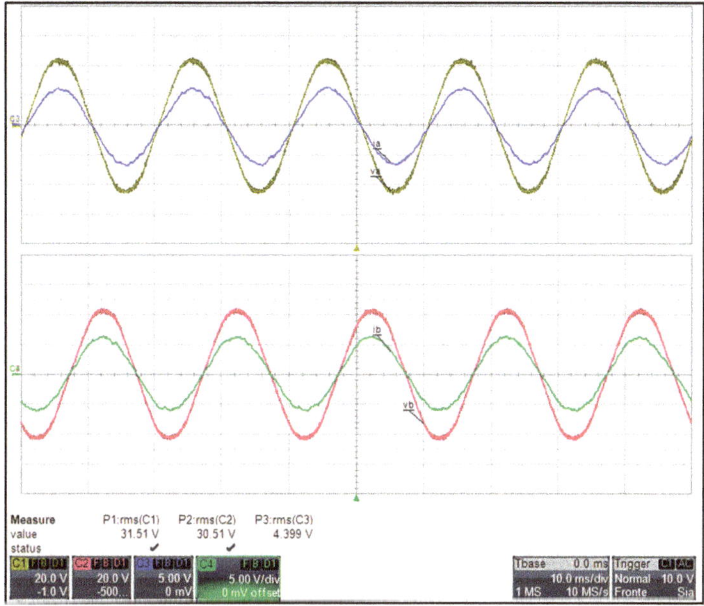

Figure 46. Measured grid voltages (20 V/div) and grid currents (5 A/div) of the phase a and b obtained with SPD at the rated power.

Figure 47 shows measured converter side current obtained with SPD at rated power while Figure 48 shows the measured grid side current in the same conditions.

First all, the differences in terms of the harmonic content between the trend of the converter side and grid side currents are evident. Moreover, a not perfect half-wave symmetry in the currents trend was observed. This phenomenon determined the present of the even-harmonics in the current.

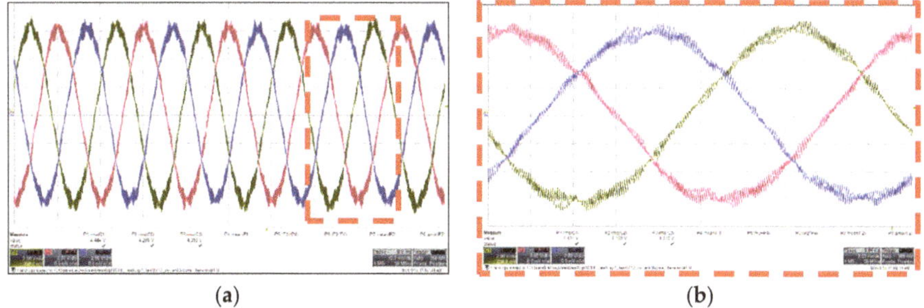

(a)   (b)

Figure 47. Measured converter side currents (2 A/div) obtained with SPD at the rated power. (**a**) Ripple in different cycles; (**b**) Magnification of ripple.

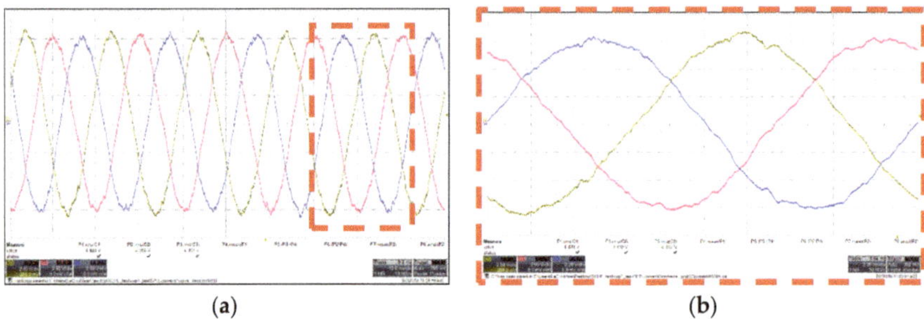

(a)   (b)

Figure 48. Measured grid side currents (2 A/div) obtained with SPD at the rated power. (**a**) Ripple in different cycles; (**b**) Magnification of ripple.

Figure 49a shows the low order harmonics spectra of the grid side current at rated power. In the first all, it interesting to note that the amplitude of the lower order harmonics are below of the standard harmonic current limits defined by IEEE 1574 and IEC 61727 at the PCC. Nevertheless, as stated above are present the even-harmonics on the harmonic spectra. Interesting comparison between the harmonic spectra centered on switching frequency of the converter side current I_a (blue bars) and grid side current I_{ga} (yellow bars) is shown in Figure 49b. The lower values of the harmonics of the grid side current demonstrate the effectiveness of the LCL filter.

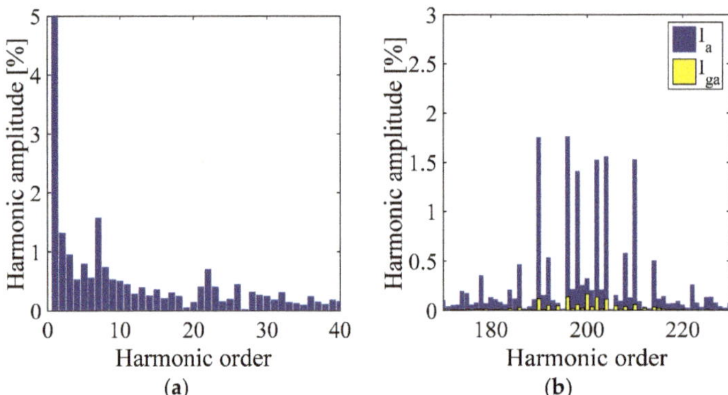

(a)   (b)

Figure 49. Calculated (**a**) low order harmonics of the grid side current and (**b**) switching frequency harmonics spectra of the converter side and grid side currents.

In Table 14 are summarized the calculated THD% for different values of the grid side current injected. It should be noted that the THD% values increase when the current injected in the grid is reduced and it is less then 5% up to $I_n/2$.

Table 14. Experimental THD% of the converter and grid side currents, obtained with SPD, for different values of the injected current into the grid.

	$I_n/3$	$I_n/2$	$2I_n/3$	I_n
Converter side current	12.17%	7.82%	6.46%	5.88%
Grid side current	7.97%	4.78%	4.26%	3.72%

Figure 50 shows the measured line voltage of the converter at rated power. It interesting to note that the line voltage presents nine-level.

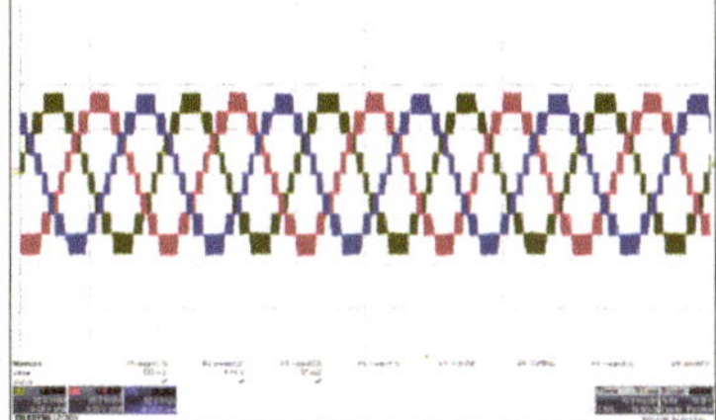

Figure 50. Measured line voltage of the converter at rated power.

Figure 51 shows the measured capacitor voltage of the LCL filter at rated power. The evident low harmonic content in the trend of the capacitor voltage demonstrate the efficacy of the LCL filter.

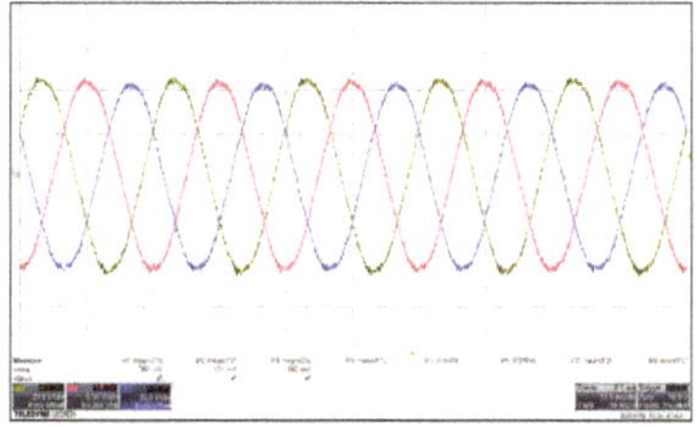

Figure 51. Measured capacitor voltage of the LCL filter at rated power.

As stated earlier, the second experimental tests have been carried out with SPS modulation techniques with the same filter used for SPD modulation techniques thank to the similar values obtained in the subsection "LCL filter Designs".

3.4.2. Phase Shifted

Figure 52 shows the measured grid phase voltages and grid currents obtained with SPS modulation technique at the rated power for each phase of the system.

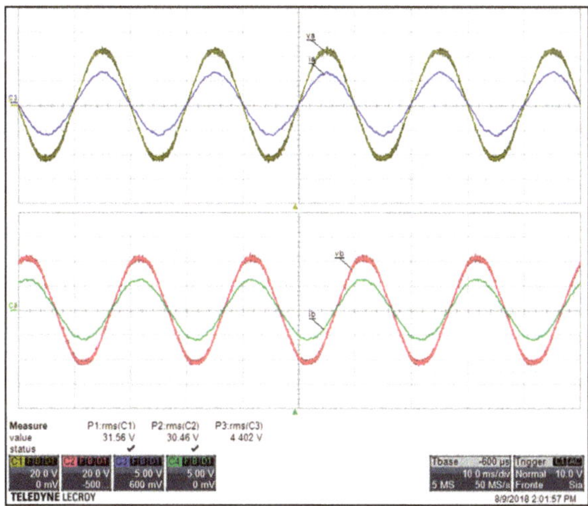

Figure 52. Measured grid voltages (20 V/div) and grid currents (5 A/div) of the phase a and b obtained with SPS at the rated power.

Also for this case, the phase angle between voltage and current of the same phase is equal to zero, thus this result demonstrates the efficacy of the control strategy.

Figures 53 and 54 show the measured converter side and grid side currents, respectively. As mentioned above, the currents trends present a not perfect half-wave symmetry and this phenomenon determined the even harmonics.

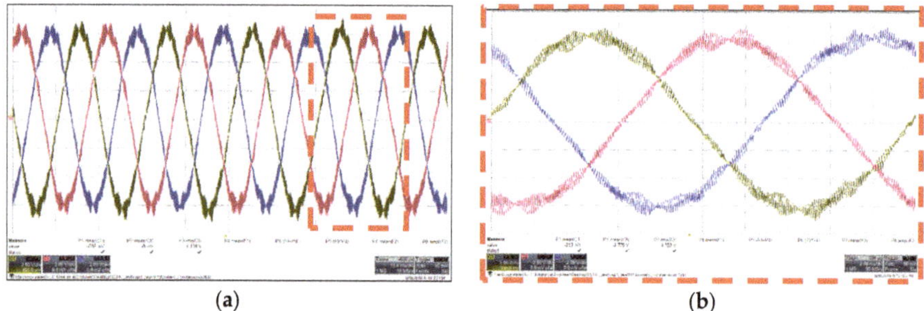

Figure 53. Measured converter side currents (2 A/div) obtained with SPS at the rated power. (**a**) Ripple in different cycles; (**b**) Magnification of ripple.

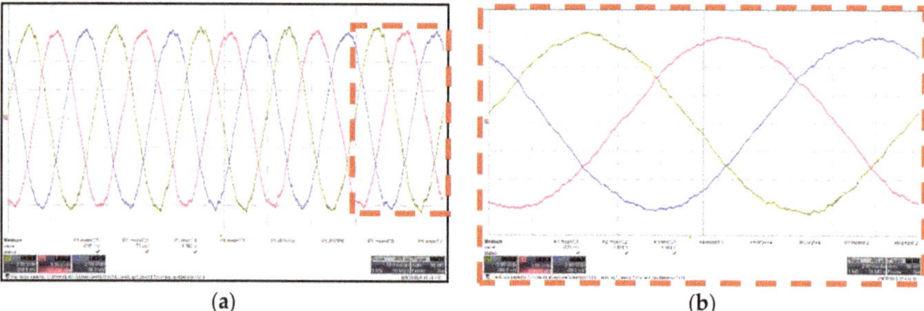

Figure 54. Measured grid side currents (2 A/div) obtained with SPS at the rated power. (**a**) Ripple in different cycles; (**b**) Magnification of ripple.

Figure 55a shows the low order harmonics of the grid current at rated power. The amplitude of the all low order harmonics are less of the current limits reported in Table 1. However, by comparing the low order harmonic spectra of SPD and SPS, it can be noted that the second order harmonic is higher in SPS modulation technique. The presence of the even harmonics also is due to absence of the DC voltage control. Moreover, in both low order harmonic spectra of the SPD and SPS is predominant a seventh harmonic with similar value.

In Figure 55b is shown the comparison between the harmonic spectra centered on switching frequency of the converter side current I_a (blue bars) and grid side current I_{ga} (yellow bars). The lower values of the harmonics of the grid side current demonstrate the effectiveness of the LCL filter.

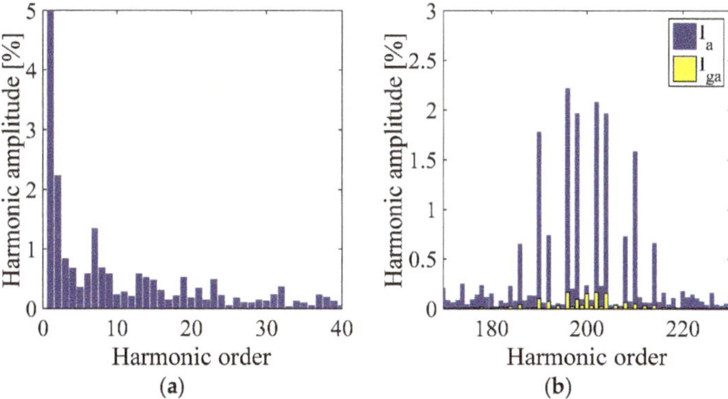

Figure 55. Calculated (**a**) low order harmonics of the grid side current and (**b**) switching frequency harmonics spectra of the converter side and grid side currents.

In Table 15 are summarized the calculated THD% for different values of side current injected in the grid. The THD% values obtained with SPS are similar respect to the previously calculated with SPD. This is an interesting result, because the modulation techniques PS based are more versatile respect to the others multicarrier modulation techniques for grid connected applications like PV systems, for example. The modulation techniques PS based allow to control each H-Bridge like a single-phase inverter and it is possible to use innovative control algorithms especially designed in dependence of the application.

Table 15. Experimental THD% of the converter and grid side currents, obtained with SPS, for different values of the injected current into the grid.

	$I_n/3$	$I_n/2$	$2I_n/3$	I_n
Converter side current	12.28%	8.41%	6.80%	5.64%
Grid side current	7.42%	4.45%	3.91%	3.33%

In addition, the line voltage build with the SPS modulation technique has nine level, as shown in Figure 56.

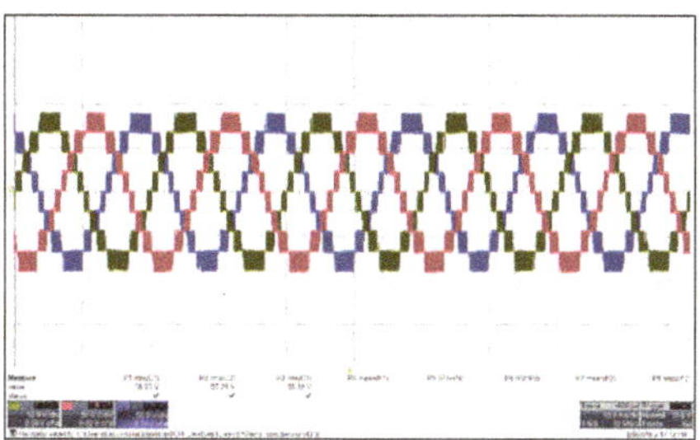

Figure 56. Measured line voltage of the converter at rated power.

Figure 57 shows the measured capacitor voltage of the LCL filter with an evident low harmonic content that also in this case demonstrate the efficacy of the LCL filter.

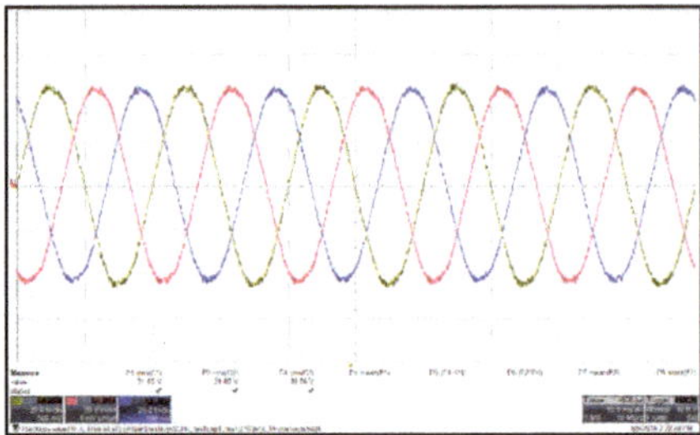

Figure 57. Measured capacitor voltage of the LCL filter at rated power.

4. Discussion

In order to face the harmonic distortion problem, two issue can be distinctly taken into account: the generation of harmonics and their suppression. Although the approach is not purely dichotomous, since a lower generation corresponds to an easier suppression, here the main results of this work can be approached with an etiological methodology.

The modulation techniques with PD as carrier signals shows a harmonic spectrum of the phase voltage with a predominant harmonic centered on the switching frequency and side band harmonics (Figure 10). By considering the modulation with POD and APOD as carriers, in the spectra, the harmonic component at switching frequency does not appear but there are only side bands (Figure 16). By considering the modulation with PS disposition, the harmonics are centered around four times the switching frequency, are present only side bands harmonics like in modulation techniques POD and APOD based (Figure 25).

In order to reduce the harmonics in the grid side, a filtering system is correctly designed. For the PD based modulation techniques, the converter side inductance L and grid side inductance L_g present the lower values (Table 4). Higher value of the converter side inductance were obtained with THIPOD modulation technique, phenomenon attributable to the higher number of the side band harmonics generated by POD carrier signals. Figure 14 (PD), 23 (POD and APOD) and 29 (PS) show the performances of the filtering by considering the different modulation techniques. For each techniques, the third harmonic of injected current is much reduced, so the comparison moves on the fifth harmonic: PD is around 1%, POD and APOD less than 0.5%, PS around 2%. Excellent performance of THIPOD for its very low values of fifth and also seventh harmonic, are remarkable, but the side inductance L is eight times the value for PD ones. By analyzing Figure 29, low order harmonics contents are present, in particular besides the fifth, seventh is predominant. Modulation techniques PS based have the higher values of the lower order harmonics compared with all modulation techniques previously described.

In conclusion, modulation techniques PD based allow obtaining good results in terms of the harmonic content on the grid current with the lower values of the LCL filter parameters. In particular, SPD represent the best solution.

The previous described good results are validated in Figure 45 with the experimental test phase; moreover, it is possible to find an experimental behavior better than the simulated one for SPD for different modulation indexes.

Finally, the approach was validated for a grid-connected system by exploiting the three-phase Variac to grid interface. Different current values were injected in the power grid, Tables 14 and 15 report that the THD% remained below the 4%, 3.72% for SPD and 3.33% for SPS techniques.

By considering the work of Colak et al. [64], as the number of levels in multilevel inverter increases, the THD in the output voltage reduces, but the drawback of increasing the levels is that that the control circuit becomes hard challenges. A comparison can be done with recent results found in literature. Kavali and Mittal obtained by MATLAB based simulation with SIMULINK environment interesting results for their single-phase five level CHBMI topology with sinusoidal pulse width modulation schemes [37]. For PD the THD was reduced to 5.69%; in case of POD it was 5.75%; for APOD it was

5.73%; the THD was 7.42% in case of PS. Results obtained in the present work follow those obtained in [37], the obtained THDs are below, and add an experimental validation to them.

5. Conclusions

As previously described, the PWM modulation techniques found large use in many industrial applications thanks their main features as easy implementation in electronic control systems, low computational cost and high flexibility. Moreover, for grid-connected applications the PWM modulation techniques are the best solution due to the lower amplitude of the low-order harmonics, reducing the filter requirements. Thus, in this work a detailed analysis taking into account all PWM modulation techniques, the LCL filter requirements and the real time implementation issues in FPGA-based prototype control board for a grid-connected three-phase five-level CHBMI, was presented.

Firstly, through a simple step-by-step procedure to design LCL filter for each modulation techniques taken into account, it was demonstrated that the lower values of the filter parameters are obtained for modulation techniques employing sinusoidal as reference signals. These interesting results were confirmed by the experimental validation of the THD% values. In particular, the SPD and SPS showed the best results in terms of the THD% values. Then, the experimental tests was focused by using the SPD and SPS modulation techniques in order to inject in the power grid different values of current through the specially designed LCL filter. Notably, the experimental tests confirmed the effectiveness of the LCL filter. The amplitude of the lower order harmonics are below of the standard harmonic current limits at the point of common coupling. Nevertheless, appeared even-harmonics on the amplitude spectra. Finally, it is possible to claim that the modulation technique SPS is the best solution for all grid-connected applications where it is necessary to control the power flow of the DC sources separately.

Funding: This research received no external funding.

Acknowledgments: This work was financially supported by MIUR-Ministero dell'Istruzione dell'Università e della Ricerca (Italian Ministry of Education, University and Research), by SDESLab (Sustainable Development and Energy Saving Laboratory), and LEAP (Laboratory of Electrical Applications) of the University of Palermo.

Conflicts of Interest: The author declares no conflict of interest.

References

1. Rashid, M.H. *Power Electronics Handbook, Devices, Circuits and Applications,* 3rd ed.; Butherworth-Heirmann is an imprint of Elsevier; Pearson Education Limited: New York, NY, USA, 2011.
2. Rodriguez, J.; Lai, J.-S.; Peng, F.Z. Multilevel inverters: A survey of topologies, controls, and applications. *IEEE Trans. Ind. Electron.* **2002**, *49*, 724–738. [CrossRef]
3. Malinowski, M.; Gopakumar, K.; Rodriguez, J.; Perez, M.A. A Survey on Cascaded Multilevel Inverters. *IEEE Trans. Ind. Electron.* **2010**, *57*, 2197–2206. [CrossRef]
4. Kouro, S.; Malinowski, M.; Gopakumar, K.; Pou, J.; Franquelo, L.G.; Wu, B.; Rodriguez, J.; Perez, M.A.; Leon, J.I. Recent Advances and Industrial Applications of Multilevel Converters. *IEEE Trans. Ind. Electron.* **2010**, *57*, 2553–2580. [CrossRef]
5. El-Hosainy, A.; Hamed, H.A.; Azazi, H.Z.; El-Kholy, E.E. A review of multilevel inverter topologies, control techniques and applications. In Proceedings of the 2017 Nineteenth International Middle East Power Systems Conference (MEPCON), Cairo, Egypt, 19–21 December 2017; pp. 1265–1275.
6. McGrath, B.P.; Holmes, D.G. Multicarrier PWM strategies for multilevel inverters. *IEEE Trans. Ind. Electron.* **2002**, *49*, 858–867. [CrossRef]
7. McGrath, B.P.; Holmes, D.G. A comparison of multicarrier PWM strategies for cascaded and neutral point clamped multilevel inverters. In Proceedings of the 2000 IEEE 31st Annual Power Electronics Specialists Conference, Conference Proceedings (Cat. No.00CH37018), Galway, Ireland, 23–23 June 2000; Volume 2, pp. 674–679.

8. Jeevananthan, S.; Madhavan, R.; Padmanabhan, T.S.; Dananjayan, P. State-of-the-art of Multi-Carrier Modulation Techniques for Seven Level Inverter: A Critical Evaluation and Novel Submissions based on Control Degree of Freedom. In Proceedings of the 2006 IEEE International Conference on Industrial Technology, Mumbai, India, 15–17 December 2006; pp. 1269–1274.

9. Holmes, D.G.; Lipo, T.A. *Pulse Width Modulation for Power Converters*; IEEE Press series on Power Engineering; Wiley Interscience: New York, NY, USA, 2004.

10. Schettino, G.; Benanti, S.; Buccella, C.; Caruso, M.; Castiglia, V.; Cecati, C.; Di Tommaso, A.O.; Miceli, R.; Romano, P.; Viola, F. Simulation and experimental validation of multicarrier PWM techniques for three-phase five-level cascaded H-bridge with FPGA controller. *Int. J. Renew. Energy Res.* **2017**, *7*, 1383–1394.

11. Schettino, G.; Buccella, C.; Caruso, M.; Cecati, C.; Castiglia, V.; Miceli, R.; Viola, F. Overview and experimental analysis of MC SPWM techniques for single-phase five level cascaded H-bridge FPGA controller-based. In Proceedings of the IECON 2016—42nd Annual Conference of the IEEE Industrial Electronics Society, Florence, Italy, 23–26 October 2016; pp. 4529–4534.

12. Benanti, S.; Buccella, C.; Caruso, M.; Castiglia, V.; Cecati, C.; Di Tommaso, A.O.; Miceli, R.; Romano, P.; Schettino, G.; Viola, F. Experimental analysis with FPGA controller-based of MC PWM techniques for three-phase five level cascaded H-bridge for PV applications. In Proceedings of the 2016 IEEE International Conference on Renewable Energy Research and Applications (ICRERA), Birmingham, UK, 20–23 November 2016; pp. 1173–1178.

13. Castiglia, V.; Livreri, P.; Miceli, R.; Raciti, A.; Schettino, G.; Viola, F. Performances of a three-phase five-level cascaded H-bridge inverter with phase shifted B-spline based modulation techniques. In Proceedings of the 6th International Conference on Renewable Energy Research and Applications (ICRERA), San Diego, CA, USA, 5–8 November 2017; pp. 1192–1197.

14. Schettino, G.; Castiglia, V.; Genduso, F.; Livreri, P.; Miceli, R.; Romano, P.; Viola, F. Simulation of a single-phase five-level cascaded H-Bridge inverter with multicarrier SPWM B-Spline based modulation techniques. In Proceedings of the 12th International Conference on Ecological Vehicles and Renewable Energies (EVER), Monte Carlo, Monaco, 11–13 April 2017.

15. Grahame, D.; Buccella, C.; Cecati, C.; Latafat, H. Digital Control of Power Converters—A Survey. *IEEE Trans. Ind. Inf.* **2012**, *8*, 437–447. [CrossRef]

16. Monmasson, E.; Cirstea, M.N. FPGA Design Methodology for Industrial Control Systems—A Review. *IEEE Trans. Ind. Electron.* **2007**, *54*, 1824–1842. [CrossRef]

17. Monmasson, E.; Idkhajine, L.; Cirstea, M.N.; Bahri, I.; Tisan, A.; Naouar, M.W. FPGAs in Industrial Control Applications. *IEEE Trans. Ind. Inf.* **2011**, *7*, 224–243. [CrossRef]

18. Rodríguez-Andina, J.J.; Valdés-Peña, M.D.; Moure, M.J. Advanced Features and Industrial Applications of FPGAs—A Review. *IEEE Trans. Ind. Inf.* **2015**, *11*, 853–864. [CrossRef]

19. Naouar, M.W.; Monmasson, E.; Naassani, A.A.; Slama-Belkhodja, I.; Patin, N. FPGA-Based Current Controllers for AC Machine Drives—A Review. *IEEE Trans. Ind. Inf.* **2007**, *54*, 1907–1925. [CrossRef]

20. Kala, P.; Arora, S. A comprehensive study of classical and hybrid multilevel inverter topologies for renewable energy applications. *Renew. Sustain. Energy Rev.* **2017**, *76*, 905–931. [CrossRef]

21. Venkataramanaiah, J.; Suresh, Y.; Panda, A.K. A review on symmetric, asymmetric, hybrid and single DC sources based multilevel inverter topologies. *Renew. Sustain. Energy Rev.* **2017**, *76*, 788–812. [CrossRef]

22. Jana, J.; Saha, H.; Bhattacharya, K.D. A review of inverter topologies for single-phase grid-connected photovoltaic systems. *Renew. Sustain. Energy Rev.* **2017**, *72*, 1256–1270. [CrossRef]

23. Monica, P.; Kowsalya, M. Control strategies of parallel operated inverters in renewable energy application: A review. *Renew. Sustain. Energy Rev.* **2016**, *65*, 885–901. [CrossRef]

24. Son, Y.; Kim, J. A Novel Phase Current Reconstruction Method for a Three-Level Neutral Point Clamped Inverter (NPCI) with a Neutral Shunt Resistor. *Energies* **2018**, *11*, 2616. [CrossRef]

25. Ye, Z.; Chen, A.; Mao, S.; Wang, T.; Yu, D.; Deng, X. A Novel Three-Level Voltage Source Converter for AC–DC–AC Conversion. *Energies* **2018**, *11*, 1147. [CrossRef]

26. Ricco, M.; Mathe, L.; Monmasson, E.; Teodorescu, R. FPGA-Based Implementation of MMC Control Based on Sorting Networks. *Energies* **2018**, *11*, 2394. [CrossRef]

27. Wu, Z.; Chu, J.; Gu, W.; Huang, Q.; Chen, L.; Yuan, X. Hybrid Modulated Model Predictive Control in a Modular Multilevel Converter for Multi-Terminal Direct Current Systems. *Energies* **2018**, *11*, 1861. [CrossRef]

28. Talon Louokdom, E.; Gavin, S.; Siemaszko, D.; Biya-Motto, F.; Essimbi Zobo, B.; Marchesoni, M.; Carpita, M. Small-Scale Modular Multilevel Converter for Multi-Terminal DC Networks Applications: System Control Validation. *Energies* **2018**, *11*, 1690. [CrossRef]
29. Kumar, N.; Saha, T.K.; Dey, J. Modeling, control and analysis of cascaded inverter based grid-connected photovoltaic system. *Int. J. Electr. Power Energy Syst.* **2016**, *78*, 165–173. [CrossRef]
30. Ravi, A.; Manoharan, P.S.; Anand, J.V. Modeling and simulation of three phase multilevel inverter for grid connected photovoltaic systems. *Sol. Energy* **2011**, *85*, 2811–2818. [CrossRef]
31. Rajkumar, M.V.; Manoharan, P.S. FPGA based multilevel cascaded inverters with SVPWM algorithm for photovoltaic system. *Sol. Energy* **2013**, *87*, 229–245. [CrossRef]
32. Prabaharan, N.; Palanisamy, K. Analysis and integration of multilevel inverter configuration with boost converters in a photovoltaic system. *Energy Convers. Manag.* **2016**, *128*, 327–342. [CrossRef]
33. Iero, D.; Carbone, R.; Carotenuto, R.; Felini, C.; Merenda, M.; Pangallo, G.; Della Corte, F.G. SPICE modelling of a complete photovoltaic system including modules, Energy storage elements and a multilevel inverter. *Sol. Energy* **2014**, *107*, 338–350. [CrossRef]
34. Javad, O.; Shayan, E.; Ali, M. Compensation of voltage sag caused by partial shading in grid-connected PV system through the three-level SVM inverter. *Sustain. Energy Technol. Assess.* **2016**, *18*, 107–118.
35. Prabaharan, N.; Palanisamy, K. A Single Phase Grid Connected Hybrid Multilevel Inverter for Interfacing Photo-voltaic System. *Energy Procedia* **2016**, *103*, 250–255. [CrossRef]
36. Rahim, N.A.; Ping, H.W.; Selvaraj, J. Elimination of harmonics in photovoltaic seven-level inverter with Newton-Raphson optimization. *Procedia Environ. Sci.* **2013**, *17*, 519–528.
37. Kavali, J.; Mittal, A. Analysis of various control schemes for minimal Total Harmonic Distortion in cascaded H-bridge multilevel inverter. *J. Electr. Syst. Inf. Technol.* **2016**, *3*, 428–441. [CrossRef]
38. Al-Othman, A.K.; Abdelhamid, T.H. Elimination of harmonics in multilevel inverters with non-equal dc sources using PSO. In Proceedings of the 2008 13th International Power Electronics and Motion Control Conference (EPE-PEMC 2008), Poznan, Poland, 1–3 September 2008; pp. 756–764.
39. Marzoughi, A.; Imaneini, H.; Moeini, A. An optimal selective harmonic mitigation technique for high power converters. *Int. J. Electr. Power Energy Syst.* **2013**, *49*, 34–39. [CrossRef]
40. Voltage Characteristics of Electricity Supplied by Public Electricity Networks. Available online: http://fs. gongkong.com/files/technicalData/201110/2011100922385600001.pdf (accessed on 15 January 2018).
41. Personen, M.A. Harmonics characteristic parameters methods of study estimates of existing values in the network. *Electra* **1981**, *77*, 35–54.
42. Electromagnetic Compatibility (EMC)–Part 3-7: LIMITS–Assessment of Emission Limits for the Connection of Fluctuating Installations to MV, HV and EHV Power Systems. Available online: https://ieeexplore.ieee. org/document/6232421/ (accessed on 15 January 2018).
43. Letha, S.S.; Thakur, T.; Kumar, J. Harmonic elimination of a photo-voltaic based cascaded H-bridge multilevel inverter using PSO (particle swarm optimization) for induction motor drive. *Energy* **2016**, *107*, 335–346. [CrossRef]
44. Rao, G.N.; Raju, P.S.; Sekhar, K.C. Harmonic elimination of cascaded H-bridge multilevel inverter based active power filter controlled by intelligent techniques. *Int. J. Electr. Power Energy Syst.* **2014**, *61*, 56–63.
45. Panda, A.K.; Patnaik, S.S. Analysis of cascaded multilevel inverters for active harmonic filtering in distribution networks. *Int. J. Electr. Power Energy Syst.* **2015**, *66*, 216–226. [CrossRef]
46. Mohan, N.; Undeland, T.M.; Robbins, W.P. *Power Electronics–Converter, Applications and Design*, 3rd ed.; John Wiley & Sons, Inc.: New York, NY, USA, 1989.
47. Miceli, R.; Schettino, G.; Viola, F. A Novel Computational Approach for Harmonic Mitigation in PV Systems with Single-Phase Five-Level CHBMI. *Energies* **2018**, *11*, 2100. [CrossRef]
48. IEEE. *IEEE Application Guide for IEEE Std 1547, IEEE Standard for Interconnecting Distributed Resources with Electric Power Systems IEEE Standard 1547.2-2008*; IEEE: Biloxi, MI, USA, 2009.
49. Crăciun, B.I.; Kerekes, T.; Séra, D.; Teodorescu, R. Overview of recent Grid Codes for PV power integration. In Proceedings of the 2012 13th International Conference on Optimization of Electrical and Electronic Equipment (OPTIM), Brasov, Romania, 24–26 May 2012; pp. 959–965.
50. Beres, R.N.; Wang, X.; Liserre, M.; Blaabjerg, F.; Bak, C.L. A Review of Passive Power Filters for Three-Phase Grid-Connected Voltage-Source Converters. *IEEE J. Emerg. Sel. Top. Power Electron.* **2016**, *4*, 54–69. [CrossRef]

51. Liserre, M.; Blaabjerg, F.; Dell'Aquila, A. Step-by-step design procedure for a grid-connected three-phase PWM Voltage Source Converter. *Int. J. Electron.* **2004**, *91*, 445–460. [CrossRef]

52. Liserre, M.; Blaabjerg, F.; Hansen, S. Design and control of an LCL-filter-based three-phase active rectifier. *IEEE Trans. Ind. Appl.* **2005**, *41*, 1281–1291. [CrossRef]

53. Bojrup, M. Advanced Control of Active Filters in a Battery Charger Application. Ph.D. Thesis, Lund University of Technology, Lund, Sweden, 1999.

54. Teodorescu, R.; Liserre, M.; Rodriguez, P. *Grid Converters for Photovoltaic and Wind Power System*; John Wiley & Sons, Ltd.: New York, NY, USA, 2011; ISBN 978-0-470-05751-3.

55. Blaabjerg, F.; Teodorescu, R.; Liserre, M.; Timbus, A.V. Overview of Control and Grid Synchronization for Distributed Power Generation Systems. *IEEE Trans. Ind. Electron.* **2006**, *53*, 1398–1409. [CrossRef]

56. Timbus, A.; Liserre, M.; Teodorescu, R.; Blaabjerg, F. Synchronization methods for three phase distributed power generation systems—An overview and evaluation. In Proceedings of the 2005 IEEE 36th Power Electronics Specialists Conference, Recife, Brazil, 16 June 2005; pp. 2474–2481.

57. International Rectifier, "Data Sheet IRFB4115PbF". Available online: www.irf.com (accessed on 15 January 2018).

58. ALTERA. *Cyclone III Device Handbook—Volume 1*; ALTERA: San Jose, CA, USA, December 2011; Available online: http://www.altera.com (accessed on 15 January 2018).

59. ALTERA. *Cyclone III Device Handbook—Volume 2*; ALTERA: San Jose, CA, USA, July 2012; Available online: http://www.altera.com (accessed on 15 January 2018).

60. ALTERA. *Quartus II Handbook Version 13.1—Volume 1: Design and Synthesis*; ALTERA: San Jose, CA, USA, November 2013; Available online: http://www.altera.com (accessed on 15 January 2018).

61. ALTERA. *Quartus II Handbook Version 13.1—Volume 2: Design Implementation and Optimization*; ALTERA: San Jose, CA, USA, November 2013; Available online: http://www.altera.com (accessed on 15 January 2018).

62. Zwoliński, M. *Digital System Design with VHDL*, 2nd ed.; Pearson-Prentice Hall: Upper Saddle River, NJ, USA, 2004.

63. Dordevic, O.; Jones, M.; Levi, E. Analytical Formulas for Phase Voltage RMS Squared and THD in PWM Multiphase Systems. *IEEE Trans. Power Electron.* **2015**, *30*, 1645–1656. [CrossRef]

64. Colak, I.; Kabalci, E.; Bayindir, R. Review of multilevel voltage source inverter topologies and control schemes. *Energy Convers. Manag.* **2011**, *52*, 1114–1128. [CrossRef]

Article

Fault Detection and Location of IGBT Short-Circuit Failure in Modular Multilevel Converters

Bin Jiang *, Yanfeng Gong and Yan Li

Department of Electrical and Electronic Engineering, North China Electric Power University,
Beijing 102206, China; yanfeng.gong@ncepu.edu.cn (Y.G.); yanli.ncepu@gmail.com (Y.L.)
* Correspondence: bin_jiang@ncepu.edu.cn; Tel.: +86-10-6177-1571

Received: 14 May 2018; Accepted: 5 June 2018; Published: 7 June 2018

Abstract: A single fault detection and location for Modular Multilevel Converter (MMC) is of great significance, as numbers of sub-modules (SMs) in MMC are connected in series. In this paper, a novel fault detection and location method is proposed for MMC in terms of the Insulated Gate Bipolar Translator (IGBT) short-circuit failure in SM. The characteristics of IGBT short-circuit failures are analyzed, based on which a Differential Comparison Low-Voltage Detection Method (DCLVDM) is proposed to detect the short-circuit fault. Lastly, the faulty IGBT is located based on the capacitor voltage of the faulty SM by Continuous Wavelet Transform (CWT). Simulations have been done in the simulation software PSCAD/EMTDC and the results confirm the validity and reliability of the proposed method.

Keywords: modular multilevel converter; IGBT short-circuit; fault detection; fault location; Differential Comparison Low-Voltage Detection Method (DCLVDM); Continuous Wavelet Transform

---

1. Introduction

Modular multilevel converters (MMCs) have attracted extensive attention and research in high-voltage and high-power applications. MMCs are composed of numbers of SMs connected in series and each SM is built up with two Insulated Gate Bipolar Translators (IGBTs), anti-paralleled diodes and a capacitor. The modular structure provides excellent features such as high output voltage quality, low harmonic distortion, low power loss, ease of construction and assembly, etc. MMC has been proved to be a valid and reliable topology for power transmission and it can be applied in many situations, i.e., interconnection with AC gird, the accessing of clean energy, the construction of DC distribution network and power supply to isolated passive loads. So far, there have been several MMC-based multi-terminal High Voltage Direct Current (MMC-MTDC) projects in operation or under construction, i.e., Tres Amigas superconductor transmission project in North American, Nan'ao MMC-MTDC and Zhoushan MMC-MTDC project in China, South-West Scheme MMC-MTDC project in Norway- Sweden, etc. [1–8].

A complete fault-tolerant strategy mainly includes fault detection, location, isolation and reconfiguration. Fault detection, as the first step in fault tolerance, should be conducted as fast as possible. A quick, reliable and precise fault detection method can gain time for the following steps which can prevent further failure and maintain steady operation [9–11]. The power semiconductor switch is one of most failure-prone components in a converter [12,13]. Considering that there are usually a large number of IGBTs in MMC, IGBT failures are more likely to take place. IGBT failures may result the converter operating abnormally, thus causing damage to other devices and might even threaten the security and reliability of power systems. Therefore, quick fault detection and isolation is vital and significant to MMCs.

For IGBT open-circuit fault detection methods, there has been extensive research. Reference [14] proposed a fault detection method by comparing the measured arm current and the expected current calculated by gate signals, capacitor voltage and phase voltage. The fault was identified if the measured current did not change as it was calculated. Reference [15] changed the voltage measurement point to the cell output terminal for control purposes to avoid extra sensors, and detected the fault based on the unconformity between the output voltage and the switching signals. However, this method required a delay unit and a memory unit in every SM, which led to high cost and complexity. The authors of [9,16] proposed a fault detection method based on a Sliding Mode Observer (SMO) for MMC. Fault occurrence is verified by comparing the observed and the measured states based on a switching model of every SM, then the SMO equations are modified to detect the faulty SM. Reference [17] proposed a fault diagnosis and tolerant control method and the fault diagnosis method detected the fault by a state observer and the knowledge of fault behaviors. In [18], a Kalman Filter (KF) was employed to detect the fault through comparing the measured state value and the estimated state value by KF and the faulty SM was located based on the voltage comparison between the faulty SM and the normal SMs. Reference [16–18] detected the fault by the changes of the circulating current, but the circulating current could also be influenced by other fault types, which may cause misjudgment. Moreover, SMO and KF require complex algorithm, huge computation and complex parameter setting.

IGBT short-circuit fault should be detected as fast as possible to protect the IGBT from destruction and to avoid another potential shoot-through fault [19,20]. Reference [21] proposed a fault detection method in multilevel converter STATCOMs based on the output DC link voltage of each phase but the method did not respond rapidly. Reference [22] proposed a quick fault identification method in cascaded H-bridge multilevel converters through the comparison between the output voltage and the reference voltage of each phase. However, this method heavily relies on the switching model of each phase. Reference [23] proposed a short-circuit fault detection method in one sub-module of cascaded H-bridge with dc link voltage but it neglected that voltage error could be generated not merely by short-circuit fault. Reference [24] proposed a short-circuit detection method through pattern recognition of IGBT gate voltage. Reference [25] proposed a self-diagnosis function for power MOSFETs and IGBTs based on monitoring of the gate charge and discharge current. However, a short-circuit fault in IGBT is usually detected by additional sensors or detection circuit such as those in [24,25]. The additional sensors and circuits add not only extra cost but also extra complexity to the systems. Comparisons are made in Table 1 to make the methods more straightforward.

Table 1. Fault detection methods.

Reference	Requires	Complexity	Detection Time
[21]	output voltage and gate signal	simple calculation	about 1 cycle
[22,23]	phase voltage	complex model and additional sensors	numbers of sampling intervals
[24,25]	gate voltage or current	additional sensors and circuits	tens of microseconds

The contributions of the paper are: (1) Differential Comparison Low-Voltage Detection Method (DCLVDM) is proposed for fault detection in terms of IGBT short-circuit. The DCLVDM is composed of two parts—the low-voltage part and the differential comparison part. (2) Continuous Wavelet Transform (CWT) is applied to locate the faulty IGBT based on the singularity of the capacitor voltage. The fault detection and location method proposed in this paper requires no additional measuring device and the algorithm is easy to realize. The simulation results also confirm the effectiveness and reliability of the proposed method.

2. Operation Principles, Fault Analysis and Calculation of MMC

2.1. Structure and Control Strategy of MMC

A three-phase MMC topology is illustrated in Figure 1 and each arm consists of n SMs, an arm inductor L_0 and an equivalent loss resistance R_0 in series. Each SM is composed of IGBT T_1, T_2, anti-paralleled diodes D_1, D_2 and a capacitor C_0. The SM is set ON/OFF under the control of a switching function S which is defined as Equation (1)

$$S = \left\{ \begin{array}{lll} 1, & g_1 = 1 \ , & g_2 = 0 \\ 0, & g_1 = 0 \ , & g_2 = 1 \end{array} \right. \tag{1}$$

where g_1 and g_2 are the gate signals for switches. When S is 1, the SM is "ON" and T_1 is conducted and T_2 is blocked. When S is 0, the SM is "OFF" and T_1 is blocked and T_2 is conducted. In normal operation, to maintain the required DC voltage, a Capacitor Voltage Balancing Method (CVBM) is applied in MMC-HVDC system [26,27]. The CVBM, SMs with higher voltages discharging ($i_{arm} < 0$) in priority and SMs with lower voltages charging ($i_{arm} > 0$) in priority, determines which SMs are ON/OFF, that is the value of S.

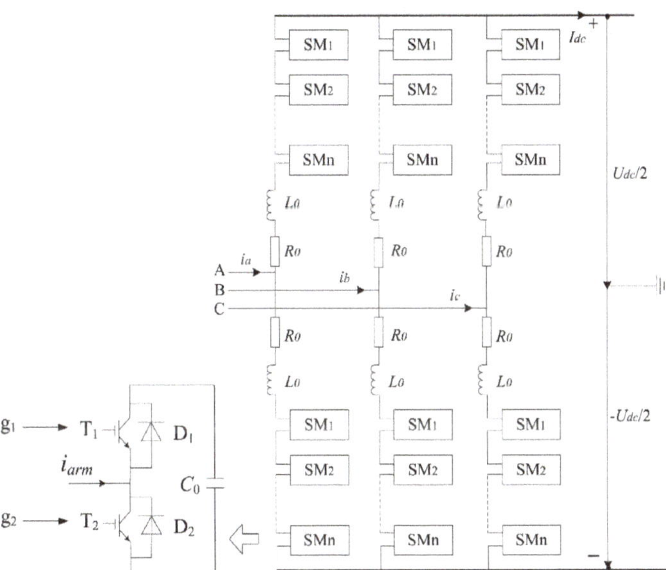

Figure 1. Three-phase MMC topology.

2.2. Fault Characteristics Analysis of SM

2.2.1. T_1 Short-Circuit

With $i_{arm} > 0$ and S = 1, T_1 short-circuit will have no impact on capacitor charging process. When S is 0 and T_2 is conducted, T_1 short-circuit will result in the capacitor's short-circuit via T_1 and T_2 and the capacitor discharges rapidly. The current path is shown in Figure 2. Similarly, with $i_{arm} < 0$ and S = 1, T_1 short-circuit will have no influence on capacitor discharging process. Only when S is 0, T_1 short-circuit will result in the capacitor short-circuit. The current path is shown in Figure 3.

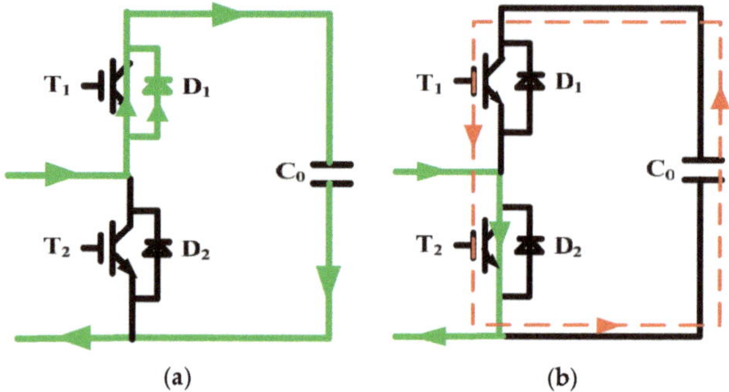

Figure 2. Current path when T_1 is short-circuit with $i_{arm} > 0$. (**a**) S = 1. (**b**) S = 0.

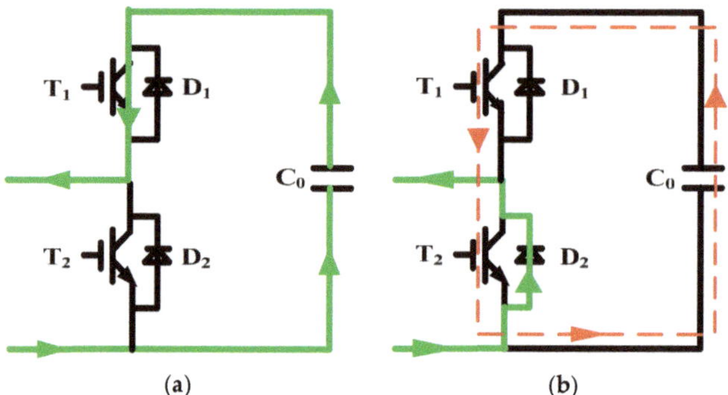

Figure 3. Current path when T_1 is short-circuit with $i_{arm} < 0$: (**a**) S = 1; (**b**) S = 0.

2.2.2. T_2 Short-Circuit

With $i_{arm} > 0$ and S = 0, T_2 short-circuit will have no impact on capacitor charging process. When S is 1, T_2 short-circuit will cause the capacitor's short-circuit via T_1 and T_2. As a result, the capacitor switches from normal charging state to fault discharging state. Similarly, with $i_{arm} < 0$ and S = 1, T_2 short-circuit will result in the capacitor switching from normal discharging state to fault discharging state. The current path is shown in Figures 4 and 5.

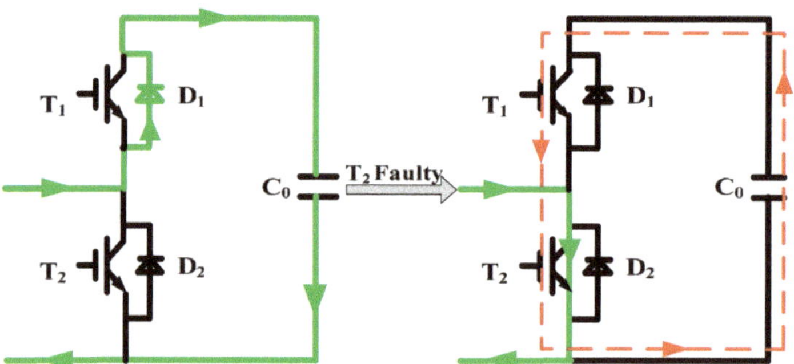

Figure 4. Current path when T_2 is short-circuit with $i_{arm} > 0$ and S = 1.

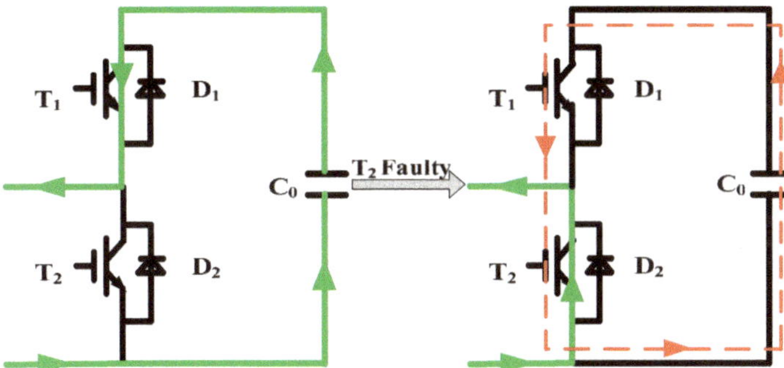

Figure 5. Current path when T_2 is short-circuit with $i_{arm} < 0$ and $S = 1$.

Based on the analysis above, state of the capacitor under different IGBTs short-circuit faults can be obtained as shown in Table 2. It can be concluded that an IGBT short-circuit will result in the capacitor's short-circuit once the other IGBT is "ON" in the same SM. Because of the small value of the time constant, the capacitor discharges rapidly. The capacitor voltage decreases and the capacitor current increases rapidly at the same time. The capacitor voltage and current saltation are the most apparent characteristics when short-circuit occurs, which can be applied to short-circuit detection method directly.

Table 2. State of the capacitor under different IGBTs short-circuit faults.

Fault IGBT	SM State	S	i_{arm}	Capacitor State
T_1	ON	1	>0	Normal
			<0	Normal
	OFF	0	>0	Short-circuit
			<0	Short-circuit
T_2	ON	1	>0	Short-circuit
			<0	Short-circuit
	OFF	0	>0	Normal
			<0	Normal

2.3. Capacitor Voltage Calculation

IGBT short-circuit will cause the capacitor to discharge via T_1 and T_2. The equivalent discharge circuit is illustrated as Figure 6. R_1 and R_2 are the equivalent on-resistance of T_1 and T_2, respectively and R equals $R_1 + R_2$. u_c is the capacitor voltage and i_c is the capacitor current.

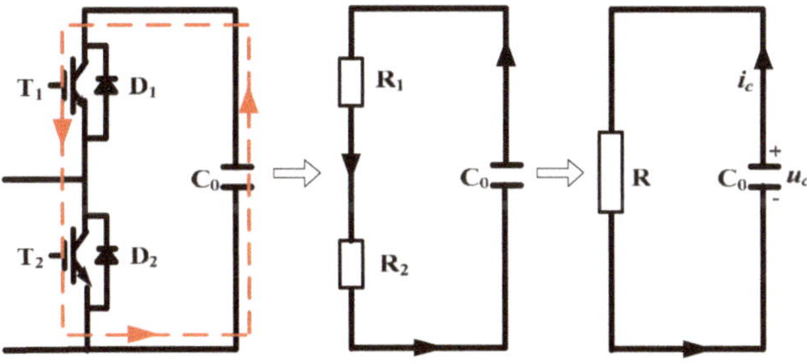

Figure 6. Equivalent discharge circuit for the capacitors.

The equivalent discharge circuit is a first-order RC circuit. Therefore, the circuit equations can be deduced as Equations (2) and (3)

$$i_c = -C_0 \frac{du_c}{dt} \tag{2}$$

$$u_c + RC_0 \frac{du_c}{dt} = 0 \tag{3}$$

With The initial state $u_c(0) = U_0$, so u_c and its derivation can be deduced as Equations (4) and (5)

$$u_c(t) = U_0 e^{-\frac{t}{RC_0}} \quad (t \geq 0) \tag{4}$$

$$\frac{du_c(t)}{dt} = -\frac{U_0}{RC_0} e^{-\frac{t}{RC_0}} = -\frac{u_c(t)}{RC_0} \quad (t \geq 0) \tag{5}$$

3. Proposed Fault Detection and Location Method for MMC

The short-circuit fault detection and location method for MMC proposed in this paper is divided into two stages. Stage 1: the fault SM detection. Stage 2: the fault IGBT location.

Stage 1: Faulty SM Detection

According to the fault characteristics analysis above, this paper proposes a novel Differential Comparison Low-Voltage Detection Method (DCLVDM) base on the capacitor voltage u_c of the fault SM. The criteria of DCLVDM is formulated as Equation (6).

$$\begin{cases} |u_c(t)| \leq \alpha \\ |du_{act}(t) - du_{cal}(t)| \leq \varepsilon \end{cases} \tag{6}$$

$|u_c(t)| \leq \alpha$ is the low-voltage part. α is the threshold value accordingly and is defined as Equation (7) where K_α is reliability coefficient and u_{ce} is the rated value of u_c under steady state. $|du_{act}(t) - du_{cal}(t)| \leq \varepsilon$ is the differential comparison part. ε is the threshold value accordingly. du_{act} and du_{cal} are the actual value and calculation value of change rate of u_c, respectively, and they are formulated as Equations (8) and (9). T_s is the sample period.

$$\alpha = K_\alpha u_{ce} \tag{7}$$

$$du_{act}(t) = -\frac{u_c(t)}{RC_0} \tag{8}$$

$$du_{cal}(t) = \frac{u_c(t) - u_c(t - T_s)}{T_s} \tag{9}$$

The criteria of DCLVDM consists of two parts: the low-voltage part and the differential comparison part. In the low-voltage part, a tentative conclusion that a short-circuit fault occurs can be reached if the capacitor voltage declines to the threshold value. However, IGBT short-circuit is not the only fault type which can cause capacitor voltage to decline, so differential comparison is introduced to confirm the fault. In differential comparison part, the calculation of du_{act} is based on the short-circuit characteristics as Equation (5), hence, the error between du_{act} and du_{cal} under fault condition can be much smaller than that under steady state, which can be used to distinguish the fault. Only if Equation (6) is proved to be true and it lasts for a certain period T_0, the IGBT short-circuit fault can be confirmed in a SM. The proposed DCLVDM requires no extra voltage measurement as the capacitor voltage is continuously measured for MMC control purpose.

Stage 2: Faulty IGBT Location Stage

As is indicated in Table 2, to locate the specific faulty IGBT, the value of Function S should be confirmed when the capacitor discharges. If there is $S = 1(S = 0)$ at the time when the capacitor voltage decreases, then T_2 (T_1) is proved to be faulty. Hence the key to locate the faulty IGBT is to locate the time spot when the capacitor voltage begins to decrease, that is, the singularity of u_c.

Continuous Wavelet Transform(CWT) is an effective time-frequency signal processing tool as it decomposes a signal in multiple scales or resolutions and retains both time and frequency domain information in the transform coefficients, which turns out to be useful in fault detection and location [28]. The CWT of a function f(t) with respect to a mother wavelet $\psi(t)$ is defined as Equations (10) and (11)

$$W(a,b) = \int_{-\infty}^{+\infty} f(t)\psi_{a,b}^*(t)dt \tag{10}$$

$$\psi_{a,b}(t) = \frac{1}{\sqrt{a}}\psi\left(\frac{t-b}{a}\right) \tag{11}$$

Here, a and b are the scale and translation factors, respectively. '*' indicates a complex conjugate. W(a,b) is the Wavelet Transform Coefficient (WTC) of f(t).

In time domain, the modulus maxima of the WTC represent the singularity of a signal and the degree of signal saltation can be characterized by the amplitude of the modulus maxima. For time-frequency analysis, a smoother continuous wavelet in time-domain is preferred as it brings better localization characteristics. In this paper, the second-order Gauss wavelet is chosen as the mother wavelet, which has excellent performance in the singularity detection [29].

To make the short-circuit fault detection and location method easier to understand, the flowchart of the method is displayed in Figure 7.

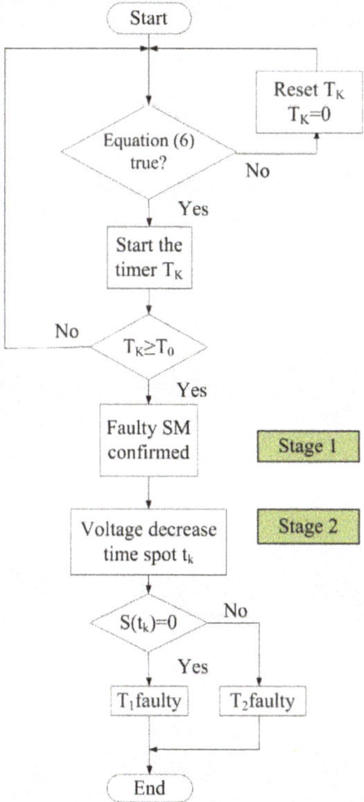

Figure 7. Flowchart of the short-circuit fault detection and location method.

4. Case Studies

In this section, to evaluate the effectiveness of the proposed fault detection and location method, a two-terminal MMC-HVDC system is constructed in the professional software PSCAD/EMTDC. The MMC circuit parameters are given in the Appendix A. Two cases are conducted:

Case 1: short-circuit fault of T_2 occurs in a SM at t = 0.02100 s.

Case 2: short-circuit fault of T_1 occurs in a SM at t = 0.03500 s.

4.1. Fault Characteristics

Figure 8a,b show the MMC performance under faulty operation of case 1 and case 2, respectively. When the short-circuit fault occurs, the capacitor voltage deceases rapidly which results in the sharp increase of the capacitor current. As previously mentioned, the capacitor current flows within the SM and does not flow to the arms, hence, the fault will have no obvious influence on the arm current.

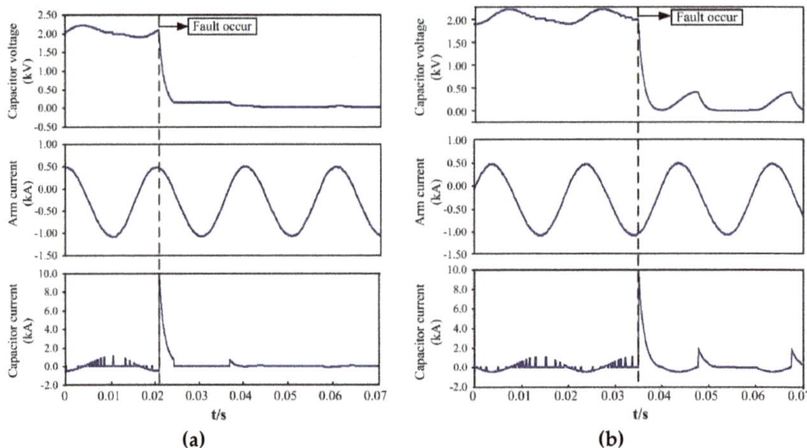

Figure 8. Performance of MMC under faulty operation: (**a**) Case 1; (**b**) Case 2.

4.2. Fault Detection

Figure 9 shows the performance of the proposed DCLVDM. To ensure the reliability and sensitivity of the detection method, the values of K_α, ε and T_0 are set as $K_\alpha = 0.8$, $\varepsilon = 0.7$, $T_0 = 1$ ms.

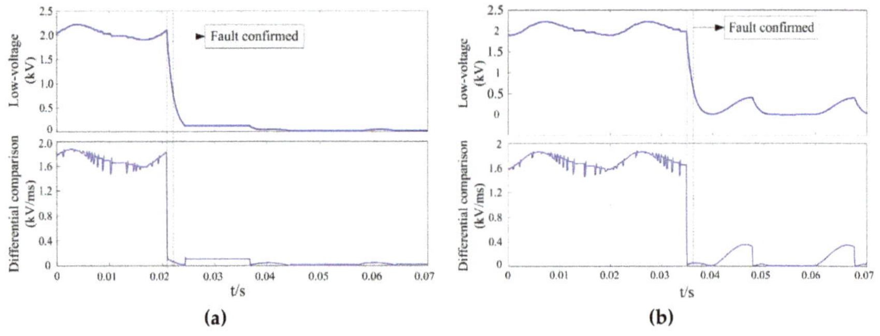

Figure 9. Performance of the proposed DCLVDM under faulty condition: (**a**) Case 1; (**b**) Case 2.

For Case 1, the short-circuit fault is detected 1.36 ms later after the fault occurs. The detection time is 1.26 ms for Case 2. The error between du_{act} and du_{cal} in steady state is much larger than that of faulty state, which is very useful in fault detection.

To prove the validity of the proposed DCLVDM, numerous simulations are carried out. The values of K_α, ε, T_0 remain the same as mentioned above. The simulation performance is similar to Figure 9. Because of space constraints, the simulation performance is not displayed. The proposed DCLVDM combined with the proper value of related parameters can detect the short-circuit fault accurately. For reliability coefficient K_α, a larger value will improve the sensitivity of the low-voltage part to voltage decline, which can shorten the detection time. However, too large a value will reduce the reliability because potential disturbance or noise, etc. which may cause minor voltage decline, can be mistaken for short-circuit fault. For the threshold, ε, the value of ε should be set between the minimum of the differential comparison under steady state and the maximum under faulty state combined with a margin. In this paper, the range of ε is about 0.35–1.48 according to the simulations. Therefore, setting ε as 0.7 is sufficient and reasonable to detect the fault. As for T_0, a larger value will improve the reliability of the fault detection process but lengthen the detection time. Because the IGBTs are of the same model in the same MMC converter and each SM has the same operation characteristics, the proposed DCLVDM and the parameter value setting are universally effective, which has been validated by numerous simulations. Actually, there is no a specification standard for the parameters setting with SM short-circuit, the parameters setting should take into consideration the sensitivity, speed, and reliability of the detection process comprehensively.

4.3. Fault Location

After the faulty SM is confirmed, the next step is to locate the faulty IGBT. As is mentioned above, the CWT of the capacitor voltage needs to be conducted. In this paper, the second-order Gauss wavelet is chosen as the mother wavelet as Equation (12). Magnitude values of the WTC are obtained for the corresponding input capacitor voltage for scale values of 1, 2, 3, 4, 5, 6, 8, 12, 16, and 20.

$$\psi(t) = \frac{2}{\sqrt{3}}\pi^{-1/4}(t^2 - 1)e^{-t^2/2} \tag{12}$$

Figure 10 shows the performance of the Gauss wavelet under various scales on the capacitor voltage. It is noticed that the Gauss wavelet under scale 1 to 8 produces the WTC modulus maxima exactly at the voltage singularity. However, a location deviation is produced under scale 12, 16, and 20. So scale 1 to 8 is sufficient to detect the singularity under this simulation condition.

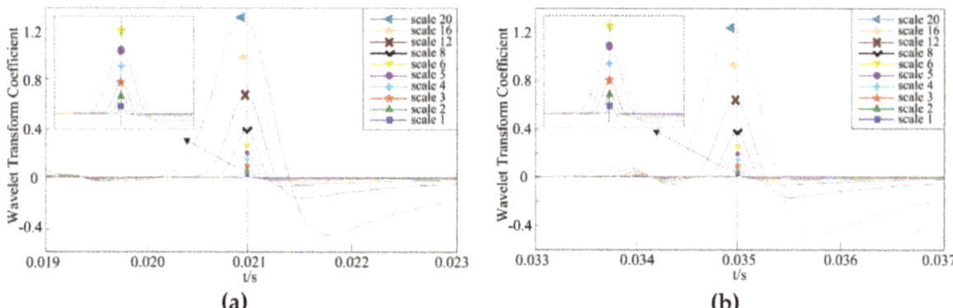

Figure 10. Wavelet Transform Coefficient under various scales: (a) Case 1; (b) Case 2.

However, noise may occur in practical operation. To investigate the influence of noise on the accuracy of the estimated singularity point, 40-db white Gaussian noise is added to the voltage signal [30,31]. As the two cases have similar characteristics, only Case 2 is conducted as an example. The simulation result is shown in Figure 11.

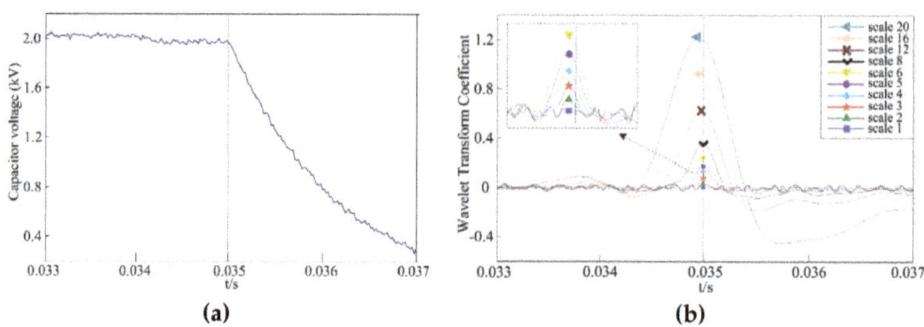

Figure 11. Fault location result with 40-db white Gaussian noise for Case 2: (**a**) Contaminated capacitor voltage; (**b**) Wavelet Transform Coefficients under various scales.

In Figure 11, firstly, a location deviation is still produced under scale 12, 16, and 20. Secondly, the fault singularity is hard to locate under scale 1 and 2 as the noise produces many more singularities in the process. Lastly, scale 3 to 8 may be chosen to locate the voltage singularity because the amplitude of the modulus maxima produced by the fault is much higher than that produced by the noise. Therefore, to locate the voltage singularity, the value of the scale cannot be too high or too low. In this paper, scale 4 is chosen to conduct the fault location as is shown in Figure 12.

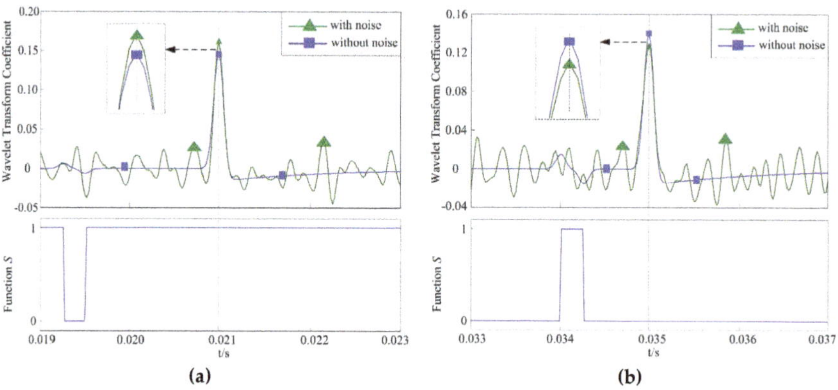

Figure 12. Faulty IGBT location under scale 4: (**a**) Case 1; (**b**) Case 2.

Figure 12 shows the detailed Wavelet Transform Coefficient (WTC) of SM voltages under scale 4 with or without noise, as well as their Function S. The singularity of the capacitor voltage caused by the fault is accurately located by the WTC modulus maxima. In Figure 12a, the singularity is located at $t = 0.02100$ s, meanwhile, the value of S is 1, So T_2 is proved to be faulty. Similarly, T_1 can be proved to be faulty according to Figure 12b at $t = 0.03500$ s. The proposed faulty IGBT location method based on CWT can accurately locate the singularity of the voltage capacitor caused by the fault, thus locating the faulty IGBT efficiently and precisely.

4.4. Comparison Analysis

As introduced in [22–25], most conventional IGBT fault detection and protection methods are hardware circuit-based and very few are algorithm-based. Hardware-based detection methods configure a detailed detection circuit for every single IGBT. In the detection circuit, gate voltage, collector-emitter voltage, collector current, etc. are been measured to detect the abnormal operation of IGBT. Therefore, hardware-based detection methods can detect the fault rapidly. However, MMC is composed of a series of SMs and each SM contains two IGBTs, thus, For an N+1 level half-bridge MMC system shown in Figure 1, at least 12N IGBTs are configured. Therefore, if the hardware detection method is adopted, for a 501-level MMC, 6000 detection circuits have to be configured besides more sensors. It absolutely adds complexity and huge cost to the MMC system. The proposed fault detection and location method in this paper is algorithm-based. The method detects the fault only by the capacitor voltage and the switch function, requiring no detection circuit or additional sensors as the capacitor voltage and switch function are originally measured for MMC control purpose. Therefore, hardware-based detection method and the algorithm-based method proposed in this paper have merits and demerits respectively, especially for IGBT of MMC, but they both provide valuable reference for MMC protection system construction in the future work.

5. Conclusions

In this paper, a novel fault detection and location method is proposed for IGBT short-circuit fault in MMC based on the SM capacitor voltage. The fault detection is carried out based on the proposed Differential Comparison Low-Voltage Detection Method (DCLVDM). No additional sensors are required, and a simple arithmetic operation is sufficient to detect the faulty SM. The fault location is fulfilled by the Continuous Wavelet Transform (CWT) based on the capacitor voltage of the faulty SM. The proposed method not only detects the faulty SM rapidly, but it can also locate the specific faulty IGBT precisely. Simulation studies conducted in PSCAD/EMTDC prove that the proposed fault detection and location method in this paper is effective. In respect of practical application, this method requires accurate voltage measurement and fast sampling frequency, both of which bring more challenges to the sensors and the calculation equipment. Both reliability and rapidity must be taken into consideration when setting the corresponding threshold values of DCLVDM. In the future, the authors will conduct deep research on the influence exerted by different disturbances and other fault location methods besides wavelet and aim to make the fault detection and location of IGBT short-circuit failure in MMC more universal, reliable and feasible.

Author Contributions: B.J. defined the problem, proposed the fault detection method and conducted the simulation. Y.G. developed the fault location method. Y.L. gathered the necessary data. B.J. and Y.G. contributed in the paper writing. B.J. and Y.L. handled the paper revisions.

Acknowledgments: This research was funded by "the Fundamental Research Funds for the Central Universities 2017XS015".

Conflicts of Interest: The authors declare no conflict of interest.

Appendix A

Table A1. Parameters of the simulated MMC system.

Quantity	Value
DC nominal voltage (U_{dc})	$\pm 20\,kV$
Number of SMs per arm (N)	20
Power transmission (P)	20 MW
Arm inductor (L_0)	15 mH
SM capacitance (C_0)	6 mF
AC line voltage (Vac)	10 kV
Loss resistance (R_0)	$0.1\,\Omega$
SM capacitor voltage	2 kV
equivalent on-resistance (R_1, R_2)	$0.01\,\Omega$

References

1. Debnath, S.; Qin, J.; Bahrani, B.; Saeedifard, M.; Barbosa, P. Operation, Control, and Applications of the Modular Multilevel Converter: A Review. *IEEE Trans. Power Electron.* **2015**, *30*, 37–53. [CrossRef]

2.	Perez, M.A.; Bernet, S.; Rodriguez, J.; Kouro, S.; Lizana, R. Circuit Topologies, Modeling, Control Schemes, and Applications of Modular Multilevel Converters. *IEEE Trans. Power Electron.* **2015**, *30*, 4–17. [CrossRef]

3.	Nami, A.; Liang, J.; Dijkhuizen, F.; Demetriades, G.D. Modular Multilevel Converters for HVDC Applications: Review on Converter Cells and Functionalities. *IEEE Trans. Power Electron.* **2015**, *30*, 18–36. [CrossRef]

4.	Harnefors, L.; Antonopoulos, A.; Norrga, S.; Angquist, L.; Nee, H.P. Dynamic Analysis of Modular Multilevel Converters. *IEEE Trans. Ind. Electron.* **2013**, *60*, 2526–2537. [CrossRef]

5.	Gebreel, A.A.; Xu, L. Power quality and total harmonic distortion response for MMC with increasing arm inductance based on closed loop–needless PID controller. *Electr. Power Syst. Res.* **2016**, *133*, 281–291. [CrossRef]

6.	Rohner, S.; Bernet, S.; Hiller, M.; Sommer, R. Modulation, Losses, and Semiconductor Requirements of Modular Multilevel Converters. *IEEE Trans. Ind. Electron.* **2010**, *57*, 2633–2642. [CrossRef]

7.	Wu, J.; Wang, Z.X.; Xu, L.; Wang, G.Q. Key technologies of VSC–HVDC and its application on offshore wind farm in China. *Renew. Sustain. Energy Rev.* **2014**, *36*, 247–255. [CrossRef]

8.	Jiang, B.; Wang, Z. The key technologies of VSC–MTDC and its application in China. *Renew. Sustain. Energy Rev.* **2016**, *62*, 297–304.

9.	Shao, S.; Watson, A.J.; Clare, J.C.; Wheeler, P.W. Robustness Analysis and Experimental Validation of a Fault Detection and Isolation Method for the Modular Multilevel Converter. *IEEE Trans. Power Electron.* **2016**, *31*, 3794–3805. [CrossRef]

10.	Sujil, A.; Choudhary, S.; Verma, J.; Kumar, R. Agent based intelligent fault detection, isolation and restoration in smart power system. In Proceedings of the 2016 IEEE Students' Conference on Electrical, Electronics and Computer Science (SCEECS), Bhopal, India, 5–6 March 2016; pp. 1–6.

11.	Hwang, I.; Kim, S.; Kim, Y.; Seah, C.E. A Survey of Fault Detection, Isolation, and Reconfiguration Methods. *IEEE Trans. Control Syst. Technol.* **2010**, *18*, 636–653. [CrossRef]

12.	Lu, B.; Sharma, S.K. A Literature Review of IGBT Fault Diagnostic and Protection Methods for Power Inverters. *IEEE Trans. Ind. Appl.* **2009**, *45*, 1770–1777.

13.	Yang, S.; Bryant, A.; Mawby, P.; Xiang, D.; Ran, L.; Tavner, P. An Industry–Based Survey of Reliability in Power Electronic Converters. *IEEE Trans. Ind. Appl.* **2011**, *47*, 1441–1451. [CrossRef]

14.	Salimian, H.; Iman–Eini, H.; Farhangi, S. Open–circuit fault detection and location in Modular Multilevel Converter. In Proceedings of the 6th Power Electronics, Drive Systems & Technologies Conference (PEDSTC2015), Tehran, Iran, 3–4 February 2015; pp. 383–388.

15.	Haghnazari, S.; Khodabandeh, M.; Zolghadri, M.R. Fast fault detection method for modular multilevel converter semiconductor power switches. *IET Power Electron.* **2016**, *9*, 165–174. [CrossRef]

16.	Shao, S.; Wheeler, P.W.; Clare, J.C.; Watson, A.J. Fault Detection for Modular Multilevel Converters Based on Sliding Mode Observer. *IEEE Trans. Power Electron.* **2013**, *28*, 4867–4872. [CrossRef]

17.	Li, B.; Shi, S.; Wang, B.; Wang, G.; Wang, W.; Xu, D. Fault Diagnosis and Tolerant Control of Single IGBT Open–Circuit Failure in Modular Multilevel Converters. *IEEE Trans. Power Electron.* **2016**, *31*, 3165–3176. [CrossRef]

18.	Deng, F.; Chen, Z.; Khan, M.R.; Zhu, R. Fault Detection and Location Method for Modular Multilevel Converters. *IEEE Trans. Power Electron.* **2015**, *30*, 2721–2732. [CrossRef]

19.	Chokhawala, R.S.; Catt, J.; Kiraly, L. A discussion on IGBT short-circuit behavior and fault protection schemes. *IEEE Trans. Ind. Appl.* **1995**, *31*, 256–263. [CrossRef]

20.	Barnes, M.J.; Blackmore, E.; Wait, G.D.; Lemire-Elmore, J.; Rablah, B.; Leyh, G.; Nguyen, M.N.; Pappas, C. Analysis of High–Power IGBT Short Circuit Failures. *IEEE Trans. Plasma Sci.* **2005**, *33*, 1252–1261. [CrossRef]

21.	Yazdani, A.; Sepahvand, H.; Crow, M.L.; Ferdowsi, M. Fault Detection and Mitigation in Multilevel Converter STATCOMs. *IEEE Trans. Ind. Electron.* **2011**, *58*, 1307–1315. [CrossRef]

22.	Brando, G.; Dannier, A.; del Pizzo, A.; Rizzo, R. Quick identification technique of fault conditions in cascaded H–Bridge multilevel converters. In Proceedings of the 2007 International Aegean Conference on Electrical Machines and Power Electronics, Bodrum, Turkey, 10–12 September 2007; pp. 491–497.

23.	Shahbazi, M.; Zolghadri, M.R.; Poure, P.; Saadate, S. Fast short circuit power switch fault detection in cascaded H–bridge multilevel converter. In Proceedings of the 2013 IEEE Power & Energy Society General Meeting, Vancouver, BC, Canada, 21–25 July 2013; pp. 1–5.

24. Lee, J.B.; Hyun, D.S. Gate Voltage Pattern Analyze for Short-circuit Protection in IGBT Inverters. In Proceedings of the 2007 IEEE Power Electronics Specialists Conference, Orlando, FL, USA, 17–21 June 2007; pp. 1913–1917.
25. Lihua, C.; Peng, F.Z.; Cao, D. A smart gate drive with self–diagnosis for power MOSFETs and IGBTs. In Proceedings of the 2008 Twenty–Third Annual IEEE Applied Power Electronics Conference and Exposition, Austin, TX, USA, 24–28 February 2008; pp. 1602–1607.
26. Yinglin, X.; Zheng, X. On the Bipolar MMC–HVDC Topology Suitable for Bulk Power Overhead Line Transmission: Configuration, Control, and DC Fault Analysis. *IEEE Trans. Power Deliv.* **2014**, *29*, 2420–2429.
27. Wang, S.S.; Zhou, X.X.; Tang, G.F.; He, Z.; Teng, L.; Bao, H.L. Analysis of submodule overcurrent caused by dc pole–to–pole fault in modular multilevel converter HVDC system. *Proc. CSEE* **2011**, *31*, 1–7. (In Chinese)
28. Silva, M.D.; Coury, D.V.; Oleskovicz, M.; Segatto, E.C. Combined solution for fault location in three–terminal lines based on wavelet transforms. *IET Gener. Trans. Dis.* **2010**, *4*, 94–103. [CrossRef]
29. Lin, S.; He, Z.Y.; Li, X.P.; Qian, Q.Q. Travelling wave time–frequency characteristic–based fault location method for transmission lines. *IET Gener. Trans. Dis.* **2012**, *6*, 764–772. [CrossRef]
30. Azizi, S.; Sanaye–Pasand, M.; Abedini, M.; Hasani, A. A Traveling–Wave–Based Methodology for Wide–Area Fault Location in Multi–terminal DC Systems. *IEEE Trans. Power Deliv.* **2014**, *29*, 2552–2560. [CrossRef]
31. Liu, X.; Osman, A.H.; Malik, O.P. Hybrid Traveling Wave/Boundary Protection for Monopolar HVDC Line. *IEEE Trans. Power Deliv.* **2009**, *24*, 569–578. [CrossRef]

Article

FPGA-Based Controller for a Permanent-Magnet Synchronous Motor Drive Based on a Four-Level Active-Clamped DC-AC Converter

Joan Nicolas-Apruzzese *, Emili Lupon, Sergio Busquets-Monge, Alfonso Conesa, Josep Bordonau and Gabriel García-Rojas

Electronic Engineering Department, Universitat Politècnica de Catalunya, 08028 Barcelona, Spain;
emili.lupon@upc.edu (E.L.); sergio.busquets@upc.edu (S.B.-M.); alfonso.conesa@upc.edu (A.C.);
josep.bordonau@upc.edu (J.B.); gabriel.garcia.rojas@alu-etsetb.upc.edu (G.G.-R.)
* Correspondence: joan.nicolas@upc.edu; Tel.: +34-93-401-7152

Received: 3 August 2018; Accepted: 28 September 2018; Published: 2 October 2018

Abstract: This paper proposes a closed-loop control implementation fully-embedded into an FPGA for a permanent-magnet synchronous motor (PMSM) drive based on a four-level active-clamped converter. The proposed FPGA controller comprises a field-oriented control to drive the PMSM, a DC-link voltage balancing closed-loop control (VBC), and a virtual-vector-based modulator for a four-level active-clamped converter. The VBC and the modulator operate in consonance to preserve the DC-link capacitor voltages balanced. The FPGA design methodology is carefully described and the main aspects to achieve an optimal FPGA implementation using low resources are discussed. Experimental results under different operating conditions are presented to demonstrate the good performance and the feasibility of the proposed controller for motor-drive applications.

Keywords: DC-link voltage balancing; field-oriented control; field-programmable gate array; multilevel active-clamped converter; motor drive

---

1. Introduction

The use of multilevel power converters for industrial applications has increased significantly in the last years thanks to their advantages compared to conventional two-level converters [1]. Some of these advantages are higher efficiency, higher power density, reduced harmonic distortion, etc. However, multilevel converters present some drawbacks, such as a higher number of switches and an increased control complexity. The higher control complexity is not just because they contain more devices. In some multilevel topologies, the DC-link bus is split into several partial voltages with the inclusion of capacitors. This implies necessary control actions to keep these capacitor voltages balanced.

This is the case, for instance, of the multilevel active-clamped (MAC) topology [2]. Figure 1 depicts the four switching states of a four-level MAC converter leg to illustrate the converter operation. The circled switches are on-state devices and the non-circled ones are off-state devices. The solid-line circled switches conduct the main current (i_o) and the dotted-line circled switches simply clamp the blocking voltage of the off-state devices to the voltage across adjacent levels. Compared to the commonly-used diode-clamped topology, which presents a lower number of switches, the MAC converter advantages are: lower conduction losses, improved switching-losses distribution, blocking voltage of a device always equal to the voltage across adjacent levels, and improved fault-tolerance capacity [3]. Motor drives, and in particular the traction inverter of electric vehicles, is one of the applications where the MAC converter appears to be of interest. Therefore, the authors propose to use a MAC converter to drive a permanent-magnet synchronous motor (PMSM).

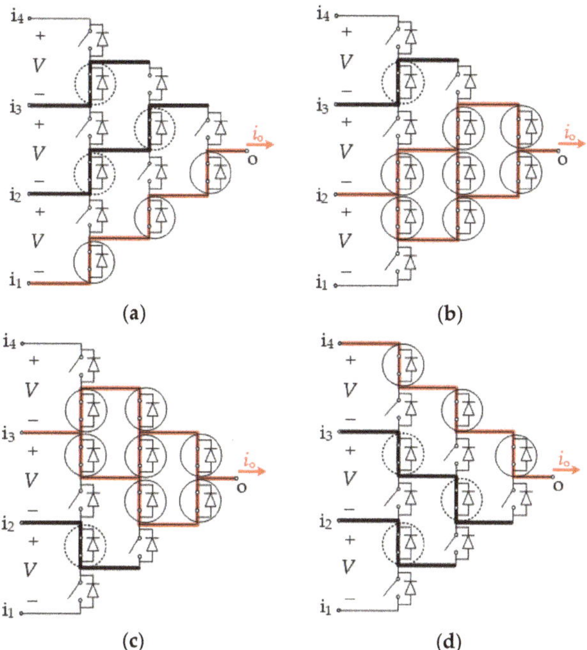

Figure 1. Four-level MAC leg switching states. (**a**) Connection to node i_1. (**b**) Connection to node i_2. (**c**) Connection to node i_3. (**d**) Connection to node i_4.

The MAC topology belongs to the family of neutral-point clamped (NPC) topologies [4]. In this family of topologies, one typical configuration consists of connecting a set of capacitors in series to passively generate the multiple voltage levels, see Figure 2. Other configurations are possible, as for instance connecting DC-voltage sources or batteries instead of capacitors across the adjacent input nodes. When DC-link capacitors are used, as in the present proposal, these topologies intrinsically present the challenge of balancing the capacitor voltages, since the classical modulation schemes lead to a voltage unbalancing. The voltage balancing problem arises from the existence of non-zero currents in the inner DC-link points (nodes i_2 and i_3 in Figure 1). This issue has been widely reported and investigated in the literature [5]. The diverse solutions proposed to solve this problem can be generally classified as hardware and software solutions. Hardware solutions introduce auxiliary circuitry to inject/draw additional current into/from the inner DC-link points to compensate the inherent converter current. Software solutions consist in defining a suitable modulation that is defined so as to maintain the average current of each inner node equal to zero over a specific period of time.

Among the software and hardware solutions, the authors propose to use a software solution since they are cheaper, present better performance, and they are simpler to implement. Among the different software solutions, the authors have selected the virtual-vector-based modulation originally introduced for three levels in [6] and extended to an arbitrary number of levels in [7], in which the average current of each inner DC-link point is maintained equal to zero over a single switching cycle. This modulation strategy enables to minimize the size of the capacitors, which leads to a higher power density. Although the applied modulation scheme is intended to preserve the balance of the DC-link capacitor voltages in every switching cycle, it is necessary to apply an additional control loop to guarantee a tight voltage balancing, since non-idealities lead to DC-link voltage unbalancing. The voltage balancing control (VBC) scheme proposed in [5] is implemented here to perform this action.

It is noteworthy that [8,9] state that the modulation scheme used here cannot work properly with dynamic loads such as motors, because the converter would not be able to keep the capacitor-voltage fluctuations low, leading to a system instability. This modulation is implemented here together with the VBC to drive a PMSM, demonstrating the feasibility of the proposed controller for motor-drive applications. Additionally, a four-level converter is used, in which the capacitor voltage balancing is much more challenging than in a three-level converter since some of the capacitor voltages may collapse [4].

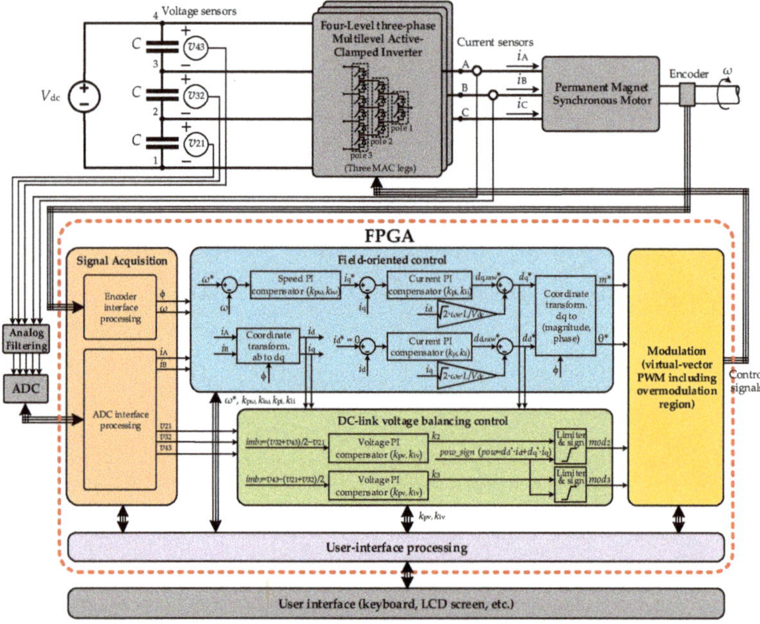

Figure 2. Global system overview.

In order to take full advantage of the traction inverter, a proper and proficient controller has to be developed. Typical power-converter digital controllers are implemented on microprocessors (µP), digital signal processors (DSP) and/or field-programmable gate arrays (FPGA). FPGA architecture permits both the parallel and sequential processing of data at high clock frequencies, which dramatically reduces the needed processing time, compared to µPs and DSPs. In addition, in cases in which the desired controller benefits from the utilization of a general-purpose processor, it can be embedded within the FPGA as many microprocessor cores are available as IPs.

The controller has to generate the 36 signals for the four-level MAC legs (12 devices per leg) at each switching cycle. Then, different automata with some duty ratios as inputs and running at high frequency (i.e., 50 MHz, which allows a time accuracy of 20 ns) have to be implemented. An FPGA implementation of such automata appears as a better solution than using a general-purpose processor with lots of timers. Additionally, for each new switching cycle, the new duty ratios have to be computed from the system input variables (currents, voltages, and rotor angle). For performing these calculations, it is desired to use measured values of the input variables as close as possible to the start of the next switching cycle, in order to maximize the control bandwidth. To this end, an ad hoc processing unit (specific purpose processing unit), implemented in the same FPGA as the above indicated automata, appears as a better solution than using an additional device, as a general-purpose processor or a DSP platform. Delay between measurement and the application of the resulting duty ratios is reduced, cost is also reduced, and synchronization between different devices is not required.

Due to the general FPGA advantages and to the specific reasons presented above in the last two paragraphs, respectively, the authors propose a full FPGA-based control implementation of a four-level three-phase MAC inverter to drive a three-phase four-pole pairs PMSM. Figure 2 presents the general overview of the electrical circuit and the proposed FPGA control structure. As it can be seen in the Figure 2, as well as the VBC closed-loop control already introduced above, a closed-loop field-oriented control (FOC) is used for driving the PMSM.

A preliminary open-loop FPGA controller implementation with the same virtual-vector modulation was presented in [10]. However, the controller in [10] did not include the closed-loop controls to operate the converter as a motor drive and to preserve the capacitor voltages balanced.

FPGAs have been employed for implementing diverse control schemes of multilevel converters ([11–21]), and also for implementing motor-drive controllers ([17–26]). References [17–21] propose FPGA-based controllers for multilevel converters operating as motor drives, as it is proposed here. However, [17–21] do not explain the FPGA controller structure, do not discuss the design methodology to obtain an efficient implementation, and generally do not deal with the voltage balancing problem.

To the best of the authors' knowledge, for the first time, a complete controller for a four-level converter of the NPC converter family [4], operating as a motor drive and including DC-link voltage balancing control, is fully-embedded into an FPGA. In addition, the controller implementation has been optimized to save FPGA resources and also to take full advantage of the FPGA potential performance capabilities.

The paper is organized as follows: Section 2 presents a summary of the used VBC, field-oriented control (FOC), and modulation scheme, presenting the equations to be implemented in the FPGA. Section 3 details the FPGA structure, and describes relevant aspects to achieve an efficient controller. In Section 4, experimental results are shown to verify the good operation of the controller under different conditions. Finally, Section 5 outlines the conclusion.

2. Closed-Loop Control and Modulation Strategy

Figure 2 presents the overall closed-loop control structure applied to the MAC converter to drive the PMSM. Two autonomous control loops are implemented: FOC and VBC.

2.1. Field-Oriented Control (FOC)

The well-known FOC is used to control the three-phase PMSM. In the blue inset of Figure 2, the FOC structure is depicted. Variables ω and φ correspond to the measured rotor angular speed and rotor electrical angle, respectively. Variables i_d and i_q are the direct and quadrature components of the three-phase currents. Variables d_d^* and d_q^* are the direct and quadrature components of the normalized reference vector required by the modulator. Command values are designated with an asterisk superscript. The control inputs are: ω^*, ω, φ, i_A, and i_B. The control scheme comprises an outer speed loop and an inner current loop. Through the PI compensators, the speed and current loops determine the reference vector polar coordinates m^* and θ^* to be the input to the modulator. Back-emf feedforward terms could be added in the current loops to improve the controller performance. They have not been implemented here for the sake of simplicity. The equations of the ab-to-dq and dq-to-(magnitude,phase) transformations are:

$$\begin{bmatrix} i_d \\ i_q \end{bmatrix} = \sqrt{2} \cdot \begin{bmatrix} \sin(\phi + 60^\circ) & \sin(\phi) \\ \cos(\phi + 60^\circ) & \cos(\phi) \end{bmatrix} \cdot \begin{bmatrix} i_A \\ i_B \end{bmatrix} \tag{1}$$

$$\begin{aligned} m^* &= \sqrt{\left(d_d^*\right)^2 + \left(d_q^*\right)^2} \\ \theta^* &= \tan^{-1}\left(d_q^*, d_d^*\right) + \phi \end{aligned} \tag{2}$$

Equation (1) assumes an isolated star point, where $i_A + i_B + i_C = 0$.

2.2. DC-Link Voltage-Balancing Closed-Loop Control (VBC)

The control scheme proposed in [5] is the one implemented here. In the green inset of Figure 2, the VBC structure is depicted. From the measured voltages v_{21}, v_{32}, and v_{43}, voltage imbalances associated to the two DC-link inner points (imb_2 and imb_3) are calculated. Then, through PI compensators, the values of variables k_2 and k_3 are determined. The sign of k_2 and k_3 depends on the direction of the converter power flow (pow_sign), calculated through the expression shown in Figure 2. Additionally, proper limits are set in k_2 and k_3 to avoid unfeasible dwell times [5]. Table 1 shows the limits and sign modification applied to k_2 and k_3. In this table, variable d_4 corresponds to an auxiliary variable that will be defined later in (4).

Table 1. Computation of variables k_2' and k_3'.

Case	k_2'	k_3'
$pow\_sign = \text{sign}(k_2)$ $pow\_sign = \text{sign}(k_3)$	$\min\left\{0.5\,;\,\lvert k_2\rvert\,;\,\frac{1-d_4}{2\cdot d_4}\right\}$	$\min\left\{0.5\,;\,\lvert k_3\rvert\,;\,\frac{3\cdot(1-d_4)}{1+6\cdot d_4}\right\}$
$pow\_sign \neq \text{sign}(k_2)$ $pow\_sign \neq \text{sign}(k_3)$	$-\min\left\{0.5\,;\,\lvert k_2\rvert\,;\,\frac{3\cdot(1-d_4)}{1+6\cdot d_4}\right\}$	$-\min\left\{0.5\,;\,\lvert k_3\rvert\,;\,\frac{1-d_4}{2\cdot d_4}\right\}$
$pow\_sign \neq \text{sign}(k_2)$ $pow\_sign = \text{sign}(k_3)$	$-\min\left\{1\,;\,\lvert k_2\rvert\,;\,\frac{1.5\cdot(1-d_4)}{1+3\cdot d_4}\right\}$	$\min\left\{1\,;\,\lvert k_3\rvert\,;\,\frac{1.5\cdot(1-d_4)}{1+3\cdot d_4}\right\}$
$pow\_sign = \text{sign}(k_2)$ $pow\_sign \neq \text{sign}(k_3)$	$\min\left\{1\,;\,\lvert k_2\rvert\,;\,\frac{1-d_4}{2\cdot d_4}\right\}$	$-\min\left\{1\,;\,\lvert k_3\rvert\,;\,\frac{1-d_4}{2\cdot d_4}\right\}$

Finally, the preliminary leg duty-ratios are modified using variables k_2' and k_3', so that the balancing of the capacitor voltages can be recovered. This part is explained below at the end of the modulation strategy subsection.

2.3. Modulation Strategy

The modulation scheme originally introduced for three levels in [6], extended to an arbitrary number of levels in [7], and extended to the overmodulation region in [27], is the one used to operate the converter. This modulation, originally defined applying the virtual-vector concept, guarantees the dc-link capacitor voltage balance in every switching cycle, provided that the phase currents remain constant over the switching cycle and that their addition is equal to zero. The modulation assumes that the switching frequency ($f_s = 1/t_s$, where t_s is the switching period) is much larger than the fundamental frequency f. This PWM allows modulation index values $m \in [0, hbc \cdot 1.1027]$, where $m = v_{ab,1,pk}/V_{dc}$, $v_{ab,1,pk}$ is the peak value of the fundamental component of the line-to-line voltage, and hbc is the overmodulation hexagonal-boundary-compression index [27]. Therefore, the PWM covers both the undermodulation (UM) and overmodulation (OM) operating modes. The OM region is further divided into two subregions (OMI and OMII), which present different reference vector trajectories [27].

A comprehensive explanation of the used modulation scheme is presented in [7,27]. A simplified description showing the final equations that have to be implemented within the FPGA is presented below.

The modulation is implemented taking advantage of the hexagonal symmetry of the space vector diagram (SVD), optimizing the FPGA resources. Therefore, the command value of the reference vector angle ($\theta^* \in [0°, 360°[$), which has been calculated previously in the FOC control algorithm, is transformed into a sextant ($sextant \in \{0, 1, 2, 3, 4, 5\}$) and an angle within a sextant ($\theta_{sext}^* \in [0°, 60°[$).

The command values of modulation index (m^*) and reference vector angle (θ_{sext}^*) are modified in case the reference vector is located in the overmodulation region to obtain corrected values of modulation index (m_c) and reference vector angle (θ_c). Tables 2 and 3 summarize these calculations. The index *hbc* is fixed to 0.98. Then, the range limits are:

$$hbc \cdot m^*_{\text{maxI}} = 0.98 \cdot 3 \ln(3)/\pi = 1.0281$$
$$hbc \cdot m^*_{\text{maxII}} = 0.98 \cdot 2\sqrt{3}/\pi = 1.0806$$

(3)

Table 2. Limiting reference vector angle θ_{lim} for overmodulation region.

Region	Application Range	θ_{lim}
UM	$0 < m^* \leq 0.98$	-
OMI	$0.98 < m^* \leq 1.0281$	$30° \cdot \frac{1.0281 - m^*}{1.0281 - 0.98}$
OMII	$1.0281 < m^* \leq 1.0806$	$30° \cdot \frac{m^* - 1.0281}{1.0806 - 1.0281}$

Table 3. Corrected values of modulation index (m_c) and reference vector angle (θ_c).

Region	Application Range	m_c	θ_c
UM	$0° \leq \theta_{\text{sext}}^* < 60°$	m^*	θ_{sxt}^*
OMI	$0° \leq \theta_{\text{sext}}^* < \theta_{\text{lim}}$	$0.98/\sin(\theta_{\text{lim}} + 60°)$	θ_{sext}^*
	$\theta_{\text{lim}} \leq \theta_{\text{sext}}^* \leq (60° - \theta_{\text{lim}})$	$0.98/\sin(\theta_{\text{sext}}^* + 60°)$	θ_{sext}^*
	$(60° - \theta_{\text{lim}}) < \theta_{\text{sext}}^* < 60°$	$0.98/\sin(\theta_{\text{lim}} + 60°)$	θ_{sext}^*
OMII	$0° \leq \theta_{\text{sext}}^* < \theta_{\text{lim}}$	$0.98/\sin(60°) = 1.1316$	$0°$
	$\theta_{\text{lim}} \leq \theta_{\text{sext}}^* \leq (60° - \theta_{\text{lim}})$	$0.98/\sin(\theta_{\text{sext}}^* + 60°)$	θ_{sext}^*
	$(60° - \theta_{\text{lim}}) < \theta_{\text{sext}}^* < 60°$	$0.98/\sin(60°) = 1.1316$	$60°$

From m_c and θ_c, the auxiliary variables d_1, d_4 and d_5 are calculated as follows:

$$d_1 = m_c \cdot \cos\left(\theta_c + 30°\right)$$
$$d_4 = m_c \cdot \cos\left(\theta_c - 30°\right)$$
$$d_5 = d_4 - d_1$$

(4)

The preliminary leg duty ratios d_{x1} and d_{x4} (indicating the duty ratio of connection of phase x to levels 1 and 4, respectively) are determined according to Table 4 from the value of the sextant of each phase *sextant_x*, and from the auxiliary variables d_1, d_4 and d_5.

Table 4. Leg duty ratios of Levels 1 and 4 depending on the sextant.

sextant_x	0	1	2	3	4	5
d_{x1}	0	d_5	d_4	d_4	d_1	0
d_{x1}	d_4	d_1	0	0	d_5	d_4

The leg duty ratios of the inner levels 2 and 3 are then calculated as follows:

$$d_{x2} = d_{x3} = (1 - d_{x1} - d_{x4})/2$$

(5)

In order to finally implement the VBC, preliminary leg duty ratios are modified according to the following equations:

$$d'_{x1} = d_{x1} \cdot (1 - k'_2 - k'_3) \cdot k_{\text{mod}}$$
$$d'_{x2} = 0.5 + k'_2 \cdot k_{\text{mod}} \cdot (d_{x1} - d_{x4}) - 0.5 \cdot d_4 \cdot k_{\text{mod}}$$
$$d'_{x4} = d_{x4} \cdot (1 + k'_2 + k'_3) \cdot k_{\text{mod}}$$
$$d'_{x3} = 1 - d'_{x1} - d'_{x2} - d'_{x4}$$

(6)

where $k_{\mathrm{mod}} = 3/(3 + k_2' - k_3')$.

3. FPGA Design and Control Implementation

Modulation and control structures presented in the previous Section have been implemented on an Altera Cyclone IV EP4CE22F17C6N FPGA device driven by a 50-MHz system clock (Altera, Intel, San Jose, CA, USA). The FPGA application has been described in VHDL. Figure 3 presents a block diagram of the FPGA application, together with its peripherals. In this figure, six different subsystems are separated in a set of six color boxes, following the same color selection as in Figure 2. Input signals are located on the left side of the boxes, while output signals are located on the right side. This criterion does not apply to the signals exchanged between the FPGA and peripherals, in which an arrow indicates the direction. For simplicity, only the main variables and constants are shown.

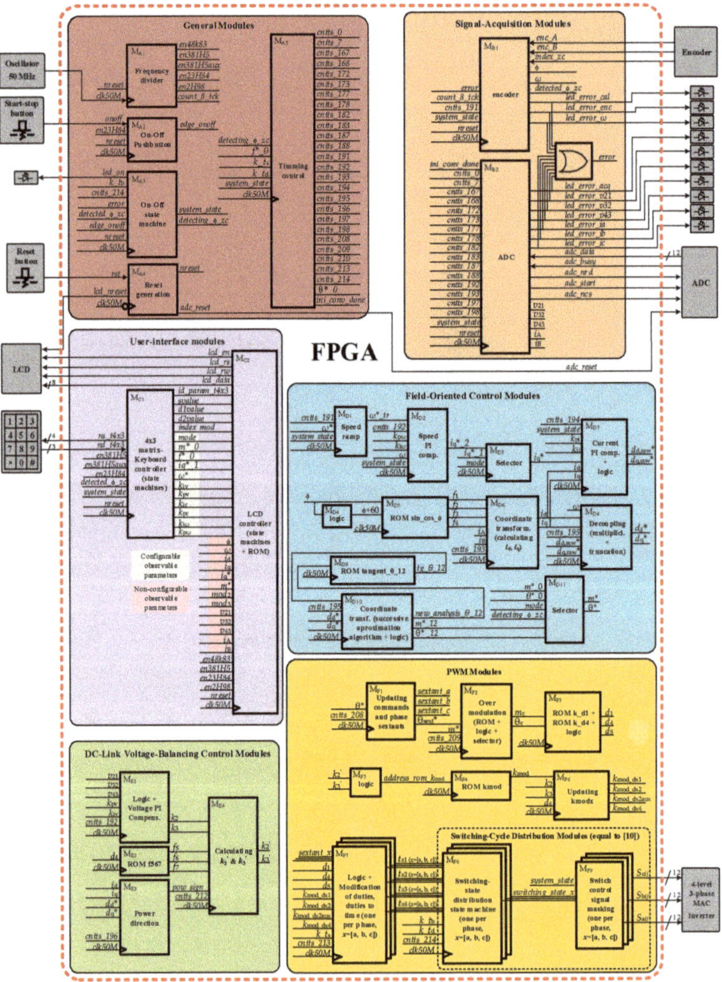

Figure 3. FPGA-controller design overview.

The FPGA has been mounted on a printed circuit board together with the sensors, the filtering circuitry, the ADC chip, and the user-interface (see Figure 3). The user-interface comprises a 4×3 matrix keyboard to introduce the values of some control variables, a three-line sixteen-character LCD screen, two pushbuttons, and eleven LEDs to easily visualize some errors and the ON-state.

Three operating modes have been defined in order to enable the use of the inverter under different system configurations, and also to bring the possibility of evaluating and tuning the different control structures individually:

-   Voltage-balancing controller mode (mode 0): in this mode, FOC is disabled, but the dc-link voltage control is enabled. With this mode, it is necessary to introduce the command values of m^* and f^* to make it operate at desired conditions.
-   Torque controller mode (mode 1): this mode enables the inner current control loop of the FOC, but the outer speed control loop is disabled. DC-link voltage control is also enabled. With this mode, it is necessary to introduce the value of i_q^*, which is proportional to the torque, to make it operate at desired conditions.
-   Speed controller mode (mode 2): this mode enables the whole closed-loop control. With this mode, it is necessary to externally introduce the command value of rotational speed ω^* to make it operate at desired conditions.

Since the control scheme requires five analog inputs (sensed i_A, i_B, v_{21}, v_{32}, and v_{43}), and the position of the rotor, an analog-to-digital converter chip and an encoder are necessary.

The AD7658 from Analog Devices has been selected as ADC chip, as it allows converting up to six analog signals synchronously. This chip is configured in parallel interface to minimize the time needed for data transmission.

Encoder RP1410 (IFM, El Prat de Llobregat, Spain) is used for obtaining the rotor position. This encoder generates three digital signals: *enc_A*, *enc_B*, and *index_zc*. Signals *enc_A* and *enc_B* present each 1024 pulses for a rotor revolution (with a phase delay of 90°). To determine the rotor angle, any single rising or falling edge of *enc_A* and *enc_B* can be counted, giving a resolution of 4096 edges per revolution, so 1024 edges per electrical cycle (the motor is a four-pole PMSM). Then, the mechanical rotor angle φ_m presents a resolution of 0.08789° (360°/4096 edges), and the electrical angle φ resolution is 0.35156°. Signal *index_zc* presents a small pulse each time a rotor revolution is accomplished (zero-crossing detection).

3.1. FPGA Basic Design Aspects

A similar design methodology to the one used in [10] has been also considered here. Variables are represented as integers. For each variable, both units and width (number of bits) have been meticulously selected to obtain the desired resolution and range avoiding underflows/overflows and to reduce FPGA used resources (truncations are often applied). Most of the variables are coded with 12 to 16 bits. Better resolution is not required as sampled variables (i.e., voltages, currents, and rotor position) are acquired just with a 12-bit resolution. Unsigned variables are usually coded in natural binary, while signed variables are coded in two's complement or sign plus magnitude when required. Divisions by constants (i.e., when changing units or calculating speed) are replaced by products followed by truncations to take advantage of the multipliers integrated in the FPGA. Other divisions are avoided as much as possible. When required, as well as for transcendental functions, they are implemented by ROMs built with the RAM blocks integrated in the FPGA. Symmetries, scaling, and offset addition are applied whenever is possible to minimize the number of bits required to achieve a certain resolution. More details are explained below in the FPGA-module-description subsection.

A 50-MHz system clock clk50M (period t_ck = 20 ns) is used to manage the FPGA. Relevant time variables, as delay time t_d = 2 t_ck, blanking time t_b = 40 t_ck, and switching period t_s = 10,000 t_ck are defined as multiples of t_ck, to optimize FPGA resources. Note that t_s = 200 μs implies f_s = 5 kHz. Note in Figure 3 the time constants k_t_b, k_t_d, and k_t_s, that are defined as $k\_t_b = t_b/t_{ck} = 2$, $k\_t_d = t_d/t_{ck} = 40$, and $k\_t_s = t_s/t_{ck} = 10,000$.

As in [10], most of the FPGA processing is synchronized with the switching cycle. A divider-by-10,000 counter *cnt_ts* is used to this purpose. Figure 4 illustrates the timing overview of a *cnt_ts* cycle and its synchronization with a switching period. A *cnt_ts* cycle starts performing all the samplings and calculations required to stablish the behavior of the next switching cycle, which starts at *cnt_ts* = 234, as soon as possible after these calculations are completed at *cnt_ts* = 214. The delay between these two events, half a blanking time, is required to properly transit from a switching cycle to the next one (d_{x4} can change from a null/non-null value to a new null/non-null value, as explained in [10]).

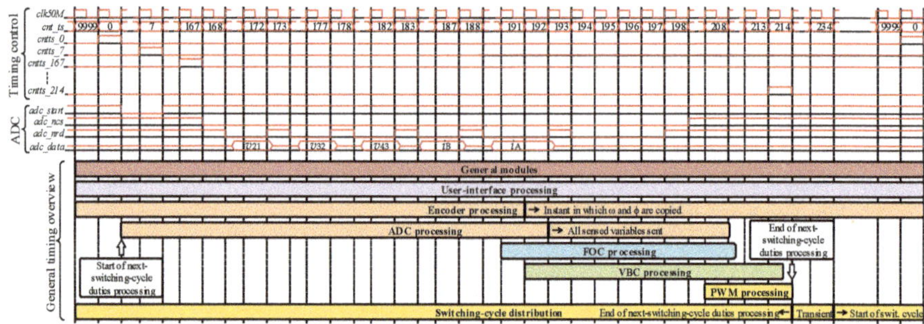

Figure 4. FPGA timing overview.

A versatile implementation has been conceived, allowing the user to modify important operational parameters through the user interface, such as the operating mode (mode), the PI compensators constants $k_{p\omega}$, $k_{i\omega}$, k_{pi}, k_{ii}, k_{pv} and k_{iv}, or the command values ω^*, i_q^*, m^* and f^*.

3.2. FPGA Module Description

3.2.1. General Modules

Module M_{A1} generates periodic enable signals of lower frequencies that are necessary to manage the periodicity of several processes. Enable signals are active high, with a pulse lasting a single t_{ck}, which is repeated at the corresponding frequency. For example, enable signal *en48k83* (48.83 kHz) is used for handling the writing process of LCD, and enable signal *en381H5* (381.5 Hz) is used for the state machine of keyboard inspection and for the state machine of the LCD screen operation. Last output signal *count_8_tck* is a 3-bit divider, which is used, together with the remaining enable signals, in other processes.

Module M_{A2} detects an active edge in signal *onoff* provided by the "ON-OFF" pushbutton, and generates an active high signal *edge_onoff*, that lasts a single t_{ck}. Pushbutton inspection is done at low frequency (23.84 Hz) for filtering possible pushbutton bounces.

Module M_{A3} implements the system ON-OFF finite state machine, which is presented in Figure 5. Binary values in each state indicate the *system_state* output of M_{A3}. From right to left, bit 0 represents the transition-and-holding bit between OFF and ON states, and bits 1, 2, and 3 represent masking bits for switches in poles 1, 2 and 3, with reference to Figure 2. In the transition from OFF state to ON state, first pole 1 is enabled, then pole 2, and lastly pole 3. This sequence is reversed in the transition from ON to OFF. When the FPGA is powered on, the motor is driven to rotate at a very low speed in mode 0 ($m^* = 0.04$ and $f^* = 1.22$ Hz) to allow an automatic initial zero-crossing detection of the rotor position. Signal *detecting_φ_zc* is activated when this automatic detection process is going on. When a filtered first pulse of encoder signal *index_zc* is detected, signal *detected_φ_zc* is activated permanently (this part of the process is carried out by module M_{B1}), the motor is stopped, and signal *detecting_φ_zc* is deactivated.

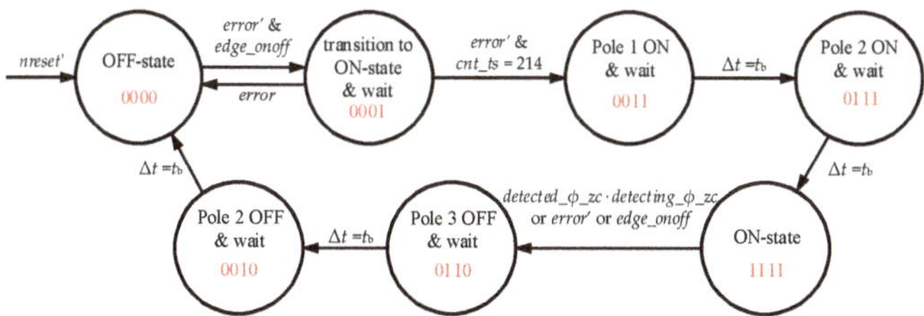

Figure 5. ON-OFF finite state machine.

Module M_{A4} generates the global reset signal *nreset* from the signal *rst* provided by the "reset" pushbutton. Signal *nreset* is active low, lasting 120 μs to guarantee minimum duration required by LCD (*lcd_nreset* and *nreset* are the same signal). Signal *adc_reset*, which lasts 10 t_{ck} and is active high, is used to initialize the ADC chip.

Module M_{A5} is the main synchronization module. It implements the divider-by-10,000 counter *cnt_ts* and generates a set of enable signals (*cntts_X*) to manage the timing of most actions performed by other modules, as indicated in Figure 4. This module also updates the value of line angle in mode 0 (θ^*_0), from the frequency command value in mode 0 (f^*_0).

3.2.2. Signal-Acquisition Modules

Module M_{B1} processes the encoder input signals *enc_A*, *enc_B*, and *index_zc* and generates the angular position φ_m of the motor, its angular speed ω, and the electrical angle φ. Input signals from the encoder are initially filtered at a sampling frequency of 6.25 MHz (50 MHz/8), discarding any value that has not remained constant for a minimum of eight consecutive samples. The angular position φ_m is updated when a new edge in any of the filtered signals derived from *enc_A* and *enc_B* is detected. A pulse of *index_zc* signal should take place every 4096 edges (1 revolution). Non-consistency produces a LED error indication. The angular speed ω is updated every 2.5 ms, and is calculated dividing the angle rotated in the last 10 ms by this time. The existence of two simultaneous edges in filtered signals derived from *enc_A* and *enc_B*, as well as an excessive speed, also result in LED error indications. Obviously, this module works asynchronously to the main synchronization module. To synchronize data, φ and ω are copied at *cnt_ts* = 191 to have convenient values for FOC and VBC processing (see Figure 4).

Module M_{B2} is in charge of handling the ADC chip according to Figure 4. Signals *adc_start*, *adc_ncs* and *adc_nrd* are generated in the FPGA to control the ADC chip. FPGA receives the acquired data through the 12-bit bus *adc_data*. Current i_c is determined from the measured ones ($i_c = -i_a - i_b$). Values out of acceptable range result in LED error indications.

3.2.3. User-Interface Modules

Module M_{C1} comprises two finite state machines that handle the 4 × 3 matrix keyboard operation. A first finite state machine handles the detection and identification of the keyboard buttons when any of them is pressed. A second finite state machine identifies the sequence of the different buttons pressed in order to determine the actions to be executed: acquisition of the identifier of the parameter to be shown in the LCD screen, modification of the value of a configurable parameter, etc.

Module M_{C2} implements three concurrent finite state machines that manage the configuration and visualization of the LCD screen. The LCD screen shows the name of the variable selected by the user to be visualized, with its current value, its identifier number and its allowed range (just for the configurable parameters). The module also includes a ROM containing the visualization format for the 24 variables that can be visualized in the LCD screen (see Figure 3).

The user-interface modules consume substantial resources of the FPGA (21.5% of the used logic elements and 27% of the used multipliers). Its explanation is simplified here because they are considered to be of less technical importance, compared to other parts of the design.

3.2.4. FOC Modules

Module M_{D1} simply generates a limiting speed ramp whenever the speed command changes. Module M_{D2} consists of the speed-error PI compensator with limited proportional, integral and total outputs. Its output value is the quadrature current command value in mode 2 (i_q*_2), which is updated when $cnt\_ts = 192$. In mode 1, the quadrature current command value (i_q*_1) is set externally. Module M_{D3} is a selector of i_q* depending on the chosen operating mode.

Modules M_{D4}, M_{D5}, and M_{D6} are in charge of implementing the coordinate transformation. Module M_{D4} implements simple logic to calculate angle $\varphi + 60°$. Module M_{D5} implements a dual-port ROM that allows obtaining the following four trigonometrical functions in a single t_{ck}, which are necessary to determine the values of i_d and i_q:

$$
\begin{aligned}
f_1 &= \sqrt{2}\ \sin(\phi) \\
f_2 &= \sqrt{2}\ \cos(\phi) \\
f_3 &= \sqrt{2}\ \sin(\phi + 60°) \\
f_4 &= \sqrt{2}\ \cos(\phi + 60°)
\end{aligned}
\tag{7}
$$

Module M_{D6} implements Equation (1) to calculate the values of i_d and i_q. Module M_{D7} contains the i_d and i_q error PI compensators with limited proportional, integral and total outputs. Output variables of PI compensators are direct and quadrature raw duties $d_{d,raw}$* and $d_{q,raw}$*, which are updated when $cnt\_ts = 194$.

Module M_{D8} implements the calculation of the decoupling factor, which is equal to $\sqrt{2} \cdot \omega_e \cdot L / V_{dc}$ (where L and V_{dc} are usually constant values), its products by currents i_d and i_q, the addition of the first product to $d_{q,raw}$* and the subtraction of the second product from $d_{d,raw}$*, to obtain duties d_d* and d_q*, respectively (see Figure 2). In general, products are done followed by a truncation (change of units), thus allowing optimizing FPGA resources.

Modules M_{D9} and M_{D10} implement the dq-to-(magnitude,phase) transformation indicated in Equation (2) to calculate the modulation index and reference vector angle in modes 1 and 2 (m*_12 and θ*_12). Both variables are obtained through a successive approximation algorithm. A ROM is used to calculate the tangent of the provisional reference vector angle used in the algorithm. Octagonal symmetry is taken into account to reduce the size of the ROM.

Module M_{D11} is just a selector for the modulation index and the reference vector angle, depending on the operating mode.

3.2.5. VBC Modules

Module M_{E1} calculates the voltage imbalances at dc-link points 2 and 3 (imb_2 and imb_3) and includes the PI compensators for the error of these variables, also calculated in the module.

Module M_{E2} consists of a ROM delivering the following auxiliary functions, which are necessary to determine the values of k_2' and k_3', as shown in Table 1:

$$
\begin{aligned}
f_5 &= \min\{1\,;\,(1 - d_4)\,/\,(2 \cdot d_4)\} \\
f_6 &= \min\{0.5\,;\,3 \cdot (1 - d_4)\,/\,(1 + 6 \cdot d_4)\} \\
f_7 &= \min\{1\,;\,1.5 \cdot (1 - d_4)\,/\,(1 + 3 \cdot d_4)\}
\end{aligned}
\tag{8}
$$

Module M_{E3} is in charge of calculating the sign of power flow, according to the expression shown in the green inset of Figure 2. Output value of *pow_sign* is updated at $cnt\_ts = 196$.

Module M_{E4} applies simple logic to obtain values of variables k'_2 and k'_3, according to Table 1.

3.2.6. Modulation Modules

Module M_{F1} adapts the value of the reference vector angle to be an angle within the first sextant ($[0°, 60°[$). The sextant values corresponding to the three phases are also determined, and given through the variables *sextant_x*. Module M_{F2} modifies θ and m^*, according to Tables 2 and 3. A ROM is used to calculate the function $m = 0.98/\sin(\theta + 60°)$. Module M_{F3} calculates the values of d_1, d_4 and d_5, according to Equation (4). Two ROMs are used to calculate d_1 and d_4.

Module M_{F4} includes a ROM for calculating the value of k_{mod}, according to Equation (6). The ROM address, which is defined as the result of expression $3 + k'_2 - k'_3$, is previously calculated through module M_{F5}. Module M_{F6} calculates the following products, which are useful to calculate modified duty ratios d'_{x1}, d'_{x2}, d'_{x3}, and d'_{x4}:

$$
\begin{aligned}
k_{mod\_dx1} &= (1 - k'_2 - k'_3) \cdot k_{mod} \\
k_{mod\_dx2} &= k'_2 \cdot k_{mod} \\
k_{mod\_dx2aux} &= 0.5 \cdot d_4 \cdot k_{mod} \\
k_{mod\_dx4} &= (1 + k'_2 + k'_3) \cdot k_{mod}
\end{aligned}
\tag{9}
$$

Module M_{F7} (one per phase) calculates d'_{x1}, d'_{x2}, d'_{x3}, and d'_{x4} according to Equation (6) and Table 4, and transforms them into time variables t_{x1}, t_{x2}, t_{x3} and t_{x4} as multiples of t_{ck}. Modules M_{F8} and M_{F9} (one per phase) are the same used in [10].

3.3. Consumed FPGA Resources

The resources used to synthesize the whole design using the Quartus II software are shown in Table 5. The complete FPGA implementation has been done making a substantial effort to save FPGA resources. As it can be seen, enough FPGA resources are still available. They can be used, for example, to include additional features in the design, to increase the switching frequency, and/or to improve the variables resolution. Furthermore, it is relevant to recall that, for practical applications, the user-interface processing is usually not required, which would increase even more the remaining FPGA resources.

Table 5. Consumed FPGA resources.

Resource	Amount Used/Total Available
Logic elements	7272/22,320 (33%)
Combinational functions	7156/22,320 (32%)
Dedicated logic registers	1183/22,320 (5%)
Pins	89/154 (58%)
Memory bits	285,056/608,256 (47%)
Embedded Multiplier 9-bit elements	74/132 (56%)
PLLs	0/4 (0%)

3.4. Control Processing Time

In order to minimize the processing time for obtaining the next-switching-cycle duties and therefore maximize the closed-loop control bandwidth, an ad hoc processing unit has been fully embedded within the FPGA, instead of using a general-purpose processor. The resulting processing time, taken from the command to sample and convert the analog signals delivered by the sensors ($cnt\_ts = 0$), until the instant when all the duties needed to generate the following switching cycle become available ($cnt\_ts = 214$), is 4.28 µs (see Figure 4). Thus, the switching frequency could be increased until approximately $1/5$ µs = 200 kHz. To maximize the control bandwidth, the processing occurs as close as possible to the start of a new switching cycle.

4. Experimental Tests

The proper operation of the FPGA controller has been tested experimentally within the system shown in Figure 6a. Main parts of the system are labelled in the figure. The shaft of the PMSM driven by the FPGA controller is coupled to an induction machine, which is not used, and also to another PMSM, which is used to set the load torque. A further detailed overview of the experimental test bed can be observed in a supplementary video attached with this paper. Figure 6b depicts FPGA control board, presenting its main parts/subcircuits.

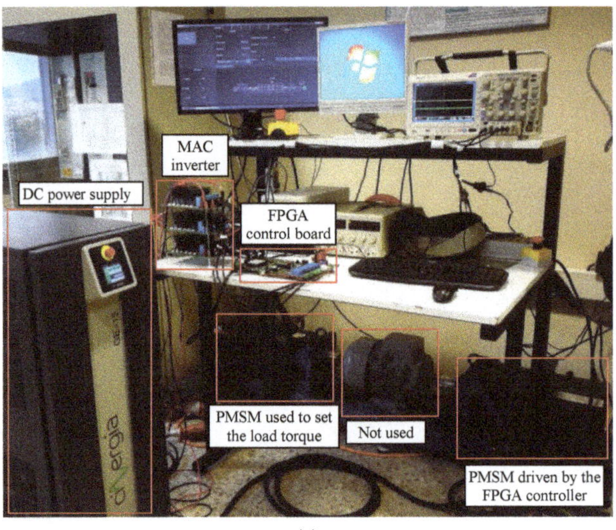

(a)

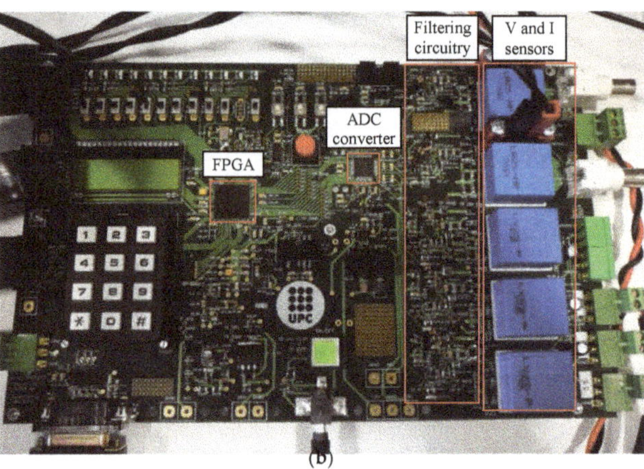

(b)

Figure 6. Experimental testbed. (**a**) Overview of the system; (**b**) FPGA control board.

The four-level three-phase MAC inverter employs 200 V STP20NF20 MOSFET devices (ST microelectronics, Amsterdam, Netherlands). The reference of the PMSM driven by the FPGA controller is 1FT6105-8SB71-2AA0 (Siemens, Berlin, Germany). This motor is a surface-magnet type PMSM, with a nominal speed of 1500 rpm and a nominal torque of 59 Nm, thus, a nominal power of 9.27 kW.

The proper system operation can be observed in Figures 7–9. In Figure 7, the good performance of the VBC under a start-up transition is depicted. Initially, in OFF state, capacitor voltages are unbalanced, but after 200 ms of operation, they become balanced. It is also remarkable the behavior of the phase current i_A. After the start-up, while the motor is accelerating, the current magnitude keeps constant at a certain level, set by the speed ramp. As soon as the motor reaches the command speed of 500 rpm, the current magnitude decreases to a lower level.

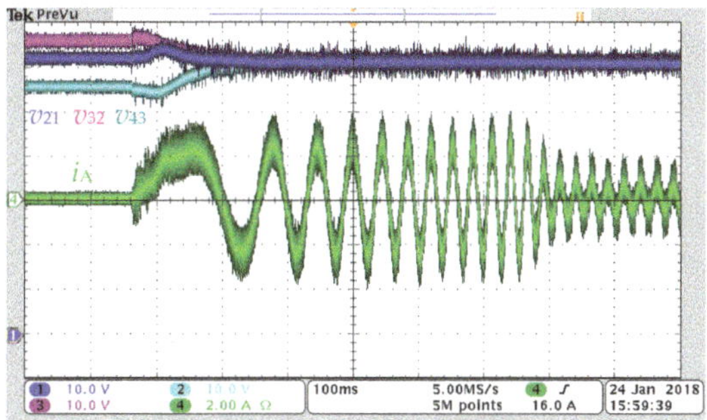

Figure 7. Experimental results of DC-link voltages and phase current i_A under a start-up transition. Conditions: $V_{dc} = 180$ V, $C = 155$ µF, $\omega^* = 500$ rpm, $k_{p\omega} = 0.1$ A/rpm, $k_{i\omega} = 0.1$ A/(rpm·s), $k_{pi} = 0.01$ A^{-1}, $k_{ii} = 1$ (A·s)$^{-1}$, $k_{pv} = 0.02$ V^{-1}, $k_{iv} = 0$ (V·s)$^{-1}$, $t_s = 100$ µs, load torque = 0 Nm.

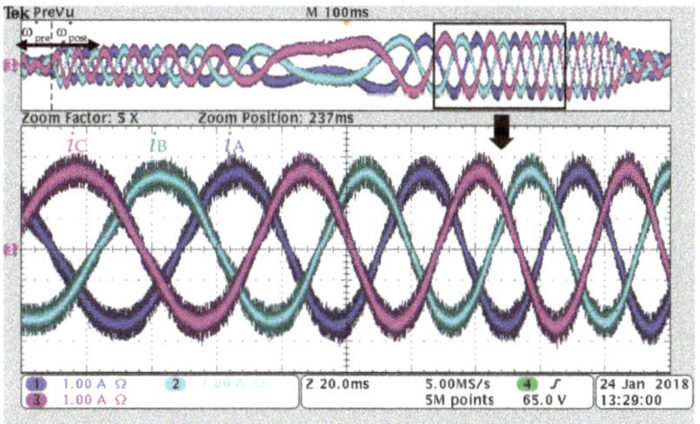

Figure 8. Experimental results of phase currents i_A, i_B, and i_C, under a change of the rotation direction. Conditions: $\omega^*_{pre} = 350$ rpm, $\omega^*_{post} = -350$ rpm (remaining conditions are the same as in Figure 7).

Figure 8 shows the phase currents under a change of rotation direction, from 350 rpm to -350 rpm. In the upper part of the figure, the whole transition is depicted. As it can be seen, the current sequence changes once the motor starts rotating in opposite direction. As in Figure 7, once the motor reaches the command speed value, the current magnitude decreases to a lower level.

Figure 9 presents the three phase currents and the phase voltage v_{A1} operating in UM and OM regions. Figure 9a shows the waveforms for the UM region, with $m^* = 0.76$. Figure 9b shows the waveforms for region OMI (0.98 < m^* < 1.028), operating with $m^* = 1.01$. Figure 9c shows the waveforms for region OMII (1.028 < m^* < 1.081), operating with $m^* = 1.03$. In UM, current waveforms are sinusoidal with almost no distortion. However, in OMI and OMII, the current waveforms present a noteworthy distortion, with a higher distortion under OMII. This is the expected behavior, since the overmodulation intrinsically introduces low-order harmonics in the phase currents [27]. It is also interesting to note that in waveforms of the phase voltage v_{A1}, the duty ratio of connection to inner levels 2 and 3 is lower when operating in OM region.

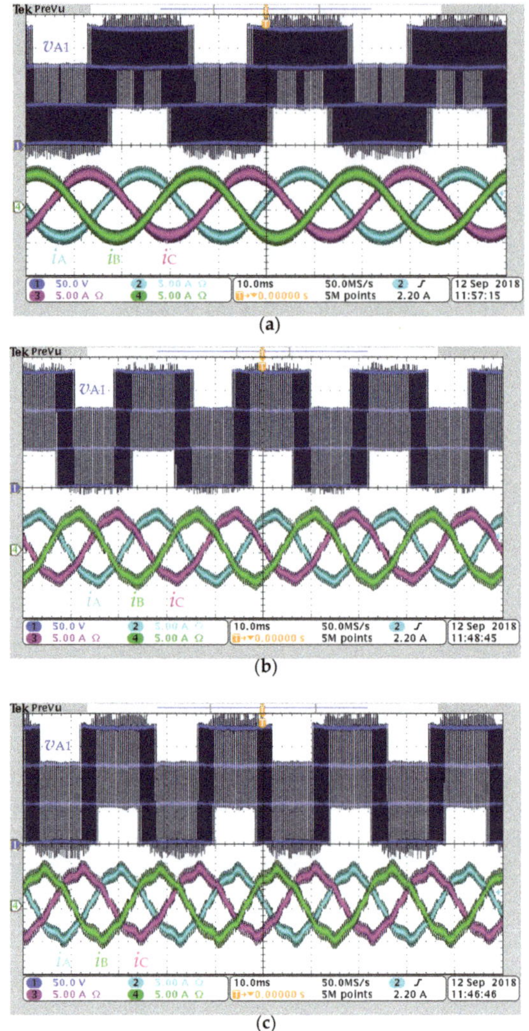

Figure 9. Experimental results of phase currents i_A, i_B, and i_C, and phase voltage v_{1A}, operating in UM and OM regions. (**a**) Operation in UM: $\omega^* = 450$ rpm, load torque = 10 Nm, $m^* = 0.76$; (**b**) Operation in OMI: $\omega^* = 610$ rpm, load torque = 10 Nm, $m^* = 1.01$; (**c**) Operation in OMII: $\omega^* = 620$ rpm, load torque = 10 Nm, $m^* = 1.03$ (remaining conditions are the same as in Figure 7 for (**a**), (**b**) and (**c**)).

As stated previously, a complementary video is included with this study. The video shows the whole system operating under conditions of Figures 7 and 8, and also under other different conditions.

5. Conclusions

A low-cost closed-loop controller fully embedded into an FPGA has been successfully implemented for a PMSM motor drive based on a four-level three-phase MAC inverter, taking full advantage of the drive potential performance capabilities. An efficient and robust implementation into a mid-range FPGA of the closed-loop VBC and FOC, together with a modulation scheme including the overmodulation region, has been achieved, consuming less than 50% of its total resources and obtaining a very low processing time. The remaining FPGA resources can be employed to increase the switching frequency, or to further improve the controller performance, including, for example, fault-tolerant controls [3], or an intelligent distribution of switching losses to better distribute the total semiconductor losses [2]. The proper operation of the whole system demonstrates the feasibility of using virtual-vector-based PWMs for neutral-point-clamped converters in motor drive applications, which had been questioned in the previous literature. In the future, the authors envision a motor drive design approach where an inexpensive switch with good performance is selected, and MAC leg structures are used to match the motor voltage rating by simply adjusting the number of levels.

Supplementary Materials: The following are available online at http://www.mdpi.com/1996-1073/11/10/2639/s1. A supplementary video included with this study shows the whole system operating under different conditions.

Author Contributions: Conceptualization, J.N.-A., E.L., S.B.-M. and J.B.; Funding acquisition, S.B.-M.; Investigation, J.N.-A., E.L. and A.C.; Methodology, J.N.-A., E.L. and S.B.-M.; Project administration, S.B.-M.; Software, E.L., A.C. and G.G.-R.; Supervision, S.B.-M., A.C. and J.B.; Validation, A.C. and G.G.-R.; Writing—original draft, J.N.-A.

Funding: This research was funded by Ministerio de Ciencia, Innovación y Universidades under grant DPI2017-89153-P.

Acknowledgments: The authors would like to thank Miquel Teixidor and his colleagues from CINERGIA, for their useful training on the usage of the GE15 grid emulator, which has been used in this study as a DC power supply.

Conflicts of Interest: The authors declare no conflict of interest.

References

1. Leon, J.I.; Vazquez, S.; Franquelo, L.G. Multilevel Converters: Control and Modulation Techniques for Their Operation and Industrial Applications. *Proc. IEEE* **2017**, *105*, 2066–2081. [CrossRef]
2. Busquets-Monge, S.; Nicolas-Apruzzese, J. A Multilevel Active-Clamped Converter Topology—Operating Principle. *IEEE Trans. Ind. Electron.* **2011**, *58*, 3868–3878. [CrossRef]
3. Nicolas-Apruzzese, J.; Busquets-Monge, S.; Bordonau, J.; Alepuz, S.; Calle-Prado, A. Analysis of the Fault-Tolerance Capacity of the Multilevel Active-Clamped Converter. *IEEE Trans. Ind. Electron.* **2013**, *60*, 4773–4783. [CrossRef]
4. Busquets-Monge, S. Neutral-Point-Clamped DC-AC Power Converters. In *Wiley Encyclopedia of Electrical and Electronics Engineering*; Webster, J.G., Ed.; John Wiley & Sons, Inc.: Hoboken, NJ, USA, 2018; pp. 1–20. ISBN 9780471346081.
5. Busquets-Monge, S.; Maheshwari, R.; Nicolas-Apruzzese, J.; Lupon, E.; Munk-Nielsen, S.; Bordonau, J. Enhanced DC-Link Capacitor Voltage Balancing Control of DC-AC Multilevel Multileg Converters. *IEEE Trans. Ind. Electron.* **2015**, *62*, 2663–2672. [CrossRef]
6. Busquets-Monge, S.; Bordonau, J.; Boroyevich, D.; Somavilla, S. The nearest three virtual space vector PWM—A modulation for the comprehensive neutral-point balancing in the three-level NPC inverter. *IEEE Power Electron. Lett.* **2004**, *2*, 11–15. [CrossRef]
7. Busquets-Monge, S.; Alepuz, S.; Rocabert, J.; Bordonau, J. Pulsewidth Modulations for the Comprehensive Capacitor Voltage Balance of n-Level Three-Leg Diode-Clamped Converters. *IEEE Trans. Power Electr.* **2009**, *24*, 1364–1375. [CrossRef]
8. Choudhury, A.; Pillay, P.; Williamson, S.S. A Hybrid PWM-Based DC-Link Voltage Balancing Algorithm for a Three-Level NPC DC/AC Traction Inverter Drive. *IEEE J. Emerg. Sel. Top. Power Electron.* **2015**, *3*, 805–816. [CrossRef]
9. Choudhury, A.; Pillay, P.; Williamson, S.S. DC-Bus Voltage Balancing Algorithm for Three-Level Neutral-Point-Clamped (NPC) Traction Inverter Drive with Modified Virtual Space Vector. *IEEE Trans. Ind. Appl.* **2016**, *52*, 3958–3967. [CrossRef]
10. Lupon, E.; Busquets-Monge, S.; Nicolas-Apruzzese, J. FPGA Implementation of a PWM for a Three-Phase DC-AC Multilevel Active-Clamped Converter. *IEEE Trans. Ind. Inform.* **2014**, *10*, 1296–1306. [CrossRef]
11. Cossutta, P.; Aguirre, M.P.; Engelhardt, M.A.; Valla, M.I. Control System to Balance Internal Currents of a Multilevel Current-Source Inverter. *IEEE Trans. Ind. Electron.* **2018**, *65*, 2280–2288. [CrossRef]

12. Zhang, Z.; Wang, F.; Wang, J.; Rodríguez, J.; Kennel, R. Nonlinear Direct Control for Three-Level NPC Back-to-Back Converter PMSG Wind Turbine Systems: Experimental Assessment With FPGA. *IEEE Trans. Ind. Inform.* **2017**, *13*, 1172–1183. [CrossRef]
13. Zhang, Z.; Hackl, C.; Kennel, R. Computationally Efficient DMPC for Three-Level NPC Back-to-Back Converters in Wind Turbine Systems With PMSG. *IEEE Trans. Power Electron.* **2017**, *32*, 8018–8034. [CrossRef]
14. Bennani-Ben Abdelghani, A.; Ben Abdelghani, H.; Richardeau, F.; Blaquière, J.M.; Mosser, F.; Slama-Belkhodja, I. Versatile Three-Level FC-NPC Converter with High Fault-Tolerance Capabilities: Switch Fault Detection and Isolation and Safe Postfault Operation. *IEEE Trans. Ind. Electron.* **2017**, *64*, 6453–6464. [CrossRef]
15. Moranchel, M.; Huerta, F.; Sanz, I.; Bueno, E.; Rodríguez, F. A Comparison of Modulation Techniques for Modular Multilevel Converters. *Energies* **2016**, *9*, 1091. [CrossRef]
16. Rodríguez-Rodríguez, J.; Venegas-Rebollar, V.; Moreno-Goytia, E. Single DC-Sourced 9-level DC/AC Topology as Transformerless Power Interface for Renewable Sources. *Energies* **2015**, *8*, 1273–1290. [CrossRef]
17. Zhang, Z.; Hackl, C.; Kennel, R. FPGA Based Direct Model Predictive Current Control of PMSM Drives with 3L-NPC Power Converters. In Proceedings of the International Exhibition and Conference for Power Electronics, Intelligent Motion, Renewable Energy and Energy Management, Nuremberg, Germany, 10–12 May 2016; pp. 1–8.
18. Umesh, B.S.; Sivakumar, K. Multilevel Inverter Scheme for Performance Improvement of Pole-Phase-Modulated Multiphase Induction Motor Drive. *IEEE Trans. Ind. Electron.* **2016**, *63*, 2036–2043. [CrossRef]
19. Kumar, P.S.; Satyanarayana, M. Comparative analysis of modulation strategies applied to seven-level diode clamped multi-level inverter fed induction motor drive. In Proceedings of the 2015 Conference on Power, Control, Communication and Computational Technologies for Sustainable Growth (PCCCTSG), Kurnool, India, 11–12 December 2015; pp. 231–237.
20. Janik, D.; Kosan, T.; Blahnik, V.; Kamenicky, P.; Peroutka, Z.; Danek, M. Complete solution of 4-level flying capacitor converter for medium-voltage drives with active voltage balancing control with phase-disposition PWM. In Proceedings of the 2014 16th European Conference on Power Electronics and Applications, Lappeenranta, Finland, 26–28 August 2014; pp. 1–8.
21. Hirani, M.; Gupta, S.; Deshpande, D.M. Comparison of performance of induction motor fed by sine pulse width modulated inverter and multi level inverter using XILINX. In Proceedings of the 2014 IEEE International Conference on Advanced Communications, Control and Computing Technologies, Ramanathapuram, India, 8–10 May 2014; pp. 264–269.
22. Chun, T.W.; Tran, Q.V.; Lee, H.H.; Kim, H.G.; Nho, E.C. A simple capacitor votage balancing scheme for the cascaded five-level inverter fed AC machine drive. In Proceedings of the 6th IET International Conference on Power Electronics, Machines and Drives (PEMD 2012), Bristol, UK, 27–29 March 2012; pp. 1–5.
23. Ponce, R.R.; Talavera, D.V. FPGA Implementation of Decimal Division for Motor Control Drives. *IEEE Lat. Am. Trans.* **2016**, *14*, 3980–3985. [CrossRef]
24. Jabeen, S.; Srinivasan, S.K.; Shuja, S.; Dubasi, M.A.L. A Formal Verification Methodology for FPGA-Based Stepper Motor Control. *IEEE Embed. Syst. Lett.* **2015**, *7*, 85–88. [CrossRef]
25. Gdaim, S.; Mtibaa, A.; Mimouni, M.F. Design and Experimental Implementation of DTC of an Induction Machine Based on Fuzzy Logic Control on FPGA. *IEEE Trans. Fuzzy Syst.* **2015**, *23*, 644–655. [CrossRef]
26. Quang, N.K.; Hieu, N.T.; Ha, Q.P. FPGA-Based Sensorless PMSM Speed Control Using Reduced-Order Extended Kalman Filters. *IEEE Trans. Ind. Electron.* **2014**, *61*, 6574–6582. [CrossRef]
27. Busquets-Monge, S.; Maheshwari, R.; Munk-Nielsen, S. Overmodulation of n-Level Three-Leg DC-AC Diode-Clamped Converters with Comprehensive Capacitor Voltage Balance. *IEEE Trans. Ind. Electron.* **2013**, *60*, 1872–1883. [CrossRef]

Article

FPGA-Based Implementation of MMC Control Based on Sorting Networks

Mattia Ricco [1,*], Laszlo Mathe [1], Eric Monmasson [2] and Remus Teodorescu [1]

[1] Department of Energy Technology, Aalborg University, 9100 Aalborg, Denmark;
    laszlo2mathe@gmail.com (L.M.); ret@et.aau.dk (R.T.)
[2] SATIE lab, Cergy-Pontoise University, 95031 Cergy Pontoise, France; eric.monmasson@u-cergy.fr
*   Correspondence: mri@et.aau.dk

Received: 6 August 2018; Accepted: 10 September 2018; Published: 11 September 2018

Abstract: In Modular Multilevel Converter (MMC) applications, the balancing of the capacitor voltages is one of the most important issues for achieving the proper behavior of the MMC. The Capacitor Voltage Balancing (CVB) control is usually based on classical sorting algorithms which consist of repetitive/recursive loops. This leads to an increase of the execution time when many Sub-Modules (SMs) are employed. When the execution time of the balancing is longer than the sampling period, the proper operation of the MMC cannot be ensured. Moreover, due to their inherent sequential operation, sorting algorithms are suitable for software implementation (microcontrollers or DSPs), but they are not appropriate for a hardware implementation. Instead, in this paper, Sorting Networks (SNs) are proposed due to their convenience for implementation in FPGA devices. The advantages and the main challenges of the Bitonic SN in MMC applications are discussed and different FPGA implementations are presented. Simulation results are provided in normal and faulty conditions. Moreover, a comparison with the widely used bubble sorting algorithm and max/min approach is made in terms of execution time and performance. Finally, hardware-in-the-loop results are shown to prove the effectiveness of the implemented SN.

Keywords: modular multilevel converters; capacitor voltage balancing; sorting networks; field-programmable gate array

---

1. Introduction

Nowadays, the Modular Multilevel Converter (MMC) has become a promising solution in different applications, such as in High Voltage Direct Current (HVDC) [1,2], high-power motor drivers [3,4] and STATic COMpensators (STATCOM) [5,6]. Thanks to several advantages, such as high modularity, scalability, low Total Harmonic Distortion (THD), high efficiency and high reliability, the interest in this topology has increased in both industry and academy [7]. However, this topology presents several challenges, such as the necessity to control the circulating current, ensure the balance of the losses among the Sub-Modules (SMs) and maintain the capacitor voltage balanced [8]. In the literature, two Capacitor Voltage Balancing (CVB) control algorithms are mainly proposed: the individual control approach [9] and the global arm control approach. The latter is commonly used in the Nearest Level Control (NLC) which requires a Sorting Algorithm (SA) [10]. Indeed, to balance the capacitor voltages (CVs), the SMs with the highest or lowest CV must be selected based on the arm current direction. Then, the SA provides a sorted list of the SMs according to their capacitor voltages.

In MMC applications, the implementation of SAs is a key challenge mainly due to the timing performances and the high computation efforts of this kind of algorithms. Low timing performances can slow down the CVB algorithm by limiting either the maximum sampling frequency or the maximum allowable number of SMs in the converter. The SAs are usually implemented in microcontrollers or in digital signal processors due to their easy implementation. However, they are based on loop/recursive operation leading to a significant increase of the execution time when the number of SMs grows. Some authors propose max/min approaches to overcome this issue [11–13]. Such methods suppose that only one SM has to be inserted/bypassed in the next sampling instant. Then, they only find the SM with the maximum (or minimum) CV by achieving a strong reduction of the execution time. However, in the case of faults or when the capacitors are approaching the maximum allowable voltage, more SMs need to be inserted or bypassed at the same time. The max/min algorithms then require more sampling periods to insert or bypass the required SMs, which leads to slow converter dynamic performance. A solution could be to run multiple times the max/min method within the same sampling instant; however, this choice increases the whole execution time by leading to the same problem highlighted for the bubble SA.

For the above-mentioned reasons, a hardware implementation of a sorting method has been proposed in this paper. Several studies have confirmed that the Field-Programmable Gate Array (FPGA) technology is really promising in industrial control applications [14]. Such platforms are more and more used in industry and in academia for achieving high timing performance, which is difficult to reach with the software counterpart. Since the FPGAs are able to exploit the inherent parallelism of the algorithm, they lead to a significant reduction of the execution time. Moreover, they are often used in MMC applications due to the possibility: to drive a huge amount of SMs [15], to implement fast protections, to interface more ADC modules [16], to implement real-time emulator [17] and for fast communication [18].

Among the different hardware implementations of sorting approaches, the Sorting Networks (SNs) have been chosen to be implemented in FPGA due to their inherent parallelism and enhanced timing performance [19]. The Bitonic SN has been considered due to its low resource requirement and its modular structure [20]. The authors in [20] compared this SN with the Odd-Even SN. However, the main aim was to provide a method for pre-evaluating the hardware resources and the execution time of the network. No simulations or hardware-in-the-loop results were provided, and no justifications were given for the use of these networks in MMC. In this work, instead, some simulation results are shown in normal and faulty conditions to highlight the benefits of adopting an algorithm that provides a complete sorted list with respect to one that only gives the SM with the highest (lowest) capacitor voltage. Another contribution of this work is the study of the achievable trade-offs between the execution time and the required resources when this kind of sorting methods are implemented for MMC applications. For this aim, three kinds of FPGA implementations are presented: a fully pipelined architecture, a hybrid structure and a fully factorized SN. In this way, designers can choose the proper solution for their requirements to achieve the best compromise between the required timing performance and the available resources. After having chosen the right solution that fits the requirements of the MMC application considered in this work, the SN architecture is compared with the classical bubble sorting algorithm and the max/min approach in terms of execution time. To demonstrate the feasibility of the proposed architecture, a Hardware-In-the-Loop (HIL) validation is also made.

The paper is divided as follows: firstly, an MMC overview is given along with its control hierarchy. Then, the Bitonic SN is presented and its peculiar aspects in MMC applications are treated. Section 4 presents the simulation results in normal and faulty conditions. After that, different FPGA implementations are proposed and compared in terms of required resources and execution time in Section 5. Then, the chosen hardware implementation is compared with the software implementation of both the classical bubble SA and the max/min approach. HIL results for a single-phase 32-SM MMC are given to demonstrate the effectiveness of the proposed SN implementation. Finally, conclusions are drawn.

2. Overview of MMC: Topology and Control

The proposed implementation of the sorting algorithm can be used in any kind of MMC. The SMs can be either half-bridge or full-bridge, without any changes in the SN. In the following, a three-phase

HVDC grid connected MMC application with half-bridge SM is considered (Figure 1). Its topology and control structure are described in the next sections.

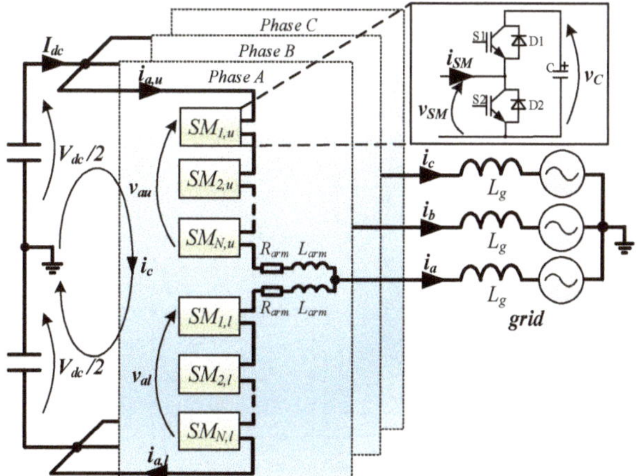

Figure 1. Schematic representation of a three-phase grid connected MMC.

2.1. MMC Topology

Each phase is composed of a leg that in turn consists of an upper and a lower arm. Each arm is composed of N series connected SMs, an arm inductor L_{arm} and the parasitic resistances in the arm, here denoted with R_{arm} [21]. The half-bridge structure for the SM is considered. It consists of two switches with two antiparallel diodes and a capacitor C, as depicted in Figure 1. The capacitor can be inserted or bypassed in the arm circuit based on the status of the two switches [8].

2.2. MMC Control Levels Hierarchy

Among the different modulation techniques, the NLC is often adopted for MMC [10]. The block diagram of a three-phase MMC controller based on such a modulation technique is displayed in Figure 2. The outer current control provides the reference voltage V_{ref} for each phase from the measured grid currents. To achieve the results presented in this paper, a classical outer current control has been implemented in the digital platform [22]. On the other hand, the circulating current control is adopted to limit the inherent circulating current ripple that is generated in the MMC. Then, the reference voltage is adjusted before the NLC. A common circulating current control based on resonant controllers has been used in this paper [22]. After that, the NLC calculates the insertion indices for the upper and lower arm for each leg, as expressed in:

$$n_{pm} = \frac{V_{ref,pm}}{V_{dc}} \tag{1}$$

where p and m represent the phase ($p = a, b, c$) and the arm ($m = u, l$), respectively. It is worth mentioning that the NLC only results the number of SMs to be inserted and it is indifferent to which SMs are selected. This task is taken in charge by the CVB control algorithm to ensure the balance between the capacitor voltages in the arm. The main elements of the CVB control algorithm are the sorting method and the selection technique. The aim of the sorting method is to put in ascendant order the SMs of an arm according to the measured capacitor voltages $V_{C_i,pm}$.

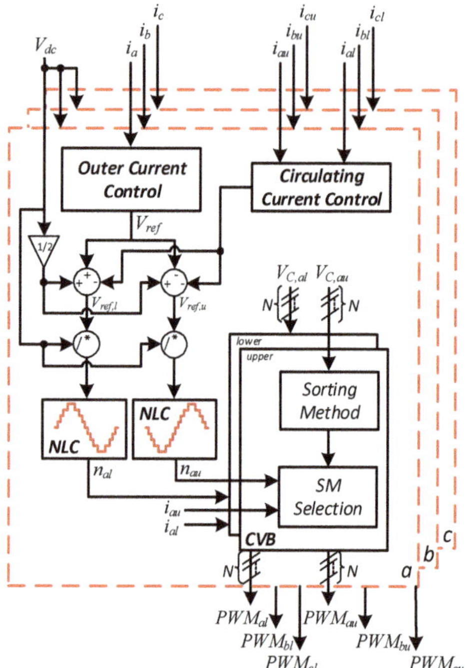

Figure 2. Block Diagram of the MMC FPGA-based controller.

The selection technique intends to select the SMs to be inserted based on the sorted list evaluated by the sorting method and the direction of the arm current i_{pm}. The best balancing performance is achieved when the sorting is executed in each sampling period and always the best SMs are selected. However, in this case, a high switching frequency f_{sw} is resulted. Different improved methods have been developed to decrease f_{sw} [23–25]. In this work, a reduction of the f_{sw} is achieved by inserting or bypassing only the difference between the required SMs and the actual ones, as shown in Figure 3. Moreover, the sorting is also performed if the capacitor voltages reach an upper or lower threshold value [23]. In this case, the SM with the highest (lowest) capacitor voltage is bypassed if the current is positive (negative) and another one is inserted in its place.

Another possible solution to reduce the f_{sw} and decrease the execution time is to adopt max/min approaches. They are based on the fact that only one SM has to be usually inserted/bypassed in one sampling period. However, this assumption is valid only during steady state operation. During faults, more SMs have to be manipulated in order to ensure fast reaction from the converter side. The max/min methods require more sampling periods to insert/bypass the required number of SMs. This leads a slower control dynamic as shown in Section 4.

2.3. Problem Statement

The aim of this paper is to deal with the efficient FPGA-implementation of the sorting method to reduce the execution time of the whole controller. To have all voltage steps equal to one level, the sampling period T_s has to satisfy [22]:

$$T_s \leq \frac{1}{\pi \cdot N \cdot f_{grid}} \tag{2}$$

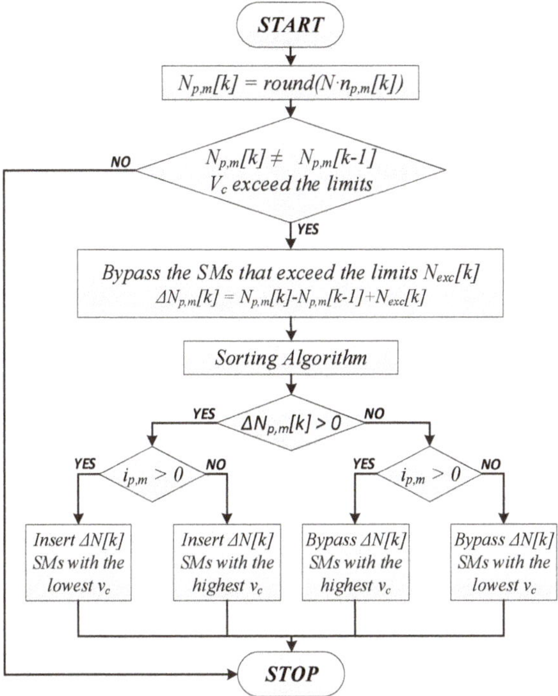

Figure 3. Flowchart of the implemented CVB algorithm in NLC.

In Equation (2), N is the number of SMs in the arm. Consequently, the controller execution time $T_{control}$ has to fulfill:

$$T_{control} \leq T_s \tag{3}$$

The common solution based on SAs has limited timing performances. They limit either the maximum allowable sampling frequency or the number of SMs in the arm. A significant reduction of $T_{control}$ can be achieved by adopting the proposed SNs. They are convenient for FPGA implementation due to their parallel structure. Moreover, they avoid the use of iterative and branch instructions.

Different factorization levels are introduced to give the possibility to choose the best tradeoff between required hardware resources and execution time. For these reasons, they are attractive for MMC applications and their detailed description is given in the next section.

3. Description of the Sorting Networks

The sorting networks are widely adopted in data processing [19] due to their timing performance. They consist of a fixed parallel structure composed of m-horizontal wires and several Compare-and-Swap (CS) operators. The latter carries out the sorting of two input elements: it compares them and ensures that the larger input value comes out from the upper output wire and the smaller input from the lower wire. Among the different SNs, the Bitonic structure is chosen in this work due to its modular aspect and to the reduced amount of CS operators [19,20].

Such a sorting network is composed of different stages which in turn are composed by several CS operators. In Figure 4, an eight-input Bitonic Sorting Network is depicted. In this case, six stages in the structure can be highlighted. The number of stages obviously depends on the number of the input. The unsorted sequence, denoted with x, is applied on the left, and the sorted list y results on the right. Only one element per wire can be applied.

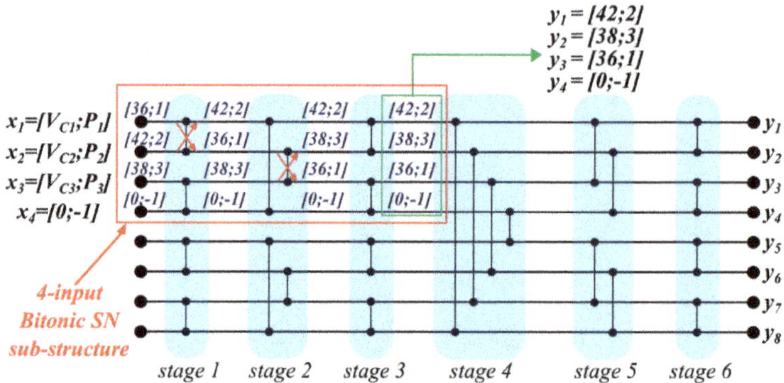

Figure 4. Eight-input Bitonic Sorting Network. It consists of six stages that in turn are composed of four CS operators. It can be seen as two four-input Bitonic SN plus Stages 4, 5 and 6. In this example, the length of the sorting list is equal to 3 to show its modularity property.

Firstly, each SM is enumerated starting from the top of each arm, as shown in Figure 1, and its position is named here $P_{i,pm}$. Each element x_i must consist of the acquired capacitor voltage $V_{C_i,pm}$ and the corresponding position $P_{i,pm}$, as depicted in Figure 4, where the subscript pm has been omitted for simplicity. A comparison of the capacitor voltages is performed and the swap operation is executed for both $V_{C_i,pm}$ and $P_{i,pm}$ if the voltages are not in the right order. The output y results in a sorted list with the voltages in ascending order along with the corresponding physical SM position. Then, if the SM with the highest voltage is needed for insertion, the first element y_1 is considered; otherwise, the last one is selected.

Another aspect that should be considered is the length of the list. Indeed, in MMC applications, the modularity is one of the main advantages [7]. It consists in the possibility to raise the power rating by adding more SMs in the arm. This means that the number of SMs, and then the length of the list, is not known a priori. However, a maximum number of SMs in an arm (named here with M) can always be presumed and then the SN is built for this worst-case scenario. In these conditions, some elements of the list can be left empty since the actual length of the list N can be different from M. To guarantee the proper behavior of the SN, these dummy elements have to be filled by setting the voltage equal to 0 and the position to -1 [20]. Therefore, these elements are kept in the last positions and the elements with positions different from -1 are selected when the SMs with lowest voltages are required.

An example of an eight-input Bitonic SN for a MMC with three SMs per arm ($M = 8$ and $N = 3$) is depicted in Figure 4. The dummy element is in Position 4 and it is kept in that position along the net. The result can be achieved before the last stage by considering only a sub-structure of the network as shown in the example. After having presented the SNs and having discussed their main peculiar aspects in MMC applications, some simulation results are shown in the next section.

4. Simulation Results

In this section, the Bitonic SN is compared with the max/min approach to show the differences between an algorithm that completely sorts the list and one that searches only the SM with the highest/lowest CV. The simulations have been performed in PLECS® power electronic simulation environment in both normal and faulty conditions. The MMC and grid parameters are given in Tables 1 and 2, respectively.

The Bitonic SN and the max/min approach have been implemented in PLECS. It is worth to note that the SN is intrinsically parallel; to simulate it its treatment has been serialized. Thus, its behavior is equal to the one achieved with a classical sorting algorithm such as the BSA. A tolerance band has also been introduced as shown in [23]. The max/min method proposed in [13] has been adopted in this work. The tolerance bands are set to 2 kV.

Table 1. MMC Parameters.

Quantity	Value
DC-link Voltage (V_{DC})	200 kV
SM Capacitor (C)	600 μF
Arm Inductance (L_{arm})	50 mH
Arm Resistance (R_{arm})	1.6 Ω
Number of SM (N)	16
Sampling frequency (f_s)	10 kHz

Table 2. Grid Parameters.

Quantity	Value
Grid frequency (f_{grid})	50 Hz
Grid Voltage (V_{grid}) line to line	121.2 kV
Grid Inductance (L_{grid})	3 mH
Grid Resistance (R_{grid})	0.1 Ω

4.1. Normal Condition

The simulation results shown in this section were achieved during steady state condition, i.e., without any faults in the system. At 0.2 s, the converter starts to deliver power to the grid. In Figure 5, the output currents, circulating currents and active power are shown when the Bitonic SN is adopted. The capacitor voltages for both the Bitonic SN and the max/min approach are depicted in Figure 6.

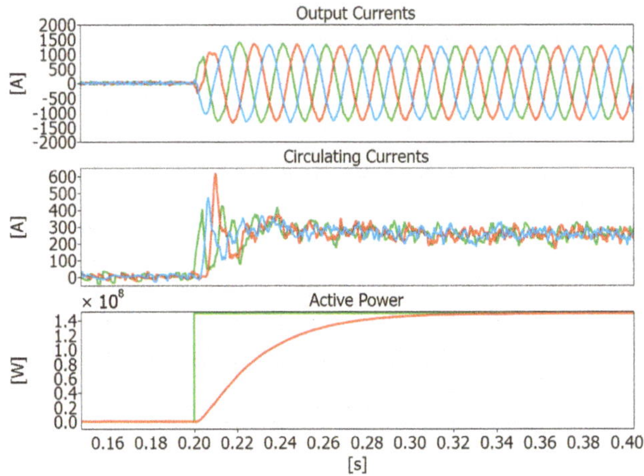

Figure 5. Output Current, Circulating Current and Active Power when the fully sorted list is adopted in the NLC.

When the best SMs are always selected, the algorithm that provides the complete sorted list achieves a better balance in comparison with the max/min approach. However, it leads to a high switching frequency, equal to 409 Hz in the studied case (Figure 6a). The max/min approach instead only selects one SM to be inserted or bypassed when a change in $N_{p,m}$ is resulted from the NLC (Figure 6b). Then, the switching frequency is intrinsically optimized and almost equal to 60 Hz.

By adopting the improved f_{sw} for the Bitonic SN, as described in Figure 3, the f_{sw} can be reduced. In this case, it is comparable to the one achieved with the max/min method, but the achieved balance is worse than the one achieved without the f_{sw} optimization, as shown in Figure 6c.

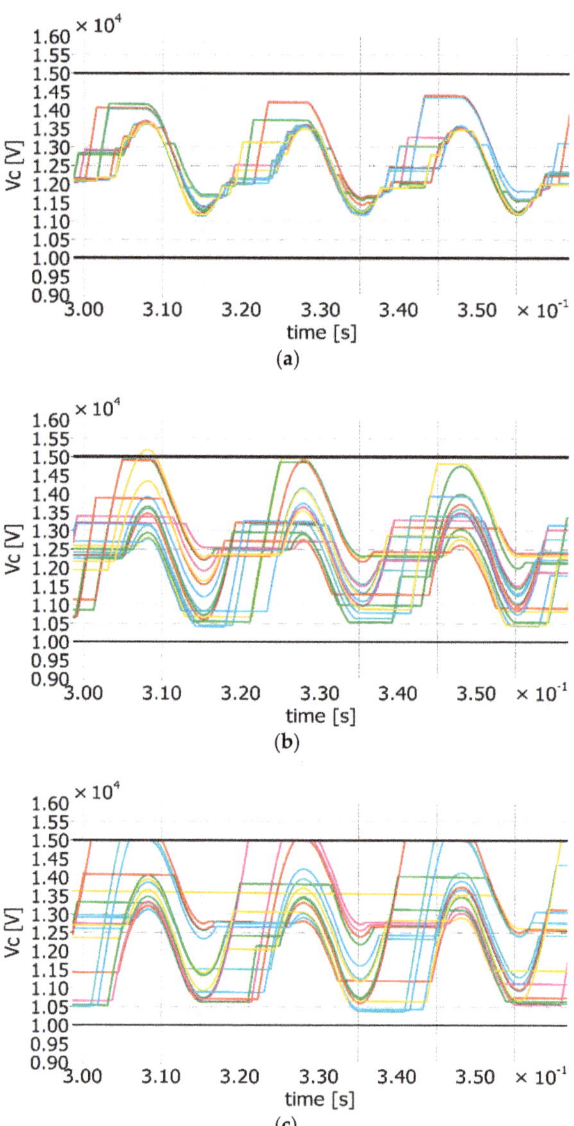

Figure 6. Capacitor Voltages of the upper arm in the phase *a*: (**a**) Sorting Method without f_{sw} optimization; (**b**) Max/Min Approach; and (**c**) Sorting Method with f_{sw} optimization as shown in Figure 3.

Finally, it can be concluded that in normal condition and when the f_{sw} optimization is active, the max/min method and the sorting algorithm give almost the same balancing results and switching frequency. From now on, the f_{sw} optimization is considered active, if not differently mentioned.

4.2. Phase-to-Ground Fault

The phase-to-ground fault is simulated by adopting the previous MMC and grid parameters. When this kind of fault appears in the system, the control algorithm demands more SMs to be inserted or bypassed in the same sampling period. The fault is applied at 0.5 s on the phase *a*. Figure 7 shows the output currents, the capacitor voltages of phase *a* and the number of switches for the same phase. When the number of switches is positive, it means that the SMs have been inserted, while, when it is negative, the SMs have been bypassed. It is worth noting that, in the case of the SN, three SMs have been bypassed at the moment of the fault. This number can relatively increase if the number of SMs in the arm is higher. The possibility to apply the required changes in one sampling period T_s enhances the dynamic performance of the controller. Indeed, at the fault instant, the algorithms that provide the fully sorted list are able to track the current reference and avoid spikes on the output current as shown in Figure 8.

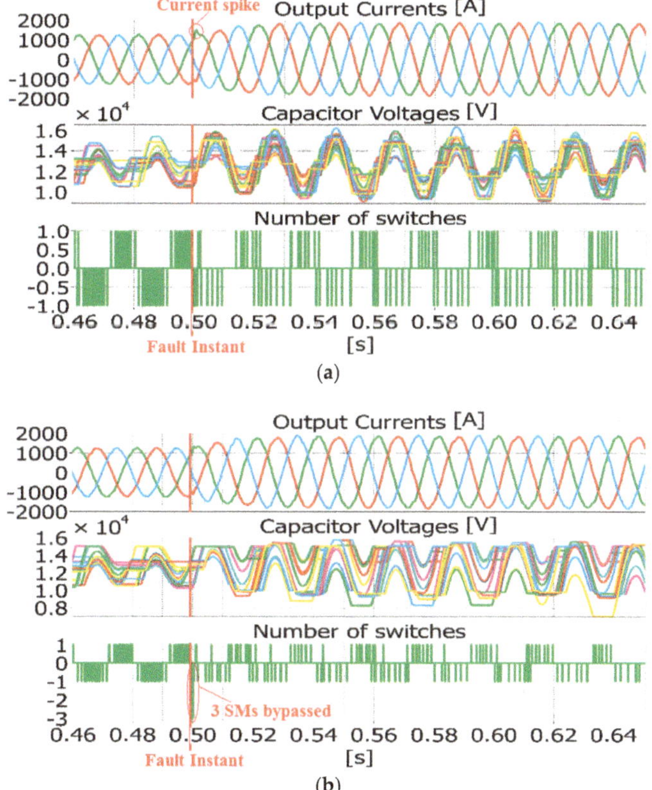

Figure 7. Output Currents, Capacitor Voltages and Number of Switches during Phase to Ground Fault at 0.5 s: (**a**) Max/Min Approach; and (**b**) Sorting Method.

On the other hand, the max/min method keeps only one switch per each T_s, requiring more sampling periods to insert/bypass the needed SMs. This leads to a lower controller dynamic that provokes a spike on the output current, as depicted in Figure 8. It is also possible to run the max/min method several times in the same sampling period to get the required SMs, but this leads to a longer execution time that can easily exceed the chosen T_s, not guaranteeing the proper controller behavior.

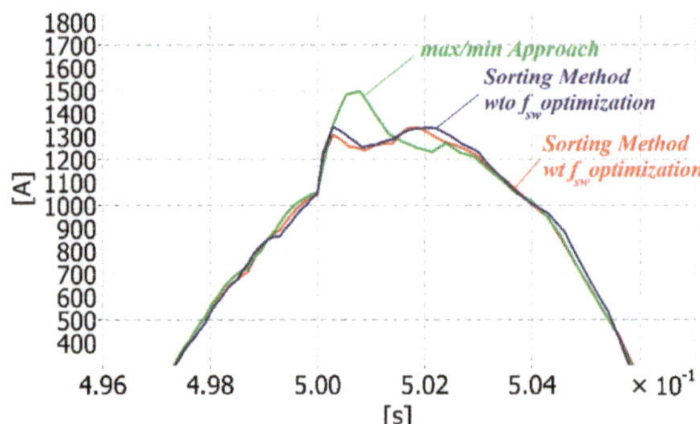

Figure 8. Zoom of the output currents on phase a during phase to ground fault. Blue line: Sorting Method without f_{sw} optimization; Green line: max/min approach; Red line: Sorting Method with f_{sw} optimization.

The advantages of the proposed SN over the min/max approach have been demonstrated and the FPGA implementation of the Bitonic SN is now dealt with.

5. FPGA Implementation of the Bitonic SN

Different FPGA-based implementations of SNs are shown and compared in this section. Firstly, the fully pipelined SN is presented. After that, a hybrid structure and a fully factorized SN are proposed for reducing the required resources. Finally, they are compared in terms of required resources and latency, i.e., the number of system clock cycles for achieving the final result.

5.1. Fully Pipelined SN

This kind of implementation of the SN allows a drastic reduction of the execution time. This architecture is driven by an external clock signal that synchronizes the data through the SN. After each stage, a bank of flip-flops is allocated for storing the results of CS operators, as shown in Figure 9. A new list of CVs can be fed to the input of the SN every clock cycle. The basic structure of the CS operator is also presented in Figure 9. It consists of a comparator and four multiplexers. The inputs are the two SM capacitor voltages V_{Ca} and V_{Cb} and the two SM positions P_a and P_b. The outputs are the sorted capacitor voltages $V_{s,Ca}$ and $V_{s,Cb}$ and their corresponding positions $P_{s,a}$ and $P_{s,b}$. The values of b_v and b_p correspond to the size of the fixed-point format of the voltage and of the position, respectively.

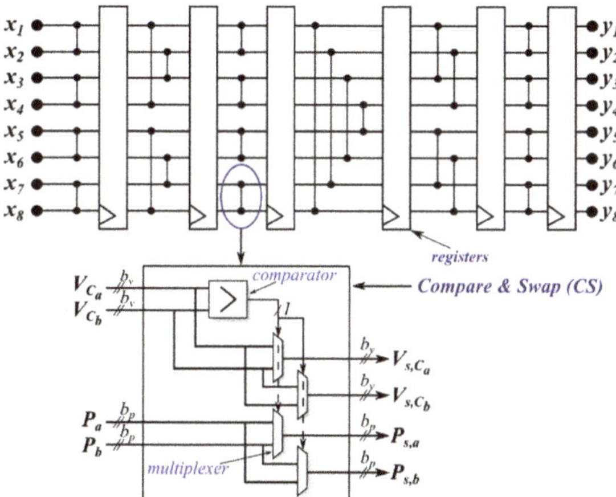

Figure 9. Fully pipelined 8-input Bitonic SN and CS operator structure.

5.2. M-Factorized SN

To reduce the required number of CS operators, the previous SN structure can be factorized. The objective is to reuse a single CS operator to perform more CS operations. However, this optimization leads to an increase of the latency in the architecture. Then, a compromise between the factorization level, i.e., which sub-structure of the Bitonic SN is factorized, and the latency is necessary. Figure 10 shows an example using $M = 8$ and the factorization level L equal to 4. This means that the four-input Bitonic SN sub-structure, highlighted in Figure 4, is factorized.

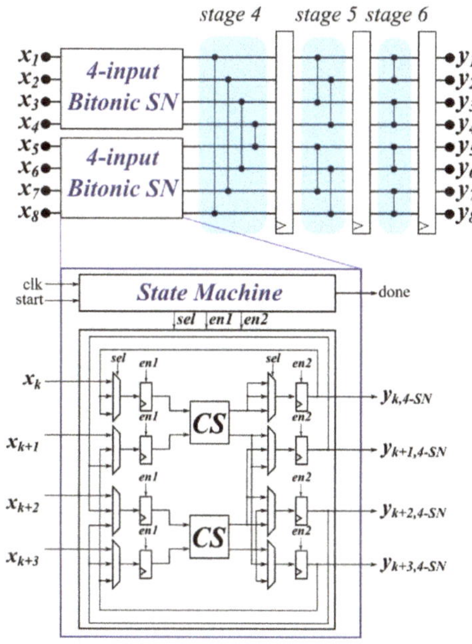

Figure 10. Hybrid factorized and pipelined synchronous eight-input SN with factorization level equal to 4.

This solution allows a reduction of the required CS operators and, therefore, the used resources. Indeed, the factorized four-input Bitonic SN only requires two CS operators instead of six. However, the latency will be 6 instead of 3 and no other list can be inserted at the input during this calculation, reducing the throughput of the architecture. It is worth noting that this solution also requires eight Multiplexers.

5.3. Fully Factorized SN

The last proposed architecture is the fully factorized SN, i.e., $L = M$. This alternative can be adopted to drastically reduce the required resources at the cost of a larger latency.

This architecture is depicted in Figure 11. The state machine generates and sends the synchronization signals to the data path which processes the input data. The Map operator implements the multiplexers and the registers needed for the factorization. The sorting done signal is raised when the sorted list is available at the output.

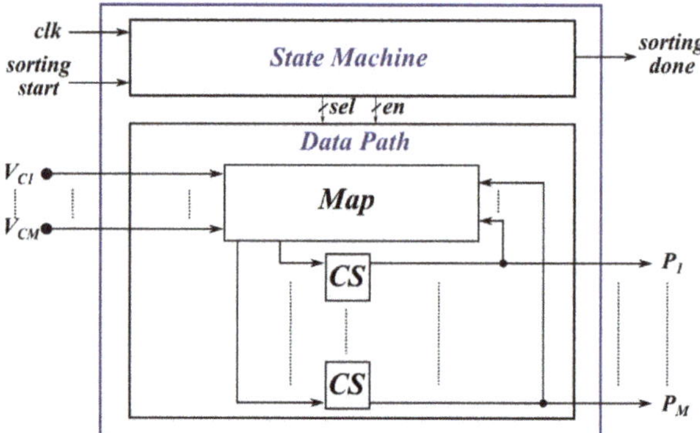

Figure 11. Fully factorized Bitonic SN implementation structure.

5.4. Comparison

In this section, the previous SN structures are compared in terms of five-input Look-Up-Tables (LUTs), Flip-Flops (FFs) and latency. The LUTs, FFs and latency for these structures can be easily pre-evaluated, as shown in [20]. Figure 12a depicts the required LUTs with different factorization level. It is shown that, by increasing the number of SMs, N, the required LUTs increase. The fully pipelined is the structure that requires the highest amount of LUTs, as expected. By increasing the factorization level, the needed LUTs can be reduced. The right y axis shows the percentage of the required LUT when the selected low-cost System-on-Chip (SoC) device is considered. Then, it is obvious that a fully pipelined structure for high numbers of N is impracticable with this kind of device. The same happen for the FFs, as depicted in Figure 12b. On the other side, a higher factorization level leads a higher latency number, as shown in Figure 12c. Thus, a compromise is required between the LUTs, FFs and latency. The designer can then pre-evaluate the required resources and the latency to choose the best solution for its requirements.

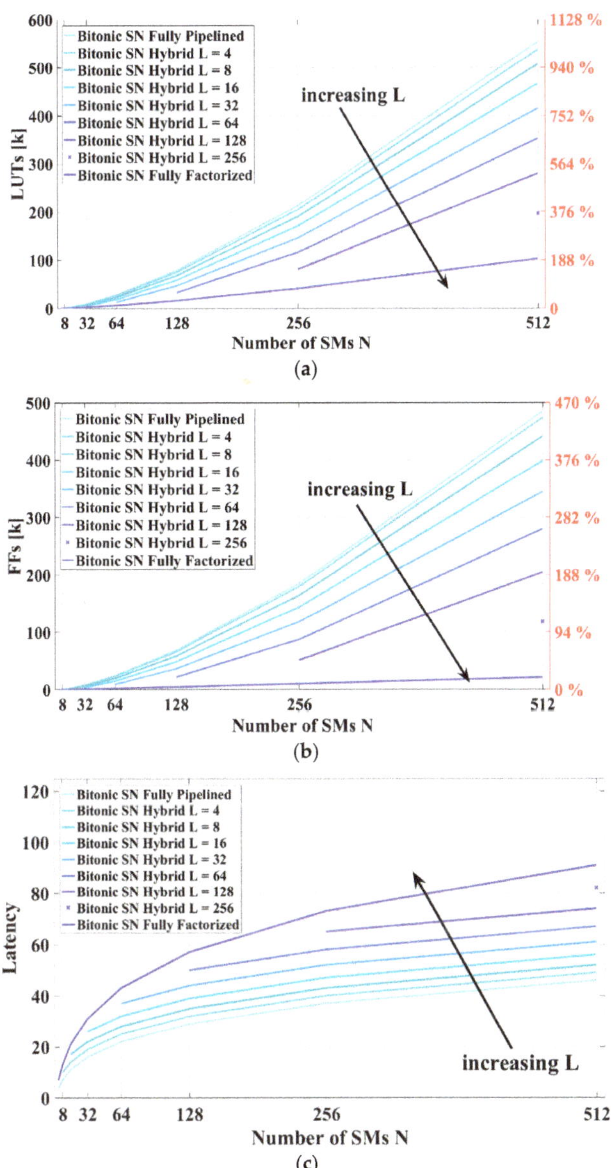

Figure 12. Evaluation of resources and timing performance of the fully pipelined Bitonic SN, the hybrid structure with different factorization level and the fully factorized SN with different input lengths (equal to the number of SMs N): (**a**) LUTs; (**b**) FFs; and (**c**) latency number.

5.5. Timing Diagram

In this section, the timing diagram for the proposed Bitonic SN is presented in Figure 13. The following control actions can be executed in parallel: the current controls (for both the output current i_s and the circulating current i_c), the upper arm sorting and the lower arm sorting. This leads a reduction of the whole control time $T_{control}$. The latter is the sum of two contributions: (a) the longer time between the current control time T_{cc} and the time needed to the Bitonic SN T_{bn}; and (b) the time $T_{selection}$ needed for selecting the SM to be inserted. From Equation (3) and Figure 13, it is possible to derive the maximum sorting time $T_{max,sort}$ that has to be guaranteed:

$$T_{max,sort} \leq T_s - T_{selection} \tag{4}$$

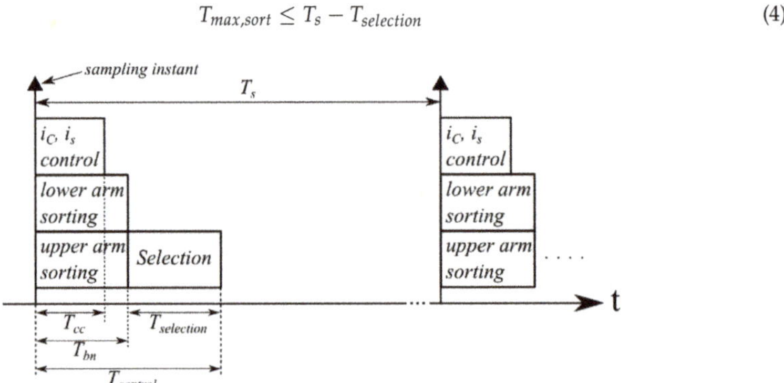

Figure 13. Timing Diagram of the MMC control based on BSN.

6. Hardware-In-the-Loop Results

A MMC system is usually composed by tens or hundreds of modules per arm. The realization of this kind of systems is very expensive, and then an usual approach is to test the control behavior using HIL approach. In this section, HIL results are provided to validate the proposed sorting method and compare it with the bubble SA and with the max/min approach in terms of the achieved sorting execution time. The Zedboard platform, equipped with a Xilinx SoC Zynq-7020 device (named here as Zynq), has been employed. This device consists of almost 85,000 logic element cells, 4.9 Mb block RAM, 220 DSP and two embedded ARM cortex A9 processors with a clock frequency equal to 667 MHz.

The MMC control, composed by the output current control and the circulating current control, has been implemented in the first ARM core of the Processor System (PS). Along with the MMC control, the bubble sorting algorithm and the max/min approach have been also carried out in the same core. On the other hand, the proposed Bitonic SN has been implemented in the Programmable Logic (PL) side by allowing the implementation of the hardware structures presented in Section 3 and the adoption of the timing diagram shown in Figure 13. The fully factorized Bitonic SN structure has been chosen to save hardware resources. In the remaining ARM core, a single-phase MMC model has been emulated based on [26]. The case of $N = 32$ has been considered, which is realistic in view of the industrial implementation of MMC-HVDC by ABB [27]. A single-phase system has been chosen considering that the performances of the Bitonic SN are identical for the single-phase and the three-phase system. Indeed, in the three-phase case, it can be easily implemented in parallel for achieving the same execution time shown in the following. The adopted capacitor value is 2.4 mF and the output energy is equal to 33 MW. The remaining MMC parameters are shown in Table 1. The whole implementation structure is depicted in Figure 14. The internal signals of the board are read and displayed on the PC through a serial communication. A sag voltage of 50% is emulated after 105 ms. The corresponding capacitor voltages, output currents, and output voltages for the max/min approach and the Bitonic SN are shown in Figure 15.

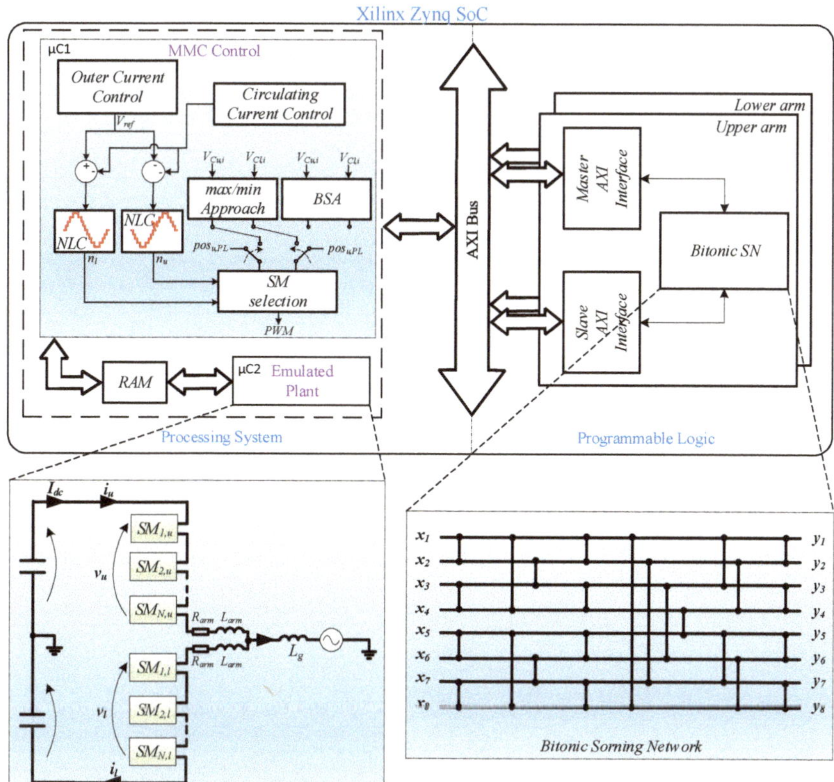

Figure 14. Hardware architecture of the developed system. The processor system and the programmable logic communicate through the AXI bus.

It is worth noting that the limited dynamic performance of the max/min approach cause an undesired overshoot in i_s of about 28%. The achieved execution times for the three techniques in both normal and fault conditions are summarized in Table 3. The sorting network has been implemented in FPGA and then some operations can be executed in parallel: the current control and the sorting algorithm in this case. Then, the control time is evaluated by summing up the longer execution time between the current control and the sorting algorithm with the selection time.

On the other hand, the bubble SA and the max/min approach have been implemented in the ARM core and then the current control, the sorting algorithm and the selection method are executed in a sequential manner. The whole execution time is then evaluated by summing up all the terms. The bubble SA achieves the worst timing performance by requiring maximum 4.56 µs in normal conditions and 6.46 µs during the emulated fault. It is worth noting that its execution time is not fixed. Thus, it is not deterministic and can be much higher than the ones obtained in this example.

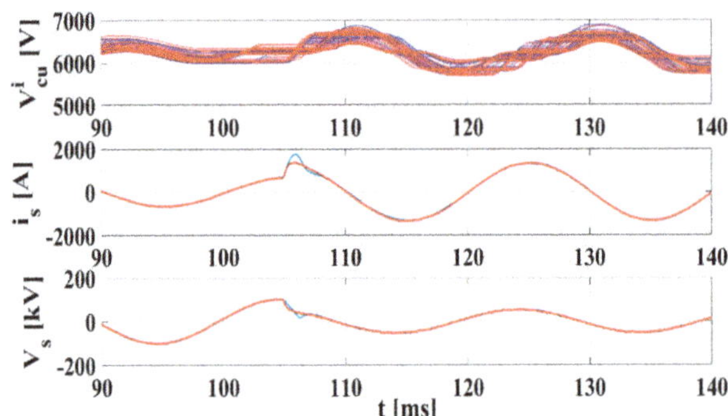

Figure 15. Capacitor Voltages, Output Current, and output voltage during HIL results with max/min approach (blue line) and Bitonic SN (red line).

Table 3. Maximum execution time of current control T_{cc}, sorting $T_{sorting}$, selection $T_{selection}$ and total control $T_{control}$ for the BSA, the max/min approach and the Bitonic SN during either normal (NC) and fault conditions (FC).

	Bubble SA		**Max/Min**		**Bitonic SN**	
	NC	**FC**	**NC**	**FC**	**NC**	**FC**
T_{cc} [µs]	0.28	0.30	0.28	0.30	0.28	0.30
$T_{sorting}$ [µs]	4.56	6.46	2.56	2.75	0.50	0.50
$T_{selection}$ [µs]	1.57	2.46	0	0	1.57	2.46
$T_{control}$	**6.41**	**9.22**	**2.84**	**3.05**	**2.07**	**2.96**

The max/min method needs 2.56 µs and 2.75 µs to be performed in normal and fault conditions, respectively. However, it does not provide a fully sorted list by causing current overshoot during faults. It is important to mention that its execution time is also not fixed. On the other hand, the proposed SN only requires a fixed execution time equal to 0.5 µs. It is also able to provide a fully sorted list useful during fault cases or when more SMs are required (like at the start-up). Its short execution time allows the proper behavior of the controller. However, if the number of SMs increases, the execution time raises and T_s decreases at the same time. In the case the execution time of the fully factorized structure is higher than the maximum allowable T_s, it is still possible to adopt either a hybrid structure or the fully pipelined SN that allow a reduction of the execution time. This choice can be done before starting the implementation thanks to the pre-evaluated resources and latency shown in Section 5 by reducing the time-to-market. Then, the right compromise between the sorting time and the required resources can be ensured. Table 4 shows the hardware resources required for the implementation of one fully factorized 64-Bitonic SN in the adopted device.

Table 4. Required resources for the implementation of one fully factorized Bitonic SN with $M = 64$ in the PL side of the Zynq device.

	Required Resources	**Utilization (%)**
LUT	5121 (53 k)	9.6
FF	1571 (106 k)	1.5

If a three-phase 64-SM MMC is of interest, six Bitonic SN have to be considered and then almost 60% of the available resources are consumed. It is also possible to use only one fully factorized Bitonic SN for accomplishing the sorting of all the arms. Obviously, this solution leads to an increment in the execution time. Another solution can be to implement a fully pipelined Bitonic SN that allows a higher throughput (a new list can be sent after the first stage has been performed) and then a lower execution time at the expense of an increment of the required resources. As already said, the right compromise has to be achieved depending on the case and the discussion in Section 5 is for this aim.

7. Conclusions

In this paper an FPGA-implementation of the Bitonic Sorting Network for MMC applications has been proposed. The main advantages of the proposed sorting method are: the fixed parallel structure and the possibility to be efficiently implemented in FPGA devices. This leads to a shorter and deterministic execution time, which results in a better performance of the CVB control. Three different architectures have been proposed to meet the different possible requirements of the application of interest: the fully pipelined architecture, the hybrid structure and the fully factorized SN. Their hardware resources and latency have been pre-evaluated and compared. This allows choosing the structure that best fits the application requirements before starting the implementation. Simulation results have been given in both normal and faulty conditions to compare the proposed sorting method with the max/min approach. The latter is considered to request the shortest execution time but it does not provide the fully sorted list by deteriorating the dynamic performance of the controller.

The fully factorized Bitonic SN has been implemented in a low-cost Xilinx Zedboard along with the MMC plant model and the corresponding control. Based on the HIL results, the proposed sorting method shows superior performance. Moreover, the execution time and the output current of the implemented SN have been compared with the ones achieved with the bubble sorting algorithm and the max/min technique in the case of $N = 32$ and a sag voltage of 50%. The output current obtained with the max/min technique presents an overshoot of about 28% in comparison with the one achieved by adopting an algorithm that provides a fully sorted list. Then, the main gain of the proposed SN compared to the max/min approach is the ability to handle fault occurrences. Moreover, the sorting and selection for the SN require 2.96 µs to be executed against the 9.22 µs and 3.05 µs needed by the bubble SA and the max/min approach, respectively. Considering the hardware resources, it requires about 10% of the available resources of the adopted board.

Author Contributions: Conceptualization, M.R. and L.M.; Software, M.R.; Validation, M.R.; Writing—original draft, M.R. and L.M.; Writing—review and editing, E.M. and R.T.

Conflicts of Interest: The authors declare no conflict of interest.

References

1. Lyu, J.; Cai, X.; Molinas, M. Frequency domain stability analysis of MMC-based HVDC for wind farm integration. *IEEE J. Emerg. Sel. Top. Power Electron.* **2016**, *4*, 141–151. [CrossRef]
2. Bergna, G.; Berne, E.; Egrot, P.; Lefranc, P.; Arzande, A.; Vannier, J.C.; Molinas, M. An energy-based controller for HVDC modular multilevel converter in decoupled double synchronous reference frame for voltage oscillation reduction. *IEEE Trans. Ind. Electron.* **2013**, *60*, 2360–2371. [CrossRef]
3. Jung, J.J.; Lee, H.J.; Sul, S.K. Control strategy for improved dynamic performance of variable-speed drives with modular multilevel converter. *IEEE J. Emerg. Sel. Top. Power Electron.* **2015**, *3*, 371–380. [CrossRef]
4. Antonopoulos, A.; Ilves, K.; Ängquist, L.; Nee, H.P. On interaction between internal converter dynamics and current control of high-performance high-power AC motor drives with modular multilevel converters. In Proceedings of the 2010 IEEE Energy Conversion Congress and Exposition, Atlanta, GA, USA, 12–16 September 2010.
5. Jafri, K.; Chen, B.C.; You, S.Z.; Chen, Y. Static and dynamic characteristics of a novel statcom based on modular multilevel converter. In Proceedings of the IET International Conference on Information Science and Control Engineering 2012 (ICISCE 2012), Shenzhen, China, 7–9 December 2012.
6. Bhesaniya, M.M.; Shukla, A. current source modular multilevel converter: Detailed analysis and statcom application. *IEEE Trans. Power Deliv.* **2016**, *31*, 323–333. [CrossRef]

7.   Debnath, S.; Qin, J.; Bahrani, B.; Saeedifard, M.; Barbosa, P. Operation, control, and applications of the modular multilevel converter: A review. *IEEE Trans. Power Electron.* **2015**, *30*, 37–53. [CrossRef]

8.   Nami, A.; Liang, J.; Dijkhuizen, F.; Demetriades, G. Modular multilevel converters for HVDC applications: Review on converter cells and functionalities. *IEEE Trans. Power Electron.* **2015**, *30*, 18–36. [CrossRef]

9.   Hagiwara, M.; Akagi, H. Control and experiment of pulsewidth-modulated modular multilevel converters. *IEEE Trans. Power Electron.* **2009**, *24*, 1737–1746. [CrossRef]

10.   Lesnicar, A.; Marquardt, R. An innovative modular multilevel converter topology suitable for a wide power range. In Proceedings of the 2003 IEEE Bologna Power Tech Conference Proceedings, Bologna, Italy, 23–26 June 2003.

11.   Siemaszko, D. Fast sorting method for balancing capacitor voltages in modular multilevel converters. *IEEE Trans. Power Electron.* **2015**, *30*, 463–470. [CrossRef]

12.   Li, W.; Gregoire, L.; Belanger, J. A modular multilevel converter pulse generation and capacitor voltage balance method optimized for FPGA implementation. *IEEE Trans. Ind. Electron.* **2015**, *62*, 2859–2867. [CrossRef]

13.   Saad, H.; Guillaud, X.; Mahseredjian, J.; Dennetière, S.; Nguefeu, S. MMC capacitor voltage decoupling and balancing controls. *IEEE Trans. Power Deliv.* **2015**, *30*, 704–712. [CrossRef]

14.   Monmasson, E.; Idkhajine, L.; Cirstea, M.N.; Bahri, I.; Tisan, A.; Naouar, M.W. FPGAs in industrial control applications. *IEEE Trans. Ind. Inform.* **2011**, *7*, 224–243. [CrossRef]

15.   Fan, B.; Li, Y.; Wang, K.; Zheng, Z.; Xu, L. Hierarchical system design and control of an MMC-based power-electronic transformer. *IEEE Trans. Ind. Inform.* **2017**, *13*, 238–247. [CrossRef]

16.   Vasiladiotis, M.; Rufer, A. Analysis and control of modular multilevel converters with integrated battery energy storage. *IEEE Trans. Power Electron.* **2015**, *30*, 163–175. [CrossRef]

17.   Shen, Z.; Dinavahi, V. Real-time device-level transient electrothermal model for modular multilevel converter on FPGA. *IEEE Trans. Power Electron.* **2016**, *31*, 6155–6168. [CrossRef]

18.   Mathe, L.; Dan Burlacu, P.; Teodorescu, R. Control of a modular multilevel converter with reduced internal data exchange. *IEEE Trans. Ind. Inform.* **2017**, *13*, 248–257. [CrossRef]

19.   Mueller, R.; Teubner, J.; Alonso, G. Sorting Networks on FPGAs. *VLDB J.* **2012**, *21*, 1–23. [CrossRef]

20.   Ricco, M.; Mathe, L.; Teodorescu, R. FPGA-based implementation of sorting networks in MMC applications. In Proceedings of the 2016 18th European Conference on Power Electronics and Applications (EPE'16 ECCE Europe), Karlsruhe, Germany, 5–9 September 2016.

21.   Perez, M.; Bernet, S.; Rodriguez, J.; Kouro, S.; Lizana, R. Circuit topologies, modeling, control schemes, and applications of modular multilevel converters. *IEEE Trans. Power Electron.* **2015**, *30*, 4–17. [CrossRef]

22.   Sharifabadi, K.; Harnefors, L.; Nee, H.P.; Norrga, S.; Teodorescu, R. *Design, Control, and Application of Modular Multilevel Converters for HVDC Transmission Systems*; Wiley: Hoboken, NJ, USA, 2016.

23.   Hassanpoor, A.; Angquist, L.; Norrga, S.; Ilves, K.; Nee, H. Tolerance band modulation methods for modular multilevel converters. *IEEE Trans. Power Electron.* **2015**, *30*, 311–326. [CrossRef]

24.   Ilves, K.; Harnefors, L.; Norrga, S.; Nee, H.P. predictive sorting algorithm for modular multilevel converters minimizing the spread in the submodule capacitor voltages. *IEEE Trans. Power Electron.* **2015**, *30*, 440–449. [CrossRef]

25.   Li, Z.; Gao, F.; Xu, F.; Ma, X.; Chu, Z.; Wang, P.; Gou, R.; Li, Y. Power module capacitor voltage balancing method for a 350-kv/1000-MW modular multilevel converter. *IEEE Trans. Power Electron.* **2016**, *31*, 3977–3984. [CrossRef]

26.   Ricco, M.; Gheorghe, M.; Mathe, L.; Teodorescu, R. System-on-chip implementation of embedded real-time simulator for modular multilevel converters. In Proceedings of the 2017 IEEE Energy Conversion Congress and Exposition (ECCE), Cincinnati, OH, USA, 1–5 October 2017.

27.   Jacobson, B.; Karlsson, P.; Asplund, G.; Harnefors, L.; Jonsson, T. VSC-HVDC transmission with cascaded two-level converters. In Proceedings of the CIGRE Conference, Paris, France, 22–27 August 2010.

Article

Hybrid Modulated Model Predictive Control in a Modular Multilevel Converter for Multi-Terminal Direct Current Systems

Zhi Wu [1], Jiawei Chu [1], Wei Gu [1,*], Qiang Huang [2], Liang Chen [2] and Xiaodong Yuan [2]

[1]   School of Electrical Engineering, Southeast University, Nanjing 210096, China; zwu@seu.edu.cn (Z.W.);
      220152251@seu.edu.cn (J.C.)
[2]   Jiangsu Electrical Power Company Research Institute, State Grid, Nanjing 211100, China;
      grassstrong@163.com (Q.H.); lancing_chen@163.com (L.C.); lannyyuan@hotmail.com (X.Y.)
*   Correspondence: wgu@seu.edu.cn; Tel.: +86-25-8779-6169

Received: 17 May 2018; Accepted: 11 July 2018; Published: 17 July 2018

Abstract: In this paper a hybrid modulated model predictive control (HM^{2}PC) strategy for modular-multilevel-converter (MMC) multi-terminal direct current (MTDC) systems is proposed for supplying power to passive networks or weak AC systems, with the control objectives of maintaining the DC voltage, voltage stability and power balance of the proposed system. The proposed strategy preserves the desired characteristics of conventional model predictive control method based on finite control set (FCS-MPC) methods, but deals with high switching frequency, circulating current and steady-state error in a superior way by introducing the calculation of the optimal output voltage level in each bridge arm and the specific duty cycle in each Sub-Module (SM), both of which are well-suited for the control of the MMC system. In addition, an improved multi-point DC voltage control strategy based on active power balanced control is proposed for an MMC-MTDC system supplying power to passive networks or weak AC systems, with the control objective of coordinating the power balance between different stations. An MMC-HVDC simulation model including four stations has been established on MATLAB/Simulink (r2014b MathWorks, Natick, MA, USA). Simulations were performed to validate the feasibility of the proposed control strategy under both steady and transient states. The simulation results prove that the strategy can suppress oscillations in the MMC-MTDC system caused by AC side faults, and that the system can continue functioning if any one of the converters are tripped from the MMC-MTDC network.

Keywords: MMC-MTDC; hybrid modulated model predictive control; optimal output voltage level; multi-point DC control

---

1. Introduction

In recent years, MMCs are gaining a lot of attention in high power/high voltage applications that involve interfacing high-voltage direct current (HVDC) systems to high voltage three-phase AC grids due to their high modularity and scalability [1,2]. Investment and research in high-voltage direct-current (HVDC) systems has been actively pursued and expanded with the aim of improving the efficiency and reliability of electric power generation, large-capacity power transmission, and linkage among different networks [1–6].

At present, research on control strategies for MMC-HVDC systems has yielded fruitful results in industry and academia. When an MMC-HVDC transmission system supplying passive networks is operating normally, its rectifier station generally uses constant current control and constant reactive control, which contains an outer power loop and an inner current loop. This control method is relatively mature and fixed, and details on the method are available in the literature [6–10]. Therefore, in this paper we focus on the control strategy used in the inverter station.

Traditional inverter station control usually involves a double closed-loop control (DCLC) strategy based on an outer voltage loop and an inner current loop. Since the response speed of the outer voltage loop is significantly slower than that of the inner current loop, the voltage quality is poor when supplying power to nonlinear loads. Furthermore, the response time of the voltage recovery increases when there are load fluctuations. Meanwhile, the overall control structure, embodying multiple proportional-integral (PI) controllers with hard-to-tune parameters, is relatively complex, and is therefore susceptible to structural parameters of the model [11,12].

Model predictive control (MPC) is gradually becoming more adopted as a control method of power converters because it is a non-linear optimization control method that can deal with nonlinear systems with complex constraints. The advantages demonstrated by this method are diverse, such as a fast response, flexibility of various goals, easy inclusion of nonlinearities, and the availability of simple modulation techniques. The MPC method based on a finite control set (FCS-MPC) constructs a multi-objective optimization function, evaluates the system's future state corresponding to the finite-switching combination of the converter, and selects the switch combination that minimizes the value of the objective function as the switching state for the next switching cycle. MPC has been applied to motor drives, high power factor rectification, and DC transmission, among others [13–21].

Reference [13] shows that the cost function may include a control target such as reduced switching frequency, reduced common-mode voltage, reduced reactive power, and reduced current ripple when controlling a power converter. Compared to proportional integral (PI) or proportional-resonant (PR) controllers, the MPC method can improve total harmonic distortion (THD) and transient characteristics [21]. References [16–18] uses FCS-MPC for the control of the MMC, which predicts the AC current, the circulating current, the sub-module capacitance voltage, and the resulting switching action of all possible switch combinations of the upper and lower arms in each control cycle. This method selects the switch combination that minimizes the objective function as the output of the next cycle and implements multi-objective optimal control.

For an MMC system with a large number of sub-modules, the existing FCS-MPC control methods involve a large amount of calculation. The Modulated Model Predictive Control (M^{2}PC) method preserves all the advantages of the FCS-MPC method and solves some problems associated with the FCS-MPC method, such as enabling variable switching frequency, delay compensation, and short sampling times [22–24]. However, little to no effort has been made to develop a M^{2}PC method for the control of MMC-HVDC systems supplying power to passive networks or weak AC systems. This paper aims to propose a new Hybrid M^{2}PC method for control of the MMC-MTDC system that addresses the key limitations faced by the FCS-MPC methods.

The contributions of this paper to the research field are:

(1) A novel voltage control strategy based on M^{2}PC is proposed for MMC-MTDC systems supplying power to passive networks or weak AC systems, which effectively regulates AC line currents and allows converters to comply with current references under severe conditions, such as severe power fluctuations or grid faults.

(2) The proposed strategy reduces the amount of calculation required compared with FCS-MPC methods when calculating optimal output voltage levels.

(3) An improved multi-point DC voltage control strategy based on active power balanced control is proposed and proved to be more applicable to MMC-MTDC systems.

2. MMC-MTDC Mathematical Model

Figure 1 shows the schematic diagram of a four-terminal MMC-MTDC system for supplying passive networks.

The rectifiers and inverters are made of three-phase MMCs. In the figure, L_s and L_p represent the AC filter inductances, while R_s and R_p denote the line equivalent resistances. U_{s1} and U_{s2} represent two independent AC power supplies. u_s and u_p represent the AC voltage of the sending end of the rectifier station and the receiving end of the inverter station, respectively. i_{dc} represents the current of the DC line. Further, C_{dc} denotes the DC capacitor, and C_1 and L_1 (C_3 and L_3) constitute the LC low-pass filter, which filters out the high-order harmonics on the inverter side, while C_2 can also provide some of the AC network reactive power support to compensate for the impact of load fluctuations on voltage stability.

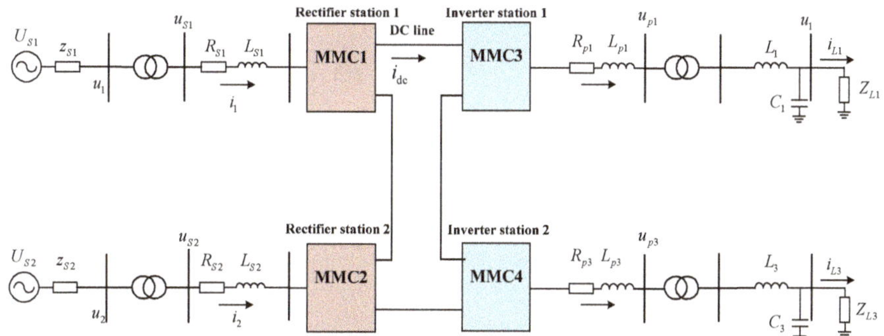

Figure 1. System configuration of a modular multilevel converter based multi-terminal direct current (MMC-MTDC) system for supplying passive networks.

Figure 2a presents the Circuit diagram of the inverter side of a three-phase MMC with six arms. Generally speaking, each arm is composed of a half-bridge submodule (SM). The MMC legs consist of three phases, a, b, and c, which are represented by the subscript j. Subscripts u and l represent the upper and lower arms of each leg. N series-connected SM, as well as the equivalent internal resistance (R) and an inductor (L), make up each leg. R_{p1} and L_{p1} denote the AC filter inductance and the line equivalent resistance of inverter station 1. Figure 2b shows the circuit diagram of SM. The switches T1 and T2 always operate in a complementary fashion.

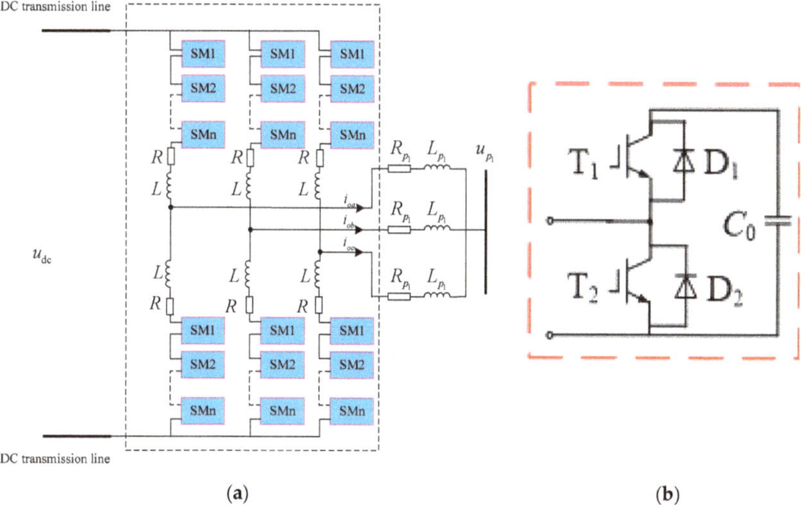

Figure 2. Circuit diagram of MMC: (**a**) Circuit diagram of the inverter side of a three-phase MMC; (**b**) Circuit diagram of Sub-Module (SM).

Using Kirchhoff's Voltage Law (KVL), the transient mathematical model of the AC side of an MMC can be obtained,

$$\frac{di_{sj}}{dt} = \frac{1}{L_{se}}\left(2u_{dj} - 2u_{p1j} - R_{se}i_{sj}\right) \tag{1}$$

$$\frac{di_{cj}}{dt} = \frac{1}{L}\left(u_{dc} - u_{cj} - Ri_{cj}\right) \tag{2}$$

where

$$\begin{cases} u_{cj} = \frac{1}{2}\left(u_{uj} + u_{lj}\right) \\ u_{dj} = -\frac{1}{2}\left(u_{uj} - u_{lj}\right) = \frac{1}{2}\left(u_{lj} - u_{uj}\right) \\ i_{cj} = \frac{1}{2}\left(i_{uj} + i_{lj}\right) \end{cases} \quad \text{and} \quad \begin{cases} L_{se} = L + 2L_{p_1} \\ R_{se} = R + 2R_{p_1} \end{cases}$$

where u_{cj} and u_{dj} denote common voltage and differential voltage, respectively, and i_{cj} represents circulation current.

Under balanced control the capacitor voltages of each sub module are equal, and the output voltage of the upper and lower bridge arm varies from 0 and N levels:

$$\begin{cases} u_{uj} = \frac{n_{uj}}{N} u_{dc}, \ n_{uj} \in [0, 1, \cdots, N] \\ u_{lj} = \frac{n_{lj}}{N} u_{dc}, \ n_{uj} \in [0, 1, \cdots, N] \end{cases} \tag{3}$$

where n_{uj} and n_{lj} represent the number of SMs for upper and lower arm inputs, respectively, and u_{uj} and u_{lj} represent the voltages of the upper and lower arms, respectively.

3. Design of the Hybrid Modulated Model Predictive Control Strategy

3.1. Design of the Inverter Side Controller

$$\Psi_{sj}(t) = \frac{1}{L_{se}}\left(2u_{dj} - 2u_{p1j} - R_{se}i_{sj}\right) \tag{4}$$

$$\Psi_{cj}(t) = \frac{1}{L}\left(u_{dc} - u_{cj} - Ri_{cj}\right) \tag{5}$$

Let T_s represent the sampling time, kT_s represent the present moment and $(k + 1)T_s$ represent the moment at the next control period. Utilizing the trapezoidal integration formula in Equation (4),

$$i_{sj}(k+1) - i_{sj}(k) = \frac{T_s}{2}\left(\Psi_{sj}(k+1) + \Psi_{sj}(k)\right) \tag{6}$$

Considering that the fluctuation of u_{sj} is negligible over one sampling period from kT_s to $(k + 1)T_s$, it can be approximated as invariant. Thus, the discretization of Equation (4) can be obtained,

$$i_{sj}(k+1) = \frac{2L_{se} - T_s R_{se}}{2L_{se} + T_s R_{se}} i_{sj}(k) + \frac{4T_s}{2L_{se} + T_s R_{se}} u_{dj}(k) - \frac{2T_s}{2L_{se} + T_s R_{se}}\left(u_{p1j}(k+1) + u_{p1j}(k)\right) \tag{7}$$

Similarly, considering that u_{dc} is constant during the sampling time, Equation (4) can be rewritten as,

$$i_{cj}(k+1) = i_{cj}(k) + \frac{T_s}{L}\left(u_{dc}(k) - u_{cj}(k)\right) \tag{8}$$

Equations (6) and (8) can be used to predict the AC line and circulating current values for all combinations of SM operations.

M^{2}PC requires a suitable modulation scheme as part of the minimization of the cost function in the MPC algorithm. In this paper we use a modulation scheme that is particularly suitable for high power converter control. In each sampling period, only one branch of one SM is allowed to switch, so as to obtain the total switching frequency of the SM, which is half of the sampling frequency [22–24].

This function becomes quite important when considering high power applications. Moreover, this switching mode helps to reduce the computational needs of the controller.

The proposed M²PC method evaluates a cost function $J(k)$ for all possible voltage levels in each leg at the start of each sampling period. This cost function with multiple prediction horizons can be defined as,

$$J_{p_1j}(k+1) = \left| u_{p_1j}(k+1) - u^*_{p_1} \right| \tag{9}$$

where

$$u^*_{p_1j} = U^*_s e^{j2\pi f^*(k+1)T_s} \tag{10}$$

where U_s^* represents the rated AC voltage; $f^* = 50$ Hz is the rated frequency [25].

By combining Equations (1)–(3) and (6), the reference voltage level of the bridge arm satisfying the AC voltage and the current control target is obtained,

$$\begin{cases} n^*_{uj} = \frac{N}{2u_{dc}}(u^*_{sj}(k+1) + u^*_{cj}(k+1)) \\ n^*_{lj} = \frac{N}{2u_{dc}}(u^*_{cj}(k+1) - u^*_{sj}(k+1)) \end{cases} \tag{11}$$

where n_{uj}^* and n_{lj}^* represent the reference voltage levels of n_{uj} and n_{lj}, respectively.

In general, the number of sub-modules applied to the MMC is large. If the traditional FCS-MPC control strategy is used, the computational load is undoubtedly enormous [26]. This paper uses Equation (11) to calculate the optimal output level of the bridge arm at the next sampling time, and then considers its neighboring $2M$ ($M \geq 1$) levels as a finite control set by selecting the appropriate M value, which can significantly reduce the amount of calculations in one control cycle. It is worth noting that in most cases n_{uj}^* and n_{lj}^*, calculated according to Equation (11), are not integers. This paper uses the method of rounding down, then selects the level combination and the upper or lower bridges. The $2M$ level combinations adjacent to the arm together constitute a new set of modulable finite controls.

A key point of voltage control in the M²PC strategy proposed in this paper is to calculate n_{uj} and n_{lj} based on the two specific voltage values of the passive network output voltage and the circulating current suppression voltage, based on Equation (11) [27–29]. The reference value thus minimizes the state performance function in Equation (7) based on the non-monotonic change characteristic of $J_{zj}(k+1)$. From the structural characteristics of the MMC it can be seen that, whether it is the upper or lower arms, increasing the number of levels in the bridge arm can increase the output value of the control target to the passive network at a future point, while reducing the number of arm levels will reduce the output voltage value to the passive network.

Therefore, the corresponding upper and lower arm level configuration parameters can be calculated according to the voltage prediction reference value at the beginning of the next period. Let the increase and decrease in the number of levels of the j phase affect the output voltage values of the next period by V_j^1 and V_j^2, respectively. In one control period described by setting a specific time node t_{p_1j} executing the control commands one after another leads u_{p_1j} $(k+1)$ to finally equal $u^*_{p_1j}$.

The ratio of time changing between V_j^1 and V_j^2 is called the duty cycle. The details of calculating this duty cycle for the proposed M²PC method are explained in detail in Reference [30]. Employing the formulation from Reference [30], the duty cycle can be obtained as,

$$d_{p_1j} = \frac{u^*_{p_1j} - u^2_{p_1j}}{u^1_{p_1j} - u^2_{p_1j}} \tag{12}$$

where $u^1_{p_1j}$ and $u^2_{p_1j}$ can be obtained from Equation (10) by substituting $u_{sj}(k)$ with V_j^1 and V_j^2 respectively. In Equation (12), d_{p_1j} represents the time ratio of $u_{sj}(k)$ needs to be set to V_j^1.

Similarly, the cost function and the duty cycle corresponding to the circulating current are,

$$J_{cj}(k+1) = \left| i_{cj}(k+1) - i^*_{cj} \right| \tag{13}$$

$$d_{cj} = \frac{i_{cj}^* - i_{cj}^2}{i_{cj}^1 - i_{cj}^2}$$

(14)

3.2. Improvement of the Model Predictive Control Algorithm

The performance cost function of Equation (13) guarantees an optimal combination of converter switches in one control cycle, but does not take into account its optimality in two or more control cycles, ignoring suboptimal switch combinations or other combinations which contain the optimal information. The algorithm relies on strong assumptions about the load behavior, and non-linear loads and load fluctuations may cause converter control system oscillation and even divergence. This paper will improve the algorithm with multi-step output predictive control to improve its robustness.

Firstly, single step prediction is performed using the discrete state equation. The number of inserted submodules (SM) is then determined to meet the requirements of multi-step model prediction. Figures 3 and 4 illustrate the principle of one step M²PC and the proposed optimized multi-step M²PC, respectively. One of the most distinctive feature of the proposed optimized multi-step M²PC is that its predicted periods become multi-step, and its control periods remain one step. Taking two-step in Figure 4 as an example, there are many paths that allow $u_{P_1 j}$ to reach the reference value at $(k + 2)T_s$, but only one path minimizes the cost function, which could be more optimized than the path in Figure 3.

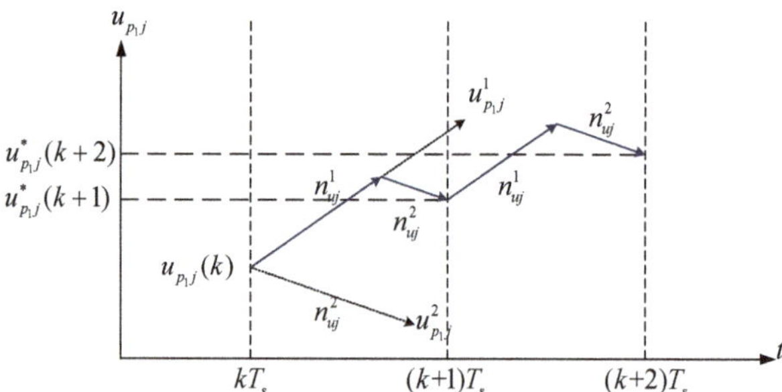

Figure 3. Principle of one step Modulated Model Predictive Control (M²PC).

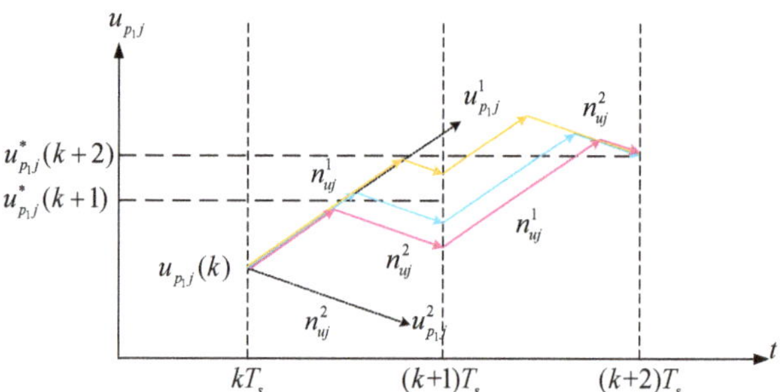

Figure 4. Principle of proposed optimized multi-step M²PC.

Therefore, the key principle of the proposed strategy is to solve for the optimal solution of the predictive model in multiple periods. In this situation, the value of the state variable in the multi-step predictive model ($X(k + p)$) needs to be calculated, and the cost function needs to make corresponding adjustments. Taking the two-step situation as an example, this paper utilizes Simpson's formula to calculate $X(k + 2)$, which can minimize the calculation error. For multi-step cases, Runge-Kutta formulae can be used.

Taking two-step in Figure 4 as an example, utilizing the Simpson integration formula in Equations (4) and (5),

$$i_{sj}(k+2) - i_{sj}(k) = \frac{T_s}{6}(\Psi_{sj}(k+2) + 4\Psi_{sj}(k+1) + \Psi_{sj}(k)) \tag{15}$$

$$i_{cj}(k+2) - i_{cj}(k) = \frac{T_s}{6}(\Psi_{cj}(k+2) + 4\Psi_{cj}(k+1) + \Psi_{cj}(k)) \tag{16}$$

Similarly, considering that u_{sj} and u_{dc} are constant during the sampling time,

$$J(k+1) = \lambda_1 \left| u_{p_1 j}(k+2) - u_{p_1}^* \right| + \lambda_2 \left| i_{cj}(k+2) - i_{cj}^* \right| \tag{17}$$

where λ_1 to λ_2 are the weighting factors of each control target.

Considering the computational complexity of the performance cost function at this time, as well as taking into account the robustness of the multi-step prediction, M can be set to 1. Meanwhile, for simplicity, it can be assumed that the load current does not change within the predicted time domain. Thus, the improved VSC-HVDC system inversion and the corresponding performance cost function are constructed using the improved Equations (15) and (16).

When a DC fault occurs, it should be noted that the fluctuation of the terminal voltage will be comparatively greater. This will undoubtedly result in instability of the controlled output voltage of the converter and the limit of the voltage range to be reached.

The DC voltage reference value of the DC voltage control link in the original converter station control system is corrected according to the amount of voltage fluctuation, and the threshold value of the modified voltage reference signal is applied to ensure that the DC voltage of the multi terminal system will not deviate from the normal operating range.

3.3. Total Cost Function and Duty Cycles

The cost function is now given by,

$$J(k+1) = \lambda_1 J_{pj}(k+2) + \lambda_2 J_{cj}(k+2) \tag{18}$$

The two weighting factors, λ_1 and λ_2 can be adjusted to achieve the desired the control performance. Since the current cost function $J_{cj}(k+2)$ already includes the amount of current necessary to charge the DC-link capacitor to the desired voltage, the importance of $J_{pj}(k+2)$ lies in its ability to reduce the steady-state error in the DC-link voltage, related to the converter losses, which is not considered in $J_{cj}(k+2)$. Therefore, the ratio λ_1/λ_2 is typically set to the minimum value that ensures zero steady-state error in the DC-link voltage [31,32].

The switching times for the two selected vectors are calculated by solving the linear system of equations in Equation (18). Once the value of K is obtained from Equation (18), the expressions for the switching times are obtained

$$\begin{cases} J_{cj}^{(1)} = \dfrac{K}{J_{cj}^{(1)}} \\[2mm] d_{cj}^{(2)} = \dfrac{K}{J_{cj}^{(2)}} \\[2mm] d_{cj}^{(1)} + d_{cj}^{(2)} = 1 \end{cases} \Rightarrow \begin{cases} d_{cj}^{(1)} = \dfrac{J_{cj}^{(2)}}{J_{cj}^{(1)} + J_{cj}^{(2)}} \\[2mm] d_{cj}^{(2)} = \dfrac{J_{cj}^{(1)}}{J_{cj}^{(1)} + J_{cj}^{(2)}} \end{cases} \tag{19}$$

Once the switching times are calculated, the M^2PC algorithm chooses the two inverter states for times $d_{cj}^{(1)}$ and $d_{cj}^{(2)}$ if they minimize the following global cost function:

$$J_{cj} = d_{cj}^{(1)} G_{cj}^{(1)} + d_{cj}^{(2)} G_{cj}^{(2)} \tag{20}$$

This solution is proposed as an alternative to an analytical duty cycle calculation. The overall M^2PC scheme is shown in Figure 5.

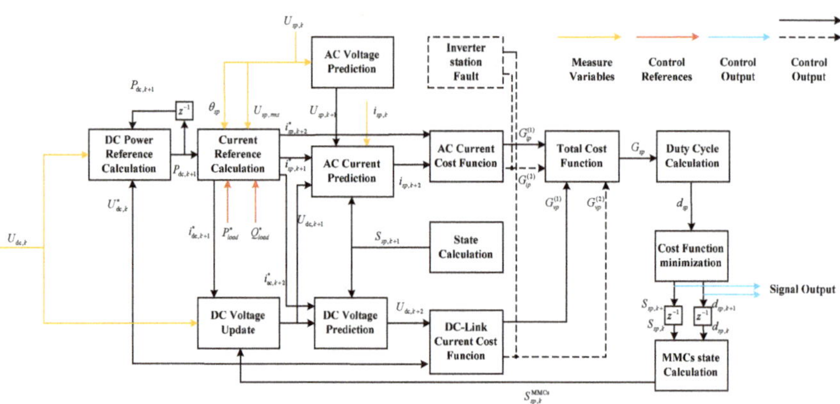

Figure 5. Overall M^2PC block scheme for the control of MMC converters.

3.4. Improved Multi-Point DC Voltage Control Strategy

The proposed multi-step M^2PC control strategy is mainly used on the inverter stations connected to the loads as a first control, while the multi-point voltage coordinated control strategy is used between these four stations to maintain the balance of active power and the stability of DC voltage as a secondary control.

Therefore, this paper further proposes a multi-point voltage coordinated control strategy based on M^2PC and power balance, which is called Hybrid M^2PC (HM^{2}PC). HM^{2}PC is integrated into the primary and secondary coordination control strategy, and its specifics are as follows:

(1) MTDC multi-point voltage coordination control requires the upper system-level controller to provide the active power reference signal to the converter station-level controller of each terminal converter station.

(2) In the upper system level controller, either the converter station with the largest converter capacity or the key converter station in the system is selected as the power balance converter station. The reference values for the active power of the remaining converter stations are taken directly from the power flow regulator system, while the reference values for the active power of the converter stations are calculated according to Equation (21), based on the reference values for the active powers of other converter stations. The variable n is the number of MTDC converter stations.

$$P_{Balance} = -\sum (P_{ref1}, P_{ref2}, \cdots\cdots, P_{refn-1}) \tag{21}$$

(3) The system level control system only updates the active power reference value to each converter station when changing the scheduling trend. At the other times, the converter stations are independently controlled according to the reference value calculated from the active power after the latest update, without much communication needs.

3.5. Implementation of the HM²PC Strategy

The execution of HM²PC includes the following steps:

(1) Sample and measure the relevant electrical parameters of the MMC-HVDC system at time kT_s, including: $u_{pj}(k), u_{sj}(k), u_{dc}(k)$, and $i_{sj}(k)$.

(2) With the prediction model, improved multi-step prediction function, and prediction correction function, combined with the time sampling value kT_s and the inverter switching state for different calculations, calculate the predicted value at the $(k + 1)T_s$ moment (or $(k + p)T_s$ moment, $p = 1, 2$).

(3) According to step (2) and the corresponding performance cost function, the predicted output of the traversing method is calculated as the inverter output in the prediction horizon from all possible combinations of the switch performance cost function and the corresponding values.

(4) The performance cost function values corresponding to each switch state are compared, and the minimum switching state g_{min} (i.e., the optimal switching state) is selected to generate switching signals, which are then driven and amplified to act on the inverter.

(5) Repeat step (1) to step (4) at the next sampling period.

4. Simulation Results and Analysis

In order to verify the effectiveness of the above predictive control algorithm, a simulation model of the VSC-MTDC system (Figure 1) in MATLAB/Simulink has been developed. The system parameters are listed in Table 1. In this study, the constant current control and constant reactive control strategy are adopted on the rectifier side of the VSC-MTDC system. On the inverter side, the traditional DCLC strategy, single-step predictive control strategy, and improved multi-step predictive control strategy (the number of predicted steps P is 3), which is added to the correction feedback, are simulated and compared.

Table 1. Parameters for the Study System of Figure 1 (Sub-Module (SM)).

Quantity	Value
AC sources system nominal Voltage	35 kV
Nominal DC voltage	±10 kV
R_s	0.1 Ω
L_s	5 mH
R_p	0.02 Ω
L_2	1.3 mH
C_2	200 μF
DC capacitance C_{dc}	4700 μF
DC line length	10 km
AC load system nominal voltage	20 kV
The number of stations	4
Arm inductance L	1 mH
Arm resistance R	0.2 Ω
SM capacitor C	1.2 mF
The number of SMs	10

4.1. MMC Converters Connected to the Load

In this simulation scenario the MMC1 stations maintain a voltage of 20 kV and the MMC3 stations maintain a power of 30 kW. Initially, the MMC4 station maintained a constant load of 40 MW. At 5 s, the load was reduced by 16 MW, and the load at 10 s increased by 16 MW, with the MMC3 station adjusting the amount of added power fluctuation.

In order to compare the static characteristics using a nonlinear load between the traditional control strategy and the adaptive modulated model predictive control strategy mentioned in this paper, several large frequency converters are connected to the low-voltage side of the passive network. The control frequency is 10 kHz and the corresponding AC current simulation results are shown in Figure 6.

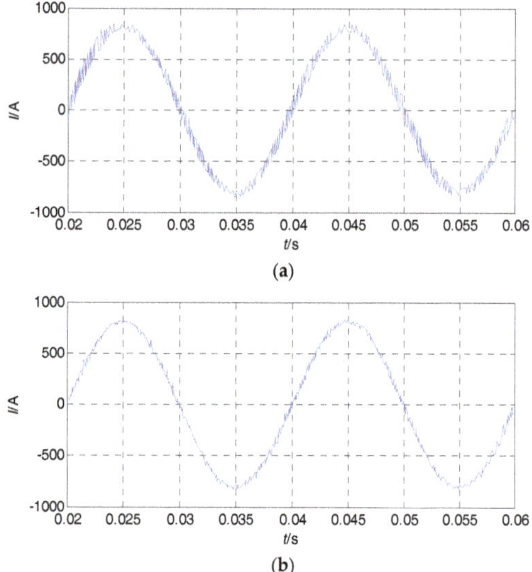

(a)

(b)

Figure 6. The simulation results of AC current: (**a**) Traditional control strategy; (**b**) The strategy proposed in this paper.

By comparing and calculating the current error of the traditional control strategy and the strategy proposed in this paper, the current signal Total Harmonic Distortion (THD) values are 9.88% and 1.83%, respectively. It can be seen that when large-capacity nonlinear loads are connected to the grid system the quality of the power supply using the traditional control strategy is significantly reduced, while the proposed method can still maintain acceptable static performance. Therefore, in the face of large-scale renewable energy integration into the distribution network, the proposed control strategy has great advantages in ensuring the quality of the power supply.

The simulation results in Figure 7a,b compare the inhibition of circulation current in a MMC converter using M^{2}PC with the improved multi-step M^{2}PC. At 0.25 s, the circulation suppression function is enabled separately. As can be seen, the circulation current contains some DC and low harmonic components. After the function is enabled, the circulation current under both control methods is rapidly reduced. The difference is that the improved multi-step M^{2}PC is able to further reduce the size of the circulation current and lower the amplitude of the pulse in the circulation current.

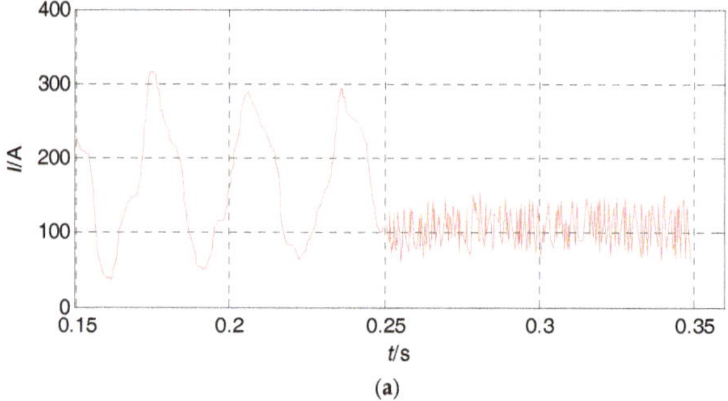

(a)

Figure 7. *Cont.*

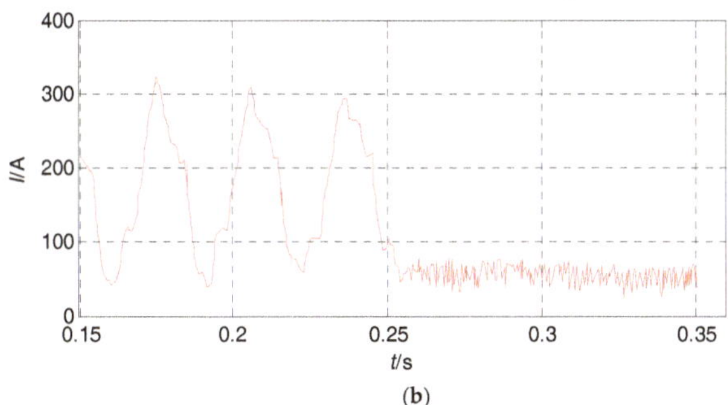

(**b**)

Figure 7. The simulation results of circulation current: (**a**) M²PC; (**b**) Improved multi-step M²PC.

4.2. Regulation Station Supplemental Power During Load Fluctuation

In this simulation scenario, the MMC1 stations maintain their voltage at 20 kV and the MMC3 stations maintain their power at 30 kW. Initially the MMC4 station maintained a constant load of 40 MW. At 5 s, the load was reduced by 16 MW, the load at 10 seconds increased by 16 MW, and the MMC3 station adjusted the amount of added power fluctuation.

The comparison of the active power transmitted by the MMC2 station and the MMC4 station in each coordinated control mode is shown in Figure 8, and the comparison of the DC voltage at each end of the MTDC system is shown in Figure 9.

In this simulation, FCS-MPC is faster than the other two control strategies. However, as the ability to adjust the load fluctuation depends entirely on the master station, the controller should balance the active power and the voltage control a ring DC grid. Furthermore, deficiencies in single point voltage control cause large voltage fluctuations, with the maximum fluctuation threshold reached 5 times. Thus, voltage and power fluctuations as well as the steady state error are the largest of the three kinds of control.

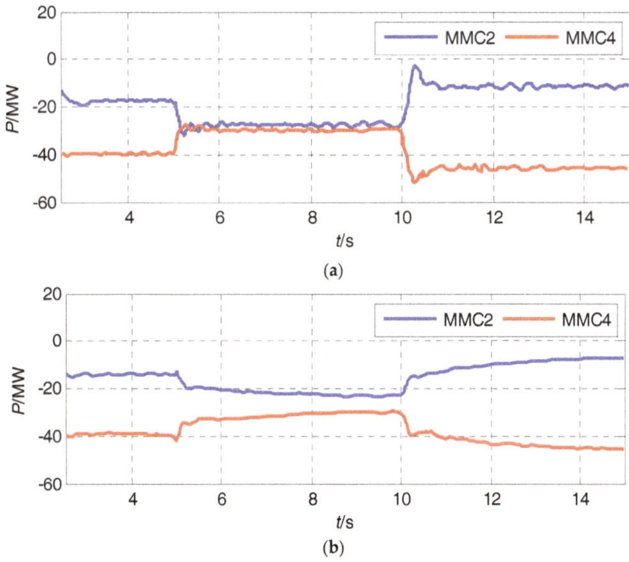

(**a**)

(**b**)

Figure 8. *Cont.*

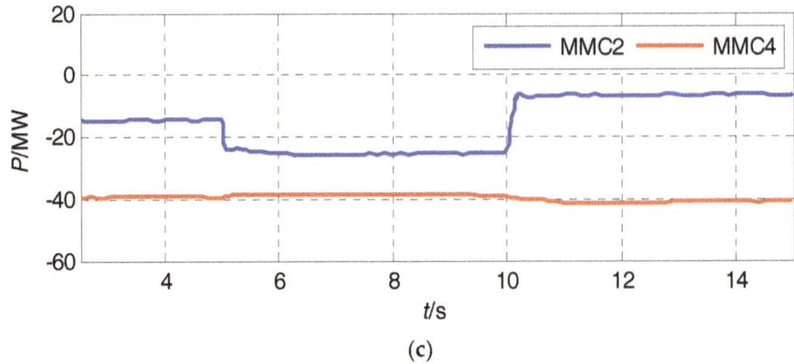

(c)

Figure 8. Simulation waveforms of load fluctuation: (**a**) The active power of the receiving end controlled by double closed-loop control (DCLC); (**b**) The active power of receiving end controlled by an MPC method based on a finite control set (FCS-MPC); (**c**) The active power of the receiving end controlled by the hybrid modulated model predictive control (HM^{2}PC).

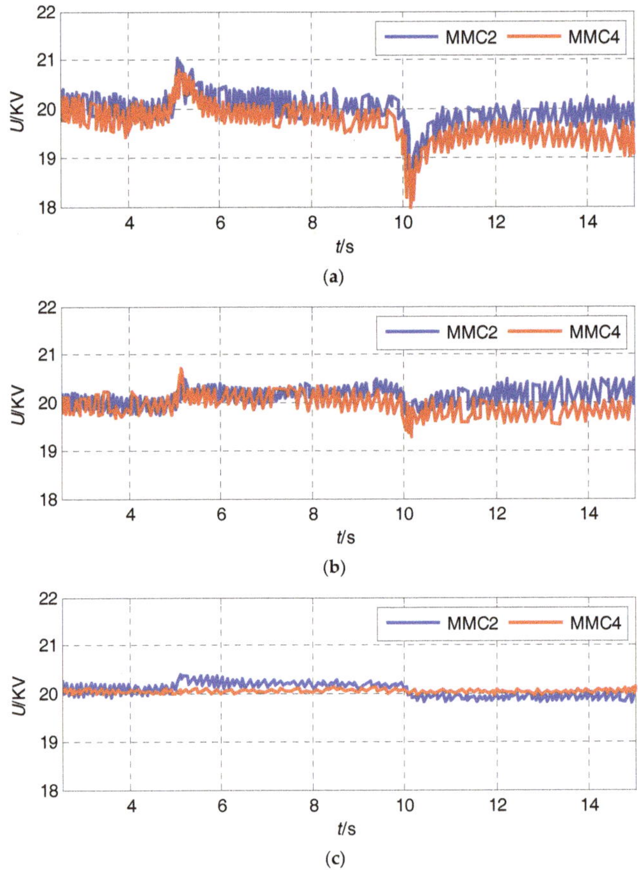

(a)

(b)

(c)

Figure 9. Simulation waveforms of load fluctuation: (**a**) The DC voltage controlled by DCLC; (**b**) the DC voltage controlled by FCS-MPC; (**c**) the DC voltage controlled by the HM^{2}PC.

The HM^{2}PC strategy proposed in this paper is faster than DCLC FCS-MPC. The DC voltage under control shows almost no fluctuation under load fluctuation, and the voltage fluctuation and power fluctuation in steady state are the smallest among the three control modes.

4.3. Aimulating the Exit of the Main Control Station Due to Failure

In this simulation scenario, the MMC1 station and the MMC4 station maintain a constant load of 10 MW and 40 MW, respectively. Initially, the VSC1 station and the VSC3 station respectively issue 20 MW and 30 MW of active power. At 5 s, the main control station MMC1 exited due to a failure. At this time, the MMC3 station became the new main control station and simultaneously delivered power to the MMC2 station and the MMC4 station.

The comparison of the active power transmitted by the MMC2 station and the MMC4 station in each coordinated control mode is shown in Figure 10, and the comparison of the DC voltage at each end of the MTDC system is shown in Figure 11.

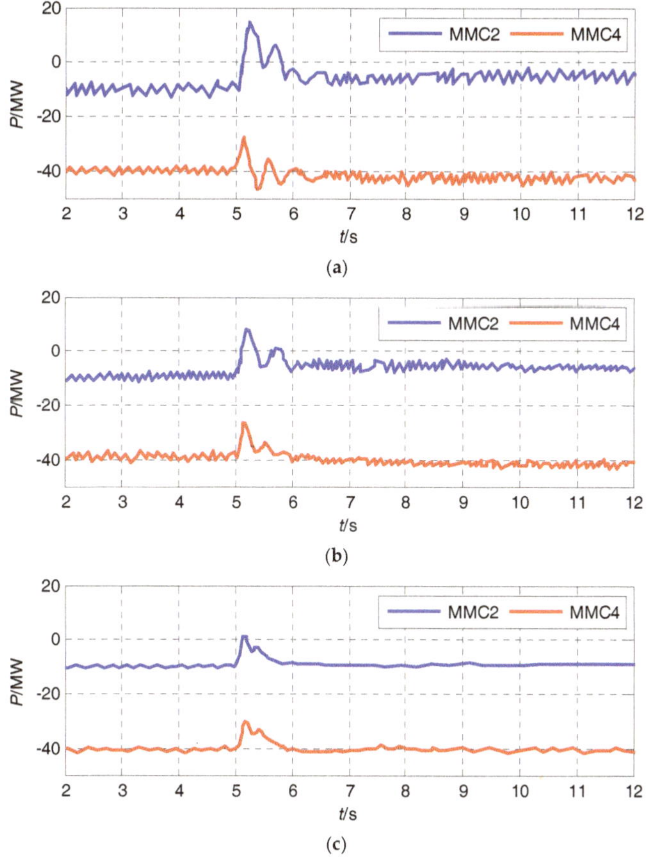

Figure 10. Simulation of the operation during an exit of an input converter station: (**a**) The active power of the receiving end controlled by DCLC; (**b**) the active power of the receiving end controlled by FCS-MPC; (**c**) the active power of the receiving end controlled by HM^{2}PC.

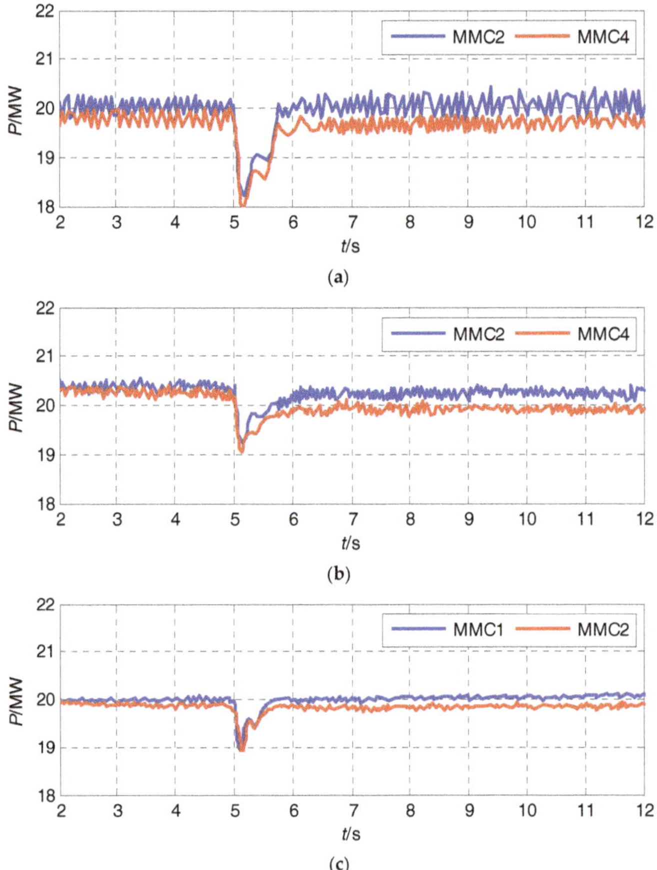

Figure 11. Simulation of the operation during an exit of the input converter station: (**a**) The DC voltage controlled by DCLC; (**b**) the DC voltage controlled by FCS-MPC; (**c**) the DC voltage controlled by HM^{2}PC.

In this simulation scenario involving a large disturbance where one end completely withdraws from the four-terminal system, all three coordinated controls can still achieve three-terminal power balance. However, the lack of voltage control capability in the master-slave control method causes an overshoot of active power to almost reach the upper limit of the master station control. If the limit is exceeded, it may cause a change in the control master station, thereby forcing the MMC4 station to maintain the DC voltage by reducing the load power. The power adjustment speed of the multi-point voltage control strategy based on an active power balance proposed in this paper performs better than voltage drop control, achieving the minimum voltage fluctuation and power fluctuation in the steady state.

5. The Scheme of Experimental Verification

In order to further verify the effectiveness of the proposed control strategy, a real-time controller hardware-in-the-loop test platform based on RT-LAB (11.1, OPAL-RT technologies Inc, Montreal, QC, Canada) can be built, as shown in Figure 12. The network frame model of MTDC system is built in the upper computer software of RT-LAB. Its main topology is put in the high performance real-time simulator OP5600 (OPAL-RT technologies Inc, Montreal, QC, Canada) and the high frequency power electronic devices are placed in the nanosecond real-time simulator OP7020 (OPAL-RT technologies Inc., Montreal, QC, Canada) containing FPGA, which are connected to the external controller through the I/O interface and communication protocol. Then, the control strategy proposed in this paper can be rewritten into the form of C language and fed into the external controller (usually Digital Signal Processor (DSP)). In addition, it can test different running scenarios by dynamically adjusting the reference values of state quantities in the control subsystem of OP5600 in real time. In a nutshell, experimental verification is the next research focus that needs to be overcome.

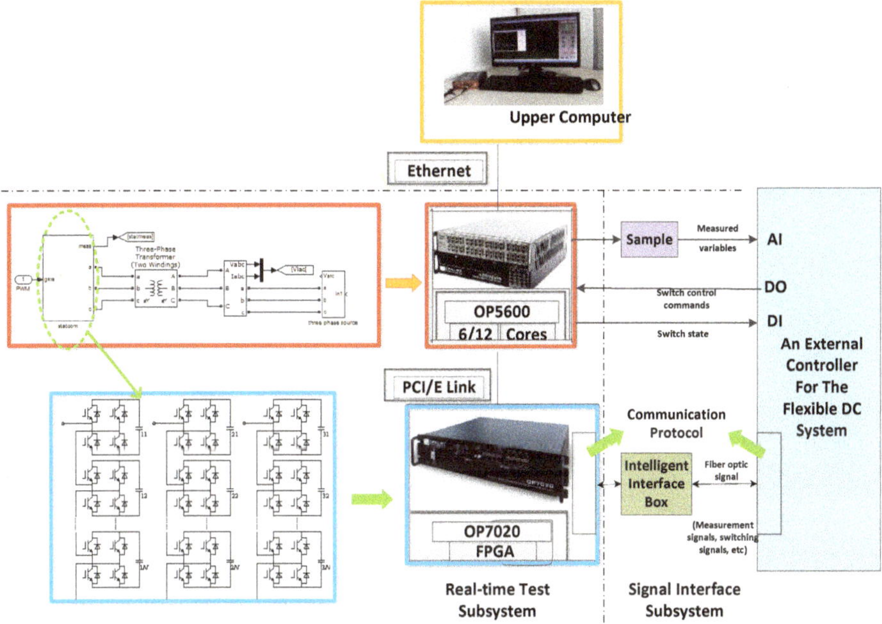

Figure 12. A schematic diagram of the real-time controller hardware-in-the-loop test platform.

6. Conclusions

A hybrid modulated model predictive control (HM^{2}PC) strategy for modular-multilevel-converter (MMC) multi-terminal direct current (MTDC) systems supplying power to passive networks or weak AC systems was developed to improve the control performance of maintaining the DC voltage and power balance of the proposed system. The proposed method, on the basis of fully preserving the superiority of the traditional model predictive control method, reduces the amount of control operations and has the characteristics of flexible structure, good robustness, and strong scalability.

Next, an improved multi-point DC voltage control strategy based on active power balanced control is proposed, which is proved to have a fast transient response and includes the control target directly in the cost function minimization algorithm. It has the further advantage of including a suitable modulation scheme inside the cost function minimization algorithm, in order to maintain a constant switching frequency equal to half the sampling frequency.

Finally, the MMC-MTDC system supplying power to passive networks or weak alternating current systems under different operating conditions is simulated to analyze and demonstrate the feasibility and effectiveness of the proposed control strategy. Besides, the scheme of experimental verification is mentioned in detail.

Energies **2018**, *11*, 1861

Author Contributions: All of the authors contributed to this research. The Authors' contributions are as follows: Writing-Original Draft, Z.W.; Writing-Review & Editing, Z.W., J.C. and W.G.; Data Curation and Investigation, Q.H.; Formal Analysis and Supervision, L.C.; Methodology and Resources, X.Y.

Funding: This work was supported by Science and Technology Project of State Grid (J2017037) and The National Natural Science Foundation of China (for youth, 51707033).

Conflicts of Interest: The authors declare no conflict of interest.

References

1. Nami, A.; Liang, J.; Dijkhuizen, F.; Demetriades, G.D. Modular Multilevel Converters for HVDC Applications: Review on Converter Cells and Functionalities. *IEEE Trans. Power Electron.* **2015**, *30*, 18–36. [CrossRef]
2. Ni, Y.X.; Vittal, V.; Kliemann, W.; Fouad, A.A. Nonlinear modal interaction in HVDC/AC power systems with DC power modulation. *IEEE Trans. Power Syst.* **2015**, *11*, 2011–2017. [CrossRef]
3. Li, X.; Song, Q.; Liu, W.; Rao, H.; Xu, S.; Li, L. Protection of Nonpermanent Faults on DC Overhead Lines in MMC-Based HVDC Systems. *IEEE Trans. Power Deliv.* **2013**, *28*, 483–490. [CrossRef]
4. Aleenejad, M.; Mahmoudi, H.; Ahmadi, R. Unbalanced Space Vector Modulation with Fundamental Phase Shift Compensation for Faulty Multilevel Converters. *IEEE Trans. Power Electron.* **2016**, *31*, 7224–7233.
5. Guo, C.; Zhao, C. Supply of an Entirely Passive AC Network through a Double-Infeed HVDC System. *IEEE Trans. Power Electron.* **2010**, *25*, 2835–2841.
6. Lesnicar, A.; Marquardt, R. An innovative modular multilevel converter topology suitable for a wide power range. In Proceedings of the 2003 IEEE Bologna Power Tech Conference Proceedings, Bologna, Italy, 23–26 June 2003; Volume 3, p. 6.
7. Gnanarathna, U.N.; Gole, A.M.; Jayasinghe, R.P. Efficient Modeling of Modular Multilevel HVDC Converters (MMC) on Electromagnetic Transient Simulation Programs. *IEEE Trans. Power Deliv.* **2015**, *26*, 316–324. [CrossRef]
8. Zhang, L.; Harnefors, L.; Nee, H.P. Interconnection of two very weak ac systems by VSC-HVDC links using power-synchronization contro. *IEEE Trans. Power Syst.* **2011**, *26*, 344–355. [CrossRef]
9. Aleenejad, M.; Mahmoudi, H.; Moamaei, P.; Ahmadi, R. A New Fault-Tolerant Strategy Based on a Modified Selective Harmonic Technique for Three-Phase Multilevel Converters with a Single Faulty Cell. *IEEE Trans. Power Electron.* **2016**, *31*, 3141–3150. [CrossRef]
10. Aleenejad, M.; Mahmoudi, H.; Ahmadi, R. A Multi-Fault Tolerance Strategy for Three Phase Cascaded H-Bridge Converters Based on HalfWave Symmetrical Selective Harmonic Elimination Technique. *IEEE Trans. Power Electron.* **2016**, *32*, 7980–7989. [CrossRef]
11. Zhang, L.; Harnefors, L.; Nee, H.P. Modeling and Control of VSC-HVDC Links Connected to Island Systems. *IEEE Trans. Power Syst.* **2011**, *26*, 783–793. [CrossRef]
12. Wang, L. *Model Predictive Control System Design and Implementation Using MATLAB*; Springer Science & Business Media: London, UK, 2009.
13. Yu, F.; Lin, W.; Wang, X.; Xie, D. Fast Voltage-Balancing Control and Fast Numerical Simulation Model for the Modular Multilevel Converter. *IEEE Trans. Power Deliv.* **2015**, *30*, 220–228. [CrossRef]
14. Saad, H.; Guillaud, X.; Mahseredjian, J.; Dennetière, S.; Nguefeu, S. MMC Capacitor Voltage Decoupling and Balancing Controls. *IEEE Trans. Power Deliv.* **2015**, *30*, 704–712. [CrossRef]
15. Moon, J.W.; Kim, C.S.; Park, J.W.; Kang, D.W.; Kim, J.M. Circulating Current Control in MMC Under the Unbalanced Voltage. *IEEE Trans. Power Deliv.* **2013**, *28*, 1952–1959. [CrossRef]
16. Rodriguez, J.; Cortes, P. *Predictive Control of Power Converters and Electrical Drives*; John Wiley & Sons: Hoboken, NJ, USA, 2012.
17. Preindl, M.; Bolognani, S. Optimal State Reference Computation with Constrained MTPA Criterion for PM Motor Drives. *IEEE Trans. Power Electron.* **2015**, *30*, 4524–4535. [CrossRef]
18. Akter, M.P.; Mekhilef, S.; Tan, N.M.; Akagi, H. Model Predictive Control of Bidirectional AC-DC Converter for Energy Storage System. *J. Electr. Eng. Technol.* **2015**, *10*, 165–175. [CrossRef]
19. Narimani, M.; Wu, B.; Yaramasu, V.; Cheng, Z.; Zargari, N.R. Finite Control-Set Model Predictive Control (FCS-MPC) of Nested Neutral Point-Clamped (NNPC) Converter. *IEEE Trans. Power Electron.* **2015**, *30*, 7262–7269. [CrossRef]

20. Kouro, S.; Cortés, P.; Vargas, R.; Ammann, U.; Rodríguez, J. Model Predictive Control-A Simple and Powerful Method to Control Power Converters. *IEEE Trans. Ind. Electron.* **2009**, *56*, 1826–1838. [CrossRef]

21. Geyer, T.; Papafotiou, G.; Morari, M. Model Predictive Direct Torque Control-PartI: Concept, Algorithm, and Analysis. *IEEE Trans. Ind. Electron.* **2009**, *56*, 1894–1905. [CrossRef]

22. Papafotiou, G.; Kley, J.; Papadopoulos, K.G.; Bohren, P.; Morari, M. Model Predictive Direct Torque Control—Part II: Implementation and Experimental Evaluation. *IEEE Trans. Ind. Electron.* **2009**, *56*, 1906–1915. [CrossRef]

23. Cortés, P.; Ortiz, G.; Yuz, J.I.; Rodríguez, J.; Vazquez, S.; Franquelo, L.G. Model Predictive Control of an Inverter With Output, Filter for UPS Applications. *IEEE Trans. Ind. Electron.* **2009**, *56*, 1875–1883. [CrossRef]

24. Qin, J.; Saeedifard, M. Predictive Control of a Modular Multilevel Converter for a Back-to-Back HVDC System. *IEEE Trans. Power Deliv.* **2012**, *27*, 1538–1547.

25. Mariethoz, S.; Fuchs, A.; Morari, M. A VSC-HVDC Decentralized Model Predictive Control Scheme for Fast Power Tracking. *IEEE Trans. Power Deliv.* **2014**, *29*, 462–471. [CrossRef]

26. Preindl, M.; Schaltz, E.; Thogersen, P. Switching Frequency Reduction Using Model Predictive Direct Current Control for High-Power Voltage Source Inverters. *IEEE Trans. Ind. Electron.* **2011**, *58*, 2826–2835. [CrossRef]

27. Xia, C.; Liu, T.; Shi, T.; Song, Z. A Simplified Finite-Control-Set Model-Predictive Control for Power Converters. *IEEE Trans. Ind. Inform.* **2014**, *10*, 991–1002.

28. Zhang, Y.; Tai, N.; Xu, B. Fault Analysis and Traveling-Wave Protection Scheme for Bipolar HVDC Lines. *IEEE Trans. Power Deliv.* **2012**, *27*, 1583–1591. [CrossRef]

29. Rashwan, A.; Sayed, M.A.; Mobarak, Y.; Shabib, G.; Senjyu, T. Predictive Controller Based on Switching State Grouping for a Modular Multilevel Converter with Reduced Computational Time. *IEEE Trans. Power Deliv.* **2017**, *32*, 2189–2198. [CrossRef]

30. Bathurst, G.N.; Watson, N.R.; Arrillaga, J. Modeling of bipolar HVDC links in the harmonic domain. *IEEE Trans. Power Deliv.* **2002**, *15*, 1034–1038.

31. Ilves, K.; Harnefors, L.; Norrga, S.; Nee, H.P. Predictive Sorting Algorithm for Modular Multilevel Converters Minimizing the Spread in the Submodule Capacitor Voltages. *IEEE Trans. Power Electron.* **2015**, *30*, 440–449. [CrossRef]

32. Mahmoudi, H.; Aleenejad, M.; Ahmadi, R. A New Multiobjective Modulated Model Predictive Control Method with Adaptive Objective Prioritization. *IEEE Trans. Ind. Appl.* **2017**, *53*, 1188–1199. [CrossRef]

Article

Low-Harmonic DC Ice-Melting Device Capable of Simultaneous Reactive Power Compensation

Jiazheng Lu, Siguo Zhu *, Bo Li, Yanjun Tan, Xiudong Zhou, Qinjun Huang, Yuan Zhu and Xinguo Mao

State Key Laboratory of Disaster Prevention and Reduction for Power Grid Transmission and Distribution Equipment, State Grid Hunan Electric Company Limited Disaster Prevention and Reduction Center, Changsha 410007, China; lujz1969@163.com (J.L.); yanting1210@126.com (B.L.); zhengyuan2017307@126.com (Y.T.); ausowny@163.com (X.Z.); dochuang@163.com (Q.H.); zhuyuan1278@163.com (Y.Z.); maoxg_0@163.com (X.M.)
* Correspondence: zhusiguo2005@163.com; Tel.: +86-0731-8633-2078

Received: 4 August 2018; Accepted: 26 September 2018; Published: 29 September 2018

Abstract: As a result of the high efficiency of ice-melting and the small power supply capacity, DC ice-melting devices are widely used in relation to transmission lines in the power grid. However, it needs to consume reactive power when ice-melting, and voltage fluctuation of the substation may be caused when the demand for reactive power is large. It also generates a large number of 5th and 7th harmonics when ice-melting. In this paper, combined with the demand for ice-melting for transmission lines and the dynamic reactive power of substations, a low-harmonic DC ice-melting device capable of simultaneous reactive power compensation is studied. The function of ice-melting and reactive power compensation can be operated simultaneously and the rectifier's main harmonics can be eliminated. The simulation and experimental research on the device was carried out in the 500 kV Chuanshan substation. The actual ice melting was carried out on the 500 kV Chuansu I line and took only 68 min to melt the ice. The 500 kV bus voltage had no negative deviation, and the positive deviation decreased from +3.09% to +1.57% within 24 h of testing. The results prove the feasibility of the proposed DC ice-melting device in this paper.

Keywords: low-harmonic DC ice-melting device; transmission line; voltage fluctuation; harmonic; dynamic reactive; substation's voltage stability

1. Introduction

Winter icing of transmission lines is one of the natural disasters of the power system. The rare long-term freezing disaster in early 2008 caused serious damage to the power grid in southern China, resulting in a large number of tower collapses, line interruptions, substation outages, and so on, which caused huge losses in many southern provinces of China.

For transmission line icing, manual deicing or improving line anti-icing designs were the main methods up until recently, but efficiency was low or investment was high. In order to cope with the impact of more and more frequent ice disasters on power system facilities, research into various ice-melting technologies has been receiving more and more attention [1–5].

Among them, DC current ice-melting has the advantages of high efficiency, a wide ice-melting range for different transmission lines, and small power supply. A previous study [6] proposes a diode-based uncontrolled rectifier DC ice-melting device, which only has the function of ice melting, and which is only used during the covering-ice period of transmission lines in winter. However, it has low utilization rate, and consumes system reactive power when ice-melting, which affects voltage stability. Another study [7] proposes a Static Var Compensator (SVC) type DC ice-melting device, which can run in two modes: DC ice-melting or SVC reactive power compensation.

However, the ice-melting and reactive power compensation cannot run at the same time and it cannot meet the reactive demand of ice melting. Further more, the harmonics of ice-melting are large, and a huge filter device is needed. Further studies [8–12] propose a new modular multi-level DC ice-melting device, which combines the dynamic reactive power compensation capability of a chain Static Var Generator (SVG) with the four-quadrant operation capability of a half-bridge modular multi-level converter (MMC). It can output DC voltage at the same time to meet the DC ice-melting demand, but the reactive power compensation and DC ice-melting need to be designed in the same capacity. The cost is high, and it is still in the theoretical research stage.

In view of the above problems associated with the ice-melting device, this paper proposes a low-harmonic DC ice-melting device that can compensate for simultaneous reactive power when ice melting. The harmonics injected into the power grid are small without a filtering device when ice melting. The device itself achieves a reactive power balance ensuring grid voltage stability. The paper is organized as follows:

Firstly, the topology of the low-harmonic DC ice-melting device with simultaneous reactive power compensation is proposed. The SVG and the ice-melting rectifier share the ice-melting transformer, which eliminates the SVG's connection reactance and reduces the SVG's output voltage and the insulation design. The design enables simultaneous operation of ice melting and reactive power compensation.

Secondly, the output voltage ripple characteristics of the low-harmonic DC ice-melting device is analyzed. The 12-pulse rectification structure is adopted, and the ripple factor value is only 0.994, which effectively reduces the output DC current fluctuation.

Thirdly, using the Fourier series, the harmonics injected into the power grid are mainly 5th and 7th harmonic currents when ice-melting. Increasing the number of power modules and the triangular carrier frequency can effectively eliminate the SVG's lower harmonic.

Fourthly, injecting low-order harmonic multi-carrier phase-shift modulation algorithm which injects 5th and 7th harmonics into a modulated wave is put forward. Using the simultaneous operation of ice melting and reactive power compensation, the algorithm can eliminate the harmonic injected into the grid when ice melting.

Finally, the simulation and experimental research on the low-harmonic DC ice-melting device which was built in the 500 kV Chuanshan substation is carried out. The results verify that DC ice-melting has low-harmonic characteristics and that the reactive power compensation function can effectively improve grid voltage stability.

2. Low-Harmonic DC Ice-Melting Device Structure

2.1. Overall Topology

The structure of the low-harmonic DC ice-melting device capable of simultaneous reactive power compensation is shown in Figure 1. It is mainly composed of an ice-melting transformer, SVG1, SVG2, Rectifier1, Rectifier2, and Isolation switch. SVG1 and SVG2 are connected to the two low-voltage winding sides of the transformer which can effectively reduce the SVG's output voltage. The SVG can absorb harmonics generated on the input side when the rectifier is running.

The SVG and the transformer realize dynamic reactive power compensation. The SVG can provide fast and flexible dynamic reactive power for the power grid, support the grid voltage especially when ice-melting, and improve the stability and power quality of the power system [13–15]. In the structure shown in Figure 1, the leakage reactance of the transformer is used as a connection reactance to replace the special reactor connected to the SVG and the grid, reducing the land occupation and cost of the SVG. The low-harmonic DC ice-melting device has the characteristics of optimized configuration of ice-melting capacity and reactive power compensation capacity. The capacity of each component of the ice-melting device can be configured according to the ice-melting and dynamic reactive power requirements of the substation avoiding the excessive design of the ice-melting or reactive capacity [16].

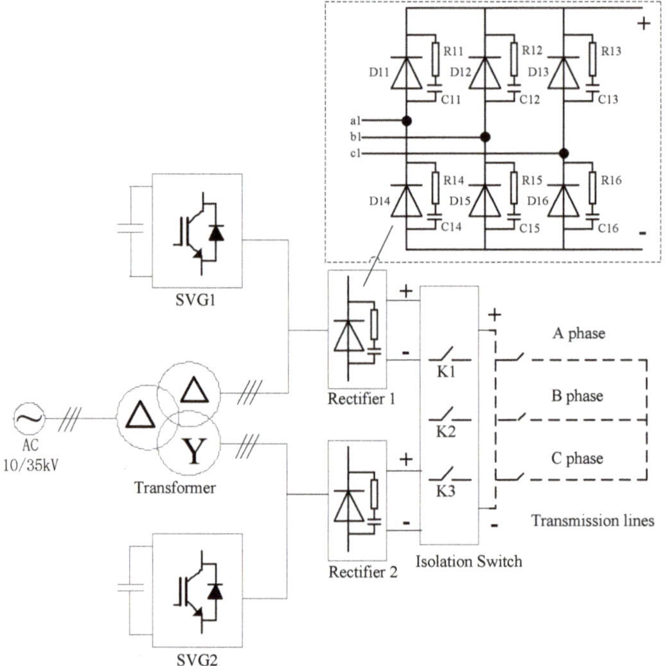

Figure 1. Structure diagram of low-harmonic DC ice-melting device.

The topology shown in Figure 1 can make full use of the transformer, providing dynamic reactive power support for the grid system during daily operation. The SVG can also eliminate harmonics and reactive power generated by the rectifier during the ice-melting operation mode.

2.2. SVG Structure

The SVG structure of the low-harmonic DC ice-melting device is shown in Figure 2. The SVG1 and SVG2 are connected to the triangular and star-shaped windings of the transformer's sub-edge without special connection reactance which is replaced by the transformer's leakage reactance.

In order to meet the requirements of the connection between SVG and power grid, the value of the transformer leakage reactance range is 8–13%. The calculation formula is as shown in Equation (1):

$$U_{kz}\% = \frac{49.6f \cdot IW\Sigma D\rho K}{etH \times 10^6},$$

(1)

where f—Frequency (Hz); I—Rated current (A); W—Coil number (Turn); et—Every potential (V); H—Average reactance height of two coils (m); ΣD—Magnetic flux leakage area (m^2); ρ—Rockwell coefficient; K—Additional reactance coefficient, generally take 1.

The SVG's output voltage is connected to the grid after being boosted by the transformer, which reduces the SVG's insulation voltage design difficulty [17,18], reduces the number of each phase power modules connected in series, and simplifies the design of the control system. In order to improve the reactive output current and meet the demand for large-capacity reactive power compensation above 100 MVar, the SVG's power modules adopt dual IGBT (Insulated Gate Bipolar Transistor) parallel structure. The parallel IGBT adopts the same control pulses.

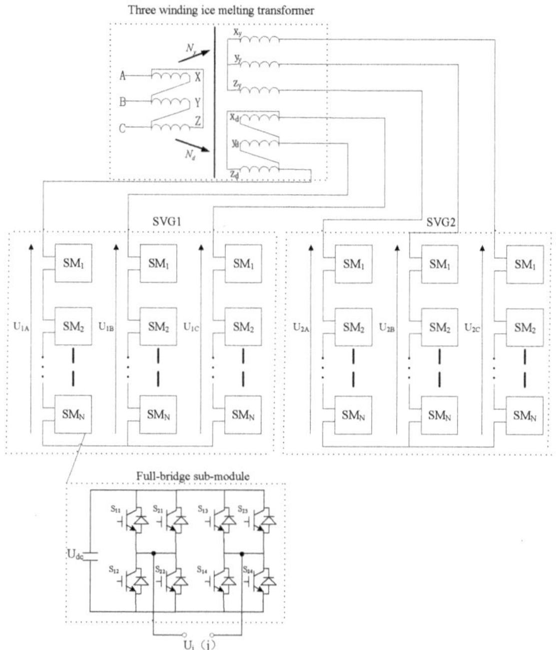

Figure 2. SVG structure.

3. Output Characteristics of Low-Harmonic DC Ice-Melting Device

3.1. Output Voltage Ripple Characteristicsof Rectifier

The low-harmonic DC ice-melting device uses two 6-pulse rectifier bridges in series or parallel configuration. Each 6-pulse rectifier bridge consists of several diodes, damping absorption resistors, and capacitors. The number of series diodes in each bridge arm is proportional to the input voltage. The two 6-pulse rectifier bridges realize series or parallel 12-pulse DC voltage output through the switch $K1$, $K2$, and $K3$. The output DC voltage is as shown in Equation (2):

$$\overline{U}_d = \frac{1}{2\pi}\int_{\frac{\pi}{2}-\frac{\pi}{m}}^{\frac{\pi}{2}+\frac{\pi}{m}} \sqrt{2}U_{2L}\sin\omega t d(\omega t) = \frac{m\sqrt{2}}{\pi}U_{2L}\sin\frac{\pi}{m}, \tag{2}$$

In Equation (2), m is the number of DC voltage pulse waves output from the rectifier. U_{2L} is the RMS (Root Meam Square) value of the AC voltage which is input to the rectifier.

The output voltage of the 12-pulse rectifier is as shown in Equation (3) where U_2 is the RMS value of the input AC voltage of the rectifier.

$$u_{d0} = (1/\frac{\pi}{12})\int_{-\frac{\pi}{24}}^{\frac{\pi}{24}} \sqrt{2}U_2\cos\omega t d(\omega t) = \sqrt{2}U_2, \tag{3}$$

Perform Fourier series decomposition on Equation (3):

$$u_{d0} = U_{d0} + \sum_{n=mk}^{\infty} b_n\cos n\omega t = U_{d0}[1 - \sum_{n=mk}^{\infty} \frac{2\cos k\pi}{n^2-1}\cos n\omega t], \tag{4}$$

where $k = 1, 2, 3 \ldots$. U_{d0} takes the form:

$$U_{d0} = \sqrt{2}U_2 \frac{m}{\pi} \sin \frac{\pi}{m}, \tag{5}$$

b_n takes the form:

$$b_n = -\frac{2\cos k\pi}{n^2 - 1} U_{d0}, \tag{6}$$

The voltage ripple factor γ_u is the ratio of the effective value of the harmonic component of the output DC voltage u_{d0} to the average value U_{d0} of the rectified voltage, namely:

$$\gamma_u = \frac{U_R}{U_{d0}}, \tag{7}$$

In Equation (7) U_R is equal:

$$\gamma_u = \frac{U_R}{U_{d0}}, \tag{8}$$

where U is equal:

$$U = \sqrt{\frac{m}{2\pi} \int_{-\frac{\pi}{m}}^{\frac{\pi}{m}} \left(\sqrt{2}U_2 \cos \omega t\right)^2 d(\omega t)} = U_2 \sqrt{1 + \frac{\sin \frac{2\pi}{m}}{\frac{2\pi}{m}}}, \tag{9}$$

According to the Equations (4)–(9) we can get the following:

$$\gamma_u = \frac{U_R}{U_{d0}} = \frac{\left[-\frac{1}{2} - \frac{m}{4\pi} \sin \frac{2\pi}{m} + \left(\frac{m}{\pi}\right)^2 \sin^2 \frac{\pi}{m}\right]^{\frac{1}{2}}}{\frac{m}{\pi} \sin \frac{\pi}{m}}, \tag{10}$$

The ripple factor values of the output voltages of different pulse wave rectifiers obtained by Equation (10) are shown in Table 1.

Table 1. The voltage ripple factor values of different pulse number m.

m	γ_u (%)
3	18.27
6	4.18
12	0.994
∞	0

It can be seen from Table 1, the more the pulse wave number of the rectifier, the smaller the output voltage ripple value. The 12-pulse rectification structure of the ice-melting device can greatly reduce the voltage ripple factor value and reduce the fluctuation of the output DC current, which is useful to facilitate line ice-melting.

3.2. Output Harmonic Characteristics of Ice-Melting

In Figure 1, since the inductance component of the impedance Z_d of the ice-transmission line is very large, the waveform of the DC output current is basically straight, and the input currents of the upper and lower 6-pulse rectifier bridges are rectangular waves, as shown in Figure 3. It is shown that each phase is turned on by 120°, and the current phases of the corresponding phases of the upper and lower bridges are different by 30°, and the amplitude is $I_d/2$.

The input currents of the two sets of 6-pulse rectifier bridges are decomposed by Fourier series, and the input current expressions of the two rectifier bridges are respectively:

$$i_{ay\varphi} = \frac{2\sqrt{3}}{\pi} I_d \left\{ \sin \omega t + \sum_{k=1,2,\ldots} (-1)^k \left[\frac{1}{6k-1} \sin(6k-1)\omega t + \frac{1}{6k+1} \sin(6k+1)\omega t \right] \right\}, \tag{11}$$

$$i_{ad\varphi} = \frac{2}{\pi} I_d \left\{ \sin \omega t + \sum_{k=1,2,\cdots} \left[\frac{1}{6k-1} \sin(6k-1)\omega t + \frac{1}{6k+1} \sin(6k+1)\omega t \right] \right\}, \tag{12}$$

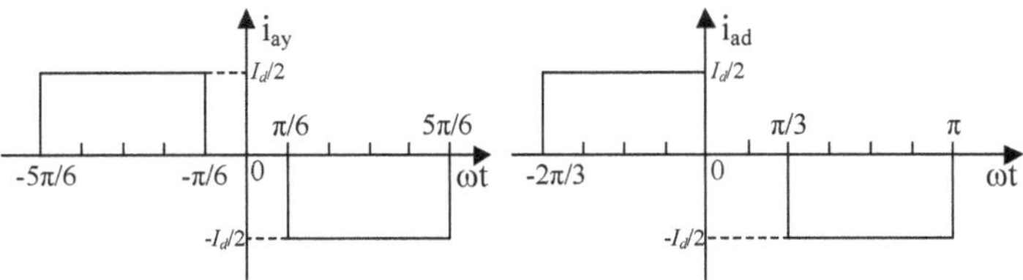

Figure 3. Input current waveform of parallel 12-pulse rectifier.

It can be obtained from Equations (11) and (12): the number of harmonics on the AC side of the rectifier of the low-harmonic DC ice-melting device is 5, 7, 11, ... , $6k \pm 1$ times, $k = 1, 2, 3, \ldots$. The RMS of the fundamental current and each harmonic are:

$$\begin{cases} I_1 = \frac{\sqrt{6}}{\pi} I_d \\ I_n = \frac{\sqrt{6}}{n\pi} I_d \end{cases}, \tag{13}$$

The RMS value of each harmonic is inversely proportional to the harmonic order. The ratio of the RMS value to the fundamental value is the reciprocal of the harmonic order. The higher the harmonic order, the smaller the harmonic amplitude, so the harmonics injected into the grid during the ice-melting are mainly the 5th and 7th harmonic currents, which are $0.156I_d$ and $0.111I_d$, respectively.

3.3. Output Harmonic Characteristics of Reactive Power Compensation

In Figure 1, the number of cascaded power modules per phase is N in SVG1 and SVG2. In order to increase the reactive output current, each power module adopts a dual IGBT parallel structure, and the output phase voltage is Fourier-decomposed:

$$\frac{U_{ug}}{U_{dc}} = Nm \sin \omega t + \sum_{n=1}^{\infty} \left(\frac{4}{n\pi} \right) \cos\left(\frac{n\pi}{2} \right) \sin\left[\frac{mn\pi}{2} \sin \omega t \right] \cdot \left[\cos n\theta_1 \ldots + \cos n\theta_N \right] \cos(n\omega_c t) + \sum_{n=1}^{\infty} \left(\frac{4}{n\pi} \right) \cos\left(\frac{n\pi}{2} \right) \cdot \sin\left[\frac{mn\pi}{2} \sin \omega t \right] \left[\sin n\theta_1 \ldots + \sin n\theta_N \right] \sin(n\omega_c t), \tag{14}$$

where U_{dc} is the power module DC bus voltage, m is the modulation degree, ω is the sinusoidal modulation wave angle frequency, ω_c is the carrier triangle wave angular frequency, and θ_k ($k = 1, 2, \ldots, N$) is the power module carrier phase-shifted angle.

When n is an odd number, $\cos\left(\frac{n\pi}{2} \right) = 0$.

When n is even, take the carrier ratio: $k_c = f_c / f$.

f_c is the triangular carrier frequency and f is the sinusoidal modulation wave frequency.

$$\sin(x \sin y) = 2 \sum_{l=1}^{\infty} J_{2l-1}(x) \sin(2l-1)y, \tag{15}$$

where J_n is an n-time Bessel function.

According to Bessel's Equation (15), Equation (14) can be expressed as:

$$\frac{U_{ug}}{U_{dc}} = Nm \sin \omega t + \sum_{n=1}^{\infty} \left(\frac{4}{n\pi} \right) \cos\left(\frac{n\pi}{2} \right) \sum_{k=1}^{\infty} J_k\left(\frac{mn\pi}{2} \right) \cdot \left[A\sin(k \pm nk_c)\omega t + 2B\sin(k\omega t)\sin(n\omega_c t) \right], \tag{16}$$

among them:

$$\begin{cases} A = \cos n\theta_1 + \cos n\theta_2 \ldots + \cos n\theta_n \\ B = \sin n\theta_1 + \sin n\theta_2 \ldots + \sin n\theta_n \end{cases}, \tag{17}$$

In the Equation (17), θ_k ($k = 1, 2, \ldots, N$) is a carrier phase-shifted angle. When the carrier phase-shifted PWM (Pulse Width Modulation) technique is adopted, the carrier angle of each power module is sequentially increased by π/N. It is not difficult to prove "$B = 0$" because "$n = 2, 4, 6, \ldots$" and "$A = 0$" when "$n < 2N$". It can be seen that the output voltage U_{ug}/U_{dc} will no longer contain "$2Nkc \pm 1$" or less harmonics. Therefore, as the number N of cascaded power modules per phase increases and the triangular carrier frequency f_c increases, the low harmonics of SVG1 and SVG2 can be effectively eliminated [19–21].

4. Injection Low-Order Harmonic Multi-Carrier Phase-Shifted Modulation Algorithm

In high-power equipment, the switching frequency of power electronics devices is so low that good control performance can't be obtained with a single converter. In order to overcome this problem, multi-carrier phase-shifted modulation algorithms were researched. A multi-carrier phase-shifted modulation algorithm is the combination of the multi-modular technique and SPWM (Sine Pulse Width Modulation) technique. The high equivalent switching frequency can be obtained with low switching frequency devices. This technique improves the equivalent switching frequency through the counteraction of lower order harmonics though not simply through processing the harmonic from lower order to higher order to get a perfectly performing harmonic feature with no filter. The American Robincon Company first invented the technology and applied for a patent (P.W.Hammond. Medium voltage PWM drive and method. U.S. Patent 5 625 545, April 1997).

In the multi-carrier phase-shifted modulation algorithm of the SVG with cascaded power module, the DC voltage utilization of the in-phase modulation algorithm is the highest, and the practical application is the most extensive. When the in-phase modulation algorithm is used, the modulation waves of the left and right bridge arms of each power module in Figure 2 are for:

$$\begin{cases} f_1(t) = m \sin \omega t \\ f_2(t) = -m \sin \omega t \end{cases},$$

(18)

The control method of each power module with three-phase AC input and the single-phase AC output adopts a sinusoidal pulse width modulation method. The power modules of the same phase have the same carrier frequency, but the phases are sequentially different by $1/N$ carrier period, as shown in Figure 4a.

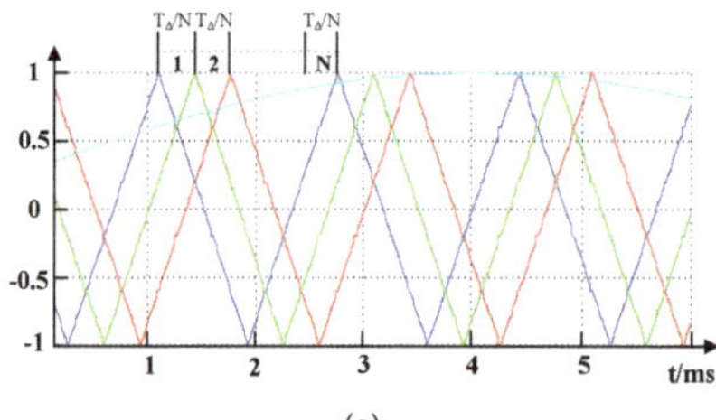
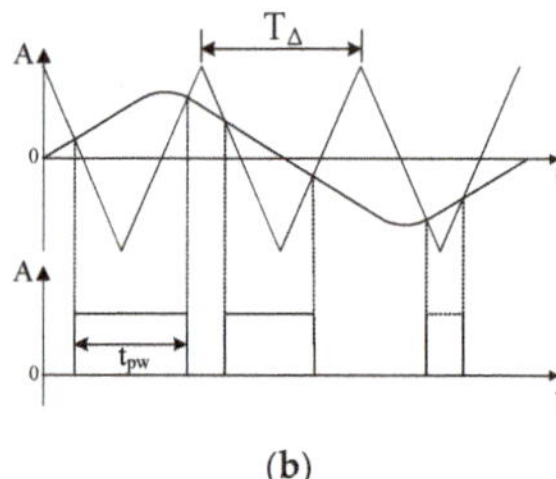

(a) (b)

Figure 4. Modulation: (**a**) Carrier phase-shifted modulation (Different colors represent different carriers); (**b**) sine pulse width modulation.

The pulse calculation method of each power module adopting sinusoidal pulse width modulation methodis shown in Figure 4b.

The formula for calculating the pulse width is:

$$t_{pw} = T_\Delta[1 + m_1 \sin \omega t]/2,$$

(19)

where T_Δ is the period of the triangular carrier, m_1 is the modulation ratio of the sine wave and the triangular wave.

In each power module of Figure 2, the power devices IGBT S_{11} and S_{21} are connected in parallel, S_{12} and S_{22} are connected in parallel, S_{13} and S_{23} are connected in parallel, and S_{14} and S_{24} are connected in parallel. The two parallel IGBT pulse control signals are the same, and the power module upper and lower arms S_{11} and S_{12}, S_{13}, and S_{14} pulse control signals are complementary. In order to avoid solving complex transcendental equations, the power switching device pulse control signal is obtained by using the regular sampling method. The calculation principle is shown in Figure 5a.

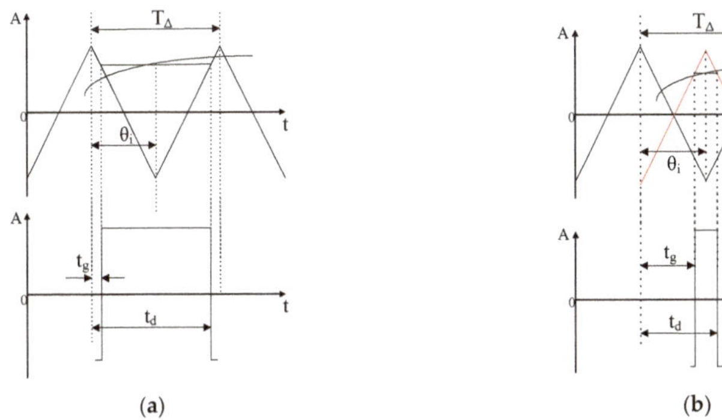

(a)   (b)

Figure 5. Pulse control signal calculation principle: (a) S_{11}; (b) S_{13}.

In Figure 5a, θ_i is the angle value of the midpoint sampling time of the triangular wave.

$$t_g = \frac{T_\Delta}{4}(1 - m_1 \sin \theta_i), \tag{20}$$

$$t_d = \frac{3T_\Delta}{4}(1 + m_1 \sin \theta_i), \tag{21}$$

The calculation results of Equations (20) and (21) are the offset time between the start and stop times of each pulse control signal of S_{11} in the power module and the corresponding triangle wave midpoint.

In Figure 5b:

$$t_g = \frac{T_\Delta}{4}(1 + m_1 \sin \theta_i), \tag{22}$$

$$t_d = \frac{3T_\Delta}{4}(1 - m_1 \sin \theta_i), \tag{23}$$

The calculation results of Equations (22) and (23) are the offset time between the start and stop times of each pulse control signal of S_{13} in the power module and the corresponding triangle wave midpoint.

In order to eliminate the 5th and 7thharmonic currents injected into the grid during the ice-melting [22–24], the 5th and 7th harmonic voltages are modulated into the power module modulation wave of SVG1 and SVG2. The harmonic amplitudes are $k1$ and $k2$, respectively, so that the generated harmonic current is equal to the amplitude of the harmonic current injected into grid when ice-melting, and the direction is opposite. After the 5th and 7th harmonic voltages are injected into the modulated wave of the left and right bridge arms of each power module, the waveform of the modulated wave is as shown in Figure 6. The calculation formulas of the modulated wave of the left and right bridge arms are as shown in Equation (24):

$$\begin{cases} f_1(t) = m \sin \omega t + k1m \sin 5\omega t + k2m \sin 7\omega t \\ f_2(t) = -m \sin \omega t - k1m \sin 5\omega t - k2m \sin 7\omega t \end{cases} \tag{24}$$

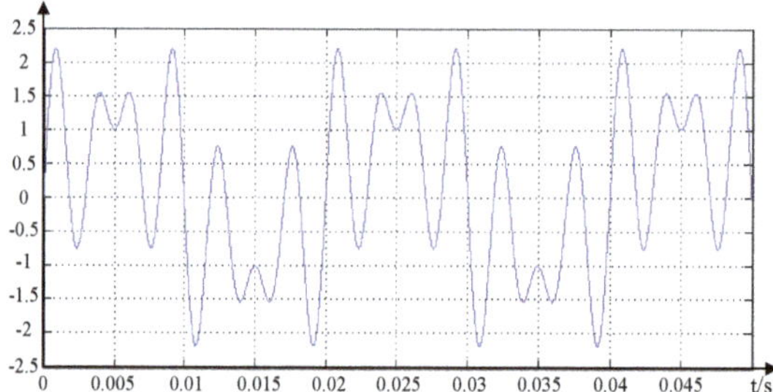

Figure 6. Modulated waveform after injection of harmonics.

According to Equations (14)–(17) and (24), after the 5th and 7th harmonics are injected, the phase voltages of each phase of SVG1 and SVG2 are:

$$U_{ug} = NmU_{dc}(\sin\omega t + k1\sin 5\omega t + k2\sin 7\omega t) \tag{25}$$

where $k1$ and $k2$ are obtained by detecting the 5th and 7th harmonic currents injected into the grid by the ice-melting rectifier and performing Fourier decomposition.

The principle of IGBT control pulse generation in each power module of the SVG is shown in Figure 7 [25–29].

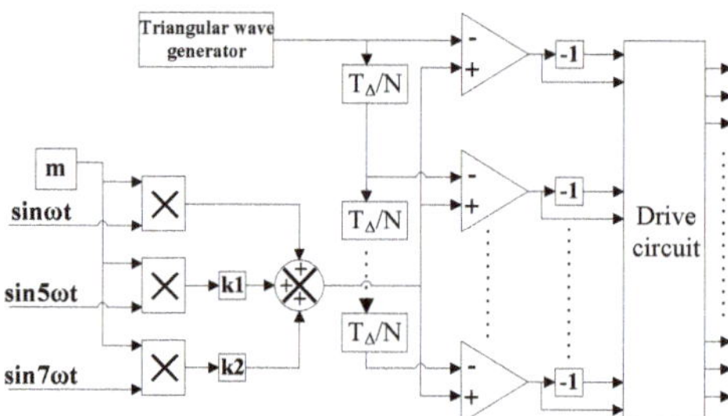

Figure 7. Schematic diagram of phase-shifted control pulse generation.

5. Simulation and Test Results

The 500 kV Chuanshan substation is located in Hengyang City, Hunan Province, China. It is an important hub substation in southern Hunan Province. In Hunan Power Grid's extremely large ice disaster in 2008, several lines of the Chuanshan substation were in a state of suspension, the icing of each line was serious, and a large area of the inverted tower and broken line appeared. The parameters of icing lines in the Chuanshan substation are shown in Table 2.

Table 2. Parameters of the icing lines in the Chuanshan substation.

Line Name	Voltage Level	Wire Type	Length (km)
Changchuan line	500 kV	4 × LGJ–400	102
Chuangu line	500 kV	4 × LGJ–500	130.3
Chuansu line	500 kV	4 × LGJ–400	127
Chuanmei line	220 kV	2 × LGJ–500	111
Chuanzhou line	220 kV	2 × LGJ–500	13.1
Chuanzhen line	220 kV	2 × LGJ–400	28
Chuangou line	220 kV	2 × LGJ–300	52

At the 500 kV voltage level of Hunan Power Grid, it can be found that the area with low voltage transient instability is concentrated in the south of the Hunan Province. The local voltage distribution map of southern Hunan is enlarged as Figure 8. A low voltage problem in a steady state and a voltage instability problem in the disturbed state exist in the southern Power Grid of Hunan Province. With high-speed rail, electric irons, and other impact loads, the dynamic reactive power demand is large in the Chuanshan substation. Therefore, it is necessary to install a dynamic reactive power compensation device of a certain capacity in southern Hunan.

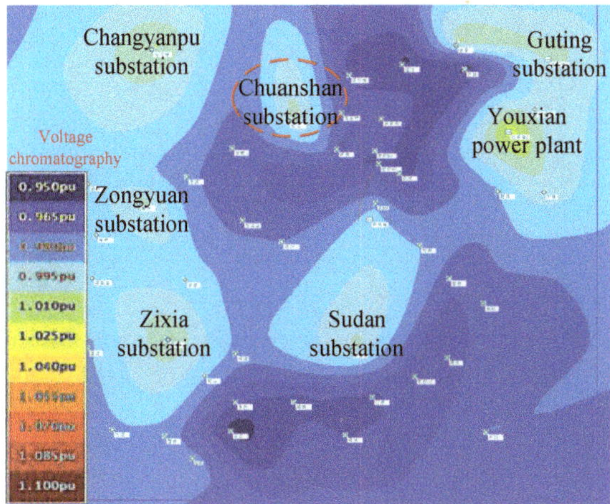

Figure 8. Local voltage distribution of 500 kV in the south of Hunan.

Combined with the demand for ice-melting of the lines in the Chuanshan substation and the dynamic reactive demand in southern Hunan, a low-harmonic DC ice-melting device with simultaneous reactive power compensation was installed in the Chuanshan substation to solve the problem of dynamic reactive power and line ice-melting at the same time. The ice-melting capacity is 120 MW, and the reactive power compensation capacity of SVG1 and SVG2 are all 50 MVar. The ice-melting transformer has a rated capacity of 120 MVA, a rated input voltage of 35 kV, a rated output voltage of 17.1 kV, and a short-circuit impedance of 15%. The ice-melting rectifier has a rated capacity of 120 MW, a rated output DC voltage of 21 kV, and a rated output DC current of 5700 A. SVG1 and SVG2 have a rated output voltage of 17.1 kV and a rated output current of 1688 A. The number of power modules per phase is 20, which is half the number of power modules compared with the 35 kV grid-connected without a transformer. This greatly improves the stability of the SVG.

5.1. Simulation Results

Using PSCAD (Power System Computer Aided Design)/EMTDC (Electro Magnetic Transient in DC System) simulation software, the ice-melting and dynamic reactive power compensation characteristics of the low-harmonic DC ice-melting device installed in the 500 kV Chuanshan substation were simulated. The transformer was connected with the 35 kV low voltage system bus in the substation. The simulation results of each function are describing in following subsections:

5.1.1. DC Ice-Melting

Taking the DC ice-melting of the 500 kV Chuangu line as an example, its DC resistance per phase was 1.88 Ω, the line length was 130.3 km, and the wire type was 2 × LGJ–500. When ice melting, the A phase and C phase are connected in a series, the output voltage of the ice-melting transformer is 17.1 kV, and the 12-pulse rectifier is connected in parallel by two bridges. During the simulation, it was possible to obtain an output DC voltage of the ice-melting device of 18.9 kV, an output DC current of 5685 A, and the voltage and current waveforms are shown in Figure 9a,b. The ripple factor value of the output voltage is small in Figure 9a and the output current waveform is smooth in Figure 9b. The actual output is consistent with the previous theoretical analysis.

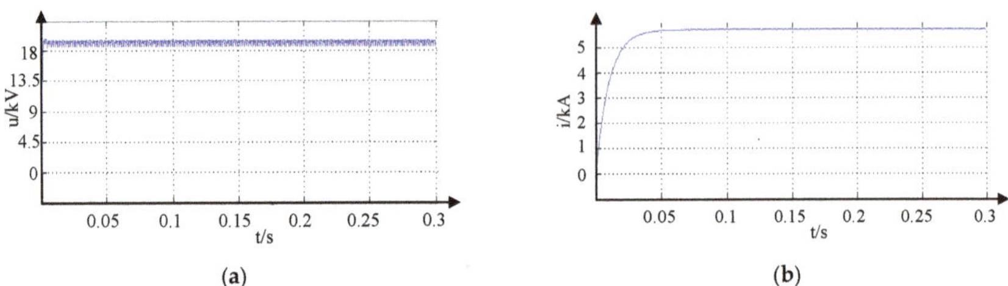

Figure 9. The output waveforms of DC ice-melting: (**a**) voltage; (**b**) current.

Figure 10 is a low-order harmonic simulation analysis of the current waveform of the input side of the ice-melting rectifier when ice-melting. When the rectifier is working, it will generate large 5th and 7th harmonics on its input side, but it can be seen from Figure 5 that the 5th and 7th harmonics in the input current are effectively eliminated by the SVG1 and SVG2 connected in parallel to the input side of the rectifier, and the content is less than 0.5%.

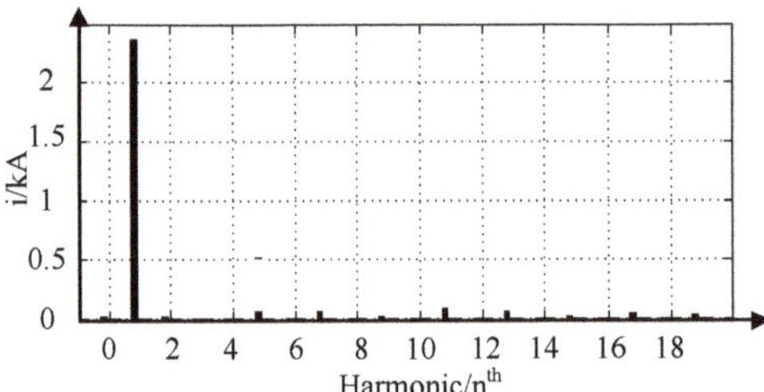

Figure 10. Low-order harmonic analysis of input current of ice-melting rectifier.

5.1.2. Output Voltage Waveform of SVG

In order to reduce the amount of calculations for the controller, the carrier frequency is generally taken as an integral multiple of the fundamental frequency. The single power module's carrier frequency is 350 Hz and the output voltage is about 494 V in SVG1 and SVG2. The output three-level voltage waveform of the single power module is as shown in Figure 11a.

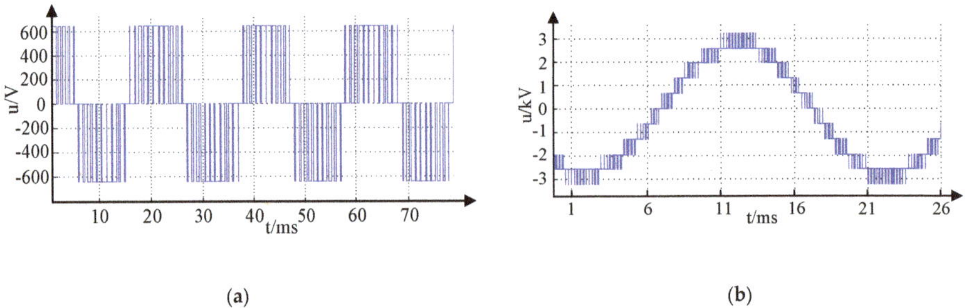

Figure 11. The output voltage waveform: (**a**) single power module; (**b**) five power modules.

Arbitrarily selecting five power modules connected in the A phase of SVG1 or SVG2 to simulate, the output of the eleven-level voltage waveform is as shown in Figure 11b. The output level of SVG is equal to "2n + 1" where n is the number of power module per phase, n = 1, 2, 3.....

5.1.3. SVG Compensation 35 kV System Reactive

If the capacitive reactive demand of the 35 kV system in the substation is 60 MVar, the reactive current in the system can be fully compensated when the SVG is put into operation at 0.2 s. Each SVG outputs reactive power 30 MVar. The system voltage, reactive load current, system reactive current, SVG1 output current, and SVG2 output current are shown in Figure 12. As can be seen from the figure, the SVG can effectively compensate the reactive load current to maintain the voltage stability of the 35 kV system voltage.

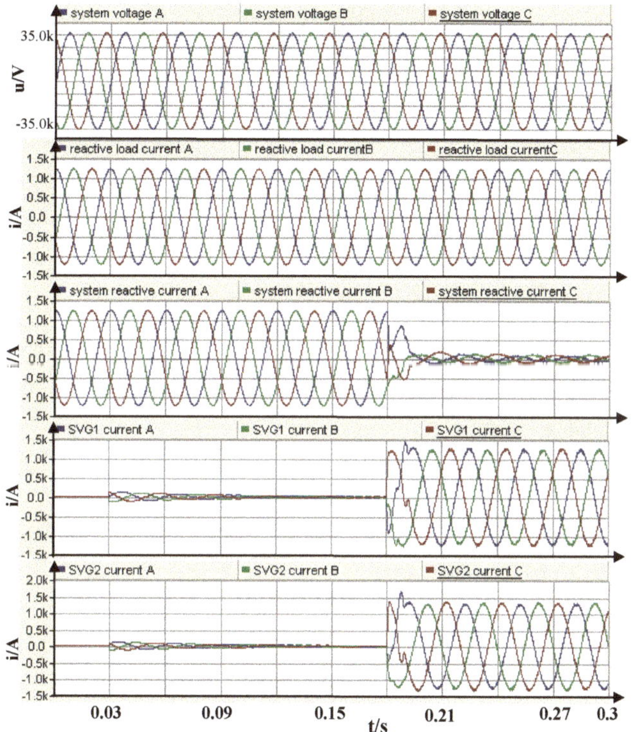

Figure 12. Compensation 35 kV system reactive power.

5.2. Test Results

The low-harmonic DC ice-melting device of the 500 kV Chuanshan substation with simultaneous reactive power compensation is shown in Figure 13. It consists of a transformer, a rectifier, SVG1, SVG2, and a cooling system.

On 28 January 2018, the 500 kV Chuansu I line had ice as thick as 50 mm. There was a risk of an inverted tower and broken line. The low-harmonic DC ice-melting device was used to melt the ice on lines. The output voltage of the transformer was 17.1 kV. The 12-pulse rectifier adopted two bridges in parallel. The DC ice-melting current was 4060 A, and the ice-melting voltage was 18.12 kV. First, the A and C phases were melted in series for about 35 min, and then the B and C phases were melted in series which took about 33 min to melt the ice. The output voltage, current and line falling ice is as shown in Figure 14a when ice-melting.

Figure 13. Low-harmonic DC ice-melting device in 35 kV Chuanshan substation.

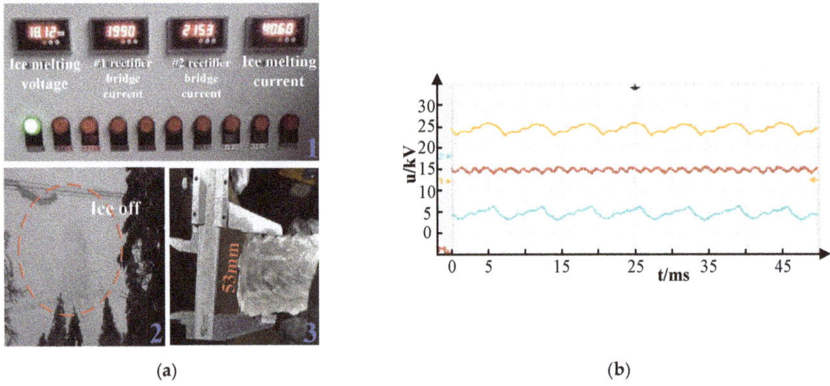

(**a**)   (**b**)

Figure 14. Actual ice melting: (**a**) (1) Ice-melting voltage; (2) current; and (3) falling ice. (**b**) Output DC voltage waveform.

The output voltage of the ice-melting is shown in Figure 14b. The voltage ripple number is 12 in each cycle, and the output voltage ripple value is about 0.99, which is consistent with the theoretical calculation. When ice-melting, SVG1 and SVG2 run in parallel with the injected low-harmonic multi-carrier phase-shift modulation algorithm. The 5th and 7th harmonic currents injected into the grid by the rectifier are effectively eliminated as shown in Figure 15.

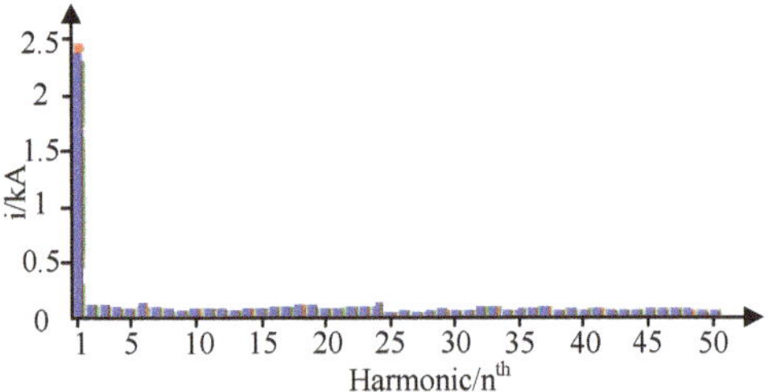

Figure 15. Low-order harmonic analysis of input current of ice-melting rectifier.

The single power module's carrier frequency is 300 Hz in SVG1 and SVG2, the output voltage is about 482 V. The output three-level voltage waveform of each power module is as shown in Figure 16a. The SVG's A phase output voltage waveform is as shown in Figure 16b. The actual test results are consistent with the previous simulation results.

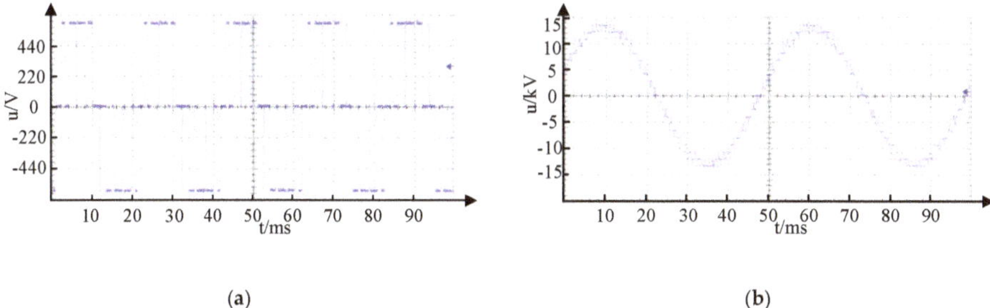

(a)

(b)

Figure 16. The output voltage waveform: (**a**) Single power module; (**b**) one-phase.

When the inductive reactive power requirement of the substation's power system is reduced from 80 MVar to 20 MVar, the single SVG output reactive power is reduced from 40 MVar to 10 MVar. The output current waveforms of the A and C phase are shown in Figure 17a. When the system inductive reactive power requirement is increased from 20 MVar to 80 MVar, the single SVG output reactive power is increased from 10 MVar to 40 MVar, and the A and C phase output current waveforms are as shown in Figure 17b.

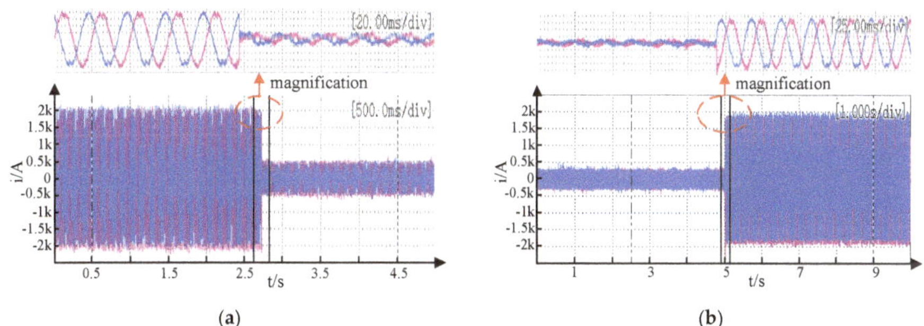

(a)

(b)

Figure 17. Output two-phase current waveform: (**a**) 40 MVar to 10 MVar; (**b**) 10 MVar to 40 MVar.

When the reactive power requirement of the power system changes from 80 MVar sensibility to −80 MVar capacitive, the single SVG output reactive power is changed from 40 MVar to −40 MVar, and the A and C phase output current waveforms are shown in Figure 18.

The SVG output reactive response time is less than 10 ms which can well meet the reactive power demand of power system according to the Figures 17 and 18.

After the low-harmonic DC ice-melting device in Chuanshan substation was put into operation, the power quality of Chuanshan substation was tested for 24 h using the power quality tester according to the test procedures. The 500 kV bus voltage had no negative deviation, and the positive deviation of 500 kV bus voltage was about 0.80–2.0% within 24 h of testing, as shown in Figure 19. The ninety-five percent probability of the positive deviation value (phase B) was about 1.57%, which was 1.52% lower than the 3.09% when there was no timely reactive power compensation at runtime.

So the low-harmonic DC ice-melting device capable of simultaneous reactive power compensation can effectively improve the voltage stability in southern Hunan Province.

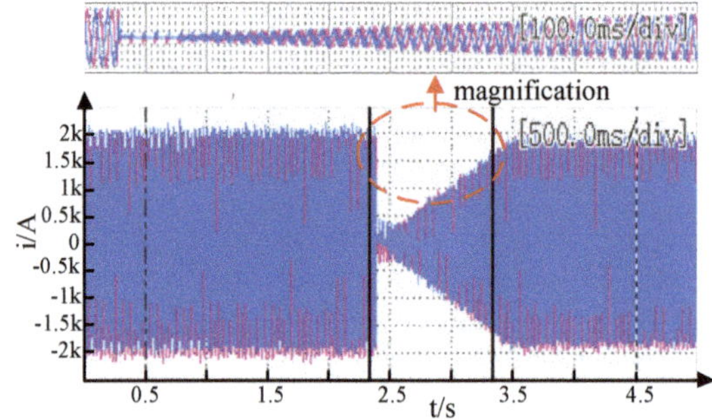

Figure 18. 40 MVar to −40 MVar output two-phase current waveform.

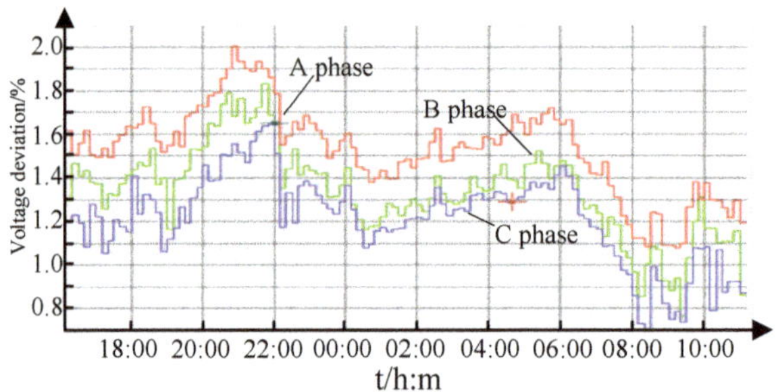

Figure 19. 500 kV bus voltage deviation.

6. Discussion

The simulated and measured results in Section 5 can be explained as follows:

The topology of a low-harmonic DC ice-melting device capable of simultaneous reactive power compensation is feasible. The SVG uses transformer leakage reactance instead of connecting reactance. It can realize the simultaneous operation of ice-melting and reactive power compensation. The running harmonics are small, and no filtering device is needed.

The twelve-pulse rectification structure in the low-harmonic DC ice-melting device can effectively reduce the fluctuation of output DC current and the ripple factor value is only 0.994.

The ice-melting and SVG reactive power compensation are simultaneously operated, and the injected low-harmonic multi-carrier phase-shift modulation algorithm can effectively eliminate the 5th and 7th harmonics mainly generated during ice melting without the need of a filter device.

After the SVG was put into operation in 500 kV Chuanshan substation, the 500 kV bus voltage had no negative deviation and the positive deviation decreased from 3.09% to 1.57% within 24 h of testing, which effectively improves the voltage stability in southern Hunan Province, China.

7. Conclusions

In this paper, a topology of a low-harmonic DC ice-melting device capable of simultaneous reactive power compensation has been proposed. The ice-melting device is mainly composed of a transformer, SVG1, SVG2, and an ice-melting rectifier. The two SVGs are connected to the two low-voltage winding sides of the transformer, which is also connected to the rectifier. The SVG can effectively absorb harmonics generated by the rectifier and improve the voltage stability.

The simulation and experimental research is carried out. The results verify the feasibility and effectiveness of the low-harmonic DC ice-melting device which provides an effective method for the research and development of the DC ice-melting device.

Author Contributions: S.Z. carried out the experiments, analyzed the test results, and wrote this paper. J.L. gave input to the analysis of test results. B.L., Y.T. and X.Z. carried out the experimental set up. Q.H., Y.Z. and X.M. carried out the simulation research.

Funding: This research was funded by State Grid Corporation of China Science and Technology Program, grant number [2016KJ0072].

Acknowledgments: The authors gratefully acknowledge the contributions of all members of the ice-melting research team in State Grid Hunan Electric Power Corporation Limited for their work on this paper.

Conflicts of Interest: The authors declare no conflict of interest.

References

1. Lu, J.; Zeng, M.; Zeng, X.; Fang, Z.; Yuan, J. Analysis of ice-covering characteristics of China Hunan power grid. *IEEE Trans. Ind. Appl.* **2015**, *51*, 1997–2002. [CrossRef]
2. Lu, J.Z.; Zhu, S.G.; Li, B.; Zhang, H.X.; Tan, Y.J.; Zhu, Y. Influence of STATCOM System on the DC-icing Rectifier Transformers' DC Magnetization and Suppression Method. *High Volt. Eng.* **2016**, *42*, 1624–1629.
3. Long, J.Z.; Zheng, L.J.; Cai, L.H.; Yong, H.F. Analysis of the causes of tower collapses in Hunan during the *2008* ice storm. *High Volt. Eng.* **2008**, *34*, 2468–2474.
4. Volat, C.; Farzaneh, M.; Leblond, A. De-icing/anti-icing techniques for power lines: Current methods and future direction. In Proceedings of the 11th International Workshop on Atmospheric Icing of Structures, Montreal, QC, Canada, 12–16 June 2005; pp. 1–11.
5. Fan, S.; Jiang, X.; Sun, C.; Zhang, Z.; Shu, L. Temperature characteristic of DC ice-melting conductor. *Cold Reg. Sci. Technol.* **2011**, *65*, 29–38. [CrossRef]
6. Tan, Y.J.; Lu, J.Z.; Fang, Z.; Li, B.; Zhang, H.X. Study of Mobile DC De-icing Equipment Based on Uncontrolled Rectification. *Cent. China Electr. Power* **2011**, *24*, 31–34.
7. Chun, L.C. Application research of SVC melting technology in 220 kV substation. *Yunnan Electr. Power* **2010**, *38*, 1–2.
8. Mei, H.; Liu, J. A novel dc ice-melting equipment based on modular multilevel cascade converter. *Autom. Elect. Power Syst.* **2013**, *37*, 96–102.
9. Zhao, Q. A Novel DC Ice-melting Device Based on MMC with Hybrid Sub-module. *South. Power Syst. Technol.* **2015**, *9*, 9–36.
10. Guo, Y.; Xu, J.; Guo, C.; Zhao, C.; Fu, C.; Zhou, Y.; Xu, S. Control of full-bridge modular multilevel converter for dc ice-melting application. In Proceedings of the11th IET International Conference on AC and DC Power Transmission, Birmingham, UK, 10–12 February 2015; pp. 1–8.
11. Fan, R.; Sun, M.; He, Z.Y.; Yan, S. Design and system test of movable DC De-icer for Jiangxi power grid. *Autom. Electr. Power Syst.* **2009**, *33*, 67–71.
12. Jin, B.H.; Nian, X.H.; Fan, R.X.; Liu, D.; Deng, M. Research of line full controlled ice-melting DC power supply with control and change strategy. *Autom. Electr. Power Syst.* **2012**, *36*, 86–91.
13. Li, S.; Wang, Y.; Li, X.; Wei, W.; Zhao, G.; He, P. Review of de-icing methods for transmission lines. In Proceedings of the 2010 International Conference on Intelligent System Design and Engineering Application, Changsha, China, 13–14 October 2010; pp. 310–313.
14. Wang, J.; Fu, C.; Chen, Y.; Rao, H.; Xu, S.; Yu, T.; Li, L. Research and application of DC de-icing technology in China southern power grid. *IEEE Trans. Power Deliv.* **2012**, *27*, 1234–1242. [CrossRef]
15. Wang, C.; Wen, J.; Li, S.; Ma, X.; Wang, J. Design on DC de-icing schemes for high voltage transmission line. In Proceedings of the 2010 5th International Conference on Critical Infrastructure (CRIS), Beijing, China, 20–22 September 2010; pp. 1–5.
16. Lesnicar, A.; Marquardt, R. An innovative modular multilevel converter topology suitable for a wide power range. In Proceedings of the 2003 IEEE Bologna Power Tech Conference Proceedings, Bologna, Italy, 23–26 June 2003.
17. Latiff, N.A.A.; Illias, H.A.; Bakar, A.H.A.; Dabbak, S.Z.A. Measurement and Modelling of Leakage Current Behaviour in ZnO Surge Arresters under Various Applied Voltage Amplitudes and Pollution Conditions. *Energies* **2018**, *4*, 875. [CrossRef]
18. Lu, J.; Xie, P.; Zhen, F.; Hu, J. Electro-Thermal Modeling of Metal-Oxide Arrester under Power Frequency Applied Voltages. *Energies* **2018**, *11*, 1610. [CrossRef]

19. Perez, M.A.; Bernet, S.; Rodriguez, J.; Kouro, S.; Lizana, R. Circuit topologies, modeling, control schemes, and applications of modular multilevel converters. *IEEE Trans. Power Electron.* **2015**, *30*, 4–14. [CrossRef]
20. Zeng, R.; Xu, L.; Yao, L.; Williams, B.W. Design and operation of a hybrid modular multilevel converter. *IEEE Trans. Power Electron.* **2015**, *30*, 1137–1146. [CrossRef]
21. Li, X.; Song, Q.; Liu, W.; Xu, S.; Zhu, Z.; Li, X. Performance analysis and optimization of circulating current control for modular multilevel converter. *IEEE Trans. Ind. Electron.* **2016**, *63*, 716–727. [CrossRef]
22. Zhang, M.; Huang, L.; Yao, W.; Lu, Z. Circulating harmonic current elimination of a CPS-PWM-based modular multilevel converter with a plug-in repetitive controller. *IEEE Trans. Power Electron.* **2014**, *29*, 2083–2097. [CrossRef]
23. Konstantinou, G.; Ciobotaru, M.; Agelidis, V. Selective harmonic elimination pulse-width modulation of modular multilevel converters. *IET Power Electron.* **2013**, *6*, 96–107. [CrossRef]
24. Tu, Q.; Xu, Z. Impact of sampling frequency on harmonic distortion for modular multilevel converter. *IEEE Trans. Power Deliv.* **2011**, *26*, 298–306. [CrossRef]
25. Li, B.; Yang, R.; Xu, D.; Wang, G.; Wang, W.; Xu, D. Analysis of the phase-shifted carrier modulation for modular multilevel converters. *IEEE Trans. Power Electron.* **2015**, *30*, 297–310. [CrossRef]
26. Townsend, C.D.; Summers, T.J.; Betz, R.E. Phase-shifted carrier modulation techniques for cascaded H-bridge multilevel converters. *IEEE Trans. Ind. Electron.* **2015**, *62*, 6684–6696. [CrossRef]
27. Hagiwara, M.; Akagi, H. Control and analysis of the modular multilevel cascade converter based on double-star chopper-cells (MMCCDSCC). *IEEE Trans. Power Electron.* **2011**, *26*, 1649–1658. [CrossRef]
28. Huneault, M.; Langheit, C.; Benny, J.; Audet, J.; Richard, J.C. A dynamic programming methodology to develop de-icing strategies during ice storms by channeling load currents in transmission networks. *IEEE Trans. Power Deliv.* **2005**, *20*, 1604–1610. [CrossRef]
29. Hagiwara, M.; Akagi, H. Control and Experiment of Pulse width-Modulated Modular Multilevel Converters. *IEEE Trans. Power Electron.* **2009**, *24*, 1737–1746. [CrossRef]

Article

Modified State-of-Charge Balancing Control of Modular Multilevel Converter with Integrated Battery Energy Storage System

Yajun Ma, Hua Lin *, Zhe Wang and Zuyao Ze

State Key Laboratory of Advanced Electromagnetic Engineering and Technology, School of Electrical and Electronic Engineering, Huazhong University of Science and Technology, Wuhan 430074, China; mayajun@mail.hust.edu.cn (Y.M.); ranger_wong@foxmail.com (Z.W.); zezuyao@163.com (Z.Z.)
* Correspondence: lhua@mail.hust.edu.cn; Tel.: +86-159-7219-3510

Received: 22 November 2018; Accepted: 24 December 2018; Published: 28 December 2018

Abstract: Modular multilevel converter with integrated battery energy storage system (MMC-BESS) has been proposed for energy storage requirements in high-voltage applications. The state-of-charge (SOC) equilibrium of batteries is essential for BESS to guarantee the capacity utilization. However, submodule voltage regulation can lead to over-modulation of individual submodules, which will limit the efficiency of SOC balancing control. Focusing on this problem, a modified SOC balancing control method with high efficiency is proposed in this paper. The tolerance for battery power unbalance is defined to quantize the convergence of SOC balancing control. Both the DC component and AC component are considered while regulating submodule voltage. The linear programming method is introduced to realize the maximum tolerance for battery power unbalance in different operation modes. Based on the analysis, by choosing appropriate submodule voltage regulation method, the efficiency of SOC balancing control is improved greatly. In addition, the SOC controller is also optimally designed to avoid over-modulation of submodules. Finally, the detailed simulation and experiment results verify the effectiveness of the analysis and proposed control strategy.

Keywords: battery energy storage system (BESS); modular multilevel converter (MMC); state-of-charge (SOC) balancing control; tolerance for battery power unbalance

---

1. Introduction

In recent years, renewable resources are becoming more and more important, but they have great impact on grid due to their stochastic nature, reducing the voltage and frequency stability [1,2]. Battery energy storage system (BESS) is necessary and effective in these applications to improve the stability [3]. As the interface between batteries and grid, many researches focus on medium and high-voltage power conversion system (PCS).

Due to the advantages in medium and high-voltage applications, modular multilevel converters (MMC) has been used in BESS (MMC-BESS). Batteries can be directly connected with the DC bus in a centralized manner [4]. However, the series connection of massive batteries reduces the reliability and increases the complexity of energy management. In [5–7], batteries are integrated into submodules of MMC, constituting a modularized and distributed PCS. Compared with the centralized structure, the reliability and flexibility of distributed structure are greatly improved. Hence, it is more applicable to medium and high-voltage applications [8]. Most importantly, MMC-BESS is not only a pure PCS, but also a three-port converter. As an example, Figure 1 is the structure diagram of wind energy system. By integrating batteries into MMC, the original PCS is eliminated, which simplifies the whole configuration and lowers the costs. In addition, BESS can coordinate the operation of AC and DC side while transferring power among them [9]. Figure 1b is the topology of MMC-BESS, consisting of three phases, each with two arms. In submodule A, batteries are directly connected with submodules [10,11]. However, low-frequency current ripple can flow into batteries, which leads to additional heat generation and temperature rise of batteries. That will accelerate the degradation of batteries, resulting in significant reduction of battery life span [12,13]. Then a DC/DC converter (isolated or non-isolated) is used to filter the low-frequency current in submodule B [14–16]. The main advantages of non-isolated DC/DC are the simple structure and low cost. The isolated DC/DC can achieve the electrical isolation between batteries and MMC, which is utilized to meet some special requirements of batteries, such as grounding. At the same time, the DC/DC converter offers an additional control degree of freedom (DOF). Therefore, the capacitor voltage can be directly controlled by the DC/DC converter with traditional dual closed-loop control structure. Compared with the methods based on traditional MMC [17,18], the capacitor voltage balancing control is greatly simplified, and this paper also prefers this control structure.

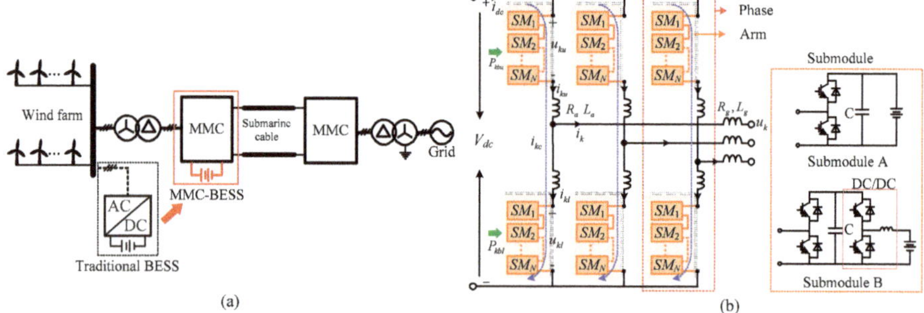

Figure 1. Wind energy generation system: (**a**) structure of offshore wind farm; (**b**) topology of MMC-BESS.

For the massive submodules of MMC-BESS, the state-of charge (SOC) of each submodule will inevitably be different. The capacity utilization of the whole BESS is limited by the submodules with highest or lowest SOC due to the over-charge or over-discharge of them. Therefore it is essential to maintain the SOC equilibrium of all submodules [19]. The SOC balancing control is usually divided into three levels, including SOC balance among phases, SOC balance between arms and SOC balance within arms. In [20], the zero-sequence voltage is injected to balance the SOC among phases. However, the calculation of zero-sequence voltage is too complex. In [21,22], the DC and fundamental circulating current are regulated to balance the SOC among phases and between arms respectively. The SOC within arms is balanced by regulating the submodule voltage, which contains both the DC component and AC component. So there are two dimensions to regulate the submodule voltage. In [23], only AC component of submodule voltage is regulated to balance SOC. In [24], the DC component is regulated to balance SOC in AC-side fault mode and the AC component is regulated to balance SOC in DC-side fault mode. In [25], a power factor is introduced to regulate both the DC component and AC components. However, the problem is that the regulation of submodule voltage can lead to over-modulation of submodules when the battery power unbalance exceeds certain limit. To avoid the over-modulation of submodules, the gain of SOC controller should be limited [17,22]. However, that will seriously limit the SOC convergence rate of the whole BESS. To guarantee appropriate SOC convergence, the key is to improve the tolerance for battery power unbalance, which is closely related to the submodule voltage regulation method. However the methods above only investigate some special cases, which are not necessarily optimal in any case. More seriously, the SOC may not converge in some case. For example in [15], the SOC will not converge when the power only transfers between DC side and batteries. Therefore, this paper aims to investigate the submodule voltage regulation method to optimize the SOC balancing control.

In this paper, capacitor voltage is controlled by the DC/DC converter with traditional dual closed-loop structure. Carrier phase shifted modulation (CPS-PWM) is used to generate switching signals. By analyzing the power flow of MMC-BESS, the factors that limit the SOC convergence rate are investigated in detail. Then the tolerance for battery power unbalance is defined to quantize the convergence of SOC balancing control. Both the DC and AC components are taken into account when regulating submodule voltage. Linear programming method is introduced to reach the maximum tolerance in different operation modes. One the basis, a modified SOC balancing control method with high efficiency is proposed by optimizing the submodule voltage regulation method. Finally, a downscaled prototype is built to verify the analysis and proposed method.

The rest of this paper is organized as follows: the power flow and principles of SOC balancing control are introduced in Section 2. The submodule voltage regulation method is investigated in Section 3. On the basis, optimized SOC balancing control strategy is proposed in Section 4. Then the detailed simulation and experiment results are given in Section 5. Finally, the conclusions are presented in Section 6.

2. Fundamental Principles of MMC-BESS

Due to the symmetry of MMC-BESS, the following analysis only takes one phase as an example and the results are also applicable to the other two phases. The subscript $k \in \{a, b, c\}$ denotes different phases, and the subscript $j \in \{u, l\}$ denotes the upper arm and lower arm respectively. $i \in \{1, 2, \ldots, N\}$ represents the number of submodule per arm.

2.1. Power Flow of MMC-BESS

In Figure 1b, u_{kj} and i_{kj} are the arm voltage and current; u_k and i_k are the AC-side voltage and current; V_{dc} and i_{dc} are the DC-side voltage and current. i_{kc} is the circulating current flowing from upper arm to lower arm. P_{kbj} is the total battery power injected into arms.

For a generalized MMC-BESS, the circulating current mainly contains DC, fundamental frequency and second harmonic component. When only considering the power flow of MMC-BESS, the second harmonic component can be ignored.

$$i_{kc} = I_{kc0} + I_{kc1} \sin(\omega\, t + \gamma_{k1}) \tag{1}$$

where I_{kc0} is the dc circulating current; I_{kc1} and γ_{k1} are the amplitude and phase of fundament circulating current.

Just like the traditional MMC, the following basic relations still exist for arm voltages and currents

$$\begin{cases} u_{ku} = \frac{1}{2}V_{dc} - v_{kc} - v_k, \quad i_{ku} = i_{kc} + \frac{1}{2}i_k \\ u_{kl} = \frac{1}{2}V_{dc} - v_{kc} + v_k, \quad i_{kl} = i_{kc} - \frac{1}{2}i_k \\ v_{kc} = R_a i_{kc} + L_a \frac{di_{kc}}{dt} \\ v_k = (R_g + \frac{1}{2}R_a)i_k + (L_g + \frac{1}{2}L_a)\frac{di_k}{dt} + u_k \end{cases} \tag{2}$$

where v_{kc} and v_k are the voltages required to drive circulating current and AC-side current respectively. In this way, the system is divided into two parts: one part is only related to circulating current and the other part is only related to AC-side current.

Ignoring power losses, the absorbed average power of arms can be calculated according to the arm voltages and currents in (1) and (2).

$$\begin{cases} P_{ku} = \underbrace{\frac{1}{2}V_{dc}I_{kc0}}_{0.5P_{kdc}} - \underbrace{\frac{1}{2}V_k I_{kc1}\cos\gamma_1}_{P_{k\Delta}} - \underbrace{\frac{1}{2} \times \frac{1}{2}V_k I_k \cos\varphi}_{0.5P_{kac}} + P_{kbu} \\ P_{kl} = \frac{1}{2}V_{dc}I_{kc0} + \frac{1}{2}V_k I_{kc1}\cos\gamma_1 - \frac{1}{2} \times \frac{1}{2}V_k I_k \cos\varphi + P_{kbl} \end{cases} \tag{3}$$

where φ is the power factor angle; V_k, I_k are the amplitude of v_k and i_k. The DC circulating current transfers the same power from DC bus to upper and lower arm, which is denoted by P_{kdc}. The fundamental circulating current transfers the same power $P_{k\Delta}$ from upper arm to lower arm. Ac-side current absorbs the same power from both upper and lower arm, which is denoted by P_{kac}.

To maintain the power balance of arms, the absorbed active power should be equal to zero, hence the total battery power injected into arms can be calculated as

$$\begin{cases} P_{kbu} = \frac{1}{2}(P_{kac} - P_{kdc}) + P_{k\Delta} \\ P_{kbl} = \frac{1}{2}(P_{kac} - P_{kdc}) - P_{k\Delta} \end{cases} \tag{4}$$

According to Equations (3) and (4), the power flow among phases and between arms is shown in Figure 2a, in which P_{dc} is the total power of dc-bus. The power flowing into AC-side for each phase is usually the same. The power absorbed from DC bus by each phase can be controlled by regulating the dc circulating current. Therefore the total battery power injected into each phase is controllable. The power flowing from upper arm to lower arm can be controlled by regulating the fundamental circulating current. Therefore, the battery power distribution between upper and lower arm can be controlled arbitrarily. In this way, the power injected into arms can be controlled independently.

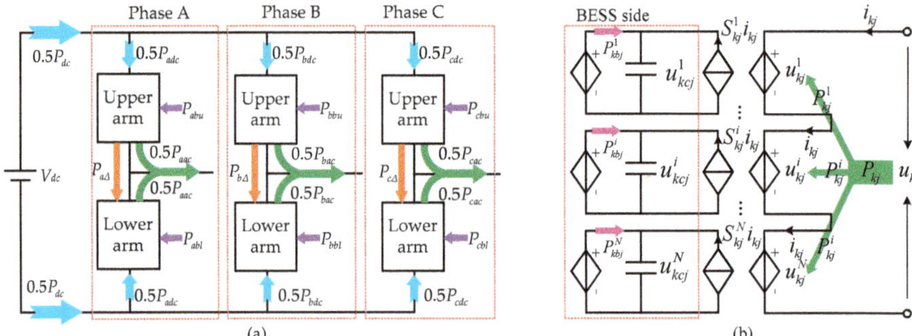

Figure 2. Power flow of MMC-BESS: (**a**) power flow among phases and between arms; (**b**) power flow among submodules within arms.

In this paper, capacitor voltage of each submodule is controlled by the DC/DC converter with traditional dual closed-loop control structure, which makes the DC/DC side of submodule as a controlled voltage source, offering stable voltage for the MMC side. Then the arm can be equivalent to Figure 2b, in which S^i_{kj}, u^i_{kcj}, and u^i_{kj} are the switching function, capacitor voltage and submodule voltage respectively. P^i_{kbj} and P^i_{kj} are the power absorbed from batteries and MMC side, which are balanced in steady state. Therefore, the battery power injected into each submodule is

$$P^i_{kbj} = -P^i_{kj} = -\frac{1}{T}\int_0^T u^i_{kj}i_{kj}dt \tag{5}$$

where T is the fundamental frequency period. For submodules connected in series within arms, the currents flowing through them are the same. The total battery power injected into arms can be flexibly allocated to submodules through regulating the DC and AC components of submodule voltage. Combination with the DC power control, AC power control and balance power control, the battery power injected into each submodule can be controlled independently.

2.2. SOC Balancing Control of MMC-BESS

The capacity utilization of the whole BESS is limited by the submodule with the highest or lowest SOC. To improve the capacity utilization of BESS, the SOC of each submodule should be maintained at the same value. The SOC of each submodule can be established by

$$SOC = \frac{Storaged\ charges}{Nominal\ capacity} \times 100\%$$
$$SOC(t) = SOC(t_0) + \frac{1}{E_{bat}} \int_{t_0}^{t} P_{bat} dt \tag{6}$$

where P_{bat} is the battery power; E_{bat} is the nominal energy, which can be calculated by multiplying the battery voltage and its capacity.

Figure 3 is the general control structure of MMC-BESS. Capacitor voltage is controlled by the DC/DC converter. The notch filter is used to filter capacitor voltage ripples, which can introduce low-frequency current ripples into batteries. The AC-side power control adopts typical control structure in dq frame just like the two-level voltage source converter [26]. The SOC balancing control is divided into three levels, including SOC balance among phases, SOC balance between arms and SOC balance within arms. SOC_{kj}^i is the SOC of the ith submodule. SOC_{kj}, SOC_k, and SOC_{BESS} are the average SOC of arms, phases, and the whole BESS respectively.

$$SOC_{BESS} = \frac{1}{3} \sum_{k=a,b,c} SOC_k, \quad SOC_k = \frac{1}{2}(SOC_{ku} + SOC_{kl}), \quad SOC_{kj} = \frac{1}{N} \sum_{i=1}^{N} SOC_{kj}^i \tag{7}$$

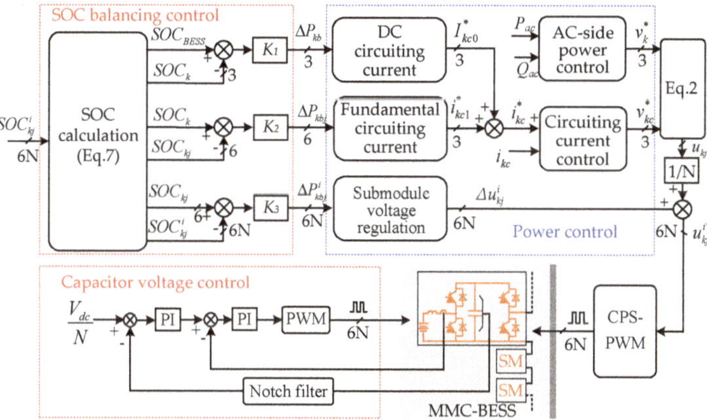

Figure 3. General control structure of MMC-BESS.

According to (6), the three-level SOC balancing control can be realized by regulating the battery power injected into phases, arms, and submodules respectively. In Figure 3, K_1, K_2, K_3 are the corresponding proportional controllers, and the battery power injected into each phase, arm, and submodule is

$$P_{kb} = \frac{1}{3} P_b + \Delta P_{kb}, \quad P_{kbj} = \frac{1}{2} P_{kb} + \Delta P_{kbj}, \quad P_{kbj}^i = \frac{1}{N} P_{kbj} + \Delta P_{kbj}^i \tag{8}$$

The battery power injected into each phase and arm is controlled by regulating the dc and fundamental circulating current. In Figure 3, the output of circulating current control v_{kc}^* is the reference of v_{kc}, which is used to drive the circulating current. The reference of dc circulating current is

$$I_{kc0}^* = \frac{1}{V_{dc}} \left(\frac{1}{3} P_{dc} - \Delta P_{kb} \right) \tag{9}$$

To prevent the fundamental circulating current from flowing into DC side, fundamental circulating current should contains no zero-sequence component, and the reference is given as [9,17]

$$
\begin{bmatrix} i^*_{ac1} \\ i^*_{bc1} \\ i^*_{cc1} \end{bmatrix} = \frac{1}{\sqrt{3}V_k} \begin{bmatrix} \sin\omega t & -\cos\omega t & \cos\omega t \\ \cos(\omega t - 2\pi/3) & \sin(\omega t - 2\pi/3) & -\cos(\omega t - 2\pi/3) \\ -\cos(\omega t + 2\pi/3) & \cos(\omega t + 2\pi/3) & \sin(\omega t + 2\pi/3) \end{bmatrix} \begin{bmatrix} \Delta P_{abu} - \Delta P_{abl} \\ \Delta P_{bbu} - \Delta P_{bbl} \\ \Delta P_{cbu} - \Delta P_{cbl} \end{bmatrix} \tag{10}
$$

In Figure 3, the battery power injected into submodules is controlled by regulating the submodule voltage. As an example, with the method in [20,25], the submodule voltage is calculated as

$$
u^i_{kj} = \frac{u_{kj}}{N} + \Delta u^i_{kj} = \frac{u_{kj}}{N}\left(1 + \frac{\Delta P^i_{kbj}}{P_{kbj}/N}\right) \tag{11}
$$

where Δu^i_{kj} is the voltage increment to regulate the battery power injected into submodules.

For half-bridge submodule used in this paper, the submodule voltage should be between 0 and V_{dc}/N. However, the voltage increment can easily result in $u^i_{kj} > V_{dc}/N$ or $u^i_{kj} < 0$, which exceeds the output range of submodules. Then over-modulation of submodules occurs, leading to massive harmonics. To solve the problem, the gain of SOC controller should be limited, and the following relation exists

$$
\frac{K_3(SOC_{kj} - SOC^i_{kj})}{P_{kbj}/N} \leq \frac{1-m}{1+m} \tag{12}
$$

where $m = 2V_k/V_{dc}$ is the modulation ratio of MMC-BESS. When the modulation ratio m is relatively large, the allowed K_3 is very small, which limits the convergence rate of SOC balancing control. More seriously, the SOC may not converge in some case, decreasing the capacity utilization of the whole BESS. Therefore, it is essential to ensure appropriate SOC convergence rate.

3. Investigation of Submodule Voltage Regulation Method

The SOC convergence rate of the whole BESS is decided by the three-level SOC balancing control. SOC balance among phases and between arms are guaranteed by regulating the DC and fundamental circulating current. In Figure 3, v^*_{kc} is injected into upper arm and lower arm to drive the circulating current. However, considering that the arm impedance of MMC-BESS is relatively small, the injection of v^*_{kc} will not lead to the over-modulation of submodules. Therefore, the SOC convergence rate of the whole BESS is mainly limited by the SOC balancing control within arms. This section mainly focuses on investigating the submodule voltage regulation method to improve the SOC convergence rate of the whole BESS.

3.1. Constraints of Submodule Voltage Regulation Method

Before the analysis, a parameter λ is defined to assess the battery power unbalance caused by SOC balancing control.

$$
\lambda^i_{kj} = \frac{P^i_{kbj} - P_{kbj}/N}{P_{kbj}/N} \times 100\%, \; \lambda^i_{kbj} \geq -1 \tag{13}
$$

The battery power unbalance degree λ reflects how much the battery power injected into submodule deviates from the average battery power of the arm. Specially, $\lambda = -1$ represents that the battery power is equal to zero, or the battery is breakdown. $\lambda < -1$ denotes that some batteries are charging while some batteries are discharging in one arm, which is usually not allowed in practice. So this paper stipulates that the battery power unbalance degree should be greater than -1.

According to Figure 3, the battery power unbalance degree is calculated as

$$
\lambda^i_{kj} = \frac{K_3}{P_{kbj}}(SOC_{kj} - SOC^i_{kj}) \tag{14}
$$

Submodule voltage contains dc component and ac component, so there are two dimensions to regulate submodule voltage. Then the voltage increment Δu^i_{kj} is given as

$$\Delta u^i_{kj} = \frac{1}{2N}\alpha^i_{kj}V_{dc} + \frac{1}{N}\beta^i_{kj}u_k \tag{15}$$

where α^i_{kj} and β^i_{kj} are defined as the DC factor and AC factor, which are used to regulate the DC and AC component of voltage respectively.

Substituting (15) into (5), the submodule power regulated by the dc factor and ac factor is

$$\Delta P^i_{kj} = \frac{1}{2N}\alpha^i_{kj}P_{kdc} - \frac{1}{2N}\beta^i_{kj}(P_{kac} + 2P_{k\Delta}) \tag{16}$$

In steady state, the total active power absorbed by submodule should be equal to zero. According to (14) and (16), the submodule voltage regulation should satisfy the relation

$$-\lambda^i_{kj}P_{kbj} = \frac{1}{2N}\alpha^i_{kj}P_{kdc} - \frac{1}{2N}\beta^i_{kj}(P_{kac} + 2P_{k\Delta}) \tag{17}$$

Increasing K_3 can improve the convergence rate of SOC, but it requires more power regulation to balance the submodule power. In different operation modes, the contribution of DC factor and AC factor to power regulation is different. Choosing appropriate DC and AC factor can increase the power regulation capability, meaning that the allowed SOC convergence rate can be improved.

To facilitate the analysis, the Equation (17) can be rewritten as

$$-\lambda^i_{kj} = \frac{1}{1-\xi_{kj}}\beta^i_{kj} - \frac{\xi_{kj}}{1-\xi_{kj}}\alpha^i_{kj} \tag{18}$$

where

$$\xi_{kj} = \begin{cases} \dfrac{P_{kdc}}{P_{kac}+2P_{k\Delta}}, & j = u \\ \dfrac{P_{kdc}}{P_{kac}-2P_{k\Delta}}, & j = l \end{cases} \tag{19}$$

ξ_{kj} is the power ratio of dc power and ac power in arms, which can denote the operation mode of MMC-BESS. To improve the SOC convergence rate, the range of allowed battery power unbalance degree should be as large as possible.

To avoid over-modulation of submodules, we find the following constraints

$$\begin{cases} (1+\alpha^i_{kj}) + \left|1+\beta^i_{kj}\right|m \leq 2 \\ (1+\alpha^i_{kj}) - \left|1+\beta^i_{kj}\right|m \geq 0 \end{cases} \tag{20}$$

In addition to constrain in (18), the dc factor and ac factor also should locate at the shadow area in Figure 4. S_1, S_2, S_3, and S_4 are the boundaries of the shadow area according to constrains in (20).

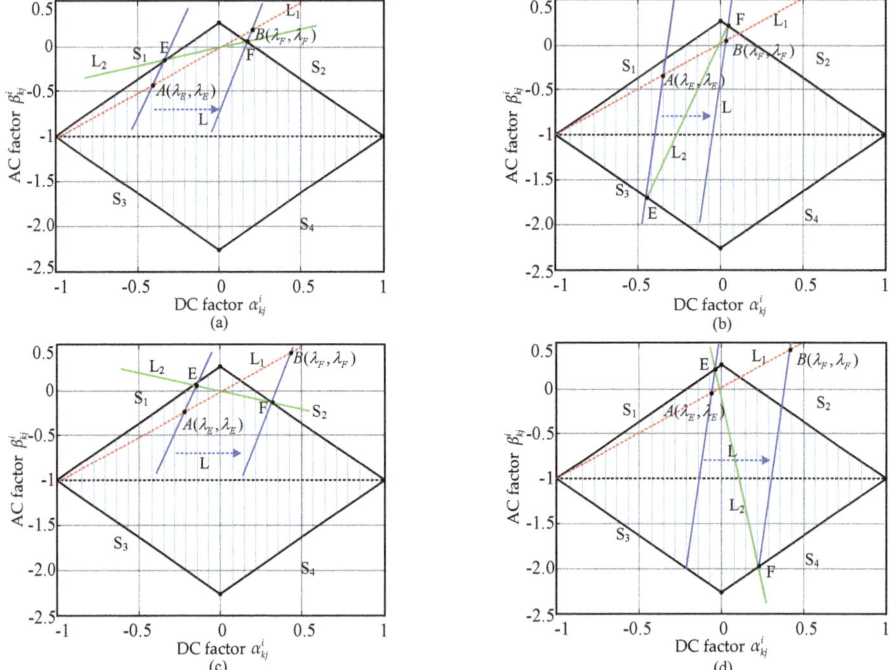

Figure 4. Choosing appropriate DC and AC factor for maximum tolerance in different conditions: (a) when $0 < \eta_{kj} \leq 1$; (b) when n (c) when $-1 < \eta_{kj} \leq 0$; (d) when $\eta_{kj} \leq -1$.

3.2. Tolearance for Battery Power Unbalance

In different operation modes, the contribution of DC factor and AC factor to power regulation is different. To facilitate the following analysis, the ratio of DC and AC factor is defined as

$$\beta_{kj}^{i} = \eta_{kj}\alpha_{kj}^{i} \tag{21}$$

By regulating η_{kj}, the allowed range of battery power unbalance degree can be improved, and the SOC convergence rate can also be improved.

Linear programming method is the typical method to study the extremum of linear objective function under linear constraints. In this paper, linear programming method is used to study the extremum of allowed unbalance degree.

According to the principles of linear programming method, the objective function is defined as

$$L: -\lambda = \frac{1}{1 - \xi_{kj}}\beta - \frac{\xi_{kj}}{1 - \xi_{kj}}\alpha \tag{22}$$

The line L_1 in Figure 4 is defined as

$$L_1: \beta = \alpha \tag{23}$$

The line L_1 intersects with L at point $(-\lambda, -\lambda)$, which can represent the unbalance degree. For a larger unbalance degree, the intersection should be far away from the origin.

The Line L_2 in Figure 4 is defined as

$$L_2: \beta = \eta_{kj}\alpha \tag{24}$$

The line L_2 denotes the ratio of DC factor and AC factor. The line segment EF is the intersection of L_2 and the shadow area. Therefore, the DC and AC factor should locate at the line segment EF.

(1)　When $0 < \eta_{kj} \leq 1$, E and F locate at S_1 and S_2 as shown in Figure 4a. The coordinates of E and F are calculated as

$$E\left(\frac{1-m}{m\eta_{kj}-1},\eta_{kj}\frac{1-m}{m\eta_{kj}-1}\right),F\left(\frac{1-m}{\eta_{kj}m+1},\eta_{kj}\frac{1-m}{\eta_{kj}m+1}\right) \tag{25}$$

(2)　When $\eta_{kj} > 1$, E and F locate at S_2 and S_3 as shown in Figure 4b. The coordinates of E and F are calculated as

$$E\left(-\frac{1+m}{m\eta_{kj}+1},-\eta_{kj}\frac{1+m}{m\eta_{kj}+1}\right),F\left(\frac{1-m}{\eta_{kj}m+1},\eta_{kj}\frac{1-m}{\eta_{kj}m+1}\right) \tag{26}$$

(3)　When $-1 < \eta_{kj} \leq 0$, point E and F still locate at S_1 and S_2 as shown in Figure 4c, and the coordinates of E and F are the same as Equation (25).

(4)　When $\eta_{kj} \leq -1$, E and F locate at S_1 and S_4 as shown in Figure 4d. The coordinates of E and F are

$$E\left(\frac{1-m}{m\eta_{kj}-1},\eta_{kj}\frac{1-m}{m\eta_{kj}-1}\right),F\left(-\frac{1+m}{m\eta_{kj}-1},-\eta_{kj}\frac{1+m}{m\eta_{kj}-1}\right) \tag{27}$$

According to the principles of linear programming method, it is obvious that λ reaches maximum or minimum values when L_1 passes through the point E or F. The point A (λ_E, λ_E) and B (λ_F, λ_F) are the corresponding intersections of the line L_1 and L. Then substituting (25)–(27) into (22), the unbalance degree λ_E and λ_F can be calculated.

In Figure 4, one of the point A and B locates at the first quadrant, and the other point locates at the third quadrant. It means that the maximum values of λ must be positive and the minimum values of λ must be negative. Therefore, we find the relations

$$\begin{cases} \lambda_p = max(\lambda_E,\lambda_F) \\ \lambda_n = min(\lambda_E,\lambda_F) \end{cases} \tag{28}$$

where λ_p denotes the tolerance for positive battery power unbalance degree, and λ_n denotes the tolerance for negative battery power unbalance degree.

Considering that the battery power unbalance degree can be negative or positive for submodules within one arm, the tolerance for battery power unbalance is defined as

$$\psi_{kj} = min\left(\left|\lambda_p\right|,\left|\lambda_n\right|\right) \tag{29}$$

If the battery power unbalance degree does not exceed the tolerance, over-modulation will not occur, which can ensure the normal operation of MMC-BESS.

According to (29), the tolerance for battery power unbalance is calculated as shown in Figure 5, which can be divided into several cases according to the power ratio.

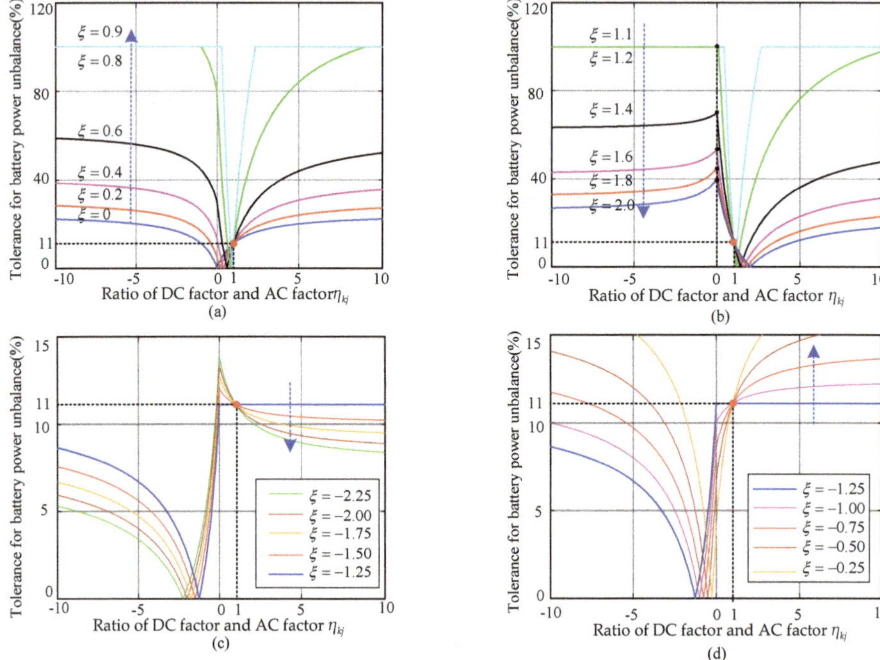

Figure 5. Power regulation capability of submodules when $m = 0.8$: (**a**) when power ratio $0 < \xi_{kj} \leq 1$; (**b**) when power ratio $\xi_{kj} > 1$; (**c**) when power ratio $\xi_{kj} \leq -1/m$; (**d**) when power ratio $-1/m < \xi_{kj} \leq 0$.

1. In Figure 5a, the power ratio $0 < \xi_{kj} \leq 1$. The power regulation capability increases when the absolute value of weighting ratio η_{kj} increases. Apparently as weighting ratio approaches infinity, the tolerance reaches the maximum value.

2. In Figure 5b, the power ratio $\xi_{kj} > 1$. The tolerance for battery power unbalance reaches the maximum value when η_{kj} is equal to zero.

According to Figure 4, when the power ratio $\xi_{kj} = -1/m$, the slope of L is the same to the boundary S_2. When L passes through point F, the intersection of L and L_1 is a fixed point, meaning that the power regulation capability is a fixed value. It can be calculated according to Equations (23) and (24).

$$\psi_{kj}\left(\xi_{kj} = -\frac{1}{m}\right) = \frac{1 - m}{1 + m} \tag{30}$$

3. For the case $\xi_{kj} < -1/m$, the tolerance for battery power unbalance reaches the maximum value when η_{kj} is equal to zero as shown in Figure 5c. For the case $-1/m < \xi_{kj} \leq 0$, the tolerance reaches the maximum value when η_{kj} approaches infinity as shown in Figure 5d.

In addition, note that all the curves in Figure 5 passes through a fixed point. The tolerance for unbalanced power has no relation with the power ratio. According to Figure 4, when the weighting ratio is equal to 1, the point B and F coincide together, so that the power regulation capability will not change, and can be calculated as

$$\psi_{kj}(\eta_{kj} = 1) = \frac{1 - m}{1 + m} \tag{31}$$

4. Modified SOC Balancing Control

The previous analysis indicates that the convergence of SOC is mainly limited by the SOC balancing control within arms. Therefore, this section mainly modifies the SOC balancing control within arms for high SOC convergence rate.

The convergence of SOC balance within arms is mainly decided by the submodule voltage regulation method, which is investigated in detail as shown in Figure 5. To maximize the SOC convergence rate, the ratio of DC and AC factor should change with the power ratio, which can be concluded as:

1. When $-1/m < \xi_{kj} \leq 1$, only ac component needs to be regulated, so that the DC factor and AC factor are

$$\begin{cases} \alpha_{kj}^i = 0 \\ \beta_{kj}^i = (1 - \xi_{kj})\lambda_{kj}^i \end{cases} . \tag{32}$$

Then according to the analysis in Section 3.2, the tolerance for battery power unbalance can be calculated as

$$\psi_{kj} = \frac{1 - m}{m(1 - \xi_{kj})} \tag{33}$$

2. When $\xi_{kj} > 1$ or $\xi_{kj} \leq -1/m$, only DC component needs to be regulated, so that the DC factor and AC factor are

$$\begin{cases} \alpha_{kj}^i = \frac{\xi_{kj}-1}{\xi_{kj}}\lambda_{kj}^i \\ \beta_{kj}^i = 0 \end{cases} \tag{34}$$

With the same method, the tolerance for battery power unbalance is

$$\psi_{kj} = \frac{\xi_{kj}}{\xi_{kj} - 1}(1 - m) \tag{35}$$

On the basis, the modified SOC balancing control within arms is shown in Figure 6. First, the unbalance degree is calculated according to (13). Then according to the power ratio, the DC and AC factor are calculated. The reference of submodule voltage is given by adding the voltage increment Δu_{kj}^i into the original submodule voltage. At last, CPS-PWM generates switching signals to control the state of submodules.

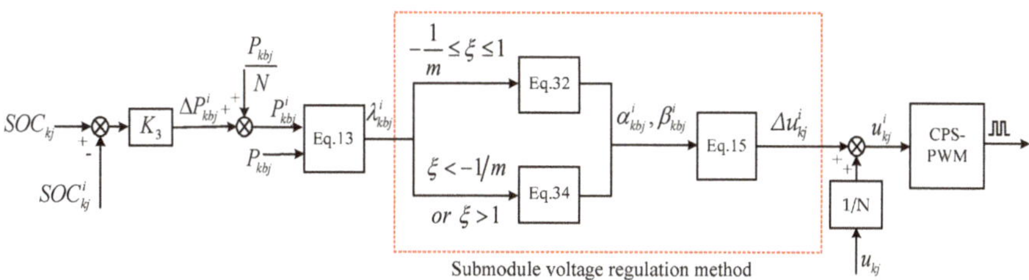

Figure 6. Modified SOC balancing control within arms.

Figure 7a shows the comparisons of tolerance for battery power unbalance when η takes some special values. In [19,25], a power factor is introduced to distribute submodule voltage, which can be equivalent to the case $\eta = 1$. The tolerance for battery power unbalance is a fixed value, but it is too small for near all power ratio, which seriously limits the convergence rate of SOC balancing control. For the case $\eta = 0$, only AC component is redistributed. However, the tolerance becomes too small when the power ratio is relatively large. For the case $1/\eta = 0$, only DC component is regulated, but the tolerance is too small when the power ratio is around zero. For the modified method, the weighting ratio changes with power ratio, and the tolerance reaches maximum for all operation modes.

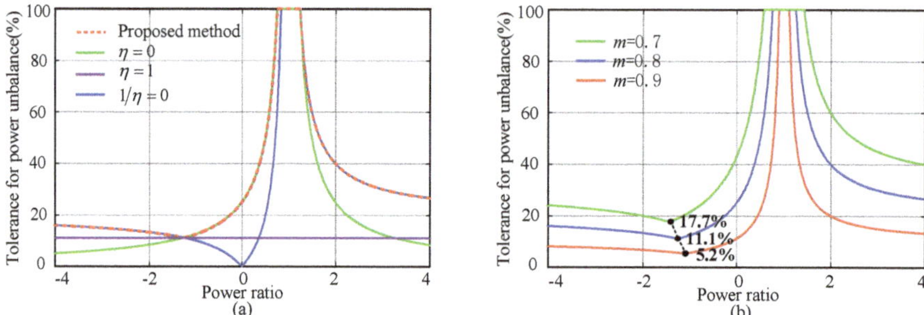

Figure 7. Tolerance for battery power unbalance: (**a**) comparison of different submodule voltage regulation methods; (**b**) influence of modulation ratio.

Figure 7b shows the relation between modulation ratio and tolerance for battery power unbalance. With the increase of modulation ratio, the tolerance sharply decreases. To avoid the over-modulation of submodules, the gain of controller should be limited as

$$K_3 \Delta \, SOC_{kj\_max} \leq \psi_{kj} \left| P_{kbj} \right| \tag{36}$$

where

$$\Delta \, SOC_{kj\_max} = max(\left| SOC_{kj} - SOC_{kj}^i \right|) \tag{37}$$

Hence the SOC controller should satisfy

$$K_3 \leq \frac{1}{\Delta \, SOC_{kj\_max}} \psi_{kj} \left| P_{kbj} \right| \tag{38}$$

Note that the power ratio and maximum SOC unbalance of six arms are different, meaning that the controller K_3 may be different for different arms. To simplify the whole SOC balancing control structure, K_3 should take the same value, which should satisfy the relations

$$K_3 \leq min(\frac{1}{\Delta \, SOC_{kj\_max}} \psi_{kj} \left| P_{kbj} \right|) \tag{39}$$

In this way, over-modulation of submodules can be avoided, and the convergence rate of SOC balancing control can be improved greatly.

5. Simulation and Experiment Results

5.1. Simulation Results

To verify the analysis and proposed model in this paper, a simulation model based on the topology shown in Figure 1 is built in MATLAB/Simulink (Mathworks, Inc., Natick, MA, USA), and the detailed parameters are shown in Table 1.

Table 1. Parameters of simulation model.

Quantity	Value	Comment
V_{dc}	400 V	DC-Link voltage
m	0.8	Modulation ratio
P_{ac}	10 KW	Nominal AC-side power
V_{sm}	100 V	Submodule capacitor voltage
N	4	Number of submodules per arm
C	5 mF	Submodule capacitance
L_a	5 mH	Arm inductor
L_{bat}	10 mH	DC/DC side inductor
V_{bat}	60 V	Nominal battery voltage
C_{bat}	1 Ah	Nominal battery capacity
f_M	2 kHz	MMC side switching frequency
f_B	10 kHz	DC/DC converter frequency

Figure 8 is the simulation results of traditional SOC balancing control method used in [25]. The power of AC side and DC side are: P_{dc} = 4.8 KW, P_{ac} = 9.6 KW. Figure 8a is the SOC of 24 submodules in MMC-BESS. At time T = 20 s, the SOC balancing control is added to the system, and then the SOC starts to converge. However the convergence rate is too small for the whole BESS. Figure 8b shows the average SOC of six arms, which can denote the SOC balancing control among phases and between arms. Figure 8c,d are the SOC of upper and lower arm in phase A. It is obvious that the convergence rate of whole BESS is mainly limited by SOC balancing control within arms. The analysis in Section 3.2 indicates that the tolerance for battery power unbalance is limited in 11% when m = 0.8. In Figure 8e, the maximum power unbalance is 10% and over-modulation almost occurs as shown in Figure 8f. It means that the SOC convergence rate has already reached the limit, however the SOC convergence of the whole BESS is still so poor.

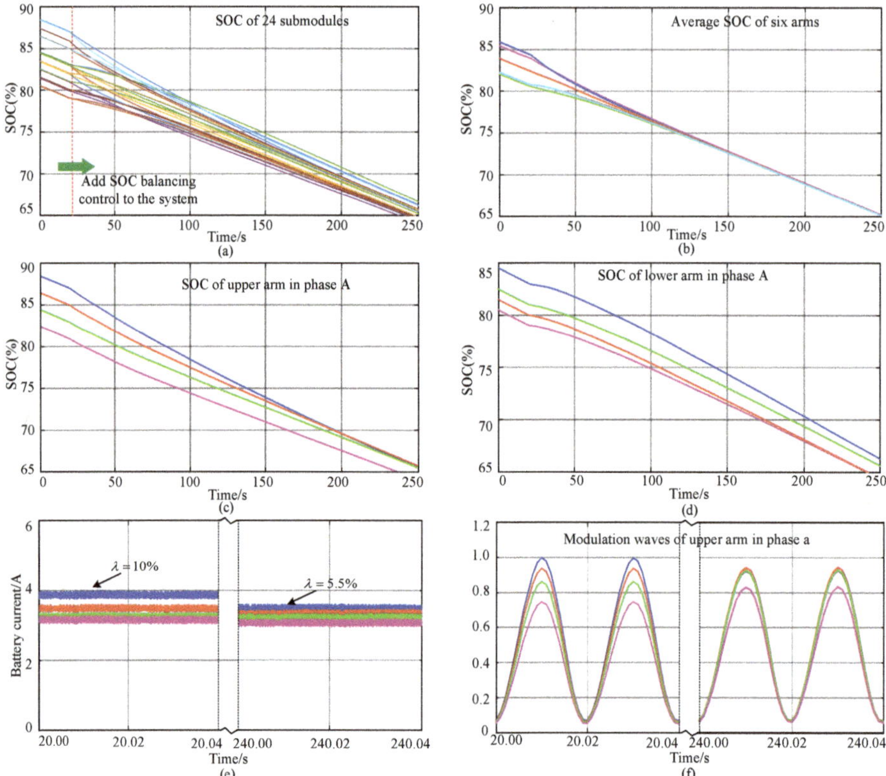

Figure 8. Simulation results of traditional SOC balancing control: (**a**) SOC of all submodules; (**b**) average SOC of six arms; (**c**) SOC of upper arm in phase a; (**d**) SOC of lower arm in phase a; (**e**) battery current of phase arm in phase a; (**f**) modulation waves of upper arm in phase A.

Figure 9 is the simulation results of the proposed SOC balancing control, and the power configuration is the same as the simulation in Figure 8. The SOC of all submodules is shown in Figure 9a. At time T = 20 s, the SOC balancing control is added to the control system, and then the SOC of the whole BESS converges to the same value at around T = 200 s. Figure 9b shows the average SOC of six arms. Figure 9c,d are the SOC of upper and lower arm in phase A. In Figure 8, the tolerance for battery power unbalance is about 50%, and the control design is according to (39). At T = 20 s, the difference of SOC among submodules is the greatest, but the battery power unbalance is still limited into the allowed values and over-modulation does not occur.

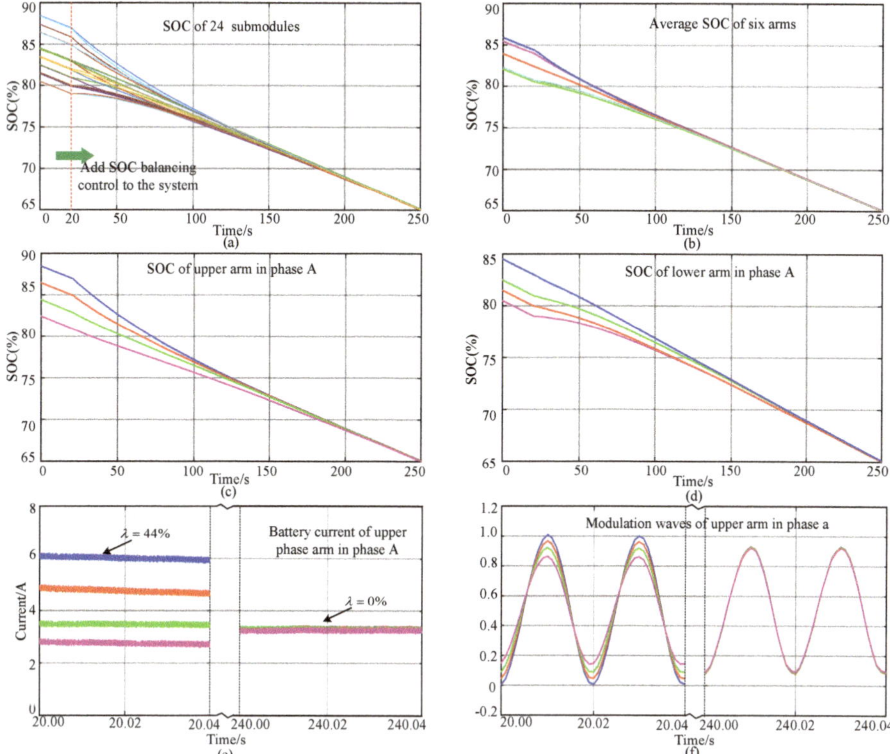

Figure 9. Simulation results of the modified SOC balancing control: (**a**) SOC of all submodules; (**b**) average SOC of all phase arms; (**c**) SOC of upper arm in phase a; (**d**) SOC of lower arm in phase a; (**e**) battery current of phase arm in phase A; (**f**) modulation waves of upper arm in phase A.

Compared with the traditional method, the proposed SOC balancing control method improves the tolerance for battery power unbalance. Therefore, the efficiency of SOC balancing control can be greatly improved and over-modulation also can be effectively avoided.

5.2. Experiment Results

The detailed simulation results above have verified the effectiveness of the analysis and proposed control strategy. For further verification, a downscaled prototype is built in this paper as shown in Figure 10. Owing to the limitation of experiment conditions. The utilized battery is the lead-acid battery, and the detailed parameters are shown in Table 2.

Figure 10. Prototype of three phase MMC-BESS.

Table 2. Parameters of experiment prototype.

Quantity	Value	Comment
V_{dc}	120 V	DC-Link voltage
m	0.8	Modulation ratio
P_{ac}	2 KW	Nominal AC-side power
V_{sm}	60 V	Submodule capacitor voltage
N	2	Number of submodules per arm
C	3 mF	Submodule capacitance
L_a	5 mH	Arm inductor
L_{bat}	10 mH	DC/DC side inductor
V_{bat}	36 V	Nominal battery voltage
C_{bat}	12 Ah	Nominal battery capacity
f_M	5 kHz	MMC side switching frequency
f_B	10 kHz	DC/DC converter frequency

Figure 11 is the steady-state waveforms of MMC-BESS, and the BESS works in charging mode. Figure 11a is the ac-side current, and Figure 11b shows the circulating current of three phases. Circulating current contains DC component and fundamental frequency component. The dc circulating currents of phase A, B, and C are different, which are controlled to balance the SOC among legs. Fundamental frequency circulating currents are injected to balance the SOC between phase arms. Figure 11c,d show the submodule capacitor voltages and battery currents of upper arm and lower arm in phase A. The ripples of capacitor voltage are filtered by the notch filter, so there is nearly no low-frequency component in the battery current. The maximum battery power unbalance degree is about 40%, which is much greater than the allowed unbalance degree in traditional method.

Figure 12 is the experiment result of optimized SOC balancing control proposed in this paper. The SOC of 12 submodules is between 76–82% at T = 0, and the SOC of all submodules converges to 70% at T = 40 min. Figure 13 is the experiment result of traditional SOC balancing control. The SOC of 12 submodules is between 55–60% at T = 0. Figures 12a and 13a shows the SOC of all submodules. it is obvious that the convergence of proposed method is much better than that of traditional method. Figures 12b and 13b denote the SOC balancing control among phases and between arms. The convergence of the two methods are basically the same. In Figure 12c,d and Figure 13c,d, the SOC balancing control within arms of the proposed method is much faster than that of traditional method. That is why the efficiency of the proposed method is much higher than that of traditional method.

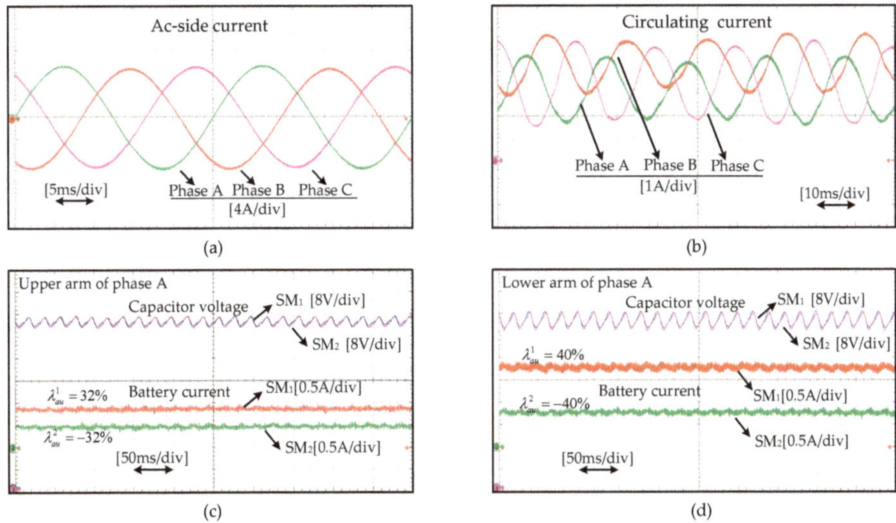

Figure 11. Steady-state waveforms of MMC-BESS with optimized SOC balancing control method: (**a**) AC-side current; (**b**) circulating current; (**c**) capacitor voltage and battery current of upper arm in phase A; (**d**) capacitor voltage and battery current of lower arm in phase A.

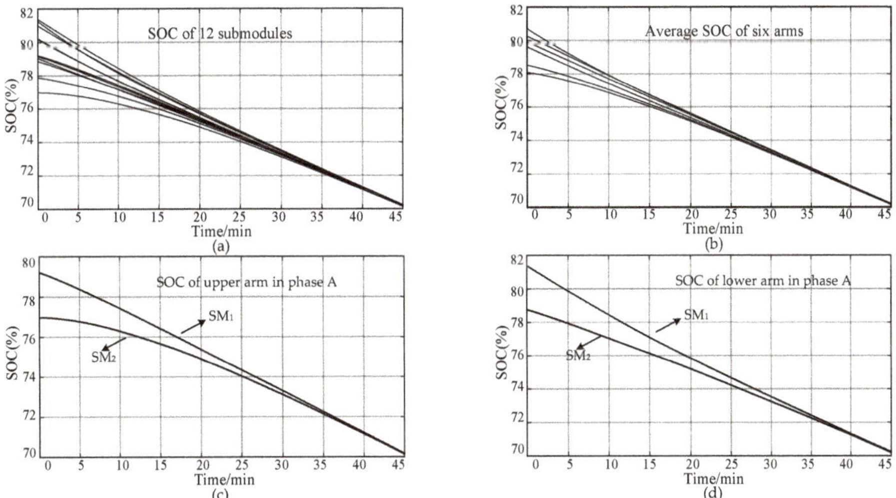

Figure 12. Experiment results of modified SOC balancing control method: (**a**) SOC of all submodules; (**b**) average SOC of phase arms; (**c**) SOC of upper arm in phase A; (**d**) SOC of lower arm in phase A.

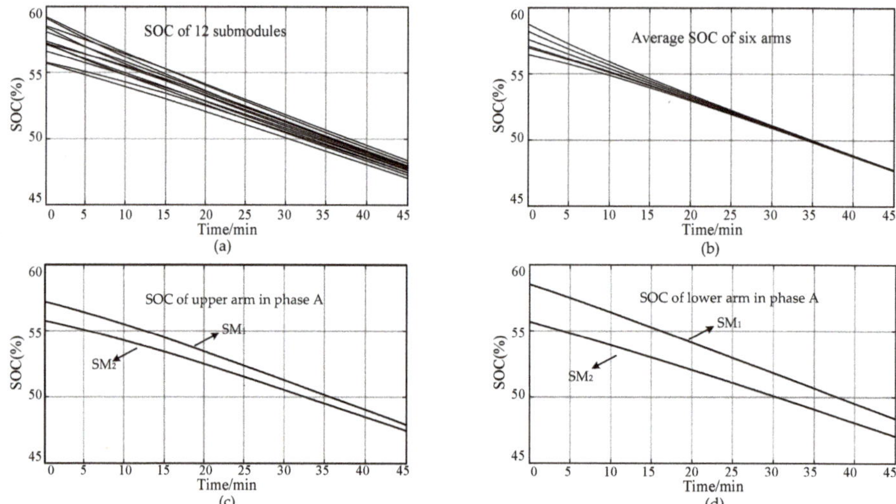

Figure 13. Experiment results of traditional SOC balancing control method: (**a**) SOC of all submodules; (**b**) average SOC of phase arms; (**c**) SOC of upper arm in phase A; (**d**) SOC of lower arm in phase A.

6. Conclusions

This paper mainly focuses on the SOC balancing control of MMC-BESS, aiming to improve the efficiency of SOC balancing control. The investigation indicates that the battery power unbalance can lead to the over-modulation of submodules, limiting the efficiency of SOC balancing control. Then the tolerance for battery power unbalance is defined to quantize the convergence of SOC balancing control. The submodule voltage regulation method is studied in detail by introducing the DC factor and AC factor. The linear programming method is introduced to reach the maximum tolerance in different operation modes. Based on the analysis, by choosing appropriate submodule voltage regulation method, the efficiency of SOC balancing control is improved greatly. To avoid over-modulation of submodules, the controller of SOC balancing control is also optimally designed. In this way, the convergence rate of SOC is greatly improved compared with traditional method. Finally, the analysis and proposed SOC balancing method are verified through detailed simulation and experiment results.

Author Contributions: Y.M. proposed the main idea, performed the theoretical analysis, and wrote the paper. H.L. contributed the experiment materials and gave some suggestions. Z.W. and Z.Z. performed the experiments.

Funding: This research was funded by National Natural Science Foundation of China, grant number 51741703.

Conflicts of Interest: The authors declare no conflict of interest.

References

1. Qian, H.; Zhang, J.; Lai, J.S.; Yu, W. A high-efficiency grid-tie battery energy storage system. *IEEE Trans. Power Electron.* **2011**, *26*, 886–896. [CrossRef]
2. Alipoor, Y.; Miura, Y.; Ise, T. Power System Stabilization Using Virtual Synchronous Generator with Alternating Moment of Inertia. *IEEE J. Emerg. Sel. Top. Power Electron.* **2015**, *3*, 451–458. [CrossRef]
3. Jiang, Q.Y.; Gong, Y.Z.; Wang, H.J. A battery energy storage system dual-layer control strategy for mitigating wind farm fluctuations. *IEEE Trans. Power Syst.* **2013**, *28*, 3263–3273. [CrossRef]
4. Baruschka, L.; Mertens, A. Comparison of cascaded H-bridge and modular multilevel converters for BESS application. In Proceedings of the Energy Conversion Congress and Exposition, Phoenix, AZ, USA, 17–22 September 2011; pp. 909–916. [CrossRef]

5. Soong, T.; Lehn, P.W. Evaluation of emerging modular multilevel converters for BESS applications. *IEEE Trans. Power Deliv.* **2014**, *29*, 2086–2094. [CrossRef]

6. Trintis, I.; Munk-Nielsen, S.; Teodorescu, R. A new modular multilevel converter with integrated energy storage. In Proceedings of the IECON 2011-37th Annual Conference on IEEE Industrial Electronics Society, Melbourne, Australia, 7–10 November 2011; pp. 1075–1080. [CrossRef]

7. Coppola, M.; Del Pizzo, A.; Iannuzzi, D. A power traction converter based on modular multilevel architecture integrated with energy storage devices. In Proceedings of the Electrical Systems for Aircraft, Railway and Ship Propulsion (ESARS), Pittsburgh, PA, USA, 14–18 September 2014; pp. 1–7. [CrossRef]

8. Hillers, A.; Stojadinovic, M.; Biela, J. Systematic comparison of modular multilevel converter topologies for battery energy storage systems based on split batteries. In Proceedings of the 17th Europe Conference Power Electronics, Geneva, Switzerland, 8–10 September 2015; pp. 1–9. [CrossRef]

9. Soong, T.; Lehn, P.W. Internal Power Flow of a Modular Multilevel Converter with Distributed Energy Resources. *IEEE J. Emerg. Sel. Top. Power Electron.* **2014**, *2*, 1127–1138. [CrossRef]

10. Gao, F.; Zhang, L.; Zhou, Q. State-of-charge balancing control strategy of battery energy storage system based on modular multilevel converter. In Proceedings of the IEEE Energy Conversion Congress and Exposition, Pittsburgh, PA, USA, 14–18 September 2014; pp. 2567–2574. [CrossRef]

11. Zhang, L.; Gao, F.; Li, N. Interlinking modular multilevel converter of hybrid AC-DC distribution system with integrated battery energy storage. In Proceedings of the IEEE Energy Conversion Congress and Exposition, Montreal, QC, Canada, 20–24 September 2015; pp. 70–77. [CrossRef]

12. Bala, S. The effect of low frequency current ripple on the performance of a Lithium Iron Phosphate (LFP) battery energy storage system. In Proceedings of the IEEE Energy Conversion Congress and Exposition, Raleigh, NC, USA, 15–20 September 2012; pp. 3485–3492. [CrossRef]

13. Puranik, I.; Zhang, L.; Qin, J. Impact of Low-Frequency Ripple on Lifetime of Battery in MMC-based Battery Storage Systems. In Proceedings of the IEEE Energy Conversion Congress and Exposition, Portland, OR, USA, 23–27 September 2018; pp. 2748–2752. [CrossRef]

14. Novakovic, B.; Nasiri, A. Modular multilevel converter for wind energy storage applications. *IEEE Trans. Ind. Electron.* **2017**, *64*, 8867–8876. [CrossRef]

15. Ma, Y.J.; Lin, H.; Wang, Z.; Wang, T. Capacitor voltage balancing control of modular multilevel converters with energy storage system by using carrier phase-shifted modulation. In Proceedings of the IEEE Applied Power Electronics Conference and Exposition, Tampa, FL, USA, 26 30 March 2017; pp. 1821–1828. [CrossRef]

16. Soong, T.; Lehn, P.W. Assessment of Fault Tolerance in Modular Multilevel Converters with Integrated Energy Storage. *IEEE Trans. Power Electron.* **2016**, *31*, 4085–4095. [CrossRef]

17. Hagiwara, M.; Akagi, H. Control and experiment of pulse width-modulated modular multilevel converters. *IEEE Trans. Power Electron.* **2009**, *24*, 1737–1746. [CrossRef]

18. Adam, G.P.; Anaya-Lara, O.; Burt, G.M. Modular multilevel inverter: Pulse width modulation and capacitor balancing technique. *IET Power Electron.* **2010**, *3*, 702–715. [CrossRef]

19. Han, W.; Zou, C.; Zhou, C. Estimation of cell SOC evolution and system performance in module-based battery charge equalization systems. *IEEE Trans. Smart Grid* **2018**. [CrossRef]

20. Vasiladiotis, M.; Rufer, A. Analysis and Control of Modular Multilevel Converters with Integrated Battery Energy Storage. *IEEE Trans. Power Electron.* **2015**, *30*, 163–175. [CrossRef]

21. Quraan, M.; Tricoli, P.D.; Arco, S. Efficiency assessment of modular multilevel converters for battery electric vehicles. *IEEE Trans. Power Electron.* **2017**, *32*, 2041–2051. [CrossRef]

22. Quraan, M.; Yeo, T.; Tricoli, P. Design and control of modular multilevel converters for battery electric vehicles. *IEEE Trans. Power Electron.* **2016**, *31*, 507–517. [CrossRef]

23. Zhang, L.; Tang, Y.; Yang, S. Decoupled Power Control for A Modular Multilevel Converter-Based Hybrid AC-DC Grid Integrated with Hybrid Energy Storage. *IEEE Trans. Ind. Electron.* **2018**. [CrossRef]

24. Chen, Q.; Li, R. Analysis and Fault Control of Hybrid Modular Multilevel Converter with Integrated Battery Energy Storage System. *IEEE J. Emerg. Sel. Top. Power Electron.* **2017**, *5*, 64–79. [CrossRef]

25. Liang, H.; Guo, L.; Song, J. State-of-Charge Balancing Control of a Modular Multilevel Converter with an Integrated Battery Energy Storage. *Energies* **2018**, *11*, 873. [CrossRef]
26. Debnath, S.; Qin, J.; Saeedifard, M. Control and stability analysis of modular multilevel converter under low-frequency operation. *IEEE Trans. Ind. Electron.* **2015**, *62*, 5329–5339. [CrossRef]

Article

Multi-Level Open End Windings Multi-Motor Drives

Salvatore Foti [1,*], **Antonio Testa** [1], **Salvatore De Caro** [1], **Tommaso Scimone** [1], **Giacomo Scelba** [2] **and Giuseppe Scarcella** [2]

[1] DI, University of Messina, 98122 Messina, Italy; atesta@unime.it (A.T.); sdecaro@unime.it (S.D.C.); tscimone@unime.it (T.S.)

[2] DIEEI, University of Catania, 95131 Catania, Italy; giacomo.scelba@dieei.unict.it (G.S.); g.scarcella@unict.it (G.S.)

* Correspondence: sfoti@unime.it

Received: 20 December 2018; Accepted: 25 February 2019; Published: 5 March 2019

Abstract: A multi-level open-end winding converter topology for multiple-motor drives is presented featuring a main multi-level inverter processing the power delivered to the motors and an active filter based on an auxiliary two-level inverter. The main inverter operates at the fundamental frequency in order to achieve low switching power losses, while the active filter is Pulse Width Modulation (PWM) operated to suitably shape the motor currents. The proposed configuration features less phase current distortion than conventional multi-level inverters operating at the fundamental frequency, while achieving a higher efficiency compared to PWM multi-level inverters. Experimental results confirm the effectiveness of such a configuration on both multiple motors-single converter and multiple motor-multiple converter drives.

Keywords: multilevel converter; multi-motor drive; harmonic mitigation; active filter; open end winding motor; high efficiency drive; high reliability applications

---

1. Introduction

A Multiple Motor Drive (MMD) is composed of some electric motors sharing the load torque [1–5]. Such a configuration costs less than a set of single motor drives, as some resources are shared between the units. Further, compared to a single drive, it is easily expandable by adding new units, moreover, the intrinsic redundancy may be used to mitigate the effects of some kinds of converter and motor faults. MMD are common in paper and textile industry and ironworks, as well as in several industry applications demanding synchronization between two or more axes, high levels of reliability and/or easy expandability. MMD systems can be grouped into two classes, namely: Multiple Motors Fed by a Single Multi-level Converter (MMSC) and Multiple Motors Fed by Multiple Multi-level Converters (MMMC). A single inverter on MMSC delivers power to all the machines, thus leading to only an approximatively proportional load sharing. On MMMC a common DC bus supplies a set of inverters, each one powering a single motor. In this case, the torque produced by each motor can be independently controlled, while retaining a common power entry and braking system.

Induction Motor (IM) based MMMCs are frequently used on rail mounted, or rubber tired, gantry cranes, equipping hoist, trolley and leg drives [6–8]. On leg drives MMMCs provide the ground to cope with different wheel diameter, unequal wheels' adhesion and slipping of transmission devices. MMDs also often equip electric locomotives, powering the axles through spur gearings [9–12]. Dual-Voltage Source Inverter (VSI)/Dual-IM and Mono-VSI/Dual-IM configurations are used with induction motors. The first, being of the MMMC type, comprises two induction motors supplied through two power converters. Such a configuration features a higher level of robustness toward inverter faults. Moreover, critical operations caused by pantograph detachment, loss of wheel adherence and stick-slip perturbation may be faced thanks to a fully independent control of the torque delivered by single machines. In recent years, MMMC has also become a viable alternative to single drives on powertrains of electrical and hybrid vehicles, featuring higher flexibility, reliability and transmission efficiency, while helping to increase the available inner space [13,14].

A straightforward way to control the speed of a MMMC is based on the common speed reference technique, providing a common speed reference signal to the speed controller of each unit. A torque follower control approach is however preferable, due to concerns regarding the control precision and system flexibility. According to such a technique, as shown in Figure 1, a single speed command is provided to a master unit, which in turns sends the torque references to the other drives [6]. Torque and flux regulation on single units is generally based on the traditional Indirect Field Oriented Control (IFOC) technique or Direct Torque Control strategy.

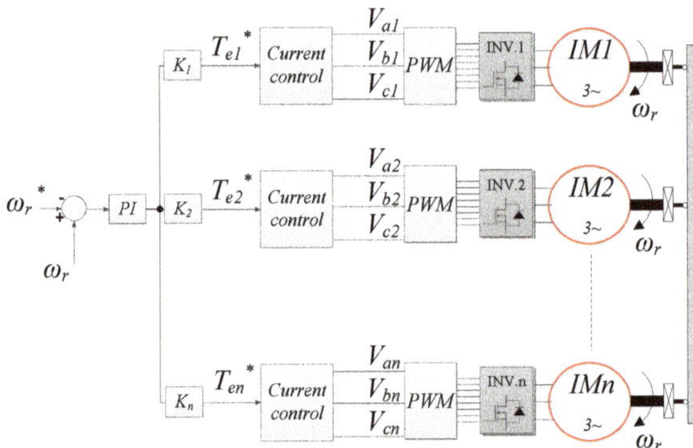

Figure 1. Common speed reference control of a Multiple Motors fed by Multiple Converters (MMMC) system.

In recent years, Multi-Level Inverters (MLIs) have been introduced on multi-motor drives. in order to generate almost sinusoidal output voltages using low frequency switching power devices, thus achieving high efficiency and electromagnetic compatibility. In addition, power devices are tasked with withstanding a fraction of the total DC input voltage, resulting in a remarkable reduction of dv/dt stresses and switches voltage ratings. However, in order to achieve a comparable phase current Total Harmonic Distortion (THD), MLI topologies working at low switching frequency could require many more power devices than conventional Pulse Width Modulation (PWM) operated bridge inverters. Cost concerns lead to limitation of the number of power devices, resulting in torque ripple and additional losses. These can be addressed by selective harmonic elimination or PWM techniques, as well as, by the addition of line reactors and tuned harmonic filters. Active power filters may also be exploited as they are able to provide a more flexible and effective attenuation of current harmonics [15–24].

An Open-end Winding (OW), multi-level, configuration for MMD applications is proposed in this paper; it features null neutral point fluctuations, low phase current ripple and improved DC bus voltage utilization [25–29]. A distinctive characteristic of the proposed configuration is the asymmetry, because that the two inverters are not of the same type and only one of the two provides power to the motors. In fact, a main multi-level inverter processes the power delivered to the motors, while an active filter, based on an auxiliary two-level inverter (TLI), shapes the phase current. Moreover, the main inverter operates at the fundamental frequency in order to achieve low switching power losses, while the active filter is PWM operated. Finally, the DC bus voltage of the two-level inverter is remarkably lower than that of the main inverter [30–36]. The proposed MMD OW configuration features a higher global efficiency and lower current THD than an equivalent PWM operated multi-level inverter for multi-motor drives [37], exploiting a specific control strategy combining low switching frequency modulation on the MLI and high frequency PWM on the TLI. Finally, it can be used either on either multiple motors-single converter or multiple motor-multiple converter drives.

2. Open Winding Multiple Motors Fed by a Single Multi-Level Converter—MMSC

Although the proposed approach can be applied to a system composed of an arbitrary number of machines, a system comprising only two identical induction motors, sharing the total load torque, is considered for simplicity. The proposed MMSC structure is shown in Figure 2; the stator windings of the two OW induction motors IM1 and IM2, are parallel connected to a MLI and a TLI, but only the first actually delivers active power to both motors, the second being operated as an active filter. This makes the proposed configuration different from most common OW motor drives, where the power supplied to the motors is evenly shared between the two inverters. The TLI, which supplies a null average power to the motors, can be supplied through a floating capacitor, avoiding the need for an additional power source and also making the two DC voltage sources V_{DC}' and V_{DC}'' independent. The motors are assumed to be speed controlled through an IFOC technique, but the load sharing between the motors cannot be managed. The output phase voltage of an MLI may take n different levels. The magnitude of each level is generally defined as the voltage across the mid-point of the DC bus and the output phase terminal. The number of possible voltage levels is generally odd, including a zero level and $n - 1$ non-zero levels and each level is identified by the value of the coefficient $l' = 0, 1, 2, \ldots, n - 1$. The *i-phase* MLI output voltage V_{MLI_stepi} is thus given by:

$$V_{MLI\_stepi} = \frac{2l' - n + 1}{2(n - 1)} V_{DC}' \tag{1}$$

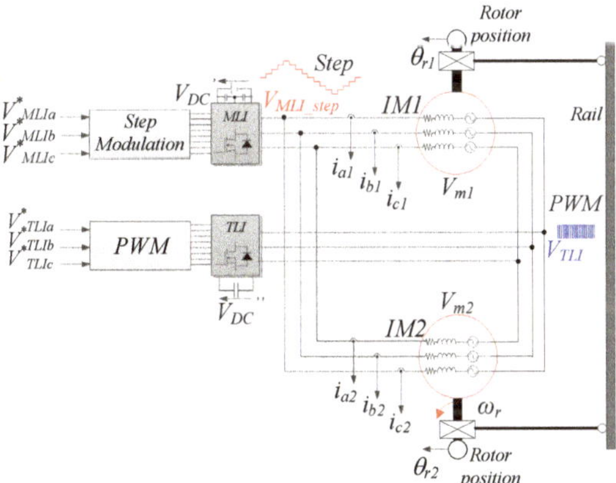

Figure 2. Proposed Open-end Winding (OW) Multi-level inverter (MLI) + Two Level Inverter (TLI) Multiple Motors Fed by a Single Multi-level Converter (MMSC) configuration.

In the same way, the *i-phase* TLI output voltage V_{TLIi} is given by:

$$V_{TLIi} = \frac{(2l'' - 1)}{2} V_{DC}'' \tag{2}$$

where $l'' = 0, 1$ is the *i-phase* TLI actual output voltage level.

The voltage V_{mi} across a generic 'i' motor phase winding, is finally given by:

$$V_{mi} = V_{MLI\_stepi} - V_{TLIi} - Vn'n'' \tag{3}$$

where $V_{n'n''}$ is the voltage across the mid points n' and n'' of the two DC buses, which is given by:

$$V_{n'n''} = \frac{1}{3}(V_{MLIa} + V_{MLIb} + V_{MLIc}) + \frac{1}{3}(V_{TLIa} + V_{TLIb} + V_{TLIc}) \tag{4}$$

The phase voltage V_{mi} takes the zero level and further $4(n + 1)$ non-zero levels when $V_{DC}'' = V_{DC}'/[2(n - 1)]$, while, some additional non zero levels become available if $V_{DC}'' < V_{DC}'/[2(n - 1)]$ [31].

The voltage levels which the proposed asymmetrical hybrid multilevel inverter configuration may take are more numerous than those of conventional Neutral Point Clamped (NPC) or Flying Capacitor (FC) MLI topologies with the same amount of power switches. In other words, the same phase voltage THD can be achieved with the proposed MMSC configuration using fewer power devices. Moreover, the proposed configuration features a switching frequency which is that of the TLI, although the main inverter switches at the fundamental frequency [37]. This leads to use power devices optimized for low switching frequency operation (i.e., featuring a low on-state voltage drop) in the MLI, and power devices suitable for high switching frequency in the TLI.

A suitable motor phase voltage modulation strategy was formulated taking into account the specific features of the proposed topology, according to the scheme in Figure 3.

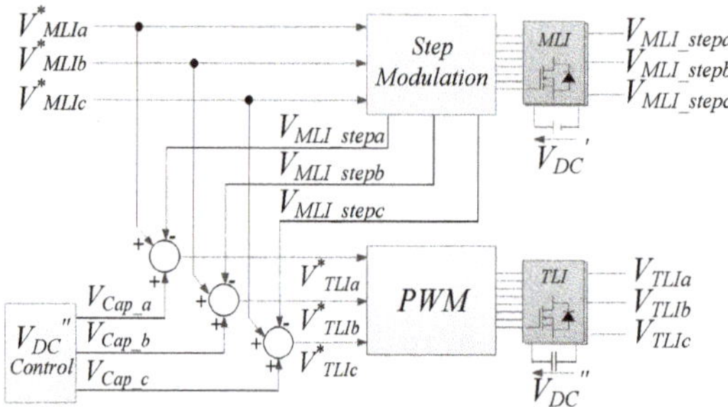

Figure 3. Block diagram of the OW MLI+TLI MMSC voltage control.

The TLI is tasked with compensating low frequency phase voltage harmonics generated by the MLI, which is step driven to reduce the switching power losses. Unwanted low frequency voltage harmonics can be suppressed by setting the voltage reference V^*_{TLIi} of the TLI to:

$$V^*_{TLIi} = V^*_{MLI\_stepi} - V^*_{MLIi} \tag{5}$$

where: V^*_{MLIi} is the fundamental harmonic of the ith motor phase reference voltage $V^*_{MLI\_stepi}$. Both quantities are depicted in Figure 4 for the case of a seven-level V_{MLI_stepi} waveform.

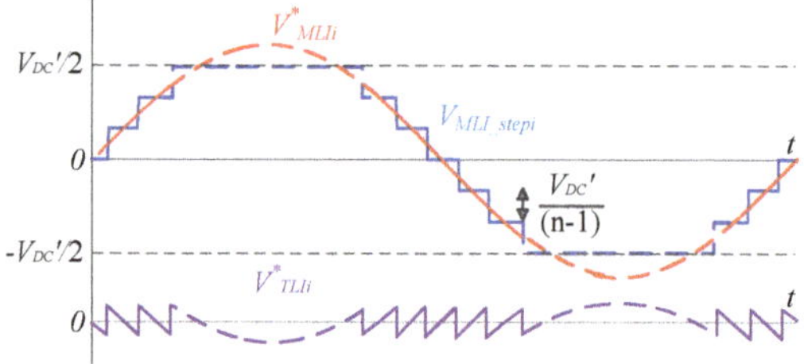

Figure 4. V_{MLI_stepi}, V^*_{MLIi} and V^*_{TLIi} waveforms.

The voltage ratio $K_V = V_{DC}''/V_{DC}'$ is a key parameter, because it impacts on the number of inverter voltage levels, on the THD of phase voltages and currents, on the maximum motor phase voltage amplitude and finally, on the ratings of TLI power devices.

The phase voltage V_{mi} is the difference between V_{MLIi} and V_{TLIi}, hence, its space vector diagram can be obtained by combination of the voltage space vector diagrams of the MLI and TLI. The simplest MMSC structure which can be obtained with the proposed approach consists of a Three-Level inverter (3LI) and a TLI (3LI+TLI).

The phase voltage of each inverter leg of the 3LI can assume three statuses namely: P, O and N. P denotes that the inverter phase voltage $V_{MLIi} = V_{DC}'$, while O indicates that $V_{MLIi} = 0$, and N that $V_{MLIi} = -V_{DC}'$. The 3LI topology is basically composed of two cells, each one supplied at $V_{DC}'/2$, hence each leg may take three switching states, resulting in $3^3 = 27$ possible inverter switching state combinations. Each space vector can be categorized into zero voltage, small voltage, medium voltage and large voltage on the basis of its magnitude. These are tabulated in Table 1. The TLI instead features $2^3 = 8$ inverter states, leading to 7 voltage vectors. As shown in Figure 5, the 3LI+TLI voltage space vector diagram is obtained by composition of the voltage space vector diagrams of the TLI and the 3LI. The number of available voltage vectors and voltage levels depends on K_V, as listed in Table 2, affecting both the THD_V and the peak amplitude V_{mpk} of the phase voltage V_{mi}. In particular, V_{mpk} increases with K_V, while a minimum THD_V is achieved for $V_{DC}'' = V_{DC}'/[(n-1)]$, [24], thus, the considered 3LI+TLI system gets the minimum THD_V for $V_{DC}'' = V_{DC}'/2$. However, V_{mpk} maximization must be also taken into consideration, thus, a useful figure of merit in determining the optimal value of K_V is $K_p = V_{mpk}/(V_{DC}' \times THD_V)$, which, according to Table 3, is maximized when $V_{DC}'' = V_{DC}'/2$. When V_{DC}'' is reduced below $V_{DC}'/2$, K_p and V_{mpk} decrease, while the number of voltage levels and voltage vectors increases, as well as THD_V. However, the voltage rating of TLI power devices is lowered. For $K_V = 1/8$, the 3LI+TLI is equivalent to conventional six-level NPC or FC inverter structures, which however require six more power devices. By setting V_{DC}'' over $V_{DC}'/2$, a higher V_{mpk} is obtained, while the number of voltage levels and voltage vectors is lowered, worsening THD_V and K_p. Moreover, when $V_{DC}'' = V_{DC}'$, the voltage rating of the TLI switches becomes equal to that of MLI devices.

Table 1. Switching states and voltage vector of Three Level Inverter (3LI).

Vector	Magnitude	Switching State
Zero vector	0	PPP, OOO, NNN
Small vector	$1/3\,V_{DC}'$	POO, PPO, OPO, OPP, POP, OOPONN, OON, NON, NOO, ONO, NNO
Medium vector	$\sqrt{3}/3\,V_{DC}'$	PON, OPN, NPO, NOP, ONP, PNO
Large vector	$2/3\,V_{DC}'$	PNN, PPN, NPN, NPP, NNP, PNP

Table 2. Power converter specifications vs. K_V for Three Level Inverter + Two Level Inverter (3LI+TLI).

K_V	Number of Inverter States	Number of Voltage Vectors	Number of Voltage Levels
1/8	216	152	8
1/4	216	91	6
1/2	216	37	4
1	216	61	5

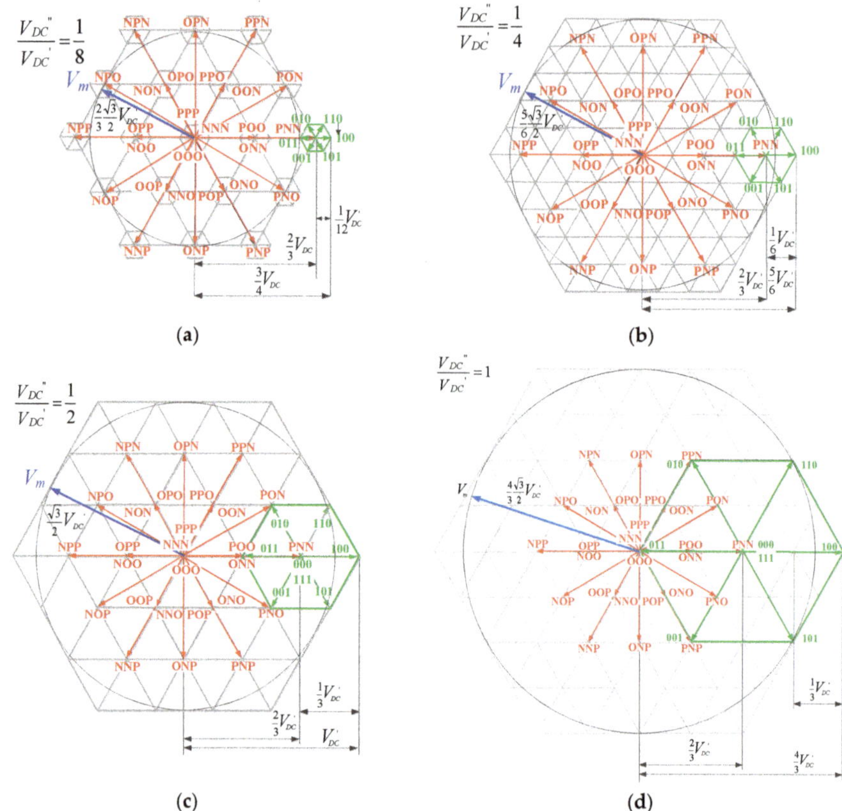

Figure 5. Phase voltage space phasor diagram (**a**) $V_{DC}''/V_{DC}' = 1/8$; (**b**) $V_{DC}''/V_{DC}' = 1/4$; (**c**) $V_{DC}''/V_{DC}' = 1/2$; (**d**) $V_{DC}''/V_{DC}' = 1$.

Table 3. V_{mpk} and THD$_V$ vs. K_V (3LI+TLI).

K_V	V_{mpk}	THD$_V$ (%)\K$_P$ $\omega_r/\omega_n = 0.1$	THD$_V$ (%)\K$_P$ $\omega_r/\omega_n = 0.3$	THD$_V$ (%)\K$_P$ $\omega_r/\omega_n = 0.7$	THD$_V$ (%)\K$_P$ $\omega_r/\omega_n = 1$
1/8	0.58 V_{DC}'	51\1.14	45\1.29	30\1.93	19\3
1/4	0.72 V_{DC}'	36\2	29\2.48	15\4.8	4.0\20.5
1/2	0.87 V_{DC}'	10\8.7	8\10.88	7.1\12.2	3.5\21.7
1	1.15 V_{DC}'	46\2.5	40\2.88	31\3.7	22.6\5.1

According to the proposed approach no active power is delivered to the two IM motors from the TLI. In practice, power devices and motors power losses in the TLI would cause a progressive discharge of the floating capacitor of the TLI DC-bus. Being floating, such a capacitor can only be charged by diverting to it a small quantity of the active power delivered by the MLI to the motors [32–44]. As shown in the control scheme of Figure 6, this is achieved by slightly modifying the TLI reference voltages. Specifically, two additional terms V_{dCap} and V_{qCap} are introduced to control the mean current flowing through the DC bus capacitor. Since the TLI is not tasked with supplying reactive power to the load, V_{dCap} can be straightforwardly related to V_{qCap}, by:

$$Q = \frac{3}{2}\left(V_{qCap}i_d - V_{dCap}i_q\right) = 0 \Rightarrow V_{dCap} = \frac{i_d}{i_q}V_{qCap} \tag{6}$$

The active power P required to hold V_{DC}'' constant is obtained by a PI controller processing the error between the reference DC bus voltage $V_{DC}''^*$ and the actual voltage V_{DC}''.

$$P = \frac{3}{2}\left(V_{qCap}i_q + V_{dCap}i_d\right) = \left(V''_{DC}{}^* - V''_{DC}\right)\left(K_{pVDC} + \frac{K_{iVDC}}{s}\right) \tag{7}$$

where s is the Laplace operator, and K_{pVDC} and K_{iVDC} are, respectively, the proportional and integral gain of the PI controller. By introducing Equation (6) into Equation (7), the dq-axes voltage reference components V_{qCap} and V_{dCap} are obtained:

$$V_{qCap} = \frac{2}{3}\frac{Pi_q}{i_q^2 + i_d^2} \qquad V_{dCap} = \frac{2}{3}\frac{Pi_d}{i_q^2 + i_d^2} \tag{8}$$

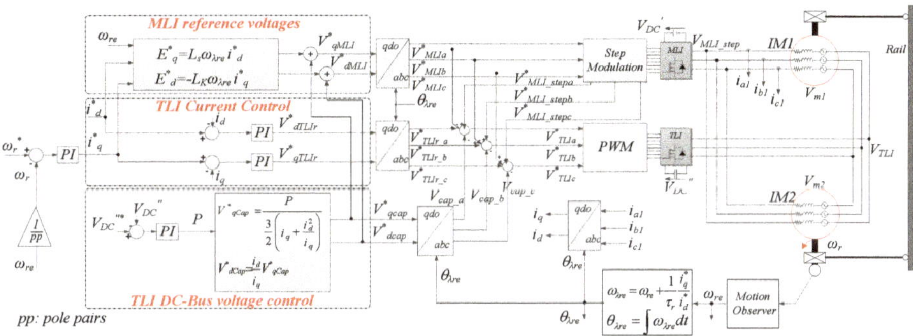

Figure 6. MLI+TLI MMSC control strategy.

The two additional terms are then transformed to the *abc* stationary frame and added to the TLI voltage references obtained from Equation (5). Due to the floating capacitor recharging, the phase current typically increases by less than 1.5% of the rated current. The considered 3LI+TLI topology requires a specific control system which is schematized in Figure 6. It consists of two subsystems, acting on the two inverters. The MLI is voltage controlled, thus, the q, d axes voltage references $V_{dqMLI}{}^*$ are made equal to the motors back EMF components $\hat{E}^*_q$ and $\hat{E}^*_d$, which in turn are estimated by

$$\begin{cases} \hat{E}^*_q = L_s\omega_{\lambda re}i_d^* \\ \hat{E}^*_d = -L_k\omega_{\lambda re}i_q^* \end{cases} \tag{9}$$

$$\begin{cases} \omega_{\lambda re} = \dfrac{R_r i_q^*}{L_r i_d^*} + \omega_r \\ L_k = -\dfrac{L_s L_r - L_m^2}{L_r} \end{cases} \tag{10}$$

where: $\omega_{\lambda re}$ is the rotor flux angular frequency, ω_r is the rotor speed of the two machines, and L_s, L_r and L_m are respectively the stator, rotor and magnetizing inductances.

As shown in Figure 6, the TLI features a feedback current control, in order to improve the shape of the current waveform and the system dynamic response. The outputs $V^*_{dqTLIr\_i}$ of the TLI current control loop are transformed to the *abc* stationary frame and then added to other voltage reference components, dealing with compensation of low order stator voltage harmonics and V_{DC}'' stabilization, this allows us to obtain the voltage references for the PWM modulator, which are given by:

$$V_{TLIi}^* = V_{TLIr\_i}^* + V_{MLI\_stepi}^* - V_{MLIi}^* + V_{cap\_i} \tag{11}$$

The currents flowing through the two stator windings cannot individually be controlled. However, even load torque sharing is obtained when considering two identical motors with a high-stiffness mechanical coupling. Whenever these conditions are not verified, torque and current unbalance may yield to a system instability. In this case, the TLI current control structure needs to be modified as suggested in references [45–51].

3. Open Winding Multiple Motors Fed by Multiple Multi-Level Converters—MMMC

The proposed OW approach can also be used on MMMC systems. An MMMC system is in general able to control the load sharing between the two motors, because providing an independent control of the stator currents of the two induction motors. However, according to the proposed approach, this would require two independent MLI and two TLI. A simpler structure was thus developed by connecting the two motors to a single MLI on one side and to a five-leg two level inverter (TLI5) on the other side, thus reducing the number of inverter switches and the associated power losses [51,52]. The circuital scheme and the control block diagram of such a topology (MLI+TLI5) are shown in Figures 7 and 8. Two different PWM strategies can be adopted on five-leg inverters [51–53]. A first one is based on the cancellation of the voltage reference of the inverter phase common to the two motors. In practice, for each set of reference voltages, the reference of the common phase is algebraically subtracted to the voltage references of the other two phases, while the common phase reference becomes equal to the difference between the two c-phase voltage references. A major drawback of this strategy is a reduction in the maximum motor phase voltage by a factor of 1/3 compared to a standard three phase motor drive. In order to overcome this drawback a second PWM strategy can be adopted, which is based on the addition, rather than subtraction, of the common phase voltage reference to the references of the other phases.

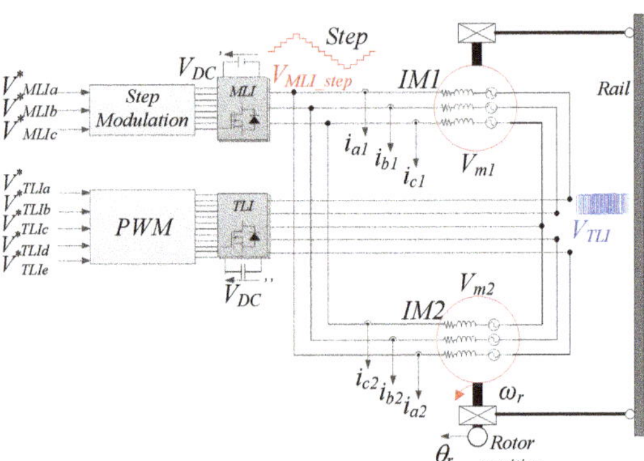

Figure 7. MLI+TLI5 OW MMMC configuration.

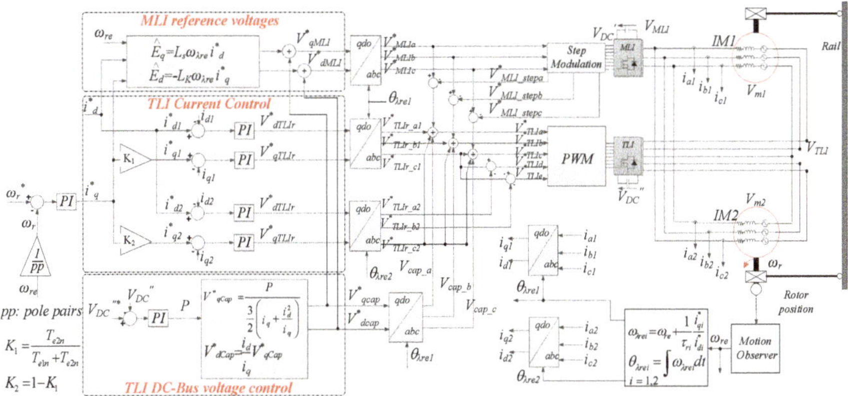

Figure 8. MLI+TLI5 OW MMMC control structure.

According to such an approach [46], the five-leg TLI voltage references are given by:

$$
\begin{aligned}
V^*_{TLIa} &= V^*_{TLIr\_a1} + V^*_{TLIr\_c2} + V^*_{MLI\_stepa} - V^*_{MLIa} \\
V^*_{TLIb} &= V^*_{TLIr\_b1} + V^*_{TLIr\_c2} + V^*_{MLI\_stepb} - V^*_{MLIb} \\
V^*_{TLIc} &= V^*_{TLIr\_c1} + V^*_{TLIr\_c2} + V^*_{MLI\_stepc} - V^*_{MLIc} \\
V^*_{TLId} &= V^*_{TLIr\_a2} + V^*_{TLIr\_c1} + V^*_{MLI\_stepa} - V^*_{MLIa} \\
V^*_{TLIe} &= V^*_{TLIr\_b2} + V^*_{TLIr\_c1} + V^*_{MLI\_stepb} - V^*_{MLIb}.
\end{aligned}
\tag{12}
$$

Since both motors are connected to a single MLI, voltage harmonic compensation is exerted by acting on the TLI *a*, *b*, *d* and *e* phase voltage references.

The average values of q-d back EMF components $\hat{E}^*_q$ and $\hat{E}^*_d$ are estimated from Equations (9) and (10), by assuming:

$$
i^*_d = i^*_{d1} = i^*_{d2}
\tag{13}
$$

$$
i^*_{q1} = K_1 i^*_q \qquad i^*_{q2} = K_2 i^*_q
\tag{14}
$$

where:

$$
K_1 = \frac{T_{e2n}}{T_{e1n} + T_{e2n}} \qquad K_2 = (1 - K_1)
\tag{15}
$$

By assuming the use of two identical motors, the q-axis current references can be imposed as:

$$
i^*_{q1} = i^*_{q2} = 0.5 i^*_q
\tag{16}
$$

4. Power Losses Assessment

The total power losses of the OW MMSC topology of Figure 2 were evaluated and compared with those of a system consisting of a single 3LI supplying two wye connected induction motors. Two different cases were investigated. In the first case, a conventional drive topology was considered, in which a single multilevel inverter was step operated and feeds two parallel connected induction motors. In the second case the inverter is operated according to a f_{swTLI} = 10 kHz space vector PWM modulation. The obtained results was then compared with those obtained with the 3LI+TLI topology, where the 3LI was driven according to a step modulation with $f_{swMLI} = \omega_{re}/2\pi$, while a f_{swTLI} = 10 kHz sinusoidal PWM strategy was used on the TLI. The TLI floating DC-bus was built around a 480 µF capacitor, while $V_{DC}{}'$ was set at 580 V and $V_{DC}{}'' = V_{DC}{}'/4$ at 145 V. This leads to the use of Insulated Gate Bipolar Transistor (IGBT) devices on the 3LI and MOSFET on the TLI.

The inverter power losses, consisting of the switching losses P_{swMLI}, P_{swTLI} and conduction losses P_{cMLI}, P_{cTLI} were calculated as in [31]:

$$P_{swMLI} = 0.5V_{ce}i_C f_{swMLI}(t_{onI} + t_{offI}) \qquad P_{cMLI} = \delta V_{ceo}i_C$$
$$P_{swTLI} = 0.5V_{DS}i_C f_{swTLI}(t_{onM} + t_{offM}) \qquad P_{cTLI} = R_{DS(on)}i_D^2 \tag{17}$$

The motor joule losses P_{joule} and motor iron losses P_{iron} are given by:

$$P_{joule} = R_S i_S^2 + R_r i_r^2 \qquad P_{iron} = \frac{E^2}{R_{fe}} \tag{18}$$

The induction motors parameters are listed in Table 4, while technical data of the IGBT used in the MLI and the MOSFET used in the TLI are summarized in Tables 5 and 6, respectively. The systems were run at ω_r = 100 rad/s.

Table 4. Technical specifications of the Induction Motors.

P_n (HP)	V_n (V)	pp	L_s (mH)	L_r (H)	L_m (H)	R_s (Ω)	R_{fe} (Ω)	R_r (Ω)	J (Kg·m²)
3	400	2	0.32	0.32	0.31	2.6	902	2.7	0.016

Table 5. MLI IGBT Data.

V_{CES} (V)	I_C (A)	T_j (°C)	V_{ceo} (V)	t_{on} (ns)	t_{off} (ns)
600	20	150	0.75	60	131

Table 6. TLI MOSFET Data.

V_{DSS} (V)	I_{DSS} (A)	T_j (°C)	$R_{DS(on)}$ (mΩ)	t_{on} (ns)	t_{off} (ns)
150	20	175	32	8.9	17.2

Figure 9 deals with power losses estimation for the three considered cases, taking into account the motor (core and winding) and inverter (switching and conduction) power losses and considering the first ninety harmonics of the stator current and voltage.

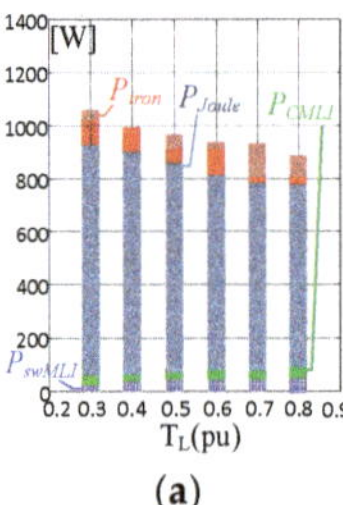
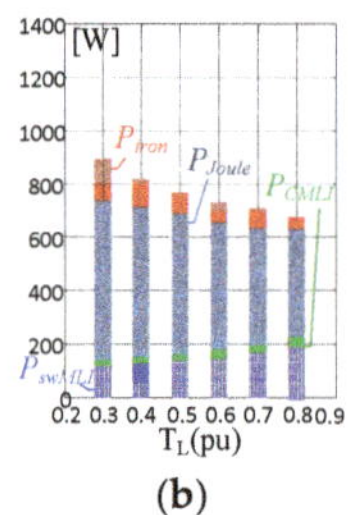
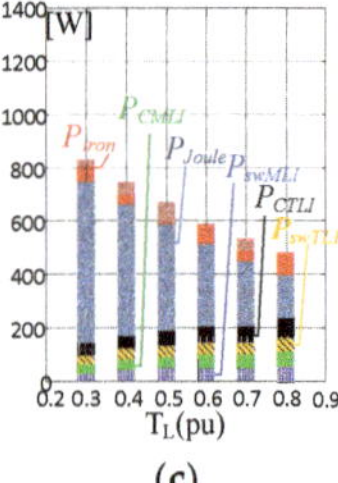

(a)   **(b)**   **(c)**

Figure 9. Motor iron losses P_{iron}, motor joule losses P_{joule}, MLI conduction losses P_{CMLI}, MLI switching losses P_{swMLI}, TLI conduction losses P_{CTLI} and TLI switching losses P_{swTLI}: (**a**) 3LI step modulated; (**b**) 3LI PWM modulated; (**c**) 3LI+TLI configuration.

When the step modulated 3LI feeds the two wye connected induction motors, as shown in Figure 9a, power losses largely consist of motor joule losses, because the motor currents are affected by low-order harmonics components. When the 3LI PWM modulated feeds the two wye connected induction motors, as shown in Figure 9b, motor joule losses are still dominant, but higher switching losses occurs. Finally, the 3LI+TLI configuration is considered. Since the MLI is step modulated and the phase motor current waveform is made close to a sinusoidal one by the active power filter, motor and MLI inverter losses are considerably reduced, especially at heavier loads.

The 3LI+TLI topology features a higher efficiency at medium and high loads, while at low loads its efficiency is comparable with that of the PWM driven 3LI. Based on computed motor and inverter losses, total efficiency was evaluated in the three considered cases, as shown in Figure 10.

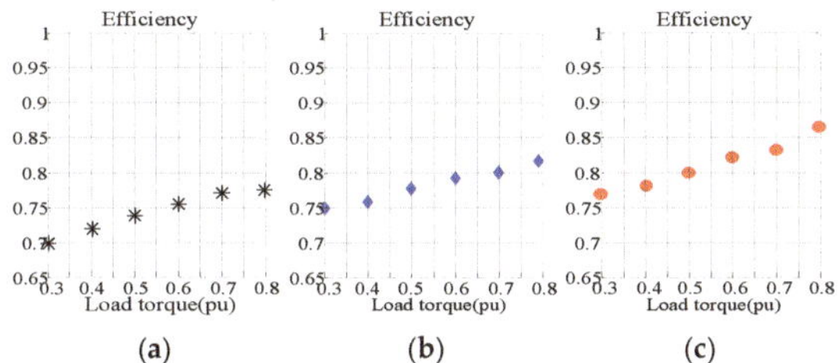

Figure 10. Total efficiency of the MMSC: (**a**) 3LI step operated; (**b**) 3LI PWM operated; (**c**) 3LI+TLI configuration.

A comparison between the performance of the 3LI+TLI5 topology and that of a conventional system consisting of two PWM operated 3LIs supplying two wye connected induction motors was also accomplished. The results shown in Figure 11, confirming that the 3LI+TLI5 MMMC is more efficient than the conventional one, especially at heavy loads.

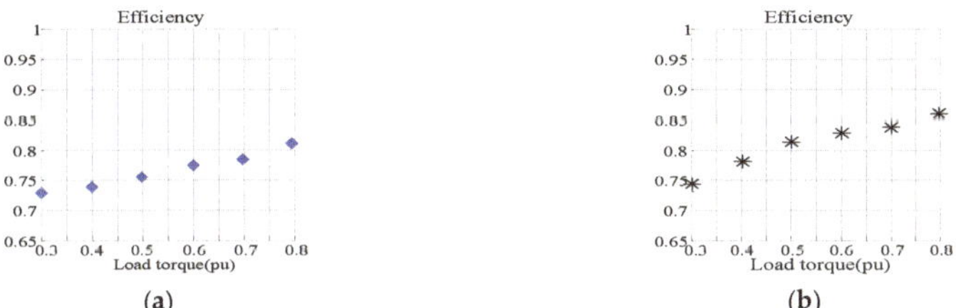

Figure 11. Total efficiency of the system: (**a**) 3LI PWM operated; (**b**) 3LI+TLI5 configuration.

5. Experimental Tests

The proposed multi-motor drive configurations were experimentally validated using two induction motors, the technical specifications of which are shown in Table 4. The MLI was equipped with an IGBT with parameters shown in Table 5 while the TLI consisted of a MOSFET, detailed in Table 6. Figure 12 shows the test rig. The first prototype featured an MMSC configuration, tailored around a five level NPC main inverter (5LI+TLI). The second prototype was still an MMSC system, but equipped with a three level NPC main inverter (3LI+TLI). The third prototype is instead a MMMC system composed of a five level NPC main inverter and a five-leg, two-level (5LI+TLI5) auxiliary inverter. All the three prototypes are field oriented controlled by a dSPACE board featuring a 100 μs update time. On each drive configuration two induction motors were present with mechanical coupling between them in order to operate them at the same rotational speed. Moreover, the TLI was PWM operated at $f_{swTLI} = 10$ kHz, with a 1 μs dead time. The DC bus voltage $V_{DC}{}'$ of the MLIs was 400 V, while $V_{DC}{}''$ was set at 50 V when a 5LI was used and at 100 V when a 3LI was used. A controllable mechanical load was realized by exploiting a conventional 3 HP vector controlled induction motor drive, detailed in Table 4.

A steady state test on a conventional step operated 5LI supplying two wye connected induction motors is shown in Figure 13. The test was performed at $\omega_r = 50$ rad/s and with no load. A remarkable stator current distortion was obtained, due to the low switching frequency. Figure 14 deals with the same test but using the 5LI+TLI MMSC topology instead of the conventional configuration. The distortion of the motors phase current was remarkably reduced, however; the efficiency was also slightly reduced, being 80% for the 5LI and 76% for the 5LI+TLI system.

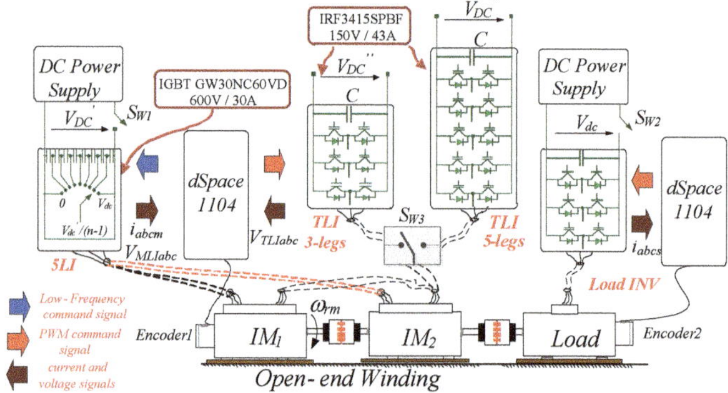

Figure 12. Experimental setup of OW MMSC.

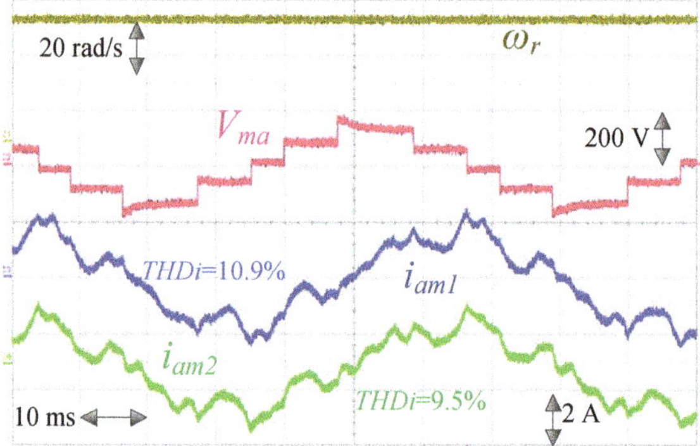

Figure 13. Conventional configuration, both motors are wye-connected and fed by a step operated 5MLI: steady state test at $\omega_r = 50$ rad/s and with no load. Motor speed ω_r, a-phase motor V_{ma}, motor currents i_{am1} and i_{am2}.

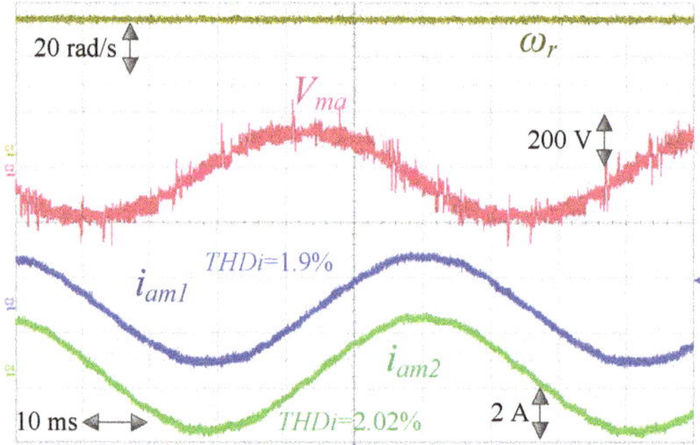

Figure 14. 5LI+TLI MMSC: steady state test at $\omega_r = 50$ rad/s. Motor speed ω_r, a-phase motor V_{ma}, motor currents i_{am1} and i_{am2}.

A comparison between the 3LI+TLI and 5LI+TLI MMSC configurations is provided in Figure 15. A steady state test ($\omega_{r1} = \omega_{r2} = 50$ rad/s) performed using the MMMC 5LI+TLI5 system is displayed in Figure 16. The c phase current of the TLI is higher than the currents of the other four phases, as it is given by the combination of the c phase currents of both motors.

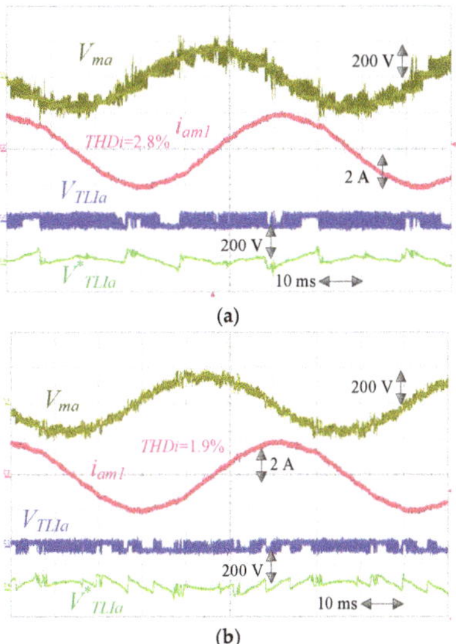

Figure 15. (a) MMSC (3LI+TLI); (b) MMSC (5LI+TLI). Steady state test, stator voltage V_{ma}, stattor current i_{am1}, TLI output voltage V_{TLIa} and TLI voltage reference V^*_{TLIa}.

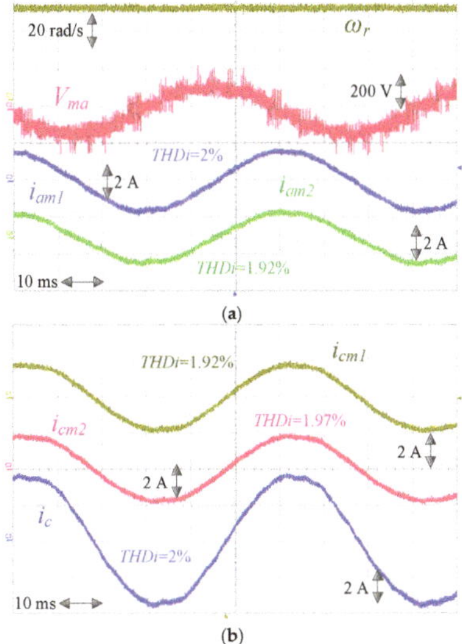

Figure 16. 5LI+TLI5 MMMC, Steady state test at $\omega_r = 50$ rad/s. (a) Motor speed ω_r, a-phase motor V_{ma}, motor currents i_{am1} and i_{am2}; (b) motor currents i_{am1} and i_{am2} and common-leg c-phase current i_c.

Speed reversals from $\omega_r = -40$ rad/s to $\omega_r = 40$ rad/s accomplished by the 5LI+TLI MMSC and the 5LI+TLI5 MMMC prototypes, are shown in Figure 17. The drives feature a good dynamic response and an effective floating capacitor voltage control. The torque load was evenly shared between the two motors.

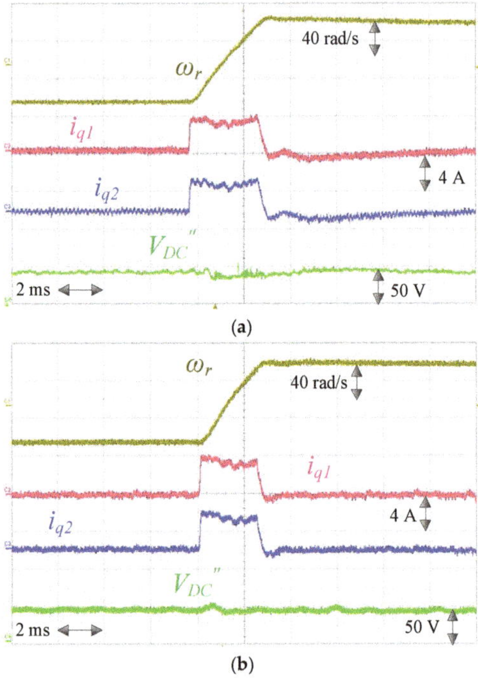

(a)

(b)

Figure 17. (**a**) 5LI+TLI5 MMSC; (**b**) 5LI+TLI5 OW MMMC: speed reversal with i_{dref} = 2 A. Motor speed ω_r, q-axes currents components i_{q1} and i_{q2} and TLI DC Bus voltage V_{DC}''.

In order to assess the capability of the 5LI+TLI5 configuration to individually control the torque produced by the two motors, the previous test was repeated with uneven loading on the two machines. As shown in Figure 18, 70% of the load torque was produced by IM1 and 30% by IM2. Figure 19 shows the phase currents of IM1 and IM2 at the steady state for the MMMC. The steady state currents and dynamical torque, as well as, and speed response are quite satisfactory.

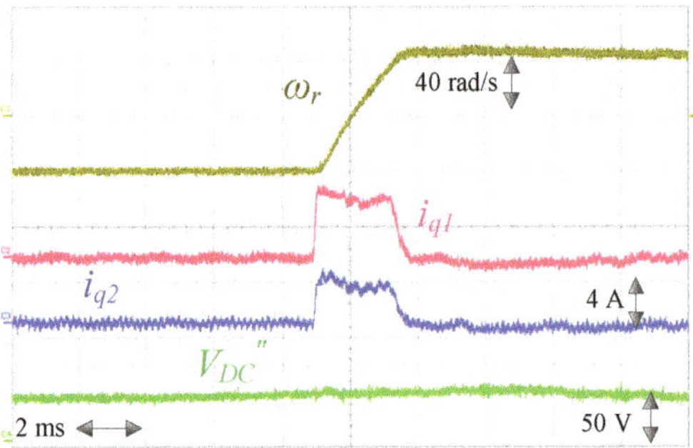

Figure 18. MMMC 5LI+TLI5: speed reversal with. i_{dref} = 2 A, i_{q1} = 0.7 i_q and i_{q2} = 0.3 i_q. Motor speed ω_r, q-axes currents components i_{q1} and i_{q2} and TLI DC Bus voltage V_{DC}''.

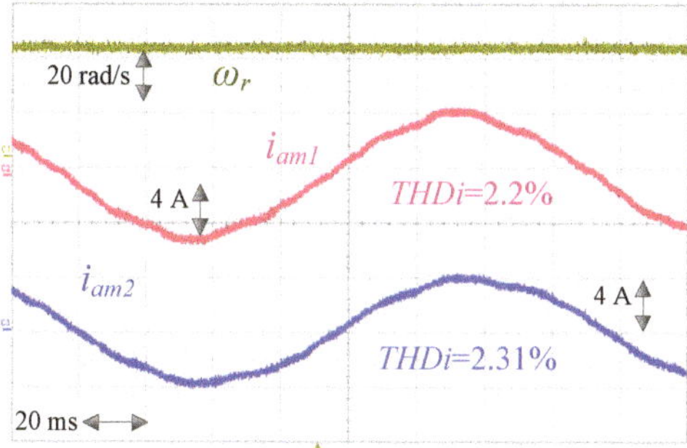

Figure 19. MMMC 5LI+TLI5: Steady state at ω_r = 50 rad/s and 50% of rated load, speed ω_r, currents i_{am1} and i_{am2}.

6. Conclusions

The paper has proposed a multi-level converter topology for multiple-motor drives based on a special open-end winding configuration. Applications of such a topology to multiple motors - single converter and multiple motor-multiple converter drives was also discussed. Specifically, two configurations, an MLI+TLI Multi Motor Single Converter system and an MLI+TLI5 Multi Motor Multi Converter system were presented and managed by purposely developed control strategies, combining a low switching frequency modulation on the MLI and high frequency PWM on the TLI. As shown in the paper, in both cases the proposed configurations produced much lower stator current distortion when compared to conventional multi-motor drives equipped with MLI switching at the fundamental frequency. Further, they generated lower power losses when compared to multi motor drives equipped with PWM operated MLI. Although the paper considers only multi motor systems comprising two identical induction machines, the proposed approach can be generalized for systems with an arbitrary amount of motors; it can also be exploited for multi motor systems using synchronous machines, and even for multi motor systems using a mix of machines of different sizes.

Author Contributions: Conceptualization, S.F., A.T., S.D.C., and G.S. (Giuseppe Scarcella); methodology, S.F., A.T., and G.S. (Giacomo Scelba); software, S.F., G.S. (Giacomo Scelba) and T.S.; validation, S.F., G.S. (Giacomo Scelba) and T.S.; supervision, A.T., G.S. (Giacomo Scelba) and G.S. (Giuseppe Scarcella).

Funding: This research received no external funding.

Nomenclature

MMD	Multiple Motor Drive
MMSC	Multiple Motors fed by a Single Converter
MMMC	Multiple Motors fed by Multiple Converters
IFOC	Indirect Field Oriented Control
TLI	Two-Level Inverter
3LI	Three-Level inverter
TLI5	Five-leg two level inverter
3LI+TLI	OW configuration including 3LI and TLI
3LI+TLI5	OW configuration including 3LI and TLI5
5LI+TLI	OW configuration including 5LI and TLI
5LI+TLI5	OW configuration including 5LI and TLI5
THD	Total Harmonic Distortion
n	MLI voltage levels
V_{MLI_stepi}	i-phase MLI output voltage with respect to n'
V_{TLIi}	i-phase TLI output voltage with respect to n''

V_{mi}	i-phase motor voltage
V_{DC}'	MLI DC Bus voltage
V_{DC}''	TLI DC Bus voltage
V_{mpk}	Peak motor voltage
i_{abc1}	abc axes IM1 phase currents
i_{abc2}	abc axes IM2 phase currents
E_{qd}	qd axes motor back EMF
i_{qd1}	qd axes IM1 phase currents
i_{qd2}	qd axes IM1 phase currents
V_{qdMLI}	qd axes MLI output voltage
R_s, R_r	Stator and rotor resistance
R_{fe}	Iron resistance
L_s, L_r	Stator and rotor inductances
L_m	Magnetizing inductance
$\omega_{\lambda re}$	Rotor flux angular frequency
K_V	Voltage ratio V_{DC}''/V_{DC}'
K_p	$V_{mpk}/(V_{DC}' * THD_V)$
ω_r	Mechanical Rotor speed
ω_{re}	Electrical Rotor speed
$\theta_{\lambda r1}, \theta_{\lambda r2}$	Rotor flux angular positions of IM1 and IM2
θ_r	Rotor angular positions of IM1 and IM2
T_{e1n}, T_{e2n}	Rated torques of IM1 and IM2
J	Total inertia of motor and load
pp	pole pairs
V_{ceo}	Collector to Emitter Saturation Voltage
t_{on}	Current Turn-On Delay Time
t_{off}	Current Turn-Off Delay Time
V_{CES}	Collector to Emitter Breakdown Voltage
I_C	Collector current
V_{DSS}	Drain-to-Source Breakdown Voltage
I_{DSS}	Drain-to-Source Leakage Current
$R_{DS(on)}$	Static Drain-to-Source On-Resistance
P_{iron}	Motor iron losses
P_{joule}	Motor joule losses
P_{CMLI}	MLI conduction losses
P_{swMLI}	MLI switching losses
P_{CTLI}	TLI conduction losses
P_{swTLI}	TLI switching losses
δ	Duty cycle of the 3LI
f_{swMLI}	Switching frequency of the 3LI
f_{swTLI}	Switching frequency of the TLI

References

1. Levi, E.; Jones, M.; Vukosavic, S.N.; Toliyat, H.A. A novel concept of a multiphase, multimotor vector controlled drive system supplied from a single voltage source inverter. *IEEE Trans. Power Electron.* **2004**, *19*, 320–335. [CrossRef]
2. Jeftenic, B.; Bebic, M.; Statkic, S. Controlled multi-motor drives. In Proceedings of the International Symposium on Power Electronics, Electrical Drives, Automation and Motion SPEEDAM, Taormina, Italy, 23–26 May 2006; pp. 1392–1398.
3. Iyer, J.; Tabarraee, K.; Chiniforoosh, S.; Jatskevich, J. An improved V/F control scheme for symmetric load sharing of multi-machine induction motor drives. In Proceedings of the Canadian Conference on Electrical and Computer Engineering (CCECE), Niagara Falls, ON, Canada, 8–11 May 2011; pp. 1487–1490.

4. Pulvirenti, M.; Scarcella, G.; Scelba, G.; Cacciato, M.; Testa, A. FaultTolerant AC Multidrive System. *IEEE J. Emerg. Sel. Top. Power Electron.* **2014**, *2*, 224–235. [CrossRef]

5. Scelba, G.; Scarcella, G.; Pulvirenti, M.; Cacciato, M.; Testa, A.; de Caro, S.; Scimone, T. Current-Sharing Strategies for Fault-Tolerant AC Multidives. *IEEE Trans. Ind. Appl.* **2015**, *51*, 3943–3953. [CrossRef]

6. Slutej, A.; Kolonic, F.; Jakopovic, Z. The new crane motion control concept with integrated drive controller for engineered crane application. In Proceedings of the IEEE International Symposium on Industrial Electronics, ISIE, Bled, Slovenia, 12–16 July 1999; pp. 1458–1461.

7. Mitrovic, N.; Petronijevic, M.; Kostic, V.; Jeftenic, B. *Electrical Drives for Crane Application, Mechanical Engineering*; Gokcek, M., Ed.; InTech: London, UK, 2012; ISBN 978-953-51-0505-3.

8. Foti, S.; Testa, A.; de Caro, S.; Scimone, T.; Pulvirenti, M. Sensorless field oriented control of multiple-motors fed by multiple-converters systems. In Proceedings of the IEEE International Symposium on Sensorless Control for Electrical Drives (SLED), Catania, Italy, 18–19 September 2017; pp. 237–242.

9. Matsumoto, Y.; Ozaki, S.; Kawamura, A. A novel vector control of single-inverter multiple-induction-motors drives for Shinkansen traction system. In Proceedings of the 16th Annual Applied Power Electronics Conference and Exposition, (APEC 2001), Anaheim, CA, USA, 4–8 March 2001; pp. 608–614.

10. Peña-Eguiluz, R.; Pietrzak-David, M.; DeFornel, B. Comparison of several control strategies for parallel connected dual induction motors. In Proceedings of the VIII IEEE International Power Electronics Congress, 2002. Technical Proceedings. CIEP 2002, Guadalajara, Mexico, 24–24 October 2002.

11. Mabrouk, W.B.; Belhadj, J.; Pietrzak-David, M. Synthesis and Analysis of one Bogie Induction Railway Traction System. In Proceedings of the on 13th European Conference on Power Electronics and Applications, Barcelona, Spain, 8–10 September 2009; pp. 1–9.

12. Achour, T.; Pietrzak-David, M. Service Continuity of an IM Distributed Railway Traction with a Speed Sensor Fault. In Proceedings of the 14th European Conference on Power Electronics and Applications, Birmingham, UK, 30 August–1 September 2011; pp. 1–8.

13. Cui, S.; Han, S.; Chan, C. Overview of multi-machine drive systems for electric and hybrid electric vehicles. In Proceedings of the IEEE Conference and Expo Transportation Electrification Asia-Pacific (ITEC Asia-Pacific), Beijing, China, 31 August–3 September 2014; pp. 1–6.

14. Arnet, B.; Jufer, M. Torque control on electric vehicles with separate wheel drives. In Proceedings of the EPE Conference, Trondheim, Norway, 8–10 September 1997; pp. 659–664.

15. Du, Z.; Tolbert, L.M.; Chiasson, J.N. Active harmonic elimination for multilevel converters. *IEEE Trans. Power Electron.* **2006**, *21*, 459–469.

16. Tolbert, L.M.; Chiasson, J.N.; Du, Z.; McKenzie, K.J. Elimination of Harmonics in a Multilevel Converter with Non equal DC Sources. *IEEE Trans. Ind. Appl.* **2005**, *41*, 75–82. [CrossRef]

17. Fong, Y.C.; Cheng, K.W.E. An adaptive modulation scheme for fundamental frequency switched multilevel inverter with unbalanced and varying voltage sources. In Proceedings of the Power Electronics Systems and Applications (PESA), Hong Kong, China, 15–17 December 2016; pp. 1–6.

18. Edpuganti, A.; Rathore, A.K. Fundamental Switching Frequency Optimal Pulsewidth Modulation of Medium-Voltage Cascaded Seven-Level Inverter. *IEEE Trans. Ind. Appl.* **2015**, *51*, 3485–3492. [CrossRef]

19. Corzine, K.A.; Sudhoff, S.D.; Whitcomb, C.A. Performance characteristics of a cascaded two-level converter. *IEEE Trans. Energy Convers.* **1999**, *14*, 433–439. [CrossRef]

20. Li, L.; Czarkowski, D.; Liu, Y.; Pillay, P. Multi-level selective harmonic elimination PWM technique in serie-connected voltage inverters. *IEEE Trans. Ind. Appl.* **2000**, *36*, 160–170. [CrossRef]

21. Napoles, J.; Leon, J.I.; Portillo, R.; Franquelo, L.G.; Aguirre, M.A. Selective Harmonic Mitigation Technique for High Power Converters. *IEEE Trans. Ind. Electron.* **2009**, *57*, 2197–2206. [CrossRef]

22. Haitham, A.R.; Holtz, J.; Rodriguez, J.; Baoming, G. Medium Voltage Multilevel Converters-State of the Art, Challenges, and Requirements in Industrial Applications. *IEEE Trans. Ind. Electron.* **2010**, *57*, 2581–2596.

23. Lai, J.-S.; Peng, F.Z. Multi-level converters-A new breed of power converters. *IEEE Trans. Ind. Appl.* **1996**, *32*, 509–517.

24. Davari, P.; Yang, Y.; Zare, F.; Blaabjerg, F. A novel harmonic elimination approach in three-phase multi-motor drives. In Proceedings of the IEEE Energy Conversion Congress and Exposition (ECCE), Montreal, QC, Canada, 20–24 September 2015; pp. 7001–7008.

25. Sivakumar, K.; Das, A.; Ramchand, R.; Patel, C. A Hybrid Multilevel Inverter Topology for an Open-End Winding Induction Motor Drive Using Two-Level Inverters in Series with a Capacitor-Fed H-Bridge Cell. *IEEE Trans. Ind. Electron.* **2010**, *57*, 3707–3714. [CrossRef]

26. Boller, T.; Holtz, J.; Rathore, A.K. Optimal pulse width modulation of a dual 3LI system operated from a single dc link. *IEEE Trans. Ind. Appl.* **2012**, *48*, 1610–1615. [CrossRef]

27. Mondal, G.; Sivakumar, K.; Ramchand, R.; Gopakumar, K.; Levi, E. A Dual Seven-Level Inverter Supply for an Open-End Winding Induction Motor Drive. *IEEE Trans. Ind. Electron.* **2008**, *56*, 1665–1673. [CrossRef]

28. Somasekhar, T.; Gopakumar, K.; Baiju, M.R.; Mohapatra, K.K.; Umanand, L. A multilevel inverter system for an inductor motor with open-end windings. *IEEE Trans. Ind. Electron.* **2005**, *52*, 824–836. [CrossRef]

29. Edpuganti, A.; Rathore, A.K. New Optimal Pulse Width Modulation for Single DC-Link Dual-Inverter Fed Open-End Stator Winding Induction Motor Drive. *IEEE Trans. Power Electron.* **2014**, *30*, 4386–4393.

30. Foti, S.; Testa, A.; Scelba, G.; de Caro, S.; Cacciato, M.; Scarcella, G.; Bazzano, D.; Scimone, T. A new approach to improve the current harmonic content on open-end winding AC motors supplied by multi-level inverters. In Proceedings of the IEEE Energy Conversion Congress and Exposition (ECCE), Montreal, QC, Canada, 20–24 September 2015; pp. 4040–4047.

31. Foti, S.; Testa, A.; Scelba, G.; de Caro, S.; Cacciato, M.; Scarcella, G.; Scimone, T. An Open End Winding Motor Approach to mitigate the phase voltage distortion on Multi-Level Inverters. *IEEE Trans. Power Electron.* **2018**, *33*, 2404–2416. [CrossRef]

32. De Caro, S.; Foti, S.; Scimone, T.; Testa, A.; Cacciato, M.; Scarcella, G.; Scelba, G. THD and efficiency improvement in multi-level inverters through an open end winding configuration. In Proceedings of the 2016 IEEE Energy Conversion Congress and Exposition (ECCE), Milwaukee, WI, USA, 18–22 September 2016; pp. 1–7.

33. Foti, S.; Scelba, G.; Testa, A.; Sciacca, A. An Averaged-Value Model of an Asymmetrical Hybrid Multi-Level Rectifier. *Energies* **2019**, *12*, 589. [CrossRef]

34. Foti, S.; de Caro, S.; Scelba, G.; Scimone, T.; Testa, A.; Cacciato, M.; Scarcella, G. An Optimal Current Control Strategy for Asymmetrical Hybrid Multilevel Inverters. *IEEE Trans. Ind. Appl.* **2018**, *54*, 4425–4436. [CrossRef]

35. De Caro, S.; Foti, S.; Scimone, T.; Testa, A.; Cacciato, M.; Pulvirenti, M.; Russo, S. Over-voltage mitigation on SiC based motor drives through an open end winding configuration. In Proceedings of the 2017 IEEE Energy Conversion Congress and Exposition (ECCE), Cincinnati, OH, USA, 1–5 October 2017; pp. 4332–4337.

36. Foti, S.; Testa, A.; Scelba, G.; Cacciato, M.; de Caro, G.S.S.; Scimone, T. Overvoltage mitigation in open-end winding AC motor drives. In Proceedings of the 2015 IEEE ICRERA, Palermo, Italy, 22–25 November 2015; pp. 235–238.

37. Scelba, G.; Scarcella, G.; Foti, S.; Testa, A.; de Caro, S.; Scimone, T. An Open-end Winding approach to the design of multi-level multi-motor drives. In Proceedings of the IECON 2016—42nd Annual Conference of the IEEE Industrial Electronics Society, Florence, Italy, 23–26 October 2016; pp. 5026–5032.

38. Haque, R.U.; Toulabi, M.S.; Knight, A.M.; Salmon, J. Wide speed range operation of PMSM using an open winding and a dual inverter drive with a floating bridge. In Proceedings of the IEEE Energy Conversion Congress and Exposition (ECCE), Denver, CO, USA, 15–19 September 2013; pp. 3784–3791.

39. Cacciato, M.; Consoli, A.; Finocchiaro, L.; Testa, A. High frequency modeling of bearing currents and shaft voltage on electrical motors. In Proceedings of the Eighth International Conference on Electrical Machines and Systems, Nanjing, China, 27–29 September 2005; pp. 2065–2070.

40. Cacciato, M.; Cavallaro, C.; Scarcella, G.; Testa, A. Effects of connection cable length on conducted EMI in electric drives. In Proceedings of the IEEE International Electric Machines and Drives Conference, Seattle, WA, USA, 9–12 May 1999; pp. 428–430.

41. Cacciato, M.; De Caro, S.; Scarcella, G.; Scelba, G.; Testa, A. Improved space-vector modulation technique for common mode currents reduction. *IET Power Electron.* **2013**, *6*, 1248–1256. [CrossRef]

42. Foti, S.; Testa, A.; Scelba, G.; Sabatini, V.; Lidozzi, A.; Solero, L. Asymmetrical hybrid unidirectional T-type rectifier for high-speed gen-set applications. In Proceedings of the IEEE Energy Conversion Congress and Exposition (ECCE), Cincinnati, OH, USA, 1–5 October 2017; pp. 4887–4893.

43. Scelba, G.; Scarcella, G.; Foti, S.; De Caro, S.; Testa, A. Self-sensing control of open-end winding PMSMs fed by an asymmetrical hybrid multilevel inverter. In Proceedings of the IEEE International Symposium on Sensorless Control for Electrical Drives (SLED), Catania, Italy, 18–19 September 2017; pp. 165–172.

44. De Caro, S.; Foti, S.; Scelba, G.; Scimone, T.; Testa, A. A six-level asymmetrical Hybrid Photovoltaic Inverter with inner MPPT capability. In Proceedings of the 6th International Conference on Clean Electrical Power (ICCEP), Santa Margherita Ligure, Italy, 27–29 June 2017; pp. 631–635.

45. Dujic, D.; Jones, M.; Vukosavic, S.N.; Levi, E. A general PWM method for a (2n+1)-leg inverter supplying n three-phase machines. *IEEE Trans. Ind. Electron.* **2009**, *56*, 4107–4118. [CrossRef]

46. Rifkin, S. Adjustable-Voltage Hoist Drives for Cranes. *IEEE Electr. Eng.* **1956**, *75*, 620–623. [CrossRef]

47. Myles, A.H.; Davies, M.C.; Srnka, L.J. DC magnetic crane hoist control for AC powered cranes. *Trans. Am. Inst. Electr. Eng. Part I Commun. Electron.* **1960**, *79*, 207–211.

48. Ruxi, W.; Yue, W.; Qiang, D.; Yanhui, H.; Zhaoan, W. Study of control methodology for single inverter parallel connected dual induction motors based on the dynamic model. In Proceedings of the 2006 37th IEEE Power Electronics Specialists Conference, Jeju, Korea, 18–22 June 2017; pp. 1–7.

49. Matsuse, K.; Kouno, Y.; Kawai, H.; Yokomizo, S. A speed-sensorless vector control method of parallel-connected dual induction motor fed by a single inverter. *IEEE Trans. Ind. Appl.* **2002**, *38*, 1566–1571. [CrossRef]

50. Kelecy, P.M.; Lorenz, R.D. Control methodology for single inverter, parallel connected dual induction motor drives for electric vehicles. In Proceedings of the Power Electronics Specialists Conference, PESC '94 Record, 25th Annual IEEE, Taipei, Taiwan, 20–25 June 1994; Volume 2, pp. 987–991.

51. Dixit, A.; Mishra, N.; Sinha, S.K.; Singh, P. A review on different PWM techniques for five leg voltage source inverter. In Proceedings of the IEEE-International Conference On Advances In Engineering, Science And Management (ICAESM), Nagapattinam, Tamil Nadu, India, 30–31 March 2012; pp. 421–428.

52. Tabbache, A.; Rebaa, K.; Marouani, K.; Benbouzid, M.E.H.; Kheloui, A. Independent control of two induction motors fed by a five legs PWM inverter for electric vehicles. In Proceedings of the 2013 4th International Conference on Power Engineering, Energy and Electrical Drives, Istanbul, Turkey, 13–17 May 2013; pp. 629–634.

53. Pradabane, S. A New SVPWM for Dual-Inverter fed Three-level Induction Motor Drive. In Proceedings of the 2013 4th International Conference on Green Computing, Communication and Conservation of Energy, Chennai, India, 12–14 December 2014; pp. 514–518.

 *energies*

Article

On PWM Strategies and Current THD for Single- and Three-Phase Cascade H-Bridge Inverters with Non-Equal DC Sources

Zhansen Akhmetov [1], Manel Hammami [2], Gabriele Grandi [2,*] and Alex Ruderman [1]

[1] Department of Electrical and Computer Engineering, Nazarbayev University, 010000 Astana, Kazakhstan; zhansen.akhmetov@nu.edu.kz (Z.A.); alexander.ruderman@nu.edu.kz (A.R.)

[2] Department of Electrical, Electronic, and Information Engineering, Univ. of Bologna, 40136 Bologna, Italy; manel.hammami2@unibo.it

* Correspondence: gabriele.grandi@unibo.it; Tel.: +39-051-20-93571

Received: 19 December 2018; Accepted: 26 January 2019; Published: 30 January 2019

Abstract: Cascade H-bridge (CHB) inverter is an attractive choice for integration of DC sources of different nature, e.g., for distributed generation with energy storage, photovoltaic generation, etc. In general, non-equal DC voltage sources can affect the total harmonic distortion (THD) of the CHB by introducing undesirable low-frequency subharmonics. This paper investigates different level-shifted (LS) and phase-shifted (PS) pulse width modulation (PWM) strategies for single- and three-phase cascade H-bridge inverters with non-equal DC sources from the load current THD minimization perspective. The best current quality is provided by LS PWM, as reported in the literature. The paper provides a simple time domain explanation of LS PWM superiority. However, PS PWM may be a preferable choice for practical applications due to fair power and loss sharing across individual H-bridges. The paper explains how to obtain the best current quality by PS PWM carriers' order arrangement (DC sources switching sequence selection).

Keywords: multilevel inverters; total harmonic distortion; level-shifted PWM; phase-shifted PWM

---

1. Introduction

Nowadays, multilevel inverters (MLIs) are widely used, since they offer improved output waveforms, smaller grid filter size, lower total harmonic distortion (THD), and reduced electromagnetic interference (EMI), compared to their two-level inverter counterparts [1–6]. In particular, MLIs are particularly suitable for medium- and high-voltage applications for both single- and three-phase systems, thanks to the possibility to work with high-voltage levels by adopting low-voltage-rated devices.

The basic multilevel converter topologies are cascade H-bridge (CHB), neutral-point-clamped (NPC), flying capacitor (FC), and modular multilevel converter (MMC) [5,6]. These kinds of converters are also adopted in photovoltaic applications due to the aforementioned advantages.

Many recently published papers deal with the estimation of voltage and current THD in multilevel inverters. In most of the cases, they are based on voltage frequency spectra numerical calculations/measurements (fast Fourier transform, FFT). The scientific community has shown a significant interest in voltage and current THD analyses for both multilevel pulse width modulation (PWM) and staircase modulation over the past years. Analytical solutions for the voltage THD of multilevel PWM single- and three-phase inverters were obtained in [7] in asymptotic approximation (high switching-to-fundamental frequency ratio). Experimental tests have been carried out in [8] to verify the calculation and analytical developments of voltage and current THDs in the case of three-, five-, and seven-level single-phase PWM inverters.

Cascade H-Bridge converter is a mature multilevel topology adopted for different applications with multiple DC sources [1–4].

A single-phase CHB inverter topology and related phase-shifted (PS) and level-shifted (LS) PWM are demonstrated in Figure 1. For LS PWM, apparent switching frequency equals the carrier one. For PS PWM, the apparent switching frequency is that of an individual H-bridge (double carrier frequency) times the number of H-bridges [9].

Current THD theoretical calculations for a single-phase multilevel PWM inverter with uniform voltage level distribution and inductance-dominated RL-load is addressed in [8]. The approach of [8] is applicable to a CHB inverter with equal DC sources.

With respect to a CHB inverter having equal DC sources, a CHB inverter with non-equal DC voltages can affect current quality by introducing undesirable subharmonic content that may violate the grid-codes [10].

This paper considers LS and PS PWM applied to single- and three-phase CHB inverters with non-equal DC sources in the context of current THD minimization. Though a converter RL-load is considered, the results are applicable to grid-connected applications as well [8].

The paper starts with a generalization of asymptotic current THD formulas for a single-phase multilevel inverter with uniform levels [8] for non-uniform voltage level distribution. The results are applicable to a single-phase CHB inverter with LS PWM and non-equal DC sources. In addition, minimal current THD requires matching LS PWM bands to DC source voltages that may be difficult to implement in real-life applications. For a single-phase CHB inverter with PS PWM and non-equal DC sources, the optimal current THD for more than three sources is achieved by a proper selection of carrier order (DC sources switching sequence), as recently shown in [11,12].

For a three-phase CHB inverter, theoretical current THD consideration becomes too complicated. Therefore, current THD is analyzed by MATLAB–Simulink simulations.

The major contributions of this paper may be formulated as follows. Theoretical current THD calculation methodology for non-uniform voltage levels allows evaluating current THD for a single-phase CHB inverter with non-equal DC sources and LS PWM. Though it is difficult to implement LS PWM in practice (PWM band matching to DC source voltages is required), the calculated current THD gives a theoretical limit.

For a three-phase CHB inverter with more than three DC sources per phase, the single-phase current THD optimal DC source switching sequences reported in [11,12] also work well. Swapping the carriers among the different phases (changing the carriers order) could be considered as an additional degree of freedom that may be used to improve the current THD.

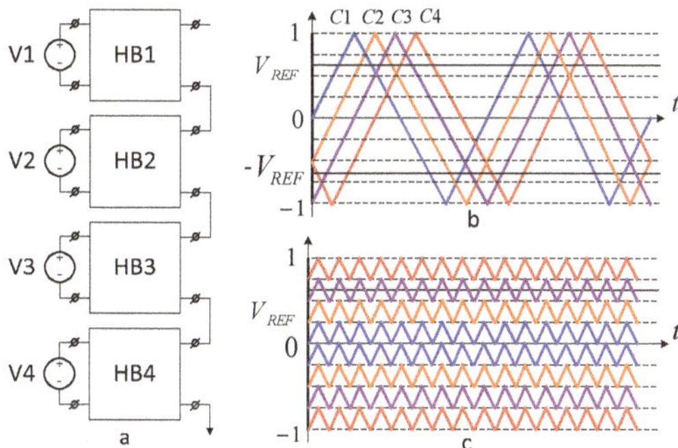

Figure 1. Single-phase cascade H-bridge (CHB) inverter: (**a**) topology example for four H-bridges; (**b**) phase-shifted (PS) pulse width modulation (PWM) carriers; (**c**) level-shifted (LS) PWM carrier arrangement.

The paper is organized as follows. Section 2 presents the current THD calculation methodology for non-uniform voltage levels applicable to a single-phase cascade inverter in the case of LS PWM and non-equal DC sources. The analysis of the current THD with PS PWM and non-equal DC sources for the cascade H-bridge inverter is presented in Section 3. The analysis of current THD has been extended to a three-phase cascade H-bridge inverter with non-equal DC sources and presented in Section 4. Section 5 presents the conclusion.

2. Current THD for a Single-Phase CHB Inverter with LS PWM and Non-Equal DC Sources

Current THD for a single-phase CHB inverter with LS PWM and equal DC sources can be calculated as suggested in [8] for uniformly distributed voltage levels. In this section, the results of [8] are generalized to acquire non-equal DC sources. For the best harmonic performance (minimal current THD), LS modulation bands must be adjusted to match non-equal DC source voltages (Figure 2).

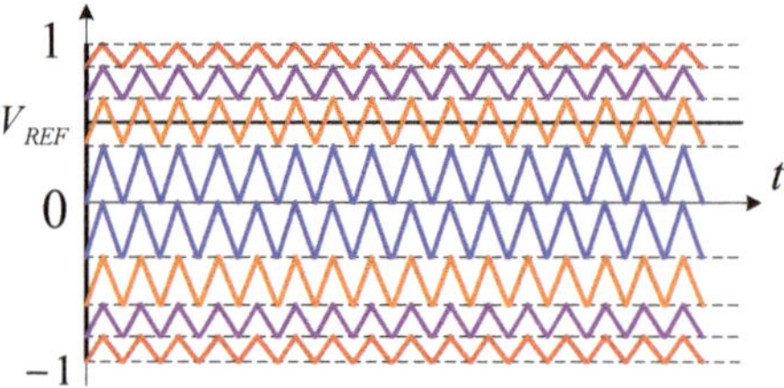

Figure 2. Four H-bridge CHB inverter LS PWM bands adjusted to match non-equal DC sources V1 > V2 > V3 > V4.

Current THD for inductance-dominated RL-load is calculated, similar to [8], as

$$THD_n(m),\% = \frac{2\pi\sqrt{2NMS^{AC}_{nNE}(m)}}{m} \cdot \frac{f_f}{f_{AS}} \cdot \sqrt{1 + \left(\frac{R}{2\pi f_f L}\right)^2} \times 100, \tag{1}$$

where L and R are load inductance and resistance; f_f and f_{AS} – fundamental frequency and apparent switching frequency; n-level count (the number of H-bridges increased by one).

For n-level CHB inverter with $(n-1)$ H-bridges with non-equal sources and LS PWM band matching (Figure 2), the current ripple normalized mean square (NMS) is found using the general formula

$$NMS^{AC}_{nNE}(m) = \frac{1}{6\pi}\left(\sum_{i=0}^{n-3}\frac{c_i}{a_{i+1}^2 - a_i^2} + \frac{b_{n-2}}{1 - a_{n-2}^2}\right), \tag{2}$$

where a_i, $0 < a_i < 1$, is "voltage level", i, $0 \le i \le n-1$, is the ratio of respective sum of DC source voltages to the total DC voltage ($a_0 = 0$, $a_{n-1} = 1$)

$$a_i = \frac{\sum_{j=1}^{i} V_j}{V_{total}}, \tag{3}$$

where $V_{total} = \sum_{j=1}^{n-1} V_j$ is the total DC voltage and m is the modulation index ($0 < m < 1$);

$$b_{n-2} = q_{n-2}\text{asin}\left(\frac{a_{n-2}}{m}\right) + d_{n-2}\left(1 - \frac{a_{n-2}^2}{m^2}\right)^{\frac{1}{2}} + m^2\left(\frac{\pi a_{n-2}^2}{4} + \pi a_{n-2} + \frac{\pi}{4}\right) + \frac{3\pi m^4}{16} + \frac{\pi a_{n-2}^2}{2}, \tag{4}$$

$$c_i = q_i\left(\text{asin}\left(\frac{a_i}{m}\right) - \text{asin}\left(\frac{a_{i+1}}{m}\right)\right) + d_i\left(1 - \frac{a_i^2}{m^2}\right)^{\frac{1}{2}} + h_i\left(1 - \frac{a_{i+1}^2}{m^2}\right)^{\frac{1}{2}}. \tag{5}$$

In Formulas (4) and (5),

$$q_i = -\frac{3}{8}m^4 - \frac{m^2}{2}\left(a_i^2 + 4a_i a_{i+1} + a_{i+1}^2\right) - a_i^2 a_{i+1}^2; \tag{6}$$

$$d_i = -m^3\left(\frac{23}{24}a_i + \frac{4}{3}a_{i+1}\right) + m\left(\frac{a_i^3}{12} - 3a_i\frac{a_{i+1}^2}{2} - 2a_i^2\frac{a_{i+1}}{3}\right); \tag{7}$$

$$h_i = m^3\left(\frac{4}{3}a_i + \frac{23}{24}a_{i+1}\right) + m\left(-\frac{a_{i+1}^3}{12} + 2a_i\frac{a_{i+1}^2}{3} + 3a_i^2\frac{a_{i+1}}{2}\right). \tag{8}$$

The current ripple NMS formula for two H-bridges (3-level)

$$NMS_{3NE}^{AC}(m) = \begin{cases} \dfrac{-\left(-\frac{3}{8}m^4 - \frac{m^2}{2}a_1^2\right)\operatorname{asin}\left(\frac{a_1}{m}\right) - \frac{4}{3}a_1 m^3 + \left(\frac{23}{24}a_1 m^3 - \frac{a_1^3}{12}m\right)\left(1 - \frac{a_1^2}{m^2}\right)^{\frac{1}{2}}}{6\pi a_1^2} \\[2em] + \dfrac{\left(-\frac{3}{8}m^4 - \frac{m^2}{2}\left(a_1^2 + 4a_1 a_2 + a_2^2\right) - a_1^2 a_2^2\right)\operatorname{asin}\left(\frac{a_1}{m}\right) + m^2\left(\frac{\pi a_1^2}{4} + \pi a_1 + \frac{\pi}{4}\right)}{6\pi\left(1 - a_1^2\right)} \\[2em] + \dfrac{\frac{3\pi m^4}{16} + \frac{\pi a_1^2}{2} + \left(-m^3\left(\frac{23}{24}a_1 + \frac{4}{3}a_2\right) + m\left(\frac{a_1^3}{12} - 3a_1\frac{a_2^2}{2} - 2a_1^2\frac{a_2}{3}\right)\right)\left(1 - \frac{a_1^2}{m^2}\right)^{\frac{1}{2}}}{6\pi\left(1 - a_1^2\right)}. \end{cases} \tag{9}$$

For a three H-bridge (4-level) CHB with non-equal DC source voltages, $V_1 \neq V_2 \neq V_3$, output voltage levels become $a_0 = 0$; $a_1 = \frac{V_1}{V_1 + V_2 + V_3}$; $a_2 = \frac{V_1 + V_2}{V_1 + V_2 + V_3}$; $a_3 = \frac{V_1 + V_2 + V_3}{V_1 + V_2 + V_3} = 1$.
Current ripple NMS for non-equal sources

$$NMS_{4NE}^{AC}(m) = \begin{cases} \dfrac{-q_0\operatorname{asin}\left(\frac{a_1}{m}\right) + d_0 + h_0\left(1 - \frac{u_1^?}{m^2}\right)^{\frac{1}{2}}}{a_1^2 * 6 * \pi} + \\[2em] + \dfrac{q_1\left(\operatorname{asin}\left(\frac{a_1}{m}\right) - \operatorname{asin}\left(\frac{a_2}{m}\right)\right) + d_1\left(1 - \frac{a_1^2}{m^2}\right)^{\frac{1}{2}} + h_1\left(1 - \frac{a_2^2}{m^2}\right)^{\frac{1}{2}}}{\left(a_2^2 - a_1^2\right) * 6 * \pi} + \\[2em] \dfrac{q_2\operatorname{asin}\left(\frac{a_2}{m}\right) + d_2\left(1 - \frac{a_2^2}{m^2}\right)^{\frac{1}{2}} + m^2\left(\pi\frac{a_2^2}{4} + \pi a_2 + \frac{\pi}{4}\right) + \frac{3\pi m^4}{16} + \frac{\pi a_2^2}{2}}{\left(1 - a_2^2\right) * 6 * \pi}, \end{cases} \tag{10}$$

where coefficients q_0, d_0, h_0, q_1, d_1 are calculated using Formulas (6)–(8). For example, for a four H-bridge (5-level) CHB inverter current ripple NMS for non-equal sources from (2),

$$NMS_{5NE}^{AC}(m) = \begin{cases} \dfrac{-q_0*\operatorname{asin}\left(\frac{a_1}{m}\right) + d_0 + h_0*\left(1 - \frac{a_1^2}{m^2}\right)^{\frac{1}{2}}}{a_1^2 * 6 * \pi} + \\[2em] \dfrac{q_1\left(\operatorname{asin}\left(\frac{a_1}{m}\right) - \operatorname{asin}\left(\frac{a_2}{m}\right)\right) + d_1\left(1 - \frac{a_1^2}{m^2}\right)^{\frac{1}{2}} + h_1*\left(1 - \frac{a_2^2}{m^2}\right)^{\frac{1}{2}}}{\left(a_2^2 - a_1^2\right) * 6 * \pi} + \\[2em] \dfrac{q_2\left(\operatorname{asin}\left(\frac{a_2}{m}\right) - \operatorname{asin}\left(\frac{a_3}{m}\right)\right) + d_2\left(1 - \frac{a_2^2}{m^2}\right)^{\frac{1}{2}} + h_2*\left(1 - \frac{a_3^2}{m^2}\right)^{\frac{1}{2}}}{\left(a_3^2 - a_2^2\right) * 6 * \pi} + \\[2em] \dfrac{q_3*\operatorname{asin}\left(\frac{a_3}{m}\right) + d_3*\left(1 - \frac{a_3^2}{m^2}\right)^{\frac{1}{2}} + m^2\left(\pi\frac{a_3^2}{4} + \pi a_3 + \frac{\pi}{4}\right) + \frac{3\pi m^4}{16} + \frac{\pi a_3^2}{2}}{\left(1 - a_3^2\right) * 6 * \pi}. \end{cases} \tag{11}$$

Theoretical current THD results obtained using (1), (2) are in good agreement with simulated ones. Calculations and simulations across this paper are carried out for the following parameters: load resistance $R = 1\,\Omega$, load inductance $L = 1$ mH, fundamental frequency $f_f = 50$ Hz, apparent switching frequency $f_{AS} = 4000$ Hz and nominal DC source voltage $V = 100$ V.

Figure 3a,b present theoretical current THD according to (1), (2), (11) for a single-phase 4-bridge CHB for three cases: non-equal sources arranged in ascending order ($a_1 = 0.2$; $a_2 = 0.45$; $a_3 = 0.7$); non-equal sources arranged in descending order ($a_1 = 0.3$; $a_2 = 0.55$; $a_3 = 0.8$); uniform voltage levels (equal sources, $a_1 = 0.25$; $a_2 = 0.5$; $a_3 = 0.75$).

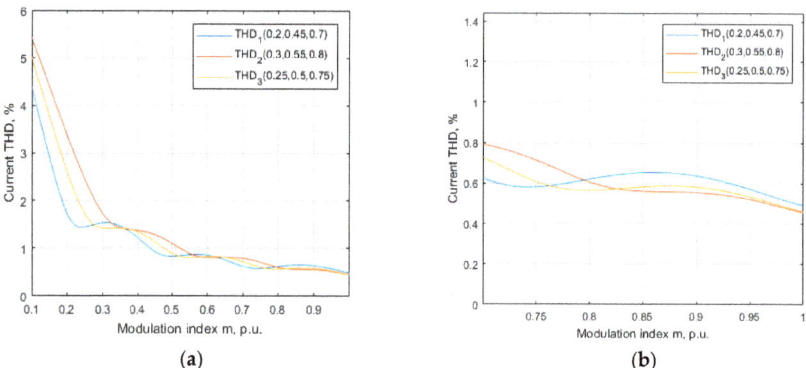

Figure 3. Total harmonic distortion (THD) for LS PWM: (**a**) $0.1 < m < 1$; (**b**) $0.7 < m < 1$.

For relatively large modulation indices $m > 0.8$, that is typical for grid-tied applications, current THDs for all three cases are close to each other. The lowest THD is achieved for non-equal sources arranged in descending order ($V_1 > V_2 > V_3 > V_4$).

For relatively small modulation indices $m < 0.3$, the lowest THD is achieved for the non-equal sources arranged in ascending order ($V_1 < V_2 < V_3 < V_4$).

Figures 4–6 show current THD simulation results for $m = 0.9$ that are in good agreement with theoretical ones (Figure 3). Theoretical current THD values are typically slightly lower than those obtained from simulation due to theoretical assumptions.

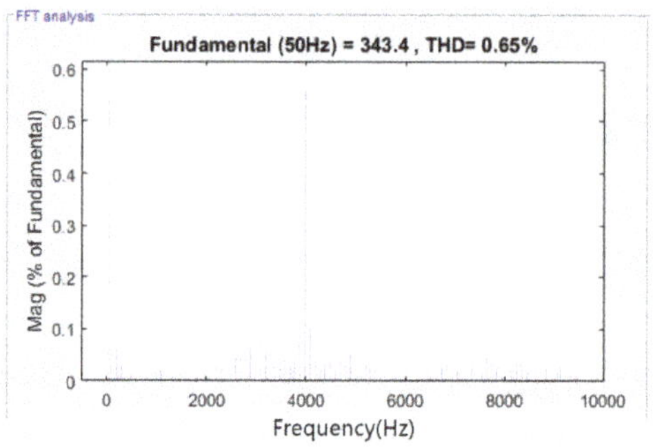

Figure 4. THD for LS PWM $a_1 = 0.2, a_2 = 0.45, a_3 = 0.7$, and $m = 0.9$.

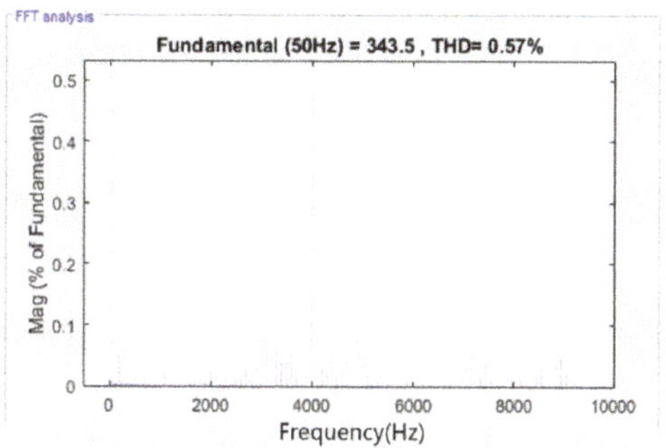

Figure 5. THD for LS PWM $a_1 = 0.3$, $a_2 = 0.55$, $a_3 = 0.8$, and $m = 0.9$.

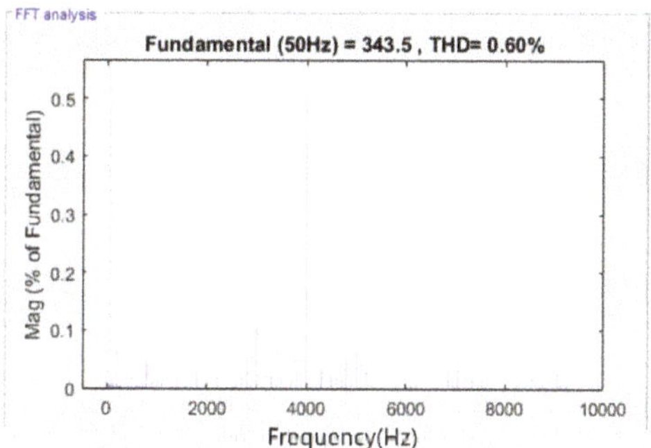

Figure 6. THD for LS PWM $a_1 = 0.25$, $a_2 = 0.5$, $a_3 = 0.75$, and $m = 0.9$.

The LS PWM considered in this section provides better current THD compared with PS PWM for single-phase and especially for three-phase CHB inverters. However, the known drawback of LS PWM is an uneven power and loss distribution across different H-bridges. For equal DC sources, the remedy may be H-bridge rotation. For non-equal DC sources, the best THD performance requires LS PWM band adjustment in accordance with true DC source voltage magnitudes (Figure 2) that may become problematic. If there is no match between non-equal DC sources and LS PWM bands, current THD is compromised, and the linearity of CHB inverter as reference signal "voltage amplifier" is violated.

3. Current THD for a Single-Phase CHB Inverter with PS PWM and Non-Equal DC Sources

For a single-phase CHB inverter with PS PWM and equal DC sources (uniform levels), current THD for inductance-dominated RL-load may be calculated according to [8]. Current THD theoretical formulas are based on asymptotic assumption, meaning that the apparent PWM frequency is much larger than the fundamental AC one.

In theory, the same formula for current THD for uniform voltage levels is applicable to both LS and PS PWM. In real life, current THD for PS PWM may be larger than that for LS PWM, that is better predicted by asymptotic formulas.

For LS PWM, voltage and current spectra show distinct apparent switching frequency (Figures 4–6). For PS PWM, the spectrum around apparent switching frequency is spread. This happens because, at reference level crossings, PWM voltage for PS PWM is shifted by half a PWM period. This effect has almost no impact on voltage THD for uniform levels—for a single-phase CHB inverter with equal sources, voltage THD is practically the same for both LS and PS PWM, and this is because voltage ripple mean square on respective PWM periods is practically the same [7].

The effect of PWM voltage half a period shift at reference level crossings, in fact, presents a low-frequency disturbance that may have a negative impact on current THD. This is because current ripple can be considered as voltage ripple integral for an inductance-dominated load [8], and it is deteriorated by the low-frequency "voltage irregularities". On the one hand, this adverse effect increases with level count increase (more reference voltage crossings). On the other hand, it reduces with apparent switching frequency increase (more switching between adjacent levels) and RL-load time constant increase (better low-frequency filtering).

Figure 7 shows current THD for a four H-bridge CHB inverter with equal sources and PS PWM. Compared with LS PWM for the same modulation index $m = 0.9$ (Figure 6), the switching frequency spectrum is spread, and the THD value is slightly larger.

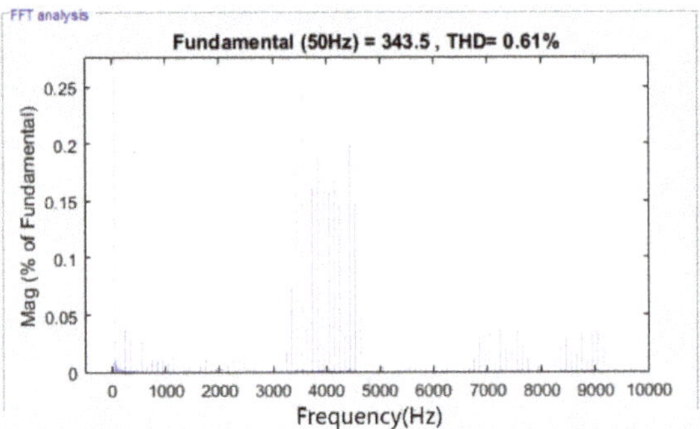

Figure 7. Four H-bridge CHB inverter current THD for PS PWM and equal sources.

In the recent papers [11] and [12], asymptotic formulas of [8] are generalized for a single-phase CHB inverter with PS PWM and non-equal DC sources. Moreover, it is shown that for more than three H-bridges, there are extreme DC source (H-bridge) switching sequences (carrier orders in Figure 1b that minimize (maximize) current THD.

Suppose DC source voltages are sorted in the ascending order

$$V_1 < V_2 < V_3 < \cdots, \tag{12}$$

and DC source switching sequences on a PWM period are denoted by number sequences like 1234 (Figure 8). In the case of three CHB cells, there are six different sequences in total, and only one sequence could be considered with respect to the THD, which is the 123. In fact, the sequences 123, 231, and 312 have the same THD due to circular permutation equivalence. Similarly, 321, 132, and 213 have the same THD due to the reversal circular permutation. In the following, "1" is always placed first to eliminate circular permutation redundancy.

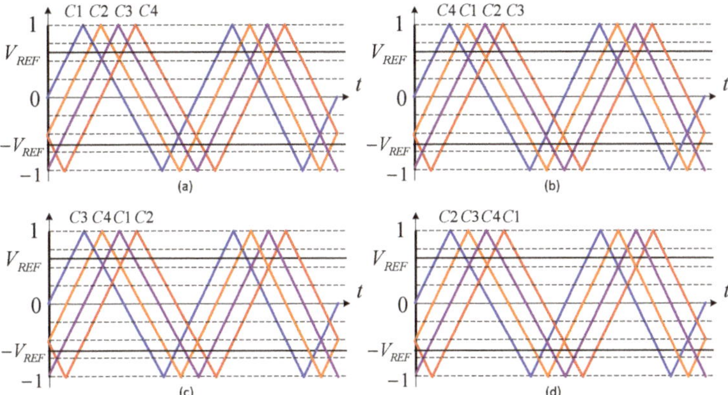

Figure 8. Four H-bridge CHB inverter switching sequences: (**a**) 1234, (**b**) 4123, (**c**) 3412 and (**d**) 2341 are equivalent under asymptotic assumption by circular permutation.

In the case of four CHB cells, only three different sequences can be considered which are 1234, 1324, and 1243. The other sequences are equivalent within circular permutation (Figure 8) and order reversal. The number of all possible DC source switching sequences for the different numbers of H-bridge cells are summarized in Table 1. It can be noticed that in the case of four and more CHB cells, the THD with PS PWM is strictly dependent on the DC sources switching sequences.

Table 1. Number of Different DC Source Switching Sequences.

Cells	4	5	6	7	8	9	10
Sequences	3	12	60	360	2520	20160	181440

As shown in [11,12], for four H-bridges, the best and worst sequences from the current THD perspective become 1423 and 1243. The switching sequences for the different numbers of H-bridge cells are given in Table 2 with respect to the best and the worst cases.

Table 2. Extreme DC Source Switching Sequences.

Cells	5	6	7	8	9
Worst	12453	124653	1246753	12468753	124689753
Best	15234	162435	1725436	18264537	192745638

Figures 9 and 10 present current THD for a single-phase CHB inverter with PS PWM for the best and worst switching sequences, respectively. Normalized non-equal voltages amount to $v_1 = 0.8$; $v_2 = 0.933$; $v_3 = 1.067$; $v_4 = 1.2$ p.u. (1.0 p.u. corresponds to 100 V). It is clearly seen that the best source switching sequence, 1423, provides the minimal current THD of 0.77% (1.04% for the worst sequence). However, the minimal THD of this PS PWM is worse than 0.6% for equal DC sources (for LS PWM with optimal bands adjustment, it is even better—0.57%).

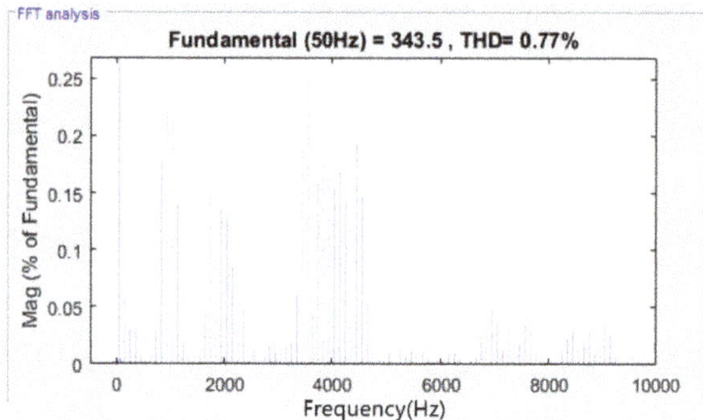

Figure 9. Four H-bridge CHB THD for PS PWM and non-equal sources best switching sequence 1423.

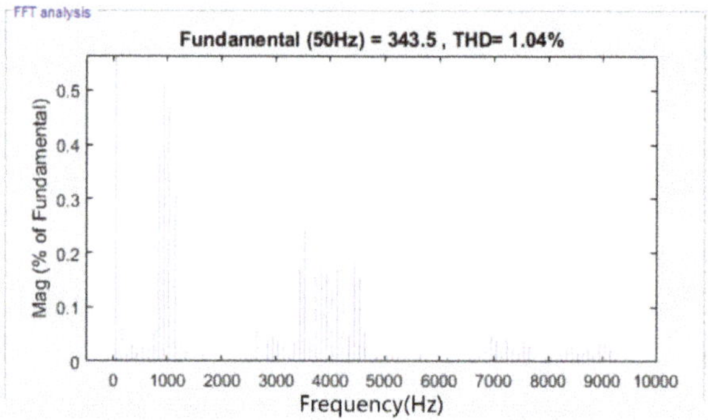

Figure 10. Four H-bridge CHB THD for PS PWM and non-equal sources worst switching sequence 1243.

Current THD was also calculated for a single-phase CHB inverter with six H-bridges and PS PWM for the best and worst switching sequences, respectively. Normalized non-equal voltages were selected as v_1 = 0.80; v_2 = 0.88; v_3 = 0.96; v_4 = 1.04; v_5 = 1.12; v_6 = 1.20 p.u. Current THD for the best switching sequence 162435 was found as 0.47%, and the same for the worst sequence—1.29% (compared with 0.4% for equal DC sources).

Current THD values for different cases of single-phase four H-Bridge CHB inverter PWM for m = 0.9, R = 1 Ω, L = 1 mH are compared in Table 3.

Table 3. Current THD for a Single-Phase Four H-Bridge CHB.

	LS PWM, Equal Sources	PS PWM, Equal Sources	LS PWM, Non-Equal Sources	PS PWM, Non-Equal Sources, Best	PS PWM, Non-Equal Sources, Worst
Calculated Current THD, %	0.59	0.59	0.56	0.76	1.03
Simulated Current THD, %	0.60	0.61	0.57	0.77	1.04

In general, there was a good agreement between theoretically calculated and simulated values. For uniform levels, simulated current THD for PS PWM was slightly larger than for LS PWM due to the unaccounted effect of PS PWM voltage half a period shift at reference level crossings (Section 2).

Current THD values for different cases of single-phase six H-Bridge CHB inverter PWM for m = 0.9, R = 1 Ω, L = 1 mH are compared in Table 4.

Table 4. Current THD for a Single-Phase Six H-Bridge CHB.

	LS PWM, Equal Sources	PS PWM, Equal Sources	LS PWM, Non-Equal Sources	PS PWM, Non-Equal Sources, Best	PS PWM, Non-Equal Sources, Worst
Calculated Current THD, %	0.25	0.25	0.24	0.29	0.82
Simulated Current THD, %	0.25	0.26	0.24	0.31	0.85

For Table 4, again, there is a good agreement between theoretically calculated and simulated current THD values. The relative difference between calculated and simulated current THD values for PS PWM is larger compared to the four H-bridge inverter because there are more unaccounted low-frequency disturbances due to PWM voltage half a period shift at reference level crossings.

4. Current THD for a Three-Phase CHB Inverter with Non-Equal DC Sources

Voltage and current quality for three-phase CHB inverters with equal DC sources for LS PWM are reported in the literature to be better than for PS PWM [13–15]. The presented frequency domain explanations seem to be complicated as they involve double Fourier series spectra calculations, sideband frequencies, etc. Presented below is an elementary time domain explanation of LS PWM superiority for three-phase converters.

The indication of modulation strategy quality for a three-phase converter is line voltage quality. For both LS and PS PWM, generated source phase voltage is of optimal nearest level switching quality. However, while for LS PWM line (phase-to-phase) voltage is still of nearest level switching type, for PS PWM, it becomes non-nearest switching that deteriorates voltage and current quality.

An elementary time domain explanation of this phenomenon is based on the effect of PS PWM voltage half a period shift at reference level crossings, as discussed in Section 2. It is about synchronization of source voltage waveforms of different phases because a line voltage is obtained as a difference between two phase voltages.

Figure 11 demonstrates the synchronization of PWM voltages of different phases for LS PWM (no matter what the specific voltage levels are). Figure 11a shows that for the pulses of the same polarity, LS PWM provides middle pulses synchronization. For the pulses of opposite polarities (Figure 11b), a middle of the pulse (peak) in one phase is synchronized with a middle of the pause (valley) in another phase. As line voltage V_{ab} is obtained by subtracting phase voltage V_b from V_a, it is the optimal nearest level switching type. Also note the line voltage frequency doubling effect.

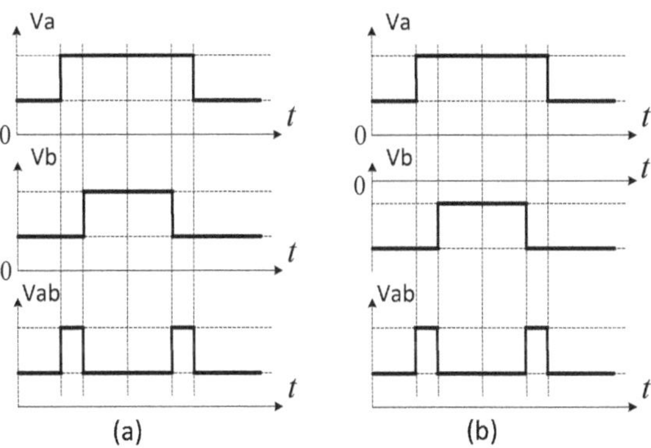

Figure 11. Different PWM phase voltages synchronization for LS PWM: (**a**) voltage pulses of the same polarity; (**b**) voltage pulses of opposite polarities.

For PS PWM, due to the effect of voltage half a PWM period shift at reference level crossings, for some portions of a fundamental period different phase, PWM voltage synchronization may be as in Figure 12 (Figure 12b is obtained from Figure 11b by half a PWM period shift). This kind of synchronization results in reduced line voltage quality due to non-nearest level switching.

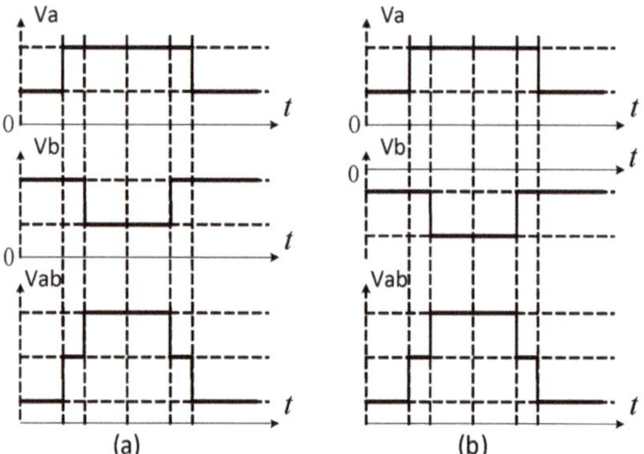

Figure 12. Different PWM phase voltages possible synchronization for PS PWM: (**a**) voltage pulses of the same polarity; (**b**) voltage pulses of opposite polarities.

Theoretical asymptotic time domain analysis of a three-phase CHB inverter voltage and current quality becomes a complicated task. Therefore, current THD values in this section were obtained by simulation for inverter Y-connected balanced RL-load and relatively large modulation index $m = 0.85$ that is characteristic for grid-tied applications. Voltage references were selected pure sinusoidal without any zero-sequence insertion.

For LS PWM and equal DC sources, line voltage and current THD are shown in Figures 13 and 14; the same for PS PWM—in Figures 15 and 16. It is clearly seen that for PS PWM, for some parts of the fundamental period, the line voltage is of the non-nearest switching type that results in current THD increase from 0.23% (in case of LS PWM) to 0.44%.

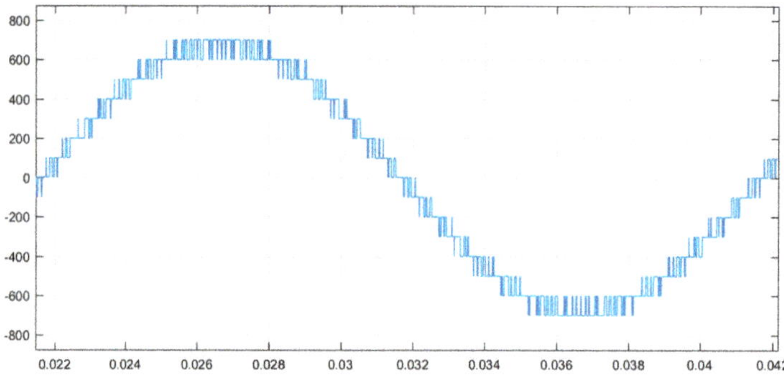

Figure 13. Three-phase four H-bridge CHB inverter line voltage for LS PWM and equal sources.

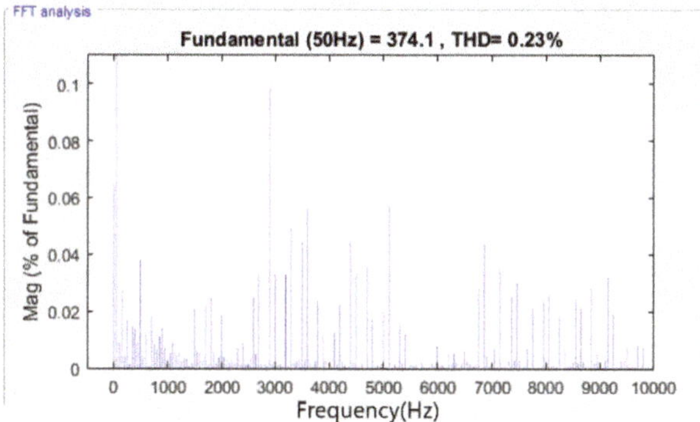

Figure 14. Three-phase four H-bridge CHB inverter current THD for LS PWM and equal sources.

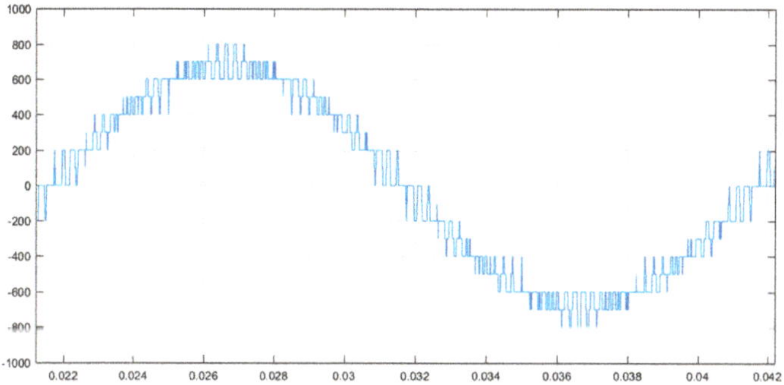

Figure 15. Three-phase four H-bridge CHB line voltage for PS PWM and equal sources.

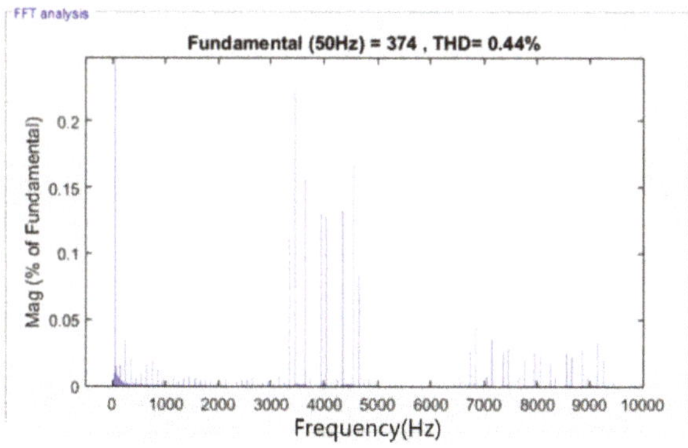

Figure 16. Three-phase four H-bridge CHB current THD for PS PWM and equal sources.

Next, consider non-equal DC sources. In this paper, for the sake of simplicity, we assume the same non-equal sources in all three phases. Specifically, for a four H-bridge CHB inverter, normalized voltages $v_1 = 0.8$; $v_2 = 0.933$; $v_3 = 1.067$; $v_4 = 1.2$ p.u. (1.0 p.u. corresponds to 100 V).

For the four H-bridge CHB inverter with LS PWM and phase DC sources arranged in the optimal descending order, line voltage and current THD are given in Figures 17 and 18. Again, this excellent current quality is achieved for ideal matching of LS PWM bands to DC sources that may be difficult to implement in practice.

The results for PS PWM for the best phase sources switching sequence are given in Figures 19 and 20, and for the worst switching one, in Figures 21 and 22.

For PS PWM, there is another degree of freedom which is carriers' synchronization in different phases. The results of Figures 19–22 were obtained assuming the same carriers for all three phases. For example, for the best sequence, the carrier sequences in three phases are (1423; 1423; 1423). However, the carriers in different phases can be shifted using circular permutation without violating phase switching sequence optimality, e.g., (1423; 3142; 2314).

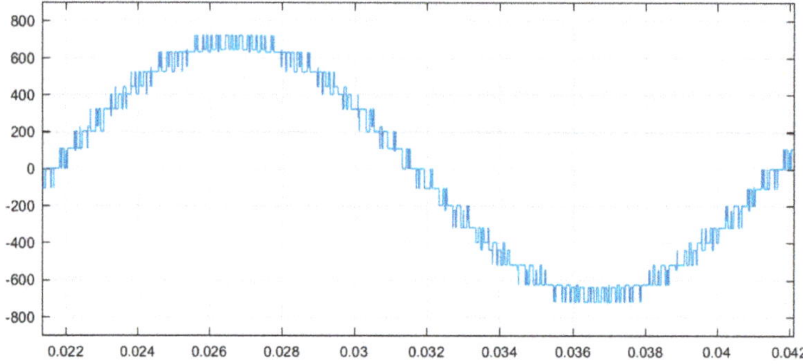

Figure 17. Three-phase four H-bridge CHB line voltage for LS PWM and non-equal sources.

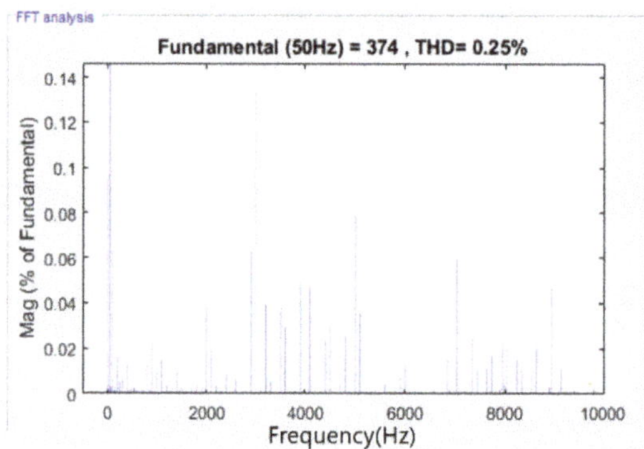

Figure 18. Three-phase four H-bridge CHB current THD for PS PWM and non-equal sources.

If the number of sources in CHB inverter phase is not a multiple of 3, then such carriers' manipulation will always make voltages and currents non-symmetric, and current THDs in different phases will be unequal. The authors have an example showing that current THD in each phase may become less than the one obtained when using the same carrier sequence in three phases.

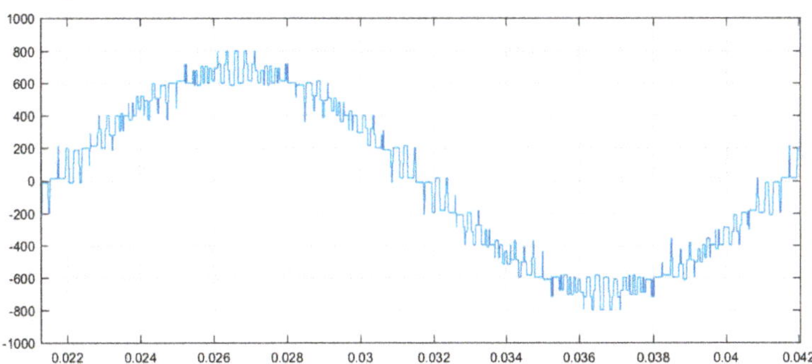

Figure 19. Three-phase four H-bridge CHB line voltage for PS PWM and best switching of non-equal sources.

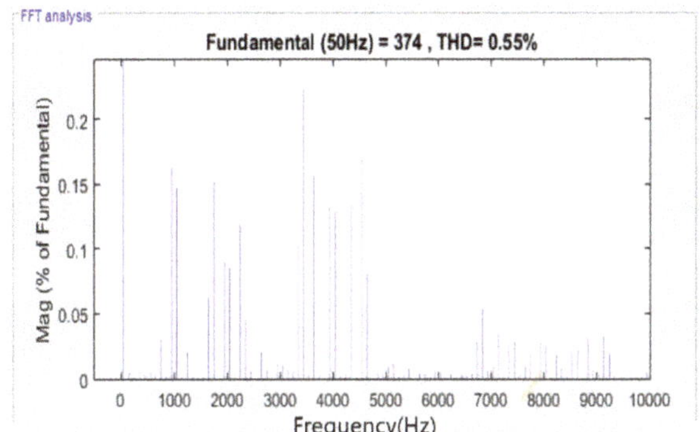

Figure 20. Three-phase four H-bridge CHB current THD for PS PWM and best switching of non-equal sources.

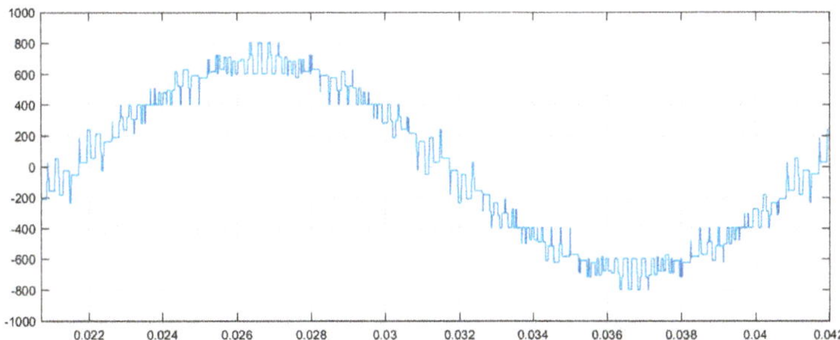

Figure 21. Three-phase four H-bridge CHB line voltage for PS PWM and worst switching of non-equal sources.

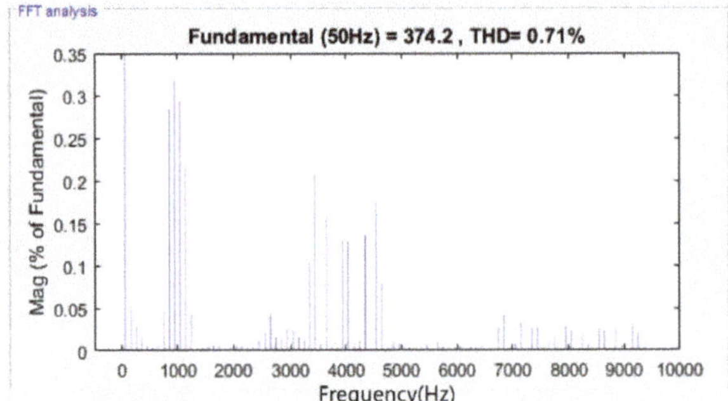

Figure 22. Three-phase four H-bridge CHB current THD for PS PWM and worst switching of non-equal sources.

5. Conclusions

This paper investigated several aspects of carrier-based PWM strategies of single- and three-phase CHB inverters with non-equal DC source voltages in the context of current THD minimization. Though the results are demonstrated for inductance-dominated RL-load, they are applicable to grid-tied applications as well.

Here are the main outcomes:

1   Asymptotic current THD formulas for a single-phase multilevel inverter with uniform voltage level distribution [8] were generalized to acquire non-uniform voltage levels. The generalized formulas are applicable to a single-phase CHB inverter with non-equal DC source voltages and LS PWM with PWM bands matching voltage values. While it may be difficult to implement LS PWM bands matching in real-life applications, obtained current THD values are theoretically minimal, and can be only compromised by PS PWM.

2   Provided is a simple time domain explanation of superiority of LS PWM over PS PWM from a voltage and current THD perspective. It is based on recognition of the effect of PS PWM voltage half a period shift at reference level crossings. While this effect has a minor impact for single-phase CHB inverters with PS PWM, it causes a significant deterioration of voltage and current THD for three-phase CHB inverters with PS PWM (non-nearest level switching) because source phase voltage synchronization becomes different from the optimal one provided by LS PWM.

3   For a three-phase CHB inverter with PS PWM and more than three unequal DC sources per phase, simulations demonstrated that the best (worst) current THD is obtained by phase switching performed according to the best (worst) DC sources switching sequences for a single-phase CHB inverter, as recently suggested in [11,12]. There is another degree of freedom to be potentially exploited, that is, different carrier sequences in three converter phases by circular permutation of optimal ones.

Future research must address the impact of zero-sequence insertion into reference voltages. At a glance, the classic third harmonic correction may deteriorate current THD in CHB inverter as opposed to its positive effect on current THD in a three-phase two-level inverter.

Author Contributions: Conceptualization, Z.A. and A.R.; Methodology and Validation, Z.A., A.R., M.H., and G.G.; Formal Analysis, A.R., and G.G.; Investigation, Z.A. and M.H.; Writing-Original Draft Preparation, Z.A. and A.R.; Writing-Review & Editing, M.H. and G.G.; Supervision, A.R., and G.G.; Funding Acquisition, G.G.

Funding: This research received no external funding.

Conflicts of Interest: The authors declare no conflict of interest.

Abbreviations

The following abbreviations and symbols are used in this manuscript:

MLI   multilevel inverters
EMI   electromagnetic interferences
CHB   cascade H-bridge
THD   total harmonic distortion
LS   level shifted
PS   phase shifted
PWM   pulse width modulation
NMS   normalized mean square
f_f fundamental frequency
f_{AS} apparent switching frequency
NPC   neutral-point-clamped
FC   flying capacitor
MMC   modular multilevel converter
FFT   fast Fourier transform

References

1. Rodríguez, J.; Bernet, S.; Wu, B.; Pontt, J.O.; Kouro, S. Multilevel voltage-source-converter topologies for industrial medium-voltage drives. *IEEE Trans. Ind. Electron.* **2007**, *54*, 2930–2945. [CrossRef]
2. Villanueva, E.; Correa, P.; Rodriguez, J.; Pacas, M. Control of a single-phase cascaded H-bridge multilevel inverter for grid-connected photovoltaic systems. *IEEE Trans. Ind. Electron.* **2009**, *56*, 4399–4406. [CrossRef]
3. Buticchi, G.; Barater, D.; Lorenzani, E.; Concari, C.; Franceschini, G. A nine-level grid-connected converter topology for single-phase transformerless PV systems. *IEEE Trans. Ind. Electron.* **2014**, *61*, 3951–3960. [CrossRef]
4. Chavarria, J.; Biel, D.; Guinjoan, F.; Meza, C.; Negroni, J.J. Energy balance control of PV cascaded multilevel grid-connected inverters under level-shifted and phase-shifted PWMs. *IEEE Trans. Ind. Electron.* **2013**, *60*, 98–111. [CrossRef]
5. Ricco, M.; Mathe, L.; Monmasson, E.; Teodorescu, R. FPGA-based implementation of MMC control based on sorting networks. *Energies* **2018**, *11*, 2394. [CrossRef]
6. Ricco, M.; Mathe, L.; Teodorescu, R. New MMC capacitor voltage balancing using sorting-less strategy in nearest level control. In Proceedings of the IEEE Energy Conversion Congress and Exposition (ECCE), Milwaukee, WI, USA, 18–22 September 2016.
7. Ruderman, A.; Reznikov, B.; Busquets-Monge, S. New MMC capacitor voltage balancing using sorting-less strategy in nearest level control. *IEEE Trans. Ind. Electron.* **2013**, *60*, 1999–2009. [CrossRef]
8. Reznikov, B.; Srndovic, M.; Familiant, Y.; Grandi, G.; Ruderman, A. Simple time averaging current quality evaluation of a single-phase multilevel PWM inverter. *IEEE Trans. Ind. Electron.* **2016**, *63*, 3605–3615. [CrossRef]
9. Malinowski, M.; Gopakumar, K.; Rodriguez, J.; Perez, M.A. A survey on cascaded multilevel inverters. *IEEE Trans. Ind. Electron.* **2010**, *57*, 2197–2206. [CrossRef]
10. Institute of Electrical and Electronics Engineers (IEEE). *IEEE Recommended Practice and Requirements for Harmonic Control in Electric Power Systems*; IEEE: Piscataway, NJ, USA, 2014.
11. Zhuldassov, N.; Ruderman, A. Single-phase PS-PWM cascaded H-bridge inverter current THD optimization by unequal DC sources switching sequence selection. In Proceedings of the IEEE 9th International Symposium on Power Electronics and Distributed Generation Systems, Charlotte, NC, USA, 25–28 June 2018.
12. Polichshuk, R.; Zhuldassov, N.; Ruderman, A.; Reznikov, B. On current THD extreme switching sequences for a single-phase cascade H-bridge inverter with phase-shifted PWM and non-equal DC sources. In Proceedings of the IEEE 18th International Conference on Power Electronics and Motion Control, Budapest, Hungary, 26–30 August 2018.
13. McGrath, B.P.; Holmes, D.G. A comparison of multicarrier PWM strategies for cascaded and neutral point clamped multilevel inverters. In Proceedings of the IEEE 31st Annual Power Electronics Specialists Conference, Galway, Ireland, 18–23 June 2000.

14. Omer, P.; Kumar, J.; Surjan, B.S. Comparison of multicarrier PWM techniques for cascaded H-bridge inverter. In Proceedings of the IEEE Students' Conference on Electrical, Electronics and Computer Science, Bhopal, India, 1–2 March 2014.
15. Sochor, P.; Akagi, H. Theoretical and experimental comparison between phase-shifted PWM and level-shifted PWM in a modular multilevel SDBC inverter for utility-scale photovoltaic applications. *IEEE Trans. Ind. Appl.* **2017**, *53*, 4695–4707. [CrossRef]

Article

Power Flow Analysis of the Advanced Co-Phase Traction Power Supply System

Xiaoqiong He [1,2], Haijun Ren [1], Jingying Lin [1,*], Pengcheng Han [1], Yi Wang [1], Xu Peng [3] and Zeliang Shu [1]

1 School of Electrical Engineering, Southwest Jiaotong University, Chengdu 611756, China; hexq@home.swjtu.edu.cn (X.H.); renhaijun@my.swjtu.edu.cn (H.R.); bk20111709@my.swjtu.edu.cn (P.H.); de_yiwang@my.swjtu.edu.cn (Y.W.); shuzeliang@swjtu.edu.cn (Z.S.)
2 National Rail Transit Electrification and Automation Engineering Technique Research Center, Southwest Jiaotong University, Chengdu 611756, China
3 Aviation Engineering Institute, Civil Aviation Flight University of China, Guanghan 611756, China; xupeng@foxmail.com
* Correspondence: linjingying@swjtu.edu.cn; Tel.: +86-028-6636-6731

Received: 18 December 2018; Accepted: 18 February 2019; Published: 24 February 2019

Abstract: The development of the traction power supply system (TPSS) is limited by the existence of the neutral section in the present system. The advanced co-phase traction power supply system (ACTPSS) can reduce the neutral section completely and becomes an important research and development direction of the railway. To ensure the stable operation of ACTPSS, it is necessary to carry out an appropriate power analysis. In this paper, the topology of advanced co-phase traction substation is mainly composed by the three-phase to single-phase cascaded converter. Then, the improved PQ decomposition algorithm is proposed to analyze the power flow. The impedance model of the traction network is calculated and established. The power flow analysis and calculation of the ACTPSS with different locations of locomotive are carried out, which theoretically illustrates that the system can maintain stable operation under various working conditions. The feasibility and operation stability of the ACTPSS are verified by the simulations and low power experiments.

Keywords: improved PQ algorithm; power flow analysis; three-phase to single-phase cascaded converter; ACTPSS

---

1. Introduction

By the end of 2018, the operation mileage of the China railway had already reached 124 thousands kilometers. Among them is the high speed railway mileage, which reached 22 thousands kilometers. There is no doubt that the China railway is of great importance to China's transportation industry. To the railway system, the traction power supply system (TPSS) is a crucial part that maintains stable and safe operation of railways [1–3]. The existing TPSS is shown in the left of Figure 1. The power supply distance of each traction substation is limited by the neutral sections, and the power supplied by each traction substation is incapable of interconnection [4–7]. As a result, when the train is running in regenerative braking mode, the power cannot be used by other locomotives because of the neutral sections. Then, the power is transmitted to a three-phase grid through the transformer, which causes the problem of large harmonic content and unbalanced voltage in three-phase power network. Because of the regenerative braking energy consumption, a large number of economic losses are caused [4]. Furthermore, the voltage range of the traction network is changed sharply, and the negative sequence, reactive power, and harmonic problems are brought into the present TPSS when the locomotives are under operation [8,9]. These problems influence the quality of the three-phase power grid, the operation efficiency, and the stability of the TPSS. Moreover, the locomotives need to decelerate and slide through the neutral section. In this way, the loss of speed and the decrease of the operation safety are problems that restrict the development of railways [10,11].

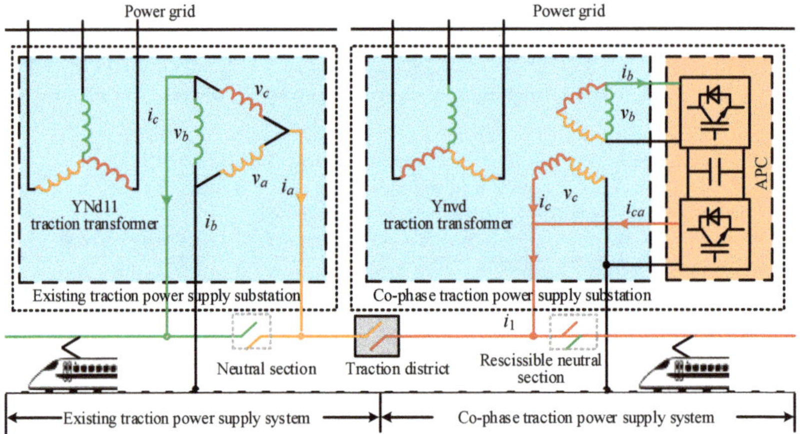

Figure 1. Traction power supply system. (**left**) The existing power supply substation, (**right**) the co-phase power supply substation based on active power compensators (APC).

To solve these problems, a great deal of research has been done in China. The co-phase traction power supply system (CTPSS) is proposed in [10], which is shown in the right of Figure 1. The active power compensators (APC) is applied in the CTPSS [4]. Half of the neutral sections can be reduced, and the power quality of the three-phase power grid can be improved to a certain extent [6]. The CTPSS has been put into operation in Meishan, Sichuan province. Voltage source converter based high voltage direct current transmission traction power supply system (VSC-Based HVDC TPSS) is proposed in [5]. Compared to TPSS, the power quality problems such as unbalance, reactive power, and harmonic distortion are greatly improved. However, the stray current is generated by the HVDC TPSS, which is harmful to the rail system, and the current supply voltage of the traction network (27.5 kV/50 Hz) is not changed by CTPSS. If there is a line using DC system, the trains will not be able to run on both systems at the same time.

In order to link all the traction networks of every different substation, the advanced traction power substation system is proposed in literature [12]. The substation based on three-phase to single-phase converter is adopted in this system [13–16]. In order to adapt to the voltage level of three-phase grid voltage and the traction network voltage, the step-down transformer and step-up transformer are added to the three-phase grid and the traction network side, respectively. The neutral sections and problems of power quality in the traditional system can be avoided. However, it is difficult to meet the existing substation requirement of capacity because the structure is limited by the power electronic devices. The advanced co-phase traction power supply system (ACTPSS) based on three-phase to single-phase cascaded converter is proposed in literature [7], as shown in Figure 2. The ACTPSS achieves the interconnection of the whole traction network, thus all the neutral sections are reduced by ACTPSS. The negative sequence, reactive power, and harmonic problems existent in the traditional TPSS can be solved effectively [12,13]. Meanwhile, the voltage of the traction network can be maintained in a relatively stable range, and regenerative braking energy can be better utilized. Thus, the ACTPSS has become essential in the research and development of future TPSS.

Due to the interconnection of the whole traction network, the energy of the locomotive is mainly supplied by the closest power supply arm, and the adjacent substations provide the remaining parts of the energy. Therefore, the system capacity of the traction power supply system can be reduced. To assess and manage the operation state of ACTPSS, the power flow analysis also needs to be researched. Criteria of voltage stability based on the existence of the power flow solvability was used in the power distribution network [17]. Another voltage stability criterion was obtained based on the volt-ampere characteristic of the branches in the system [18]. This power analysis research of the public three-phase power grid provides a good theoretical basis for the power flow analysis of the traction network. In the power supply model of auto transformer (AT) traction substation, the power flow and voltage distribution in different working conditions of high-speed railway are analyzed by establishing a general multi-conductor chain circuit model [19,20]. The influence of traction load on power quality of the three-phase power grid was illustrated. These laid a good foundation for building the model of the traction network [21,22]. In ACTPSS, no node voltage of the traction network is allowed to exceed the prescribed voltage standard [23]. In this way, each traction substation supplies the power equitably and stably. Furthermore, the voltage range of the traction network maintains steadily. Therefore, it is necessary to carry out the power flow analysis of ACTPSS and assess the voltage variation degree of the traction network so that the dispatchers can take correspondent measures.

In this paper, firstly, the structure of ACTPSS based on a three-phase to single-phase cascaded converter is illustrated, and the corresponding control strategy and modulation strategy are given. Secondly, on the basis of ACTPSS, the mathematical model of traction network impedance is established by multi-conductor model, and the power flow analysis of the traction network is carried out by the improved PQ decomposition algorithm. Thirdly, the stability performance of ACTPSS is verified through the analysis of the power flow under different locomotive operation conditions. Finally, the correctness of the theory analysis is verified through the simulation, and the operation reliability of ACTPSS is verified through the low power experiment.

2. Configurations and Strategy

2.1. The Structure of ACTPSS

According to the structure of ACTPSS, as shown in Figure 2, the traction substation consists of the multi-winding step-down transformer and the three-phase to single-phase cascaded converter. The primary side of the multi-winding step-down transformer is connected with the three-phase power grid, and the secondary sides of it are connected with the three-phase to single-phase converters. To improve the withstand voltage of the traction substation, the cascaded topology is utilized in each traction substation. In order to improve the withstand voltage of the single model, the three-level neutral point clamped (3L-NPC) topology is adopted in this paper.

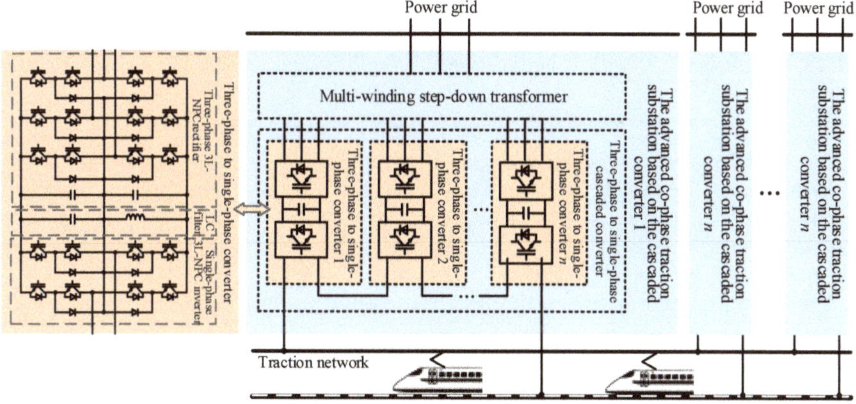

Figure 2. The structure of the advanced co-phase traction power supply system (ACTPSS).

The three-phase to single-phase converter is comprised of the three-phase 3L-NPC rectifier and the single-phase 3L-NPC inverter. The three-phase 3L-NPC rectifier converts the power from three-phase alternating current (AC) to direct current (DC), then the single-phase 3L-NPC inverter converts the power from DC to single-phase AC. The output of the three-phase to single-phase cascaded converter as the output of the traction substation is connected with the traction network. The amplitude, the frequency, and the phase of the traction substation output voltage are controlled to be in accordance with the traction network. The traction substations are paralleled with the traction network and achieve the power interconnection of the whole system.

2.2. The Control Strategy and Modulation Strategy of Three-Phase 3L-NPC Rectifier

In order to ensure that the three-phase 3L-NPC rectifiers operate under unit power factor, P-Q transformation, voltage, and current double closed loop control strategy is adopted [24]. The control strategy and modulation strategy of the three-phase 3L-NPC rectifier are illustrated by Figure 3. Firstly, the phase of input voltage v_{abc} is extracted through phase lock loop (PLL). The locked phase is used as the reference phase of input current i_{abc} to achieve coordinate transformation from the three-phase static coordinate (abc) to the two-phase rotating coordinate (dq). Under the two-phase rotating coordinate, the active component and reactive component of the current are controlled, respectively. The reference of the active component of the current is relative with the output voltage v_{dc}, while the reference of the reactive component of the current is 0. The Space Vector Pulse Width Modulation (SVPWM) strategy is adopted as the modulation strategy for the rectifier.

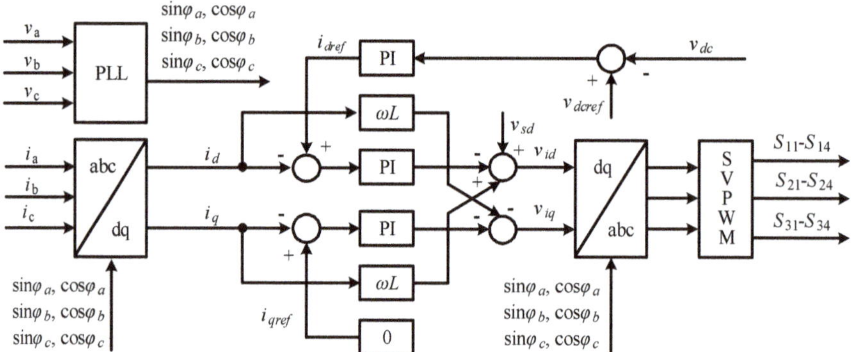

Figure 3. The control strategy and modulation strategy of the three-phase three-level neutral point clamped (3L-NPC) rectifier.

2.3. The Control Strategy and Modulation Strategy of Single-Phase 3L-NPC Cascaded Inverter

The control strategy and modulation strategy of the single-phase 3L-NPC cascaded inverter are shown in Figure 4. The double closed loop control strategy is applied in this paper. The outer loop control strategy is designed to steady the output voltage. The inner loop control strategy is designed to adjust the dynamic response of the system. The value of output voltage and inductance current are detected to involve the control strategy.

The carrier phase shift strategy is used as the modulation strategy [25]. Every inverter model has the same modulation wave produced by the controller, while the triangle angle waves are different. The phase angle difference of two adjacent triangle angle waves is π/n, where the n represents the number of the inverter models. The single-phase SVPWM strategy is used as the modulation strategy to drive each inverter model.

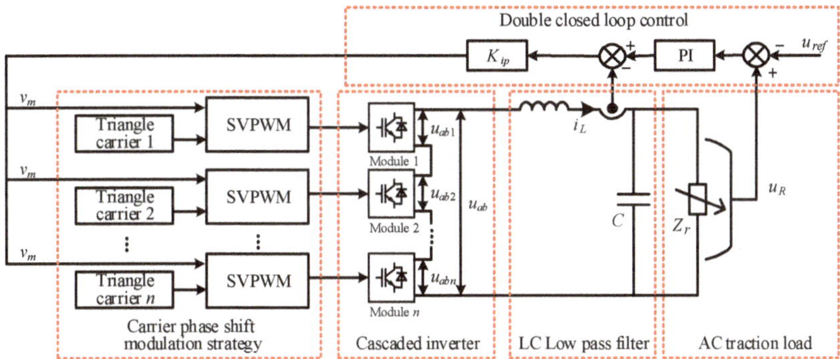

Figure 4. The control strategy and modulation strategy of single-phase 3L-NPC cascaded inverter.

3. The Traction Network Impedance Model of ACTPSS

Because the steel rail and ground belong to ferromagnetic materials, the current of the circuit changes with the operation state change of the traction load, which has great influence on the equivalent inductance and impedance of the steel rail. Generally, the length of steel rail is recognized as infinity, and the distribution parameters between the steel rail and ground are nonlinear. Furthermore, the traction impedance is also recognized as nonlinear. Compared with the impedance calculation of the power system, the impedance calculation of the traction network is much more complicated. In this paper, the impedance model of the traction network based on ACTPSS is built by the method of the simplified equivalent model.

3.1. The Simplified Equivalent Model of Traction Network

The original circuit model of the traction network is shown in Figure 5. Viewed from the output port of the traction substation, the impedance of the traction network consists of several equivalent impedances of the conductor-ground circuit [26].

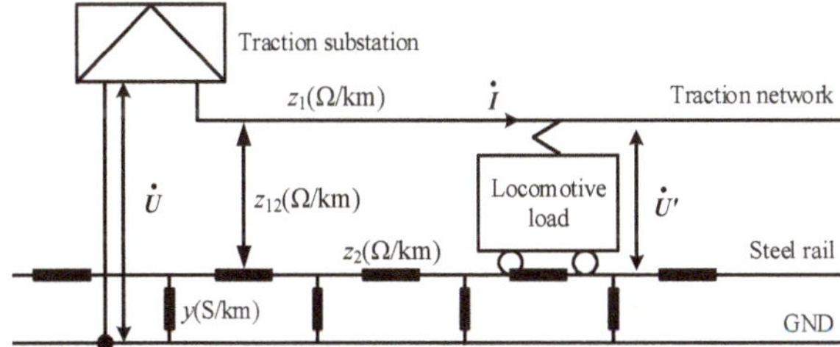

Figure 5. The original circuit model of the traction network.

According to Figure 5, it is known that the unit length impedance of traction network z can be calculated by Equation (1):

$$z = \frac{\dot{U} - \dot{U}'}{\dot{I}l} \tag{1}$$

In Equation (1), the voltage between the output port of the traction substation and ground is $\dot{U}$, the voltage of the locomotive load is $\dot{U}'$, the length of the traction network is l, and the current of the traction network is $\dot{I}$.

It is known from Equation (1) that the z is nonlinear and cannot be calculated by formula flexibly. It is assumed that the current of steel rail can flow into the ground instantly, or the current of the ground can flow into the steel rail instantly, and the current can flow into the output port of the traction substation with the same amplitude as well. Then, the admittance γ between the steel rail and the ground can be considered to approach infinity, and the conduction current of steel rail can be considered to tend to 0. Taking these assumptions as the simplification condition, Equation (2) can be expressed as:

$$\lim_{\gamma \to \infty} \frac{1}{2}(1 - k_z)\dot{I}\left[e^{-\gamma(l-x)} + e^{-\gamma x}\right] \approx 0 \tag{2}$$

In Equation (2), $k_z = z_{12}/z_2$. The mutual impedance is z_{12}, the self-impedances are z_1 and z_2, and the distance between the locomotive and the traction substation is x.

Then, the current of the steel rail I_T and the ground I_G can be recognized as:

$$\begin{cases} \dot{I}_T(x) = \dot{I}_T = k_z\dot{I} \\ \dot{I}_G(x) = \dot{I}_G = (1 - k_z)\dot{I} \end{cases} \tag{3}$$

The simplified impedance model of the traction network is illustrated by Figure 6 [26]. The model is composed of two current loops. In the first loop, the current flows through the traction network and the ground. In the second loop, the current flows through the steel rail and the ground. The second current loop just reflects the induced current, which is far bigger than the conduction current. According to the light of this simplified impedance model, the unit length impedance of the traction network z can be figured out by the calculation of the self-impedance of the first and second loop.

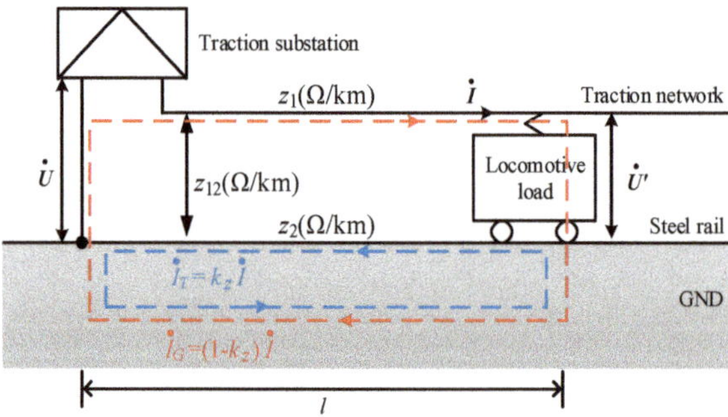

Figure 6. The simplified impedance model of the traction network.

According to the Carson theory [27], the ground can be recognized as virtual wire. Then, Equation (4) can be expressed as:

$$\begin{cases} z_1 = r_1 + 0.05 + j0.1446g\dfrac{D_g}{R_{e1}} & \Omega/\text{km} \\[2mm] z_2 = r_2 + 0.05 + j0.1446g\dfrac{D_g}{R_{e2}} & \Omega/\text{km} \\[2mm] z_{12} = 0.05 + j0.1446g\dfrac{D_g}{d_{12}} & \Omega/\text{km} \end{cases} \tag{4}$$

In Equation (4), the r_1 and R_{g1} represent the effective resistance and equivalent radius of the first loop, respectively; r_2 and R_{g2} represent the effective resistance and equivalent radius of the second loop, respectively; the D_g represents the depth of the equivalent earth-return circuit; d_{12} is the center distance between the conductor in the first and second loop.

Therefore, the z can be calculated as follow:

$$z = z_1 - \frac{z_{12}^2}{z_2} \tag{5}$$

According to Equation (5), the impedance of the traction network z_l with the length l can be expressed as follows:

$$z_l = z \times l \tag{6}$$

3.2. The Impedance Calculation for the Traction Network of ACTPSS

In ACTPSS, the induced current is far bigger than the conduction current of the second loop. When the distance between locomotive and traction substation is longer than 5 km, the conduction current can be ignored, and the impedance of the traction network can only be calculated by the inducted current. Generally, the error between the calculation result and the actual model is within 5%. Therefore, the simplified equivalent model of the traction network is always adopted. The wire type of GLCA-100/215 was selected as the traction network wire in this paper. Then, $r_1 = 0.184$ $\Omega/$km, $r_2 = 0.295$ $\Omega/$km, $R_{g1} = 8.56$ mm, $R_{g2} = 12.2$ mm, $d_{12} = 5850$ mm. According to the simplified equivalent model of the traction network, the impedance can be calculated as shown in Table 1.

Table 1. The impedance calculation results of the traction network.

The Type of Impedance	Value
The self-impedance of traction network	$z_1 = 0.234 + j0.728$ $(\Omega/$km$)$
The self-impedance of steel rail	$z_2 = 0.345 + j0.706$ $(\Omega/$km$)$
The mutual impedance	$z_{12} = 0.05 + j0.318$ $(\Omega/$km$)$
The unit length impedance of traction network	$z = 0.2527 + j0.598$ $(\Omega/$km$)$

4. The Power Flow Analysis for Traction Network of ACTPSS

4.1. The Improved PQ Decomposition Algorithm

The PQ decomposition algorithm, which can be used to calculate the power flow, is based on the simplified power flow calculation progress of the polar coordinate formula in the Newton-Raphson algorithm [28]. Generally, in the transmission line equivalent model of the AC high voltage power grid, it is considered that $x \gg r$, where the x is the line reactance and r is the line impedance. Therefore, the active power is mainly affected by the voltage phase, while the reactive power is mainly affected by the voltage amplitude. If the effect of voltage phase change on reactive power distribution and the effect of voltage amplitude change on active power distribution are ignored, Equations (7) and (8) can be obtained as follows:

$$\Delta P = H\Delta\theta \tag{7}$$

$$\Delta Q = LU^{-1}\Delta U \tag{8}$$

From Equations (7) and (8), it is known that the correction equation of the active power ΔP and the reactive power ΔQ can be iterated, respectively. Compared with the $2n$-order linear equations of the Newton-Raphson algorithm, the PQ decomposition algorithm transforms it into two n-order linear equations. The calculation time can be reduced sharply. However, in the iteration, the coefficient matrix H and L, which belong to the asymmetric matrix, are changing constantly. Thus, it is necessary to change coefficient matrices into the symmetric matrices so that the calculation can be further simplified. Because of the relatively little voltage phase difference between two ports of the circuit, it can be supposed as follows:

$$\begin{cases} \cos \delta_{ij} \approx 1 \\ G_{ij} \sin \delta_{ij} \ll B_{ij} \\ Q_i \ll U_i^2 B_{ii} \end{cases} \tag{9}$$

Then, the expression of the Jacobian matrix can be expressed as:

$$H_{ij} = L_{ij} = U_i U_j B_{ij} \tag{10}$$

$$H_{ii} = L_{ii} = U_i^2 B_{ii} \tag{11}$$

Then, Equations (7) and (8) can be rewritten as:

$$\Delta P = UBU\Delta\theta \tag{12}$$

$$\Delta Q = UBU\left(U^{-1}\Delta U\right) = UB\Delta U \tag{13}$$

In Equations (12) and (13), the U is the diagonal matrix of the effective value of node voltage, and the B represents the susceptance matrix.

By multiplying Equations (12) and (13) with U^{-1}, Equations (14) and (15) can be obtained as follows:

$$U^{-1}\Delta P = B'U\Delta\theta \tag{14}$$

$$U^{-1}\Delta Q = B''\Delta U \tag{15}$$

The matrix forms are shown as follows:

$$\begin{bmatrix} \Delta P_1/U_1 \\ \Delta P_2/U_2 \\ \vdots \\ \Delta P_n/U_n \end{bmatrix} = B' \begin{bmatrix} V_1\Delta\theta_1 \\ V_2\Delta\theta_2 \\ \vdots \\ V_n\Delta\theta_n \end{bmatrix} \tag{16}$$

$$\begin{bmatrix} \Delta Q_1/U_1 \\ \Delta Q_2/U_2 \\ \vdots \\ \Delta Q_m/U_m \end{bmatrix} = -B'' \begin{bmatrix} \Delta U_1 \\ \Delta U_2 \\ \vdots \\ \Delta U_m \end{bmatrix} \tag{17}$$

In Equations (16) and (17), the coefficient matrices B' and B'' have the same form. The B' is the n order matrix and the B'' is the $n - m - 1$ order matrix, where the m represents the number of the PV node. Both the B' and B'' are imaginary parts of the admittance matrix, and both are symmetric matrices.

However, the transmission line reactance of the traction network in ACTPSS is not much larger than the resistance, thus the PQ decomposition algorithm cannot be applied directly. Based on the PQ decomposition algorithm, the improved PQ decomposition algorithm is proposed to calculate the power flow of the traction network.

The impedance model of the traction network is shown in Figure 7a. The model after simplifying the impedance between node i and node j into the two parallel impedances model is shown in Figure 7b. The value of the simplified model is identical to the original mode. The value of the simplified model can be expressed as Equation (18):

$$\begin{cases} Y' = -j\frac{1}{X} \\ Y'' = \frac{R}{R^2+X^2} + j\left(\frac{1}{X} - \frac{X}{R^2+X^2}\right) \end{cases} \tag{18}$$

In Equation (18), the X is the line reactance of the original model, the R is the line resistance of the original model, and the unit of them is Ω. If we suppose that the real part of the Y'' is 0, the transmission line reactance of the traction network can be sufficiently larger than the resistance, and the condition $G_{ij}\sin\delta_{ij} \ll B_{ij}$ can be satisfied. To make sure that the power flow calculation will not be wrong and the system can keep balance after the change of the impedance, the lacking power part will be added by the other power sources in node i and node j. The way to add the power sources can be determined by Figure 7c. Combined with the node voltage and the modified Y'', the power of node i and node j can be illustrated as:

$$\begin{cases} P_i = U_i\left[G''\left(U_j\cos\theta_{ij} - U_i\right) + B''U_j\sin\theta_{ij}\right] \\ Q_i = U_i\left[G''U_j\sin\theta_{ij} + B''\left(U_j\cos\theta_{ij} - U_i\right)\right] \\ P_j = U_j\left[G''\left(U_i\cos\theta_{ij} - U_j\right) + B''U_i\sin\theta_{ij}\right] \\ Q_j = -U_j\left[G''U_i\sin\theta_{ij} + B''\left(U_i\cos\theta_{ij} - U_j\right)\right] \end{cases} \tag{19}$$

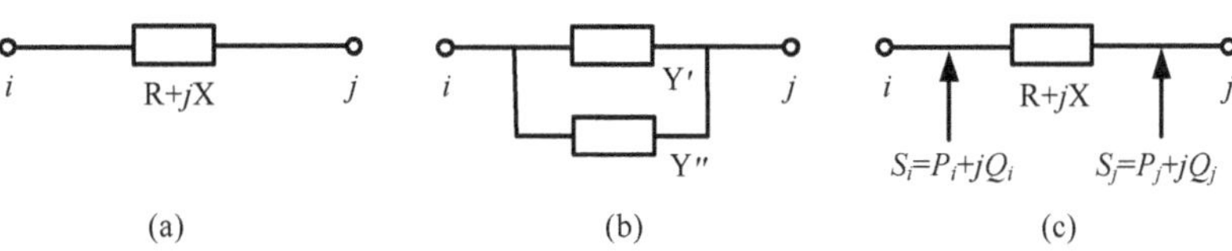

Figure 7. The models of the traction network. (**a**) The impedance model of the traction network; (**b**) the parallel equivalent impedance model of the traction network; (**c**) the simplified equivalent impedance model of the traction network.

In Equation (19), the G'' is the conductance of the modified model, the B'' is the susceptance of the modified model, and the unit of them is S. When the condition $G_{ij}\sin\delta_{ij} \ll B_{ij}$ has been satisfied, the subsequent calculation of the improved PQ decomposition algorithm is same as the PQ decomposition algorithm. In the power flow calculation of the improved PQ decomposition algorithm to the traction network, the locomotives are considered as the power sources, and the traction substations are recognized as the equivalent node models of voltage source. The equivalent mathematic model of ACTPSS, which can apply the improved PQ decomposition algorithm, is exhibited in Figure 8.

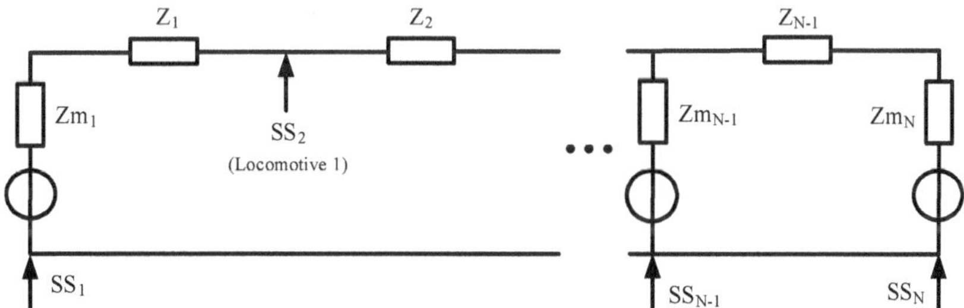

Figure 8. The equivalent mathematic model of ACTPSS.

4.2. The Power Flow Calculation of ACTPSS

In this section, the improved PQ decomposition algorithm is applied to calculate the power flow and to verify the operation stability of the system. The software PAST is used to build the model shown by Figure 9. The model consists of three traction substations and seven locomotives and simulates the condition in which several locomotives run in a certain section. The type of locomotive is CRH380A, the active power is 9600 kW, and the power factor is 0.99.

Under the ideal state, the output voltages of traction substation SS1, SS2, and SS3 are coincidental, thus they can be recognized as the balanced node. The Z_{SS1}, Z_{SS2}, and Z_{SS3} represent the resistances of the three traction substations, respectively. The value of them is set as $0.5 + j3.375\ \Omega$ in this model. The Z_i ($i = 1, 2 \ldots 6$) represents the resistance of the traction network, and the value of unit length is set as $0.2527 + j0.298\ \Omega$. The L_i ($i = 1, 2, \ldots 7$) represents the seven locomotives, respectively, and the locomotive can be seen as a PQ node because the power of locomotive is constant. In conclusion, the system model is composed of 10 nodes.

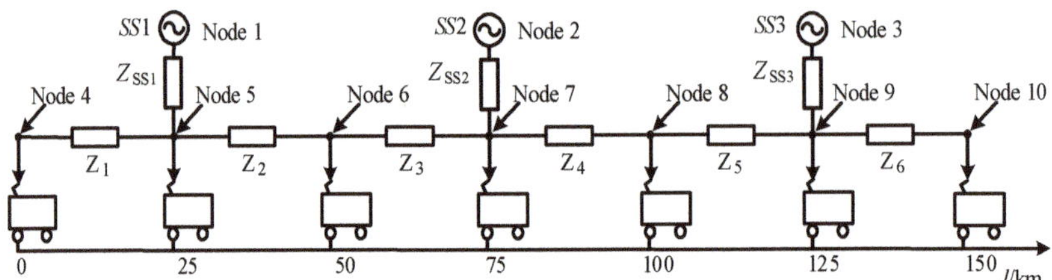

Figure 9. The equivalent model of ACTPSS.

Power flow calculation results of ACTPSS are listed in Table 2. In this paper, the voltage values of node 4 and node 10 in the model are the lowest, which are 21.3758 kV. Obviously, 21.3758 kV is still higher than 19 kV. At this condition, the locomotive can continue to run, which indicates that ACTPSS proposed in this paper can ensure the locomotive operates normally.

Table 2. The power flow calculation results of ACTPSS.

Node	Voltage Amplitude U (kV)	Voltage Phase α (rad)	Complex Power S (MVA)
1	27.500	0.000	$25.484 + j10.218$
2	27.500	0.000	$21.159 + j6.625$
3	27.500	0.000	$25.484 + j10.218$
4	21.3758	−0.35922	$-9.6 + j(-1.37)$
5	25.9499	−0.11363	$-9.6 + j(-1.37)$
6	24.3719	−0.20953	$-9.6 + j(-1.37)$
7	26.4185	−0.09387	$-9.6 + j(-1.37)$
8	24.3719	−0.20953	$-9.6 + j(-1.37)$
9	25.9499	−0.11363	$-9.6 + j(-1.37)$
10	21.3758	−0.35922	$-9.6 + j(-1.37)$

The line impedance exists in the traction network model. By using the node voltage and node admittance matrix, it can be calculated that the total power of the system is $72.135 + j27.062$ MVA, the total consumption power of the system load is $67.2 + j9.595$ MVA, and the total power loss of the system is $4.935 + j17.472$ MVA. The power loss of every line between two nodes is illustrated in Table 3. The power losses are expressed in the form of complex power with the inductive reactive power. Because the distance between the locomotive and the node is different, the power loss of the line between the nodes is different.

Table 3. The power losses of ACTPSS.

Line	Power Losses	Line	Power Losses
1–5 line	$0.499 + j3.365$	6–7 line	$0.307 + j0.726$
2–7 line	$0.325 + j2.194$	7–8 line	$0.307 + j0.726$
3–9 line	$0.499 + j3.365$	8–9 line	$0.199 + j0.471$
4–5 line	$1.3 + j3.077$	9–10 line	$1.3 + j3.077$
5–6 line	$0.199 + j0.471$	Total	$4.935 + j17.472$

4.3. The Calculation of Output Power Under Different Locomotive Location

Since the different location of the locomotive directly affects the output power of each traction substation, the influence of different locomotive locations on the output power of substation with two traction substations is analyzed as an example. The model is established in Figure 10.

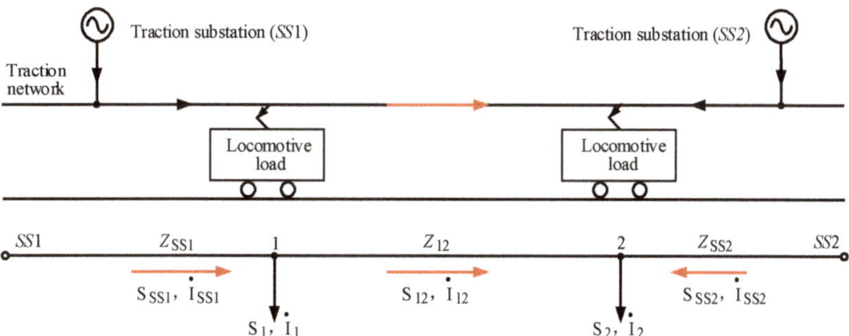

Figure 10. The model of ACTPSS with two traction substation.

Two locomotives are located at point 1 and 2 in Figure 10. The locomotive is considered as a single-particle model in the analysis. The output current of traction substations SS1 and SS2 are i_{SS1} and i_{SS2}, respectively. The line impedances between the traction substations SS1 and SS2 to the locomotive are Z_{SS1} and Z_{SS2}, respectively. The line impedance between point 1 and point 2 is Z_{12}. The currents of locomotives 1 and 2 are i_1 and i_2, respectively.

According to the Kirchhoff's Voltage Law (KVL) and Kirchhoff's Current Law (KCL):

$$\dot{V}_{SS1} - \dot{V}_{SS2} = Z_{SS1}\dot{I}_{SS1} + Z_{12}\dot{I}_{12} - Z_{SS2}\dot{I}_{SS2} \tag{20}$$

$$\dot{I}_{SS1} - \dot{I}_{12} = \dot{I}_1 \tag{21}$$

$$\dot{I}_{12} - \dot{I}_{SS2} = \dot{I}_2 \tag{??}$$

Assuming that the output voltage of the SS1 and SS2 are V_{SS1} and V_{SS2}, and the currents I_1, I_2 of the locomotive are known, then Equations (21) and (22) can be simplified as:

$$\dot{I}_{SS1} = \frac{(Z_{12} + Z_{SS2})\dot{I}_1 + Z_{SS2}\dot{I}_2}{Z_{SS1} + Z_{12} + Z_{SS2}} + \frac{\dot{V}_{SS1} - \dot{V}_{SS2}}{Z_{SS1} + Z_{12} + Z_{SS2}} \tag{23}$$

$$\dot{I}_{SS2} = \frac{Z_{SS1}\dot{I}_1 + (Z_{SS1} + Z_{12})\dot{I}_2}{Z_{SS1} + Z_{12} + Z_{SS2}} - \frac{\dot{V}_{SS1} - \dot{V}_{SS2}}{Z_{SS1} + Z_{12} + Z_{SS2}} \tag{24}$$

In the actual condition, there must exist the voltage drop along the traction network. Therefore, the power flow calculation of the traction network also needs a similar method. Generally, the power loss of the traction network is ignored, and the power is also calculated by the voltage V. If we suppose that $V = V_N\angle 0°$ and $S = V_N I$, then Equations (25) and (26) can be acquired after multiplying the conjugate value of Equations (23) and (24) with the V_N.

$$S_{SS1} = \frac{\left(\dot{Z}_{12} + \dot{Z}_{SS2}\right)S_1 + \dot{Z}_{SS2}S_2}{\dot{Z}_{SS1} + \dot{Z}_{12} + \dot{Z}_{SS2}} + \frac{\left(\dot{V}_{SS1} - \dot{V}_{SS2}\right)V_N}{\dot{Z}_{SS1} + \dot{Z}_{12} + \dot{Z}_{SS2}} \tag{25}$$

$$S_{SS2} = \frac{\dot{Z}_{SS1}S_1 + \left(\dot{Z}_{SS1} + \dot{Z}_{12}\right)S_2}{\dot{Z}_{SS1} + \dot{Z}_{12} + \dot{Z}_{SS2}} - \frac{\left(\dot{V}_{SS1} - \dot{V}_{SS2}\right)V_N}{\dot{Z}_{SS1} + \dot{Z}_{12} + \dot{Z}_{SS2}} \tag{26}$$

Based on Equations (25) and (26), it can be obtained that the output power of the traction substation contains two parts. One part is decided by the parameter of the locomotive itself; the other is unrelated to the parameter of itself, which is called the cyclic power.

The simulation that sets three traction substations in the system is also achieved. In this simulation, the locomotive runs from the start (0 km) to the end (150 km). When the locomotive is at seven different locations, the output power of the traction substations are illustrated as shown in Table 4.

Table 4. The output power of traction substations.

Distance	0 km	25 km	50 km	75 km	100 km	125 km	150 km
Traction substation 1	9.848	8.803	5.007	0.789	0.458	0.073	0.091
Traction substation 2	0.93	0.79	4.627	8.088	4.627	0.79	0.93
Traction substation 3	0.091	0.073	0.0458	0.789	5.007	8.803	9.848
Total	10.869	9.666	9.6798	9.666	10.092	9.666	10.869

Fitting the data of Table 4, the relationship between the distance and the output power ratio of each traction substation is expressed as Figure 11. It can be known by Figure 11 that the output power is varied with the change of the distance between the locomotive and the substation. Compared with the present traction substation, 10 percentages of the output power can be reduced by the traction substation of ACTPSS (approximately), which could have significant economic influence on the China railway system.

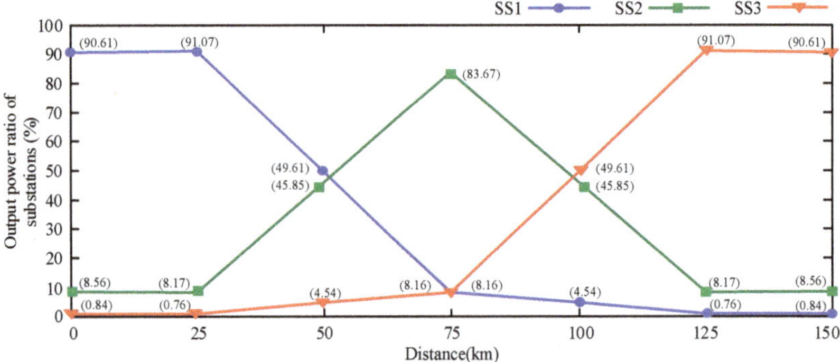

Figure 11. The relationship between the location of the locomotive and the output power ratio of the traction substation.

5. The Simulation Analysis and Experimental Verification

5.1. The Simulation Analysis of Three-Model Three-Phase to Single-Phase Cascaded Converter Based on ACTPSS

The simulation of the three-model three-phase to single-phase cascaded converter based on ACTPSS was built, and the feasibility of the control strategy and modulation strategy was verified. The simulation parameters are exhibited in Table 5.

Table 5. The parameters of the three-model cascaded converter simulation.

Parameters	Value
The number of the cascaded model	3
The DC voltage of the single model	15 kV
The reference voltage of the traction network	27.5 kV
Power of the traction network	9.6 MW
The value of the capacitor in DC side	1 mF
The value of the fliter inductance	10 mH
The value of the fliter capacitor	10 μF
The carrier wave frequence of the inverter	1 kHz

The output voltage waves of the three-phase 3L-NPC rectifiers are shown in Figure 12, and it can be known that the voltages could achieve stability after about 0.17 s. As the output waves appeared, the second-order ripple was suppressed by the LC filter. By the rectifiers, the steady and reliable DC voltages could be obtained.

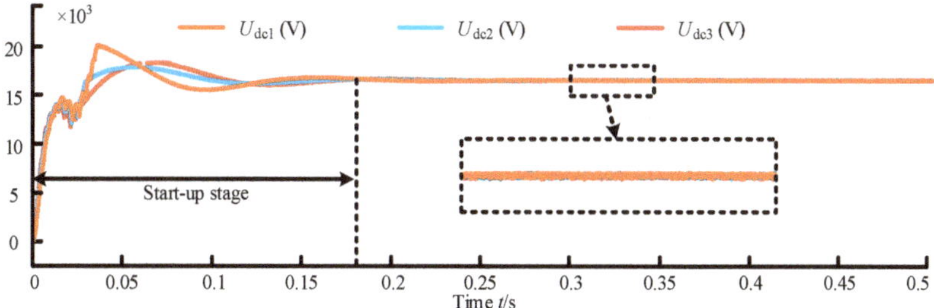

Figure 12. The output voltage waves of three three-phase rectifiers.

The output voltage and current waves of the three-model 3L-NPC cascaded inverter after filter are illustrated by Figure 13a. The u_R and i_R are the output voltage and current after the filter, respectively. In order to display the voltage and the current wave of output load properly and in the same scope, the current waveform was expanded by 50 times, the output voltage and current waveform were closed to the standard sine waves, and the output voltage that had little harmonic component and high power factor could be provided by the cascaded inverter. The 5-level output voltage waves of the three-inverter and the output voltage of the cascaded inverter are shown in Figure 13b. The u_{ab} is the 13-level output voltage of the three-model cascaded inverter, of which the effective value is 27.5 kV. The feasibility and availability of the cascaded inverter's control strategy and modulation strategy have been verified by Figure 13.

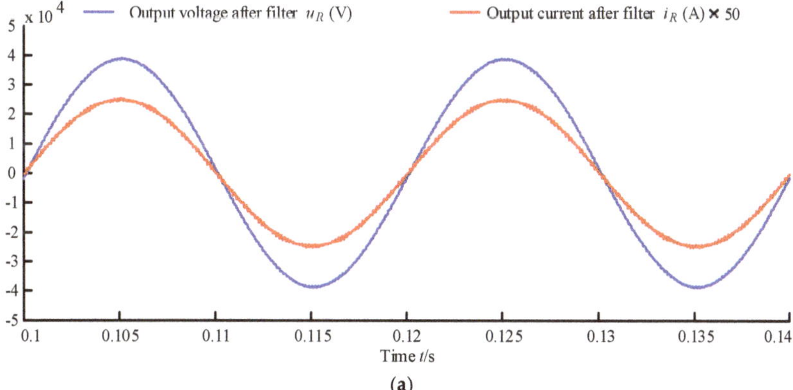

Figure 13. *Cont.*

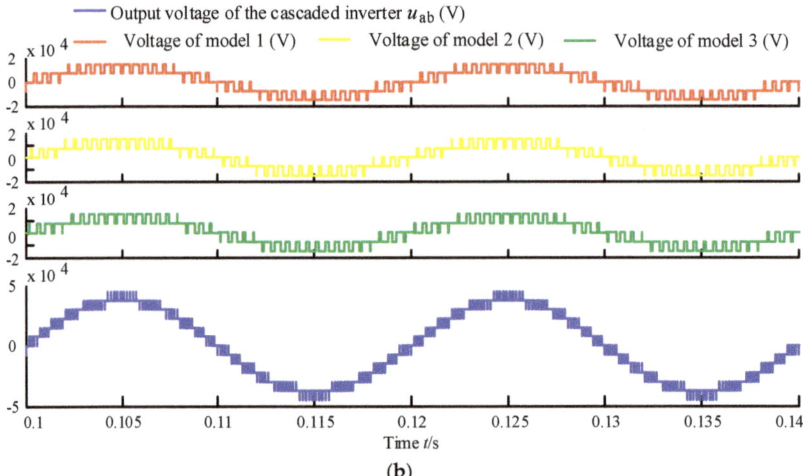

Figure 13. The output voltage and current waves of 3L-NPC cascaded inverter. (**a**) The output voltage and current wave of the three-model 3L-NPC cascaded inverter after filter; (**b**) the output voltage waves of the three single-phase 3L-NPC inverter and the output voltage of 3L-NPC cascaded inverter.

5.2. The Simulation Analysis of ACTPSS

The simulation of ACTPSS was built, and three traction substations and three locomotives were set in this simulation. The locomotives worked as the traction load. The parameters of the system are illustrated in Table 6.

Table 6. The parameters of ACTPSS.

Parameters	Value
The number of the traction substation	3
The reference voltage of the traction network	27.5 kV
The value of the fliter inductance in network side	10 mH
The value of the fliter capacitor in network side	10 μF
The carrier wave frequence of the inverter	1 kHz

The locomotive 1 (R_1) was set to run in 0.5–1.5 s, 2.5–3.5 s, and 4.5–7 s in the simulation, the locomotive 2 (R_2) was set to run in 1.5–2.5 s and 3.5–6.0 s, while the locomotive 3 (R_3) was set to run in 2.5–5.0 s. The output current waves of the three traction substations and the current waves of the three locomotives are shown in Figure 14.

The traction substation 1 (SS1) was connected with the traction network at 0.5 s, the start of the system operation. The traction substation 2 (SS2) was connected with the traction network at 1.0 s, and the traction substation 3 (SS3) at 2.0 s. In 0.5–1.0 s, only the R_1 was under operation, and the power was provided to the R_1 only by the SS1, thus the current of the R_1 was the same as the current of the SS1 at this period. In 1.0–1.5 s, the power was supplied to the R_1 by both the SS1 and the SS2. In 1.5–2.0 s, only the R_2 was under operation, and the power was supplied to the R_2 by both the SS1 and the SS2. Because the distance between the R_2 and the SS2 was shorter, the power supplied by the SS2 was more than that supplied by the SS1. In 2.0–2.5 s, the power was supplied to the R_2 by the SS1, the SS2, and the SS3. Because the location of R_2 was between the SS2 and the SS3, the power provided by the SS1 was the least. In 2.5–3.0 s, the R_1 and the R_3 were under operation, and the power of them was supplied by the three traction substations together. At this period, the R_3 was on the right side of the SS3, and the power provided by the SS3 was more than that supplied by the SS1 and the SS2.

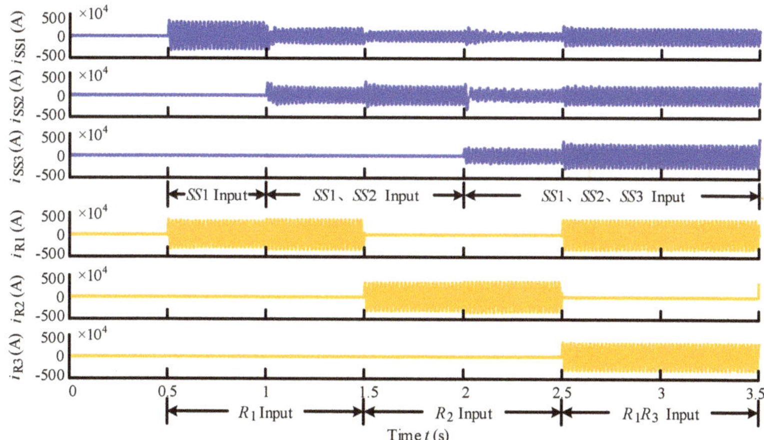

Figure 14. The current waves of the traction substations and locomotives in 0–3.5 s.

The output voltage waves of the traction substations and the voltage waves of locomotives in 0–3.5 s are exhibited in Figure 15. From the simulation results, it is clear that the output voltages of the traction substations and the voltages of locomotives were adjusted properly when the traction loads were input or excision. The voltages were maintained within the international prescribed scope, from 19 kV to 29 kV, ensuring that the locomotives could be under the normal operation.

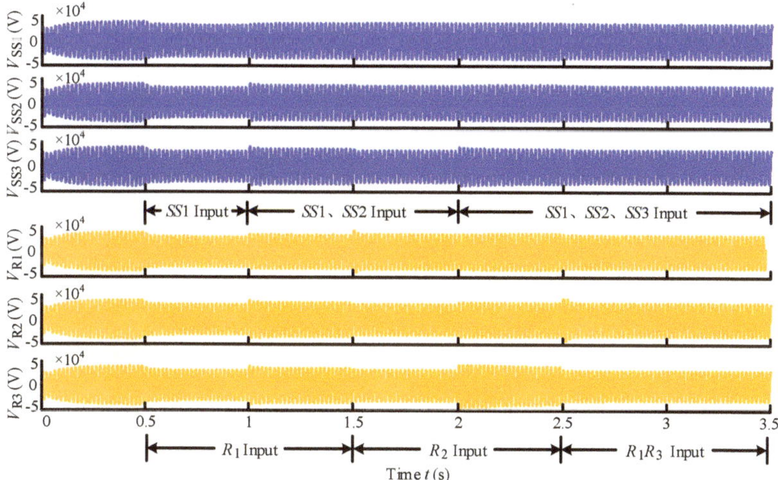

Figure 15. The voltage waves of the traction substations and the locomotives in 0–3.5 s.

The output current waves of the traction substations and the current waves of locomotives in 3.5–7.0 s are exhibited in Figure 16. In 3.5–4.5 s, the R_2 and the R_3 were under operation, and the power of them was supplied by the three traction substation together. Because the distance between the R_2, the R_3, and the SS1 was longer, the power supplied by the SS1 was the least. In 4.5–5.0 s, the R_1, the R_2, and the R_3 were under operation, and the power of them was also provided by the three traction substations together. Because the R_3 was farther from the SS1 and the SS2 than from the SS3, the power supplied from the SS3 to the R_3 was the most. In 5.0–6.0 s, when the SS3 was excision because of fault, the R_1 and the R_2 were still under operation and obtained the power from the SS1 and the SS2.

In 6.0–7.0 s, the R_1 was under operation and located between the SS1 and the SS2. The power was supplied to the R_1 by the SS1 and the SS2 equally.

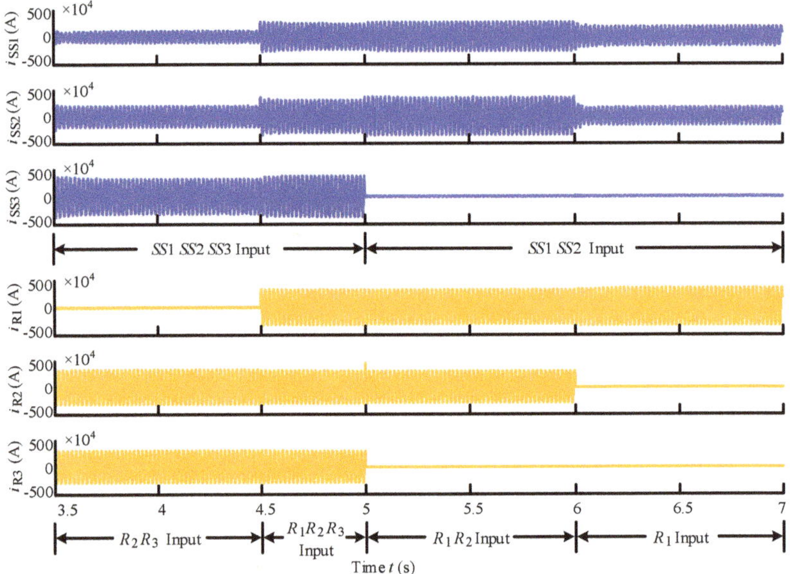

Figure 16. The current waves of the traction substations and locomotives in 3.5–7 s.

The output voltage waves of the traction substations and the voltage waves of locomotives in 3.5–7.0 s are exhibited in Figure 17. From the simulation results, it is clear that the output voltages of the traction substations and the voltages of locomotives were adjusted properly when the traction loads were input or excision, and the voltages were maintained within the international prescribed scope, from 19 kV to 29 kV, ensuring that the locomotives could be under the normal operation.

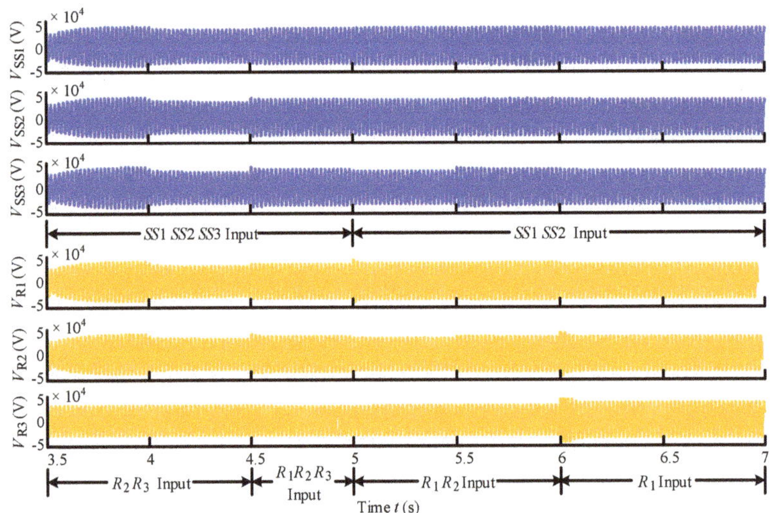

Figure 17. The voltage waves of the traction substations and locomotives in 3.5–7 s.

The output active and reactive power of the traction substations are illustrated by Figure 18. The converters inside the locomotives were controlled to operate with the unity power factor, thus the locomotives could be considered the pure resistance load. Therefore, under the steady operation of the locomotive, the reactive power was almost zero. When the locomotive started up or broke, the reactive power was outputted to achieve the power balance of the traction network.

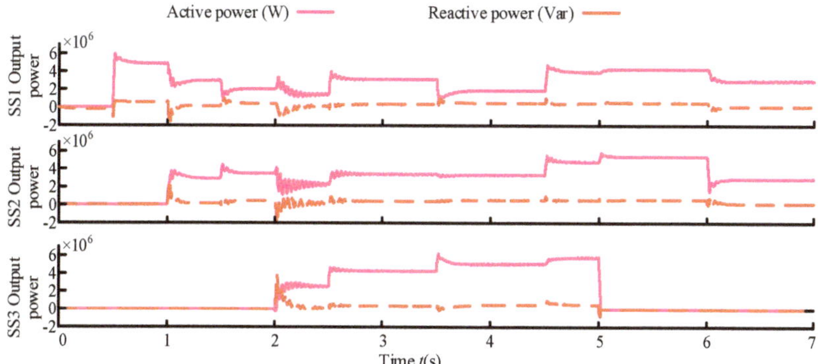

Figure 18. The output active and reactive power of three traction substations.

5.3. The Exprimental Verification

The experimental platform of low power was built to verify the feasibility of the structure, control strategy, and modulation strategy. The platform consists of three parallel substations, which were composed of the three-model 3L-NPC cascaded converter.

The output voltage waves of the three-model 3L-NPC cascaded converter are shown in Figure 19. Three 3L-NPC models were cascaded, thus the output voltage wave of each inverter was 5-level and the wave of the 3L-NPC cascaded converter was 13-level. The output voltage waves of the experiment were stable with the theoretical analysis.

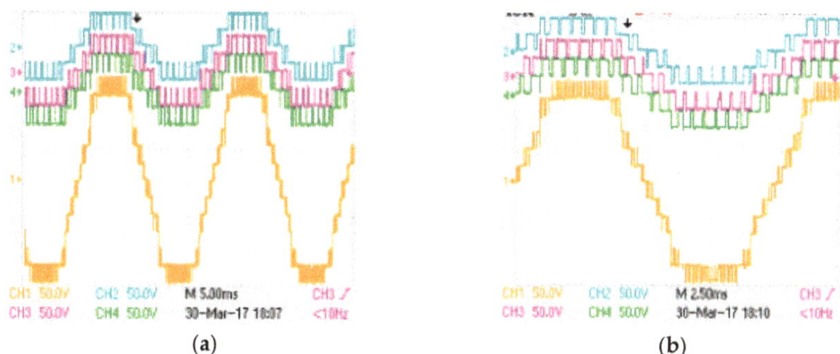

(a) (b)

Figure 19. The output voltage waves of the three-model 3L-NPC cascaded converter. (a) The relationship of the output voltage waves; (b) the output voltage waves in about one period (CH1: the output voltage of cascaded converter; CH2: the voltage of the inverter 1; CH3: the voltage of the inverter 2; CH4: the voltage of the inverter 3).

The output current waves of the three parallel traction substations and the load voltage wave are exhibited in Figure 20. The same amplitude and phase of three current waves were obtained after about 50 ms to adjust. The current could be outputted by three traction substations equally. The feasibility of the theoretical analysis and simulation could be verified by the results of the low power experiment.

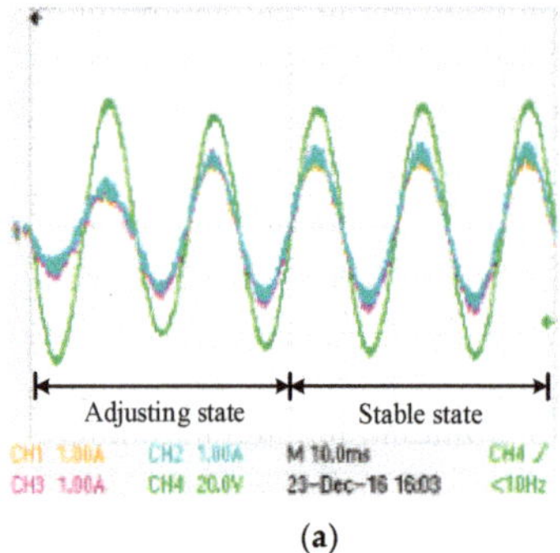

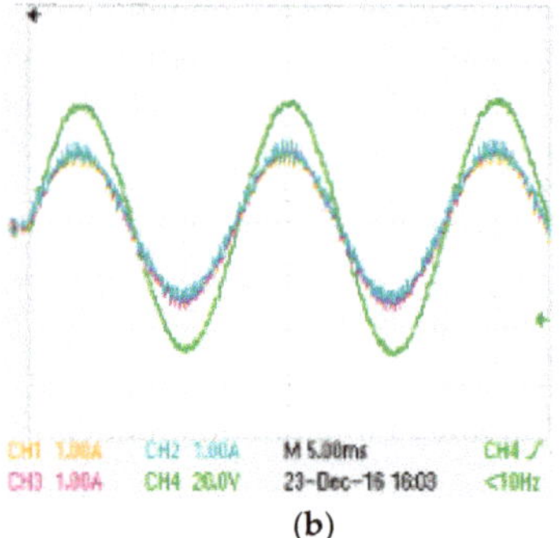

Figure 20. The output current waves of the three parallel traction substations and the load voltage wave. (**a**) The waves under the adjusting state and stable state; (**b**) the relationship between the current waves and the load voltage wave under the stable state (CH1: the output current of the SS1; CH2: the output current of the SS2; CH3: the output current of the SS3; CH4: the voltage of load.).

6. Conclusions

In this paper, ACTPSS based on the three-phase to single-phase cascaded converter was studied to solve the problems of power qualities and to cancel all the neutral sections in traditional TPSS. Based on the research of ACTPSS, an improved PQ decomposition algorithm was proposed to calculate the power flow of ACTPSS. As a result, the impedance model of the traction network was built and analyzed. Then, the power flow situations under different operation conditions were analyzed. Finally, simulation and experimental results were given to validate the proposed improved PQ decomposition algorithm strategy. According to the aforementioned analysis and implementation, the following advantages could be obtained:

(1) The induced current in the rail-earth loop was far greater than that in the conduction circuit. As a result, when the distance between the locomotive and the substation was greater than 5 km, the resulting impedance calculation error was less than 5%. Therefore, the simplified equivalent model of the traction network was valid in the traction network impedance model of ACTPSS.

(2) The results showed that the minimum voltage of the traction network was higher than 19 kV, which met the running requirements of locomotive load. The analyses indicated that the capacity of the individual substation could be reduced by nearly 10%. The theoretical fundament of ACTPSS was laid in this paper.

(3) The power losses of the ACTPSS between the nodes were calculated. According to the calculation of the output power, the calculation results illustrated that the output power was affected by the distance between the locomotive and the traction substation. The closer the distance between the locomotive and the substation, the more power would output by the substation.

(4) On the basis of theoretical analysis, the correctness and validity of the proposed improved PQ decomposition algorithm strategy were proven by the simulation and the experimental, which laid a foundation for the application of ACTPSS.

Author Contributions: X.H. and H.R. analyzed the strategy and conceived the experiment; J.L. collected and analyzed the data; P.H. performed the experiment; Y.W. and X.P. wrote the paper; and Z.S. contributed the experiment prototype.

Acknowledgments: This work was supported by the National Natural Science Foundation of China (Grant Nos. 51477144) and the National Rail Transit Electrification and Automation Engineering Technique Research Center Open Project (Grant No. NEEC-2017-A01).

Conflicts of Interest: The authors declare no conflict of interest.

References

1. Feng, C.; Zhu, Q.; Yu, B.; Zhang, Y. Complexity and vulnerability of high-speed rail network in China. In Proceeddings of the 2017 36th Chinese Control Conference (CCC), Dalian, China, 26–28 July 2017; pp. 10034–10039.
2. Zhang, J.; Wu, M.; Liu, Q. A Novel Power Flow Algorithm for Traction Power Supply Systems Based on the Thévenin Equivalent. *Energies* **2018**, *11*, 126. [CrossRef]
3. Zhao, N.; Roberts, C.; Hillmansen, S.; Nicholson, G. A Multiple Train Trajectory Optimization to Minimize Energy Consumption and Delay. *IEEE Trans. Intell. Transp. Syst.* **2015**, *16*, 2363–2372. [CrossRef]
4. Wang, H.; Li, Y.; Liu, H.; Wu, L.; Sun, Y. Transmission characteristics of harmonics and negative sequence components of electrified railway in power system. In Proceeddings of the 2016 International Conference on Smart Grid and Clean Energy Technologies (ICSGCE), Chengdu, China, 19–22 October 2016; pp. 301–306.
5. Shu, Z.; Xie, S.; Lu, K.; Zhao, Y.; Nan, X.; Qiu, D.; Zhou, F.; Gao, S.; Li, Q. Digital detection, control, and distribution system for co-phase traction power supply application. *IEEE Trans. Ind. Electron.* **2013**, *60*, 1831–1839. [CrossRef]
6. Yang, X.; Hu, H.; Ge, Y.; Aatif, S.; He, Z.; Gao, S. An Improved Droop Control Strategy for VSC-Based MVDC Traction Power Supply System. *IEEE Trans. Ind. Appl.* **2018**, *54*, 5173–5186. [CrossRef]
7. He, X.; Guo, A.; Peng, X.; Zhou, Y.; Shi, Z.; Shu, Z. A Traction Three-Phase to Single-Phase Cascade Converter Substation in an Advanced Traction Power Supply System. *Energies* **2015**, *8*, 9915–9929. [CrossRef]
8. Xie, B.; Zhang, Z.; Li, Y.; Hu, S.; Luo, L.; Rehtanz, C.; Krause, O. Reactive Power Compensation and Negative-Sequence Current Suppression System for Electrical Railways with YNvd-Connected Balance Transformer—Part II: Implementation and Verification. *IEEE Trans. Power Electron.* **2017**, *32*, 9031–9042. [CrossRef]
9. Zhang, D.; Zhang, Z.; Wang, W.; Yang, Y. Negative Sequence Current Optimizing Control Based on Railway Static Power Conditioner in V/v Traction Power Supply System. *IEEE Trans. Power Electron.* **2016**, *31*, 200–212. [CrossRef]
10. Li, Q.; Zhang, J.; He, W. Study of a new power supply system for heavy haul electric traction. *J. China Railw. Soc.* **1988**, *10*, 23–31.
11. Zhang, R.; Lin, F.; Yang, Z.; Cao, H.; Liu, Y. A Harmonic Resonance Suppression Strategy for a High-Speed Railway Traction Power Supply System with a SHE-PWM Four-Quadrant Converter Based on Active-Set Secondary Optimization. *Energies* **2017**, *10*, 1567. [CrossRef]
12. He, X.; Shu, Z.; Peng, X.; Zhou, Q.; Zhou, Y.; Zhou, Q.; Gao, S. Advanced Cophase Traction Power Supply System Based on Three-Phase to Single-Phase Converter. *IEEE Trans. Power Electron.* **2014**, *29*, 5323–5333. [CrossRef]
13. Han, P.; He, X.; Wang, Yi.; Ren, H.; Peng, X.; Shu, Z. Harmonic Analysis of Single-Phase Neutral-Point-Clamped Cascaded Inverter in Advanced co-phase traction Power Supply System Based on the Big Triangular Carrier Equivalence Method. *Energies* **2017**, *11*, 1–16.
14. Townsend, C.D.; Baraciarte, R.A.; Yu, Y.; Tormo, D.; Parra, H.Z.; Demetriades, G.D.; Agelidis, V.G. Heuristic Model Predictive Modulation for High-Power Cascaded Multilevel Converters. *IEEE Trans. Ind. Electron.* **2016**, *63*, 5263–5275.
15. Haridas, K.; Khandelwal, S.; Das, A. Three phase to single phase modular multilevel converter using full bridge cells. In Proceedings of the 2016 IEEE International Conference on Power Electronics, Drives and Energy Systems (PEDES), Trivandrum, India, 14–17 December 2016; pp. 1–5.
16. Lin, H.; Shu, Z.; He, X.; Liu, M. N-D SVPWM With DC Voltage Balancing and Vector Smooth Transition Algorithm for a Cascaded Multilevel Converter. *IEEE Trans. Ind. Electron.* **2018**, *65*, 3837–3847. [CrossRef]
17. Plattner, G.; Semlali, H.F.; Kong, N. Analysis of probabilistic load flow using point estimation method to evaluate the quantiles of electrical networks state variables. *CIRED Open Access Proc. J.* **2017**, *2017*, 2087–2091. [CrossRef]
18. Liu, H.; Tang, C.; Han, J.; Li, T.; Li, J.; Zhang, K. Probabilistic load flow analysis of active distribution network adopting improved sequence operation methodology. *IET Gener. Trans. Distrib.* **2017**, *11*, 2147–2153. [CrossRef]

19. He, Z.; Hu, H.; Zhang, Y.; Gao, S. Harmonic Resonance Assessment to Traction Power-Supply System Considering Train Model in China High-Speed Railway. *IEEE Trans. Power Deliv.* **2014**, *29*, 1735–1743. [CrossRef]

20. Hu, H.; Gao, S.; Shao, Y.; Wang, K.; He, Z.; Chen, L. Harmonic Resonance Evaluation for Hub Traction Substation Consisting of Multiple High-Speed Railways. *IEEE Trans. Power Deliv.* **2017**, *32*, 910–920. [CrossRef]

21. Xu, S.; Miao, S. Calculation of TTC for multi-area power systems based on improved Ward-PV equivalents. *IET Gen. Trans. Distrib.* **2017**, *11*, 987–994. [CrossRef]

22. Hu, S.; Xie, B.; Li, Y.; Gao, X.; Zhang, Z.; Luo, L.; Krause, O.; Cao, Y. A Power Factor-Oriented Railway Power Flow Controller for Power Quality Improvement in Electrical Railway Power System. *IEEE Trans. Ind. Electron.* **2017**, *64*, 1167–1177. [CrossRef]

23. Madhusoodhanan, S.; Tripathi, A.; Patel, D.; Mainali, K.; Kadavelugu, A.; Hazra, S. Solid-State Transformer and MV Grid Tie Applications Enabled by 15 kV SiC IGBTs and 10 kV SiC MOSFETs Based Multilevel Converters. *IEEE Trans. Ind. Appl.* **2015**, *51*, 3343–3360. [CrossRef]

24. Wang, C.; Xiao, L.; Zheng, X.; Lv, L.; Xu, Z.; Jiang, X. Analysis, Measurement, and Compensation of the System Time Delay in a Three-Phase Voltage Source Rectifier. *IEEE Trans. Power Electron.* **2016**, *31*, 6031–6043. [CrossRef]

25. Jie, S.; Stefan, S.; Qu, B.; Zhang, Y.; Chen, K.; Zhang, R. Modulation Schemes for a 30-MVA IGCT Converter Using NPC H-Bridges. *IEEE Trans. Ind. Appl.* **2015**, *51*, 4028–4040.

26. Li, Q.; He, J. *Analysis of Traction Power Supply System*; Southwest Jiaotong University Press: Chendu, China, 2012.

27. Olsen, R.G.; Pankaskie, T.A. On the Exact, Carson and Image Theories for Wires at or Above the Earth's Interface. *IEEE Trans. Power Appar. Sys.* **1983**, *PAS-102*, 769–778. [CrossRef]

28. Rao, L.; He, R.; Wang, Y.; Yan, W.; Bai, J.; Ye, D. An efficient improvement of modified Newton-Raphson algorithm for electrical impedance tomography. *IEEE Trans. Magn.* **1999**, *35*, 1562–1565.

Article

Small-Scale Modular Multilevel Converter for Multi-Terminal DC Networks Applications: System Control Validation

Elie Talon Louokdom [1], Serge Gavin [2], Daniel Siemaszko [3], Frédéric Biya-Motto [1], Bernard Essimbi Zobo [1], Mario Marchesoni [4] and Mauro Carpita [2,*]

[1]   Laboratory of Electronics, Department of Physics, Faculty of Science, University of Yaoundé I, P.O. Box 812 Yaoundé, Cameroon; talonelie@gmail.com (E.T.L.); biyamotto@yahoo.fr (F.B.-M.); bessimb@yahoo.fr (B.E.Z.)
[2]   Département des Technologies Industrielles (TIN), Haute école Spécialisée de Suisse Occidentale (HES-SO), University of Applied Sciences of Western Switzerland, Route de Cheseaux 1, 1401 Yverdon-les-Bains, Switzerland; serge.gavin@heig-vd.ch
[3]   Power Electronics and Systems Consultancy, Rue de Lyon 27, CH-1201 Geneva, Switzerland; daniel.siemaszko@pesc.ch
[4]   Department of Electrical, Electronic, Telecommunications Engineering and Naval Architecture, University of Genova, Via all'Opera Pia 11A, 16145 Genova, Italy; marchesoni@unige.it
*   Correspondence: mauro.carpita@heig-vd.ch; Tel.: +41-(0)24-5576306

Received: 16 May 2018; Accepted: 22 June 2018; Published: 28 June 2018

Abstract: This paper presents the design and implementation of a digital control system for modular multilevel converters (MMC) and its use in a 5 kW small-scale prototype. To achieve higher system control reliability and multi-functionality, the proposed architecture has been built with an effective split of the control tasks between a master controller and six slave controllers, one for each of the six arms of the converter. The MMC prototype has been used for testing both converter and system-level controls in a reduced-scale laboratory set up of a Multi-Terminal DC transmission network (MTDC). The whole control has been tested to validate the proposed control strategies. The tests performed at system level allowed exploration of the advantages of using an MMC in a MTDC system.

Keywords: digital controller; digital signal processors (DSP); modular multilevel converters (MMC); multi-terminal DC network (MTDC)

---

1. Introduction

Due to the energy challenges the world is facing today, the interest in the integration into the utility grids of renewable energy sources has significantly increased in recent years. In this context, high voltage direct current (HVDC) systems are considered one of the best options to achieve high reliability in future long-distance offshore grids that are needed to interconnect offshore wind farms, loads and large-scale storage facilities [1].

Among the different power converter topologies proposed in the literature on HVDC applications, Modular Multilevel Converters (MMC) [2] have emerged in recent years due to their attractive properties such as low harmonic distortion, scalability and flexibility. The use of MMC for HVDC systems has been largely studied by the scientific community and some commercials products have been developed by constructors such as ABB, Siemens and Alstom [3,4].

HVDC systems using MMC installed in the last decade are mostly point-to-point systems [5–7]. However, the needs to strengthen the existing AC transmission grids and to balance the intermittent power of offshore wind farms over a wider area by interconnecting multiple neighboring HVDC systems have increased the interest on Multi-terminal HVDC systems (MTDC) [8,9]. The use of MMC in MTDC systems instead of conventional two-level voltage source converters (VSC) has the already cited advantages of being redundant and scalable, to reduce or eliminate bulky harmonics filters on the AC side and to avoid DC link capacitors banks. Despite these advantages, many technical challenges must be tackled to allow their large-scale use. Among these challenges, the lack of a standardized grid code for interconnecting adjacent HVDC systems [10], the DC faults protection management [11,12], the experimental validation of system control strategies to ensure power flow and DC voltage control can be cited.

Concerning this last issue, MMC-based MTDC system control can be split among a high-level (or system-level) control and a low-level (or converter-level) control [13–18]. To test and verify these control levels under realistic conditions, as well as their interactions, it is useful to make use of small-scale laboratory prototypes, able to handle the various operational modes. Some small-scale prototypes have been proposed in the scientific literature since the introduction of MMC. For instance, a prototype of MMC is built in [19], with 44 submodules (SM) per arm, each SM capacitor with a nominal voltage at 10 V. In [20], a 20 kW back-to-back MMC-based system with 3 SM per arm is presented, while [21] presents a 25 kW six-level MMC prototype. In [22], a hybrid small-scale prototype is proposed for Alternate Arm Converter (AAC) and MMC.

Considering all these previous studies, it is obvious that, given the large number of submodules used to form the whole structure of an MMC [6], the complex command and control schemes require efficient architectures. This can only be implemented by digital control techniques, making use of advanced FPGA (Field Programmable Gate Array) or DSP (Digital Signal Processors) for fast calculation and accurate timing of the switching signals of the multiple power semiconductors. An FPGA-based control of an MMC has been proposed in [23], making use of a lookup table for the generation of the output references. This approach is not well suited for closed loop control of both the external and internal dynamic behavior of the MMC. it is obvious that a single microcontroller will have difficulties in effectively performing the complex control schemes and the communication with external peripherals. Most of previous works propose a combination of DSP and FPGA boards to handle the control and communication. In [24], the combination of both processors (DSP and FPGA) is used as a central supervision unit. The DSP performs the analog to digital conversion and all the high-level control tasks, while the FPGA manages the modulation, the capacitor voltage balancing and the communication with the submodules tasks. This allows the obtaining of the advantage of a central modulator, which generates all naturally synchronized PWM signals for all switches. However, the capacitor voltage balancing in an MMC usually uses a sorting algorithm which is in itself a sequential process, not leading to an efficient use of parallel logical resources.

Distributed controller architecture can meet these challenges. A distributed architecture fits very well with the scalability of the converter's structure and the computational load is shared between several microprocessors. Nevertheless, a distributed architecture of the controller requires an effective synchronization of all arm controllers to ensure that the gate signals of all submodules are synchronous. A robust and fast communication link between the master controller and arm controllers must also be assured.

In this paper, a small-scale MMC prototype with a controller architecture designed with only DSPs is presented. The master controller uses a dual core processor which combines a C2000 Texas Instruments (TI) MCU for real time control tasks and an Arm Cortex-M3 processor for the communication purposes. This master controller interacts with six slave controllers, one for each arm, implemented with another C2000 Texas Instruments MCU. This MMC prototype is used in a reduced-scale laboratory setup of a MTDC [25] to implement both system-level and converter-level controls, and to study MMC interactions with other VSC in an existing MTDC system.

Novelty of the paper concerns the experimental verification of the usefulness of the proposed MMC control structure to ensure power flow and DC voltage control in MTDC systems. Testing on a reduced-scale mock-up is, in the opinion of the authors, a step further in comparison to the various HIL (Hardware In a Loop) real-time simulation systems options available on the market. Moreover, it will give to the reader a deep description of several implementation details, concerning both hardware and software choices made.

The paper is organized as follows: Section 2 presents the topology of MMC and a brief review of its control and MTDC control strategies. Section 3 describes the controller structure, the task partitioning as well as the communication protocols. The actual implementation and the experimental validation of the proposed architecture are presented in Section 4. A discussion on how this architecture may be practically extended to be used in MMC systems utilizing more submodules is done in Section 5. Finally, conclusion is given in Section 6.

2. Structure of MMC and Control Strategies

2.1. Structure

Figure 1 presents the typical structure of an MMC converter [1]. It consists of six arms, working as voltage sources. Each arm is made of N power submodules. A submodule consists of a storage capacitor and a half or full bridge converter. The bridge is used to insert or bypass the capacitor. The three-phase AC voltages and the total arm voltage are respectively composed of 2N + 1 and N + 1 levels per signal cycle, corresponding to different insertions or removal of capacitors. This minimizes voltage harmonic distortions and consequently allows reduction/elimination of the filter on the AC side. Furthermore, a large and bulky capacitor on the DC side is no longer necessary. The modularity of the converter allows the use of power semiconductors with reduced voltage to achieve high AC and DC voltages.

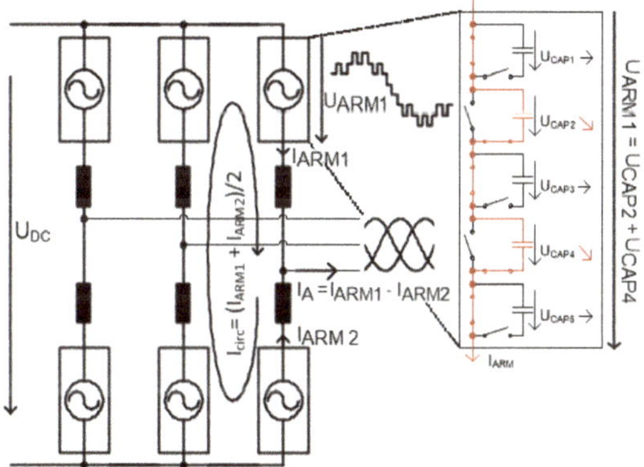

Figure 1. Structure of the Modular Multilevel Converter.

2.2. Converter-Level Control

MMCs need controllers fulfilling several purposes: the control of external voltages and/or currents loops, the control of internal current loops and the control and balancing of the capacitor voltages [4].

2.2.1. Modulation Strategies of MMC

Several modulation techniques have been proposed in the literature [4,26–30]. In this paper, phase-shifted carrier pulse width modulation (PS-PWM) has been chosen, given the low number of submodules used in the prototype [26]. For a full-scale application with an increasing number of modules, other modulation schemes should be preferred, e.g., the Nearest Level Control [1].

2.2.2. Output Current and Energy Stored Control

The output current control in the MMC is similar to the control used in conventional 2-level VSC. For the internal control of MMC, two approaches can be used: Non-energy-based approach, where the output DC current is uncontrolled [17] and Energy-based control, where the energy stored in converter is controlled making use of the circulating currents [31], which are an intrinsic feature of the MMC converters. In this paper, an Energy-based control, already presented by the authors in [32], has been used to decouple the capacitor voltages from the DC bus. Figure 2a illustrates its working principle.

2.2.3. Capacitor Voltage Balancing

The sub-module voltage balancing is implemented to prevent the divergence of capacitor voltages, which would eventually result in the collapse of the entire system. Much research has dealt with SM voltage balancing techniques [26,27]. The strategy retained in this paper is based on what was presented in [32]. Figure 2b summarizes this balancing strategy; the meaning of all the involved signals is explained in the caption of Figure 2. In principle, each sub-module capacitor voltage is measured and compared with the mean value of all arm capacitor voltage. The difference resulting from this comparison, with a sign given by the arm current direction, is used by a proportional controller to provide a correction value which is added to the set point voltage given to the sub-module PWM generation.

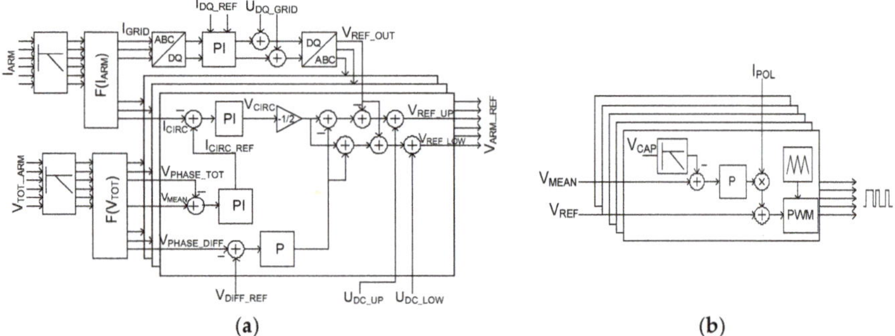

(a)   (b)

Figure 2. (a) MMC control strategy diagram. I_{ARM} are the six arm currents of the MMC, I_{GRID} are the three line currents of the utility grid computed using the block $F(I_{ARM})$ from the arm's currents. I_{CIRC} are the three-circulating currents in each phase of the MMC computed by the block $F(I_{ARM})$. I_{CIRC_REF} is the circulating current reference in each phase. I_{DQ_REF} are the AC line current references in the *dq* frame. U_{DQ_GRID} are the three AC grid voltages in the *dq* frame. V_{REF_OUT} are the three AC grid voltage references given by PI controllers. V_{TOT_ARM} are the six total arms' voltages. V_{PHASE_TOT} is the total voltage of each phase of the MMC. V_{PHASE_DIF} is the voltage difference between the upper and lower arms of the same phase. V_{DIF_REF} is the voltage difference reference between the upper and lower arms of the same phase. U_{DC_LOW} and U_{DC_UP} are the half of DC grid voltage. V_{MEAN} is the mean voltage of each phase. V_{ARM_REF} are the six arms' voltage references given by the controller. P is a proportional controller; (b) Submodules capacitors voltage balancing strategy: V_{MEAN} is the mean voltage of each arm. V_{CAP} is the capacitor voltage of each submodule. V_{REF} is the voltage reference of each submodule capacitor. I_{POL} gives the sign of the arm current. PWM is the block generating gates signals.

2.3. MTDC Control Strategies

The control of a MTDC system focused on DC line voltage control and the control of the power exchange among HVDC stations. Several control strategies have been proposed in the literature.

These include voltage margin control, voltage droop control, dead-band voltage droop control and non-dead band voltage control [33].

The voltage margin control method is an extension of the point-to-point HVDC transmission systems control. One of the terminals (also called master terminal or "slack-bus") controls the DC voltage and the other terminals (slaves terminals) can arbitrarily (or based on available resources) inject or draw power. Controlling the MTDC network voltage at a single terminal has the drawbacks that the master converter is the only one to participate in the regulation of the DC voltage. It is, therefore, necessary that the AC network associated with the master converter can absorb or provide all the power variations necessary for the balance of the MTDC system, in particular in case of a fault. Moreover, the single terminal used as balanced terminal must be sized to cope with all power variations situations. This result is a weakness of the entire system since if the master converter is lost, the MTDC system will be no longer regulated and collapses [8].

In the voltage droop control method, the DC voltage variation is used as a common signal by all converters that adjust their power based on this DC voltage. Thus, the task of controlling the DC voltage is shared among all the converters. To stabilize the MTDC system, a dead zone can be added to the droop control setting, also called dead-band voltage droop control. This allows discrimination between normal and disturbed operation of the MTDC system. However, the control activity of the converters within the band (normal operation) is fully lost. This has the disadvantage that some of the droop control parameters are set to zero and infinity, which does not give any degree of freedom for optimization [34]. In the non-dead band voltage droop control, the dead zone is replaced by a real power-voltage characteristic slightly inclined. Therefore, different droop constants can be used, depending on the deviation between DC voltage set-point and measure. Several control characteristics of a non-dead band voltage control scheme are shown in Figure 3: a reference voltage and power are set for each converter, and the balancing of the system occurs by increasing the DC voltage in case of excess of power, while decreasing it otherwise. The slopes of the different sections of the characteristic are chosen according to the connected loads or sources (e.g., wind generators cannot usually provide large amounts of additional power). This method is effective to ensure stable operation of the DC network, also in case of a failure of one or even several converters.

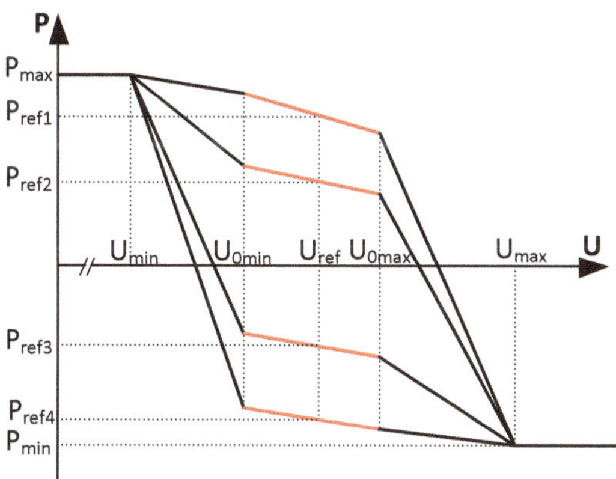

Figure 3. Power-voltage control characteristics of a non-dead band voltage control [33].

3. MMC Control System Design

3.1. Overall Controller Operation

The MMC controller architecture is presented in Figure 4. A distributed controller architecture has been chosen, given the computational limitations of a centralized controller [35]. Compared to the distributed architecture presented in [35], a dedicated synchronization signal has been used here for an effective synchronization of all arm controllers. More details about the synchronization method are given in Section 3.3.4.

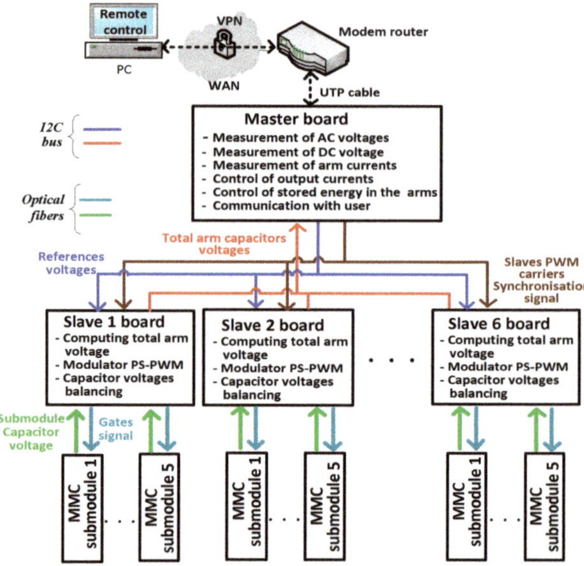

Figure 4. Signals flows between controllers and power submodules.

All control and communication tasks are distributed between the "master" controller and the arms "slave" controllers. The master controller manages the real and reactive power flow, and the communications tasks with the external world. The master controller communicates to the six slave controllers the voltage references, the arm current signs, the synchronization signal of the arm controllers and other commands such as start driving submodules. Each slave controller periodically sends back to the master the total arm voltage and performs the balancing of its arm capacitor voltages using the arm current sign received from master.

3.2. Slave Controllers and Power Modules

The main tasks of the slave controllers consist of receiving each capacitor voltage, performing capacitor voltage balancing and generating command signals for each submodule. The slave controllers communicate with the master board using inter-integrated circuit bus (I2C), and with submodules using optical fibers. This ensure the insulation requirements between the control board and the power boards. The advantages of I2C bus are the reduced number of wires and the quite simple implementation in a multi-slave environment. The I2C fast mode has been chosen with a clock frequency of 350 kHz. Each arm controller is implemented using a TMS320F28335 TI C2000 DSP family. The frequency sent by the power submodule for the capacitor voltage measurement is measured by the enhanced capture (eCAP) peripheral of the DSP and then converted into voltage. The power submodules are driven by the slave controller through optical fibers. To prevent faults and generate the dead-times, the signals are pretreated by a programmable logic device (CPLD) before being sent to the gates. The capacitor voltages are measured using a voltage-controlled oscillator (VCO) which converts the capacitor voltage into a variable frequency signal. It is then sent to the corresponding arm controller through an optical fiber link. In this application, for a capacitor voltage range between 1 and 250 V, the VCO frequency varies linearly from 3 to 450 kHz. Each module has DC bus connectors and AC output connectors (middle points of H-bridge arms). Two PWM input signals are available to control the two IGBT legs. A 3.2 mF capacitor is connected between the DC bus for local power storage. In this paper, two IGBTs only (half-bridge operation) are mounted in the power module, requiring just one PWM input. The nominal ratings of the module are 10 A and 200 V, with some 1 kV isolation voltage. The low power supply for the sub-module is provided by a medium-frequency current-fed system, consisting of an isolated power supply and a cable passing through thirty small torus from the secondary of each torus, one for each sub-module [36]. An inverter fed a nominal current of 5 A in the cable at 45 kHz. Figure 5 presents the sub-module schematic and its hardware implementation. Table 1 give some specifications of the main power submodules devices.

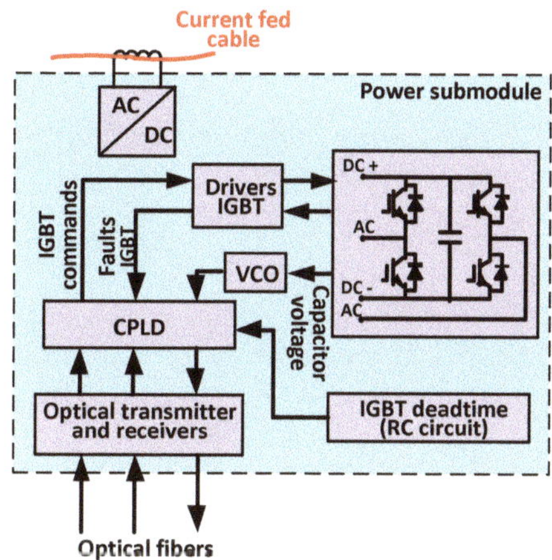

Figure 5. Power submodule schematic (principle).

Table 1. Main power submodule devices.

Main Devices	References
CPLD	XILINX XC9536XL-10VQG44C
VCO	AD654JRZ
IGBT	IXBH16N170
Optical transmitter	AVAGO HFBR-1522Z
Optical receiver	AVAGO HFBR-2522Z
IGBT gate driver	ST TD350E
Electrolytic capacitors	ESMH451VND102MB63T
Film Capacitor	MKP1848C63012JY5

3.3. Master Controller

The master controller board is designed using a TI DSP Concerto F28M35x. The main tasks of the master board are the communication with the user or host, the digital conversion of measurements, the energy flow control, the start and stop tasks and the communication with the slave boards.

3.3.1. User Communication

The master communicates with operator through a User Interface (UI). It interacts with the M3 core of the master board using Ethernet protocol allowing, for instance, the converter to be compliant with IEC61850. Information such as real power, reactive power, nominal current, start and stop command are send to the master by the user through the UI. This feature allows command of the converter remotely in a Wide Area Network (WAN).

3.3.2. Start and Stop Tasks

Once the master controller receives start command from UI, switches connect the converter to the grid through pre-charge resistors. Flowcharts of the pre-charge and the shutdown of the MMC are presented in Figure 6.

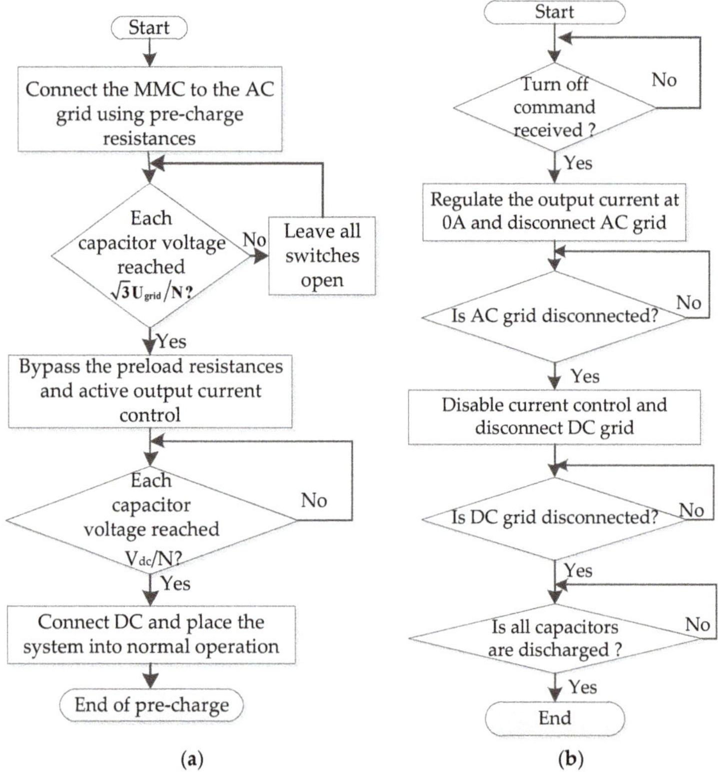

(a)   (b)

Figure 6. Flowchart of the pre-charge and turn off strategy. (**a**) Pre-charge; (**b**) Turn off.

3.3.3. Current and Voltage Measurements and Energy Flow Control

ADC operations are performed by the C2000 core. Afterwards, PLL-based grid synchronization, AC current PI controllers and energy balancing are executed. The master controller is then able to send the six voltage references to each individual arm controller. To increase the bandwidth of the communication between master and slaves, the switching frequency has been chosen as one half of the sampling frequency. The arm voltage references are then updated every half period of the switching frequency; see details next section.

3.3.4. Communication and Synchronization with Slave Boards

Figure 7a shows the master controller data transfer process to the slaves in six sampling periods. Figure 7b,c show the communication protocol developed to meet the data transfer needs. The voltage references are coded with 12 bits.

The four remaining bits are used to send the corresponding arm current sign and other commands to each slave such as the commands to start and stop the driving of submodules. All sending and receiving operations are performed within one sampling period. This protocol ensures that the voltage references and arm current sign, which are key information for the control, are updated and sent to slaves every sampling cycle. After three switching periods, the master has received the six arm voltages. This reception delay, caused by the low speed of I2C, is not so critical for the control since the capacitor voltage balancing is performed by the arm controllers. This delay has been taken into account in the controller design.

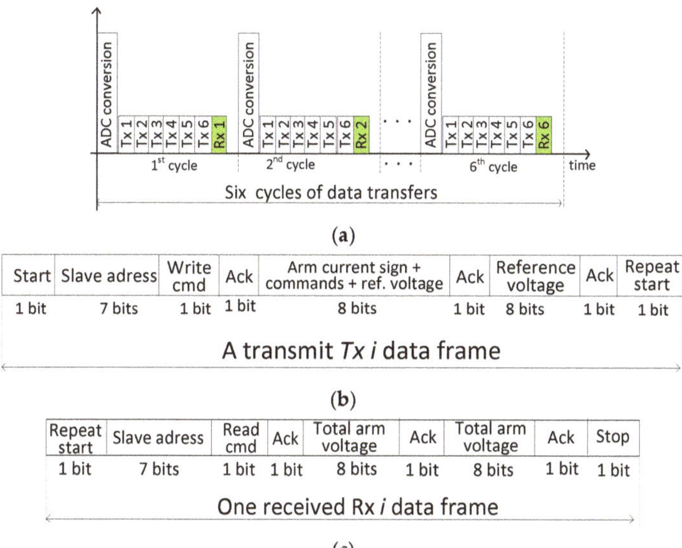

Figure 7. Master to slaves communication protocols. (**a**) Master controller data transfer with slaves per ADC cycle of conversion. $Tx\ i$ = transmit to slave i; $Rx\ i$ = receipt from slave i; (**b**) data protocol used when sending messages to each slave; (**c**) data protocol used to receive message from each slave.

A distributed architecture of the controller requires an effective synchronization of all arm controllers, to ensure that the gates signals of all submodules are synchronized. To manage this issue, a dedicated synchronization signal is used as event trigger to adjust the phase of the first PWM on each of the six slave controllers. The other carriers of the same slave controller are referenced to the first carrier.

3.4. Protection Functions

Protection functions are included at various levels of the control system. The master controller redundantly checks for possible failures by analyzing the measurements received from ADC module. If a failure is detected, it sends the appropriate command to slave controllers to stop the submodules. The master also checks if a slave controller fails to communicate. In that case, it will also send the suitable command to the other slave boards to stop the whole converter. Submodule protections concerning power section faults, overvoltage and undervoltage are directly performed by the slave controllers. Their thresholds values are included in the capacitor's voltage balancing scheme. If a non-critical failure occurs on a submodule, slave controller sends the appropriate command to bypass the faulty submodule. If the occurred failure is critical for the safe operation of the converter, the slave controller sends the appropriate command to the master and stops the entire arm. The last protection layer is performed by the CPLD on the submodule board. The CPLD uses a logic to protect the semiconductors in case of improper commands or incompatible capacitor voltage.

4. Experimental Validation

The proposed architecture has been implemented on a 5 kW reduced-scale MMC demonstrator. Figure 8 shows the whole MMC demonstrator. The main system parameters are shown in Table 2. Tests have been carried out with MMC working as inverter on a three-phase passive load with a DC link of 800 V and in a reduced-scale laboratory setup of a MTDC transmission network [25,37].

The experimental validation of the control architecture and the overall functionality of the MMC includes several aspects. The communication between the master board and an UI, the data flow between the master and slave boards together with the synchronization of the generated PWM signals, and the operation of the modules driven by the slave boards have been tested.

The current and voltage waveforms as seen from the AC side, the MMC output current and energy stored control and the power flow control in a MTDC network have been tested too.

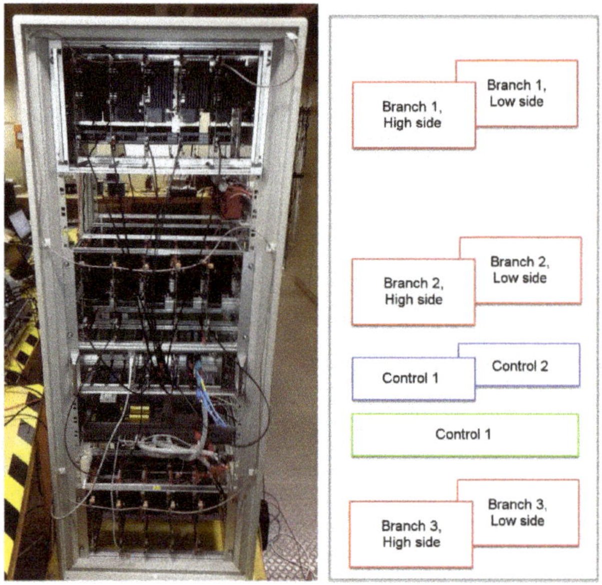

Figure 8. Whole converter reduced-scale prototype.

Table 2. Experimental setup.

Symbol	Quantity	Value
U_{grid}	RMS grid AC voltage	230 V
I_{nom}	RMS nominal current	7 A
U_{DC}	DC voltage	800 V
L	Arm inductance	6 mH
f_s	Switching frequency	1 kHz
f_{ech}	Sampling frequency	2 kHz
f_{grid}	Grid frequency	50 Hz
C	Submodule capacitor	3.3 mF
Vc	Submodule nominal voltage	160 V
N	Number of submodule per arm	5
R_{load}	AC passive load resistance	100 Ω
L_{load}	AC load Inductance	3 mH

4.1. Communications Tests

To check the communication between master board and the UI, a hypertext transfer protocol (HTTP) has been used, allowing the master board to host a web page.

The correct flow of data between master board and slave boards has been tested, checking the information timing on the serial data line (SDA) and serial clock line (SCL) of the I2C bus. Figure 9 shows communication signals in compliance with the protocol presented in Figure 7a. The communication between master and slaves is effectively performed at each sampling period (half of the switching period). Data transfer duration between boards at each sampling cycle is about 400 µs, i.e., 80% of the period. The synchronization of PWM signals between two slave boards is also highlighted in Figure 9.

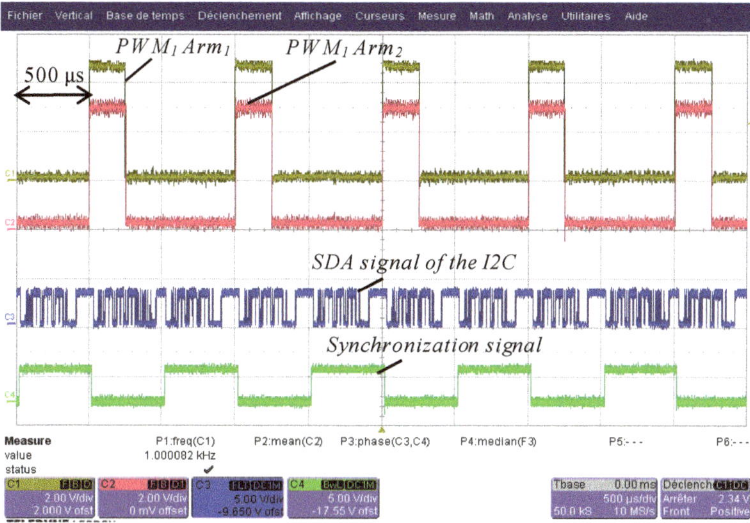

Figure 9. Signals on the converter: PWM 1 signals of two arms controllers (duty cycle 25%), Serial data (SDA) line of the I2C (blue trace) and arms' synchronization signal (green trace).

4.2. MMC Control Tests

4.2.1. Pre-Charge Test

The pre-charge is performed directly from the AC side without using any external power supply, in accordance with the flowchart presented in Figure 6a and to the fact that the module's control is supplied independently of the capacitor voltage. The transient behavior of the voltage of a single SM capacitor during the pre-charge is presented in Figure 10.

During the uncontrolled phase, this evolution corresponds to the exponential charge waveform of a capacitor as expected. Bypassing the pre-charging resistors creates a voltage jump of approximately 15 V, due to the direct connection of the MMC to the AC network. The evolution of the voltage of each capacitor during the second phase is approximately linear, thanks to the output current controller that limits the current withdrawn from the AC side. With the chosen parameters, the entire pre-charge takes less than one minute.

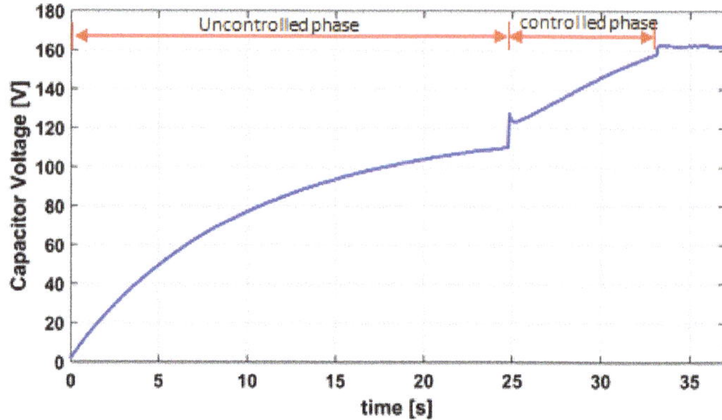

Figure 10. Evolution over time of a submodule capacitor voltage during pre-charge.

4.2.2. Capacitor Voltage Balancing Test

During this test, the capacitor voltage balancing algorithm has simply been disabled and enabled for short periods the evaluation of its impact on the operation of the MMC. This test has been performed with the DC bus at 150 V to prevent any accidental capacitors failures. To better illustrate the dynamics of the balancing strategy at nominal voltage, some simulations were performed where all capacitor voltages were unbalanced with a large spread. Their initial values are: $V_{C1} = 240$ V, $V_{C2} = 80$ V, $V_{C3} = 192$ V, $V_{C4} = 128$ V and $V_{C5} = 272$ V.

Figure 11 presents the evolution over time of the capacitor voltages on the same arm in the two situations. It can be observed in Figure 11a that, when the algorithm is disabled, the SM capacitor voltages will not converge to their average value. Figure 11b shows that, when the balancing algorithm is activated, after about 0.2 s, all SM capacitor voltages converge to their average value. This result confirms the effectiveness of the SM capacitor balancing algorithm presented in Section 2.2.3.

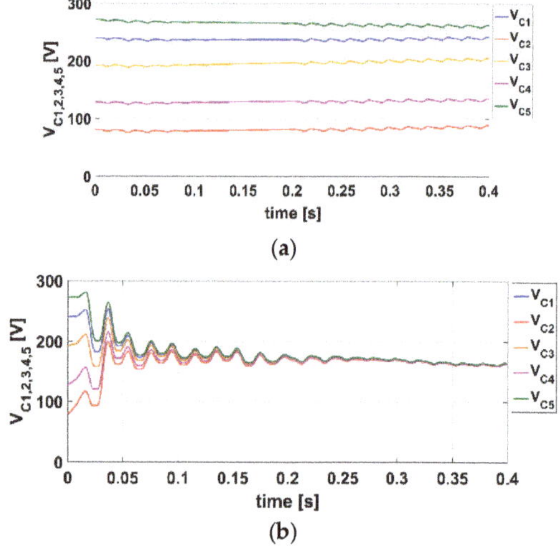

Figure 11. Capacitor balancing test on the MMC, (**a**) With the balancing algorithm disabled; (**b**) With the balancing algorithm enabled.

4.2.3. Output Current and Energy Stored Control Tests

The output current controller has been tested with the MMC working as an inverter, supplied by a DC voltage of 800 V and injecting power into the AC utility grid. Figure 12 presents the dynamics of the controller during a change of the reference of the output current from 1 A to 3 A. The new reference value is reached in less than half a period of the output current denoting thereby a good dynamic of the regulator. The control of the energy stored in the converter is performed as presented in Section 2.2.2. Figure 13 presents the control of total arm voltages. In Figure 13a, the total arm voltages are not controlled. When the DC voltage changes, the total voltages of the upper and lower arms follow the DC voltage.

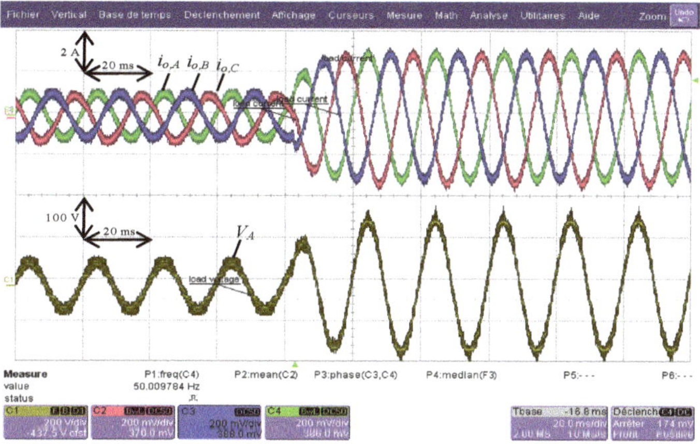

Figure 12. Output current control: output currents and output voltage of the phase A.

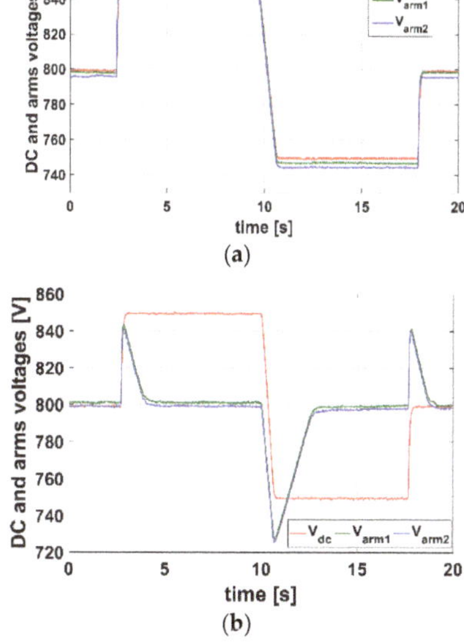

(a)

(b)

Figure 13. *Cont.*

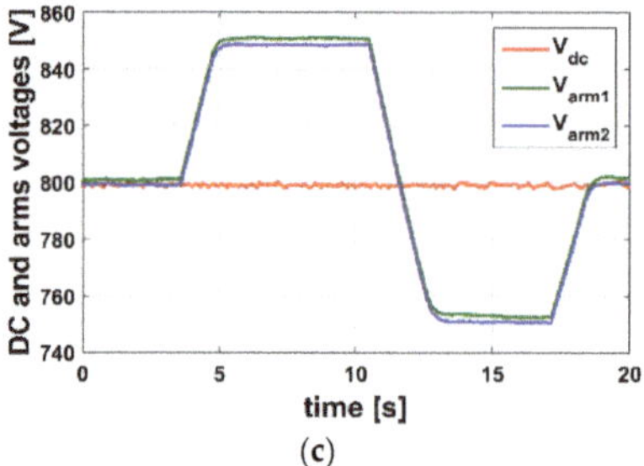

Figure 13. Energy control tests results, (**a**) DC voltage, upper and lower arms total voltages without an energy control; (**b**) DC voltage, upper and lower arms total voltages with an energy control and step change in the DC voltage; (**c**) DC voltage, upper and lower arms total voltages with an energy control and step change of total arm voltages.

Hence the DC voltage is not decoupled from capacitor voltages. If a short time brake occurs in the DC bus, this will directly affect the capacitor voltages. In Figure 13b,c, the energy control is enabled. In Figure 13b, the total arm voltages are controlled at 800 V and the DC voltage steps successively from 800 V to 850 V, from 850 V to 750 V and from 750 V to 800 V. After a short transient phase, the total arm voltages remain at their reference value. In Figure 13c, the DC voltage is held at 800 V and the total arm voltages are set to step successively from 800 V to 850 V, from 850 V to 750 V and from 750 V to 800 V. The energy control allows a decoupling between capacitor voltages and the DC bus voltage. The dynamics of the total arm voltage control loop has been intentionally reduced to avoid overvoltage on the capacitors.

4.3. Power Flow Control in a MTDC Network

The MMC has been implemented in a reduced-scale laboratory setup to test system-level control algorithms. Figure 14 shows the MTDC mock-up previously developed and described in [1]. The laboratory set-up presented in Figure 14b simulates the following imaginary situation: Three converters, namely VSC$_{\text{DELTA}}$, VSC$_{\text{GAMMA}}$, and the MMC, simulate an MTDC connection between three different AC grids. This could be the case as an example of the connection between Italy, Sardinia, and Corsica, actually made with thyristors technology. All converters are implemented with independent controllers to test the stability of the DC grid. A non-dead band voltage control [37] has been implemented on VSCDELTA and VSCGAMMA converters while the MMC controls the power injected in the AC grid. Several tests have been performed to verify the functionalities of the MMC when connected to a MTDC transmission network. The DC lines have been (roughly) simulated with a lumped parameters π equivalent circuit.

In a first test, different power profiles have been imposed on the AC side and corresponding powers are absorbed by the MMC on the DC side, as illustrated in Figure 15a. On the DC side, the stability of the DC network is secured, showing the effectiveness of the non-dead band voltage control. On the MMC side, the energy control allows maintaining of a stable voltage in the converter's sub-modules, decoupling them from the DC-link voltage variations as illustrated in Figure 15b.

A second test was performed to observe the whole MTDC system behavior when the SM of the MMC are used to store and restore energy in the system. The power transmitted between the DC and the AC1 grid remains constant. The DC voltage variation and two arms of the MMC voltages are shown in Figure 16a, and the power variations on the three VSC are shown in Figure 16b. It can be seen in Figure 16 that the DC bus is not affected by these energy variations on the MMC and that the transients due to these charges and discharges are well absorbed.

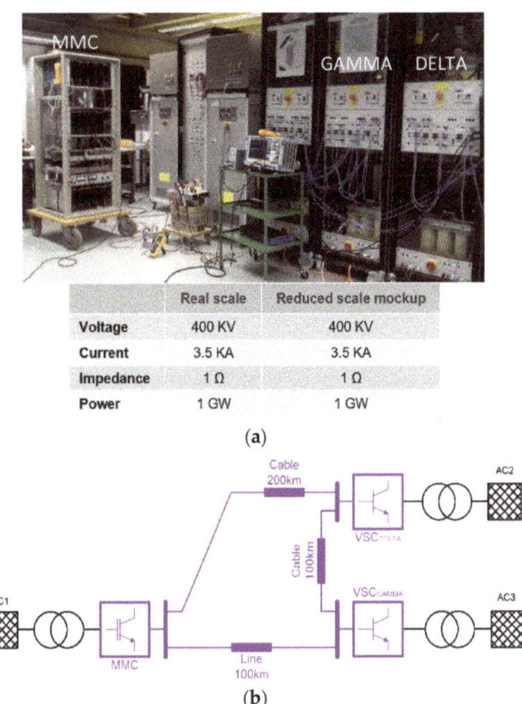

	Real scale	Reduced scale mockup
Voltage	400 KV	400 KV
Current	3.5 KA	3.5 KA
Impedance	1 Ω	1 Ω
Power	1 GW	1 GW

(**a**)

(**b**)

Figure 14. MTDC mockup, (**a**) MTDC mockup on laboratory; (**b**) Single line diagram of the laboratory setup.

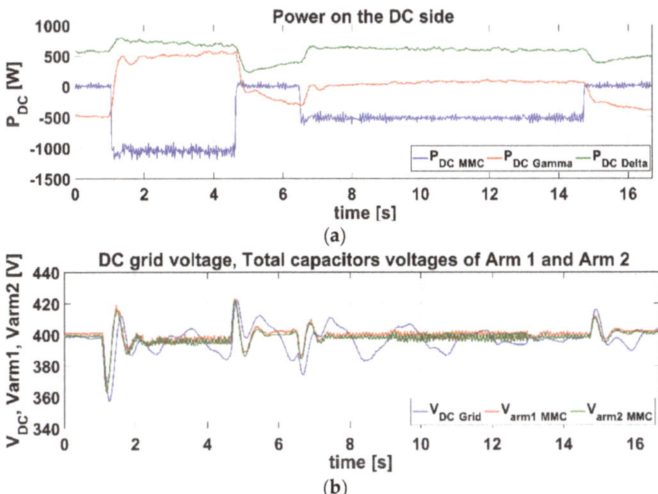

Figure 15. MTDC results with a variation of power transmitted in the onshore AC grid, (**a**) Power variations in the experimental set-up with controlled MMC and two inverters implemented with non-dead band voltage control; (**b**) DC voltage variation and two arms of the MMC voltages during power flow.

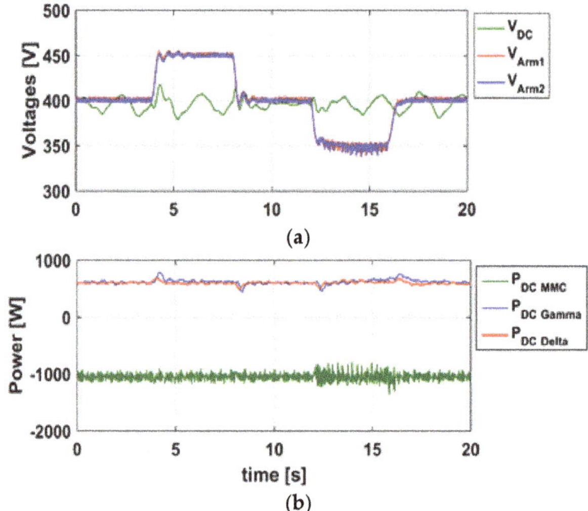

(a)

(b)

Figure 16. MTDC results when the stored energy in the MMC varies, (**a**) DC voltage variation and two arms of the MMC voltages; (**b**) Power variations in the experimental set-up with controlled MMC and two inverters implemented with non-dead band voltage control.

A third test was run to test the MTDC system with all VSC implementing the non-dead band droop control. Figure 17 presents the DC link voltage and power variation over the three converters when slight power variations are provoked. It can be seen on Figure 17 that the DC bus is well controlled, and the transients are well absorbed by the three converters when the entire system is facing power variations. This third scenario also highlights the contribution of each converter to the stability of the DC bus.

Those results confirm the correct behavior of the MMC in a MTDC system and against power variation.

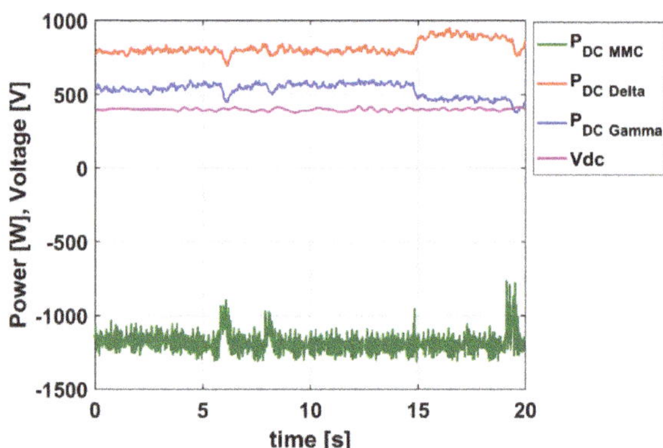

Figure 17. MTDC results when all VSC implement Non-dead band droop control.

5. Discussion and Further Improvements

Concerning the controller architecture presented in this paper, the goal was to propose a light architecture with well-known DSP that could be quickly built and that could successfully be implemented in a small-scale prototype to test the control strategies developed. One critical point of this control structure is its scalability in case of a higher number of modules. Due to its decentralized architecture, in principle it is suitable for a higher number of SM in each arm, because all the operations assigned to the master controller and the communication between the master and the slave controllers do not depend on the number of SM to control. The local management of each arm is performed by an independent controller sending the reference set-points to all PWM modulators. However, the DSP used has twelve independent PWM outputs, so a maximum number of 12 modules can be controlled, without acting on the slave hardware.

For more modules, the slave controller could be combined, or even replaced by a low-cost FPGA that would produce as many PWM outputs as required. In case of combination of the slave DSP and an FPGA, the capacitor balancing and sorting algorithms which are sequential operations remain the task of the DSP. These aspects are under development and will not be discussed further in this paper.

The proposed hardware structure will be used for a lower demanding application in terms of voltage rating, i.e., a medium-voltage application such as a MV-Statcom or MV-SOP (Soft Open Point). In this case, the number of modules to be managed is about 20–24, which is feasible with the proposed multi-DSP approach.

Another critical point is the robustness of the I2C bus. It appeared a few times during tests in a highly EM perturbed environment that the I2C bus was perturbed when the bus was not shielded enough. However, all the EMC problems in the prototype have been solved with suitable shielding and accurate state-machine and communication tuning to avoid critical situations, such as the switching of the main contactors.

6. Conclusions

This paper presents the design and the implementation of a small-scale modular multilevel converter. The proposed MMC prototype has been used to test both converter and system-level controls in a reduced-scale MMC-based MTDC network. Several MMC control levels have been tested and the results obtained validated the control strategies implemented and the controller architecture designed. The tests performed at system level have enabled exploration of the advantages of using an MMC-based MTDC system, which include the capability of the MMC to store and restore the energy in the system, its scalability which allows the extension of the DC link voltage as desired, and improvement to the AC side voltage harmonics contents.

Author Contributions: Conceptualization, E.T.L., S.G., D.S. and M.C.; Formal analysis, all authors; Funding acquisition, M.C.; Software, E.T.L., S.G. and D.S.; Supervision, M.C.; Validation, all authors; Writing—original draft, E.T.L.; Writing—review & editing, E.T.L. and M.C.

Funding: The authors acknowledge EOS-Holding (CH) for the funding of the research presented in this paper. This research is part of the activities of the Swiss Centre for Competence in Energy Research on the Future Swiss Electrical Infrastructure (SCCER-FURIES), which is financially supported by the Swiss Innovation Agency (Innosuisse-SCCER program).

Acknowledgments: The authors would like to thank P. Favre-Perrod, D. Leu, H. Parisod (HES-SO), J. Braun (CERN) for their contributions and useful suggestions.

Conflicts of Interest: The authors declare no conflict of interest.

References

1. Siemaszko, D.; Carpita, M.; Favre-Perrod, P. Conception of a modular multilevel converter in a multi-terminal DC/AC transmission network. In Proceedings of the Power Electronics and Applications (EPE'15 ECCE-Europe), Geneva, Switzerland, 8–10 September 2015. [CrossRef]
2. Lesnicar, A.; Marquardt, R. An innovative modular multilevel converter topology suitable for a wide power range. In Proceedings of the IEEE Bologna Power Tech Conference, Bologna, Italy, 23–26 June 2003. [CrossRef]

3. Sharifabadi, K.; Harnefors, L.; Nee, H.P.; Norrga, S.; Teodorescu, R. *Design, Control, and Application of Modular Multilevel Converters for HVDC Transmission Systems*; Wiley-IEEE Press: Hoboken, NJ, USA, 2016.

4. Perez, M.A.; Bernet, S.; Rodriguez, J.; Kouro, S.; Lizana, R. Circuit topologies, modelling, control schemes and applications of modular multilevel converters. *IEEE Trans. Power Electron.* **2015**, *30*, 4–17. [CrossRef]

5. Dorn, J.; Gambach, H.; Strauss, J.; Westerweller, T.; Alligan, J. Transbay cable-A breakthrough of VSC multilevel converters in HVDC transmission. In Proceedings of the CIGRE San Francisco Colloquium, San Francisco, CA, USA, 7–9 March 2012.

6. Francos, P.L.; Verdugo, S.; Alvarez, H.F.; Guyomarch, S.; Loncle, J. INELFE—Europe's first integrated onshore HVDC interconnection. In Proceedings of the IEEE Power and Energy Society General Meeting, San Diego, CA, USA, 22–26 July 2012.

7. Hussennether, V.; Rittiger, J.; Barth, A.; Worthington, D.; Rapetti, D.G.M.; Huhnerbein, B.; Siebert, M. Project borwin2 and helwin1-large scale multilevel voltage sources converter technology for bundling of offshore windpower. In Proceedings of the CIGRE, Paris, France, 26–31 August 2012.

8. Zhang, L.; Zou, Y.; Yu, J.; Qin, J.; Vittal, V.; Karady, G.; Shi, D.; Wang, Z. Modeling, control, and protection of modular multilevel converter-based multi-terminal HVDC systems: A review. *CSEE J. Power Energy Syst.* **2017**, *3*, 340–352. [CrossRef]

9. Ronanki, D.; Williamson, S. Modular Multilevel Converters for transportation electrification: Challenges and opportunities. *IEEE Trans. Transport. Electrification* **2018**, *4*, 399–407. [CrossRef]

10. Van Hertem, D.; Ghandhari, M. Multi-terminal VSC HVDC for the European supergrid: Obstacles. *Renew. Sustain. Energy Rev.* **2010**, *14*, 3156–3163. [CrossRef]

11. Bucher, M.K.; Franck, C.M. Fault current interruption in multiterminal HVDC networks. *IEEE Trans. Power Deliv.* **2015**, *31*, 87–95. [CrossRef]

12. Li, R.; Xu, L.; Yao, L.; Williams, B.W. Active control of DC fault currents in DC solid-state transformers during ride-through operation of multi-terminal HVDC systems. *IEEE Trans. Energy Convers.* **2016**, *31*, 1336–1346. [CrossRef]

13. Egea-Alvarez, A.; Bianchi, F.; Junyent-Ferre, A.; Gross, G.; Gomis-Bellmunt, O. Voltage control of multiterminal VSC-HVDC transmission systems for offshore wind power plants: Design and implementation in a scaled platform. *IEEE Trans. Ind. Electron.* **2013**, *60*, 2381–2391. [CrossRef]

14. Aragüés-Peñalba, M.; Egea-Àlvarez, A.; Galceran Arellano, S.; Gomis-Bellmunt, O. Droop control for loss minimization in HVDC multi-terminal transmission systems for large offshore wind farms. *Electr. Power Syst. Res.* **2014**, *112*, 48–55. [CrossRef]

15. Debnath, S.; Qin, J.; Bahrani, B.; Saeedifard, M.; Barbosa, P. Operation, control, and applications of the modular multilevel converter: A review. *IEEE Trans. Power Electron.* **2015**, *30*, 37–53. [CrossRef]

16. Belhaouane, M.; Freytes, J.; Ayari, M.; Colas, F.; Gruson, F.; Braiek, N.; Guillaud, X. Optimal control design for Modular Multilevel Converters operating on multi-terminal DC Grid. In Proceedings of the 2016 Power Systems Computation Conference (PSCC), Genova, Italy, 20–24 June 2016; pp. 1–7. [CrossRef]

17. Ji, H.; Chen, A.; Liu, Q.; Zhang, C. A new circulating current suppressing control strategy for modular multilevel converters. In Proceedings of the 36th Chinese Control Conference (CCC), Dalian, China, 26–28 July 2017; pp. 9151–9156. [CrossRef]

18. Zama, A.; Benchaib, A.; Bacha, S.; Frey, D.; Silvant, S. High dynamics control for MMC based on exact discrete-time model with experimental validation. *IEEE Trans. Power Deliv.* **2018**, *33*, 477–488. [CrossRef]

19. Zhou, Y.; Jiang, D.; Hu, P.; Guo, J.; Liang, Y.; Lin, Z. A prototype of modular multilevel converters. *IEEE Trans. Power Electron.* **2014**, *29*, 3267–3278. [CrossRef]

20. Binbin, L.; Dandan, X.; Dianguo, X.; Rongfeng, Y. Prototype design and experimental verification of modular multilevel converter based back-to-back system. In Proceedings of the International Symposium on Industrial Electronics, Istanbul, Turkey, 1–4 June 2014; pp. 626–630.

21. Moranchel, M.; Sanchez, F.M.; Bueno, E.J.; Rodriguez, F.J.; Sanz, I. Six-level modular multilevel converter prototype with centralized hardware platform controller. In Proceedings of the IECON 2015, Yokohama, Japan, 9–12 November 2015; pp. 863–868.

22. Jasim, O.F.; Moreno, F.J.; Trainer, D.R.; Feldman, R.; Farr, E.M.; Claree, J.C. Hybrid experimental setup for alternate arm converter and modular multilevel converter. In Proceedings of the 13th IET international conference on AC and DC Power Transmission, Manchester, UK, 14–16 February 2017; pp. 1–6.

23. Islam, M.R.; Guo, Y.; Zhu, J. FPGA-based control of modular multilevel converters: Modeling and experimental evaluation. In Proceedings of the International Conference on Electrical & Electronic Engineering (ICEEE), Rajshahi, Bangladesh, 4–6 November 2015; pp. 89–92.
24. Lago, J.; Sousa, G.J.M.; Heldwein, M.L. Digital control/modulation platform for a modular multilevel converter system. In Proceedings of the COBEP, Gramado, Brazil, 27–31 October 2013; pp. 271–277.
25. Luginbühl, M.; Pidancier, T.; Favre-Perrod, P. Transmission d'énergie par réseaux à courant continu multiterminaux. *Bull. AES/Electrosuisse* **2014**, *12*, 70–74.
26. Hagiwara, M.; Akagi, H. Control and experiment of pulse width modulated modular multilevel converters. *IEEE Trans. Power Electron.* **2009**, *24*, 1737–1746. [CrossRef]
27. Siemaszko, D. Fast sorting method for balancing capacitor voltages in modular multilevel converters. *IEEE Trans. Power Electron.* **2015**, *30*, 463–470. [CrossRef]
28. Ängquist, L.; Antonopoulos, A.; Siemaszko, D.; Ilves, K.; Vasiladiotis, M.; Nee, H.P. Open-loop control of modular multilevel converters using estimation of stored energy. *IEEE Trans. Ind. Appl.* **2011**, *47*, 2516–2524. [CrossRef]
29. Edpuganti, A.; Rathore, A.K. Optimal pulsewidth modulation of medium-voltage modular multilevel converter. *IEEE Trans. Ind. Appl.* **2016**, *52*, 3435–3442. [CrossRef]
30. Franquelo, L.G.; Rodriguez, J.; Leon, J.I.; Kouro, S.; Portillo, R.; Prats, M.M. The age of multilevel converters arrives. *IEEE Trans. Ind. Electron. Mag.* **2008**, *2*, 28–39. [CrossRef]
31. Delarue, P.; Gruson, F.; Guillaud, X. Energetic macroscopic representation and inversion based control of a modular multilevel converter. In Proceedings of the 15th European Conference on Power Electronics and Applications (EPE'13 ECCE Europe), Lille, France, 3–5 September 2013; pp. 1–10.
32. Siemaszko, D.; Talon Louokdom, E.; Parisod, H.; Braun, J.; Gavin, S.; Eggenschwiler, L.; Favre-Perrod, P.; Carpita, M. Implementation and experimental set-up of a Modular Multilevel Converter in a Multi Terminal DC/AC transmission network. In Proceedings of the 18th European Conference on Power Electronics and Applications (EPE'16 ECCE Europe), Karlsruhe, Germany, 5–9 September 2016; pp. 1–12.
33. Luginbühl, M.; Lalou, M.J. Contrôle coordonné des convertisseurs de réseaux MTDC. *Bull. AES/Electrosuisse* **2013**, *12*, 36–39.
34. Vrana, T.K.; Zeni, L.; Fosso, O.B. Active power control with undead-band voltage & frequency droop for HVDC converters in large meshed DC grids. In Proceedings of the EWEA Conference, Copenhagen, Denmark, 16–19 April 2012.
35. Talon, E.L.; Gavin, S.; Siemaszko, D.; Biya-Motto, F.; Essimbi, B.Z.; Carpita, M. Design and implementation of a multi-dsp based digital control system architecture for modular multilevel converters. In Proceedings of the IEEE-PEMC, Varna, Bulgaria, 25–30 September 2016; pp. 1182–1187.
36. Wen, H.; Xiao, W.; Lu, Z. Current-fed high-frequency AC distributed power system for medium–high-voltage gate driving applications. *IEEE Trans. Ind. Electron.* **2013**, *60*, 3736–3751. [CrossRef]
37. Talon Louokdom, E.; Siemaszko, D.; Gavin, S.; Leu, D.; Favre-Perrod, P.; Carpita, M. Use of Modular Multilevel Converter in multi-terminal DC transmission network: Reduced scale set-up and experimental results. In Proceedings of the 19th European Conference on Power Electronics and Applications (EPE'17 ECCE Europe), Warsaw, Poland, 11–14 September 2017; pp. P.1–P.9.

 *energies*

Article

Theoretical and Experimental Investigation of the Voltage Ripple across Flying Capacitors in the Interleaved Buck Converter with Extended Duty Cycle

Peter Zajec * and Mitja Nemec

Faculty of Electrical Engineering, Department of Mechatronics, University of Ljubljana, Trzaska 25, 1000 Ljubljana, Slovenia; mitja.nemec@fe.uni-lj.si
* Correspondence: peter.zajec@fe.uni-lj.si; Tel.: +386-1-4768-479

Received: 25 March 2018; Accepted: 18 April 2018; Published: 21 April 2018

Abstract: The interleaved buck converter with an extended duty cycle is analyzed in terms of unexplored parasitic switching states that diminish the switch utilization and its safety due to high-magnitude charging and discharging currents. The analysis explains the origin of the states and their effects and demonstrates their correlation with the existing voltage ripple on flying capacitors. The article further demonstrates that the voltage ripple can no longer be arbitrarily chosen as parasitic states emerge whenever the ripple exceeds an identified critical value being equal to the twofold voltage drop on the diode. A simple design criterion for flying capacitance is proposed. For a limited set of battery-powered DC–DC converters, a solution permitting the use of smaller capacitance by adding an extra switch is proposed. The derived findings are verified using experimental and simulation results.

Keywords: DC–DC conversion; interleaved buck; parasitic switching states

1. Introduction

DC–DC converters, capable to operate at a high voltage conversion ratio between the input V_{in} and output voltages V_{out}, are gaining noticeable consideration in different applications [1,2], ranging from point-of-load converters to converters in several hybrid vehicle configurations. In these applications, power switches are commonly poorly utilized. Switch utilization is, in general, defined as the ratio between the output power consumed on a load and the product of the maximum voltage and current on the switch, thus being proportional to the duty cycle [3,4]. In the past, low switch utilization was successfully addressed in inductor-tapped solutions; with the only drawback of the increased blocking voltage [5–8]. In recent years, the voltage rating of the switch is commonly reduced by applying multi-level DC–DC converters, such as the one with a flying capacitor [9–14]. In this case, the switch utilization is increased at the price of increased control complexity. On the other hand, the switches with lower voltage ratings exhibit lower conduction and switching losses, consequently increasing the overall efficiency.

Traditionally, switch utilization can be enhanced by sharing the total power among converters operating in parallel. Such a converter is designated as a multi-phase converter [15–20]. Additional benefits can be further gained by interleaving as the current ripple through the output smoothing capacitor is decreased. In order not to exceed the current rating of the individual switch, the complexity of control becomes problematic not only due to the increased number of current transducers but mostly as the current has to be evenly shared both in the steady state and during transients.

The advantages of both concepts, i.e. multi-level and multi-phase, have been successfully combined in the interleaved multi-phase Buck converter with an extended duty cycle proposed in [19,20]. There, two different topologies are presented. Both feature almost the same benefits, differing in the number of switches per phase. In a topology with two metal oxide semiconductor field effect transistors (MOSFET) per output phase (analyzed in this paper—Figure 1), all MOSFETs (T_1–T_4) are subjected to the same current (I_{load}/n, where n denotes the number of phases) and voltage ($V_{in}/2$) stress. Most importantly, the current sharing among phases takes place spontaneously. Accordingly, only the load current needs to be measured, thus make the analyzed topology an ideal example for high-current applications.

Regardless of the topology, flying capacitors are exposed to large current and thermal stress, which leads to a more rapid ageing of the component [21]. Its consequences are reduced capacitance and increased equivalent series resistance (ESR). Accordingly, when selecting capacitors, a wider design margin is needed in order to mitigate the capacitor's degradation and to guarantee error-free operation.

Commonly, when analyzing DC–DC converters, bulky capacitors are implied, assuming the voltage ripple on the capacitor is small enough compared to its average voltage. This is certain for output smoothing capacitor, which has the direct impact on the output voltage quality and on the electromagnetic interference (EMI). On the other hand, in practice, a larger voltage ripple is usually allowed across flying capacitors in order to decrease the required volume in high-density designs. No serious impacts on the basic operation of the converter owed to a higher voltage ripple on flying capacitors have been reported [9–14].

This paper offers an in-depth analysis of interleaved buck topology. The paper also derives a condition referring to the minimum capacitance that separates error-free operation from the appearance of the parasitic switching states, and discuss the effects if the condition is not met. The origin of the parasitic switching states is analyzed in detail as not yet reported in references dealing with this topology [19,20]. Furthermore, a simple solution that prevents the appearance of parasitic states and their effects is proposed and commented.

2. The Operating Principle of the Interleaved Buck Converter

Figure 1 depicts the original scheme of the two-phase interleaved buck consisting of four MOSFETs and freewheeling diodes D_{11} and D_{22} [19]. Consistently with the original paper, only the continuous conduction mode (CCM) operation is assumed.

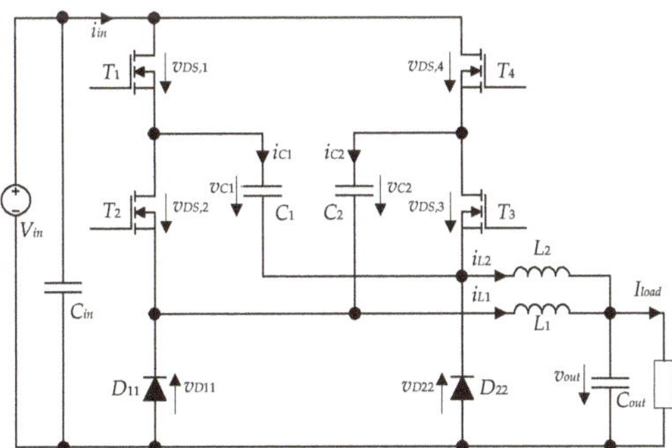

Figure 1. The interleaved two-phase buck converter with an extended duty cycle.

Accordingly, the converter enters four active switching intervals (identified by the conduction of a particular switch) that take part in a predefined sequence—either clockwise (CW sequence: T_4-T_3-T_2-T_1) or counter-clockwise (CCW sequence: T_1-T_2-T_3-T_4). Between two consecutive active switching intervals, the freewheeling interval (FW) occurs. In that interval, inductors' currents flow through D_{11} and D_{22}. All active switching intervals have the same duty cycle (d_{sw}) which never exceeds 0.25. In the same paper, it is further demonstrated that, assuming the circuit symmetry, the average voltage across the flying capacitor (C_1, C_2) equals $V_{in}/2$, and the load current is equally shared among inductors. As a result, in steady-state operation, the output voltage is proportional to $V_{in}\cdot d_{sw}/2$. To sum up, all results and conclusions provided in [19,20] apply only to highly idealized cases due to the assumed bulkiness of C_1 and C_2.

2.1. Deriving the Voltage Ripple on Flying Capacitors

During the operation, C_1 and C_2 are charged and discharged interchangeably by i_{L1} and i_{L2} as seen in Figures 2 and 3. As a result, they are inherently subjected to short current pulses of high magnitude that equal I_{L1} and I_{L2} respectively.

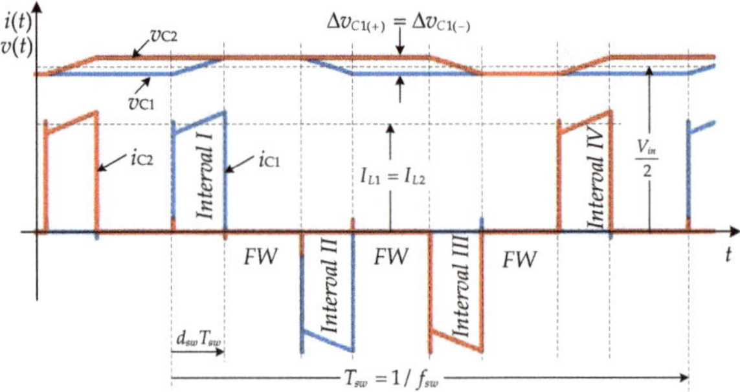

Figure 2. Simulation results showing voltages and currents that flow through flying capacitors (details regarding the simulation models built in LTspiceXVII can be found in Section 3).

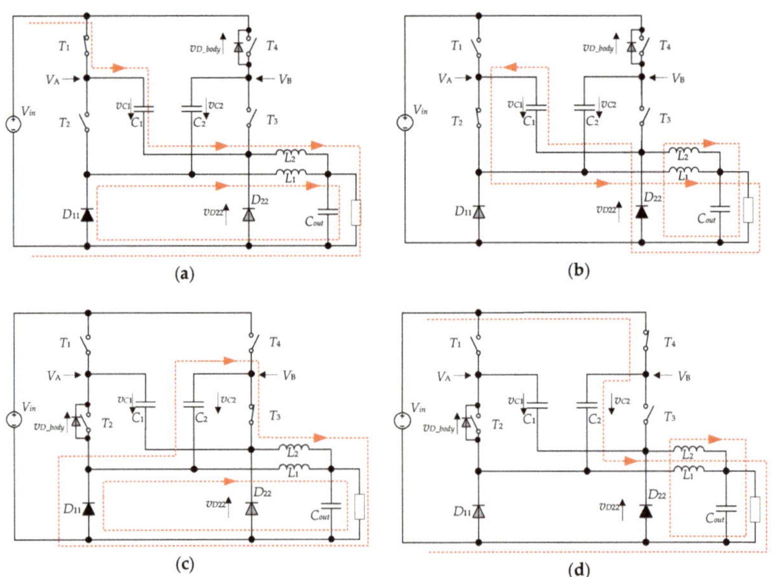

Figure 3. Equivalent circuits with indicated current paths (in red) when Δv_C (see text) is below a critical value in: (**a**) *Interval I*; (**b**) *Interval II*; (**c**) *Interval III*; (**d**) *Interval IV*.

In *Interval I*, the capacitor C_1 is charged by the inductor current i_{L2}. In *Interval II*, when T_2 is switched ON, C_1 is discharged by current i_{L1}.

Assuming a lossless capacitor v_{C1} remains unchanged during the FW interval, as well as in consecutive intervals when v_{C2} at first decays in *Interval III* (due to i_{L2}) and then in *Interval IV* rises back (due to i_{L1}). In the steady-state, the voltage increase and decrease on the individual flying capacitor are in equilibrium:

$$\begin{aligned}
\Delta v_{C1(+)} &= \frac{I_{L2} \cdot d_{sw}}{C_1 \cdot f_{sw}} = \frac{I_{L1} \cdot d_{sw}}{C_1 \cdot f_{sw}} = \Delta v_{C1(-)} \\
\Delta v_{C2(+)} &= \frac{I_{L1} \cdot d_{sw}}{C_2 \cdot f_{sw}} = \frac{I_{L2} \cdot d_{sw}}{C_2 \cdot f_{sw}} = \Delta v_{C2(-)}
\end{aligned}. \tag{1}$$

Assuming symmetrical circuit the voltage ripples, expressed as a peak–peak value, are equal ($\Delta v_{C1} = \Delta v_{C2} = \Delta v_C$) on both flying capacitors.

In *Interval II* (Figure 3b), the diode D_{22} is forward-biased. Therefore, the potential in node A is equal to the sum:

$$V_{A,II} = -v_{D22} + v_{C1}. \tag{2}$$

By neglecting the voltage drop on MOSFET (T_2), the potential in node B is:

$$V_{B,II} = -v_{D22} + v_{C1} + v_{C2}. \tag{3}$$

Figure 2 demonstrates that at the end of *Interval I*, the capacitor voltages v_{C1} and v_{C2} tend to reach their maximum values ($V_{in}/2 + \Delta v_C/2$) and remain unchanged until the start of *Interval II*. Consequently, at the start of *Interval II*, V_B reaches its maximum as well:

$$V_{B,IImax} = -v_{D22} + V_{in} + \Delta v_C. \tag{4}$$

Providing that $V_{B,II\,max}$ is lower than $V_{in} + v_{D\_body}$, the body diode in MOSFET (T_4) remains reverse-biased. Thus, if an equal voltage drop across the body diode and the freewheeling diode is assumed, the voltage sum across flying capacitors ($V_{in} + \Delta v_C$) should be kept below $V_{in} + 2v_D$. The latest can be rephrased into the condition:

$$\Delta v_C \leq 2v_D, \tag{5}$$

where $2v_D$ is recognized as a critical value of the voltage ripple Δv_C.

In a similar way, at the start of *Interval IV* (Figure 3d), the body diode in MOSFET (T_2) could conduct only if the sum of flying capacitor voltages drops below its minimum $V_{in} - \Delta v_C$. In the meantime, node B is fastened to the positive supply, causing:

$$V_{B,IV} = V_{in}, \tag{6}$$

whereas the potential in node A remains unchanged compared to *Interval II*:

$$V_{A,IV} = -v_{D22} + v_{C1}. \tag{7}$$

However, as voltages on both flying capacitors have already reached their minimum ($V_{in} - \Delta v_C/2$), $V_{A,IV}$ drops to:

$$V_{A,IVmin} = -v_{D22} + \frac{V_{in} - \Delta v_{C1}}{2}. \tag{8}$$

By assuming equal voltage drops across the body diode and the freewheeling diode, the same condition already stated in Equation (5) determines whether the body diode in MOSFET (T_2) remains biased in reverse or it turns into conduction.

2.2. The Origin of the Parasitic Switching States

Referring to the derived Equations (2)–(8), it is evident that the maximum voltage ripple permitted on flying capacitors cannot be chosen arbitrarily during the design process. If the ripple exceeds $2v_D$, the parasitic switching state emerges as the body diode of the inactive MOSFET is forward-biased. Figure 4a shows an existing current path (dashed red line) during *Interval II* and an extra path (solid blue line) that appears through the body diode in T_4.

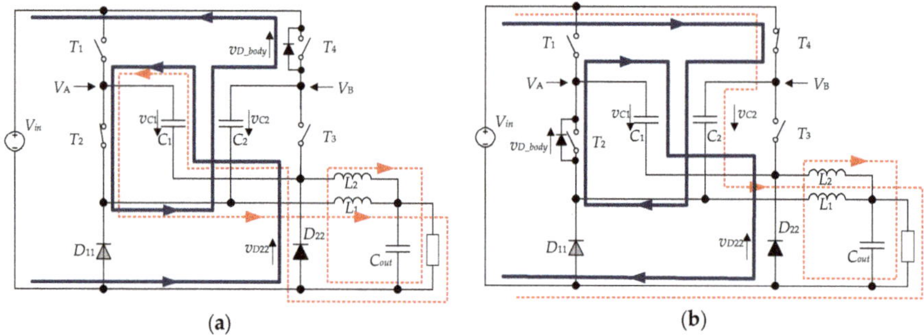

(a) (b)

Figure 4. Equivalent circuits showing extra current paths (in blue) that occur when Δv_C exceeds the critical value: (**a**) in *Interval II**; (**b**) in *Interval IV**.

The extra current path emerges at the beginning of *Interval II* and exists only for a limited time, being denoted as *Interval II**. Figure 5 shows this phenomenon in detail. During this interval, a surplus charge, which has accumulated on flying capacitors, is abruptly discharged back to the voltage source.

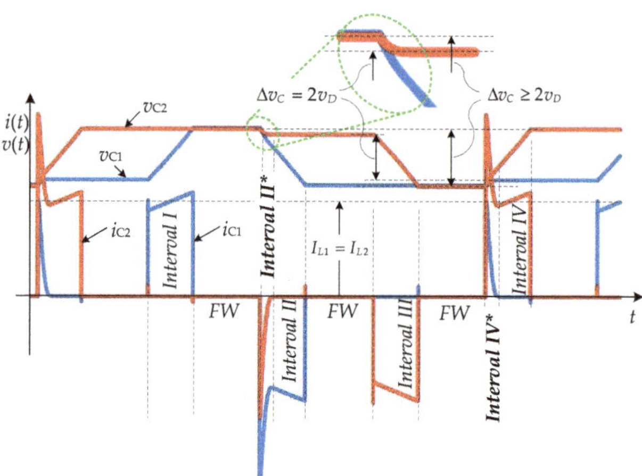

Figure 5. Simulation results showing flying capacitor currents and voltages with zoomed voltage waveforms in parasitic *Interval II**.

This overcharging does not jeopardize the validity of the derived Equations (2–8) as these define electric potentials in nodes A and B only in *Interval II* and *Interval IV*. In fact, at the end of *Interval I*, the flying capacitors could be charged to a higher voltage as shown in the zoomed part in Figure 5, depending on the parameters involved in Equation (1), thus forcing the voltage ripple over the critical value. If this is the case, the capacitors discharge until their ripples drop below $2v_D$. If the ripples remain inside the boundaries, the error-free operation depicted in Figure 2 takes place.

The shape of the discharging current, which flows simultaneously through both flying capacitors, follows:

$$i(t) = \frac{V}{\omega_r L_p} e^{-\alpha t} \cdot \sin(\omega_r t); \ \alpha = \frac{R}{2L_p}; \ \omega_r^2 = \frac{1}{L_p C} - \left(\frac{R}{2L_p}\right)^2, \tag{9}$$

where V stands for the voltage difference seen in the zoomed section in Figure 5. The shape is defined by *R-L-C* parameters found in the depicted path (Figure 4a, blue line) where small resistances (R) and inductances (L_p), both contributed by parasitic components of circuit and flying capacitors (C), have a dominant impact on the magnitude and period of the signal. Its period could be as short as the conduction interval of power switches, whereas its magnitude can easily reach or even surpass the inductor current.

Figure 4b indicates a similar parasitic switching state which could occur when T_4 turns ON if in that instant the sum of voltages across C_1 and C_2 is too low to maintain the body diode of T_2 reverse-biased. In that case, the capacitors are abruptly charged from the power source through the forward-biased body diode of MOSFET T_2. As D_{22} is forward biased by the freewheeling current forced by L_2, the surge current (blue line) can be considered to flow in opposite direction thus decreasing the current through D_{22}.

It is apparent that such an operation is not desirable as conduction losses may increase considerably, but also since semiconductors and flying capacitors could be overstressed by the current. In addition, if the front side of the converter includes an overcurrent protection, its level should be high enough to prevent an unintentional tripping. Therefore, to prevent the occurrence of these states and their impacts, the capacitance should be set higher than the critical one:

$$C_{crit} = \frac{I_{load}}{2v_D \cdot f_{sw}} \cdot \frac{V_{out}}{V_{in}}. \tag{10}$$

Equation (10) is derived from Equation (1) by taking into account that the load current is equally shared among inductors and the voltage ratio V_{out}/V_{in} is proportional to $d_{sw}/2$. It is important to point out that the form of Equation (10) is not significantly different from equations that can be derived for an arbitrary converter with flying capacitors [9,11,13,14]. The critical capacitance in this particular converter is inherently limited by topology itself, as the ripple must not exceed the twofold voltage drop on the diodes ($2v_D$). This finding is the essential contribution to the original papers [19,20] in order to preserve an error-free operation.

3. Simulation and Experimental Results

Simulations have been performed using LTspiceXVII. The simulation model did not include any additional inductances besides those already present in the models of transistors and diodes.

Experimental verifications have been carried out in order to prove the theoretical reasoning and simulation results already partially presented when introducing the parasitic switching states and their effects in the previous section. The experimental setup including a custom-made converter with a TMS320F2806 DSP (Texas Instruments, Dallas, TX, USA) is presented in Figure 6. Table 1 summarizes the main parameters of the experimental setup.

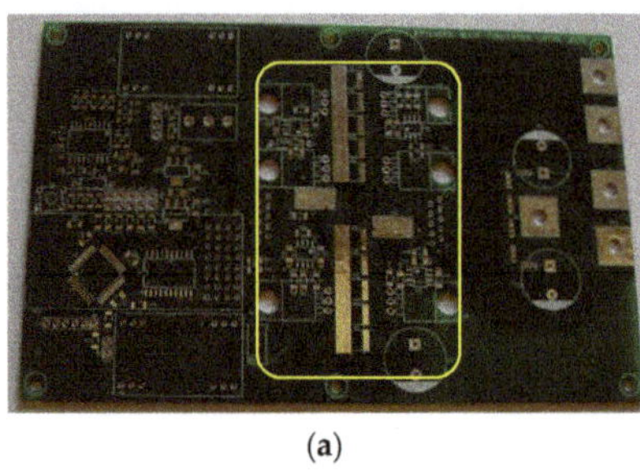

(a)

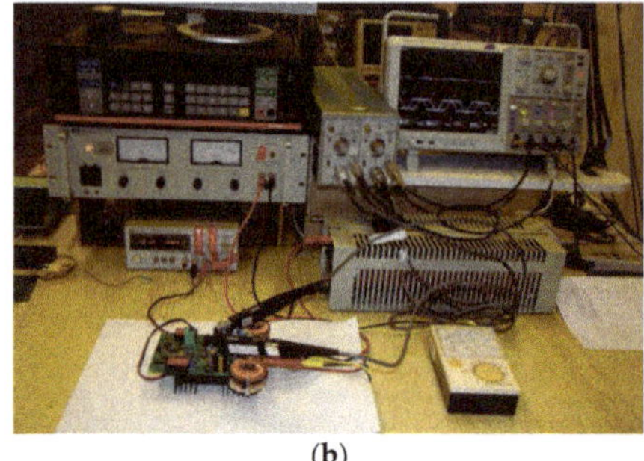

(b)

Figure 6. Pictures of the experimental setup: (**a**) an unpopulated printed circuit board (PCB) with a marked (yellow) section consisting of six MOSFETs and flying capacitors; (**b**) converter at the test bench.

Table 1. Parameters of the prototype converter and its simulation model.

Label	Description	Value
V_{in}	Input voltage	30 V
V_{out}	Output voltage	3 V
$I_{load,max}$	Load current	50 A
C_1, C_2	Flying capacitance	50 µF
L_1, L_2	Inductance	55 µH
f_{sw}	Switching frequency	33 kHz

The converter is deployed on a two-layer PCB. The topology is the same as in Figure 1 but the D_{11} and D_{22} were replaced with MOSFETs in order to increase the flexibility of converter thus enabling additional research. For measurements, these MOSFETs were permanently OFF to emulate the freewheeling diodes D_{11} and D_{22}. In order to reduce EMI to the lowest possible level, flying capacitors are placed in a close proximity of switching nodes with a high *dv/dt*, thus keeping the current loops that are subjected to pulses with a high *di/dt* short as well. In order to enable the current measurement through flying capacitors, the PCB was not equipped with multilayer ceramic capacitors (MLCC). Instead, they are placed on dedicated holders which are connected to the rest of the circuit by solid wires where a current probe (A6302 from Tektronix, Beaverton, OR, USA) can be clamped on. To simplify the tests, each MLCC holder was populated with three 15 µF/50 V capacitors. According to the known current that flows through parallel capacitors, their capacitance was estimated at 30 µF at an average capacitor voltage (15 V). For tests requiring a critical or higher capacitance, the difference was realized by fastening a temporary capacitor of an adequate capacitance.

The first test was carried out in order to verify the steady-state waveforms of the converter and its numerical model built in LTspiceXVII. Due to the additional inductance introduced by the aforesaid capacitor placement, tests have been performed at a reduced current capability $(I_{load,\,max*} = 20\text{ A})$ in order to reduce overvoltage on MOSFETs. In accordance with Equation (10), the critical capacitance amounts to 44 µF. Figure 7a,b depict i_{C1} (blue) and i_{C2} (red) at different load currents and with a capacitance greater than C_{crit} specified in Equation (10). The shape and magnitude of both currents closely match simulation traces seen in the top part of Figure 7c,d, but there is slight asymmetry of magnitudes between currents due to mismatch of inductance and capacitor values.

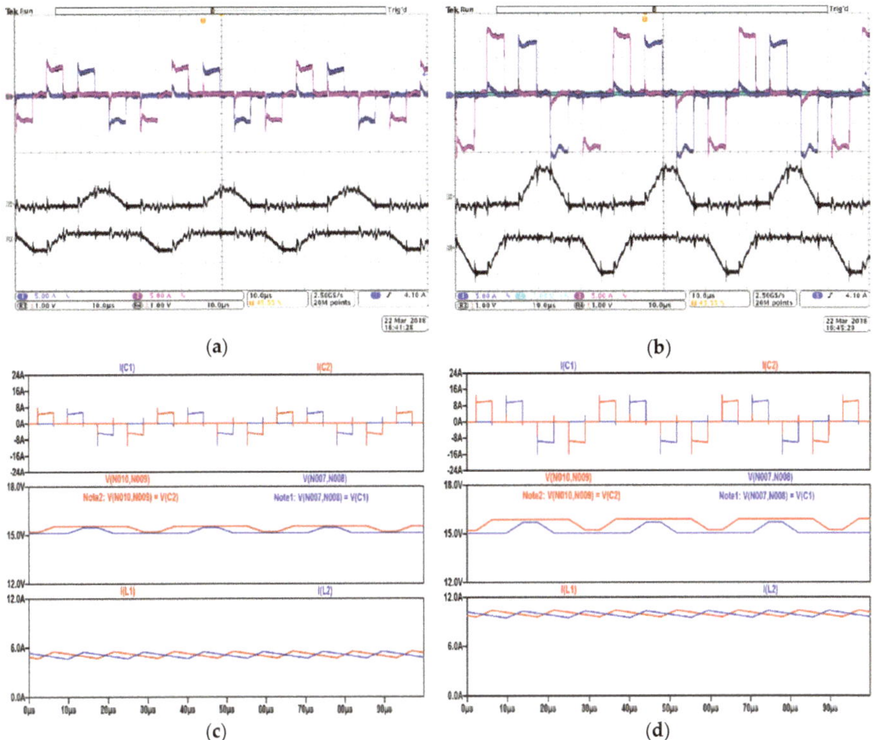

Figure 7. Experimental and simulation results obtained at $C_1 = C_2 = 60\ \mu F$, ($C_{crit} = 44\ \mu F\ @\ I_{load,max^*} = 20\ A$). Flying capacitors' currents (blue: i_{C1}; red: i_{C2}) and voltages (AC components only) on flying capacitors (top black: v_{C1}; bottom black: v_{C2}), both measured at an average inductor current: (**a**) 5 A; (**b**) 10 A. Simulated results at an average inductor current: (**c**) 5 A and (**d**) 10 A: the upper part, flying capacitors' currents: the middle part, voltages on flying capacitors: the lower part, inductors' currents.

Current spikes, which appear in experimental traces whenever an individual MOSFET turns ON, are caused by the reverse recovery charge Q_{rr} of the freewheeling diodes. Owing to the increased inductance, the trailing edges of these spikes are additionally prolonged. In addition, a capacitive coupling due to the close proximity between switching nodes with a high dv/dt and the current probe rendered the shape of current measurements, too. When voltages—especially those referenced against one of the switching node—are measured with passive voltage probes, the quality of captured traces on oscilloscope (Tektronix–DPO 4034B) usually worsen substantially. As a result, voltages shown on oscilloscope window were not measured simultaneously with current traces. Instead, they were measured separately and then recalled from internal memories.

In addition, voltage ripples in the middle part of Figure 7c,d exhibit their dependency on the load current (traces at the bottom of the same figures) as identified in the analysis. As it can be noticed, the experimental voltage ripples (at the bottom part of the Figure 7a,b) match the simulated ones quite faithfully, both in the magnitude and the shape. The average value of the flying capacitor voltages differs from $V_{in}/2$, being theoretically derived with neglected voltage drops on MOSFETs and freewheeling diodes.

The effectiveness of the proposed analytical developments given by Equations (2)–(10) is further confirmed in Figure 8, where an experimental verification was carried out considering the violation of the critical value of flying capacitance. Compared to the simulation results in Figure 8a, it can be noticed that the experimental currents in Figure 8b exhibit a longer duration of parasitic intervals. Referring to Equation (9), this deviation can be explained as simulation did not include any additional parasitic inductances which were present in the circuit (PCB traces, transistor terminals, equivalent series inductance of capacitors, capacitor holder).

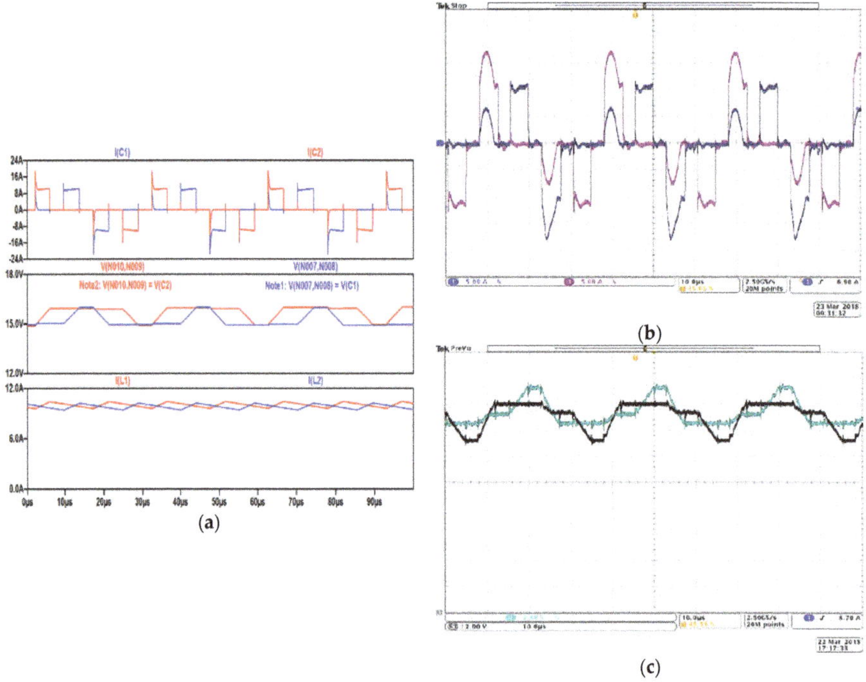

Figure 8. Simulation and experimental results obtained at an average inductor current 10 A and at $C_1 = C_2 = 35\ \mu$F: (**a**) simulated results: the upper part, flying capacitors' currents: the middle part, voltages on the flying capacitors: the lower part, inductors' currents; (**b**) measured results: flying capacitors' currents (blue: i_{C1}; red: i_{C2}); (**c**) voltages on the flying capacitors (blue: v_{C1}; black: v_{C2}).

4. Discussion

All in all, the benefit of knowing the critical capacitance can be of higher importance in the converter with n larger than 2, particularly if the converter works at higher operational temperatures. There, the selection and placement of flying capacitors in a confined volume are strengthened in order to satisfy thermal and EMI specifications. Commonly, MLCCs of type X8R, X7R or X5R are implemented [22,23]. Nowadays, they can be produced with high-temperature grades (150 °C) and with a high rated capacitance (22 μF/100 V) in a relatively small package of size 2220. On the other hand, MLCC faces a considerable voltage dependency. As a result, its rated capacitance could be met only at a reduced voltage. For high volumetric parts, the decrease of C with voltage can be greater than 50% of its rated value [23]. Furthermore, a current derating has to be taken into account in order to reduce the dissipating power inside of the flying capacitors. All this increases the chance that in space-confined designs the flying capacitance has to be chosen close to its critical value, thus increasing the probability to violate Equation (10). Taking into account the ageing effects of MLCC as well [24–26], the aforesaid becomes even more likely. Furthermore, the situation becomes even worse if diodes D_{11} and D_{22} are replaced with a synchronous MOSFET, in order to boost the efficiency even further. In this case, the peak–peak voltage ripple across C_1 and C_2 falls under the voltage drop of a single body diode, basically halving the voltage ripple given in Equation (5).

As an additional remark, it was found out that, if the sequence changes from CCW to CW, parasitic switching states occur at the beginning of *Interval I* and *Interval III*. Since the body diodes of T_1 and T_3 are biased in the forward direction, the flying capacitors are again exposed to charging and discharging currents. All together, tests show that despite high expectations, the applicability of the converter could be significantly limited due to Equation (10).

Nevertheless, simulation results additionally exhibit that in the case of battery-powered converters—equipped with a MOSFET for battery reversal protection—the parasitic switching states vanish completely. Figure 9a,b show results obtained when extra MOSFETs were added in series with the top positioned MOSFETs T_1 and T_4, forming a back-to-back switch.

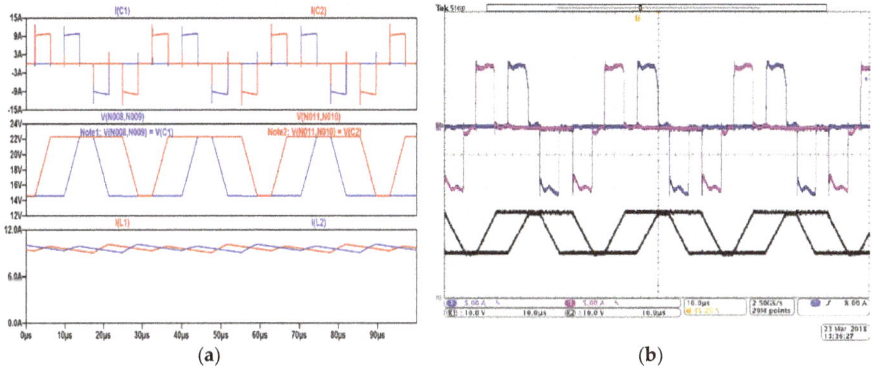

Figure 9. Comparison of results obtained at an average inductor current 10 A, with inserted extra switches and with $C_1 = C_2 = 5\,\mu\text{F}$: (**a**) simulation results: the upper part, flying capacitors' currents: the middle part, voltages on flying capacitors: the lower part, inductors' currents: (**b**) measured results: the upper part, flying capacitors' currents (blue: i_{C1}; red: i_{C2}): the lower part, voltages on flying capacitors (bottom black: v_{C1}; top black: v_{C2}).

The extra MOSFETs are switched simultaneously with T_1 and T_4. As a result, no discharging current could flow back to the voltage supply irrespective of the CW or CCW sequence. In this case, the flying capacitor is not required to fulfil Equation (5). Thus, a larger voltage ripple may be accepted. In fact, traces on the left and right side in Figure 9 correspond to a flying capacitance of just $5\,\mu\text{F}$. Traces i_{C1} and i_{C2} are similar to those in Figure 7d obtained with a much larger capacitance ($60\,\mu\text{F}$). Although the voltage ripples in Figure 9a,b are more than five times larger than the ones in Figure 7d, the load current is ideally shared among both inductors. In Figure 10, current sharing is proved even in the case when the resistance of one inductor has been intentionally increased to fivefold.

As it can be noticed in Figures 9 and 10, not only the ripple voltages increase, the average voltage across flying capacitors tends to increase as well. In this particular case, the converter remains functional at the price of an uneven voltage stress of the individual MOSFET (T_1–T_4). Furthermore, as extra MOSFETs are switched simultaneously with T_1 and T_4 and at the zero-current condition, the total losses of the switches remain more or less unaffected. The control itself remains simple and straightforward. The additional MOSFETs are placed at the bottom side of PCB in order to be efficiently cooled and to keep the parasitic inductance as low as possible.

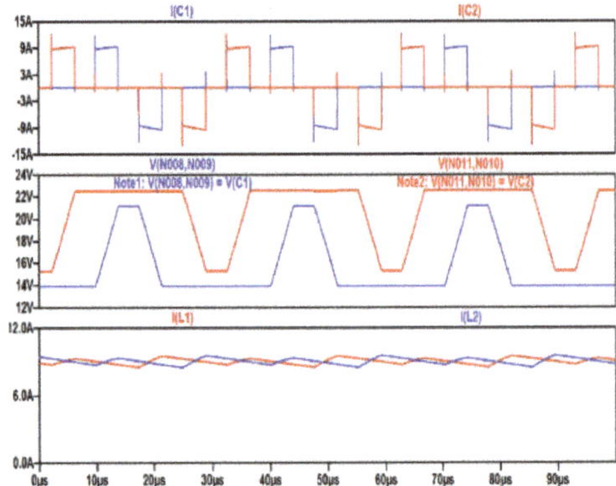

Figure 10. Simulation results obtained at an average inductor current 10 A, with inserted extra switches, with $C_1 = C_2 = 5\ \mu F$ and with unsymmetrical inductors' resistances $R_{L2} = 5R_{L1}$: the upper part, flying capacitors' currents: the middle part, voltages on flying capacitors: the lower part, inductors' currents.

5. Conclusions

The voltage ripple in the interleaved buck converter with an extended duty cycle has been analyzed in this paper, as the converter has been recognized as an ideal candidate for high-current DC–DC applications from many points of view. Specifically, in contrast to other converters being referenced, it provides an equal current and voltage stress on all MOSFETs. And most importantly, it requires less current transducers as the current sharing in output inductors takes place automatically.

The analysis presented in this paper focuses on unexplored switching states which occur under specific conditions in the two-phase interleaved converter. As their occurrence has a significant negative impact on the converter operation, a critical capacitance was derived in order to avoid them. This has been verified with a simulation model and confirmed by the measurements performed on the prototype converter. Furthermore, a simple mitigation solution based on the additional switch per converter's phase is proposed and verified. The solution is justified for battery-powered converters enabling to install a smaller flying capacitance as required by Equation (10).

Author Contributions: Both authors contributed to all of the aspects of the manuscript. To emphasize a particular specific contribution, Peter Zajec wrote the paper, designed and performed the experiments. Mitja Nemec wrote the code for the DSP and analyzed the simulation and experimental results.

Conflicts of Interest: The authors declare no conflict of interest.

References

1. Forouzesh, M.; Siwakoti, Y.P.; Gorji, S.A.; Blaabjerg, F.; Lehman, B. Step-Up DC-DC Converters: A Comprehensive Review of Voltage-Boosting Techniques, Topologies, and Applications. *IEEE Trans. Power Electron.* **2017**, *32*, 9143–9178. [CrossRef]
2. Du, S.; Wu, B. A Transformerless Bipolar Modular Multilevel DC-DC Converter With Wide Voltage Ratios. *IEEE Trans. Power Electron.* **2017**, *32*, 8312–8321. [CrossRef]
3. Erickson, R.W.; Maksimovic, D. *Fundamentals of Power Electronics*; Springer Science & Business Media: Berlin, Germany, 2007; ISBN 978-0-306-48048-5.
4. Schwingal, N.; Bernet, S. Power Semiconductor Loss Comparison of Low-Voltage, High-Power Two-Level and Three-Level Voltage Source Converters. In Proceedings of the PCIM Europe 2015, International Exhibition and Conference for Power Electronics, Intelligent Motion, Renewable Energy and Energy Management, Nuremberg, Germany, 19–20 May 2015; pp. 1–9.

5. Park, J.H.; Cho, B.H. Nonisolation Soft-Switching Buck Converter With Tapped-Inductor for Wide-Input Extreme Step-Down Applications. *IEEE Trans. Circuits Syst. Regul. Pap.* **2007**, *54*, 1809–1818. [CrossRef]

6. Yao, K.; Ye, M.; Xu, M.; Lee, F.C. Tapped-Inductor Buck Converter for High-Step-Down DC-DC Conversion. *IEEE Trans. Power Electron.* **2005**, *20*, 775–780. [CrossRef]

7. Park, J.H.; Cho, B.H. The Zero Voltage Switching (ZVS) Critical Conduction Mode (CRM) Buck Converter With Tapped-Inductor. *IEEE Trans. Power Electron.* **2005**, *20*, 762–774. [CrossRef]

8. Williams, B.W. Unified Synthesis of Tapped-Inductor DC-to-DC Converters. *IEEE Trans. Power Electron.* **2014**, *29*, 5370–5383. [CrossRef]

9. Zhang, F.; Du, L.; Peng, F.Z.; Qian, Z. A new design method for high efficiency DC-DC converters with flying capacitor technology. In Proceedings of the Twenty-First Annual IEEE Applied Power Electronics Conference and Exposition, APEC '06, Dallas, TX, USA, 19–23 March 2006.

10. Lei, Y.; Liu, W.C.; Pilawa-Podgurski, R.C.N. An analytical method to evaluate flying capacitor multilevel converters and hybrid switched-capacitor converters for large voltage conversion ratios. In Proceedings of the 2015 IEEE 16th Workshop on Control and Modeling for Power Electronics (COMPEL), Vancouver, BC, Canada, 12–15 July 2015; pp. 1–7.

11. Shenoy, P.S.; Lazaro, O.; Amaro, M.; Ramani, R.; Wiktor, W.; Lynch, B.; Khayat, J. Automatic current sharing mechanism in the series capacitor buck converter. In Proceedings of the 2015 IEEE Energy Conversion Congress and Exposition (ECCE), Montreal, QC, Canada, 20–24 September 2015; pp. 2003–2009.

12. Zhang, F.; Peng, F.Z.; Qian, Z. Study of the multilevel converters in DC-DC applications. In Proceedings of the 2004 IEEE 35th Annual Power Electronics Specialists Conference (IEEE Cat. No.04CH37551), Aachen, Germany, 20–25 June 2004; pp. 1702–1706.

13. Parastar, A.; Gandomkar, A.; Seok, J.K. High-Efficiency Multilevel Flying-Capacitor DC/DC Converter for Distributed Renewable Energy Systems. *IEEE Trans. Ind. Electron.* **2015**, *62*, 7620–7630. [CrossRef]

14. Jin, K.; Yang, M.; Ruan, X.; Xu, M. Three-Level Bidirectional Converter for Fuel-Cell/Battery Hybrid Power System. *IEEE Trans. Ind. Electron.* **2010**, *57*, 1976–1986. [CrossRef]

15. Chuang, C.F.; Pan, C.T.; Cheng, H.C. A Novel Transformer-less Interleaved Four-Phase Step-Down DC Converter With Low Switch Voltage Stress and Automatic Uniform Current-Sharing Characteristics. *IEEE Trans. Power Electron.* **2016**, *31*, 406–417. [CrossRef]

16. Pattnaik, S.; Panda, A.K.; Mahapatra, K.K. An Improved Multiphase converter for New Generation Microprocessr. In Proceedings of the Industrial and Information Systems, 2008, ICIIS 2008. IEEE Region 10 and the Third international Conference on, Kharagpur, India, 8–10 December 2008; pp. 1–5.

17. Harada, K.; Nishijima, K.; Nakano, T.; Nabeshima, T.; Sato, T. Analysis of Double Step-Down Two-Phase Buck Converter for VRM. In *Zuk. Durch Informationstechnik Schnell-Mobil-Intell. Informationstechnik Für Menschen-50 Jahre ITG Vorträge Jubil. Am 26 27 April 2004 Johann-Wolfgang-Goethe-Univ. Frankf. Am Main Mit CD-ROM*; VDE Verlag: Berlin/Offenbach, Germany, 2004; p. 497.

18. Gleissner, M.; Bakran, M.M. Design and Control of Fault-Tolerant Nonisolated Multiphase Multilevel DC-DC Converters for Automotive Power Systems. *IEEE Trans. Ind. Appl.* **2016**, *52*, 1785–1795. [CrossRef]

19. Jang, Y.; Jovanovic, M.M.; Panov, Y. Multiphase buck converters with extended duty cycle. In Proceedings of the IEEE Applied Power Electronics Conference (APEC'06), Dallas, TX, USA, 19–23 March 2006; pp. 38–44.

20. Jang, Y.; Jovanović, M. Non-isolated Power Conversion System Having Multiple Switching Power Converters. U.S. Patent 7,230,405, 12 June 2007.

21. Rating and Derating for Low-Voltage Multilayer Ceramic Capacitors (MLCCs). Available online: https://nepp.nasa.gov/files/25843/2013-Teverovsky-pres-MLCCsBME-n263.pdf (accessed on 19 March 2018).

22. Stewart, J.; Neely, J.; Delhotal, J.; Flicker, J. DC link bus design for high frequency, high temperature converters. In Proceedings of the 2017 IEEE Applied Power Electronics Conference and Exposition (APEC), Tampa, FL, USA, 26–30 March 2017; pp. 52–55.

23. Choi, D.H.; Randall, C.; Furman, E.; Ma, B.; Balachandran, U.B.; Zhang, S.; Lanagan, M. Energy and power densities of capacitors and dielectrics. In Proceedings of the 2015 IEEE International Workshop on Integrated Power Packaging (IWIPP), Chicago, IL, USA, 3–6 May 2015; pp. 52–55.

24. Biba, D.R.; Musuroi, S.; Svoboda, M. A new approach to lifetime and loading stress level estimation for Multilayer Ceramic Capacitors in Electronic Control Units. In Proceedings of the 2017 International Conference on Optimization of Electrical and Electronic Equipment (OPTIM) 2017 Intl Aegean Conference on Electrical Machines and Power Electronics (ACEMP), Brasov, Romania, 25–27 May 2017; pp. 37–42.

25. Jiang, W.; Hu, Y.; Bao, S.; Lijie, S.; Yuwei, Z.; Cui, Y.; Qiang, L. Analysis on the causes of decline of MLCC insulation resistance. In Proceedings of the 2015 16th International Conference on Electronic Packaging Technology (ICEPT), Changsha, China, 11–14 August 2015; pp. 1255–1257.
26. Scarpulla, J.; Ayvazian, T.; Buell, W.; Campbell, M.; Dubitsky, A.; Lin, R.; Martin, W.; Moss, S.; Nuccio, S.; Yarbrough, A.; et al. Thin MLCC (multi-layer ceramic capacitor) reliability evaluation using an accelerated ramp voltage test. In Proceedings of the IEEE Accelerated Stress Testing Reliability Conference (ASTR), Pensacola Beach, FL, USA, 28–30 September 2016; pp. 1–12.

 *energies*

Article

Voltage Balance Control Analysis of Three-Level Boost DC-DC Converters: Theoretical Analysis and DSP-Based Real Time Implementation

Driss Oulad-Abbou [1,2,*], Said Doubabi [1] and Ahmed Rachid [2]

[1] Laboratory of Electric Systems and Telecommunications, Cadi-Ayyad University, BP 549, Av Abdelkarim Elkhattabi, Gueliz, 4000 Marrakesh, Morocco; s.doubabi@uca.ma

[2] Laboratory of Innovative Technologies, University of Picardie Jules Verne, 80025 Amiens, France; ahmed.rachid@u-picardie.fr

* Correspondence: driss.ouladabbou@gmail.com; Tel.: +212-651-260-941

Received: 28 September 2018; Accepted: 26 October 2018; Published: 8 November 2018

Abstract: In this paper, a step-by-step description to get a unique three-level boost DC–DC converter (TLBDC) (DC—direct current) small signal model is first presented and validated through simulations and experiments. This model allows for overcoming the usage of two sub-models as in the conventional modeling approach. Based on this model, voltage balance (VB) controllers are designed and VB control analysis is presented. Two VB controllers, namely Proportional Integral (PI) and Fuzzy, were analyzed when the VB control was applied on both TLBDC switches or only one. According to the obtained simulation and experimental results, the proposed model gives an accurate approximation in dynamic, small perturbations around an operating point and steady state modes. Moreover, it has been shown that VB is achieved in a reduced time when VB control is applied on both the TLBDC's switches. Furthermore, the Fuzzy controller performs better than PI controller for VB control.

Keywords: three-level boost DC-DC converter; small signal modeling; voltage balance control

---

1. Introduction

In recent decades, modeling and control of DC–DC (DC—direct current) converters have gained much attention. This is due to their increased uses in various applications, such as voltage regulation [1–4], renewable energy interfacing [5–7], electric vehicle charging [8–10], etc. The conventional boost and buck converters are the basic topologies that are shown in Figure 1a,b, respectively. Due to their simplicity and high efficiency, they are the most used DC–DC converters. However, because of high voltage stress on their switching components, these conventional converters are not recommended for medium- and high-voltage ratings that require more powerful switching devices, which increase the cost, the volume, and the system complexity.

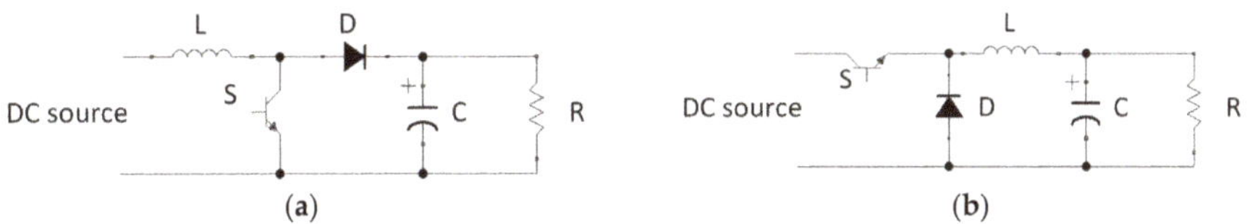

Figure 1. Conventional two-level schemes: (**a**) a boost converter, and (**b**) a buck converter.

Multilevel DC–DC converters are a suitable solution to overcome the aforementioned limitations. This is due to their ability to operate at high power ratings with higher efficiencies compared to conventional two-level topologies. They also provide other advantages such as low distortion of the output voltage and lower switching losses [2–4,11–14]. The three-level DC-DC boost converter (TLBDC) depicted in Figure 2a, has been widely discussed [14–19]. The converter fundamentals and design considerations were presented in Reference [19], where it has been shown, for instance, that the converter inductance and capacitors can be significantly reduced when compared to the two-level boost DC–DC converter.

Based on the state-space modeling approach, several TLBDC models were presented [17,18,20–22], where two sub-models were used: the first one used for a duty ratio (DR) less than 50%, and the second one is used for a DR greater than 50%. Hence, a selection parameter is required to distinguish between these two sub-models. Using a state space averaged modeling (SSAM) approach and a small signal model (SSM), introduced in Reference [23] and discussed in detail in Reference [3], the transfer functions around a corresponding operating point could be extracted. A discrete-time approach is another way for TLBDC modeling [11,24]. However, it requires long and complex calculations when compared to the previous SSAM method. In Reference [25], a DSP-based implementation of a self-tuning Fuzzy controller for TLBDC has been presented. The converter was modeled using SSAM and three cases based on the DR values were presented: for a DR less than 50%, for a DR higher than 50%, and for any DR. The main objective was the output voltage controller synthesis. However, neither the modeling procedure has been described in detail, nor the simulation and practical model validation were carried out. Moreover, comparison between the single model and the conventional modeling approach was not addressed.

The proper operation of TLBDC needs the balance of the output capacitors' voltages. Different voltage balance (VB) control methods were presented [15,21,26–34]. In References [21,29], and referring to Figure 2, the VB control was achieved by delaying forward or backward SW2 switch control signals of the TLBDC. Another method using an existing energy storage system to ensure the VB was presented in Reference [26]. In References [30–34], the VB control was performed by a PI controller. The controller output was added to the DR of the switch SW1 and subtracted from the DR of switch SW2. A sensor-less VB control method was also proposed in Reference [15] using a PI controller whose output was added to the DR of switch SW2. Finally, in Reference [25], the output capacitors' voltages were sensed, and a PI-controller was used for VB control. The controller output was added to the SW2 switch DR.

Through this literature review, it is clear that the main VB control methods consist in the following: add a small perturbation to one (or both) converter's switch(es) DR(s), or adjust the delay between the switches control signals. However, the method to choose the TLBDC switch(es) on which VB control should be applied was not addressed.

Based on these motivations, and unlike Reference [25], where the main goal was the output voltage controller synthesis, this paper adds further contributions to the state of the art by giving a step-by-step description of the followed method to get a unique model for a TLBDC working in continuous conduction mode (CCM), with a non-zero inductor equivalent series resistor. The unique model allows for avoiding the usage of two sub-models as in the conventional modeling approach, and facilitates synthesizing a convenient VB controller. This model has been validated using simulation and experimental tests, and a comparison with the conventional modeling approach is addressed. On the other hand, a technique is presented to best ensure the VB of the TLBDC. The analysis is carried out using two different VB methods and controllers, namely PI and Fuzzy controllers. This allows for figuring out the convenient controller and the adequate way for the VB control of the TLBDC.

The rest of paper is organized as follows. Section 2 describes the TLBDC operation and the developed small signal model (SSM). The VB control of the TLBDC is analyzed in Section 3, followed by the conclusion. Each of Sections 2 and 3 gives theoretical developments as well as simulation and experimental results.

2. Three-Level Boost DC–DC Converter Small Signal Modeling

The electrical scheme of the TLBDC under study is shown in Figure 2a. It is composed of an inductor L, two power switches SW1 and SW2, two switching diodes D1 and D2, and finally two output capacitors C1 and C2. $u_1(t)$ and $u_2(t)$ are the SW1 and SW2 control signals, respectively. These control signals are phase-shifted by 180°, and two operating modes could be distinguished: a DR less than 50% and a DR higher than 50%. The control signals for these two cases are shown in Figure 2b,c, respectively [15,17–19,25].

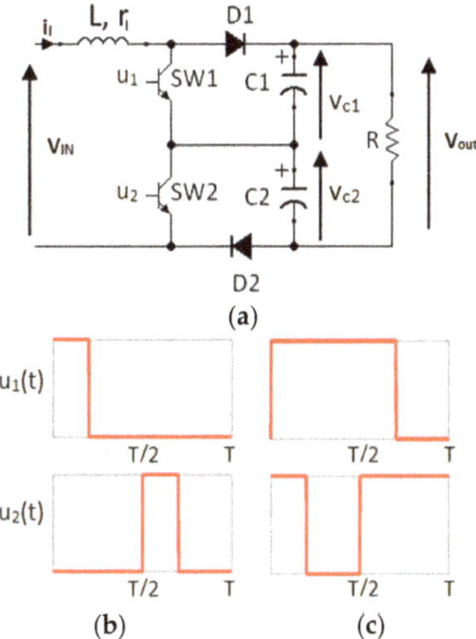

Figure 2. (a) The electrical scheme of the TLBDC under study, (b) TLBDC control signals for a DR less than 50%, and (c) TLBDC control signals for a DR higher than 50%.

Under CCM, the TLBDC is described by a set of equations and equivalent electrical schemes. These are summarized in Table 1 and Figure 3, respectively. i_l, r_l, v_{c1}, v_{c2}, v_{IN}, and v_{out} are the inductor current, inductor equivalent series resistor (ESR) (that equals 0.1 Ω in our case), capacitor C1 voltage, capacitor C2 voltage, and input and output voltages, respectively.

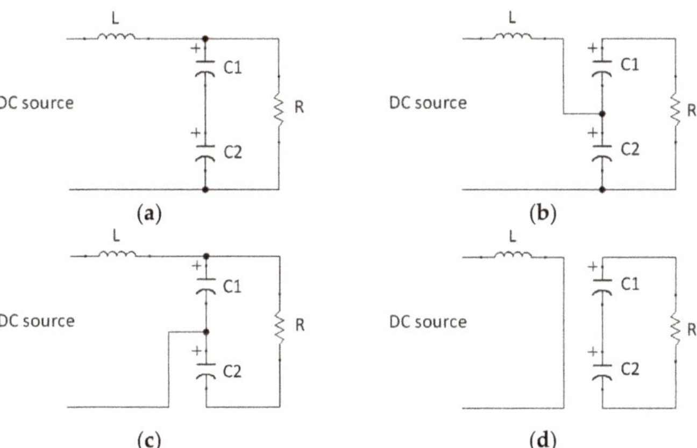

Figure 3. TLBDC equivalent electrical schemes for control signals $u_1(t)$-$u_2(t)$ sequence: (a) 0-0, (b) 1-0, (c) 0-1 and (d) 1-1.

Table 1. Differential equations for each control signals sequence of TLBDC working in CCM.

State of the Control Signals ($u_1(t)$-$u_2(t)$)	Differential Equations
0-0	$\dfrac{d}{dt}i_l = -\dfrac{1}{L}v_{c1} - \dfrac{1}{L}v_{c2} + \dfrac{1}{L}v_{IN} - \dfrac{r_l}{L}i_l,$ (1) $\dfrac{d}{dt}v_{c1} = \dfrac{1}{C1}i_l - \dfrac{1}{R\cdot C1}v_{c1} - \dfrac{1}{R\cdot C1}v_{c2},$ (2) $\dfrac{d}{dt}v_{c2} = \dfrac{1}{C2}i_l - \dfrac{1}{R\cdot C2}v_{c1} - \dfrac{1}{R\cdot C2}v_{c2},$ (3) $v_{out} = v_{c1} + v_{c2},$ (4)
1-0	$\dfrac{d}{dt}i_l = -\dfrac{1}{L}v_{c2} + \dfrac{1}{L}v_{IN} - \dfrac{r_l}{L}i_l,$ (5) $\dfrac{d}{dt}v_{c1} = -\dfrac{1}{R\cdot C1}v_{c1} - \dfrac{1}{R\cdot C1}v_{c2},$ (6) $\dfrac{d}{dt}v_{c2} = \dfrac{1}{C2}i_l - \dfrac{1}{R\cdot C2}v_{c1} - \dfrac{1}{R\cdot C2}v_{c2},$ (7) $v_{out} = v_{c1} + v_{c2},$ (8)
0-1	$\dfrac{d}{dt}i_l = -\dfrac{1}{L}v_{c1} + \dfrac{1}{L}v_{IN} - \dfrac{r_l}{L}i_l,$ (9) $\dfrac{d}{dt}v_{c1} = \dfrac{1}{C1}i_l - \dfrac{1}{R\cdot C1}v_{c1} - \dfrac{1}{R\cdot C1}v_{c2},$ (10) $\dfrac{d}{dt}v_{c2} = -\dfrac{1}{R\cdot C2}v_{c1} - \dfrac{1}{R\cdot C2}v_{c2},$ (11) $v_{out} = v_{c1} + v_{c2},$ (12)
1-1	$\dfrac{d}{dt}i_l = \dfrac{1}{L}v_{IN} - \dfrac{r_l}{L}i_l,$ (13) $\dfrac{d}{dt}v_{c1} = -\dfrac{1}{R\cdot C1}v_{c1} - \dfrac{1}{R\cdot C1}v_{c2},$ (14) $\dfrac{d}{dt}v_{c2} = -\dfrac{1}{R\cdot C2}v_{c1} - \dfrac{1}{R\cdot C2}v_{c2},$ (15) $v_{out} = v_{c1} + v_{c2},$ (16)

Based on the differential Equations (1)–(16), the TLBDC state space equations for the four control signals sequences are given by Equations (17)–(24), where Equations (17) and (18) correspond to the state space equations for 0-0 control signals state, Equations (19) and (20) correspond to the state space equations for 0-1 control signals state, Equations (21) and (22) correspond to the state space equations for 1-0 control signals state, and Equations (23) and (24) correspond to the state space equations for 0-0 control signals state.

$$\frac{d}{dt}\begin{vmatrix} i_l \\ v_{c1} \\ v_{c2} \end{vmatrix} = \begin{vmatrix} -\dfrac{r_l}{L} & -\dfrac{1}{L} & -\dfrac{1}{L} \\ \dfrac{1}{C1} & -\dfrac{1}{R\cdot C1} & -\dfrac{1}{R\cdot C1} \\ \dfrac{1}{C2} & -\dfrac{1}{R\cdot C2} & -\dfrac{1}{R\cdot C2} \end{vmatrix} \cdot \begin{vmatrix} i_l \\ v_{c1} \\ v_{c2} \end{vmatrix} + \begin{vmatrix} \dfrac{1}{L} \\ 0 \\ 0 \end{vmatrix} \cdot v_{IN}, \tag{17}$$

$$v_{out} = \begin{vmatrix} 0 & 1 & 1 \end{vmatrix} \cdot \begin{vmatrix} i_l \\ v_{c1} \\ v_{c2} \end{vmatrix}, \tag{18}$$

$$\frac{d}{dt}\begin{vmatrix} i_l \\ v_{c1} \\ v_{c2} \end{vmatrix} = \begin{vmatrix} -\frac{r_l}{L} & 0 & -\frac{1}{L} \\ 0 & -\frac{1}{R \cdot C1} & -\frac{1}{R \cdot C1} \\ \frac{1}{C2} & -\frac{1}{R \cdot C2} & -\frac{1}{R \cdot C2} \end{vmatrix} \cdot \begin{vmatrix} i_l \\ v_{c1} \\ v_{c2} \end{vmatrix} + \begin{vmatrix} \frac{1}{L} \\ 0 \\ 0 \end{vmatrix} \cdot v_{IN}, \tag{19}$$

$$v_{out} = \begin{vmatrix} 0 & 1 & 1 \end{vmatrix} \cdot \begin{vmatrix} i_l \\ v_{c1} \\ v_{c2} \end{vmatrix}, \tag{20}$$

$$\frac{d}{dt}\begin{vmatrix} i_l \\ v_{c1} \\ v_{c2} \end{vmatrix} = \begin{vmatrix} -\frac{r_l}{L} & 0 & -\frac{1}{L} \\ 0 & -\frac{1}{R \cdot C1} & -\frac{1}{R \cdot C1} \\ \frac{1}{C2} & -\frac{1}{R \cdot C2} & -\frac{1}{R \cdot C2} \end{vmatrix} \cdot \begin{vmatrix} i_l \\ v_{c1} \\ v_{c2} \end{vmatrix} + \begin{vmatrix} \frac{1}{L} \\ 0 \\ 0 \end{vmatrix} \cdot v_{IN}, \tag{21}$$

$$v_{out} = \begin{vmatrix} 0 & 1 & 1 \end{vmatrix} \cdot \begin{vmatrix} i_l \\ v_{c1} \\ v_{c2} \end{vmatrix}, \tag{22}$$

$$\frac{d}{dt}\begin{vmatrix} i_l \\ v_{c1} \\ v_{c2} \end{vmatrix} = \begin{vmatrix} -\frac{r_l}{L} & -\frac{1}{L} & 0 \\ \frac{1}{C1} & -\frac{1}{R \cdot C1} & -\frac{1}{R \cdot C1} \\ 0 & -\frac{1}{R \cdot C2} & -\frac{1}{R \cdot C2} \end{vmatrix} \cdot \begin{vmatrix} i_l \\ v_{c1} \\ v_{c2} \end{vmatrix} + \begin{vmatrix} \frac{1}{L} \\ 0 \\ 0 \end{vmatrix} \cdot v_{IN}, \tag{23}$$

$$v_{out} = \begin{vmatrix} 0 & 1 & 1 \end{vmatrix} \cdot \begin{vmatrix} i_l \\ v_{c1} \\ v_{c2} \end{vmatrix}, \tag{24}$$

Using discrete variables $u_1(t)$ and $u_2(t)$, Equations (17)–(24) could be assembled into one equation. The obtained TLBDC model is given by Equations (25) and (26).

$$\frac{d}{dt}\begin{vmatrix} i_l \\ v_{c1} \\ v_{c2} \end{vmatrix} = \begin{vmatrix} -\frac{r_l}{L} & -\frac{1-u_1(t)}{L} & -\frac{1-u_2(t)}{L} \\ \frac{1-u_1(t)}{C1} & -\frac{1}{R \cdot C1} & -\frac{1}{R \cdot C1} \\ \frac{1-u_2(t)}{C2} & -\frac{1}{R \cdot C2} & -\frac{1}{R \cdot C2} \end{vmatrix} \cdot \begin{vmatrix} i_l \\ v_{c1} \\ v_{c2} \end{vmatrix} + \begin{vmatrix} \frac{1}{L} \\ 0 \\ 0 \end{vmatrix} \cdot v_{IN}, \tag{25}$$

$$v_{out} = \begin{vmatrix} 0 & 1 & 1 \end{vmatrix} \cdot \begin{vmatrix} i_l \\ v_{c1} \\ v_{c2} \end{vmatrix}, \tag{26}$$

Equations (25) and (26) could be written as:

$$\frac{d}{dt}\begin{vmatrix} i_l \\ v_{c1} \\ v_{c2} \end{vmatrix} = \left[\begin{vmatrix} -\frac{r_l}{L} & -\frac{1}{L} & -\frac{1}{L} \\ \frac{1}{C1} & -\frac{1}{R \cdot C1} & -\frac{1}{R \cdot C1} \\ \frac{1}{C2} & -\frac{1}{R \cdot C2} & -\frac{1}{R \cdot C2} \end{vmatrix} + u_1(t) \cdot \begin{vmatrix} 0 & \frac{1}{L} & 0 \\ -\frac{1}{C1} & 0 & 0 \\ 0 & 0 & 0 \end{vmatrix} + u_2(t) \cdot \begin{vmatrix} 0 & 0 & \frac{1}{L} \\ 0 & 0 & 0 \\ -\frac{1}{C2} & 0 & 0 \end{vmatrix} \right] \cdot \begin{vmatrix} i_l \\ v_{c1} \\ v_{c2} \end{vmatrix} + \begin{vmatrix} \frac{1}{L} \\ 0 \\ 0 \end{vmatrix} \cdot v_{IN}, \tag{27}$$

$$v_{out} = \begin{vmatrix} 0 & 1 & 1 \end{vmatrix} \cdot \begin{vmatrix} i_l \\ v_{c1} \\ v_{c2} \end{vmatrix}, \tag{28}$$

Let us denote $\bar{x} = \begin{vmatrix} \bar{i_l} \\ \bar{v_{c1}} \\ \bar{v_{c2}} \end{vmatrix}$, $\bar{v_{IN}}$, $\bar{v_{out}}$, $\bar{u_1}$, and $\bar{u_2}$ as the average values of the state vector

$x = \begin{vmatrix} i_l \\ v_{c1} \\ v_{c2} \end{vmatrix}$, v_{IN}, v_{out}, u_1, and u_2, respectively. Using this notation, the obtained SSAM is given by Equations (29) and (30):

$$\frac{d}{dt}\begin{vmatrix} \overline{i_l} \\ \overline{v_{c1}} \\ \overline{v_{c2}} \end{vmatrix} = \left[\begin{vmatrix} -\frac{r_l}{L} & -\frac{1}{L} & -\frac{1}{L} \\ \frac{1}{C1} & -\frac{1}{R\cdot C1} & -\frac{1}{R\cdot C1} \\ \frac{1}{C2} & -\frac{1}{R\cdot C2} & -\frac{1}{R\cdot C2} \end{vmatrix} + \overline{u_1}\cdot\begin{vmatrix} 0 & \frac{1}{L} & 0 \\ -\frac{1}{C1} & 0 & 0 \\ 0 & 0 & 0 \end{vmatrix} + \overline{u_2}\cdot\begin{vmatrix} 0 & 0 & \frac{1}{L} \\ 0 & 0 & 0 \\ -\frac{1}{C2} & 0 & 0 \end{vmatrix} \right]\cdot\begin{vmatrix} \overline{i_l} \\ \overline{v_{c1}} \\ \overline{v_{c2}} \end{vmatrix} + \begin{vmatrix} \frac{1}{L} \\ 0 \\ 0 \end{vmatrix}\cdot\overline{v_{IN}}, \tag{29}$$

$$\overline{v_{out}} = \begin{vmatrix} 0 & 1 & 1 \end{vmatrix}\cdot\begin{vmatrix} \overline{i_l} \\ \overline{v_{c1}} \\ \overline{v_{c2}} \end{vmatrix}, \tag{30}$$

Each variable can be written as the sum of small alternating current (AC) variations and DC steady-state quantities as follows:

$$\overline{x} = X + \tilde{x}, \tag{31}$$

$$\overline{u_1} = U_1 + \tilde{u}_1, \tag{32}$$

$$\overline{u_2} = U_2 + \tilde{u}_2, \tag{33}$$

$$\overline{v_{IN}} = V_{IN} + \widetilde{v_{IN}}, \tag{34}$$

$$\overline{v_{out}} = V_{out} + \widetilde{v_{out}}, \tag{35}$$

Using this decomposition, Equations (31)–(35), Equations (29) and (30) become:

$$\dot{\tilde{x}} = [\widetilde{u_1}A_1 + \widetilde{u_2}A_2]\cdot\tilde{x} + B\widetilde{v_{IN}} + [A_0 + U_1A_1 + U_2A_2]\cdot\tilde{x} + \widetilde{u_1}A_1X + \widetilde{u_2}A_2X + [A_0 + U_1A_1 + U_2A_2].X + BV_{IN}, \tag{36}$$

$$\overline{v_{out}} = \begin{vmatrix} 0 & 1 & 1 \end{vmatrix}\cdot\begin{vmatrix} \overline{i_l} \\ \overline{v_{c1}} \\ \overline{v_{c2}} \end{vmatrix}, \tag{37}$$

where: $A_0 = \begin{vmatrix} -\frac{r_l}{L} & -\frac{1}{L} & -\frac{1}{L} \\ \frac{1}{c1} & -\frac{1}{R\cdot C1} & -\frac{1}{R\cdot C1} \\ \frac{1}{C2} & -\frac{1}{R\cdot C2} & -\frac{1}{R\cdot C2} \end{vmatrix}$, $A_1 = \begin{vmatrix} 0 & \frac{1}{L} & 0 \\ \frac{-1}{C1} & 0 & 0 \\ 0 & 0 & 0 \end{vmatrix}$, and $A_2 = \begin{vmatrix} 0 & 0 & \frac{1}{L} \\ 0 & 0 & 0 \\ \frac{-1}{C2} & 0 & 0 \end{vmatrix}$.

Neglecting the higher-order terms, steady-state terms are null, and supposing that the supply voltage is constant, terms written in bold, italic, and bold italic in Equation (36), respectively [3]. The SSM of the TLBDC is given by Equations (38) and (39):

$$\dot{\tilde{x}} = [A_0 + U_1A_1 + U_2A_2].\tilde{x} + \widetilde{u_1}A_1X + \widetilde{u_2}A_2X, \tag{38}$$

$$\widetilde{v_{out}} = \begin{vmatrix} 0 & 1 & 1 \end{vmatrix}\cdot\begin{vmatrix} \widetilde{i_l} \\ \widetilde{v_{c1}} \\ \widetilde{v_{c2}} \end{vmatrix}, \tag{39}$$

The proposed model is validated through simulation and experimental results. Simulations were performed on MATLAB software (Matworks, Natick, MA, USA) using the ode23 function, while the experimental tests were carried out on the experimental setup depicted in Section 3. The TLBDC parameters used for these tests are listed in Table 2.

The simulated and experimental output voltage curves for the switched model and the SSM around 30% and 60% DRs are respectively illustrated in Figures 4 and 5, where 4% positive and negative perturbations were introduced around those DR values.

Based on the results reported in Figure 4, it can be seen that the SSM behavior was in accordance with the switched one. In addition, the presented experimental results in Figure 5 were closely matching those obtained from the proposed SSM. By analyzing these results, one can see that the proposed SSM gave an averaged behavior of the TLBDC for both DR cases.

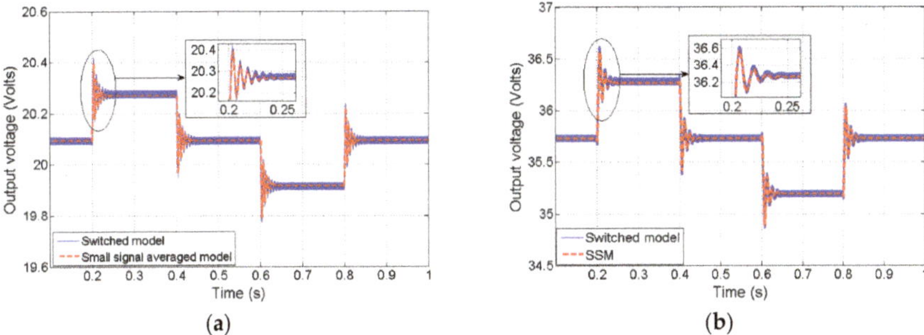

Figure 4. Simulated SSM and switched model output voltage curves in the case of 4% DR perturbation: (**a**) around 30%, and (**b**) around 60% DRs.

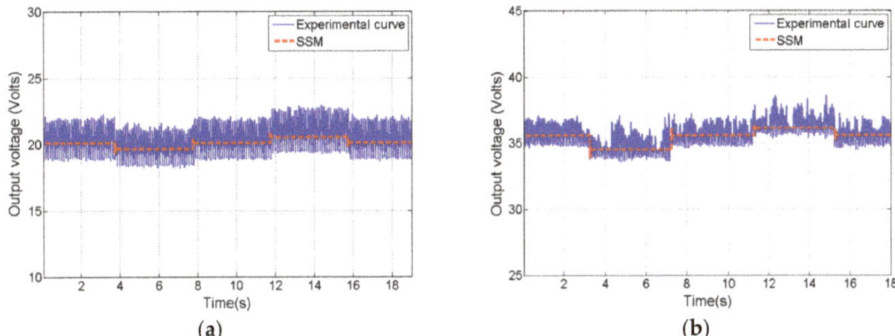

Figure 5. Experimental and SSM output voltages curves with 4% perturbation width: (**a**) around thirty percent and (**b**) around sixty percent DRs.

The results for a DR transition from a value less than 50%, namely 30%, to another one higher than 50%, namely 60%, are shown in Figure 6. The proposed SSAM and experimental output voltage curves are shown in Figure 6a, while Figure 6b illustrates the conventional SSAM and experimental output voltage curves. Finally, Figure 6c presents a comparison between the conventional approach and the proposed one.

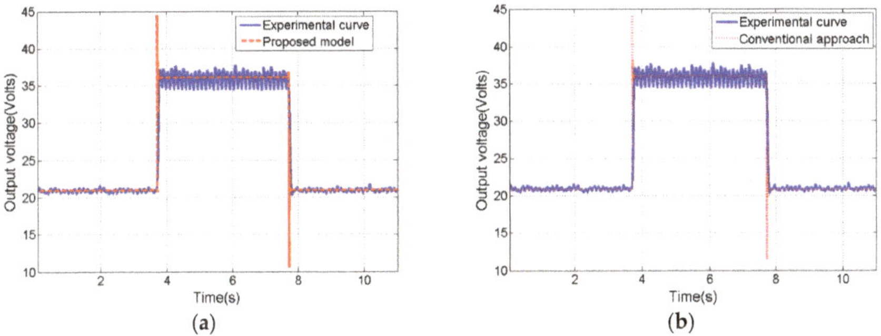

Figure 6. *Cont.*

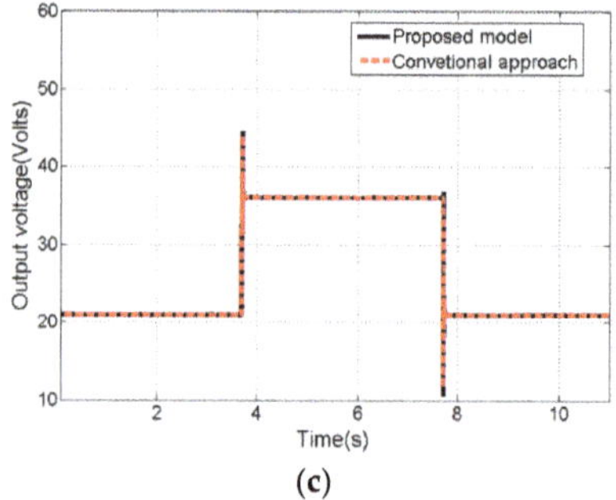

Figure 6. Curves for a DR value change from 30% to 60% to 30%: (**a**) proposed model and experimental output voltages, (**b**) experimental and conventional approach output voltages, and (**c**) proposed and conventional approach output voltages.

By analyzing the results depicted in Figure 6, the model behavior when the DR was changed from a value less than 50% to another one higher than 50% is similar to the conventional one. Both SSAM approaches had identical output voltage curves, but the conventional approach used two different sub-models for the two duty ratio ranges, 0–50% and 50–100%, which required an additional selection parameter that allowed for choosing the convenient model [17,18,20–22]. Additionally, unlike the previous works [17,18,20–22,25], the followed procedure for TLBDC modeling was described in step-by-step detail.

Applying Laplace transforms with zero initial conditions and using the superposition theorem, the small-signal duty-cycles $\widetilde{u_1}$ and $\widetilde{u_2}$ to state vector $\widetilde{x}$ transfer functions are as follows [3,24,35]:

$$\frac{\widetilde{x}(s)}{\widetilde{u_1}(s)} = |sI - [A_0 + U_1 A_1 + U_2 A_2]|^{-1} . A_1 X, . \tag{40}$$

$$\frac{\widetilde{x}(s)}{\widetilde{u_2}(s)} = |sI - [A_0 + U_1 A_1 + U_2 A_2]|^{-1} . A_2 X, \tag{41}$$

The transfer functions $\frac{\widetilde{V_{c1}}(s)}{\widetilde{u_1}(s)}$, $\frac{\widetilde{V_{c1}}(s)}{\widetilde{u_1}(s)}$, $\frac{\widetilde{V_{c2}}(s)}{\widetilde{u_1}(s)}$, $\frac{\widetilde{V_{c2}}(s)}{\widetilde{u_2}(s)}$, and $\frac{\widetilde{V_{c1}}(s)}{\widetilde{V_{c2}}(s)}$ can be deduced, and the required VB controllers are then designed.

3. Three-Level Boost DC–DC Converter Voltage Balance Control (VBC) Analysis

In order to assess the suitable method/controller for the VB control of the TLBDC, a comparison between two different methods using two different controllers, PI and Fuzzy, is carried out. The DR is set by an outer control loop, and the PI/Fuzzy controller ensures the VB control. The VB controllers' parameters are illustrated in Figure 7. The VB control is applied on both switches by subtracting the VB controller output from SW1 DR and adding it to SW2 DR, or on the lower switch only, by adding it to SW2 DR as illustrated in Figure 7. The duty cycles u_1 and u_2 are then used to generate the SW1 and SW2 control signals, respectively. SW1 and SW2 control signals are phase shifted by 180° as previously shown in Figure 2b,c.

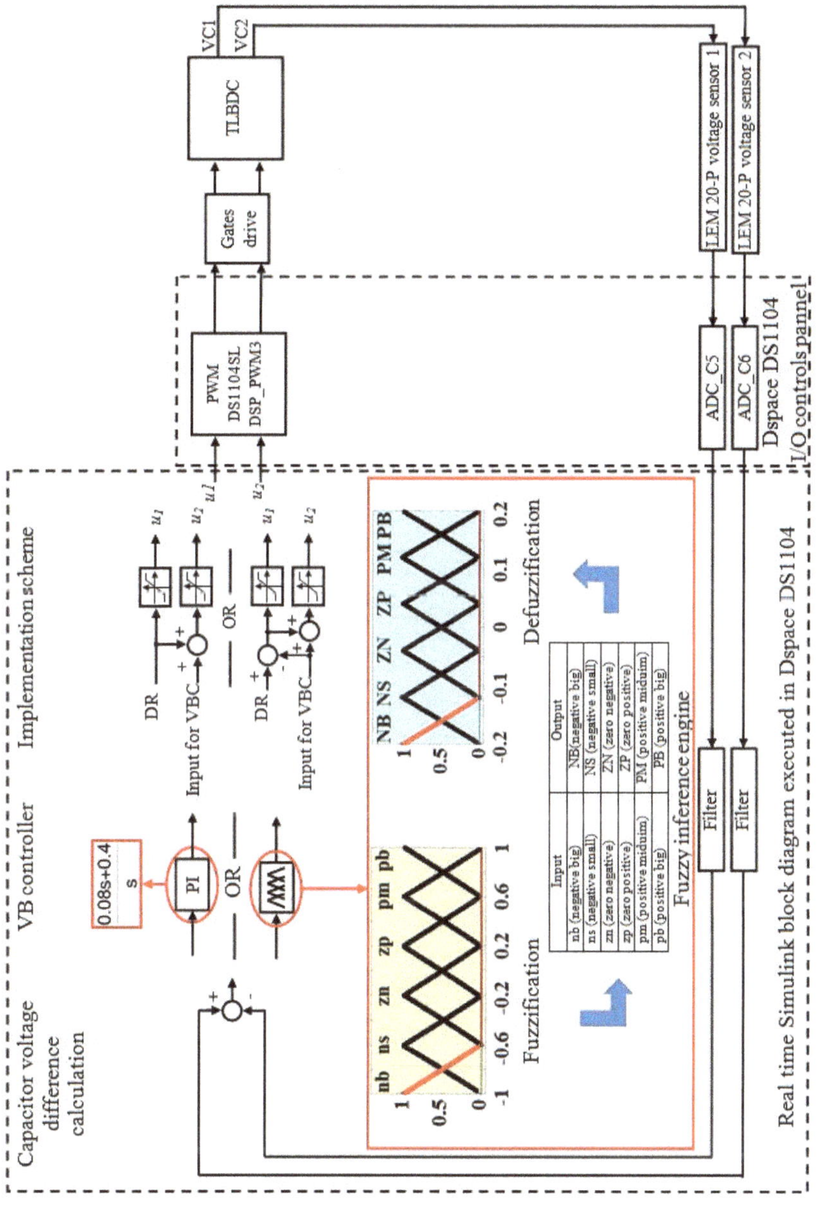

Figure 7. Block diagram of voltage balance control (VBC) schemes.

The aforementioned comparisons were examined via simulations performed in Matlab/Simulink software (Matworks, Natick, MA, USA), while the experimental tests were performed on the TLBDC prototype shown in Figure 8. The simplified scheme of the experimental setup and the TLBDC parameters are shown in Figure 9 and Table 2, respectively.

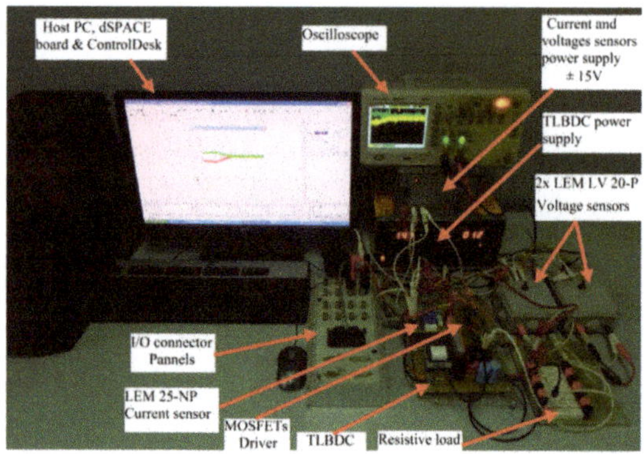

Figure 8. TLBDC Experimental setup.

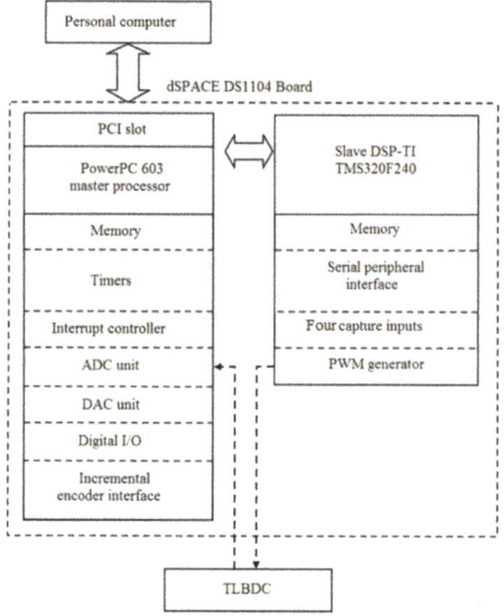

Figure 9. Block-diagram of the dSPACE DS1104 controller board.

Table 2. TLBDC parameters.

Parameter	Value
Switching frequency	12.5 kHz
Inductance, ESR	9 mH, 0.1 Ω
Output capacitors	100 uF
Input voltage	15 Volts
Load	82 Ω
Diode's forward voltage	0.5 Volts

VB control was implemented using the dSPACE 1104. After building the TLBDC VBC based on real-time Simulink-blocks, including the dSPACE 1104 slave-PWM generator and analog to digital (A/D) converters, the C code was automatically generated, downloaded and executed on the dSPACE board. The 180° phase-shifted control signals were generated using the dSPACE 1104. The logic signals were provided to an IR2110 gate driver that allowed for controlling the two TLBDC's MOSFETs (Metal–Oxide–Semiconductor Field-Effect Transistors). The dSPACE DS1104 ControlDesk monitor software was used to visualize and save the experimental data. The implemented Matlab/Simulink models on the dSPACE DS1104 board are shown in Figure 10, where Figure 10a illustrates the implemented model when the VB was applied on the lower switch of TLBDC, and Figure 10b shows the implemented model when the VB control was applied on both TLBDC switches. The VB controller, indicated in Figure 10, was either a Fuzzy or PI controller whose parameters are shown in Figure 7.

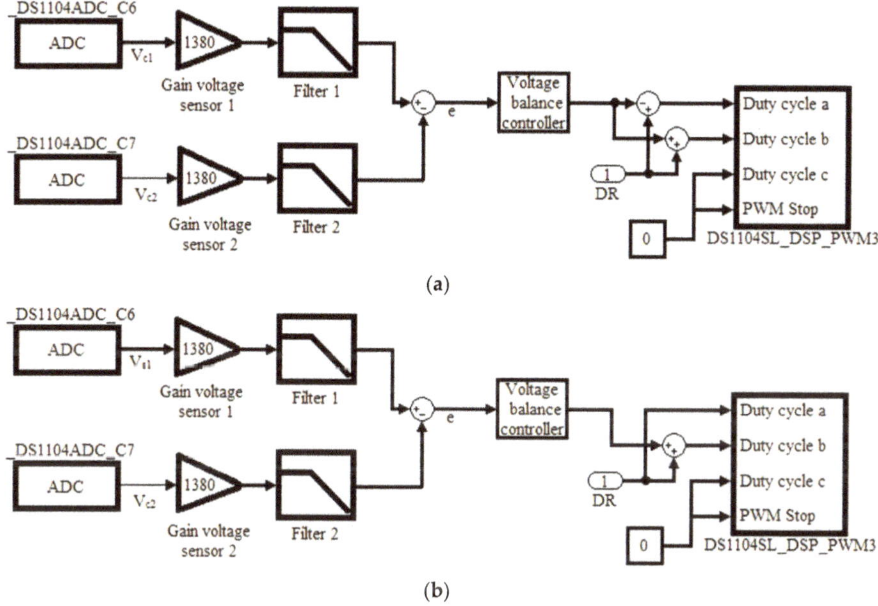

Figure 10. Matlab/Simulink implemented model on the dSPACE DS1104 board: (**a**) VB control applied on the TLBDC lower switch, and (**b**) VB control applied on either switche of the TLBDC.

Simulation and experimental results are depicted in Figures 11 and 12, respectively. Simulated output capacitors' voltages before and after applying the VB control at t = 0.025 s are presented in Figure 11. Using a PI controller, the VB was approximately achieved in 5 ms and 15 ms when the VBC was applied on both TLBDC switches, or only one, respectively. While the Fuzzy controller ensured a VB within 3 ms and 10 ms when the VBC was applied on both switches or on the lower switch, respectively. The same results could be deduced from experimental results presented in Figure 12, where the VB control was applied at t = 0.05 s. The VB, using a PI controller, was achieved in approximately 0.1 s and 0.3 s, when applying the VB control on both switches or one switch, respectively. While it was approximately achieved, using a Fuzzy controller, within 0.08 s and 0.15 s when the VB control was applied on both switches or on the lower switch, respectively.

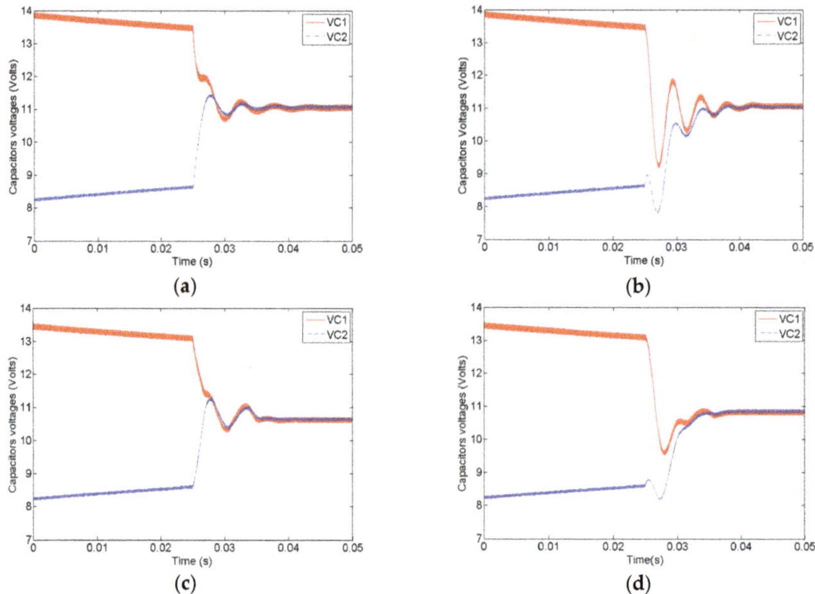

Figure 11. Simulated output capacitors' voltage curves after applying a balancing control at t = 0.025 s:
(**a**) on both switches using a PI VB controller, and (**b**) on the lower switch using a PI VB controller, (**c**) on
both switches using a Fuzzy VB controller, and (**d**) on the lower switch using a Fuzzy VB controller.

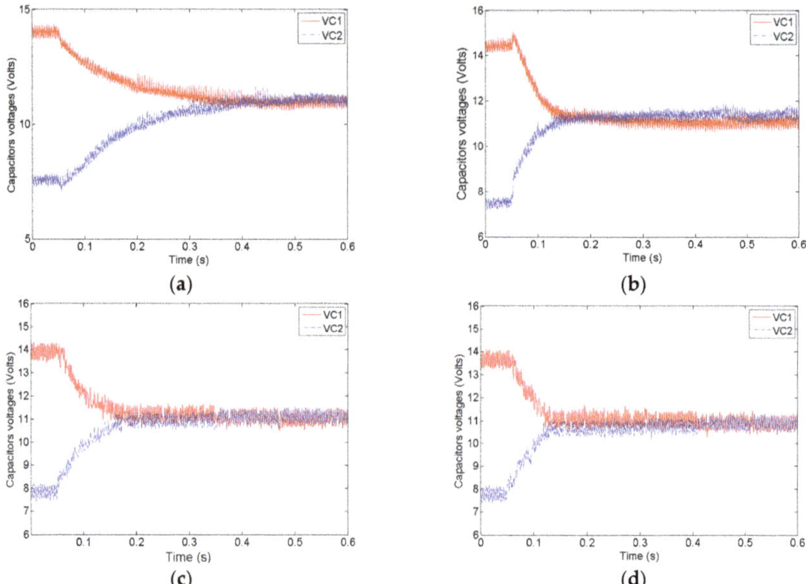

Figure 12. Experimental output capacitors' voltage curves: (**a**) on both switches using a PI VB controller,
and (**b**) on the lower switch using a PI VB controller, (**c**) on both switches using a Fuzzy VB controller,
and (**d**) on the lower switch using a Fuzzy VB controller.

According to the previous results, static and dynamic behaviors of the proposed model are in agreement with the experiments. The slight observed differences were mainly caused by the simplified assumptions made in the analysis, the slight errors introduced by measuring instruments, etc.

By analyzing the obtained results from the VB control analysis, one can see that the experimental results were in good agreement with the simulated ones. The differences observed in the VB controller's response times, in simulations and experiments, were mainly due to delays included by the digital to analogue (D/A) conversions, the processing time for real time implementation, and the needed time for the voltage average value calculation loop.

The analysis has shown that a VB was ensured in all cases. However, for both controllers, applying the VB control on either of the TLBDC's switches allows achieving the VB within a reduced time compared to applying it on one switch only. This showed that the works presented in References [29–33], where a VB control was applied on both TLBDC switches, have used an efficient way to ensure a VB control. In addition, the Fuzzy VB controller showed better performances compared to the PI controller, in terms of the requested time to ensure a VB for both cases as indicated in Table 3.

Table 3. Time to ensure VBC.

VBC		Requested Time to Ensure VBC Applied on One Switch (ms)	Requested Time to Ensure VBC Applied on Either Switch (ms)
PI	Simulation	15	5
	Experiments	300	100
Fuzzy	Simulation	10	3
	Experiments	150	80

4. Summary and Conclusions

The results of this study present a significant advance in the modeling and control of TLBDCs. This research also fills the gap in the related literature concerning this topic and provides new findings. The TLBDC unique model that describes the converter behavior for all DR values was first described in details. Based on the TLBDC switches' states and their equivalent electrical schemes, the state-space modeling of a non-zero inductor ESR TLBDC was carried out, and its SSM was then derived and validated using a TLBDC prototype. In a second stage, a VB control analysis was presented. Two VB controllers, PI and Fuzzy types, were used and their outputs were applied on both or one TLBDC switch(es), respectively. This allowed for choosing the efficient way and convenient controller for the TLBDC VB control.

The obtained results showed a good agreement between simulations and experiments. They also demonstrated that the developed model gave an accurate estimation of the TLBDC behavior. Generally, the presented results reflected an accurate approximation of the real results in dynamic, small perturbations around a corresponding operating point, and steady-state modes. These results have also shown that VB was achieved in all cases. However, applying the VB control on both switches allowed for achieving a VB in a reduced time compared to applying it on one switch. In addition, the Fuzzy controller presented good results, in terms of required time to ensure a VB control, when compared to the PI VB controller.

Author Contributions: Conceptualization, D.O.-A., S.D. and A.R.; Methodology, D.O.-A., S.D. and A.R.; Software, D.O.-A., S.D. and A.R.; Validation, D.O.-A., S.D. and A.R.; Writing—Original Draft Preparation, D.O.-A.; Writing—Review and Editing, D.O.-A., A.R. and S.D.; Supervision, S.D. and A.R.; Funding Acquisition, S.D. and A.R.

Funding: This work is performed in the framework of VERES Project funded by the Research Institute in Solar Energy and New Energies (IRESEN).

Conflicts of Interest: The authors declare no conflict of interest.

References

1. Sun, Y.; Ma, L.; Zhao, D.; Ding, S. A Compound Controller Design for a Buck Converter. *Energies* **2018**, *11*, 2354. [CrossRef]
2. Rashid, M.H. *Power Electronics Handbook*, 3rd ed.; Butterworth-Heinemann: Oxford, UK, 2011.
3. Monmasson, E. *Power Electronic Converters-PWM Strategies and Current Control Techniques*; ISTE Ltd.: London, UK, 2011.
4. Gerard, V.P.; Eduard, A. *CMOS Integrated Switching Power Converters*; Springer: New York, NY, USA, 2009.
5. Reza Tousi, S.M.; Moradi, M.H.; Basir, N.S.; Nemati, M. A function-based maximum power point tracking method for photovoltaic systems. *IEEE Trans. Power Electron.* **2016**, *31*, 2120–2128. [CrossRef]
6. Hossain, M.Z.; Rahim, N.A.; Selvaraj, J. A/L Recent progress and development on power DC-DC converter topology, control, design and applications: A review. *Renew. Sustain. Energy Rev.* **2018**, *81*, 205–230. [CrossRef]
7. Patel, H.; Agarwal, V. Maximum power point tracking scheme for PV systems operating under partially shaded conditions. *IEEE Trans. Ind. Electron.* **2008**, *55*, 1689–1698. [CrossRef]
8. Miao, K.; Ramachandaramurthy, V.K.; Yong, J.Y. Integration of electric vehicles in smart grid: A review on vehicle to grid technologies and optimization techniques. *Renew. Sustain. Energy Rev.* **2016**, *53*, 720–732.
9. Goli, P.; Shireen, W. PV powered smart charging station for PHEVs. *Renew. Energy* **2014**, *66*, 280–287. [CrossRef]
10. Torreglosa, J.P.; Fern, L.M.; García-trivi, P.; Jurado, F. Control and operation of power sources in a medium-voltage direct- current microgrid for an electric vehicle fast charging station with a photovoltaic and a battery energy storage system. *Energy* **2016**, *115*, 38–48.
11. El Aroudi, A.; Robert, B.; Martinez-Salamero, L. Modelling and analysis of multi-cell converters using discrete time models. In Proceedings of the IEEE International Symposium on Circuits and Systems, Island of Kos, Greece, 21–24 May 2006; pp. 2161–2164.
12. Zhang, Y.; Shi, J.; Fu, C.; Zhang, W.; Wang, P.; Li, J.; Sumner, M. An Enhanced Hybrid Switching-Frequency Modulation Strategy for Fuel Cell Vehicle Three-Level DC-DC Converters with Quasi-Z Source. *Energies* **2018**, *11*, 1026. [CrossRef]
13. Hafez, A.A.A. Multi-level cascaded DC/DC converters for PV applications. *Alex. Eng. J.* **2015**, *54*, 1135–1146. [CrossRef]
14. Zajec, P.; Nemec, M. Theoretical and Experimental Investigation of the Voltage Ripple across Flying Capacitors in the Interleaved Buck Converter with Extended Duty Cycle. *Energies* **2018**, *11*, 1017. [CrossRef]
15. Chen, H.C.; Lin, W.J. MPPT and voltage balancing control with sensing only inductor current for photovoltaic-fed, three-level, boost-type converters. *IEEE Trans. Power Electron.* **2014**, *29*, 29–35. [CrossRef]
16. Zhang, Y.; Sun, J.T.; Wang, Y.F. Hybrid boost three-level DC-DC converter with high voltage gain for photovoltaic generation systems. *IEEE Trans. Power Electron.* **2013**, *28*, 3659–3664. [CrossRef]
17. Bougrine, M.D.; Benalia, A.; Benbouzid, M.H. Simple sliding mode applied to the three-level boost converter for fuel cell applications. In Proceedings of the International Conference on Control, Engineering and Information Technology, Tlemcen, Algeria, 25–27 May 2015; pp. 1–6.
18. Yaramasu, V.; Wu, B. Predictive control of a three-level boost converter and an NPC inverter for high-power PMSG-based medium voltage wind energy conversion systems. *IEEE Trans. Power Electron.* **2014**, *29*, 5308–5322. [CrossRef]
19. Ruan, X.; Li, B.; Chen, Q.; Tan, S.C.; Tse, C.K. Fundamental considerations of three-level DC-DC converters: Topologies, analyses, and control. *IEEE Trans. Circuits Syst. I Regul. Pap.* **2008**, *55*, 3733–3743. [CrossRef]
20. Ma, H.; Yang, C.; Zhang, Y.Y. Analysis and design for single-phase three-level boost PFC converter with quasi-static model. In Proceedings of the 37th Annual Conference of the IEEE Industrial Electronics Society, Melbourne, VIC, Australia, 7–10 November 2011; pp. 4385–4390.
21. Krishna, R.; Soman, D.E.; Kottayil, S.K.; Leijon, M. Pulse delay control for capacitor voltage balancing in a three-level boost neutral point clamped inverter. *IET Power Electron.* **2015**, *8*, 268–277. [CrossRef]
22. Meleshin, V.; Sachkov, S.; Khukhtikov, S. Three-level boost converters. Modes, sub-modes and asymmetrical regime of operation. In Proceedings of the 16th European Conf. on Power Electronics and Applications, Lappeenranta, Finland, 26–28 August 2014; pp. 1–10.

23. Middlebrook, R.D.; Cuk, S. A general unified approach to modelling switching-converter power stages. In Proceedings of the Power Electronics Specialists Conference, Cleveland, OH, USA, 8–10 June 1976; pp. 18–34.
24. Khaldi, H.S.; Ammari, A.C. Fractional-order control of three level boost DC/DC converter used in hybrid energy storage system for electric vehicles. In Proceedings of the 6th International Renewable Energy Congress, Sousse, Tunisia, 24–26 March 2015; pp. 1–7.
25. Nouri, A.; Salhi, I.; Elwarraki, E.; El Beid, S.; Essounbouli, N. DSP-based implementation of a self-tuning fuzzy controller for three-level boost converter. *Electr. Power Syst. Res.* **2017**, *146*, 286–297. [CrossRef]
26. Rivera, S.; Wu, B. Electric Vehicle Charging Station with an Energy Storage Stage for Split-DC Bus Voltage Balancing. *IEEE Trans. Power Electron.* **2016**, *32*, 2376–2386. [CrossRef]
27. Tan, L.; Zhu, N.; Wu, B. An Integrated Inductor for Eliminating Circulating Current of Parallel Three-Level DC–DC Converter-Based EV Fast Charger. *IEEE Trans. Ind. Electron.* **2016**, *63*, 1362–1371. [CrossRef]
28. Tan, L.; Wu, B.; Yaramasu, V.; Rivera, S.; Guo, X. Effective Voltage Balance Control for Bipolar-DC-Bus-Fed EV Charging Station With Three-Level DC–DC Fast Charger. *IEEE Trans. Ind. Electron.* **2016**, *63*, 4031–4041. [CrossRef]
29. Vitoi, L.A.; Krishna, R.; Soman, D.E.; Leijon, M.; Kottayil, S.K. Control and implementation of three level boost converter for load voltage regulation. In Proceedings of the Industrial Electronics Society Conference, Vienna, Austria, 10–13 November 2013; pp. 561–565.
30. Tan, L.; Wu, B.; Rivera, S.; Yaramasu, V. Comprehensive DC Power Balance Management in High-Power Three-Level DC–DC Converter for Electric Vehicle Fast Charging. *IEEE Trans. Power Electron.* **2016**, *31*, 89–100. [CrossRef]
31. Xia, C.; Gu, X.; Shi, T.; Yan, Y. Neutral-Point Potential Balancing of Three-Level Inverters in Direct-Driven Wind Energy Conversion System. *IEEE Trans. Energy Convers.* **2011**, *26*, 18–29. [CrossRef]
32. Oulad-Abbou, D.; Doubabi, S.; Rachid, A.; García-Triviño, P.; Fernández-Ramírez, L.M.; García-Vázquez, C.A.; Sarrias-Mena, R. Combined control of MPPT, output voltage regulation and capacitors voltage balance for three-level DC/DC boostconverter in PV-EV charging stations. In Proceedings of the International Symposium on Power Electronics, Electrical Drives, Automation and Motion (SPEEDAM), Amalfi, Italy, 20–22 June 2018; pp. 372–376.
33. Costa, L.F.; Mussa, S.A.; Barbi, I. Capacitor voltage balancing control of multilevel DC-DC converter. In Proceedings of the Brazilian Power Electronics Conference, Gramado, Brazil, 27–31 October 2013; pp. 332–338.
34. Zhao, Q.; Fang, Y.; Ma, M.; Wang, J.; Xie, Y. Study on a Fuzzy Controller for the Balance of Capacitor Voltages of Three-Level Boost Dc-Dc Converter. In Proceedings of the International Power Electronics and Application Conference and Exposition (PEAC), Shanghai, China, 5–8 November 2014; pp. 993–996.
35. Mobarrez, M.; Ghanbari, N. Subhashish Bhattacharya Control Hardware-in-the-Loop Demonstration of a Building-Scale DC Microgrid Utilizing Distributed Control Algorithm. In Proceedings of the PES General Meeting 2018, Portland, OR, USA, 5–9 August 2018.

MDPI

St. Alban-Anlage 66

4052 Basel

Switzerland

Tel. +41 61 683 77 34

Fax +41 61 302 89 18

www.mdpi.com

Energies Editorial Office

E-mail: energies@mdpi.com

www.mdpi.com/journal/energies